W9-BRU-220

Essentials of Mathematics

Essentials of Mathematics

Fourth Edition

Russell V. Person, D.Sc.
Professor of Mathematics,
Capitol Institute of Technology
Kensington, Maryland

JOHN WILEY & SONS, New York • Chichester • Brisbane • Toronto • Singapore

Dedication:

This Fourth Edition of my book ESSENTIALS OF MATHEMATICS is lovingly dedicated to the memory of my wife, Emily Rose, who died on January 4, 1980, and without whose patience and devotion this book could not have been written. May this book be a lasting memorial to her.

Library of Congress Cataloging in Publication Data:

Person, Russell V
Essentials of mathematics.

Includes indexes.
1. Mathematics—1961– I. Title.
QA39.2.P473 1979 510 79–10708
ISBN 0-471-05184-5

Printed in the United States of America

10 9 8 7 6 5

Preface

This fourth edition contains basically the same material and coverage as the third edition, with a few minor changes and additions. The purpose of this edition, like that of former editions, is to provide the necessary mathematics for students entering any field of technology and for anyone desiring a good working knowledge of arithmetic, algebra, geometry, logarithms, trigonometry, with an introduction to calculus.

In this edition there is a slight change in the arrangement of topics:

1. The explanation of the metric system now appears in Chapter 32 rather than in Chapter 5. This change has been made because it seems reasonable that the geometric measurements of length, area, and volume should be explained first in terms of our common English system, and then in terms of the metric system. The explanation of the relation between the two systems has been enlarged somewhat, with emphasis on changes from the English system to the metric system. Chapter 32 also includes an explanation of dimensional analysis.
2. The chapter on number bases has been moved from Chapter 6 to Chapter 53, near the end of the book. This chapter has been enlarged to include a more complete explanation of the binary, the octal, and the sexagesimal systems.
3. A section on the use of electronic calculators has been added in Chapter 54.

Other minor topics have been added in this edition:

1. Proportional division has been included in Chapter 23, Ratio Proportion, Variation.
2. A special method of solving certain higher degree equations has been included in Chapter 20.
3. An explanation of the derivation of Pythagorean triples has been added in Chapter 45, Vectors.

A change has been made in the placement of quizzes in the book. One quiz, Form A, has been included at the end of each chapter, instead of a quiz only after several chapters, as in the third edition. This change was suggested by some users of previous editions. A second quiz, Form B, similar to Form A for each chapter, is included in the Instructor's Manual.

As a result of suggestions by users of previous editions, some minor but important changes have been made in this fourth edition:

1. Wordage has been reduced in some places where such reduction will not interfere with good understanding.

2. In some instances, the number of problems in an exercise has been reduced where previously there had been an unnecessary repetition of the same kinds of problems. This decrease in the number of problems in an exercise should be helpful to instructors in making assignment selections easier.
3. Assignment exercises have been revised in some cases and more problems of a practical nature have been added.
4. More worked-out illustrative examples have been added to help students' understanding, for example, in connection with trigonometric identities.
5. More illustrations of the graphs of the trigonometric and the inverse trigonometric functions have been added.

An Instructor's Manual, published separately, contains the following:

1. A second quiz, Form B, similar to Form A, for each chapter of the book.
2. Answers for even-numbered exercise problems and answers for all quizzes.
3. Some suggestions for the presentation of some topics.

I express my appreciation to users of previous editions for their many excellent comments and suggestions for improving the usefulness of the book. I especially express my appreciation to Dr. Irving L. Kosow of John Wiley, for his suggestions, help, and encouragement in preparing this fourth edition.

Russell V. Person

Preface to the Third Edition

The purpose of this third edition, as in the first and second editions, is to provide the elementary mathematical background needed by students preparing to enter a field of technology. The primary emphasis is on the practical use of mathematics. *Essentials of Mathematics* is designed to meet the need for a text covering arithmetic, algebra, geometry, logarithms, trigonometry, and the elements of calculus for use in technical institutes, junior colleges, and in first-year college general mathematics courses.

The book follows a classroom-tested organization. The separate subjects and topics are discussed in a logical order that facilitates learning. Each topic is based on the material that has been taken up before, and no topic depends on one that follows.

The approach in the presentation is the same as in previous editions:

1. The book is written for students' understanding.

2. It assumes almost no previous mathematical background except elementary arithmetic.

3. Explanations of mathematical concepts and processes are given in simple, easily understood language on an adult level.

4. Emphasis is on understanding and thinking rather than on the memorization of rules.

5. Reasoning is inductive. In most instances students themselves learn to formulate and state the rules and principles of mathematics.

6. Students learn the "Why" as well as the "How" of mathematical operations.

7. Numerous worked-out illustrative examples show students exactly what to do and how to do it.

8. Each topic presents the necessary essentials in an orderly, logical, and interesting manner so as to be most useful to students.

9. Presentation follows the basic psychological laws of learning:
(a) Motivation is established by showing the need.
(b) Learning proceeds from the known to the unknown, from the specific to the general, and from the simple to the complex.

PREFACE TO THIRD EDITION

What is new in this edition?
The following changes have been made in this third edition:

1. The most significant change is the addition of a section on calculus. This final section of the book covers the elements of differential and integral calculus of algebraic and some transcendental functions. The topics are introduced in a nonrigorous manner with just enough theory for a useful understanding of the subject for technical students. This section on calculus is based partially on material as presented in my text, *Calculus with Analytic Geometry*, published by Rinehart Press, Inc., 1970.
2. Another major change is the use of four-place instead of five-place logarithms.
3. The slope formula and the writing of equations of straight lines are included in Chapter 17, on graphing of equations.
4. Chapter 23, on systems of quadratics, is rewritten to form a more logical and coherent presentation of the topics.
5. Irrational equations have been included in Chapter 20 on quadratic equations, instead of in a separate chapter.
6. Inequalities are presented in Chapter 24.
7. Polar coordinates are presented as a section in Chapter 46 on vectors.
8. In many instances explanations have been made slightly more concise by reducing wordage where understanding is not jeopardized.
9. Approximately sixty percent of the exercise problems for practice are new. Moreover, they have been rearranged so that they progressively increase in difficulty in each exercise. The result is a better selection and a more practical arrangement of problems. Exercises also contain more problems of a strictly technical nature.

The above additions and changes are the result of many valuable suggestions by users of previous editions, and for those suggestions I am deeply grateful.

R.V.P.

Preface to the First Edition

This book was prepared as a textbook in the essentials of elementary mathematics. It consists chiefly of material I have used for the last six years in a preparatory mathematics course covering the five subjects, arithmetic, geometry, algebra, logarithms, and trigonometry.

The book presupposes very little or no previous knowledge in the fields covered, with the exception of elementary arithmetic. It contains more detailed explanation than is usually found in textbooks of this kind. One of the most difficult features of many mathematics texts, especially for the beginner, is the overabundance of symbols. It is my contention that the ideas and concepts of mathematics are not difficult. They are made difficult only when they are surrounded by and buried in a heap of symbols. In many instances a word can be used just as conveniently as a symbol and with more understanding. Even the word *equals* is sometimes preferable to the equality sign. Whenever a new concept is introduced, it is explained and defined in ordinary everyday language. Definitions are accurate without being complicated by difficult mathematical terminology.

Although written from material used in residence classes and planned for such use, the book is equally well suited to home-study courses. It is designed to enable a student to make good progress in learning mathematics without the aid of a personal instructor. It is addressed to the student earnestly and diligently seeking knowledge. Explanations are carefully given, and each step in a new process is explained and illustrated with examples. At the same time there is ample opportunity for the instructor in residence classes to vary the presentation to suit his own personal taste. The book contains an adequate supply of exercises so that selections can be made for class assignments.

The subject matter is entirely general and is equally suitable for technical and nontechnical schools. The book can be used as a text in general mathematics in any school or institute offering a course in the fields it covers. It might well form the basis for a college one-year terminal course in general mathematics for nonscience majors.

Reasoning is inductive rather than deductive. In most instances the students formulate their own rules. The emphasis is always on understanding before memorizing, on thinking rather than on the mere application of rules. Even axioms must appear reasonable.

PREFACE TO THE FIRST EDITION

Mathematics is not presented as something final and complete but as a fascinating field that contains opportunity for growth. The attempt is made to have students realize that there may be more than one approach to a problem and several ways of arriving at a solution. The emphasis is upon procedure and correct reasoning, not simply upon getting the right answer. The purpose is to have the student get some understanding of the power and the beauty as well as the practical utility of mathematics.

I wish to express my gratitude to all of my colleagues and to others who have given helpful comments and suggestions in the writing of this book, especially to Mr. Lattie M. Upchurch, Jr., for his encouragement, and to John Wiley & Sons for many helpful suggestions in the course of editing the manuscript.

R. V. P.

Contents

PART ONE ARITHMETIC

CHAPTER ONE Whole Numbers

1-1 OUR COMMON NUMBER SYSTEM

In order to get a good understanding of the various operations used in arithmetic, one should have a clear idea of the formation of our common number system. Our system of numbers makes use of only ten symbols, called *digits*: 0, 1, 2, 3, 4, 5, 6, 7, 8, and 9. By means of these ten symbols we can write a number of any size.

Each of these symbols was invented by early man to stand for a certain number of objects. The symbols were first used for counting. For instance, at one time in the prehistoric past a person may have found this many pebbles: o, o. After many years someone decided to call this many *two* and to represent them by the symbol 2. After many more years someone who had one more than two (that is, this many, o, o, o,) decided to call them *three* and to represent them by the symbol 3. He might just as well have called them *seven* and represented them by the symbol 5. Later on, this many objects, o, o, o, o, were called *four* and written 4. The other symbols up to nine (9) were invented in the same way. The zero (0) was the last of the ten symbols invented. Of course, it must be understood that the names and symbols were not exactly the same as we know them in English.

Then, when counting continued to one more than nine (such as this many pebbles, o, o, o, o, o, o, o, o, o, o, which we now call ten), there was no symbol for the number. It was decided to lump all of these objects together as one bunch and to write the digit for the bunch. This 1 was placed in a position a little farther to the left. A zero was placed to the right of 1: thus 10. This is the way we now represent *ten* objects. In other words, 1 written in the second place to the left indicates a bunch of ten.

This *place value* of the digits was one of the greatest inventions in mathematics. When we write 27, we mean 2 bunches of ten each, and 7 single objects more, or 20 + 7. In the number 27 the 7 is in the position we call *units*' place and the 2 is in the position we call *tens*' place. By means of this place value of a digit, we can represent a number of any size, no matter how large.

When we count to 99, we have 9 bunches of ten each and 9 single objects more, or 90 + 9. Now, if we have one more object, we get *ten* bunches, which we call one hundred objects. Since we have no symbol for ten, we lump the ten bunches together and call it a large bunch. We might call it a bunch of the *second order.* We write 1 and follow it with two zeros: 100. The 1 is said to be in *hundreds'* place. When we write a number such as 358 for a number of objects, we mean 3 large bunches of one hundred each, 5 small bunches of ten each, and 8 single objects more; that is, 300 + 50 + 8.

As we get larger and larger bunches, or bunches of higher orders, we move the digits for these bunches farther and farther to the left. We might say 1000000 (one million) represents 1 bunch of the sixth order. In a larger number, such as 92,000,000, the only purpose of the zeros is to place the 92 far enough to the left so that it will indicate 92 million. If the numbers are as large as ten thousand or more, we usually separate the digits by commas into groups of three, starting at the right, thus:

5,870,000,000,000

In our common number system the first place starting at the right represents single units; the next place to the left represents tens; the third place, hundreds; the fourth place, thousands; the fifth place, ten-thousands; the sixth place, hundred-thousands; the seventh place, millions. In the United States a billion is represented in the tenth place. That is, 1 billion is a thousand millions. However, in England and in some other countries 1 billion means a million millions.

1-2 KINDS OF NUMBERS

Numbers may be classified as *integers* and *fractions.* An integer is a whole number. The positive integers, 1, 2, 3, 4, and so on, are used in counting and, are therefore, sometimes called *counting* numbers.

A *fraction* indicates a part or parts of a whole number, such as $\frac{1}{2}$, $\frac{3}{5}$, $\frac{4}{9}$, and so on. Fractions will be studied in Chapter 2.

An *even* number is a number that can be exactly divided by 2, such as the numbers, 8, 14, 376, 9530, and so on. Numbers that are not exactly divisible by 2 are called *odd* numbers, such as 3, 7, 19, 61, and 847. Whether a number is odd or even is determined by the right-hand digit. In any number, if the right-hand digit is even, such as in 7936, the number itself is even and is divisible by 2. If the right-hand digit is odd, as in the number 4865, then the number itself is odd and is not divisible by 2.

1-3 ADDITION OF NUMBERS

When numbers are added, the result is called the *sum* of the numbers. The sum of 5, 7, and 4 is 16. The numbers added together are called *addends.* Addition is usually indicated by the plus sign (+).

In the addition of numbers, two laws are useful. We probably use these laws constantly without being aware of their names. The *commutative law* for addition states that two numbers may be added in either order. For example, to add 5 and 3, we may write

$$5 + 3 \quad \text{or} \quad 3 + 5$$

Now we may begin with 5 objects and add 3 to this number. However, we may begin with 3 objects and add 5 to them. The result is 8 in both instances. In general, if a and b represent any two numbers, respectively, then, by the *commutative law*,

$$a + b = b + a$$

Another law useful in adding several addends is the *associative law* for addition. The associative law states that numbers to be added may be associated in any order. For example, suppose we have the addition

$$16 + 3 + 7$$

We might first say $16 + 3 = 19$. Then we add 7 to 19 and get 26. However, we know it is convenient to add 10 to any number. Now we might recognize that $3 + 7 = 10$. Then we can say $16 + 10 = 26$. The two ways of associating addends can be shown by parentheses. We enclose in parentheses the numbers to be added first. Then, by the *associative law*

$$(16 + 3) + 7 = 16 + (3 + 7)$$

In general, if a, b, and c represent any three addends, respectively, then by the *associative law*,

$$(a + b) + c = a + (b + c)$$

The commutative and associative laws enable us to pair up any combination of addends for easy calculation. For example, note the pairs of addends that make 10 in the following example:

$$6 + 7 + 3 + 5 + 8 + 5 + 2 = 36$$

10   10   10

This pairing of addends that make 10 is very useful in column addition.

To add several numbers, we usually arrange them in column form, placing units under units, tens under tens, and so on. The sum is written below the numbers. In adding each column it is helpful to note any combinations of digits that make 10.

Example

Add $697 + 7486 + 29354 + 78$.

Solution

The form for vertical addition is shown at the right. We begin by adding the digits in the units' column. The sum is 25, which is 2 tens and 5 units. The 5 is placed under the units' column, and the 2 tens are carried to the top of the tens' column. We continue in similar manner, adding each column, moving toward the left. The carrying process should be done mentally, although in long columns, the carried digits are sometimes written at the top of the proper column.

$$\begin{array}{r} 697 \\ 7486 \\ 29354 \\ 78 \\ \hline 37615 \end{array}$$

To be sure that your work in mathematics is correct, you should develop the habit of checking it in some way. Probably the best way to check addition is simply to add the numbers a second time. Another way is to reverse the direction of adding. If you add the columns downward the first time, then check by adding upward. However, the best way to avoid errors in addition is to be sure you know instantly the sum of each of the following sets of numbers. They are called the 45 addition combinations.

$$\begin{array}{ccccccccccccccc} 2 & 3 & 4 & 3 & 9 & 5 & 7 & 5 & 3 & 4 & 8 & 2 & 1 & 9 & 4 \\ 1 & 2 & 5 & 4 & 9 & 2 & 1 & 3 & 9 & 7 & 3 & 9 & 8 & 1 & 4 \\ \hline \\ 6 & 2 & 6 & 7 & 9 & 1 & 8 & 7 & 6 & 9 & 5 & 7 & 4 & 1 & 5 \\ 5 & 4 & 6 & 3 & 7 & 6 & 8 & 5 & 2 & 5 & 1 & 6 & 8 & 4 & 8 \\ \hline \\ 3 & 6 & 2 & 7 & 9 & 6 & 1 & 7 & 4 & 9 & 3 & 5 & 7 & 2 & 6 \\ 3 & 9 & 2 & 8 & 4 & 8 & 1 & 2 & 6 & 8 & 1 & 5 & 7 & 8 & 3 \\ \hline \end{array}$$

1-4 SUBTRACTION OF INTEGERS

When one number is subtracted from another, the result is called the *difference* or *remainder.* The number subtracted is called the *subtrahend.* The number from which the subtrahend is subtracted is called the *minuend.*

The symbol for subtraction is the minus sign, a short horizontal line (−). To indicate 7 subtracted from 15, we write, $15 - 7$. This is read "15 minus 7." The minus sign means that the second number is to be subtracted from the first, not the reverse. For this reason, in subtraction the numbers cannot be reversed. In addition $a + b$ equals the same as $b + a$. But in subtraction $a - b$ does not mean the same as $b - a$.

In subtracting numbers we begin with the units, as in addition.

Example 1

Subtract 435 from 897.

Solution

The work is shown at the right. We begin by subtracting 5 units from 7 units, and place the result, 2, in units' place. Next, 3 from 9 leaves 6, in tens' place. Finally, 4 from 8 leaves 4, in hundreds' place.	897 435 462

Subtraction may be checked by addition. If the remainder is added to the subtrahend, the result should be the minuend. Checking is done mentally.

In subtraction, it is often necessary to use a trick called *borrowing.* This is the opposite of *carrying* in addition. There are other methods of subtraction, but the borrowing method is probably most common. The actual borrowing is done mentally, although in some instances it may be shown.

Example 2

Subtract 279 from 863.

Solution

$$\begin{array}{r} 863 \\ \underline{279} \\ 584 \end{array}$$

The form is shown at the right but the borrowing is not shown. Since we cannot subtract 9 from 3, we borrow 1 ten from the 6 tens, leaving 5 tens in the minuend. The 1 ten is changed to 10 units, which is combined with the 3 units to make 13. Now we subtract 9 from 13 and get 4. As a second step, we borrow 1 hundred from 8, leaving 7 hundreds. The 1 hundred is changed to 10 tens and combined with the 5 tens, making 15 tens. We take 7 tens from 15 tens and get 8 tens in second place. Finally, we take 2 hundreds from 7 hundreds and get 5 hundreds.

Example 3

Subtract 269 from 704.

Solution

$$\begin{array}{r} 704 \\ \underline{269} \\ 435 \end{array}$$

The work, except for the borrowing, is shown at the right. We cannot borrow 1 from zero. Then we go to the hundreds column and borrow 1 hundred from 7 hundreds, leaving 6 hundreds. This 1 hundred is changed to 10 tens. One of these 10 tens is changed to 10 units, which together with the 4 units makes 14 units. Now we take 9 units from 14 units, and get 5 units. As a second step, we take 6 tens from 9 tens, leaving 3 tens. Finally, we take 2 hundreds from 6 hundreds, leaving 4 hundreds.

In a series of horizontal terms involving addition and subtraction the operations are performed in the order in which they occur. If any quantity is enclosed in parentheses, brackets, or braces, or some form of grouping, the quantity within the parentheses is to be considered as a single quantity. In such a case the operations indicated within the parentheses are to be performed first.

$$15 - 13 + 9 = 11; \qquad 24 + (6 - 5 + 2) - 6 - (5 + 6) = 10$$

Exercise 1-1

Add the following: (They may be worked several times for practice).

1.	368	958	739	386	976	459	738
	284	517	871	627	249	349	493

2.	746	568	847	643	798	867	828
	956	479	348	727	558	598	679

3.	864	475	536	728	693	349	977
	247	739	478	859	385	695	361
	953	351	854	381	418	738	183
	158	156	632	256	725	415	746
	632	994	259	523	572	872	528
	471	218	361	864	241	549	274
	349	872	743	547	539	261	532

4. 498 + 1534 + 23 + 43,576 **5.** 9164 + 37 + 12,986 + 423

6. 2056 + 304 + 74,607 + 53 **7.** 43 + 3862 + 298 + 14,317

8. 24 + 5986 + 307 + 63,364 **9.** 96 + 453 + 7 + 347,208

Subtract the bottom number from the top:

10.	5684	1923	8488	5635	3785
	4309	1042	619	978	3159

11.	7043	9804	4132	8423	9006
	3897	2973	3901	7138	5842

12.	5904	2973	3601	4826	7015
	4917	584	2073	2067	3960

13.	6500	1745	1032	4000	7005
	2759	796	579	1809	6738

14. Combine $24 + (8 - 5) - (7 + 2) - (6 - 4)$

15. Combine $30 - (8 + 3) - (9 - 4) + (4 - 1)$

16. Combine $27 - (16 - 9) + (3 + 2) - (6 + 8)$

17. Combine $58 + 21 - 5 - 6 - 1 + 8 - 7$

18. A family on an automobile tour takes a trip of 1620 miles to be completed in five days, Monday through Friday. Each day they traveled the following number of miles: Monday: 348; Tuesday: 355; Wednesday: 298; Thursday: 346. How many miles must be driven on Friday?

19. Four voltages in series in an electric circuit have a combined voltage of 320 volts. If three of the voltages are; respectively, 85 volts, 48 volts, and 124 volts, find the fourth voltage. (In an electric circuit, if voltages are arranged in series, the combined voltage is equal to the sum of the voltages.)

20. In a certain electric circuit, three resistors in series have the following ratings, respectively: 245 ohms; 186 ohms; and 362 ohms. If the circuit requires a total resistance of 1000 ohms, find the amount of resistance that must be added to the circuit. (If resistors are arranged in series, the total resistance is found by adding the resistances.)

21. In a certain electric circuit, four capacitors arranged in parallel have a total capacitance of 200 microfarads. If three of the capacitors have ratings of 67, 56, and 45 microfarads, find the rating of the fourth capacitor. (If capacitors are arranged in parallel, the total capacitance is found by adding the capacitances of the separate capacitors.)

1-5 MULTIPLICATION

Multiplication is the process of taking one number two or more times. Multiplication can be considered as a shortened form of addition. For instance, if we wish to add three 7's, we can say 3×7. The symbol for multiplication is the sign $\times$. The symbol is read "times." The expression 3×7 is read, "3 times 7" and means the same as adding three 7's: thus $7 + 7 + 7$. In column form these expressions can be written

Addition:	Multiplication:
7	7
7	× 3
7	
21	21

Multiplication can be indicated in other ways. It is sometimes indicated by a raised dot placed between the numbers: thus $3 \cdot 7$ means 3×7. We may also enclose in parentheses each of the numbers to be multiplied and then place one next to the other without any sign between them: (3)(7) = 21. Whenever any two numbers are written in this form, they are meant to be multiplied together.

In the multiplication of two numbers, one number is called the *multiplicand* and the other is called the *multiplier.* The *multiplicand* is the number that is to be taken a given number of times. The multiplier is the number that tells how many times the multiplicand is to be taken. In many examples it makes little difference which number is taken as the multiplier and which is taken as the multiplicand.

The answer obtained in multiplication is called the *product.* A product always implies multiplication. The numbers multiplied together are called the *factors* of the product. As an example, the factors of 21 are 3 and 7 because 3 times 7 equals 21. The product of the three numbers, 2, 3, and 5, is 30. Therefore the factors of 30 are 2, 3, and 5.

A *composite* number is a number composed of integral factors other than 1 and the number itself. For example, 15 is a composite number because it is composed of the two factors, 3 and 5. A *prime* number is a number that cannot be separated into any integral factors other than itself and 1, such as the number 17.

In addition we have already noted two laws of numbers: the *commutative* and the *associative* laws. In the multiplication of numbers we have the corresponding laws. For example,

$$3 \times 7 = 7 \times 3$$

The *commutative law* for *multiplication* states that the multiplication of two numbers may be done in either order. In general terms, if a and b represent any two numbers, respectively, then by the commutative law,

$$a \times b = b \times a$$

We often make use of the commutative law to check multiplication by reversing the order of multiplication. For example,

$$327 \times 453 = 453 \times 327$$

The *associative law for multiplication* enables us to associate certain factors in such a way as to simplify computation. For example, suppose we wish to multiply the following:

$$17 \times 5 \times 2$$

We might first say $17 \times 5 = 85$. Then we multiply 85 by 2 and get 170. However, we know that it is convenient to multiply any number by 10. Now we might first recognize that $5 \times 2 = 10$. Then we can say

$$17 \times 10 = 170$$

The two ways of associating the factors can be shown by parentheses.

We enclose in parentheses the numbers to be multiplied first. Then, by the associative law,

$$(17 \times 5) \times 2 = 17 \times (5 \times 2)$$

The answer is 170 in both instances. The *associative law for multiplication* may be indicated in general, using the numbers a, b, and c:

$$(a \times b) \times c = a \times (b \times c)$$

The associative and commutative laws for multiplication together enable us to pair up any combinations of numbers that lead to easy calculation. For example, note the multiplication by 10's in the following:

$$9 \times \underbrace{2 \times 5}_{10} \times 7 \times \underbrace{5 \times 2}_{10} = 6300$$

To get the answer we first take 9×7 and then multiply the result by 100.

In multiplying numbers containing more than one digit each, we usually place one number above the other with the right-hand digits in line. The larger number is usually placed above and is considered as the multiplicand.

Example 1

Multiply 394 by 7.

Solution

394	394	394	394	multiplicand
× 7	× 7	× 7	× 7	multiplier
	8	58	2758	product

Explanation

We begin the multiplication with the units' digits. That is, $7 \times 4 = 28$. The product, 28, means 2 tens and 8 units. The 8 units are placed in the units' place, and the 2 tens are carried over to the tens' position. Any digit carried over in this manner is sometimes written above the next digit at the left in the multiplicand. Multiplying 7×9 tens, we get 63 tens, which, added to the 2 tens carried, makes 65 tens, or 6 hundreds and 5 tens. The 5 is placed in the tens' place in the answer, and the 6 hundreds are carried over to the hundreds' place. Multiplying 7×3 hundreds, we get 21 hundreds. This 21 hundreds, added to the 6 hundreds carried, makes 27 hundreds, or 2 thousands and 7 hundreds.

Example 2

Multiply 543 by 96.

Solution

543	multiplicand
96	multiplier
3258	multiplication by 6
4887	multiplication by 90
52128	product

Explanation

First, we multiply the multiplicand, 543, by 6, the same as with a single-digit multiplier. Next, we multiply by the 9 in the multiplier. However, this 9 represents tens. Therefore, when we say $9 \times 3 = 27$, we really mean 90×3. The product of $90 \times 3 = 270$, which is 27 tens, or 2 hundreds and 7 tens. In order to make adjustment for the fact that the 27 represents tens, we place 7 in the tens' place, and carry the 2 over the hundreds' place. After multiplying by each digit in the multiplier, we add the products.

Example 3

Multiply 2987 by 405.

Solution

2987	multiplicand
405	multiplier
14935	multiplication by 5
11948	multiplication by 400
1209735	product

If the multiplier contains a zero digit, the multiplication by this digit need not be shown, since the product of any number by zero is zero.

Example 4

Multiply 967 by 840.

Solution

967	multiplicand
840	multiplier
38680	multiplication by 40
7736	multiplication by 800
812280	product

Here the multiplier ends in zero. In such cases the multiplier is often written farther to the right, as shown here. Although the multiplication by zero is zero, we must write the zero in units' place to make the answer correct.

Example 5

Multiply 80436 by 7050.

Solution

```
    80436    multiplicand
     7050    multiplier
  -------
  4021800    multiplication by 50
563052       multiplication by 7000
---------
567073800    product
```

In this example notice especially the zeros in multiplicand and multiplier. Such zeros often cause students much trouble.

1-6 DIVISION

Division is the process of determining how many times one number is contained in another. The symbol for division is ÷. The expression 28 ÷ 4 is read "28 divided by 4." This indicates that we are to find the number of times 4 is contained in 28. Division may also be shown as a fraction. Thus 28 ÷ 4 can be written $\frac{28}{4}$. The number divided by another is called the *dividend.* The number divided into the dividend is called the *divisor.*

We see that 28 ÷ 4 = 7. The answer obtained by division is called the *quotient.* A quotient always implies division, just as the word *product* always implies multiplication. In the example, 28 ÷ 4 = 7, the dividend is 28, the divisor is 4, and the quotient is 7. In most instances there is a *remainder* after the division. For example, 29 ÷ 4 = 7 with a remainder of 1.

Division may be checked by multiplication. The quotient times the divisor should equal the dividend. If there is a remainder, the division can be checked by multiplying the quotient by the divisor and adding the remainder. The result should be the dividend.

Example 1

Divide 37,538 by 7.

In this example, the dividend is 37,538 and the divisor is 7. When the divisor is a single digit, we use a form called *short division.* The usual form is shown here:

$$\begin{array}{l} \quad 5\ \ 3\ \ 6\ \ 2, \text{ remainder} = 4 \\ 7\overline{)3\,7\,{}^{2}5\,{}^{4}3\,{}^{1}8} \end{array}$$

We begin by dividing 7 into 37, which is 5, with a remainder of 2. Here we are really dividing 7 into 37,000, which is 5000, with a remainder of 2000. The quotient, 5, is written directly above the 7 in 37, and the 2 is carried over to the 5; that is, the 2000 remaining after the first division represents 20 hundreds and is added to the 5 hundreds in the dividend. The result is 25 hundreds to be divided by 7. From here on the division follows the same procedure. Whenever there is a remainder, it is carried over and placed next to the following digit. We divide 7 into 25, which is 3 with a remainder of 4. The 4 is carried over and placed next to the 3, making 43. Then we divide 7 into 43, which is 6, with a remainder of 1. The 1 is placed next to the 8, making 18. Finally, we divide 7 into 18, which is 2, with a remainder of 4.

The answer can be checked by multiplication. If we multiply the quotient by the divisor and add to this product the remainder, the result should be the dividend.

In short division the separate remainders are found mentally and are usually not shown. For instance, in the foregoing example the student should perform subtractions mentally. The work then appears as follows:

$$\begin{array}{l} \quad 5\,3\,6\,2, \text{ remainder} = 4 \\ 7\overline{)3\,7\,5\,3\,8} \end{array}$$

In division a digit is placed in the quotient for each division or attempted division. Zeros will sometimes appear in the quotient.

Example 2

Divide 423,543 by 6.

$$\begin{array}{l} \quad\ 7\,0\,5\,9\,0, \text{ remainder} = 3 \\ 6\overline{)4\,2\,3\,5\,4\,3} \end{array}$$

In this example notice that a zero is placed in the quotient when each step in the division involves the division of 6 into a smaller number. Of course, in this example, it is not necessary to write a zero for the division of 6 into 4, the very first digit of the dividend, since this would not change the value of the answer.

Long division is the procedure of showing the subtraction and the remainder in each step of the division. The following examples show a common form for long division. Each digit of the quotient should be placed directly above the last digit of the dividend used in each separate step. In Example 3 notice that 6, the first digit of the quotient, is placed directly above the 5 in the dividend, since the first division is 47 into 295.

Example 3

Divide 29,508 by 47.

Solution

```
      627
 47)29508
    282
    ---
     130
      94
     ---
     368
     329
     ---
      39
```

Step 1. Divide 47 into 295. Place the 6 in the quotient *directly above the 5.*
Step 2. Multiply the divisor, 47, by the 6 and write the product, 282, below the 295.
Step 3. Subtract, leaving 13.
Step 4. Bring down the 0, the next digit of the dividend.
Step 5. Divide 47 into 130 and write the result, 2, as the second digit of the quotient.
Step 6. Multiply the divisor, 47, by the 2 and write the product, 94, below the 130.
Step 7. Subtract, leaving 36.
Step 8. Bring down the 8, the next digit of the dividend, making 368.
Step 9. Divide 47 into 368 and write the result, 7, in the quotient.
Step 10. Multiply the divisor, 47, by the 7 of the quotient and write the product, 329, under the 368.
Step 11. Subtract, leaving a remainder of 39.

Long division is used mainly when the divisor contains two or more digits. When the divisor contains only a single digit, as in Examples 1 and 2, the remainders are carried along mentally. Although long division in arithmetic is often performed correctly by habit, it is well to consider each of the steps, since the same procedure is followed in long division in algebra.

Example 4

Divide 113,673 by 37.

Solution

```
        3072
 37)113673
    111
    ---
      267
      259
      ---
        83
        74
        --
remainder 9
```

In this example notice that after the first division by the divisor, 37, and the subtraction, the first remainder is only 2. When the next digit is brought down, the new dividend becomes only 26. The divisor can not be divided into the 26. Since one digit of the dividend has already been brought down at this point, a digit must be written in the quotient for this division—in this case, zero.

Warning. An error that is sometimes made is to write too small a number in the quotient for a particular division. In the following example an error of this kind has been purposely made to point out the danger.

Example 5

Divide 58,692 by 73.

In the division shown here the first digit in the quotient has been called 7. However, when the subtraction is made, the remainder is 75, which is greater than the divisor. Therefore, the first digit in the quotient is not correct. It should be 8. The example is worked out correctly as shown.

```
     7     (wrong)
73)58692
   511
    75

     804   (correct division)
73)58692
   584
     292
     292
```

Exercise 1-2

Multiply the following:

1.	2873 × 49	3516 × 98	4867 × 59	6295 × 68	5923 × 76
2.	7784 × 76	4678 × 916	5308 × 897	4971 × 872	3158 × 936
3.	4036 × 902	6814 × 807	7459 × 609	8916 × 705	5789 × 801
4.	3148 × 936	4864 × 824	9617 × 193	3976 × 829	9580 × 705
5.	5291 × 42	10582 × 84	3737 × 333	9999 × 4444	2927 × 352

Divide as indicated:

6. 419,382 ÷ 6 **7.** 210,938 ÷ 7 **8.** 816,320 ÷ 8

9. 31,416 ÷ 3 **10.** 14,142 ÷ 2 **11.** 602,064 ÷ 6

12. 80,001 ÷ 9	**13.** 419,320 ÷ 4	**14.** 35,217 ÷ 43
15. 10,488 ÷ 24	**16.** 29,664 ÷ 32	**17.** 55,091 ÷ 89
18. 67,744 ÷ 73	**19.** 64,008 ÷ 84	**20.** 17,777 ÷ 29
21. 29,792 ÷ 38	**22.** 39,292 ÷ 47	**23.** 33,512 ÷ 59
24. 26,101 ÷ 43	**25.** 34,036 ÷ 67	**26.** 57,816 ÷ 72
27. 60,716 ÷ 86	**28.** 53,336 ÷ 59	**29.** 81,463 ÷ 28
30. 444,444,444 ÷ 36	**31.** 777,777,777 ÷ 63	**32.** 69,506 ÷ 89

33. If you take a circle tour plane trip from Washington, D.C. by way of the following cities back to Washington, what is the total length of your flight: Washington to Chicago, 597 miles; Chicago to Minneapolis, 355 miles; Minneapolis to Los Angeles, 1524 miles; Los Angeles to Dallas, 1240 miles; Dallas to St. Louis, 547 miles; St. Louis to Miami, 1061 miles; and Miami to Washington, 923 miles?

34. On a particular construction project 35 tons of concrete are needed. The first six truck loads weighed, respectively, 8710; 11,600; 11,120; 10,460; 9070; and 9830 pounds. How many more pounds are needed? (1 ton = 2000 pounds.)

35. Sound travels approximately 1080 feet per second. If the sound of thunder is heard 12 seconds after the flash was seen, how far away was the lightning?

36. Light travels 186,000 miles per second. How long does it take the light from the sun, about 93,000,000 miles away, to reach the earth?

37. Find the length of a light-year, the distance light travels in one year.

38. In 22 hours of driving time the odometer on a car changed from a reading of 35,479 to 36,722 miles. Find the average rate of travel in miles per hour.

39. For the Apollo 15 flight, the J-1 rocket, first stage of V-5, had five engines, each with a thrust of 1,500,000 pounds. Find the total thrust.

40. The J-2 rocket, second stage of V-5, Apollo 15, had five engines each with a thrust of 200,000 pounds. Find the total thrust of these engines.

41. At one point in the flight of Apollo 15, the fuel consumption was

28,000 pounds per second. If the burn continued for 150 seconds at this rate, what was the total fuel consumption, in tons, for the 150 seconds?

42. Four voltages in series have a combined voltage of 1200 volts. If three of the voltages are, respectively, 150 volts, 345 volts, and 475 volts, find the fourth voltage.

43. Three resistors have the following ratings, respectively: 462 ohms, 573 ohms, and 614 ohms. What additional resistance is required if the necessary resistance, connected in series, is 2000 ohms?

44. Three capacitors in parallel have a total capacitance of 85 microfarads. If two of them have ratings of 28 and 37 microfarads, respectively, find the third.

45. During one month a family used the following appliances, with indicated wattage, for the time indicated. Find the total number of *kwh* (kilowatt-hours) for the month:

frying pan, 1150 watts, 12 hours; toaster, 850 watts, 8 hours;
grill, 1350 watts, 6 hours; lights, 370 watts, 90 hours.

QUIZ ON CHAPTER 1. FORM A.

(Do not use any form of calculator for this quiz.)

1. Add the following numbers:

$$487 + 396 + 623 + 578 + 354 =$$

2. Add the following numbers:

$$398 + 4763 + 79 + 5127 + 6089 =$$

3. Subtract the bottom number from the top in each set:

694	873	635	807	703	1042	9000	3502
273	419	486	364	247	963	3496	2467

4. Multiply the two numbers as indicated:
(a) (637)(92) **(b)** (783)(56); **(c)** (807)(79); **(d)** (5069)(86);
(e) (4276)(354) **(f)** (6004)(897) **(g)** (7968)(2970) **(h)** (8593)(389)

5. Divide as indicated; show remainders:
(a) $47569 \div 6$ **(b)** $68085 \div 7$ **(c)** $30795 \div 8$ **(d)** $41056 \div 9$
(e) $25239 \div 7$ **(f)** $7506 \div 23$ **(g)** $22593 \div 56$ **(h)** $7295 \div 28$
(i) $229600 \div 79$ **(j)** $26170 \div 87$

6. Four pieces, each 18 in. long, are cut from a board 7 ft long. Find the length of the remaining piece.

7. The following deposits were made in a bank: \$485; \$890; \$562; \$96; \$1730. Find the total amount of the deposits. If a withdrawal of \$968 was then made, how much of the deposits was left in the bank?

8. On a trip of 1300 miles made in four days, the average for the first three days was 358 miles per day. How many miles were traveled the fourth day?

9. How many pieces each 15 in. long can be cut from a board 14 ft long, and how much remains?

CHAPTER TWO Fractions

2-1 NEED FOR FRACTIONS

As long as people used numbers only for *counting*, they had no need for fractions. Whether he was counting pebbles or his sheep, early man did not require fractions. Counting his sheep, he could not say 1, 2, $2\frac{1}{2}$, etc.

Numbers also serve in another capacity besides *counting*. They are necessary in *measuring*. When we measure anything, whether length, weight, time, area, or electric current, we use a *unit of measure*. The *unit* is simply a small definite amount of the same kind of thing as the quantity that is to be measured. The measurement is made by noting how many times the quantity measured contains the unit.

When early man wished to measure a particular length (for example, the length of his hut), he may have picked up a stick of convenient length to use as a unit of measure. As he laid off the unit of length (the stick) along the wall of his hut, he found that the number of times he laid off the unit was not a whole number. That is, he found that he laid off the unit, perhaps five times, and then had about a half unit left over. In this way fractions and approximations came into arithmetic. If arithmetic had been used only for counting "discrete" objects, fractions would never have been needed. In all measurement we are faced with a *continuum* rather than separate and *discrete* objects.

Counting is exact; measurement is only approximate.

It is well to distinguish between the two uses of numbers: (*a*) for counting and (*b*) for measuring.

Exercise 2-1

Tell which of the two uses of numbers is employed in each of the following situations:

1. The number of hairs on your head.

2. The length of a given hair.

3. The number of leaves on a tree.

4. The number of leaves on all the trees in the world.

5. The weight of a cupful of water.

6. The number of drops in a pail of water.

7. The number of automobiles in the United States.

8. The length of a certain automobile.

9. The weight of an automobile.

10. The length of the diameter of a particular circle.

11. The length of the circumference of a circle.

12. The number of whole days in a year.

13. The time length of a year.

14. The number of pages in a particular book.

15. The number of printed characters in a book.

16. The amount of ink required to print a book.

2-2 DEFINITION OF A FRACTION

There are two ways of looking at a fraction. In arithmetic probably our first understanding of a fraction was about as follows:

Let us consider the meaning of the fraction $\frac{3}{4}$. Suppose we divide a circle into four equal parts (Fig. 2-1). Each part is called a "fourth" and is denoted by the fraction $\frac{1}{4}$. The number below the line is called the denominator because it *denominates*, or *names*, the part. It may be called the *namer.* The word *denominate* means to *name.* If we take three of these parts, or "fourths," we show this amount by the fraction $\frac{3}{4}$. The number, 3, above the line is called the *numerator* because it *enumerates*, or *counts*, the number of parts taken. It may be called the *counter.* The three fourths are shaded in the figure.

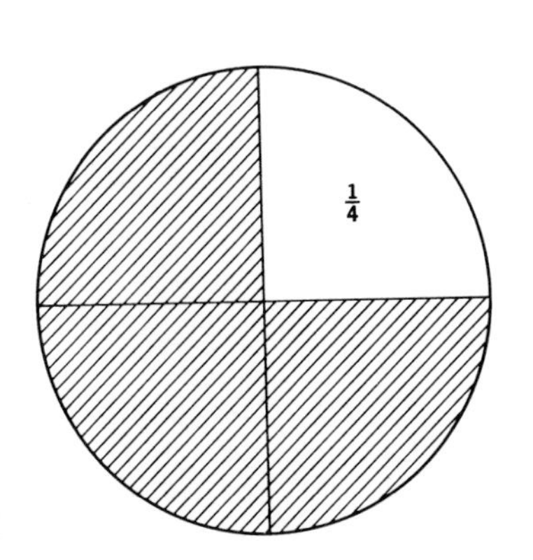

Figure 2-1. Three "fourths."

The foregoing is the most common explanation of the meaning of a fraction. However, there is another way of looking at a fraction. A fraction can be considered as an *indicated division.* The fraction $\frac{3}{4}$ can be taken to mean 3 divided by 4, or $3 \div 4$. The horizontal line between the numerator and the denominator can be taken as a symbol of division.

As an example showing the two approaches, consider the meaning of $\frac{3}{4}$ of a dollar. We can take the denominator, 4, to mean that one dollar is first divided into 4 parts, each part called a "fourth." The value is 25 cents. Then the numerator, 3, means that we take 3 of these "fourths." The value is 75 cents.

As the second approach, we look upon the fraction, $\frac{3}{4}$, as $3 \div 4$; that is, 3 dollars divided by 4. Then we have $\$3.00 \div 4 = \$0.75$. The value is 75 cents, the same as before, but the approach is different.

As another example, consider the meaning of $\frac{2}{3}$ of a foot. One third of a foot is 4 inches; two thirds is 8 inches. On the other hand, the fraction $\frac{2}{3}$ can mean 2 feet divided by 3; that is, 24 inches $\div$ 3 = 8 inches. Again, the answer is the same but the approach is different.

There is often an advantage in considering a fraction as an indicated division. The fraction $\frac{7}{9}$ means $7 \div 9$. Moreover, division may be written as a fraction. For example, $13 \div 4$ can be written as the fraction $\frac{13}{4}$. This view of a fraction has many advantages, especially in algebra.

The numerator and the denominator are called the *terms* of a fraction.

A *proper fraction* is a fraction in which the numerator is less than the denominator; that is, the fraction has a value less than 1. The following are proper fractions: $\frac{1}{4}, \frac{5}{8}, \frac{5}{6}$.

An *improper fraction* is a fraction in which the numerator is equal to, or greater than, the denominator; that is, an improper fraction has a value equal to, or more than, 1. The following are improper fractions: $\frac{5}{5}, \frac{8}{7}, \frac{5}{2}, \frac{261}{43}$.

A *mixed number* is a number consisting of a whole number and a fraction. The following are mixed numbers:

$$4\tfrac{2}{3};\quad 15\tfrac{2}{7};\quad 596\tfrac{39}{56}$$

It should be noted that a mixed number, such as $12\frac{4}{9}$, is really the sum of a whole number and a fraction.

$$12\tfrac{4}{9} \quad \text{means} \quad 12 + \frac{4}{9}$$

2-3 CHANGING THE FORM OF NUMBERS CONTAINING FRACTIONS

Before we work problems involving fractions, we must devise certain rules for operating with such numbers.

It is often necessary to change the form of a fraction. An improper fraction can be changed to a whole number or to a mixed number. For instance, the improper fraction $\frac{23}{6}$ can be changed to a mixed number by dividing the numerator by the denominator. We know that there are 6 sixths in each whole unit. To find the number of whole units in 23 sixths, we divide 23 by 6, and get 3 whole units and 5 sixths over. Therefore, $\frac{23}{6} = 3 + \frac{5}{6} = 3\frac{5}{6}$.

Rule. *To change an improper fraction to a whole number or a mixed number, divide the numerator by the denominator. The result is the whole number. Any remainder is placed over the denominator as a fraction.*

Sometimes it is desirable to change a whole number or a mixed number to an improper fraction. For instance, we might wish to change the whole number 7 into fourths. Since there are 4 fourths in each whole unit, then in 7 whole units there are (7)(4), or 28, fourths; that is,

$$7 = \frac{28}{4}$$

Rule. *To change a whole number into an improper fraction with any desired denominator, multiply the whole number by the desired denominator and place the result over the denominator.*

To change a mixed number, such as $8\frac{2}{3}$, to an improper fraction, we first change the 8 to thirds: $8 = \frac{24}{3}$. Then we add the $\frac{2}{3}$ to the $\frac{24}{3}$.

$$8\tfrac{2}{3} = \frac{26}{3}$$

Rule. *To change a mixed number to an improper fraction, multiply the whole number by the denominator and to this result add the numerator of the given fraction. Place the total over the denominator.*

Exercise 2-2

Change each of the following mixed numbers to improper fractions:

1. $5\frac{2}{3}$ **2.** $7\frac{1}{4}$ **3.** $9\frac{1}{2}$ **4.** $14\frac{1}{3}$ **5.** $16\frac{3}{5}$

6. $21\frac{4}{5}$ **7.** $15\frac{5}{6}$ **8.** $26\frac{3}{4}$ **9.** $48\frac{2}{3}$ **10.** $31\frac{3}{8}$

11. $14\frac{2}{7}$ **12.** $7\frac{4}{15}$ **13.** $16\frac{2}{3}$ **14.** $33\frac{1}{3}$ **15.** $66\frac{2}{3}$

16. $11\frac{1}{9}$ **17.** $8\frac{8}{9}$ **18.** $4\frac{13}{16}$ **19.** $5\frac{23}{32}$ **20.** $17\frac{3}{10}$

Reduce the following improper fractions to whole numbers or mixed numbers:

21. $\frac{19}{4}$ **22.** $\frac{18}{3}$ **23.** $\frac{43}{5}$ **24.** $\frac{30}{7}$ **25.** $\frac{41}{6}$

26. $\frac{75}{8}$ **27.** $\frac{65}{11}$ **28.** $\frac{78}{3}$ **29.** $\frac{67}{9}$ **30.** $\frac{97}{10}$

31. $\frac{86}{13}$ **32.** $\frac{68}{17}$ **33.** $\frac{319}{6}$ **34.** $\frac{251}{8}$ **35.** $\frac{468}{7}$

36. $\frac{702}{9}$ **37.** $\frac{396}{12}$ **38.** $\frac{983}{15}$ **39.** $\frac{4997}{16}$ **40.** $\frac{1755}{17}$

2-4 FUNDAMENTAL PRINCIPLE OF FRACTIONS

Before we consider the various operations with fractions, the following principle must be thoroughly understood.

Fundamental Principle of Fractions. *If the numerator and the denominator of any fraction are multiplied or divided by the same quantity (other than zero), the value of the fraction will not be changed.*

To illustrate this principle, suppose we start with the fraction $\frac{12}{18}$. The fundamental principle states that if we *multiply* the numerator and the denominator by any number, the fraction will still have the same value. Suppose we *multiply both terms* of the fraction by the number 7. Then we get the new fraction $\frac{84}{126}$. This new fraction has the same value as $\frac{12}{18}$.

You may wonder why this is true. If we multiply the fraction $\frac{12}{18}$ by 1, the value of the fraction will not change. However, the number 1 may be expressed in some form such as $\frac{3}{3}$, $\frac{7}{7}$, or $\frac{29}{29}$. Now let us multiply the fraction $\frac{12}{18}$ by $\frac{7}{7}$:

$$\frac{7}{7} \cdot \frac{12}{18} = \frac{84}{126}$$

When we multiply both numerator and denominator of the fraction by 7, we are only multiplying the fraction by 1. Therefore, the value of the fraction is not changed.

The fundamental principle also says that if we *divide* numerator and denominator of a fraction by the same number, the fraction will still have the same value. Suppose we *divide both terms of the fraction* $\frac{12}{18}$ by the number 6. Then we get the fraction $\frac{2}{3}$. The new fraction has the same value as $\frac{12}{18}$. Dividing the numerator and the denominator by the same number does not change the value of the fraction.

It should be noted that if the same number (not zero) is *added to*, or *subtracted from*, both numerator and denominator of a fraction, the value of the fraction *will be changed*. For example, the fraction $\frac{2}{3}$ will have a different value if we add some number such as 4 to both numerator and denominator. If we add 4 to both terms of the fraction, we get a new fraction, $\frac{6}{7}$, which has a different value from $\frac{2}{3}$.

2-5 REDUCING FRACTIONS

By use of the fundamental principle, it is possible to change the terms of a fraction without changing the value. Such a change is called *reducing* the fraction. Work with fractions is often simplified if the numbers are small. For example, the fraction $\frac{15}{20}$ can be changed in form by dividing both terms by 5. Then

$$\frac{15}{20} = \frac{15 \div 5}{20 \div 5} = \frac{3}{4}$$

The value of the fraction has not been changed. We say the fraction is reduced to lower terms because the numbers have been made smaller.

In reducing fractions to lower terms, we divide numerator and denominator by any number contained in both an even number of times. Now the question arises: How can we tell what number can be divided into numerator and denominator? Sometimes we can tell easily by inspection. In many cases the problem is more difficult. At this point we need to know how to tell the divisibility of numbers. Here are some tests: let us test the number 131,736.

(a) A number is divisible by 2 if the last digit at the right is divisible by 2; that is, if the number is even; in the example, 6 is divisible by 2.

(b) A number is divisible by 4 if the number represented by the last two digits is divisible by 4; in the example, 36 is divisible by 4.

(c) A number is divisible by 8 if the number represented by the last three digits is divisible by 8; in the example, 736 is divisible by 8.

(d) A number is divisible by 3 if the sum of its digits is divisible by 3; in the example, $1 + 3 + 1 + 7 + 3 + 6 = 21$, divisible by 3.

(e) A number is divisible by 9 if the sum of its digits is divisible by 9; in the example, 21 is not divisible by 9, therefore the number itself is not.

(f) A number is divisible by 10 if it ends in zero; this number does not.

(g) A number is divisible by 5 if it ends in 5 or zero; this number does not.

(h) Divisibility by 7: there is no simple test; try it out.

(i) A number is divisible by 11 if the *difference* between the *sums* of the *alternate digits* is zero or divisible by 11; in the example, for one set of digits we have $1 + 1 + 3 = 5$; for the alternate digits, $3 + 7 + 6 = 16$. The difference is 11; therefore the number is divisible by 11.

Note. For divisibility by 3 or 9, the digits may be added until the result is only one digit. First we get 21. Then we can add $2 + 1 = 3$.

A fraction is said to be reduced to lowest terms when the numerator and the denominator do not contain a common factor other than 1. For example, $\frac{48}{60}$ can be changed to $\frac{12}{15}$ by dividing both terms by 4. However, the fraction can be reduced still further to $\frac{4}{5}$ by dividing both terms by 3. The fraction is now in its lowest terms, since the numerator and denominator do not contain a common factor other than 1.

The expression *reducing a fraction* sometimes also refers to changing a fraction to *higher* terms. This procedure is often necessary. For instance, the fraction $\frac{3}{4}$ can be changed to higher terms by multiplying numerator and denominator by some number.

$$\frac{3}{4} = \frac{(8)(3)}{(8)(4)} = \frac{24}{32}$$

Exercise 2-3

Reduce the following fractions to lowest terms:

1. $\frac{18}{54}$ **2.** $\frac{27}{45}$ **3.** $\frac{30}{42}$ **4.** $\frac{54}{66}$

5. $\frac{36}{84}$ **6.** $\frac{36}{78}$ **7.** $\frac{84}{132}$ **8.** $\frac{75}{105}$

9. $\frac{72}{168}$ **10.** $\frac{108}{144}$ **11.** $\frac{105}{135}$ **12.** $\frac{72}{192}$

13. $\frac{140}{224}$ **14.** $\frac{105}{189}$ **15.** $\frac{180}{252}$ **16.** $\frac{192}{288}$

17. $\frac{168}{378}$ **18.** $\frac{135}{360}$ **19.** $\frac{54}{234}$ **20.** $\frac{288}{1056}$

21. $\frac{462}{616}$ **22.** $\frac{168}{840}$ **23.** $\frac{210}{672}$ **24.** $\frac{168}{1155}$

25. $\frac{1260}{1512}$ **26.** $\frac{1782}{2673}$ **27.** $\frac{3003}{8008}$ **28.** $\frac{2310}{3234}$

29. $\frac{3780}{9072}$ **30.** $\frac{7182}{7938}$ **31.** $\frac{6174}{6468}$ **32.** $\frac{4165}{7595}$

Change the fractions in each of the following sets to new fractions with higher terms and with the indicated denominator.

33. $\frac{3}{4}, \frac{2}{3}, \frac{3}{8}, \frac{7}{12}$; denominator: 24

34. $\frac{3}{5}, \frac{7}{8}, \frac{5}{12}, \frac{11}{40}$; denominator: 120

35. $\frac{2}{3}, \frac{6}{7}, \frac{4}{5}, \frac{8}{15}$; denominator: 105

36. $\frac{4}{9}, \frac{8}{27}, \frac{7}{12}, \frac{31}{36}$; denominator: 108

2-6 ADDITION AND SUBTRACTION OF FRACTIONS

If two or more fractions have the same denominators, they are added or subtracted by adding or subtracting their numerators. The result is placed over the denominator.

Example 1.

$$\frac{7}{8} + \frac{3}{8} = \frac{7 + 3}{8} = \frac{10}{8}.$$ The sum $\frac{10}{8}$ should be reduced to $\frac{5}{4}$.

Example 2.

$$\frac{7}{9} + \frac{8}{9} - \frac{2}{9} = \frac{7 + 8 - 2}{9} = \frac{13}{9} = 1\tfrac{4}{9}$$

If the denominators are different, we must first change the form of some or all of the fractions, so that all have the same denominator.

First we must find the *lowest common denominator* (LCD), that is, the smallest number that can be divided by all the denominators. Then we change each fraction to a new fraction having the LCD as its denominator. After making this change, we proceed as usual for fractions with the *same denominator.*

Example 3.

Add $\frac{3}{4} + \frac{2}{3} + \frac{5}{8} - \frac{7}{12}$.

The denominators are 4, 3, 8, and 12. By inspection, we find the LCD is 24. Now, each fraction must be expressed as a fraction with a denominator of 24. To change the form of the fractions we use the *fundamental principle.* Let us first write down the denominator of the fractions:

$$\frac{\quad}{24} + \frac{\quad}{24} + \frac{\quad}{24} - \frac{\quad}{24}$$

The $\frac{3}{4}$ is changed to $\frac{18}{24}$ by multiplying numerator and denominator by 6. The fraction $\frac{2}{3}$ is changed to $\frac{16}{24}$ by multiplying both terms by 8. The fraction $\frac{5}{8}$ is changed to $\frac{15}{24}$ by multiplying both terms by 3. The fraction $\frac{7}{12}$ is changed to $\frac{14}{24}$ by multiplying both terms by 2. Now the fractions can be added:

$$\frac{18}{24} + \frac{16}{24} + \frac{15}{24} - \frac{14}{24} = \frac{18 + 16 + 15 - 14}{24} = \frac{35}{24} = 1\tfrac{11}{24}$$

In adding mixed numbers, we might use any one of several methods. Probably the most practical way is to place the numbers in column form. Then we add whole numbers and fractions separately. To add

$43\frac{1}{2}$, $59\frac{1}{4}$, $127\frac{4}{5}$, and $413\frac{7}{10}$, we place them in column form as follows:

$$\begin{aligned} 43\tfrac{1}{2} &= 43\tfrac{10}{20} \\ 59\tfrac{1}{4} &= 59\tfrac{5}{20} \\ 127\tfrac{4}{5} &= 127\tfrac{16}{20} \\ 413\tfrac{7}{10} &= 413\tfrac{14}{20} \\ \hline 642 \quad & + \frac{45}{20} = 644\tfrac{5}{20} = 644\tfrac{1}{4} \end{aligned}$$

Short Form		or	20 (LCD)	
$43\frac{1}{2}$	$\frac{10}{20}$		$43\frac{1}{2}$	10
$59\frac{1}{4}$	$\frac{5}{20}$		$59\frac{1}{4}$	5
$127\frac{4}{5}$	$\frac{16}{20}$		$127\frac{4}{5}$	16
$413\frac{7}{10}$	$\frac{14}{20}$		$413\frac{7}{10}$	14

Each fraction is expressed as a new fraction with the lowest common denominator. Then the fractions and whole numbers are added separately. The sum of the whole numbers is 642. The sum of the fractions is $\frac{45}{20}$, which, reduced to a mixed number in lowest terms, becomes $2\frac{1}{4}$. This quantity is combined with 642. The total is $644\frac{1}{4}$.

In some instances it is convenient, when adding fractions, to change mixed numbers to improper fractions. This method is not practical if the whole numbers are large. However, when whole numbers are small, this method is often simple.

Example 4

Add $6 + \frac{4}{5} + 4\frac{3}{5} + 3\frac{1}{2}$.
We can change all numbers to tenths. Then we have

$$\frac{60}{10} + \frac{8}{10} + \frac{46}{10} + \frac{35}{10} = \frac{149}{10} = 14\tfrac{9}{10}$$

Subtraction of numbers involving fractions and mixed numbers is usually done by placing the subtrahend below the minuend as with whole numbers. However, in subtraction we often run into some difficulties. The following examples show the subtraction of fractions.

Example 5

Subtract $492\frac{7}{8} - 154\frac{1}{8}$.
This problem presents no difficulty.

$$\begin{array}{r} 492\tfrac{7}{8} \\ 154\tfrac{1}{8} \\ \hline \end{array}$$

Subtracting, $338\frac{6}{8} = 338\frac{3}{4}$

We subtract $\frac{1}{8}$ from $\frac{7}{8}$. The result is $\frac{6}{8}$. Then we subtract the whole numbers.

Example 6

Subtract $536\frac{5}{12} - 142\frac{5}{12}$.

$$\begin{array}{r} 536\frac{5}{12} \\ \underline{142\frac{5}{12}} \\ 394 \end{array}$$

Subtracting,

In this example we subtract $\frac{5}{12}$ from $\frac{5}{12}$, which is zero. However, this zero (0) must *not* be written down below the fractions. If it is written, the entire answer would appear to be 3940, which is not correct.

Example 7

Subtract $436\frac{5}{8} - 124$.

$$\begin{array}{r} 436\frac{5}{8} \\ \underline{124\phantom{\frac{5}{8}}} \\ 312\frac{5}{8} \end{array}$$

Subtracting,

In this example we subtract nothing from $\frac{5}{8}$. We simply bring down the $\frac{5}{8}$ as the fractional part of the remainder.

Example 8

Subtract $436 - 124\frac{5}{8}$.

$$\begin{array}{r} 5\phantom{6\frac{8}{8}} \\ 43\not{6}\frac{8}{8} \\ \underline{124\frac{5}{8}} \end{array} \qquad \begin{array}{r} 435\frac{8}{8} \\ \underline{124\frac{5}{8}} \\ 311\frac{3}{8} \end{array}$$

In this example, we must take one whole unit from the 6 and change it to $\frac{8}{8}$ before we can subtract the $\frac{5}{8}$.

Example 9

Subtract $832\frac{4}{7} - 158\frac{6}{7}$.

$$\begin{array}{r} 832\frac{4}{7} \\ \underline{158\frac{6}{7}} \end{array} \qquad \begin{array}{r} 831\frac{11}{7} \\ \underline{158\frac{6}{7}} \\ 673\frac{5}{7} \end{array}$$

In this example we cannot subtract $\frac{6}{7}$ from $\frac{4}{7}$. Therefore, we take one whole unit from the 2 in the minuend and change it to $\frac{7}{7}$. This quantity, $\frac{7}{7}$, combined with the $\frac{4}{7}$ already in the minuend, makes $\frac{11}{7}$. Now we subtract $\frac{6}{7}$ from $\frac{11}{7}$, which leaves $\frac{5}{7}$, as the fractional part of the answer.

2-7 FINDING THE LOWEST COMMON DENOMINATOR

We have seen that if fractions are to be added or subtracted, they must have the same denominator. If the denominators are diffierent, then the form of the fractions must be changed so that they have a common denominator. The first step in changing fractions to higher terms is to determine some number that is divisible by all of the given denominators.

In some examples the LCD can be determined by inspection, which is simply a good guess. However, in many instances, we need a more systematic method of determining the LCD of two or more fractions.

It is first necessary to understand what is meant by a *multiple* of a number. A multiple of any given number is some number that is exactly divisible by the given number. The following numbers are all multiples of 8 because they are all exactly divisible by 8 : 8, 16, 24, 32, 40, 48, 56, 64, 72, 80, etc. The following numbers are all multiples of 12 because they are all exactly divisible by 12 : 12, 24, 36, 48, 60, 72, etc.

A *common multiple* of two or more numbers is some number that is exactly divisible by each of the given numbers. Thus the following numbers are common multiples of 8 and 12 because all can be divided exactly by 8 and 12 : 24, 48, 72, 96, etc. Notice that the number 24 is the smallest number exactly divisible by both 8 and 12. Such a number is called the *lowest common multiple* (LCM) of the two numbers 8 and 12. The lowest common multiple of two or more numbers is the smallest number that is exactly divisible by each of the given numbers. When we add or subtract fractions, we first find the lowest common multiple of the denominators. This number is the lowest common denominator of the fractions.

There are several ways of determining the lowest common multiple of two or more numbers. One method is shown in the following example.

Example

Find the lowest common multiple of these numbers: 24, 45, 75, and 210.

Step 1. We write each number as a product of its prime factors. To find the prime factors of a number, we begin dividing the number by the smallest number contained in the given number. For instance, to find the prime factors of 24, we divide 24 by 2, then the result by 2, the next result by 2, and so on until the quotient is not divisible by any other factor. The result of the divisions shows that the prime factors of 24 are 2, 2, 2, and 3. We do the same with the other numbers and then write each as a product of its factors:

$$24 = 2 \cdot 2 \cdot 2 \cdot 3$$
$$45 = 3 \cdot 3 \cdot 5$$
$$75 = 3 \cdot 5 \cdot 5$$
$$210 = 2 \cdot 3 \cdot 5 \cdot 7$$

Step 2. Now, for the LCM, we write down each of the prime factors as many times as it is contained in any one of the given numbers. The factor 2 is contained three times in 24, which is the greatest number of times we find the factor 2 in any one number. Therefore, we set down $2 \cdot 2 \cdot 2$, as part of the LCM. The factor 3 is contained twice in 45. Therefore, we write $3 \cdot 3$ as part of the LCM. The greatest number of times we find 5 in any one of the given numbers is twice. Therefore, we write down two 5's. The factor 7 is found only once in any one of the given numbers. Therefore, one 7 must be used in the LCM. The lowest common multiple is therefore the product of the following factors:

$$2 \cdot 2 \cdot 2 \cdot 3 \cdot 3 \cdot 5 \cdot 5 \cdot 7$$

The product, 12,600, is the LCM, that is, the smallest number that will contain the numbers 24, 45, 75, and 210.

Exercise 2-4

Find the lowest common multiple of each of the following sets of numbers:

1. 6, 8, 12 **2.** 4, 6, 3 **3.** 2, 3, 4, 5

4. 4, 6, 9, 24 **5.** 10, 15, 20 **6.** 3, 5, 7, 11

7. 6, 9, 18, 36 **8.** 4, 5, 6, 10 **9.** 30, 42, 60

10. 24, 54, 60, 90 **11.** 84, 90, 180, 630 **12.** 180, 210, 315

13. 120, 180, 240, 300 **14.** 60, 168, 200, 270

15. 45, 48, 180, 252 **16.** 120, 180, 250, 360

17. 36, 120, 270, 600 **18.** 54, 324, 900, 225

19. (a) Add each of the following sets of mixed numbers:

$48\frac{5}{8}$	$32\frac{1}{8}$	$41\frac{1}{7}$	83	$54\frac{4}{9}$	$62\frac{3}{5}$	$87\frac{4}{7}$
$19\frac{3}{8}$	$17\frac{3}{16}$	$23\frac{5}{14}$	$25\frac{4}{5}$	$24\frac{5}{6}$	28	$25\frac{4}{7}$

(b) In each of the examples in No. 19(a), subtract the bottom number from the top.

Add and subtract the following as indicated. Reduce result if possible:

20. $7\frac{2}{5} + 4\frac{7}{8}$ **21.** $4\frac{5}{6} + 8\frac{4}{7}$ **22.** $6\frac{3}{4} + 5\frac{5}{9}$

23. $2\frac{7}{12} + 6\frac{5}{16}$ **24.** $6\frac{3}{8} + 8\frac{5}{12}$ **25.** $5\frac{7}{24} + 3\frac{13}{36}$

26. $\frac{7}{8} - \frac{5}{6}$ **27.** $\frac{5}{8} - \frac{3}{10}$ **28.** $\frac{11}{15} - \frac{7}{10}$

29. $12\frac{5}{16} - 8$ **30.** $9 - 5\frac{3}{8}$ **31.** $23 - 14\frac{7}{12}$

32. $38\frac{7}{10} - 26\frac{4}{15}$ **33.** $53\frac{5}{12} - 13\frac{8}{9}$ **34.** $91\frac{5}{24} - 48\frac{9}{16}$

35. $52\frac{3}{20} - 8\frac{11}{15}$ **36.** $11\frac{5}{13} - 10\frac{5}{13}$ **37.** $7\frac{4}{15} - 2\frac{8}{35}$

38. $5\frac{4}{9} + 2\frac{7}{12} - \frac{5}{8}$ **39.** $7\frac{4}{15} - 4\frac{5}{6} + 6\frac{7}{10}$ **40.** $3\frac{11}{15} + 6\frac{9}{10} - 5\frac{3}{8}$

41. $6\frac{4}{15} + 4\frac{2}{21} - 7\frac{3}{5}$ **42.** $8\frac{4}{45} - 3\frac{11}{30} - 2\frac{5}{36}$ **43.** $9\frac{8}{45} - 5\frac{7}{30} - 1\frac{5}{24}$

44. $9\frac{4}{15} - 1\frac{5}{12} - 2\frac{7}{8}$ **45.** $9\frac{4}{9} - 3\frac{8}{15} - 4\frac{11}{18}$ **46.** $8\frac{7}{24} - 2\frac{4}{15} - 3\frac{3}{20}$

2-8 MULTIPLICATION OF FRACTIONS

Probably everyone who reads this knows the rule for multiplying fractions.

Rule. *In multiplying fractions, multiply the numerators together for the numerator of the product and multiply the denominators together for the denominator of the product.*

This rule can be shown to hold for all multiplication involving fractions.

Example 1

Find $\frac{7}{8} \cdot \frac{4}{9}$.

Solution

If the numerators are multiplied together and the denominators are multiplied together, we find that a factor "4" can be divided into both terms of the answer. This "4" can be divided into a numerator and a denominator before the terms are multiplied; thus

$$\frac{7}{\underset{2}{\cancel{8}}} \cdot \frac{\cancel{4}}{9} = \frac{7}{18}$$

Example 2

Find $\frac{8}{15} \cdot \frac{5}{12} \cdot \frac{6}{7}$.

Solution

In this example three numbers can be divided into numerators and denominators; namely, 3, 4, and 5.

$$\frac{\overset{2}{\cancel{8}}}{\underset{3}{\cancel{15}}} \cdot \frac{\overset{2}{\cancel{6}}}{7} \cdot \frac{\cancel{5}}{\underset{3}{\cancel{12}}} = \frac{4}{21}$$

Example 3

Find $(3\frac{4}{7}) \cdot (6\frac{3}{10})$.

Solution

The best way to multiply mixed numbers is to change the mixed numbers to improper fractions.

$$\left(3\tfrac{4}{7}\right)\cdot\left(6\tfrac{3}{10}\right) = \frac{\overset{5}{\cancel{25}}}{\cancel{7}}\cdot\frac{\overset{9}{\cancel{63}}}{\underset{2}{\cancel{10}}} = \frac{45}{2} = 22\tfrac{1}{2}$$

Example 4

Find $\left(\frac{5}{8}\right)\cdot\left(496\frac{3}{7}\right)$.

Solution

Changing a mixed number to an improper fraction sometimes leads to a large improper fraction. In some examples the multiplication may instead be done in parts, although this method is probably more susceptible to error. First we arrange the numbers to be multiplied in column form. We call the $\frac{5}{8}$ the multiplier and the $496\frac{3}{7}$ the multiplicand.

	$496\frac{3}{7}$
	$\frac{5}{8}$
Multiply $\frac{3}{7}$ by $\frac{5}{8}$	$\frac{15}{56}$
Multiply 496 by 5	2480
Divide by 8	310
Add the fraction to 310	$310\frac{15}{56}$

2-9 RECIPROCAL OF A NUMBER

Before beginning the division of fractions, we need to understand what is meant by the reciprocal of a number. The *reciprocal* of any number is defined as the result of 1 divided by the number. The reciprocal of 5 is $\frac{1}{5}$. The reciprocal of 7 is $\frac{1}{7}$. The reciprocal of 1 is $\frac{1}{1}$, or 1. The reciprocal of 487 is $\frac{1}{487}$. The reciprocal of the fraction $\frac{3}{5}$ is

$$1 \div \frac{3}{5} \quad \text{or} \quad \frac{1}{\frac{3}{5}}$$

Let us determine the result of this division. By using the fundamental principle of fractions, we can change the expression $\frac{1}{\frac{3}{5}}$ into a simpler form. As the fraction is written, the numerator is 1. The denominator is $\frac{3}{5}$. Now we multiply both numerator and denominator by 5.

$$\frac{(5)(1)}{(5)\left(\frac{3}{5}\right)} = \frac{5}{3}$$

Therefore, the reciprocal of $\frac{3}{5}$ is $\frac{5}{3}$. We have the following rule:

Rule. *The reciprocal of a fraction can be obtained simply by inverting the fraction.* Thus the reciprocal of $\frac{4}{7}$ is $\frac{7}{4}$.

By definition, the product of any number and its reciprocal is equal to 1.

2-10 DIVISION INVOLVING FRACTIONS

Perhaps most people who read this know the following rule:

Rule. *In division involving fractions, multiply the dividend by the reciprocal of the divisor. In other words, invert the divisor and then multiply.*

Example 1

$$\frac{4}{7} \div \frac{2}{3} = \frac{4}{7} \cdot \frac{3}{2} = \frac{6}{7}.$$

Let us see why this rule is true. The rule depends on two facts: (1) the product of any number and its reciprocal is 1; and (2) any number divided by 1 is equal to the number itself.

To see why the rule is true, let us write the foregoing problem with the dividend $\frac{4}{7}$ as the numerator and the divisor $\frac{2}{3}$ as the denominator of a new fraction. The result is called a complex fraction:

$$\frac{\frac{4}{7}}{\frac{2}{3}}$$

Now let us multiply the numerator and the denominator of this complex fraction by $\frac{3}{2}$, the reciprocal of the denominator $\frac{2}{3}$. We get

$$\frac{\frac{4}{7} \cdot \frac{3}{2}}{\frac{2}{3} \cdot \frac{3}{2}} = \frac{\frac{6}{7}}{1} = \frac{6}{7}$$

The following examples show the application of this rule:

$$\text{(a) } 5 \div \tfrac{3}{4} = 5 \cdot \tfrac{4}{3} = \tfrac{20}{3} = 6\tfrac{2}{3} \qquad \text{(b) } \tfrac{5}{8} \div 3 = \tfrac{5}{8} \cdot \tfrac{1}{3} = \tfrac{5}{24}$$

$$\text{(c) } 4\tfrac{2}{3} \div 2\tfrac{3}{5} = \tfrac{14}{3} \div \tfrac{13}{5} = \tfrac{14}{3} \cdot \tfrac{5}{13} = \tfrac{70}{39} = 1\tfrac{31}{39}$$

The rules for the multiplication and division of fractions make it possible to simplify the work in many problems involving several multiplications and divisions. For example, in some problems we find it necessary to multiply two or more numbers together and then to divide the product by one or more other numbers. The result may be a problem such as the following:

$$\frac{48 \times 35 \times 36}{63 \times 50 \times 32}$$

If we perform the multiplications as indicated, we first get the result

$$\frac{60480}{100800}$$

This fraction can be reduced by dividing numerator and denominator by any factor common to both terms.

The work can be greatly simplified by first dividing the numerators and the denominators of the original form by common factors before multiplying. We see at once that we can divide the following factors into both numerator and denominator: 16, 7, 9, 5, 2, 2.

$$\frac{\overset{3}{\cancel{48}} \times \overset{\cancel{5}}{\cancel{35}} \times \overset{\overset{\cancel{2}}{\cancel{4}}}{\cancel{36}}}{\underset{\cancel{9}}{\cancel{63}} \times \underset{\underset{5}{\cancel{10}}}{\cancel{50}} \times \underset{\cancel{2}}{\cancel{32}}} = \frac{3}{5}$$

The student is warned: This method of simplifying an expression cannot be used when additions or subtractions appear in numerator or denominator or both.

In the following example, the numerator and the denominator of the expression must be expanded as indicated.

$$\frac{6 \times 8 \times 5 + 1}{12 \times 4 \times 10} = \frac{241}{480}$$

A problem can often be set up in a form suitable for reducing. All numbers that are to be multiplied together are placed as factors in the numerator. All numbers to be used as divisors can be written in the denominator separated by the *multiplication* sign.

Example 2

A man drives 480 miles in 12 hr. What is his average speed in feet per second?

Solution

A distance of 480 miles is equal to (480)(5280) ft. The total number of feet is not expanded but is left in factored form. The number of feet is now divided by the number of seconds in 12 hr: 12 hr are equal to (12)(60)(60) sec. Now we set up the problem.

$$\frac{480 \times 5280}{12 \times 60 \times 60} = 58\tfrac{2}{3} \text{ ft per sec}$$

The answer is obtained easily by dividing numerator and denominator by any numbers contained in both.

Exercise 2-5

Multiply or divide as indicated:

1. $\frac{2}{5} \times \frac{4}{7}$	**2.** $(9\frac{3}{7})(2\frac{1}{2})$	**3.** $(2\frac{6}{7})(\frac{3}{5})$	**4.** $(\frac{12}{35})(\frac{24}{37})$
5. $(18)(\frac{4}{9})$	**6.** $(6\frac{3}{4})(70)$	**7.** $(9\frac{3}{4})(6\frac{1}{2})$	**8.** $(17\frac{3}{5})(2\frac{5}{8})$
9. $(33)(\frac{4}{15})$	**10.** $(4\frac{5}{8})(36)$	**11.** $(26\frac{4}{7})(1\frac{3}{4})$	**12.** $(5\frac{5}{9})(514)$
13. $(15\frac{6}{7})(5\frac{4}{9})$	**14.** $(\frac{3}{8})(4\frac{9}{20})$	**15.** $(12\frac{16}{21})(4\frac{7}{8})$	**16.** $(24)(5\frac{3}{16})$
17. $\frac{3}{4} \div \frac{5}{8}$	**18.** $\frac{4}{9} \div 2\frac{2}{3}$	**19.** $\frac{7}{8} \div \frac{3}{4}$	**20.** $\frac{16}{25} \div 1\frac{4}{5}$
21. $9\frac{2}{7} \div 1\frac{3}{7}$	**22.** $2\frac{5}{8} \div \frac{3}{10}$	**23.** $2\frac{4}{15} \div \frac{8}{11}$	**24.** $72 \div 1\frac{3}{7}$
25. $2\frac{8}{21} \div 4$	**26.** $15 \div \frac{9}{10}$	**27.** $27 \div \frac{18}{85}$	**28.** $5\frac{2}{5} \div 16$
29. $13\frac{1}{15} \div 1\frac{8}{25}$	**30.** $3\frac{13}{24} \div 7\frac{3}{16}$	**31.** $\frac{28}{43} \div 6$	**32.** $11\frac{2}{3} \div 15$

Simplify

33. $(15\frac{2}{3})(8\frac{5}{16})(13\frac{4}{15})$	**34.** $(6\frac{4}{5})(8\frac{3}{4})(10\frac{4}{15})$	**35.** $(4\frac{8}{25})(3\frac{10}{21})(5\frac{7}{16})$
36. $(3\frac{48}{91})(5\frac{7}{9})(4\frac{27}{32})$	**37.** $(5\frac{35}{81})(4\frac{2}{7})(2\frac{4}{15})$	**38.** $(8\frac{18}{25})(6\frac{3}{16})(7\frac{20}{27})$
39. $(36\frac{3}{5})(5\frac{10}{27})(9\frac{3}{8})$	**40.** $(4\frac{5}{18})(3\frac{3}{14})(6)$	**41.** $(3)(\frac{5}{20}) \div (11\frac{2}{3})$
42. $\dfrac{28 \times 75 \times 66}{125 \times 44 \times 27}$	**43.** $\dfrac{54 \times 55 \times 98}{35 \times 18 \times 88}$	**44.** $\dfrac{18 \times 30 \times 77}{22 \times 42 \times 45}$
45. $\dfrac{12 \times 60 + 1}{48 \times 30 - 2}$	**46.** $\dfrac{2 + 18 \times 45}{5 + 27 \times 50}$	**47.** $\dfrac{40 \times 50 \times 60}{40 + 50 + 60}$

48. Five pieces, measuring, respectively, $3\frac{3}{8}$, $5\frac{1}{4}$, $4\frac{5}{16}$, $2\frac{9}{32}$, and $3\frac{15}{16}$ in., are cut from a strip of brass 2 ft, 6 in. long. How much of the strip remains if the amount of waste is $\frac{1}{32}$ in. per cut?

49. A strip of aluminum is cut into seven pieces measuring, respectively, $2\frac{3}{8}$, $3\frac{1}{16}$, $3\frac{5}{8}$, $4\frac{5}{64}$, $2\frac{17}{64}$, $3\frac{3}{4}$, and $2\frac{1}{2}$ in. If the waste for each cut is $\frac{1}{32}$ in., find the length of the original strip.

50. How many pieces, each $3\frac{5}{16}$ in. long, can be cut from a piece of metal 28 in. long if $\frac{1}{16}$ in. is allowed for each cut as waste?

51. Twelve pieces, each $1\frac{9}{16}$ in. long are cut from a strip of copper 24 in. long. How much of the strip remains if $\frac{1}{32}$ in. is allowed per cut for waste?

52. Two holes are to be drilled in a metal plate so that the distance between the holes is $1\frac{7}{8}$ in. The diameters of the holes are $\frac{3}{8}$ and $\frac{7}{32}$ in., respectively. Find the distance between the centers.

53. A rectangle is $5\frac{7}{8}$ in. long and $3\frac{5}{16}$ in. wide. Find the distance around the rectangle.

54. If a man works at a job for 10 weeks, $5\frac{1}{2}$ days per week, and $7\frac{3}{4}$ hr per day, how many hours does he work at the job?

55. The gasoline tank on a car holds $12\frac{3}{4}$ gal. If the car travels at an average of $16\frac{5}{8}$ miles on 1 gal, how many times will the tank have to be filled to travel 1000 miles?

56. On a blueprint of a particular house $\frac{1}{2}$ in. represents 1 ft. Find the lengths on the blueprint that will represent the dimensions of the following rooms: $11\frac{1}{2} \times 13$ ft; $18\frac{1}{4} \times 14\frac{3}{8}$ ft; $12\frac{1}{3} \times 15\frac{1}{2}$ ft; and $16\frac{3}{4} \times 12\frac{5}{8}$ ft.

QUIZ ON CHAPTER 2, FRACTIONS. FORM A.

1. Reduce to lowest terms:

(a) $\frac{36}{84}$ **(b)** $\frac{108}{180}$ **(c)** $\frac{51}{68}$ **(d)** $\frac{63}{91}$

2. Combine each of these into a single fraction or mixed number:

(a) $\frac{7}{8} + \frac{3}{4} - \frac{5}{12} =$ **(b)** $5\frac{4}{9} + 2\frac{5}{6} - 3\frac{7}{36} =$

3. Find the lowest common multiple of these numbers:

36; 40; 75; 100

4. Add the numbers in each set:

(a) $18\frac{3}{8}$, $13\frac{4}{5}$ **(b)** $17\frac{4}{9}$, $12\frac{5}{9}$ **(c)** $19\frac{5}{6}$, $14\frac{5}{6}$

5. In No. 4, subtract the bottom number from the top in each set.

6. Multiply as indicated in each set:

(a) $\frac{3}{8} \times \frac{5}{6} \times \frac{4}{7}$ **(b)** $3\frac{3}{4} \times 1\frac{2}{7} \times 4\frac{2}{3}$ **(c)** $270 \times \frac{5}{12} \times 18$

7. Divide as indicated in each set:

(a) $\frac{7}{16} \div \frac{3}{8}$ **(b)** $\frac{8}{15} \div \frac{12}{25}$ **(c)** $2\frac{4}{9} \div 2\frac{14}{15}$ **(d)** $2\frac{4}{7} \div 1\frac{4}{5}$

8. Simplify each of these expressions:

(a) $4\frac{2}{3} \times 2\frac{1}{4} \div 3\frac{1}{5}$ **(b)** $3\frac{3}{4} \div 4\frac{1}{6} \div 3\frac{3}{5}$

9. The weights of three chickens were, respectively: $2\frac{3}{4}$ lb, $3\frac{1}{2}$ lb, and $3\frac{1}{6}$ lb. Find the total weight.

10. The following lengths were cut from an iron rod 2 ft long: $5\frac{3}{4}$ in., $8\frac{1}{2}$ in., and $6\frac{5}{8}$ in. If the waste in cutting was $\frac{1}{16}$ in. for each cut, find the length of the remaining piece of the rod.

11. How many pieces each $3\frac{3}{4}$ in. can be cut from a 2-ft strip of copper and how much of the strip remains? Disregard waste in cutting.

12. A car traveled 275 miles in $6\frac{1}{4}$ hours. Find the average rate per hour.

CHAPTER THREE Decimal Fractions

3-1 DEFINITION

The decimal fraction was one of the major advances in mathematics. For a long time people had used the place value of the digits to indicate the number 10 and numbers larger than 10. For instance, when we write a number such as 4444, the 4 at the extreme right, as we have said, indicates units. The 4 at the left of units place indicates *tens;* that is, the second 4 from the right has a value equal to ten times as much as the first 4. The third 4 from the right has a value ten times as much as the second 4, and so on.

Placing a digit in the second, third, fourth, or fifth place from the right multiplies its value by 10, 100, 1,000, or 10,000. This place value of digits was early recognized as one of the most important ideas in arithmetic.

However, for a long time no one ever thought of placing digits at the *right* of units place. Then, less than 400 years ago, decimal fractions were invented. It was seen that a digit could be written one place to the right of units place with some mark, such as a period, between them. In this position a digit could represent one tenth as much as in units place. Thus a 4 written one place to the right of units place, as 0.4, means $\frac{4}{10}$.

The point separating the units place from the fractional part is called a *decimal point.* A fraction written with digits at the right of the decimal point is called a *decimal fraction.* If no decimal point is shown in a number, the number is understood to be an integer and the decimal point is understood to be at the right of the number. When we write the number 63, the decimal point is understood to be just at the right of the 3.

Notice the difference in value of each of the 1's in these numbers.

1000	one thousand
100	one hundred
10	one ten
1	one unit
0.1	one tenth
0.01	one one-hundredth
0.001	one one-thousandth

For each place a digit is moved to the right, the digit represents another division by 10, or one tenth as much as in the preceding place.

Annexing zeros to the right of a decimal fraction does not change the value. These fractions all have the same value:

$$0.3 = \frac{3}{10}, \qquad 0.30 = \frac{30}{100}; \qquad 0.300 = \frac{300}{1000}$$

In the same way, annexing zeros to the left of a whole number does not change the value. The value of 934.5 is not changed by writing it 000934.5000.

A number consisting of a whole number and a decimal fraction is sometimes called a *mixed decimal.* Thus

the number 0.352 is a decimal fraction
the number 614.54 is a mixed decimal

The number of decimal places in a number means the number of digits to the right of the decimal point. The number 0.352 is said to have three decimal places. The number 0.0086 has four decimal places. The number 614.54 has two decimal places.

To summarize, whenever a digit is moved one place farther to the left, its value is multiplied by 10. Whenever a digit is moved one place farther to the right, its value is divided by 10. The number 3333.333 means $3000 + 300 + 30 + 3 + \frac{3}{10} + \frac{3}{100} + \frac{3}{1000}$. The number 72508.6439 means

7 ten-thousands
2 thousands
5 hundreds
0 tens
8 units
6 tenths
4 hundredths
3 thousandths
9 ten-thousandths

In reading a number the word "and" is used *only* at the decimal point. The foregoing number is read "seventy-two thousand five hundred eight *and* six thousand four hundred thirty-nine ten-thousandths."

3-2 CHANGING A COMMON FRACTION TO A DECIMAL

A common fraction such as $\frac{3}{4}$ can easily be changed to a decimal if we look upon a fraction as an indicated division. The fraction $\frac{3}{4}$ means $3 \div 4$. To find the decimal fraction equal to $\frac{3}{4}$, we perform the division $4\overline{)3}$. We place a decimal point after the 3 and add zeros at the right of the decimal point, as we divide. We place the decimal point for the answer directly above the decimal point in the dividend. After two divisions the remainder is zero.

$$\begin{array}{r} 0.75 \\ 4\overline{)3.0} \\ \underline{2\,8} \\ 20 \\ \underline{20} \\ 0 \end{array}$$

The answer can easily be checked by writing 0.75 as a common fraction, $\frac{75}{100}$, and then reducing the fraction. It becomes $\frac{3}{4}$.

Some common fractions do not come out even as a decimal fraction. For instance,

$$\frac{1}{3} = 0.3333\ldots \qquad \frac{3}{7} = 0.4285714285714\ldots$$

Whenever a common fraction is changed to a decimal fraction, the result is either an even decimal or repeating decimal, as shown in the following examples:

$$\frac{2}{5} = 0.4; \quad \frac{1}{4} = 0.25; \quad \frac{3}{8} = 0.375; \quad \frac{4}{25} = 0.16;$$

$$\frac{7}{16} = 0.4375; \quad \frac{2}{3} = 0.666666\ldots; \quad \frac{5}{11} = 0.45454545\ldots$$

By a repeating decimal, we mean a decimal in which the number after a certain point has the same set of consecutive digits. In the fraction $\frac{1}{3}$ the decimal will consist of a repetition of 3's. In the fraction $\frac{5}{11}$ the decimal will consist of a repetition of the two digits, 45.

Since a decimal fraction always indicates the denominators, 10, 100, 1000, etc., it follows that many common fractions cannot be stated as exact decimals. If a common fraction has a denominator that can be changed to 10, 100, 1000, etc., then the fraction can be stated as an exact decimal. For instance, $\frac{3}{5}$ can be changed to $\frac{6}{10}$ and can therefore be written 0.6; $\frac{1}{2}$ can be stated as $\frac{5}{10}$ and can be written 0.5. However, a fraction such as $\frac{4}{7}$ cannot be changed to tenths or hundredths, etc., and cannot be written as an exact decimal.

3-3 CHANGING A DECIMAL FRACTION TO A COMMON FRACTION

It is sometimes desirable to change a fraction from decimal form to common fraction form. This is easily done by writing the entire decimal as a common fraction and then reducing it to lowest terms.

Example 1

$$\text{(a) } 0.25 = \frac{25}{100} = \frac{1}{4} \qquad \text{(b) } 0.625 = \frac{625}{1000} = \frac{5}{8}$$

To change a mixed decimal to a mixed number, change only the fraction part.

Example 2

$$16.24 = 16\tfrac{24}{100} = 16\tfrac{6}{25}$$

Example 3

$$0.87\tfrac{1}{2} = 0.875 = \frac{875}{1000} = \frac{7}{8}$$

Example 4

Change to common fraction form: $0.16\frac{2}{3}$.

This can first be written as a common fraction: $\dfrac{16\frac{2}{3}}{100}$

Now we multiply numerator and denominator by 3: $\dfrac{3 \times 16\frac{2}{3}}{3 \times 100} = \dfrac{50}{300} = \dfrac{1}{6}$

Exercise 3-1

Reduce the following common fractions and mixed numbers to decimals:

1. $\frac{1}{2}$	**2.** $\frac{3}{4}$	**3.** $\frac{5}{8}$	**4.** $\frac{9}{16}$	**5.** $\frac{4}{11}$
6. $1\frac{1}{4}$	**7.** $3\frac{3}{16}$	**8.** $5\frac{3}{7}$	**9.** $6\frac{7}{8}$	**10.** $4\frac{5}{13}$
11. $8\frac{2}{5}$	**12.** $4\frac{1}{3}$	**13.** $7\frac{5}{6}$	**14.** $3\frac{7}{16}$	**15.** $1\frac{3}{32}$
16. $4\frac{7}{20}$	**17.** $9\frac{8}{9}$	**18.** $\frac{31}{15}$	**19.** $\frac{26}{3}$	**20.** $5\frac{3}{40}$
21. $9\frac{5}{32}$	**22.** $\frac{27}{8}$	**23.** $\frac{41}{32}$	**24.** $\frac{185}{9}$	**25.** $\frac{135}{16}$

Reduce the following decimal fractions to common fraction form:

26. 0.6	**27.** 0.25	**28.** 0.333	**29.** 0.16
30. 0.45	**31.** 1.2	**32.** 0.375	**33.** $0.33\frac{1}{3}$
34. 4.75	**35.** 8.05	**36.** 7.048	**37.** $3.02\frac{1}{2}$
38. 13.64	**39.** 15.075	**40.** 4.0625	**41.** 5.0375
42. 3.002	**43.** 8.005	**44.** $9.87\frac{1}{2}$	**45.** $0.300\frac{1}{2}$
46. $3.3\frac{1}{3}$	**47.** 11.08	**48.** $2.66\frac{2}{3}$	**49.** $0.00\frac{1}{7}$

50. Show that any common fraction can be expressed as repeating decimal.

3-4 ADDITION OF DECIMALS

In adding decimal fractions or mixed decimals, we place the numbers in column form with the decimal points in line.

Example

Add 25 + 48.7 + 5.724 + 864.1 + 0.0372.

```
 25
 48.7
  5.724
864.1
  0.0372
---------
943.5612; sum
```

3-5 SUBTRACTION OF DECIMALS

In subtracting decimal fractions, or mixed decimals, we place one number below the other with decimal points in line.

Example 1

Subtract 145.397 − 31.729.

```
145.397
 31.729
-------
113.668; remainder
```

Example 2

Subtract 56.4 − 13.6537.

```
56.4000
13.6537
-------
42.7463; remainder
```

Example 3

Subtract 2 − 0.53728.

```
2.00000
0.53728 \
-------  > add to check
1.46272 /
```

Remember that subtraction can be checked by addition.

Exercise 3-2

Perform the following additions and subtractions.

1. 567.43 + 79.96 + 869.58

2. 573.9 + 958 + 1957.2

3. 7.084 + 287.6 + 0.851

4. 495 + 98.3 + 0.04074

5. 678.4 + 4.78 + 80000

6. 9.3 + 0.187 + 384.7 + 0.0937

7. 3.7 + 58.53 + 28 + 98.357

8. 78.82 + 3.8 + 0.0367 + 690

9. 3.65 + 54.28 + 0.0538 + 701

10. 42.6 + 74 + 0.0036 + 2.36794

11. From 85.73 subtract 44.67

12. From 9827 subtract 427.9

13. From 3580 subtract 29.47

14. From 780.5 subtract 31.83

15. From 719.3 subtract 9.485

16. From 576.3 take 27.486

17. From 377.4 take 188.432

18. From 80.8 take 41.9032

19. From 3.96 take 2.83914

20. From 86.4 take 0.06081

21. From 3 subtract 0.25

22. From 7 subtract 1.4963

23. From 2 subtract 0.38658

24. From 5 subtract 0.89798

25. From 1 subtract 0.06027

26. Find 5 − 1.39782

27. Find 6 − 5.42697

28. Find 3 − 1.58314

29. Find 2 − 0.63825

30. Find 1 − 0.71345

31. A family goes on a tour and drives the indicated distance during each of the first four days: Monday, 234.5 miles; Tuesday, 315.2 miles; Wednesday, 342.6 miles; Thursday, 286.1 miles. If the trip is 1400 miles in all, how many miles must they drive the fifth day to complete the trip in five days?

32. The following lengths were cut from a strip of steel (in inches): 3.75, 4, 4.25, 4.5, and 4.75. If the original strip was 24 in. and 0.05 in. was wasted in cutting each piece, what was the length of the piece remaining?

3-6 MULTIPLICATION OF DECIMAL FRACTIONS OR MIXED DECIMALS

In multiplying two numbers involving decimals (1) we place one above the other just as with whole numbers; (2) we perform the actual computation without regard to decimal points, just as though the numbers were whole numbers; (3) we place the decimal point in the answer. To place the decimal point correctly, we point off as many places from the right as there are decimal places in the multiplier and multiplicand combined.

Example 1

Multiply (2.95)(3.1).

```
  2.9 5
    3.1
 ------
  2 9 5
8 8 5
 ------
9 1 4 5
```

The decimal point is now placed in proper position by counting three places from the right. The product is 9.145.

To see why the decimal point is determined by this rule, suppose we write the numbers as improper common fractions. Then (2.95)(3.1) is the same as

$$\frac{295}{100} \times \frac{31}{10} = \frac{9145}{1000} = 9.145$$

Notice that the number 2.95 indicates a denominator of 100 and the number 3.1 indicates a denominator of 10. When the denominators are multiplied together, the denominator of the product is 100×10, or 1000. The product of the numerators is then divided by 1000, which means that we point off three decimal places.

In some cases it is necessary to add zeros at the left of the answer in multiplication.

Example 2

Multiply (0.034)(0.02).

```
0.034
 0.02
-----
   68
```

In order to get the necessary decimal places in the product (that is, *five*), we must annex three zeros at the left of 68. The product is 0.00068.

3-7 ROUNDING OFF NUMBERS

When we come to division involving decimal fractions, we are first faced with the problem of rounding off numbers. For instance, when we change the common fraction $\frac{3}{7}$ to a decimal fraction, we get the following:

```
   0.428571428571428571 . . .
7)3.0000000000
```

The answer is an unending, repeating decimal. We are then faced with the problem of when to stop. Usually we stop the division at some point so that the number we have is convenient to use. In a problem such as this we would probably drop all the digits after the first four or five. The numbers we keep are called *significant* digits.

It is important to understand exactly what is meant by a *significant* digit. In any given number the first significant digit is the first digit other than zero starting at the left of the number. In a decimal fraction, such as 0.00062049, the first significant digit is 6. The first three zeros at the left are not significant digits. The decimal 0.00062049 has five significant digits. The 0 between the 2 and the 4 is a significant digit. In a whole number, such as 93,407,000, the zeros at the right do not count as significant digits unless definitely stated as such. The number 93,407,000 has five significant digits.

In the division of reducing $\frac{3}{7}$ to a decimal, as shown above, when some digits at the right are dropped, we say the number is "rounded off." In the number 0.428571428571 . . . , if we keep the first five digits, we have 0.42857. In some cases the last digit kept is increased by 1, and in other cases it remains the same.

Suppose we say the distance from the earth to the sun at a particular instant is 93,284,716 miles. In most cases we are not interested in such a high degree of accuracy. Usually, we would say the distance is aproximately 93,000,000 miles. In other words, we round off the number 93,284,716 and use only the two important (or significant) digits, 93. We sometimes say a number such as 93,000,000 is stated in round numbers. The term "rounding off" comes from the words "round numbers," which refer to zeros. When we round off any number, we imply that the number we are using is only approximate, yet accurate enough for practical purposes.

In order to understand clearly the method of rounding off numbers, let us take the number 647.2. Suppose we wish to round off this number so that we shall have only three significant digits. In this case we drop the 2 and retain the number 647. What we mean is that 647 is approximately equal to 647.2.

However, if we wish to round off the number 647.9 to three significant digits, we can say that the number 648 is approximately equal to 647.9. In this case the 7 is increased to 8 because the digit we drop (that is, the 9) is more than 5.

For rounding off numbers, we have the following rules:

1. If the first digit dropped is less than 5, then the last digit kept is left as it is. Then the number is rounded *downward.*

2. If the first digit dropped is larger than 5, then the last digit kept is increased by 1. Then the number is rounded *upward.*

3. If the digit dropped is exactly 5, followed only by zeros, then we need a special rule. When the 5 is dropped, the last digit kept is not changed if it is *even*. If the last digit kept is *odd*, it is changed to the next higher *even* digit. That is, the last digit kept is always made an *even* digit. Remember that this special rule applies only when the digit dropped is 5 followed only by zeros.

As a result of this last rule, if we round off many numbers ending in 5, the net effect will be that we shall round off such numbers *upward* about as many times as we round them *downward*.

Numbers are rounded off not only in division but also in multiplication and, in fact, in any computation when we wish to retain only a certain number of significant digits.

The following examples show the application of the special rule for rounding off numbers ending in 5 followed by only zeros:

To four significant digits, the number 43275 becomes 43280.
To four significant digits, the number 98365 becomes 98360.
To three significant digits, the number 248503 becomes 249000.
To three significant digits, the number 627450 becomes 627000.

Exercise 3-3

Round off each of the numbers listed below to five, four, three, and two significant digits.

1. 473529	**2.** 29.6725	**3.** 5081362	**4.** 183.935
5. 852.9217	**6.** 357261.4	**7.** 49.25004	**8.** 0.0728476
9. 0.1739651	**10.** 9172.653	**11.** 488535	**12.** 4265312
13. 9270403	**14.** 0.2610851	**15.** 3.258135	**16.** 0.000429965
17. 0.00584036	**18.** 0.860106	**19.** 57.0046	**20.** 0.03074545

3-8 DIVISION INVOLVING DECIMAL FRACTIONS

In the division of decimal fractions, most people use the following procedure.

Example 1

Divide

$$0.23\overline{)97.635}$$

First we mentally move the decimal point in the dividend and divisor as many places to the right as there are decimal places in the divisor. This makes the divisor a whole number. Then the decimal point is placed in the answer directly above the new position of the decimal point in the dividend. This should be done before division is

```
          4
0.23)97.635
```

started. In order to be sure of the proper place for the decimal point in the answer, we place each number in the quotient directly above the number used in the dividend for each division. For instance, 23 is contained in 97 only four times. The 4 is written above the 7. Then the division is performed just as with whole numbers.

Example 2

Divide 45.837 by 3.12 and carry the answer to four significant digits.

```
         14.691
3.12)45.837
       312
       1463
       1248
        2157
        1872
         2850
         2808
           420
```

The answer, rounded off to four digits, is 14.69.

Exercise 3-4

Perform the following multiplications:

1. 91.7 × 4.8 **2.** 63.9 × 0.58 **3.** 0.948 × 8.6

4. 8.56 × 3.8 **5.** 7.85 × 0.75 **6.** 9.28 × 59

7. 7.36 × 0.089 **8.** 84.9 × 0.58 **9.** 0.748 × 3.9

10. 0.0863 × 0.074 **11.** 0.0293 × 0.054 **12.** 0.00628 × 0.39

13. 98.5 × 0.0087 **14.** 0.0279 × 0.067 **15.** 8.963 × 3.21

16. 0.09378 × 0.516 **17.** 697.82 × 0.408 **18.** 0.75806 × 0.083

19. 0.05796 × 1.007 **20.** 1087.4 × 0.816 **21.** 0.8507 × 0.0513

22. 0.00126 × 1.034 **23.** 4965.2 × 0.204 **24.** 0.00769 × 3500

Perform the following divisions:

25. 2.1044 ÷ 26 **26.** 3621.5 ÷ 54 **27.** 0.11862 ÷ 46

28. 141.38 ÷ 3.9 **29.** 68.83 ÷ 0.019 **30.** 234.91 ÷ 0.28

31. $0.03301 \div 3.9$ **32.** $6.1392 \div 0.76$ **33.** $48613 \div 0.065$

34. $0.47602 \div 0.492$ **35.** $0.07814 \div 0.057$ **36.** $118.6 \div 0.0019$

37. $55.413 \div 1.87$ **38.** $36435 \div 38.2$ **39.** $458.21 \div 0.673$

40. $4394.2 \div 0.048$ **41.** $0.00402 \div 58.1$ **42.** $0.02002 \div 0.037$

3-9 A FEW SHORTCUTS

There are many so-called *short cuts* for fast computation in arithmetic. The following five rules cover a few of the most useful in multiplication and division.

Rule 1. *To multiply any number by 10, 100, 1000, etc., move the decimal point toward the right as many places as there are zeros in the multiplier.*

Examples

$$10 \times 48.63 = 486.3$$
$$100 \times 48.63 = 4863$$
$$1000 \times 48.63 = 48630$$
$$1{,}000{,}000 \times 0.0003724 = 372.4$$

Rule 2. *To divide any number by 10, 100, 1000, etc., move the decimal point toward the left as many places as there are zeros in the divisor.*

Examples

$$48.63 \div 10 = 4.863$$
$$48.63 \div 100 = 0.4863$$
$$48.63 \div 1000 = 0.04863$$
$$295.37 \div 1{,}000{,}000 = 0.00029537$$

Rule 3. *To multiply a number by 5, move the decimal point one place to the right and divide by 2.*

Examples

$$5 \times 8 = \frac{1}{2} \text{ of } 80 = 40$$

$$5 \times 367 = \frac{1}{2} \text{ of } 3670 = 1835$$

Rule 3 means that we actually multiply by 10 and then divide by 2, which is equivalent to multiplying by 5. The rule can be extended to multiplication by 50, 500, etc.

Rule 4. *To multiply a number by 25, move the decimal point two places to the right and divide by 4.*

Examples

$$25 \times 36.8 = \frac{1}{4} \text{ of } 3680 = 920$$

$$25 \times 893 = \frac{1}{4} \text{ of } 89300 = 22325$$

Rule 4 means that we actually multiply by 100 and then divide by 4, which is equivalent to multiplying by 25. The rule can be extended to multiplication by 250, 2500, etc.

Rule 5. *To multiply a number ending in $\frac{1}{2}$ by itself, multiply the whole number by one more than itself and then annex the fraction $\frac{1}{4}$.*

This is one of the most useful rules in mathematics for rapid multiplication. It is useful especially in engineering but also in other fields of mathematics, such as statistics.

Example 1

$$(6\tfrac{1}{2})(6\tfrac{1}{2}).$$

Take 6×7 and annex the fraction $\frac{1}{4}$. *Answer:* $42\frac{1}{4}$.

Example 2

$$(19\tfrac{1}{2})(19\tfrac{1}{2}).$$

Take 19×20 and annex the fraction $\frac{1}{4}$. *Answer:* $380\frac{1}{4}$.

Rule 5 can also be used when $\frac{1}{2}$ is written as a decimal, 0.5.

Example 3

$$(8.5)(8.5)$$

Take 8×9 and annex the fraction 0.25. *Answer:* 72.25.

Example 4

$$(24.5)(24.5)$$

Take 24×25 and annex the fraction 0.25. *Answer:* 600.25.

Rule 5 can also be used for any number ending in 5 even though it is not a decimal.

Example 5

(245)(245).
Take 24×25 and annex the two digits 25. *Answer:* 60,025.

Example 6

(125)(125).
Take 12×13 and annex 25. *Answer:* 15,625.

If a problem requires several multiplications and divisions involving decimal fractions, it can often be simplified in the same manner as with whole numbers. We have seen how the following example can be reduced:

$$\frac{24 \times 25 \times 49}{35 \times 28 \times 45}$$

In this example the numerator and the denominator can be divided successively by 3, 4, 5, 5, 7, and 7, and the fraction reduces to $\frac{2}{3}$.

If the numbers in the numerator and/or the denominator contain decimals, the computation can be simplified in the same way. However, in that case, it is best first to multiply the numerator and the denominator by 10, 100, 1000, or by some other number of 10's so that the decimals disappear. This is done simply by *moving the decimal point the same number of places in the numerator and the denominator.*

Suppose, in arranging the work in a problem, we get the following fractional expression:

$$\frac{0.6 \times 175 \times 4.8 \times 0.12}{1.25 \times 4 \times 3.6 \times 0.028}$$

The decimals can be eliminated by multiplying numerator and denominator by 10 a sufficient number of times. In this case we multiply numerator and denominator by 1000000, so that all decimals disappear. To multiply both terms by 1000000, we simply move the decimal point a total of six places in the numerator and a total of six places in the denominator. The problem then becomes:

$$\frac{6 \times 175 \times 48 \times 1200}{125 \times 4 \times 36 \times 28}$$

All the numbers are now integers, and the fraction can be reduced to 120.

Exercise 3-5

Multiply each of the following numbers by 10, by 100, and by 1000:

1. 48.6 **2.** 5.38 **3.** 0.0084 **4.** 0.0526

5. 62 **6.** 2700 **7.** 0.762 **8.** 0.00389

9. 2.4 **10.** 183 **11.** 0.00005 **12.** 0.00049

13–24. Divide each of the numbers in No. 1–12 by 10, by 100, and by 1000.

Multiply each of the following numbers by 5 and by 25:

25. 456 **26.** 23720 **27.** 827 **28.** 63.2

29. 34.78 **30.** 13.61 **31.** 0.796 **32.** 0.005201

Multiply by a short method:

33. $(7\frac{1}{2})(7\frac{1}{2})$ **34.** $(11\frac{1}{2})(11\frac{1}{2})$ **35.** (8.5)(8.5) **36.** (23.5)(23.5)

37. $(5\frac{1}{2})(5\frac{1}{2})$ **38.** (9.5)(9.5) **39.** (465)(465) **40.** (31.5)(31.5)

41. (625)(625) **42.** (750)(750) **43.** $(18\frac{1}{2})(18\frac{1}{2})$ **44.** (1335)(1335)

Simplify each of the following:

45. $\dfrac{240 \times 7.2 \times 0.36 \times 8.4}{4.8 \times 0.15 \times 168.96}$ **46.** $\dfrac{42 \times 62.5 \times 2.43 \times 0.7854}{2.64 \times 0.175 \times 270 \times 432}$

47. $\dfrac{3.1416 \times 6.45 \times 62.5 \times 8.8}{2.2 \times 1728 \times 31.5 \times 2.54}$ **48.** $\dfrac{6.28 \times 377 \times 60 \times 0.0128}{0.7854 \times 39.37 \times 62.4 \times 60}$

Exercise 3-6

Assume that measurements are approximate in all cases.

1. On an automobile trip a stop was made at each of the following towns, *A*, *B*, *C*, *D*, and *E*. The odometer showed the following readings:

at beginning of trip:	34,782.7	at town *C*: 34,884.5
at town *A*:	34,817.3	at town *D*: 34,942.8
at town *B*:	34,858.6	at town *E*: 35,006.1

Find each of the distances between the stops. What was the length of the entire trip? If the entire trip took 14.3 gal of gasoline, what was the average mileage per gallon? Find the cost of the gasoline used at 69.6 cents per gal.

2. A car travels the following distances on four successive days: Monday, 285.3 miles; Tuesday, 318.4 miles; Wednesday, 364.8 miles; and Thursday, 197.2 miles. Find the total number of miles traveled on the four days. What was the average number of miles per day? On the entire trip 76.2 gal of gasoline were used. What was the average mileage per gallon?
   Find the total cost of gasoline for the trip at 71.9 cents per gal.

3. The sides of a triangle are 7.3, 8.25, and 9.63 in., respectively. Find the perimeter of the triangle. (The perimeter is the distance around a figure.)

4. Find the area and perimeter of a rectangle 2.43 in. long and 1.82 in. wide.

5. A rectangular room is 21.35 ft long and 14.8 ft wide. What is the length of the molding required to reach around the room? How many square feet are there in the floor?

6. A square is 31.5 in. on a side. Find its perimeter and area.

7. Find the area and the perimeter of a room 18.8 ft long and 13.5 ft wide.

8. What must be the length of a rectangle 2.71 in. wide if it is to contain approximately 12 sq in?

9. Two holes, one of them $\frac{7}{16}$ in. in diameter and the other $\frac{11}{32}$ in. in diameter, are to be drilled in a metal plate so that the distance between the holes is $4\frac{1}{4}$ in. Find the distance between the centers.

10. What is the total thickness of a pile of 12 metal sheets of iron, if each sheet is 0.045 in. thick?

11. A pile of 15 sheets of metal has a total thickness of 2.14 in. What is the approximate thickness of each sheet?

12. A copper wire has a diameter of 0.032 in. How many turns of the wire can be wound on a coil that is 3 in. long?

13. A piece of aluminum 4.625 in. long is cut from a strip 20 in. long. If the cut itself wastes 0.04 in., how long is the remaining strip?

14. Five bars of iron weigh, respectively, $7\frac{3}{4}$, $5\frac{1}{8}$, $4\frac{9}{16}$, $6\frac{1}{2}$, and $5\frac{1}{5}$ lb. Change weights to decimal fractions, and then find the total weight and the average weight. Round off final answer to three significant digits.

15. Six strips of brass measuring, respectively, 3.275, 4.35, 2.625, 0.875, 5.125, and 4.6 in. are cut from a strip 32 in. long. How long a strip is left if 0.045 in. is wasted in each cut?

16. A strip of metal is cut into five strips of the following lengths, respectively: 4.25, 3.95, 7.3, 6.485, and 5.74 in. The waste for each cut is 0.035 in. How long was the original piece?

17. Five pieces, each measuring 1.375 in. long, are cut from a strip of silver alloy 12 in. long. If each cut wastes 0.018 in., what is the length of the remaining piece?

18. How many pieces, each 3.125 in. long, can be cut from a strip of brass if the waste for each cut is $\frac{1}{16}$ in. and the strip is 25 in. long?

19. A bar of copper is cut into 8 pieces, each $2\frac{7}{16}$ in. long. If each cut wastes $\frac{1}{32}$ in., find the length of the original bar.

20. Six pieces measuring, respectively, $2\frac{7}{16}$, $5\frac{1}{4}$, $6\frac{1}{2}$, $7\frac{5}{8}$, $3\frac{1}{8}$, and $3\frac{5}{16}$ in. are cut from a piece of aluminium, 30 in. long. If the waste is $\frac{1}{16}$ in. per cut, find the length of the remaining strip. Work first by common fractions and then by decimal fractions.

21. How many pieces, each 2.225 in. long, can be cut from a strip of brass 1 yd long if the waste is 0.04 in. per cut?

22. Most substances expand when heated. The rate of change per degree of change in temperature (Celsius scale) is called the *coefficient of expansion.* The coefficient of expansion of steel is about 0.000011. This means that steel expands 0.000011 of its length for each degree (C°) rise in temperature. If a bar of steel measures 56.8 ft at 10° C, find its length at 35° C. State the increase in inches.

23. The coefficient of expansion of aluminum is about 0.000023. If a bar of aluminium measures 113.6 in. at 20° C, find its length at 35° C.

24. The coefficient of expansion of ordinary glass is about 0.0000083, and of Pyrex Ⓣ it is about 0.0000032. If a rod of each kind of glass is measured as 24.0 in. at a temperature of 20°, find the length of each at 100° C. Find the difference in the expansion of each. What is the significance in the use of each in dishes for cooking and baking?

25. The resistance in an electrical circuit usually increases with an increase in temperature. The rate of increase in resistance per degree of change in temperature is called the *temperature coefficient of resistance.* For copper this coefficient is about 0.004. If a copper wire of a certain size and length has a measured resistance of 48.0 ohms at 20° C, find the resistance at 35° C.

26. A certain resistor has a measured resistance of 150 ohms at 20° C.

At 25° C it is 152.4 ohms. Find the temperature coefficient of resistance.

27. The resistance of the alloy *constantan* is about 27.5 times the resistance of copper. If the resistance of a particular length of copper wire is 11.2 ohms, find the resistance of a similar wire of the material *constantan.*

28. The temperature coefficient of constantan is about 0.00001. If a copper wire of a particular length and size has a resistance of 4.80 ohms at 20° C, find the resistance of a similar wire of constantan at the same temperature. Find the resistance of each at 80° C, and find the difference in the increases in resistance. What is the advantage of using an alloy like constantan in electrical instruments?

QUIZ ON CHAPTER 3, DECIMAL FRACTIONS. FORM A.

1. Change these decimal fractions to common fraction form:

0.36; 0.85; 0.42; 0.275; 1.45; 0.008

2. Change these common fractions to decimal form:

$\frac{3}{8}$; $\frac{9}{16}$; $\frac{7}{12}$; $\frac{15}{32}$; $3\frac{7}{8}$; $4\frac{3}{4}$; $2\frac{5}{16}$; $\frac{7}{200}$

3. Add the following numbers: 53.8 + 147.36 + 4.953 + 0.079 + 367 + 0.006

4. Subtract as indicated:

(a) 73.5 − 16.72; **(b)** 8 − 2.5714;
(c) 4 − 0.0081; **(d)** 9.1 − 8.007

5. Multiply as indicated:

(a) (62.8)(0.39); **(b)** (84.3)(1.406); **(c)** (0.287)(0.032)

6. Divide as indicated:

(a) 405 ÷ 5.6 **(b)** 342.9 ÷ 4.5
(c) 474 ÷ 0.075 **(d)** 0.00481 ÷ 7.4

7. Twelve sheets of metal of uniform thickness have a total thickness of 16.2 in. What is the thickness of each sheet?

8. An automobile trip of 357.2 miles required 23.5 gal of fuel. What was the average number of miles per gallon?

9. A trip of 316.1 miles was made in 7 hours 15 minutes. What was the average number of miles per hour?

10. How many pieces each 6.5 in. long can be cut from a strip of copper 3 ft long if the waste for each cut is 0.04 in.? How much remains?

11. A strip of brass is cut into 9 pieces, each 2.6 in. long. If the waste per cut was 0.05 in., what was the length of the original strip?

CHAPTER FOUR Percentage

4-1 DEFINITION

The expression *percent* is so common in everyday speech that probably most adults know what is meant by "one hundred percent," "50 percent," or some other similar expression. The statement, "You are one hundred percent right" means "You are completely right." When we say, "He lost 50 percent of his money," we mean he lost half of it.

The word "percent" means "hundredths" or "by the hundred." The expression 25 percent means "25 hundredths" or "25 out of every hundred." The symbol for percent is %. When we say 25%, we mean "25 out of every hundred," $\frac{25}{100}$ as a common fraction, or 0.25 as a decimal.

Note, especially in the example, 25% that the number 25 by itself is a *whole* number, or integer; 25 alone is *not* a fraction. The expression becomes a fraction when we attach the percent sign (%), as in 25%; when we place the denominator 100 below the 25: $\frac{25}{100}$; or when we place the decimal point before the 25: 0.25.

When we use the expression *percent*, such as 25%, we always mean a part of *some quantity*. The expression 25% is equivalent to the common fraction $\frac{1}{4}$. For instance, 25% of \$48 means $\frac{25}{100}$ of \$48, or $\frac{1}{4}$ of \$48, which is \$12.

4-2 RELATION BETWEEN PERCENTS AND FRACTIONS

In multiplication or division involving percents it is first necessary to change the percent to a decimal or common fraction. On the other hand, when we have found an answer in the form of a decimal, we often wish to state it as a percent. The following rules and examples show how these changes are made.

Rule 1. *To change a percent to a common fraction, omit the percent sign (%) and write* 100 *below the number of percent. Then reduce the fraction if possible.*

For instance, the expression 15% can be reduced as follows: $15\% = \frac{15}{100} = \frac{3}{20}$. The expression 20% is equal to the fraction $\frac{20}{100}$ or to 0.20. The value can be reduced to lower terms as in any fraction: $20\% = \frac{20}{100} = \frac{1}{5}$.

Examples

$$25\% = \frac{25}{100} = \frac{1}{4} \qquad 0.5\% = \frac{0.5}{100} = \frac{5}{1000} = \frac{1}{200}$$

$$63\% = \frac{63}{100} \qquad 125\% = \frac{125}{100} = 1\tfrac{1}{4}$$

$$5\% = \frac{5}{100} = \frac{1}{20} \qquad 200\% = \frac{200}{100} = 2$$

$$1\% = \frac{1}{100} \qquad 12\tfrac{1}{2}\% = \frac{12\frac{1}{2}}{100} = \frac{25}{200} = \frac{1}{8}$$

Rule 2. *To change a percent to a decimal, move the decimal point two places toward the left and then omit the percent sign.*

It is easy to see that this rule is reasonable if we first change the percent to a common fraction.

Examples

$$47\% = \frac{47}{100} = 0.47 \qquad 142\% = \frac{142}{100} = 1.42$$

$$8\% = \frac{8}{100} = 0.08 \qquad 3.25\% = \frac{3.25}{100} = \frac{325}{10000} = 0.0325$$

$$3.4\% = \frac{3.4}{100} = \frac{34}{1000} = 0.034 \qquad 0.02\% = \frac{0.02}{100} = \frac{2}{10000} = 0.0002$$

Rule 3. *To change a decimal fraction to percent, move the decimal point two places to the right and annex the percent sign* (%). A common fraction can be changed first to a decimal fraction and the result changed to percent.

Examples

$0.32 = 32\%$ $\qquad 0.0625 = 6.25\%$ $\qquad 7.5 = 750\%$

$0.415 = 41.5\%$ $\qquad 0.0001 = 0.01\%$ $\qquad 10 = 1000\%$

$0.007 = 0.7\%$ $\qquad 1.75 = 175\%$ $\qquad 0.14\tfrac{2}{7} = 14\tfrac{2}{7}\%$

$\frac{3}{4} = 0.75 = 75\%$ $\qquad \frac{5}{16} = 0.3125 = 31.25\%$ $\qquad \frac{1}{32} = 0.03125 = 3.125\%$

Exercise 4-1

Change the following percents to common fractions or mixed numbers and to decimal form:

1. 35% **2.** 48% **3.** $4\tfrac{3}{8}\%$ **4.** 7%

5. $\frac{1}{8}$%	**6.** 10%	**7.** $37\frac{1}{2}$%	**8.** $162\frac{3}{4}$%
9. $14\frac{2}{7}$%	**10.** 450%	**11.** 175%	**12.** 95%
13. 1000%	**14.** $3\frac{1}{5}$%	**15.** $33\frac{1}{3}$%	**16.** $\frac{1}{4}$%
17. $1\frac{7}{8}$%	**18.** $\frac{1}{16}$%	**19.** $1\frac{1}{3}$%	**20.** $5\frac{1}{4}$%

Change the following common fractions, mixed numbers, and decimals to percents:

21. $\frac{4}{5}$	**22.** $\frac{3}{16}$	**23.** $\frac{7}{64}$	**24.** $\frac{15}{32}$
25. $5\frac{5}{8}$	**26.** $1\frac{3}{25}$	**27.** 0.025	**28.** 0.0015
29. 1.453	**30.** 2.3	**31.** $\frac{2}{3}$	**32.** $\frac{4}{11}$
33. $3\frac{5}{9}$	**34.** 1.04	**35.** 5.1	**36.** $1\frac{35}{128}$
37. 0.0005	**38.** 0.075	**39.** 0.6	**40.** 2

4-3 RATE, BASE, PERCENTAGE

In a problem involving percent three quantities must always be considered. These three quantities are called the *rate*, the *base*, and the *percentage*, respectively. In order to see the relation between these three quantities, consider the example 25% of \$48 = \$12. In this statement, 25% is called the *rate*, \$48 is called the *base*, and \$12 is called the *percentage*.

In any percentage statement the rate is the indicated fractional part of a particular quantity. As such, the rate is equivalent to a common or decimal fraction. The base is the number of which a fractional part is taken. The base usually has a name or denomination of some kind, such as dollars, objects, or measurements, such as feet or pounds.

The foregoing example can be stated by using a decimal fraction in place of *percent* for the rate. Thus the statement 25% of \$48 = \$12 can be written

$$0.25 \times \$48 = \$12$$

From this example notice that the general percentage statement may be given as

$$\text{rate} \times \text{base} = \text{percentage}$$

Note especially that the *percentage* is a *product* and that the *rate* and the *base* are *factors* multiplied together. Whenever we know the two factors in any problem in multiplication, we multiply them together to form the product. Whenever we know the product and one of the factors, we can find the other factor by dividing the product by the known factor. For instance, if we know the two factors are 5 and 7, we can find the product by multiplying 5 and 7 together. The product is

35. If we know the product of two factors is 91 and one of the factors is 7, we find the other factor by dividing the product by the known factor. The other factor is 91 ÷ 7, which is 13.

4-4 FINDING THE PERCENTAGE WHEN THE RATE AND THE BASE ARE KNOWN

If the rate and the base are given, we have a problem in which the two factors are known and the product (the percentage) is to be found. If two factors are known, their product is found by multiplying the factors together. Therefore, we have the following rule:

Rule 4. *To find the percentage when the rate and the base are given, multiply the base by the rate. The rule may be stated as*

$$\textit{percentage} = \textit{rate} \times \textit{base}$$

The percentage has the same name or denomination as the base. We have said that 100% means $\frac{100}{100}$, or all of a particular quantity. If the rate is equal to 1 (that is, 100%), then the percentage is equal to the base. If the rate is less than 1 (that is, less than 100%), then the percentage is less than the base. If the rate is more than 1 (that is, more than 100%), then the percentage is more than the base. These three different conditions are illustrated in the following examples. Remember, rate × base = percentage.

Example 1

$$100\% \text{ of } 80 \text{ lb} = 1.00 \times 80 \text{ lb} = 80 \text{ lb}$$

Example 2

$$25\% \text{ of } 80 \text{ ft} = 0.25 \times 80 \text{ ft} = 20 \text{ ft}$$

Example 3

$$150\% \text{ of } \$80 = 1.5 \times \$80 = \$120$$

Exercise 4-2

Find the percentage in each of the following problems:

1. 24% of $3500 **2.** 12.5% of $720

3. 64.5% of 520 miles **4.** 87.5% of 10,560 ft

5. 40.4% of $385 **6.** 7.5% of $36.80

7. 3.4% of $16,500 **8.** 1.6% of $24,500

9. 0.5% of $15,840 **10.** 0.02% of $18,000

11. 0.005% of $7200

12. $\frac{1}{3}$% of 9600 lb

13. 120% of 800 lb

14. 135% of 460 bushels

15. 140% of $32,500

16. 106% of $26500

17. 200% of $600

18. 325% of $480

19. One year the price of a house was $48,500. The next year it was 115% as much as the price of the previous year. What was the new price?

20. A man bought a house for $56,800 and paid 12.5% as a down payment. How much was the down payment and how much was left to be paid?

21. A woman has $84,000 invested as follows: 17.5% in stocks, 24.5% in bonds, and the rest in savings and loan companies. How much has she invested in each type of investment?

22. A certain metal alloy contains the following: copper: 80%; tin: 10%; zinc: 2.5%. How much of each of these metals is contained in 2000 lb of the alloy?

23. One year a farmer raised 6500 bushels of corn. The next year his crop was 116% as much as the previous year. What was the crop the second year?

24. A family receiving a yearly income of $16,400 spent the following amounts:

Food: 21%   Housing: 16%;   Clothing: 9.5%
Utilities: 13%   Medical: 10.5%   Taxes: 11.5%

Insurance and savings: 14%; Miscellaneous: the remainder. Find the amount spent for each item.

4-5 FINDING THE RATE WHEN THE BASE AND THE PERCENTAGE ARE KNOWN

If we know the base and the percentage, we have a problem in which the product of two factors is known. The product is the percentage. One of the factors is the base. The part to be found is the other factor, which is the rate. We must then divide the product (the percentage) by the known factor (the base). Therefore, we have the following rule.

Rule 5. *To find the rate when the base and the percentage are given, divide the percentage by the base.* The rule may be stated as

$$\text{rate} = \text{percentage} \div \text{base}$$

As an example showing the use of this rule, suppose we have the following problem: a man earns $6000 a year and spends $1170 of it for food. What part or percent of his income does he spend for food?

In this problem the rate is unknown. We have given the percentage, $1170, which is the *product* of two factors. One of the factors is the base, $6000. Therefore, in this problem we divide the product, $1170, by the known factor, $6000.

$$\$1170 \div \$6000 = 0.195$$

The quotient, 0.195, is the other factor, or the rate. We can check the answer by the question: does $0.195 \times \$6000 = \$1170$? Since the problem calls for the answer in percent, we change the decimal 0.195 to 19.5%.

Exercise 4-3

Find the rate (in percent) in each of the following problems:

1. $1600 is___percent of $6000?
2. $123 is what percent of $8200?
3. 240 ft is what percent of 750 ft?
4. $52.50 is what percent of $420?
5. $1344 is what percent of $7680?
6. 162 lb is what percent of 3600 lb?
7. $14.20 is what percent of $648?
8. $5.40 is what percent of $450?
9. $42 is what percent of $7500?
10. 32 miles is what percent of 580 miles?
11. $480 is what percent of $320?
12. 60 ft is what percent of 48 ft?
13. $240 is what percent of $200?
14. 414 lb is what percent of 360 lb?
15. $84 is what percent of $35?
16. $171.40 is what percent of $130?
17. A family with a yearly income of $14,500 spent the following amounts for the items listed. What percent of their income was spent for each? (Round off answers to nearest tenth of 1 percent.)

Food: $3030 Housing: $2500
Clothing: $1480 Utilities: $2230
Taxes: $1580 Medical: $1540
Insurance and savings: $1840 Miscellaneous: $300

4-6 FINDING THE BASE WHEN THE RATE AND THE PERCENTAGE ARE KNOWN

A third type of problem involves finding the base when the rate and the percentage are given in the problem. Here, again, we have the product, as well as one of the factors. The other factor, the base, is to be found. The problem requires that we divide the known product (percentage) by the known factor (rate). Therefore, we have the following rule:

Rule 6. *To find the base when the rate and the percentage are given, divide the percentage by the rate.* The rule may be stated as

base = percentage ÷ rate

As an example showing the use of this rule, suppose we have the following problem: a family spends $2940 a year, or 18.6 percent of its yearly income, for rent. What is the family's yearly income?

In this problem, the base is unknown. In the regular percentage statement, we have

18.6% of income = $2940

Then

(0.186) × income = $2940

To find the unknown factor, the income, we divide the percentage, $2940, by the known factor, the rate, which is 0.186. We get

$2940 ÷ 0.186 = $15,806.45

The answer is their income, or the base.

In a problem of this kind, the answer is rounded off. The answer, therefore, is approximate. However, it should be remembered that it is not the income that is approximate. The income is an exact number of dollars and cents. The same is true with regard to the amount spent for rent. The part that is approximate is the rate, 0.186.

Exercise 4-4

Find the base in each of the following problems:

1. $15.85 is 25% of what?

2. $16.80 is 32% of what?

3. $35.36 is 52% of what?

4. $24.70 is 76% of what?

5. $124.20 is 45% of what?

6. $85.50 is 18% of what?

7. 42.5 lb is 12.5% of what?

8. $34.65 is 7.5% of what?

9. $133.70 is 17.5% of what?

10. $105.30 is 23.4% of what?

11. $10.44 is 2.4% of what?

12. 2.108 is 3.25% of what number?

Supply the correct number for each blank:

13. 135% of ____ = $1674

14. 32.5% of ____ = $53.30

15. 2.4% of ____ = $4.44

16. $8.40 = 1.25% of ____

17. A man bought a car for $3480, which was 21.3% of his yearly income. How much was his yearly income? (Round off answer to nearest dollar.)

In percentage problems, perhaps the greatest difficulty is in identifying correctly the *rate*, the *base*, and the *percentage*. Here are three statements that might be remembered as a help in working problems in percentage:

1. The percentage is a product of two factors. Therefore, whenever the percentage is given, the process will be division; that is, the percentage must be divided by the given base or the given rate.
2. The base and the percentage have the same denomination or name, such as dollars, feet, miles, pounds, objects, or quantities of any kind.
3. The rate has no name or denomination but is simply an indicated fractional part of some quantity. The rate is usually less than 100%, but it may be over 100%.

4-7 INCREASES AND DECREASES

In some problems a particular quantity is to be increased or decreased by a given percent of the quantity. In such problems, we first find the amount of increase or decrease. The original amount is the base. Therefore, we multiply the rate times the base to get the increase or decrease. Then we add or subtract the increase or decrease to find the new amount.

Example 1

One year the population of a town was 13,560. The next year it had increased by 7.5%. Find the population after the increase.

Solution

For the increase we have

$$0.075 \times 13{,}560 = 1017$$

Adding the increase,

$$13{,}560 + 1017 = 14{,}577; \text{ population after increase.}$$

Example 2

A man's suit was marked to sell for $124. At a sale it was sold at a discount of 22.5%. Find the selling price.

Solution

For the discount, we have

$$0.225 \times \$124 = \$27.90$$

Subtracting, we get

$$\$124 - \$27.90 = \$96.10; \text{ selling price}$$

In some problems, the original quantity is given together with the amount after an increase or decrease. Then our problem is to find the rate of increase or decrease. In such problems, the base is taken as the original number, that is, the first number in time.

Example 3

One year a house was priced at \$47,500. The following year the price was \$56,000. What was the rate of increase in the price?

Solution

We first find the increase:

$$\$56{,}000 - \$47{,}500 = \$8{,}500$$

The increase represents the percentage. To find the rate of increase, we use the first quantity as the base. For the rate of increase, we get

$$8500 \div 47{,}500 = 0.178947\ldots$$

We round off the answer to the nearest tenth of 1 percent and get

$$\text{rate of increase} = 17.9\%$$

Example 4

One year the population of a certain city was 87,200. The following year it was only 76,300. Find the rate of decrease during the year.

Solution

First we find the decrease:

$$87{,}200 - 76{,}300 = 10{,}900$$

To find the rate of decrease, we take the first number as the base

$$10{,}900 \div 87{,}200 = 0.125$$

The rate of decrease was therefore 12.5%.

In another type of problem, we are required to find the original amount after it has been increased or decreased by a given rate.

Example 5

After an increase of 20%, the price of an article was \$144. What was the original price?

Solution

In this problem, the base is the original price, which is unknown. Here we take the original price to be 100% of itself. Then, when 20% is added, the new price is 120% of the original price. That is,

$$\$144 = 120\% \qquad \text{of original price}$$

Now we know the rate and the percentage. Then we divide:

$$\$144 \div 1.20 = \$120, \qquad \text{original price}$$

To check the problem, we can take 20% of $120 and get $24, which is the increase. Adding the increase, we have

$$\$120 + \$24 = \$144$$

Example 6

After a discount of 22.5%, the price of an article was $207.70. What was the original price?

Solution

Again we take the original price as 100% of itself. Then, when 22.5% is subtracted, the new price is 77.5% of the original price. That is,

$$\$207.70 = 77.5\% \qquad \text{of original price}$$

In this problem, the percentage is 207.70; the rate is 77.5%. Dividing, we get

$$\$207.70 \div 0.775 = \$268.00, \qquad \text{original price}$$

As a check on the work, we can take

$$22.5\% \text{ of } \$268 = \$60.30, \text{ discount}$$

Subtracting, we get the new price:

$$\$268 - \$60.30 = \$207.70$$

Exercise 4-5

Find the correct number for the blank in each of the following statements:

1. 12.5% of $68 = ____

2. 67.5% of ____ = $205

3. 87.5% of 360 miles = ____

4. 135% of $24 = ____

5. ____% of $720 = $108

6. 351 cm is ____% of 5560 cm

7. 135% of ____ = $837

8. 32.5% of ____ = $53.3

9. $44 is ____% of $32

10. 0.015 in. is ____% of 0.24 in.

11. During one year a family received an income of $12,600 and spent the following portions for each of the items shown. Find the amount spent for each item.

Food	19%	Medical expenses	8.5%
Housing	18%	Taxes	9.7%
Utilities	15%	Insurance and savings	12%
Clothing	10.5%	Miscellaneous	7.3%

12. A family receiving an income of $9200 one year spent the amount indicated for each item shown. Find the rate of total income spent on each item.

Food	$1850	Medical expense	$790
Housing	1680	Taxes	920
Utilities	1360	Insurance and savings	870
Clothing	980	Miscellaneous	750

13. One year the population of a town was 36,800. The following year the population had increased 8.5%. What was the population after the increase?

14. One year the population of a town was 165,000. A year later it had decreased by 5.2%. What was the population after the decrease?

15. The population of a town was 48,000. A year later it was 50,640. What was the rate of increase?

16. During one year the population of a town decreased from 64,500 to 61,700. What was the rate of decrease?

17. Find the rate of increase or decrease:
(a) When population increases from 30,000 to 35,000.
(b) When population decreases from 35,000 to 30,000.

18. Find the gain or loss and the rate of gain or loss on the following:
(a) When an article is bought for $120 and sold for $140.
(b) When an article is bought for $140 and sold for $120.

19. Find the gain or loss and the selling price for each of the following:
(a) When an article is bought for $72 and sold at a gain of 25%.
(b) When an article is bought for $90 and sold at a loss of 20%.

20. A baseball player's batting average one year was 0.294. How many hits did he make if he had 483 official times at bat?

21. Which of the following players had a higher batting average: *A*, who had 148 hits in 512 times at bat; or B, who had 121 hits in 423 times at bat?

22. After a man's weight had increased by 8%, he weighed 167.4 lb. What was his weight before the increase?

23. A bar of iron expanded in length from 87.5 in. to 87.5028 in. What was the percent of expansion?

24. After an increase of 7.5% in his salary, a man received a salary of $9030 a year. What was his salary before the increase?

25. In an AC electric circuit the effective value of current or voltage is approximately 70.7% of the maximum or peak value. The effective value is that shown on an ammeter or voltmeter. If the maximum voltage in a circuit is 320 V, what will be the meter reading?

26. If the maximum current in an AC circuit is 420 milliamps, find the effective value.

27. If a meter shows a reading of 110 volts in an AC circuit, find the peak voltage.

28. If an ammeter shows 6.5 amperes in an AC circuit, find the maximum current.

29. The average value of current or voltage in an AC circuit is approximately 63.7% of the maximum value. Find the average voltage if the maximum is 165 volts.

30. If the average current in an AC circuit is 9 amperes, find the peak current.

31. A voltmeter shows a reading of 146 volts in an AC circuit. Find the maximum voltage, and then find the average voltage.

32. Sound has a velocity of approximately 1080 ft/sec at a temperature of 0° C. The velocity increases about 2 ft/sec for each rise of 1° C. Find the velocity when the temperature is 25° C; and find the percent of increase.

33. A certain resistor is rated at 3500 ohms, with a tolerance or possible error of ±8%. What are the top and bottom limits of its resistance?

34. The efficiency of a motor is given by the ratio between the power output and the power input. That is, e = (power out)/(power in). If a motor produces an output of 360 watts of power and requires an input of 500 watts, what is the efficiency of the motor (in percent)?

QUIZ ON CHAPTER 4, PERCENTAGE. FORM A.

1. Change the following percents to decimal fractions and then to common fraction form or mixed numbers:

45%; 6%; 80%; 225%; 0.4%; 0.02%; 320%; 600%; 23.6%; 37.5%

2. Change these fractions to decimal form and then to percents:

$\frac{3}{4}$; $\frac{5}{8}$; $\frac{3}{5}$; $\frac{5}{16}$; $1\frac{4}{5}$; $2\frac{1}{2}$; $\frac{15}{32}$; $\frac{2}{3}$

3. Find the percentage in each of the following:

(a) 24% of $65 **(b)** 3% of $840 **(c)** 0.5% of $480
(d) 1.6% of 700 **(e)** 250% of 68 **(f)** 140% of $160

4. In each of the following, find the rate, stated as a percent:

(a) 24 is what percent of 96? **(b)** 36 is what percent of 800?
(c) 32.4 is what percent of 450? **(d)** $6 is what percent of $900?
(e) 1.2 is what percent of 2400? **(f)** 90 is what percent of 75?

5. Find the base in each of the following:

(a) $24 is 15% of what? **(b)** $16.80 is 3.5% of what?

6. Find the selling price of each of the following:

(a) An article cost $45 and sold at a gain of 16%.
(b) An article cost $240 and sold at a loss of 15%.

7. Find the rate of gain or loss for each of the following:

(a) Cost: $750, sold for $840. **(b)** Cost: $2800, sold for $2320.

8. After a discount of 16%, an article was sold for $63. Find the original price.

9. For a gain of 12%, an article was sold for $504. What was the cost?

10. A voltmeter shows a voltage of 120 volts in a circuit. What is the maximum voltage? (A voltmeter shows 0.707 of the maximum voltage.)

CHAPTER FIVE Square Roots of Numbers

5-1 DEFINITION If we multiply a number by itself, we call the answer the *square*, or the *second power*, of the number. For instance, in the expression

$$7 \times 7 = 49$$

we call 49 the square of 7 or the second power of 7.

In order to indicate the multiplication of a number by itself, we often use a small number called an *exponent*.

An exponent is a small number placed at the right and a little above another number, called the base, to show how many times the base is to be used as a factor.

Thus

$$7^2 \text{ means } 7 \times 7 = 49$$

The small 2 placed near the 7 means that two 7's are to be *multiplied* together. In this example 7 is the *base* and 2 is the *exponent* of the power.

In the same way

$$5^3 \text{ means } 5 \times 5 \times 5 = 125$$

In this example 5 is the base. The 3 is the exponent of the power and means three 5's are to be multiplied together. The 125 is called the *third power* of 5.

The second power of a number is called the *square* of the number. The third power of a number is called the *cube* of the number. If the exponent is larger than 3, we name the power by the exponent. Thus

5^4, or 625, is called the *fourth power* of 5

7^5 is called the *fifth power* of 7

10^6 is called the *sixth power* of 10

Note. The exponent is usually shown in smaller type than the type used for the base.

5-2 ROOTS OF NUMBERS

If a number is used twice as a factor to form a product, the number is called the *square root* of the product. As an example, 7 is the square root of 49 because $7 \times 7 = 49$. Stated in another way, *the square root of a number is one of the two equal factors of the number.*

If a number is used three times as a factor to form a product, the number is called the *cube root* of the product. As an example, 7 is the cube root of 343 because $7 \times 7 \times 7 = 343$. *The cube root of a number is one of the three equal factors of a number.*

Other examples:

9 is the square root of 81 because $9 \times 9 = 81$
4 is the cube root of 64 because $4 \times 4 \times 4 = 64$
2 is the fifth root of 32 because $2^5 = 32$
3 is the fourth root of 81 because $3^4 = 81$.

Finding a root of a number is the opposite of finding a power. To find a power of a number, we multiply the number by itself. For instance, the third power of 6, that is, 6^3, is 216. The process of finding a power is called *involution.* The opposite process is called *evolution.* In the process of *evolution* we find the *root* of a number; that is, we try to find the number that has been multiplied by itself a certain number of times to produce the given number.

The symbol for a root is the sign $\sqrt{\ }$, called the *radical* sign. It means that a certain root is to be found. The number under the radical sign is called the *radicand* and is the number of which the root is to be found. A small number, called the *index*, is often placed in the notch of the radical sign to indicate the root to be found. Thus $\sqrt[2]{49}$ means the square root of 49, which is 7. The index 2 is usually omitted. Then we write

$$\sqrt{49} = 7$$

In order to indicate the cube root, we use the index number, 3. Thus $\sqrt[3]{125}$ means that the cube root of 125 is to be found, and we write

$$\sqrt[3]{125} = 5$$

Other examples:

$$\sqrt[4]{81} = 3 \qquad \sqrt[5]{32} = 2 \qquad \sqrt[3]{512} = 8$$

The only roots with which we shall be concerned in this chapter are *square* roots. Our problem is how to find the square roots of numbers.

We find the square roots of some numbers by inspection. For instance, if we square all the integers from 1 to 9, we get the following squares: 1, 4, 9, 16, 25, 36, 49, 64, 81. The square roots of these numbers can be determined from memory; that is,

$$\sqrt{1} = 1 \qquad \sqrt{16} = 4 \qquad \sqrt{49} = 7$$
$$\sqrt{4} = 2 \qquad \sqrt{25} = 5 \qquad \sqrt{64} = 8$$
$$\sqrt{9} = 3 \qquad \sqrt{36} = 6 \qquad \sqrt{81} = 9$$

A few others are quite generally known and can be determined by inspection:

$$\sqrt{100} \qquad \sqrt{144} \qquad \sqrt{225} \qquad \sqrt{121} \qquad \sqrt{169} \qquad \sqrt{400}$$

However, when we get to larger numbers, we cannot tell the square root from memory. We need some systematic method for finding the square root.

First, we need to consider what is meant by a *perfect square.* A number such as 49 is a perfect square because the square root of 49 is *exactly* 7. The square root of 169 is exactly 13, because 13×13 makes exactly 169. The number $6\frac{1}{4}$ is a perfect square. The square root of $6\frac{1}{4}$ is exactly $2\frac{1}{2}$ because $(2\frac{1}{2})(2\frac{1}{2})$ is exactly $6\frac{1}{4}$. A perfect square is a number whose square root can be stated *exactly* as a fraction or a whole number.

On the other hand, consider the number 19. There is no whole number or fraction that can be multiplied by itself to make exactly 19. The square root of 19 does not come out even as a decimal or common fraction. We can say it is approximately 4.359. If we multiply 4.359 by itself, we get a number slightly larger than 19. If we use 4.358, we get a number slightly smaller than 19. A number that cannot be expressed as a whole number or as a common fraction is called an *irrational* number. The square roots of many numbers are irrational numbers.

If the square root of a number is irrational, we carry out the answer to as many decimal places as we need for any particular degree of accuracy. Just where to stop is a matter of good judgment.

In most cases in actual practice we find the square roots of numbers by looking them up in a table. The engineer uses a slide rule, an electronic calculator, or logarithms. There is, however, a way to work out the square root of any number. As a student of mathematics, you should understand the process and be able to use it when the need arises. The steps in the process are now shown by examples.

Example 1

Find the square root of 1918.44.

Step 1. Separate the number into groups of digits, two digits in a group, starting at the decimal point. Use a mark such as a prime mark, ′, not a decimal point. Draw a line over the

$$\overline{19'\ 18'.44}$$

number and place the decimal point for the answer directly above the decimal point in the number. The answer will have one digit for each *group* of digits in the number itself. If the right group has only one digit, annex a zero. If the left group has only one digit, it is considered as one group.

Step 2. Find the largest perfect square that can be subtracted from the first group at the left. In this case the largest perfect square in 19 is 16. The first digit in the answer will always be the square root of this perfect square; in this case it is 4. Place the 4 above the 19 for the first digit of the answer.

```
   4    .
 ---------
 19'18.'44
```

Step 3. Place the square, 16, below the 19 and subtract, leaving 3.

Step 4. Bring down the next group of digits, 18, and place them next to the 3, making 318, which we call the first remainder.

```
   4    .
 ---------
 19'18.'44
 16
 --
  3 18
```

At this point we know that the square root will be some number between 40 and 50, since the original number lies between 40^2 and 50^2.

Step 5. Now we double the partial answer with a zero attached. The result is 80. Divide 80 into the remainder, 318. The result is 3. Then "3" *may* be the next digit in the answer. However, this number must be considered only as a trial answer. In order to determine whether the 3 is satisfactory, we must add the 3 to 80 and then multiply the result by 3; thus

```
  80              4  3.
 + 3            ---------
 ---            1918.44
  83             16
 × 3             --
 ---              318
 249              249
                  ---
                   69
```

Since 249 is less than the remainder, 318, we know that 3 is the next digit in the answer. It is placed above the second group of digits, 18. The answer is therefore 43 plus some decimal fraction. The 249 is placed below the 318 and subtracted, leaving 69.

Step 6. Bring down the next group of digits, 44.

Step 7. Now we proceed in the same way as in Step 5. We double the partial answer with a zero attached. The result is 860. Divide 860 into the remainder, 6944. The result is 8. This may be the next digit in the answer. However, to be sure, we first add the 8 and then multiply by 8; thus

```
                  4  3. 8
                ---------
                1918.44
                16
                --
                 318
                 249
                 ---
                   6944
                   6944
                   ----
```

```
  860
  + 8
  ---
  868
  × 8
 ----
 6944
```

Since the number 6944 is not larger than the remainder, we know that 8 is the third digit in the answer. Moreover, the remainder is now zero. Therefore, the square root is exactly 43.8. To check the answer, we multiply 43.8 by itself.

Example 2

Find the square root of 13.472 to five significant digits.
In this problem, at one point, when a group of two digits is brought down, the resulting number is too small to be divided by the necessary divisor. Then we place a zero in the answer in the proper position and bring down the next two digits. The work is left to the student.

5-3 A SHORT METHOD FOR FINDING SOME ROOTS

There is an important principle that enables us to find the roots of some numbers by a short method.

Principle. *If a radicand can be factored, the radical can be written as the product of two radicals.*

To see the meaning of this principle, consider the example

$$\sqrt{36}$$

Now, we know that $\sqrt{36} = 6$. However, suppose we separate the radicand into two factors and then write the radical as the product of two radicals:

$$\sqrt{36} = \sqrt{4 \cdot 9} = \sqrt{4} \cdot \sqrt{9}$$

The result reduces to the product $(2)(3) = 6$. The answer is correct for this example. In a more complete study of radicals, we shall find that the principle is always true for all radicals.

The foregoing principle can be used to find the square root of some numbers.

Example 1

Find $\sqrt{32}$.

Solution

We separate the radicand, 32, into two factors such that one of them is a perfect square. The largest perfect square contained as a factor in 32 is 16. Then we have

$$\sqrt{32} = \sqrt{16 \cdot 2}$$

Now we write the expression as the product of two radicals and simplify:

$$\sqrt{16 \cdot 2} = \sqrt{16} \cdot \sqrt{2} = 4\sqrt{2}$$

If we now remember that the square root of 2 is approximately 1.4142, we have

$$\sqrt{32} = \sqrt{16 \cdot 2} = \sqrt{16} \cdot \sqrt{2} = 4\sqrt{2} = 4(1.4142) = 5.6568 \text{ (approx.)}$$

The short method of finding roots depends on remembering the approximate values of a few irrational roots, such as the following (approximate values):

$$\sqrt{2} = 1.4142; \quad \sqrt{3} = 1.732; \quad \sqrt{5} = 2.236;$$
$$\sqrt{6} = 2.449; \quad \sqrt{7} = 2.646$$

Warning. *The square root of a sum or difference cannot be found by taking the square root of each part, as in the example*: $\sqrt{9 + 4} = \sqrt{13}$.

If one side of a square is given, then we find the area of the square by finding the square of the side.

Example 1

One side of a square measures 13.5 in. Find the number of square inches in the area of the square.

Solution

To find the area, we take $(13.5)^2 = 182.25$ (sq in.)

If the area of a square is known, we find the length of one side by taking the square root of the area.

Example 2

A square room has an area of 702.25 sq ft. Find the length of a side of the room.

Solution

Taking the square root of the area, we get

$$\sqrt{702.25} = 26.5;$$

the length of one side is 26.5 ft.

Example 3

A square has an area of 520 sq in. Find the approximate length of one side of the square.

Solution

Taking the square root of the area, we get

$$\sqrt{520} = 22.8 \text{ (approximately)}$$

Exercise 5-1

Find the square root of each of the following numbers. If the root is irrational, carry the answer out to five significant digits and then round off the answer to four significant digits.

1. 55,225	**2.** 70,225	**3.** 5522.5	**4.** 7022.5
5. 1823.29	**6.** 2162.25	**7.** 2.6569	**8.** 2.9584
9. 0.162409	**10.** 0.258064	**11.** 12.54	**12.** 14.6
13. 388.4	**14.** 346.3	**15.** 8686	**16.** 4802
17. 69.89	**18.** 91.4	**19.** 4252	**20.** 3238
21. 30.7	**22.** 84.1	**23.** 0.875	**24.** 0.683
25. 0.00608	**26.** 0.00409	**27.** 0.00328	**28.** 0.00477
29. 5084	**30.** 5021	**31.** 936.4	**32.** 50.21

Find the following by a simple method if possible (approximate values if irrational):

33. $\sqrt{18}$	**34.** $\sqrt{50}$	**35.** $\sqrt{12}$	**36.** $\sqrt{72}$
37. $\sqrt{108}$	**38.** $\sqrt{200}$	**39.** $\sqrt{24}$	**40.** $\sqrt{45}$
41. $\sqrt{100 + 25}$	**42.** $\sqrt{(100)(25)}$	**43.** $\sqrt{36 + 64}$	**44.** $\sqrt{(36)(64)}$

45. A square room has an area of 525 sq ft. Find the approximate length of a side of the room.

46. A square field contains 870 square rods. Find the approximate length of a side of the field.

47. A rectangle is 36 in. long and 20 in. wide. Find the side of a square that has the same area.

48. Find the size of a single ventilating pipe that has the same cross-sectional area as the combined area of two square pipes, one 12 in. on a side and the other 8 in. on a side.

49. A building lot is a rectangle 110 ft long and 52 ft wide. A square lot has an area 430 sq ft less than the rectangular lot. Find the size of the square lot.

50. A cement floor is a square 30 ft on a side. Find the length of a square floor that has twice as much area.

QUIZ ON CHAPTER 5. SQUARE ROOTS OF NUMBERS. FORM A.

Find the square root of each of the following numbers. If the answer is irrational, round off the answers to three significant digits.

1. 812.25 **2.** 2540.16 **3.** 0.00328

4. 4770 **5.** 3880 **6.** 14.9

Find the approximate value of each of the following by means of the short method. Use the following approximate values:

$\sqrt{2} = 1.4142$; $\sqrt{3} = 1.732$; $\sqrt{5} = 2.236$; $\sqrt{6} = 2.449$

7. $\sqrt{98}$; **8.** $\sqrt{48}$; **9.** $\sqrt{20}$; **10.** $\sqrt{54}$

11. Find the length of one side of a square whose area is 1024 sq ft.

12. A room contains approximately 342.25 sq ft. Find the length of one side of the room.

PART TWO ALGEBRA

CHAPTER SIX Operations with Signed Numbers

6-1 MEANING OF ALGEBRA

Algebra may be called *generalized arithmetic.* To understand this statement, suppose we ask: what is the cost of 7 books at $4 each? Here we are using arithmetic. The numbers 7 and 4 are *arithmetic* numbers. They are specific, definite numbers. The total cost, $28, is also an *arithmetic* number.

Now, suppose we ask a general question: how do you find the total cost of some books if you know the number of books and the cost of each book? The answer is: you multiply the cost of one book by the number of books. To state the rule a little more simply, we can say

$$\text{(number of books)} \times \text{(cost of each)} = \text{total cost}$$

Actually, the statement is simply a rule for finding the total cost. If we wish to state the rule in a still shorter form, we may use the letter N for the number of books, the letter C for the cost of each book, and the letter T for total cost. The rule becomes

$$N \times C = T$$

Consider another example. If a rectangle is 8 inches long and 5 inches wide, its area is 40 square inches. We find the area by multiplication: $8 \times 5 = 40$. All the numbers here used are *arithmetic* numbers.

Now, suppose we ask the general question: how do you find the area of any rectangle if you know its length and its width? The answer is: you multiply the number of linear units in the length by the number of linear units in the width. The result gives us the number of corresponding square units in the area. Stated as a rule, this statement becomes

$$\text{length} \times \text{width} = \text{area}$$

The statement is really simply a rule that tells how to find the area. If we wish to state the rule in a still shorter form, we may use the letter L for length, the letter W for width, and the letter A for area. The rule then becomes

$$L \times W = A \qquad \text{or} \qquad LW = A$$

In algebra, when two or more letters representing numerical values are written next to each other without a symbol between them, their values are meant to be multiplied together.

6-2 LITERAL NUMBERS

When letters of the alphabet are used to represent numbers, as we used the letters N, C, T, L, W, and A, such letters are called *literal* numbers. *Arithmetic* numbers are numbers used only in arithmetic, such as 3, 4, 7, 12, 473, and so on. Literal numbers are letters of the alphabet that are used to represent arithmetic numbers. They are such numbers as x, y, z, n, t, w, a, and so on.

Literal numbers may be called *general* numbers because they represent general values rather than specific values. Where we say $A = L \times W$, the letters represent the area, length, and width, respectively, of any rectangle; that is, a general rectangle. It is for this reason that algebra may be called *generalized arithmetic.*

Literal numbers cannot actually be added, subtracted, multiplied, or divided in the same way as arithmetic numbers. If we wish to find the sum of 5 and 3, we add them and get 8. If we wish to combine literal numbers in some way, such as the numbers x and y, all we can do is to indicate or express the operation and result. For instance, if we wish to add the numbers x and y, we show the sum as $x + y$. The difference between x and y is shown as $x - y$. The product is xy. The quotient of x divided by y is $x \div y$ or x/y.

In arithmetic we generally indicate multiplication by the sign $\times$. However, since the letter x is often used as an algebraic number, the multiplication sign is usually omitted. Instead, in algebra we sometimes indicate multiplication by placing a dot between the numbers to be multiplied; thus $3 \cdot 4$. If there is no danger of confusion, we omit the multiplication sign entirely. Thus xy means x times y; $2a$ means 2 times a. Multiplication of algebraic numbers can also be indicated by placing parentheses around each of the numbers without any sign between them; thus $(x)(y)$ means x times y; $(3)(4)$ means 3 times 4. If two literal numbers, or a literal number and an arithmetic number, are placed next to each other without a sign between them, they are always meant to be multiplied together.

It often happens that we wish to indicate that the sum of two numbers, or their difference, is to be considered as a single quantity. In such cases we enclose the quantity in parentheses. Any set of numbers enclosed in parentheses must always be considered as a single quantity and is treated as such. The sum of x and y can be shown as $x + y$. However, to show three times this quantity, we enclose the sum $x + y$ in parentheses and write a "3" before it: $3(x + y)$.

Exercise 6-1

Express the following:

1. The sum of a, b, and c

2. The product of 3 and x

3. The sum of $3x$ and $4y$

4. Seven more than x

5. A number x more than 7

6. m more than n

7. Eight less than y

8. y less than 8

9. 15 increased by n

10. c increased by x

11. 10 decreased by x

12. t decreased by y

13. The difference between a and b (if a is greater than b)

14. The difference between a and b (if b is greater than a)

15. The result of m divided by n

16. The quotient of $3n$ divided by $5x$

17. The sum of x and y divided by their product

18. The product of R and r divided by their sum

19. The sum of a and b decreased by the sum of x and y

20. Five times the sum of m and n

21. Three times the difference between h and k

22. The product of x and y added to the quotient of m divided by n

23. The product of $4a$ and $7b$

24. The sum of $5x$ and $8y$ divided by the difference between $10x$ and $4y$

6-3 NEGATIVE AND POSITIVE NUMBERS

In algebra we make use of what we call *negative* numbers. Negative numbers are numbers that are less than zero.

In arithmetic we did not use negative numbers. The problem of subtracting a number from a smaller number is considered impossible in arithmetic. For that reason, for a long time negative numbers were considered to be impossible. They were called *absurd, ridiculous,* and "fictitious." Yet they continued to force themselves into mathematics.

We shall see that such numbers are neither impossible nor fictitious. One of the most common uses is in connection with a thermometer. When the temperature drops down to zero, it can drop still lower. Suppose we say that we shall call numbers below zero *negative* numbers. Then, when the temperature is ten degrees below zero, we can call this a *negative ten degrees.* A negative number is usually indicated by placing a minus (−) before the number. Ten degrees below zero can then be called −10°. If you go outdoors some day when the temperature is −50°, you will realize that such a temperature is not "fictitious."

Numbers that are more than zero are called *positive* numbers to distinguish them from negative numbers. A positive number can be indicated by placing a plus sign (+) before the number. Ten degrees above zero can be indicated by +10°.

When numbers are considered as positive or negative numbers, they are called *signed* numbers. If a number has no sign before it, it is considered positive. In arithmetic we work with only positive numbers.

In arithmetic we use the signs + and − to indicate addition and subtraction, respectively. As *signs of operation,* they tell what mathematical operation is to be performed. However, when these signs are used to indicate positive and negative numbers, they are *signs of quality.* They indicate the *kind* of number, such as +3, −2, +7, +10, −5.

We have seen that temperatures below zero may be indicated by negative numbers, such as −10°, −15°, and so on. Negative numbers can be used in many other situations. If we indicate a distance of 3 miles *east by a* +3, then we may indicate a distance of 3 miles *west* by a −3. If we wish to show distances north by positive numbers, we may show distances south by negative numbers. If +3 (amperes) indicates a flow of electric current in a particular direction, then −3 (amperes) indicates a flow in the opposite direction.

If we represent elevation *above* sea level by a positive number, we may represent elevation *below* sea level by a negative number. If assets are indicated by positive numbers, then liabilities may be indicated by negative numbers. If a man has a debt of 15 dollars, we may say he has −15 dollars.

In a game we may sometimes make a score that takes us backward. We say we "get set'." Such a score may be called a negative score. A score that takes us forward may be called a positive score.

6-4 A NUMBER SCALE

Positive and negative numbers may be shown on a number scale represented by a horizontal line, as shown here. Positive numbers or distances are laid off to the right of zero, negative numbers or distances to the left, as shown in Figure 6-1.

Basically, negative numbers are counted in a direction opposite that for positive numbers. Any direction may be chosen for positive numbers, but the usual practice is to count them to the right of zero. The

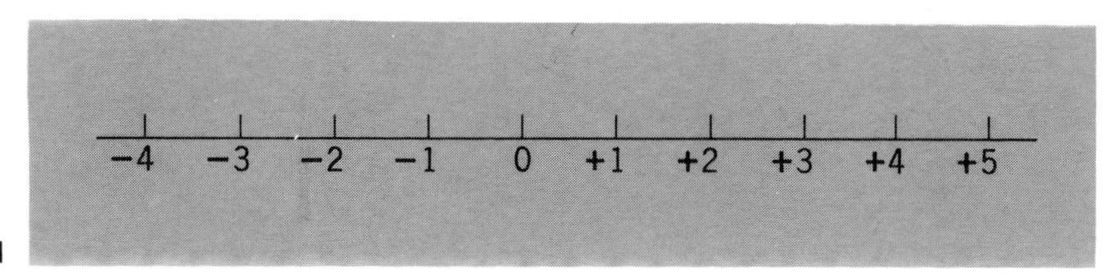

Figure 6-1

zero point is then the dividing point between the positive and negative numbers.

6-5 ABSOLUTE VALUE OF A NUMBER

The *absolute* value (sometimes called *numerical* value) of a number is its value without regard to the sign. The absolute value of $+10$ is the same as the absolute value of -10. The absolute value of a number is usually indicated by a pair of vertical lines, one on each side of the number, as $|-10|$. Then we can say $|+5|$ has the same value as $|-5|$.

Notice these two statements:

This statement is not true: $10 = -10$

This statement is true: $|10| = |-10|$

We know that a $+10$ is not the same as a -10. However, their absolute values are equal. Actually, the absolute value refers to the distance from zero. The two numbers, $+10$ and -10, are the same distance from zero.

6-6 OPERATIONS WITH SIGNED NUMBERS: ADDITION

Since algebra deals with positive and negative numbers, our first problem is to formulate rules for operating with these numbers. Then we must be sure to follow these rules when operating with signed numbers.

First, let us consider addition. One of the easiest and best ways to learn how to operate with signed numbers is to consider how we add scores in some games. In some games you may make a score that increases your total or a score that decreases your total; that is, you may make a score that takes you backward, in which case we say you "get set." If you use a positive number to indicate a "good" score that increases your total, then you may use a negative number to indicate a "bad" score that decreases your total. If you make a score of $+20$ and then later make a score of -15, your net, or total, score is $+5$.

Suppose, at the beginning of a game, you "get set," or go back 30 points. Your total score at that point is actually 30 points less than nothing. In a game this is sometimes called "going in the hole." You may call this a score of -30. It means that if you have a score of -30

and someone else has a score of zero (0), then he is still ahead of you by 30 points. It means also that you must make 30 points before you will have a score of zero.

Suppose the following sets of scores are made. Let us find the net total score in each set. This is called "adding" the scores. If a score has no sign, it is to be considered positive. The total score is shown for each set.

+20	+20	−15	−20	5	−10	−15
+15	−6	+10	−10	−15	25	4
+35	+14	−5	−30	−10	+15	−11

From these examples we can formulate the rule for adding signed numbers:

Rule. *To add two numbers with like signs, find the sum of their absolute values and prefix the common sign.*

To add two numbers with unlike signs, find the difference between their absolute values and then prefix the sign of the number having the greater absolute value.

If several scores are to be added, the positive scores may be combined first, then the negative scores, and finally the two combined to form the net total score.

Example

Find the sum of the signed numbers in the column at the right: −10, +7, −9, +14, −18, +4; sum −12

Solution

Adding the negative numbers:	−37
Adding the positive numbers, we get	+25
Now we add these signed numbers and get	−12

Exercise 6-2

Add algebraically:

1.	+17	−28	−43	+45	−72	28
	+42	+69	−19	−13	+47	−87
2.	59	−27	71	−92	0	−62
	−32	+61	0	36	−59	62

3.	20	−70	−16	−18	23	−15
	−15	+43	48	−4	−41	32
	−10	−35	−19	42	16	−19
	27	+19	24	−37	−56	−12
	−18	11	−21	16	15	+14

4.	−415	+513	−695	+591	392	−473
	−107	+624	+862	−376	−505	882

5.	−57.8	+0.596	−81.3	−47.2	−42.9	−0.038
	0.196	+2.68	+23.7	−5.81	1.38	+0.103

6.	131	−213	−48	153	136	−1.26
	−175	407	193	−215	−249	3.78
	−243	−316	−418	−84	−73	6.54
	56	526	−53	423	152	−4.08

6-7 SUBTRACTION OF SIGNED NUMBERS

Subtraction of signed numbers presents a little extra difficulty. Suppose you make the following scores in five trials in a game: +20, −15, +12, −8, +4. Your total score is 13.

Now, suppose you are told that you should have had only four trials. You are told to "erase" or "subtract" the last score. Here *subtract* means to *erase* the score.

Your total score was	+13	or by adding	+13
You subtract	+4	and	−4
Your net score is now	+9	we have	+9

Subtracting the +4 from the +13 is the same as adding a −4 to your score.

In another game suppose you make the following scores: +20, −15, +12, +9, −5. Your total score is 21.

Now, suppose you are told to subtract the last score (subtract means to *erase*). If you subtract or erase the last score, which is −5, your net score becomes 26, instead of 21. Actually, subtracting the −5 increases your score by 5.

Your first score was	+21	or by adding	+21
You subtract	−5	and	+5
Your net score now is	26	we have	+26

Subtracting the −5 from the 21 is the same as adding a + 5 to your score.

From the foregoing examples we formulate the rule for subtraction of signed numbers:

Rule. *To subtract two signed numbers, mentally change the sign of the subtrahend and then proceed as in algebraic addition.*

Note. Algebraic subtraction may be checked by addition, the same as subtraction in arithmetic. The remainder added to the subtrahend should equal the minuend.

Exercise 6-3

Subtract the bottom number from the top number:

1.	+9 −5	−6 −9	+23 +18	+43 −21	−56 +17	−62 −35
2.	32 74	−21 54	+73 −49	−37 56	−41 −93	−62 −75
3.	38 −81	−63 28	−79 54	−23 −23	0 −25	0 74
4.	98 −98	726 −45	−523 127	−416 −324	−917 −520	−144 −300
5.	−32 40	−635 −847	−322 0	0 −322	−315 83	+558 −442
6.	17 28	−46 −24	35 −16	37 −69	−51 18	14 48
7.	−3.28 −5.	6. 1.872	−0.831 −7.12	+0.904 +9.801	−7. −0.537	−1. 4.5322

8. Add the two numbers in each of the examples in No. 7.

6-8 HORIZONTAL ADDITION AND SUBTRACTION

Suppose we have this problem in addition:

$$\begin{array}{r} +6 \\ -3 \\ +7 \\ -2 \\ \hline \end{array}$$

If we place these signed numbers in a horizontal line, we can indicate addition by enclosing each number with its sign in parentheses and then placing the addition sign (+) between them. First, let us write

$$(+6) \text{ plus } (-3) \text{ plus } (+7) \text{ plus } (-2) = 8$$

Now, we replace the word *plus* with the sign for addition (+). This sign is then the sign of *operation.*

$$(+6) + (-3) + (+7) + (-2) = 8$$

Here, the plus signs between the numbers are signs of operation, that is, addition. The signs inside the parentheses are the signs of positive and negative numbers.

If we drop all the plus signs of operation and omit the parentheses, we have

$$+6 - 3 + 7 - 2 = 8$$

The answer can be obtained simply by combining the signed numbers as we would in arithmetic. For horizontal addition, we have this rule:

Rule 1. *To add signed numbers horizontally, simply write down each number with its proper sign and omit the signs of addition. Then combine the numbers.*

To indicate subtraction in horizontal form, we enclose each number with its sign in parentheses and then place the sign of operation, a minus sign, between the numbers. For example, suppose we have this problem: from (+12) subtract (−7). We first write

$$(+12) - (-7)$$

We have seen that subtracting any number is equivalent to adding the same number with its sign changed. Therefore,

$$(+12) - (-7) \text{ means the same as } (+12) + (+7)$$

Now we apply the rule for horizontal addition and get

$$+12 + 7 = +19$$

For horizontal subtraction we have this rule:

Rule 2. *In horizontal subtraction, omit the signs of subtraction and simply write down each number to be subtracted with its sign changed. Then combine the terms as in horizontal addition.*

Examples

(a) $(-5) - (+4) = -5 - 4 = -9$

(b) $(+3) - (-7) = +3 + 7 = +10$

(c) $(+8) - (+3) = +8 - 3 = +5$

When horizontal additions and subtractions occur in the same problem, the signs of operation may be dropped, provided the rules for horizontal addition and subtraction are observed.

Example

$(+8) + (+7) + (-3) - (+5) - (-2) + (4) =$

$+8 \quad +7 \quad -3 \quad -5 \quad +2 \quad +4 = 13$

Exercise 6-4

Combine each of the following into a single answer:

1. $(+8) + (-3) - (-4) - (+6) + (+9) =$

2. $(+5) + (+12) - (-7) + (-2) - (+8) =$

3. $(-7) - (+6) + (-13) + (+9) - (-18) =$

4. $(-12) + (11) - (-30) + (-17) - (+26) =$

5. $(+13) - (-15) - (8) + (-16) + (-20) =$

6. $(+14) - (+9) + (-7) - (-3) + (+6) - (-3) - (-8) =$

7. $(-16) - (-3) + (9) - (-1) + (-5) - (11) - (-2) =$

8. $(-9) + (14) + (-23) - (15) - (+12) + (13) + (-8) =$

9. $(16) - (18) + (-12) - (-41) + (-9) - (14) - (-5) =$

10. $(-15) - (13) + (-7) - (32) + (-2) - (-20) + (-4) =$

11. $(12) - (-15) - (24) + (-13) - (+6) - (16) + (-3) =$

12. $(-17) - (11) + (-12) - (+5) + (18) + (3) - (-16) =$

13. $(18) + (-21) - (-73) + (+4) + (31) - (+45) - (14) =$

14. $(-312) - (+415) + (247) - (-763) - (658) - (-1234) =$

15. $(139) - (-254) - (267) + (980) - (114) + (-844) =$

6-9 MULTIPLICATION OF SIGNED NUMBERS

In multiplying signed numbers, we multiply the absolute numerical values in the same way as in arithmetic. Our main problem, then, is to determine the sign of the product.

Multiplication can be considered a short form of addition. In a problem in arithmetic, such as (3×482), the 3 is the multiplier. It shows how many times the multiplicand is to be added. As addition, the problem may be written

$$\begin{array}{r} 482 \\ 482 \\ \underline{482} \end{array}$$

Now suppose we have this problem in the multiplication of signed numbers: $(+3) \cdot (+4)$. Let us write one number above the other.

Example 1

Multiply

$$\begin{array}{r} +4 \\ \underline{+3} \end{array}$$

The absolute numerical value of the product is 12. Our problem now is to determine the sign of the answer.

Let us call the +3 the multiplier. Suppose we consider the plus sign (+) before the 3 as the sign of operation (addition). Then, since multiplication is a short form of addition, the problem means

add three "+4's"

If we look upon the +4's as scores in a game, the problem means
adding three +4's to your score changes your total by +12

Example 2

Multiply

$$\begin{array}{r} -4 \\ \underline{+3} \\ -12 \end{array}$$

Again, the absolute numerical value of the answer is 12. To determine the sign of the product, consider the plus sign (+) before the multiplier 3, as the sign of the operation (addition). The problem then means

add three "−4's"

Adding three −4's to your score changes your total by −12. Therefore, $(+3)(-4) = -12$.

Example 3

Multiply $(+4)(-3)$.
Or in column form, multiply

$$\begin{array}{r} +4 \\ \underline{-3} \\ -12 \end{array}$$

Again, the absolute numerical value of the answer is 12. To determine the sign of the product, consider the minus sign ($-$) before the multiplier, 3, as the sign of operation (subtraction). The problem then means

subtract three " +4's"

Subtracting three +4's from your score changes your total by -12. Therefore, $(-3)(+4) = -12$.

This product can be shown in another way. By the *commutative law*, the problem can be taken to mean:

add four "-3's"

Then, when the problem is written in column form, we get the same answer.

Example 4

Multiply $(-3)(-4)$.
Or in column form, multiply

$$\begin{array}{r} -4 \\ \underline{-3} \\ +12 \end{array}$$

Again, the absolute numerical value of the answer is 12. To determine the sign of the product, consider the minus sign ($-$) before the multiplier, 3, as the sign of operation (subtraction). Then the problem means

subtract three "-4's"

Subtracting three -4's from your score changes your total by +12. Therefore, $(-3)(-4) = +12$.

From the foregoing examples we can formulate the rule for multiplying signed numbers.

Rule. *The product of two numbers with like signs is a positive number. The product of two numbers with unlike signs is negative.*

In symbols, the rule for the signs of a product may be stated in this way:

$$(+)(+) = + \qquad (+)(-) = -$$
$$(-)(-) = + \qquad (-)(+) = -:$$

However, remember that this way of showing the rule for the multiplication of signed numbers is only symbolic and the signs themselves are not to be considered as factors.

The rule for the product of signed numbers can be applied to the product of several factors, as in the following examples.

Example 5

Find the product of $(-5)(-4)(+3)(+2)$

Solution

The product of the first two factors, $(-5)(-4)$, is $+20$. Then we have for the next products: $(+20)(+3) = +60$; and $(+60)(+2) = +120$

Example 6

Find the product: $(-3)(-7)(-1)(+2)(+4)$

Solution

The product of the first two factors, $(-3)(-7)$, is $+21$.
For the next product, we have: $(+21)(-1) = -21$.
For the next product, we have: $(-21)(+2) = -42$.
For the final product, we get: $(-42)(+4) = -168$

From these and other examples, we can formulate the rule:

Rule. *For an even number of negative factors, the product is positive. For an odd number of negative factors, the product is negative.*

Note. When we come to the multiplication of signed numbers under radical signs, we shall find it necessary to modify slightly the statement: $(-)(-) = +$.

Exercise 6-5

Multiply the following:

1. $(-3)(-7)(-4) =$

2. $(-5)(+4)(-6) =$

3. $(+8)(-1)(-5) =$

4. $(12)(-2)(+8) =$

5. $(-1.2)(3.5)(4.3) =$

6. $(-\frac{1}{2})(-24)(-\frac{2}{3}) =$

7. $(+36)(-0.04)(-2) =$

8. $(-7.2)(+0.13)(-3.5) =$

9. $(6.3)(-2.4)(-1)(-2) =$

10. $(-\frac{3}{5})(\frac{3}{7})(-\frac{5}{8})(+4) =$

11. $(-1)(-2)(+3)(-4)(+4) =$

12. $(-3)(-1)(-2)(-2)(+3) =$

13. $(+1)(-3)(+1)(-3)(-5) =$

14. $(-2)(+3)(-5)(-2)(-6)(+2) =$

15. $(3)(-4)(-1)(+5)(-1) =$

16. $(7)(+\frac{1}{2})(-3)(-\frac{2}{3})(-\frac{3}{8}) =$

17. $(-1)(-1)(4)(2)(5)(-1)(\frac{4}{15}) =$

18. $(-2)(-5)(-\frac{1}{4})(-\frac{1}{15})(10) =$

19. $(-4)(-3)(-3)(-5)(-1) =$

20. $(-4)(+3)(7)(-1)(-2)(+8) =$

21. $\begin{array}{r}-81\\-42\\\hline\end{array}$ $\quad$ $\begin{array}{r}+124\\+31\\\hline\end{array}$ $\quad$ $\begin{array}{r}-98\\+23\\\hline\end{array}$ $\quad$ $\begin{array}{r}-76\\-12\\\hline\end{array}$ $\quad$ $\begin{array}{r}-.072\\+.13\\\hline\end{array}$ $\quad$ $\begin{array}{r}-.472\\5.2\\\hline\end{array}$ $\quad$ $\begin{array}{r}0.0425\\-0.362\\\hline\end{array}$

6-10 DIVISION OF SIGNED NUMBERS

In division we have the same rules for signs as in multiplication.

Rule. *The division of numbers with like signs gives a positive quotient. The division of numbers with unlike signs gives a negative quotient.*

Examples

$(+12) \div (+4) = +3 \qquad (-12) \div (+4) = -3$

$(-12) \div (-4) = +3 \qquad (+12) \div (-4) = -3$

Division may be checked by multiplication in the same way as in arithmetic; that is, the quotient multiplied by the divisor should equal the dividend. This method of checking division also applies to signs as well as to the numerical values.

In symbols, the rule for the sign of the quotient in division may be stated in this way:

$$(+) \div (+) = + \qquad (+) \div (-) = -$$

$$(-) \div (-) = + \qquad (-) \div (+) = -$$

Exercise 6-6

Divide as indicated:

1. $(+15) \div (+3) =$ **2.** $(-16) \div (+2) =$ **3.** $(-24) \div (+3) =$

4. $(36) \div (-9) =$ **5.** $(-40) \div (-8) =$ **6.** $(30) \div (5) =$

7. $(-128) \div (8) =$ **8.** $(144) \div (-6) =$ **9.** $(180) \div (-12) =$

10. $(-152) \div (-4) =$ **11.** $(-8) \div (-12) =$ **12.** $(-17) \div (+4) =$

13. $(13) \div (-5) =$ **14.** $(-20) \div (-8) =$ **15.** $(16) \div (12) =$

16. $(-\frac{5}{12}) \div (-\frac{3}{4}) =$ **17.** $(+\frac{3}{7}) \div (-\frac{6}{11}) =$ **18.** $(-\frac{4}{7}) \div (\frac{3}{5}) =$

19. $\dfrac{-936}{-8} = \quad \dfrac{-322}{+23} = \quad \dfrac{+966}{-42} = \quad \dfrac{-527}{-31} = \quad \dfrac{+0.16}{8} =$

20. $\dfrac{-0.72}{+1.2} = \quad \dfrac{-5.6}{1.4} = \quad \dfrac{-0.3794}{-1.4} = \quad \dfrac{8.4}{-0.7} = \quad \dfrac{-0.0028}{7} =$

21. $\dfrac{-7}{3} = \quad \dfrac{+15}{-10} = \quad \dfrac{+0.3798}{+180} = \quad \dfrac{48}{-60} = \quad \dfrac{-5376}{24} =$

22. A football halfback carrying the ball 7 times during a game made the following yardages for the times he carried the ball: loss 4 yards, gain 13 yards, gain 6 yards, gain 8 yards, loss 3 yards, gain 0 yards, gain 2 yards. Show his gains and losses by signed numbers and then find his total yardage. What was his average gain or loss per carry during the game?

23. A football player made the following gains and losses in a game: gain 2 yards, loss 5 yards, gain 4 yards, loss 7 yards, gain 3 yards, loss 8 yards, loss 3 yards. Show his gains and losses by signed numbers and find his total yardage gained. What was the average gain per carry?

24. An elevator operator made the following moves starting at the ground floor: up 3 floors, up 5 floors, down 2 floors, down 4 floors, up 8 floors, up 1 floor, down 9 floors, down 3 floors. Show the moves by signed numbers. Where was the operator at the end of the last move mentioned?

QUIZ ON CHAPTER 6, SIGNED NUMBERS. FORM A.

1. Add the signed numbers in each set:

+25	−17	−47	−35	23	−27	0	−28	+36	−16
+23	−32	−13	+14	−42	0	−34	+28	+36	−16

2. In No. 1, subtract the bottom number from the top.

3. Add each set:

−8	+7	−12	+16	−13	+6	−31.4
+6	−3	−7	−23	+14	−15	+15.8
−13	−18	+4	−14	−7	−8	−14.7
24	14	9	−8	−9	+6	+21.2

4. In each set, subtract the bottom from the top:

−4.52	+3.6	−6.47	5.24	−8.36
−7.	−4.35	+3.12	7.86	−2.43

5. Remove parentheses and combine the terms:

(a) $(5) - (+6) + (+7) - (+4) - (-9)$;

(b) $(-3) - (-8) + (-5) - (-6) + (+4)$

6. Multiply the numbers in each set:

+35	−42	−53	+28	480	−12	−3.25	+4.96
+8	+7	−9	−6	−3.5	−23	−5.62	−1.25

7. Multiply as indicated:

(a) $(-5)(-6)(3)$;

(b) $(-4)(-6)(-2)(3)(-1)$; **(c)** $(-3\frac{3}{4})(-4\frac{4}{5})(-2\frac{1}{3})$

8. Divide as indicated:

(a) $(-63) \div (-7)$; **(b)** $(84) \div (-6)$; **(c)** $(-420) \div (14)$

(d) $(-\frac{8}{15}) \div (-\frac{4}{5})$ **(e)** $(4\frac{3}{8}) \div (-5\frac{1}{4})$; **(f)** $(-0.3906) \div (-1.26)$

CHAPTER SEVEN Operations with Algebraic Expressions

7-1 DEFINITIONS

A *product* is the answer obtained by multiplying two or more quantities together. For instance, in the multiplication $3 \cdot 5 = 15$, the 15 is the product of the two numbers 3 and 5. The product of 2, 2, 3, and 7 is 84. The product of $5x$ and y is $5xy$. The word *product* always implies multiplication.

The quantities multiplied together to form a product are called the *factors* of the product. The numbers 3 and 5 are factors of 15. One set of factors of $5xy$ is $5x$ and y.

Sometimes a factor can itself be further split up into factors. For instance, we can say that the factors of 60 are 6 and 10. However, the 6 and the 10 can be separated into other factors. The 6 is a product of 2 and 3. The 10 can be separated into two factors, 2 and 5. A factor that cannot be further separated into other factors except itself and 1 is called a *prime* factor. The prime factors of 60 are 2, 2, 3, and 5. In like manner, the prime factors of $3axy$ are 3, a, x, and y.

If two numbers are multiplied together, such as in the product $3 \cdot 7$, either factor is called the *coefficient*, of the other. The word *coefficient* means that the two numbers are *efficient together* in forming the product 21. In this example, 3 is the coefficient of 7 and 7 is the coefficient of 3. However, both numbers are factors of 21.

It is important that you understand exactly what is meant by the words *coefficient, factor*, and *product*. In the product $5xy$

the coefficient of x is $5y$
the coefficient of y is $5x$
the coefficient of 5 is xy
the coefficient of xy is 5
the coefficient of $5y$ is x
the coefficient of $5x$ is y

The numbers 5, x, and y are the *factors* of the product $5xy$.

In the product $5xy$ the number "5" is the *numerical* coefficient. By *numerical* coefficient we mean the *arithmetic* factor. If a literal number is shown without the numerical coefficient, its coefficient is always understood to be 1. The numerical coefficient of xy is 1. The numerical coefficient of $-xy$ is -1.

A *term* is any algebraic expression not separated within itself by a plus or minus sign. A term consists only of factors. The following expressions are single terms:

$$3xy, \quad 4x^2y^3, \quad -7abcd, \quad -2x, \quad y, \quad 2a^2b^2cx^4y^5z$$

A term indicates a *product,* not a sum or difference. The sign of a term is the sign preceding it. If no sign is expressed, the term is always *positive.*

Like, or *similar,* terms are terms that are exactly alike in their *letter parts.* The terms shown here are *like terms,* since the letter part is x^2y in all:

$$4x^2y, \quad -7x^2y, \quad -5x^2y, \quad 15x^2y, \quad -x^2y$$

Unlike, or *dissimilar,* terms are terms that are not exactly alike in their letter parts. The following are *unlike, or dissimilar, terms:*

$$4x^2y, \quad -7xyz, \quad -5xy^2, \quad 3x^2z, \quad 2x, \quad -5y^2$$

A *monomial* is an algebraic expression consisting of only one term, such as $5xy$. A *polynomial* is an algebraic expression of more than one term. The following expressions are polynomials:

$$5x^2 + 3xy, \quad 4xy + 7ab - x^2y, \quad 7x^2 - 2xy + 3y^2 - 5x + 2y$$

A polynomial indicates the sum or difference of two or more terms.

A polynomial of two terms is usually called a *binomial.* A polynomial of three terms is often called a *trinomial.* The expression $5x^2 + 3xy$ is a binomial. The expression $4xy + 7ab - x^2y$ is a trinomial.

An *exponent* is a mathematical notation whose meaning must be thoroughly understood by a student of algebra. To avoid errors in the use of exponents, we must understand the exact meaning of the word.

It often happens that we wish to multiply a number by itself, such as $5 \cdot 5$. To indicate this multiplication, we can write 5 with a small 2 placed at the right and a little above the 5; thus 5^2. The 2 here indicates that two 5's are to be multiplied. The 2 so placed and having this meaning is called an *exponent.* It indicates the *power* to which the 5 is to be raised. The number 5 on which the exponent is placed is called the *base.*

The expression 7^2 is called "7 squared," or "7 raised to the second power." The expression, 7^3, is read "7 cubed," or "7 raised to the third power." The exponent 3 placed on the base 7 means that the base is to be used three times as a factor; that is, $7^3 = 7 \cdot 7 \cdot 7 = 343$. If we wish to indicate the multiplication $x \cdot x \cdot x \cdot x$, we can write it x^4.

This expression is read "x raised to the fourth power," or simply "x fourth."

If any number is shown without an exponent, its exponent is always understood to be 1 (unity). The number x means x^1. Also, 5^1 means 5.

From the several examples given we summarize the definition of an exponent:

An exponent is a number placed at the right and slightly above another number, the base, to show how many times the base is to be used as a factor.

7-2 ADDITION AND SUBTRACTION OF MONOMIALS

We add or subtract like terms simply by adding or subtracting their numerical coefficients. Notice that the letter part is not changed. The following examples show *addition* of monomials:

+7 miles	$8x^2y$	$-9x^3y^2$	$-3abc$	$6mn$	$-5xy$
+2 miles	$2x^2y$	$5x^3y^2$	$-2abc$	$-mn$	xy
+9 miles	$10x^2y$	$-4x^3y^2$	$-5abc$	$5mn$	$-4xy$

In each of the following examples the bottom monomial is subtracted from the top (recall the rule for subtraction):

$+5xy^2$	$-5a^3b$	$-3cd$	$-6xyz$	$+rst$	$-7h^2k^3$
$+2xy^2$	$2a^3b$	$-8cd$	$-7xyz$	$-4rst$	$-6h^2k^3$
$+3xy^2$	$-7a^3b$	$+5cd$	xyz	$+5rst$	$-h^2k^3$

If two terms are unlike, their sum or difference can only be indicated or expressed. Add the following:

7 dollars	$+8x$	$-6a$	$5x^2$
3 pesos	$+2y$	$-4b$	$-3x$
7 dollars + 3 pesos	$8x + 2y$	$-6a - 4b$	$5x^2 - 3x$

In each of the following examples the bottom monomial is subtracted from the top (recall the rule for subtraction):

$+7x$	$-4a$	$3cd$	$-5x^2y$	u
$+3y$	$+9b$	$-5mn$	$2xy^2$	v
$+7x - 3y$	$-4a - 9b$	$3cd + 5mn$	$-5x^2y - 2xy^2$	$u - v$

In horizontal addition or subtraction of monomials like terms may be combined. For example,

$$3x^2y - 4xy^2 - 3x + x^2y + 2xy^2 + 4x = 4x^2y - 2xy^2 + x$$

7-3 ADDITION AND SUBTRACTION OF POLYNOMIALS

To add or subtract polynomials, we write one polynomial under the other so that like terms fall in the same column. We then add or subtract each column separately, just as we add or subtract monomials.

Example 1

Add the following polynomials:

$$3x^2 - 4xy + 5y^2;\ 3xy - 3y^2 - x^2 + ab;\ 3x + 4x^2 - 2y^2 - 2ab.$$

Solution

$$\begin{array}{rrrrr}
3x^2 & -\ 4xy & +\ 5y^2 & & \\
-x^2 & +\ 3xy & -\ 3y^2 & +\ ab & \\
4x^2 & & -\ 2y^2 & -\ 2ab & +\ 3x \\
\hline
6x^2 & -\ xy & & -\ ab & +\ 3x
\end{array}$$

None of the terms in the sum can be combined further, since all are unlike terms.

Example 2

From $5x^2 - 3xy + 2y^2 - 2x$ subtract the polynomial $3y^2 - 2x^2 - 5y - 3xy$.

Solution

$$\begin{array}{rrrrr}
5x^2 & -\ 3xy & +\ 2y^2 & -\ 2x & \\
-2x^2 & -\ 3xy & +\ 3y^2 & & -\ 5y \\
\hline
7x^2 & & -\ y^2 & -\ 2x & +\ 5y
\end{array}$$

7-4 HORIZONTAL ADDITION AND SUBTRACTION OF POLYNOMIALS

The addition and subtraction of polynomials may be indicated horizontally by use of parentheses. Any terms enclosed in a set of parentheses must be considered as a single quantity. The addition of two polynomials, such as $4x^2 - 3xy - 2y^2$ and $2x^2 + 3xy - 5y^2$, can be shown by enclosing each in a set of parentheses and placing the addition sign between them:

$$(4x^2 - 3xy - 2y^2) + (2x^2 + 3xy - 5y^2)$$

You recall that when we add like terms we simply combine the terms according to the rules of algebraic addition. Since none of the signs is changed in addition, we can remove the parentheses without changing any signs and then combine like terms; thus

$$4x^2 - 3xy - 2y^2 + 2x^2 + 3xy - 5y^2 = 6x^2 - 7y^2$$

Horizontal subtraction of polynomials can be indicated by enclosing each polynomial in a set of parentheses and placing the subtraction sign, $-$, between the two quantities. For instance, if we wish to indicate the subtraction of the polynomial $2x^2 - 3x + 5$ from the polynomial $5x^2 + 2x + 5$, we can indicate the subtraction in this way:

$$(5x^2 + 2x + 5) - (2x^2 - 3x + 5)$$

Before like terms can be combined, the parentheses must be removed. Remember that in subtraction we change the sign of the number subtracted and then add algebraically. The problem can be changed to

$$(5x^2 + 2x + 5) + (-2x^2 + 3x - 5)$$

Now, if the parentheses and the sign of operation, $+$, are dropped, like terms may be combined; thus

$$5x^2 + 2x + 5 - 2x^2 + 3x - 5 = 3x^2 + 5x$$

From the foregoing examples we can formulate a simple rule for horizontal addition and subtraction of polynomials.

Rule. *If a quantity within parentheses is preceded by a plus sign (+), this plus sign and the parentheses may be omitted without changing the sign of any term within the parentheses.*

If a quantity within parentheses is preceded by a minus sign (−), this minus sign and the parentheses may be omitted provided that the sign of each term within the parentheses is changed.

Exercise 7-1

First add each of the following sets of monomials; then subtract the bottom monomial from top in each set.

1.	$3x$	$-4x^2$	$-5st$	$+5n$	0	$4xy$	$-7x^2y$
	$5x$	$-7x^2$	$+4st$	$-3n$	$-3xy$	$-3x$	$-xy$
2.	$-5b$	$-5n$	$3a^3$	$-6xy^2$	$-7x^3y$	$8xy^3$	$-9c^2$
	$8b$	0	$-9a^3$	$-7x^2y$	$4x^3y$	$-5x$	$-2c$
3.	xy^3	$-3st^2$	$-7xy$	$-3xyz$	$-4m^2n$	$-9x^2y$	$-12xy$
	$7xy^3$	$-8st^2$	$+6xy$	$+3xyz$	$3m^2n$	$-9x^2y$	$-12xy^2$
4.	$5a^2b^3$	$5bx^3$	$-17ax$	$6m^2n$	$-4x^2y^3$	$-12yz^3$	$-5a^2c^3$
	$-4a^2b^3$	$-7bx^3$	$+15ax$	$15m^2n$	$-13x^2y^3$	$+8yz^3$	$6a^2c^2$

Add the following:

5.	**6.**	**7.**	**8.**
$-3x^2y$	ab^3	$-4x^2y^3$	$-bc^3d^2$
$-5x^2y$	$-8ab^3$	$-x^2y^3$	$6bc^3d^2$
$7x^2y$	$4ab^3$	$+9x^2y^3$	$-5bc^3d^2$
$+x^2y$	$2ab^3$	$-5x^2y^3$	$-bc^3d^2$

Horizontal addition and subtraction. Combine the following:

9. $2a + 4b + 3c - 3b + 5a - c + a - ab + 5c + 6b - 2a - b =$

10. $3x - (2y) - (-5x) + (-4z) - 3y +$
$(6z) - (-2x) - (+3z) + 4y + z =$

11. $4x + 3y + 6 - (+3x) - (-7y) -$
$2 + 5z + 5x - (+4y) + (-6z) + 5 - x =$

12. $ab + 4b^2 - 5a^2 - 3ab + 3a + 2b -$
$2a^2 - b^2 - (-a) - (+ab) + 7 =$

Add the following polynomials:

13.
$$\begin{array}{r} 3x^2 - 5xy + 6y^2 \\ x^2 + 4xy - 3y^2 \\ -4x^2 - xy - 2y^2 \\ \hline \end{array}$$

14.
$$\begin{array}{r} 5x^3 - 3x^2y + 4y^3 \\ -2x^3 + 4x^2y - 2xy^2 \\ -3x^3 + x^2y - 3y^2 + 3xy^2 \\ \hline \end{array}$$

Subtract the following polynomials (bottom from the top):

15.
$$\begin{array}{r} 5x^2 - 4xy + 6y^2 \\ 3x^2 + xy - 5y^2 \\ \hline \end{array}$$

16.
$$\begin{array}{r} -x^2 - 4xy + 3y^2 - 2x \\ -3x^2 - xy - 3x + 4y \\ \hline \end{array}$$

17. Add $5x^2 + 2xy - 3y^2$ to $x^2 + 6y^2 - 4xy$.

18. Add $3x^2 - 7xy + 2y^2$ to $2xy - y^2 - x^2$.

19. Add $3nx - 5n^2 + x^2$ to $3x^2 - n^2 + 2nx$.

20. Add $y + 2xy - 4x^2 - y^2$ to $x^2 + 3y^2 - 4x - 2xy$.

21. Add $5x^2 - 3xy + y^2$; $5xy - 3x^2 + 2xy$; $5x + 4xy - 3x^2$.

22. Subtract $4x^2 - 2xy - 3y^2$ from $2x^2 - 5xy - y^2$.

23. Subtract $3x^2 - y^2 - 5xy - 3x$ from $x^2 - 2xy - 3y^2$.

24. From $5x^2y + 2x^2y^2 - 3xy^2$ take $2x^2y - 4xy^2 + 3x^2y^2$.

25. From $8x^3 - 3x^2 + 4x$ take $3x^3 - 2x^2 - 5x$.

26. Take $-2x^2 + xy + y^2$ from $x^2 + y^2 + 3xy$.

7-5 CHECKING THE WORK IN ALGEBRA

Since the letters we use in algebra represent arithmetic numbers, we can check our work by replacing each letter with some particular value. For example, suppose we have this example in addition:

$$\begin{array}{r} 3x^2 - 4xy + 5y^2 \\ x^2 + 5xy - 4y^2 \\ \hline 4x^2 + xy + y^2 \end{array}$$

The sum is $4x^2 + xy + y^2$

Now, if we let x equal some value, such as 2, and y equal some other value, such as 3, we then find that the first polynomial is equal to 33, the second is equal to -2, and the sum is 31, which checks with the sum.

In checking your work with particular values, you may use any arithmetic numbers for x or y. However, the numbers 1 or 0 should not be selected, since such a check would be worthless. The *best check* on your work consists of these two steps:

(a) *First, know how to do the work correctly.*
(b) *Then, do the work over again.*

QUIZ ON CHAPTER 7, OPERATIONS WITH ALGEBRAIC EXPRESSIONS. FORM A.

1. Add the following expressions:

$4x$	$-6xy$	$-5x^2y$	$+8xy^2$	$3ab$
$-7x$	$-2xy$	$+9x^2y$	$+4xy^2$	$2cd$

In Nos. 2–7, add the terms in each set:

(2)	(3)	(4)	(5)	(6)	(7)
$-5x^2y$	$-ab^2$	$+4x^2y^2$	$3abc$	$-2cd$	$6xy^2z$
$3x^2y$	$7ab^2$	$-3x^2y^2$	$5abc$	$-6cd$	$-xy^2z$
$-7x^2y$	$-3ab^2$	$-8x^2y^2$	$-abc$	$8cd$	$-4xy^2z$
x^2y	$-2ab^2$	$7x^2y^2$	$-6abc$	$-cd$	$2xy^2z$

8. Combine like terms:

$$3a + 2b - 3c + a - b + 4c - 6ab + 2a - 3ab + c$$

9. Combine:

$$2x - (+3y) - (-2x) - (-4y) + (-3z) + (5x) - (-2z)$$

10. Add:

$$(3x^2 + 2xy - 4y^2 + 3x) + (4x^2 + 2y^2 - y - 3xy)$$

11. Add:

$$(4x^2 - xy + 3y + 2y^2 + 3) + (4xy - 5x^2 - 7 - 3y^2 + 2x)$$

12. In No. 10, subtract the first polynomial from the second.

13. In No. 11, subtract the second polynomial from the first.

14. Simplify and combine:

$$(4x^2 - 3xy - 2y^2 - 3x) - (3xy - 3y^2 + y - x^2)$$

CHAPTER EIGHT Multiplication and Division

8-1 MEANING OF AN EXPONENT

In the multiplication and division of algebraic quantities, it is essential that we understand clearly the meaning of an exponent. In Chapter 2 we defined the word. However, at this point we need to emphasize the meaning. We repeat the definition of an exponent.

Definition. *An exponent is a number placed at the right and a little above another number, called the base, to show how many times the base is to be used as a factor.*

8-2 EXPONENTS IN MULTIPLICATION

Let us see what we mean by multiplying two powers, such as $x^4 \cdot x^3$:

$$x^4 \text{ means } x \cdot x \cdot x \cdot x$$
$$x^3 \text{ means } x \cdot x \cdot x$$

Therefore, $x^4 \cdot x^3$ means $(x \cdot x \cdot x \cdot x) \cdot (x \cdot x \cdot x)$. By using another exponent, we can write the product as x^7. Therefore, we can say

$$x^4 \cdot x^3 = x^{4+3} = x^7$$

If we analyze in the same way the multiplication of two other powers, such as $n^5 \cdot n^8$, we find that the product is n^{13}. From these examples and many others we can formulate the rule for exponents in the multiplication of algebraic quantities:

Rule. *In multiplying algebraic quantities, expressed as powers, we add exponents of the same base.* (We are multiplying the *quantities*, not the exponents.)

As a formula:

$$(x^a)(x^b) = x^{a+b}$$

This rule holds true for arithmetic numbers as well as for literal numbers.

$$2^5 \cdot 2^3 = 2^8 \text{ means } 32 \cdot 8 = 256$$
$$10^5 \cdot 10 = 10^6 \text{ means } 100{,}000 \cdot 10 = 1{,}000{,}000$$
$$5^3 \cdot 5^4 = 5^7 \text{ means } 125 \cdot 625 = 78{,}125$$

We state without proof the following facts:

1. The rule for adding exponents in multiplication holds true for fractional exponents: (add exponents.)

$$x^{\frac{1}{4}} \cdot x^{\frac{3}{8}} = x^{\frac{5}{8}}$$

2. The rule holds true for negative and zero exponents, as well as for positive exponents: (add exponents.)

$$(x^6)(x^{-2}) = x^4$$
$$(x^{-5})(x^{-2}) = x^{-7}$$
$$(x^5)(x^0) = x^5$$

3. The rule holds true also for literal exponents: (add exponents.)

$$(x^n)(x) = x^{n+1} \qquad (x^a)(x^b) = x^{a+b}$$
$$(x^{2c})(x^c) = x^{3c} \qquad (2^n)(2) = 2^{n+1}$$
$$(x^{a^2-3a+5})(x^{2a^2+a-2}) = x^{3a^2-2a+3}$$

If we have the multiplication of two quantities that include different bases, we add exponents of *each* base, separately.

Example

Multiply $(x^3y^2)(x^4y^3)$.
The quantity $x^3y^2 = x \cdot x \cdot x \cdot y \cdot y$
The quantity $x^4y^3 = x \cdot x \cdot x \cdot x \cdot y \cdot y \cdot y$
Therefore, $(x^3y^2)(x^4y^3) = x \cdot x \cdot x \cdot y \cdot y \cdot x \cdot x \cdot x \cdot x \cdot y \cdot y \cdot y$
Since multiplication of factors may be done in any order, the foregoing factors may be written simply as x^7y^5. Therefore, $(x^3y^2)(x^4y^3) = x^7y^5$.

When we multiply the two quantities together, we can get the product quickly by adding separately the exponents of x and the exponents of y. The exponent of x in the product is equal to the *sum* of the exponents of x in the factors. The exponent of y in the product is equal to the *sum* of the exponents of y in the factors.

8-3 MULTIPLICATION OF MONOMIALS

Example 1

Suppose we have this problem: $(-2x^3y^2z)(-5axy^4)$. To formulate a rule for multiplying monomials, let us first separate these monomials into prime factors:

$$-2x^3y^2z = (-2) \cdot x \cdot x \cdot x \cdot y \cdot y \cdot z$$
$$-5axy^4 = (-5) \cdot a \cdot x \cdot y \cdot y \cdot y \cdot y$$

The product, after rearranging factors, is

$$(-2) \cdot (-5) \cdot a \cdot x \cdot x \cdot x \cdot x \cdot y \cdot y \cdot y \cdot y \cdot y \cdot y \cdot z$$

By multiplying the factors, we get the product:

$$(-2x^3y^2z)(-5axy^4) = +10ax^4y^6z$$

In the multiplication of monomials we have the following steps:

Step 1. *Determine the sign of the product by the rule for the multiplication of signed numbers.*
Step 2. *Multiply the numerical coefficients of the factors to get the numerical coefficient of the product.*
Step 3. *Multiply the literal parts by writing down all literal factors and adding the exponents of like literal factors.*

Example 2

Multiply $(-3ab^2x^3)(-4a^2cx^2)(-5b^4cxy^2)$.
The product of three negative numbers is negative. The product of the numerical coefficients is $(3)(4)(5) = 60$. Adding the exponents of like literal factors, we get the product: $-60a^3b^6c^2x^6y^2$

8-4 EXPONENTS IN DIVISION

Since division is the inverse of multiplication, we should expect that division would require a procedure opposite to that of multiplication in the handling of exponents. In multiplication we *add* exponents of the same base. *In division we subtract exponents of the same base.* (The student should be careful *not* to say simply, "When we divide, we subtract." Remember we are dividing the *quantities*, not the exponents.)

The rule for exponents in division can be seen in examples. Suppose we wish to divide the quantity x^6 by the quantity x^2; that is, $x^6 \div x^2$. Let us first write the division as a fraction and then separate the quantities into prime factors:

$$\frac{x^6}{x^2} = \frac{x \cdot x \cdot x \cdot x \cdot x \cdot x}{x \cdot x}$$

If we now divide numerator and denominator by $x \cdot x$, we get

$$\frac{x^6}{x^2} = \frac{\cancel{x} \cdot \cancel{x} \cdot x \cdot x \cdot x \cdot x}{\cancel{x} \cdot \cancel{x}} = x^4$$

Stated in horizontal form, $x^6 \div x^2 = x^4$.

From the foregoing example we can formulate the rule for division:

Rule. *In dividing algebraic quantities expressed as powers, we subtract the exponent of the divisor from the exponent of the dividend.*

As a formula:

$$x^a \div x^b = x^{a-b}$$

Examples

$$\frac{n^9}{n^3} = n^6; \qquad a^{20} \div a^4 = a^{16}; \qquad x^5 \div x = x^4.$$

Note. *Division may be checked by multiplication.*

We state without proof the following facts:

1. The rule for subtracting exponents in division holds true also for *fractional exponents* (subtract exponents):

$$x^{\frac{3}{4}} \div x^{\frac{1}{8}} = x^{\frac{3}{4}-\frac{1}{8}} = x^{\frac{5}{8}}$$

2. The rule holds true for *negative and zero exponents,* as well as for positive exponents (subtract exponents):

$$x^5 \div x^{-2} = x^7 \qquad x^6 \div x^0 = x^6 \qquad x^{-3} \div x^{-7} = x^4$$

$$x^3 \div x^5 = x^{-2} \qquad x^5 \div x^5 = x^0$$

3. The rule holds true also for *literal exponents* (subtract exponents):

$$x^n \div x = x^{n-1} \qquad x^p \div x^q = x^{p-q} \qquad x^{4n} \div x^n = x^{3n}$$

$$(x^{2a^2-3a+4}) \div (x^{a^2+a+5}) = x^{a^2-4a-1}$$

8-5 DIVISION OF MONOMIALS

In division the number divided by another is called the *dividend.* The number that is divided into another is called the *divisor.* The answer obtained is called the *quotient.* After the division is completed as far as possible, any part of the dividend that is left over is called the *remainder.*

Suppose we have the following problem in division:

$$(-20x^5y^4z) \div (-5x^4y^2)$$

To formulate a rule for dividing monomials, let us first separate the monomials into prime factors and write the division as a fraction:

$$\frac{-20x^5y^4z}{-5x^4y^2} = \frac{(-)2 \cdot 2 \cdot 5 \cdot x \cdot x \cdot x \cdot x \cdot x \cdot y \cdot y \cdot y \cdot y \cdot z}{(-)5 \cdot x \cdot x \cdot x \cdot x \cdot y \cdot y}$$

If we divide numerator and denominator by equal factors, we get the answer. Remember, $(-) \div (-) = +$.

$$\frac{(-) \cdot 2 \cdot 2 \cdot \cancel{5} \cdot \cancel{x} \cdot \cancel{x} \cdot \cancel{x} \cdot \cancel{x} \cdot x \cdot \cancel{y} \cdot \cancel{y} \cdot y \cdot y \cdot z}{(-) \cdot \cancel{5} \cdot \cancel{x} \cdot \cancel{x} \cdot \cancel{x} \cdot \cancel{x} \cdot \cancel{y} \cdot \cancel{y}} = +4xy^2z$$

In division of monomials we have the following steps:

Step 1. *Determine the sign of the quotient by the rule for division of signed numbers.*
Step 2. *Divide the numerical coefficients to get the numerical coefficient of the quotient.*
Step 3. *Divide the literal parts by writing down all the literal factors of the dividend and subtracting the exponents of like factors in the divisor.*

Example

Divide $-36a^2bx^3y^4z$ by $4abxy^4$.

Solution

The sign of the quotient is equal to $(-) \div (+) = -$.
Dividing the numerical coefficients, we get 9.
Subtracting the exponents of the literal factors, we get ax^2z.
Therefore, the complete quotient is $-9ax^2z$.

Note. The division of monomials can be performed without writing out the separate prime factors.

8-6 MULTIPLICATION OF A POLYNOMIAL BY A MONOMIAL

To indicate the multiplication of a polynomial by a monomial, we can place the expressions next to each other, but *the polynomial must be enclosed in parentheses.* For example, to indicate the multiplication of the polynomial, $5x^2 - 7xy + 4y^2$, by the monomial 3, we enclose the polynomial in parentheses and then place the 3 either before or after the polynomial:

$$3(5x^2 - 7xy + 4y^2)$$

The expression then means that the entire polynomial is to be multiplied by the monomial. For the product we multiply *each* term of the polynomial by 3 and get

$$3(5x^2 - 7xy + 4y^2) = 15x^2 - 21xy + 12y^2$$

That the product is correct can be seen if we recall that multiplication is a shortened form of addition. In addition form, the problem becomes:

$$\begin{array}{rl} & 5x^2 - 7xy + 4y^2 \\ & 5x^2 - 7xy + 4y^2 \\ & 5x^2 - 7xy + 4y^2 \\ \hline \text{Adding like terms,} & 15x^2 - 21xy + 12y^2 \end{array}$$

The foregoing example is an illustration of the following rule:

Rule. *To multiply a polynomial by a monomial, multiply each term of the polynomial by the monomial.*

This rule is the application of the so-called *Distributive Law*, which states that *multiplication is distributive with respect to addition.* In general terms, the law states that $a(b + c) = ab + ac$.

Note. Be careful to observe all the rules for signs, coefficients, and letters with their exponents.

Example 1

Find the indicated product: $-3x^2y(4x^2 - 5xy - 7y^2)$.

Solution

The first term of the product is $(-3x^2y)(4x^2) = -12x^4y$. The second term of the product is $(-3x^2y)(-5xy) = +15x^3y^2$. The third term of the product is $(-3x^2y)(-7y^2) = +21x^2y^3$. The complete product is $-12x^4y + 15x^3y^2 + 21x^2y^3$.

Example 2

Find the indicated product: $x(x^3 - 4x^2 + 5x + 1)$.
By the foregoing rule, the product is $x^4 - 4x^3 + 5x^2 + x$.

Example 3

Perform the indicated multiplication: $2x(3x^2 - 5xy + 8y^2)(-4y)$.

Solution

In this example the polynomial $3x^2 - 5xy + 8y^2$ is to be multiplied by $2x$ and also by $(-4y)$. Notice that $-4y$ must be enclosed in parentheses, or the meaning will not be clear.

The actual multiplying may be done in two different ways. We may first multiply the monomial $2x$ by the monomial $(-4y)$. The monomial multiplier then becomes $-8xy$. In this case we have

$$-8xy(3x^2 - 5xy + 8y^2) = -24x^3y + 40x^2y^2 - 64xy^3$$

The product could also be found by multiplying the polynomial first by $2x$. However, in an example such as the foregoing the simplest way to find the product is to consider the entire monomial multiplier as $(2x)(-4y)$ or $-8xy$.

8-7 DIVISION OF A POLYNOMIAL BY A MONOMIAL

Since division and multiplication are inverse processes, we have the following rule for division:

Rule. *To divide a polynomial by a monomial, divide each term of the polynomial by the monomial.*

This rule is simply a restatement of the *Distributive Law*, this time in reverse. In general terms,

$$ab + ac = a(b + c); \qquad \text{then} \qquad \frac{ab + ac}{a} = b + c$$

As in the division of monomials, be careful to observe the rules for signs, coefficients, and exponents.

Example 1

Divide the polynomial $(18x^4y - 15x^3y^2 - 12x^2y^3)$ by $(-3x^2y)$.

This division problem may also be written as a fraction:

$$\frac{18x^4y - 15x^3y^2 - 12x^2y^3}{-3x^2y}$$

Here we divide *each* term of the polynomial by the monomial $-3x^2y$ and get the quotient

$$-6x^2 + 5xy + 4y^2$$

The answer can be checked by multiplication.

Example 2

Divide as indicated: $(5x^3 - 4x^2 + x) \div (x)$.

When each term of the polynomial is divided by the divisor, x, the quotient is $5x^2 - 4x + 1$. The correctness of the answer may be checked by multiplication.

Notice that the 1 must not be omitted in this example.

Example 3

Divide $(12ax^2 - 6a^2x^3 + 3ax) \div (3ax)$.

In this example the quotient is $4x - 2ax^2 + 1$. Whenever we find 1 as a factor with other factors, then the 1 may be omitted.

Exercise 8-1

Rewrite each of the following by the use of exponents:

1. $2 \cdot 2 \cdot 2 \cdot x \cdot x \cdot x \cdot x \cdot y \cdot y$

2. $3 \cdot a \cdot a \cdot b \cdot b \cdot b \cdot c \cdot c \cdot c \cdot c \cdot c$

3. $5 \cdot 5 \cdot 5 \cdot 5 \cdot x \cdot x \cdot x \cdot y \cdot z$

4. $7 \cdot 7 \cdot 7 \cdot a \cdot a \cdot a \cdot a \cdot a \cdot a \cdot b \cdot b \cdot c \cdot d$

5. $10 \cdot 10 \cdot 10 \cdot 10 \cdot x \cdot x \cdot y \cdot y \cdot z \cdot z \cdot z \cdot z \cdot z$

6. $3 \cdot 3 \cdot 3 \cdot 3 \cdot 5 \cdot 5 \cdot m \cdot m \cdot m \cdot m \cdot m \cdot n \cdot n \cdot n$

Multiply the following:

7. $(3x)(2x)$

8. $(-4x)(-5x^2)$

9. $(-2x^2y)(3xy^3)$

10. $(-3y^2)(y^5)$

11. $(7ax^3y)(-xy^2)$

12. $(4x^4)(5x^5)$

13. $(-3x^3)(-6x^6)$

14. $(5n^5)(n)$

15. $(5ab)(3cd)$

16. $(-c^2x^3y)(-cxy)$

17. $(-3x^2y)(-4x^3z^2)(-5xy^3z^4)$

18. $(-2ab^2c)(7a^3c^2d^4)(-3b^3cd)$

19. $(4mn^2)(-6m^2n^3)(2amn)(-2a^2)$

20. $(5r^2st)(-9rs^2)(-11rs^4t^2)$

21. $(-9xy^2z)(-7x^2z^3)(3x^3yz^4)$

22. $(-3ab)(-2a^2c)(4b^2cd)(-ab)$

23. $(-a^2b)(-a^2b^2c)(-bc^2)(+a)$

24. $-(-2x)(-4x^3y)(-3xy^2)(-y)$

25. $(-1)(-2a)(-3b)(4c)(-5d)(6e)$

26. $-(-t)(u)(-v)(-w)(x)(-y)(-z)$

27. $(3x)(4x^2 + 5xy - 2y^2)$

28. $-2x^2y(x^2y - 4x^3y^2 + x^4y^3)$

29. $2x(4x^2 - 5xy - 7y^2)yz$

30. $3a^2(4a^2 + 4ab - 5b^2)(-2b)$

31. $-4ac^2(-2a^3 + 3a^2bc - 5ab^2c^2 + 3b^3c^4)$

32. $-5xy^2(5x^3 - 3x^2y - xy^2z + y^3z^2)2z$

33. $2xy(4x^3 - 5x^2y + 4xy^2 - y^3 + 2x + 1)$

34. $xy^2z(1 - 4xy^3 - 6y^2z + 3xyz^2 + 7x - 3y)$

Divide as indicated:

35. $(20x^6y^2z^5) \div (-2xyz)$

36. $(-32a^4b^5c) \div (8a^4bc)$

37. $(14m^2n^3xy) \div (7mn^3x)$

38. $(+9ab^2c^3d) \div (6abcd)$

39. $(5x^2yz^3) \div (-5x^2yz^3)$

40. $(+50rs^2t^4) \div (-10rs^2t^4)$

41. $(42x^3yz) \div (-6xy)$

42. $(-36x^3y^4z) \div (+9x^2z)$

43. $(12x^3y^4 - 10xy^3 + 8x^2y^2) \div (-2xy)$

44. $(24ax^2y - 36a^2xy^2 + 12axy) \div (+12axy)$

45. $(8r^2s^3t^4 - 6rs^2t + 4st^3 - 5r^3s^4t^5) \div (2st)$

46. $(6x^6 - 8x^8 + 10x^{10} - 12x^{12} - 4x^4 + 2x^2) \div (-2x^2)$

8-8 MULTIPLICATION OF A POLYNOMIAL BY A POLYNOMIAL

The multiplication of a polynomial by another polynomial in algebra is somewhat similar in form to the multiplication in arithmetic of a two-or-more-digit number by another similar number. As an example, consider the following multiplication problem in arithmetic: (32)(42). We usually write one number above the other:

$$\begin{array}{r} 32 \\ 42 \end{array} \qquad\qquad \begin{array}{ccc} 3 & & 2 \\ \uparrow & \times & \uparrow \\ 4 & & 2 \\ \hline \end{array}$$

In the multiplication of two numbers, one is called the multiplicand and the other is the multiplier. The number taken a certain number of times is called the *multiplicand.* The number that indicates how many times the multiplicand is to be taken is called the *multiplier.* In arranging the numbers for multiplication, we usually consider the top number the multiplicand and the bottom number the multiplier.

In the foregoing example in arithmetic, notice that we must make *four* separate multiplications. They are $2 \cdot 2$, $2 \cdot 3$, $4 \cdot 2$, and $4 \cdot 3$. The results are placed in their proper positions and added. Each digit in the multiplicand is multiplied by both digits in the multiplier.

In algebra, if we wish to multiply one polynomial by another, we place one below the other, just as in arithmetic. Suppose we have this problem:

Example 1

Multiply $(3x - 4y)(2x + 5y)$.

We use parentheses around each binomial to indicate that it is to be considered as one quantity. For multiplication, we place one binomial below the other:

$$\begin{array}{r} 3x - 4y \\ \underline{2x + 5y} \end{array}$$

Here we have four separate multiplications, just as we had in the arithmetic example. The four separate multiplications are shown by arrows. *Each* term of one binomial must be multiplied by *both* terms of the other.

$$\begin{array}{l} 3x - 4y \\ \uparrow \\ \underline{2x + 5y} \\ 6x^2 \end{array}$$

In algebra we begin multiplication at the *left* side instead of at the right as in arithmetic. The first step is to multiply $(2x)(3x)$. The first product, $6x^2$, is placed at the left just below the $2x$ of the multiplier. The *next* step is the multiplication $(2x)(-4y) = -8xy$. The result is written as the second term of the product.

$$\begin{array}{l} 3x \quad - 4y \\ \uparrow \\ \underline{2x \quad + 5y} \\ 6x^2 - 8xy \end{array}$$

The third step is the multiplication $(5y)(3x) = 15xy$. Since the product is a term similar to the term $-8xy$, it may be placed in the same column for convenience in adding. The fourth multiplication is $(5y)(-4y)$, which equals $-20y^2$, placed at the right. The partial products are finally combined by addition, as in arithmetic. The complete product is shown.

$$\begin{array}{llll} 3x & - & 4y & \\ & & \uparrow & \\ 2x & + & 5y & \\ \hline 6x^2 & - & 8xy & \\ & + & 15xy & - 20y^2 \\ \hline 6x^2 & + & 7xy & - 20y^2 \end{array}$$

Example 2

Multiply $(4x - 7y)$ by $(3x + 5y)$.

Solution

$$\begin{array}{llll} 4x & - & 7y & \\ 3x & + & 5y & \\ \hline 12x^2 & - & 21xy & \\ & + & 20xy & - 35y^2 \\ \hline 12x^2 & - & xy & - 35y^2 \end{array}$$

Example 3

Expand $(7 - 3x^2 + 4x^3 - 5x)(2 + 3x)$.

Solution

Before beginning the actual multiplication, we shall often find it desirable to rearrange the terms of the multiplicand and the multiplier. Although the multiplication could be done without making any changes, it is usually better to arrange the terms in a descending order of powers of some letter. Therefore, we arrange the terms of multiplicand and multiplier so that each starts with the highest power of x.

Rearranging terms,	$4x^3 - 3x^2 - 5x + 7$
	$3x + 2$
Multiplying by $3x$,	$12x^4 - 9x^3 - 15x^2 + 21x$
Multiplying by 2,	$8x^3 - 6x^2 - 10x + 14$
The product is	$12x^4 - x^3 - 21x^2 + 11x + 14$

In the foregoing multiplication you will notice that each term of the multiplicand, $4x^3 - 3x^2 - 5x + 7$, is first multiplied by $3x$. Then each term of the multiplicand is multiplied by +2. If like terms appear in these multiplications, they are placed in the same columns and then added. If no like terms appear, each term is placed by itself since it cannot be combined with other terms. Note that like terms are added *algebraically*.

For multiplication involving two polynomials we have the following rule.

Rule. *Multiply each term of one polynomial by each term of the other, and combine any like terms that appear.*

In general terms, the rule says

$$(a + b)(c + d) = ac + ad + bc + bd$$

If like terms appear in the multiplication, they can be aligned in the same column to simplify their addition.

Exercise 8-2

Perform the indicated multiplications:

1. $(3x - 5y)(2x + 3y)$

2. $(3x^2 - 5 + 4x)(x - 3)$

3. $(2x^2 - 3xy + y^2)(3x + y)$

4. $(8 - 2x^3 + 3x)(3x - 2)$

5. $(2x^2 - 2xy - y^2)(x - 7)$

6. $(3x^2 - 4x - 5)^2$

7. $(x^2 - 2xy + y^2)^2$

8. $(x - 2)(x^2 + 4x + 8)$

9. $(2x + 1)(3x^2 - 4x - 5)$

10. $(x^2 + 3x - 7)(4 - x)$

11. $(4x^3 - xy^2 - y^3 - 3x^2y)(x - y)$

12. $(2x^3 - 5x - 2 + 4x^2)(2x - 1)$

13. $(2x - 3 + x^3 - 5x^2)(2 + 3x)$

14. $(3x - x^2 + 4 - x^3)(2x - 3)$

15. $(2x^2 - 2 - 4x^3 + 3x)(4 - 3x^2)$

16. $(3x - 2)^3$

17. $(3 + x^2 - 2x)(x^3 - 1 - x - 2x^2)$

18. $(2x^3 - 5 - x^2)(x^2 + 1 - 3x)$

19. $(x - 3)(2x + 5)(x - 4)$ **20.** $(x + 4)(2x - 1)(4 - x)(2x + 1)$

21. $(4x^2 - 6xy + 9y^2)(2x + 3y)$ **22.** $(x - 3)^4$

23. $(a^2 + b^2 + ab)(a^2 + ab - b^2)$ **24.** $(x - 2)(x - 3)(x + 2)(x + 3)$

25. $(x - 2)(x + 3)(x - 4)(x + 1)(x - 6)$

26. $(3x^2y + 4axy^2 - 7aby^3)(2a - 3b)$

27. $(x^4 - x^3y + x^2y^2 - xy^3 + y^4)(x + y)$

28. $(2r^3 - r^2t - 5rt^2 + 6t^3)(3r - 2t)$

29. $(2x^2 - 3x + 5)(4x^3 + 2x^2 - 7x - 3)$

30. $3x(2x - 5)(2x + 5)(x^2 + 2x - 3)(2y)$

8-9 DIVISION OF A POLYNOMIAL BY A POLYNOMIAL

The division of one polynomial by another polynomial in algebra is similar in form to long division in arithmetic.

Suppose we have the following problem:

$$(6x^4 + 5x^3 - 12x^2 - 8x + 3) \div (3x - 2)$$

Let us now recall the steps in a problem in long division in arithmetic, for example, $(7685) \div (28)$. We write the two problems side by side and note the steps in each:

Arithmetic

$$\begin{array}{r} 2 \\ 28\overline{)7685} \\ 56 \\ \hline 208 \end{array}$$

Algebra

$$\begin{array}{r} 2x^3 \\ 3x - 2\overline{)6x^4 + 5x^3 - 12x^2 - 8x + 3} \\ 6x^4 - 4x^3 \\ \hline + 9x^3 - 12x^2 \end{array}$$

Step 1. In arithmetic we divide 28 into 76. We first consider 2 into 7. The answer is 3. However, this is too large for the first number in the quotient, so we write 2 as the quotient's first digit.

In algebra we divide $3x$ into $6x^4$. The answer is $2x^3$, which is the first term in the quotient. In algebra we do not need to "try" this answer. There is no guess work here in algebra.

The proper division of the first term of the divisor into the first term of the dividend will always result in the correct first term of the quotient.

Step 2. In arithmetic we multiply the entire divisor by the first digit of the quotient; that is, we "multiply back," taking 2 times the divisor, 28, and placing the product, 56, under the 76. Subtracting, we get 20 as a first remainder.

In algebra we multiply the entire divisor by the first *term* of the quotient; that is, we "multiply back," taking $2x^3$ times the divisor $3x - 2$. The product, $6x^4 - 4x^3$, is placed under the first part of the dividend. Subtracting, we get $9x^3$ as a remainder.

Step 3. In arithmetic we bring down 8, the next digit. In algebra we bring down $-12x^2$, the next term.

Step 4. From this point on, we continue to divide, repeating the same steps as in arithmetic. In algebra we continue until we get a remainder of zero or a remainder whose highest power of x is lower than the highest power of x in the divisor. If there is some remainder other than zero, it may be placed above the divisor as a fraction, as in arithmetic, and then added to the quotient as a fraction.

In algebra, the complete steps are as follows:

1. Divide $3x$ into $6x^4$. The result is $2x^3$.
2. Place the $2x^3$ as the first term in the quotient.
3. Multiply back, taking the $2x^3$ in the quotient times the *entire* divisor. This makes $6x^4 - 4x^3$. This quantity is placed below the first part of the dividend, as in arithmetic.
4. Subtract. The remainder is $+9x^3$ (algebraic subtraction).
5. Bring down $-12x^2$, the next term.
6. Divide $3x$, the *first* term of the divisor, into the $+9x^3$. The result is $+3x^2$.
7. Place $3x^2$ as the second term of the quotient.
8. Multiply back, taking $+3x^2$ times the entire divisor. The result, $9x^3 - 6x^2$, is placed below the new dividend, as in arithmetic.
9. Subtract. The remainder is $-6x^2$.
10. Bring down $-8x$, the next term of the dividend.
11. Continue to divide in the same way until the highest power of x in the dividend is less than the highest power in the divisor.
12. The remainder is -5.

The problems in arithmetic and algebra are shown here complete.

$$
\begin{array}{r}
274 + \frac{13}{28} \\
28\overline{)7685} \\
56 \\
\hline
208 \\
196 \\
\hline
125 \\
112 \\
\hline
13
\end{array}
\qquad
\begin{array}{r}
2x^3 + 3x^2 - 2x - 4 - 5/(3x - 2) \\
3x - 2\overline{)6x^4 + 5x^3 - 12x^2 - 8x + 3} \\
6x^4 - 4x^3 \qquad\qquad\qquad\quad \\
\hline
+ 9x^3 - 12x^2 \qquad\qquad \\
9x^3 - 6x^2 \qquad\qquad \\
\hline
- 6x^2 - 8x \qquad \\
- 6x^2 + 4x \qquad \\
\hline
- 12x + 3 \\
- 12x + 8 \\
\hline
- 5
\end{array}
$$

Note 1. In division it is necessary to arrange the terms of both dividend and divisor in a *descending order of powers* before beginning the division. Start with the highest power of x, or whatever the letter. If there are two letters with various powers, then select one of them for the arrangement according to descending powers.
Note 2. If one of the powers is missing, it means that the coefficient of that power is zero (0). A space should be allowed in the dividend for each missing term. This vacancy may be conveniently shown by placing "+0" for each of the missing terms.

Example

Divide $(3x^4 - 7 + 8x - 5x^2) \div (x - 2)$

Before beginning the division, we rearrange the terms of the dividend and divisor in a descending order of powers of x, starting with the highest power. We make allowance for any missing terms by placing +0 in the place of each missing term.

$$x - 2\overline{)3x^4 + 0 - 5x^2 + 8x - 7}$$

The completion of the division is left as an exercise for the student. The quotient is $(3x^3 + 6x^2 + 7x + 22)$, with a remainder of 37.

Exercise 8-3

Divide as indicated:

1. $(x^2 + 8x + 15) \div (x + 3)$

2. $(x^2 - 9x + 18) \div (x - 3)$

3. $(3x^2 + 10x - 6) \div (x + 4)$

4. $(3x^2 - 7x - 6) \div (x + 2)$

5. $(2x^3 - 3 - 9x^2) \div (2x - 1)$

6. $(x + 8x^3 - 14x^2) \div (2x - 3)$

7. $(7 - 5x + 3x^3) \div (x + 2)$

8. $(5 - 14x + 2x^3) \div (x + 3)$

9. $(x + 8x^3 - 14x^2) \div (2x - 3)$

10. $(6x^3 - 4x - 7x^2) \div (3x - 2)$

11. $(12x^3 - 11x - 5x^2) \div (4x - 3)$

12. $(6x^3 - 13x - 5x^2) \div (3x - 4)$

13. $(6c^2 - 5cd - 9d^2) \div (2c - 3d)$

14. $(6x^2 + 5xy - 3y^2) \div (3x - 2y)$

15. $(v^2 - 9v - 8v^3 + 6v^4 - 10) \div (3v + 2)$

16. $(4x^2 - 9) \div (2x + 3)$

17. $(6x - x^3 + 12x^4 - 23x^2 - 5) \div (3x - 4)$

18. $((9a^2 - 16) \div (3a - 4)$

19. $(13x^2 - 23x + 18x^4 - 3) \div (3x - 2)$

20. $(t^3 + 64) \div (t + 4)$

21. $(32x^4 + 3 - 58x^2 - 2x) \div (4x - 5)$

22. $(n^3 - 27) \div (n - 3)$

23. $(11 + 6x^3 + 8x^4 - 37x) \div (4x - 5)$

24. $(x^4 - 81) \div (x + 3)$

25. $(9x^3 - 12x^4 + 16x^2 - 7) \div (3 - 4x)$

26. $(x^5 - 32) \div (x - 2)$

27. $(34x^3 - 24x^4 + x - 2) \div (2 - 3x)$

28. $(x^6 - 1) \div (x + 1)$

29. $(6x^3 - 3x^2y - 2xy^2 + 12y^3) \div (2x - 3y)$

30. $(6a^4 + 6b^4 + 22a^3b + ab^3) \div (3a + 2b)$

31. $(2c^4 + 7c^3 + 13c^2 + 11c + 3) \div (c^2 + 2c + 3)$

32. $(6x^4 + 5x^3 + 17 - 7x - 8x^2) \div (3x + 2x^2 - 4)$

33. $(x^2y^2 - 7x^3y - 31xy^3 + 12x^4 - 15y^4) \div (2xy + 5y^2 + 3x^2)$

34. $(5y^5 - 3x^3y^2 + 6x^5 - 2xy^4) \div (y^2 + 2x^2 + 2xy)$

35. $(4x^4 - 5x^2 - 6) \div (x^2 - 2)$ **36.** $(3x^6 - 5x^3 - 2) \div (3x^3 + 1)$

8-10 SYNTHETIC DIVISION

Synthetic division is a way of making long division short. This can be done by omitting much of the writing, especially the needless repetition of certain terms. Consider the following problem:

$$(3x^4 - 10x^3 + 6x^2 + 9x - 16) \div (x - 2)$$

First, let us work the problem in the usual way and note the repetition of certain terms: In the division shown, the starred terms need not be written. Note that they are simply repetitions. In each subtraction, the first remainder is always zero, so we might omit repeating the term. Also, the terms, $6x^2$, $9x$, and -16, need not be brought down. The subtraction can be done just as the terms stand.

$$
\begin{array}{rrrrrr}
 & 3x^3 & -\ 4x^2 & -\ 2x & +\ 5 & \\
\cline{2-6}
x-2) & 3x^4 & -\ 10x^3 & +\ 6x^2 & +\ 9x & -\ 16 \\
 & {*3x^4} & -\ 6x^3 & & & \\
\cline{2-3}
 & & -\ 4x^3 & +\ {*6x^2} & & \\
 & & -\ {*4x^3} & +\ 8x^2 & & \\
\cline{3-4}
 & & & -\ 2x^2 & +\ {*9x} & \\
 & & & -\ {*2x^2} & +\ 4x & \\
\cline{4-5}
 & & & & +\ 5x & -\ {*16} \\
 & & & & +\ {*5x} & -\ 10 \\
\cline{5-6}
 & & & & & -\ 6
\end{array}
$$

We may therefore omit the unnecessary repetition of terms. Moreover, we may even omit the letter x itself and write only the coefficients. Then we have the form shown at the right. In the result the starred numbers here show the quotient coefficients. To make the arrangement more compact, we move all the numbers up nearer the dividend and get

$$\begin{array}{rrrrrr}
 & 3 & -4 & -2 & +5 & \\ \hline
-2) & 3^* & -10 & +6 & +9 & -16 \\
 & & -6 & & & \\ \hline
 & & -4^* & & & \\
 & & & +8 & & \\ \hline
 & & & -2^* & & \\
 & & & & +4 & \\ \hline
 & & & & +5^* & \\
 & & & & & -10 \\ \hline
 & & & & & -6
\end{array}$$

$$\begin{array}{rrrrrr}
 & 3 & -4 & -2 & +5 & \\ \hline
-2) & 3 & -10 & +6 & +9 & -16 \\
 & & -6 & +8 & +4 & -10 \\ \hline
 & & -4 & -2 & +5 & -6
\end{array}$$

Note that the numbers just below the dividend came about through multiplying the divisor by each term of the quotient. Now, if we bring the first coefficient of the dividend down in line with the remainders, these numbers in the bottom row will show the coefficients in the quotient, the last number being the final remainder. Therefore, we need not write the quotient. We get

$$\begin{array}{r|rrrrr}
-2 & 3 & -10 & +6 & +9 & -16 \\
 & \downarrow & -6 & +8 & +4 & -10 \\ \hline
 & 3 & -4 & -2 & +5 & -6
\end{array}$$

The numbers in the bottom row came about through subtraction. Now we make one more change. To avoid subtraction (and use addition instead), we change the sign of the divisor from -2 to $+2$.

Now let us summarize the steps in synthetic division, using the foregoing example:

1. With the terms of the dividend arranged in a descending order of powers, write the coefficients, including a zero for each missing power: $3\ -10\ +6\ +9\ -16$

2. Change the sign of the constant in the divisor and write this number at the *left* (some books say *right*):

$$\begin{array}{r|rrrrr}
2 & 3 & -10 & +6 & +9 & -16 \\
 & & & & & \\ \hline
\end{array}$$

3. Leaving a working space under the dividend, draw a horizontal line below.

4. Bring down the first coefficient of the dividend:

$$\begin{array}{r|rrrrr}
2 & 3 & -10 & +6 & +9 & -16 \\
 & \downarrow & & & & \\ \hline
 & 3 & & & &
\end{array}$$

5. Multiply this number by the divisor, place the product under the second term of the dividend, and *add*:

$$\begin{array}{r|rrrrr} 2 & 3 & -10 & +6 & +9 & -16 \\ & \downarrow & 6 & & & \\ \hline & 3 & -4 & & & \end{array}$$

6. Multiply the result by the divisor, place the product under the third term of the dividend, and *add*:

$$\begin{array}{r|rrrrr} 2 & 3 & -10 & +6 & +9 & -16 \\ & & 6 & -8 & & \\ \hline & 3 & -4 & -2 & & \end{array}$$

7. Proceed in the same way to the end of the polynomial. We get

$$\begin{array}{r|rrrrr} 2 & 3 & -10 & +6 & +9 & -16 \\ & & 6 & -8 & -4 & 10 \\ \hline & 3 & -4 & -2 & +5 & -6 \end{array}$$

The last term in the bottom row is the remainder, and the other terms are the coefficients of the quotient in descending powers of x. The degree of the quotient is one less than the degree of the dividend. The quotient is, therefore,

$$3x^3 - 4x^2 - 2x + 5$$

with a remainder of -6.

If the final remainder is zero, the division is even, and we can say the dividend can be factored into divisor and quotient.

Example 1

Divide $(2x^4 - 11x^3 - x + 17 + 14x^2) \div (x - 3)$

Solution

$$\begin{array}{r|rrrrr} 3 & 2 & -11 & +14 & -1 & +17 \\ & & 6 & -15 & -3 & -12 \\ \hline & 2 & -5 & -1 & -4 & +5 \end{array}$$

Quotient: $2x^3 - 5x^2 - x - 4$; remainder $= 5$

Example 2

Divide: $(6 + 9x^3 - x + 4x^4) \div (x + 2)$

Solution

$$\begin{array}{r|rrrrr} -2 & 4 & 9 & 0 & -1 & 6 \\ & & -8 & -2 & 4 & -6 \\ \hline & 4 & 1 & -2 & 3 & 0 \end{array}$$

Quotient: $4x^3 + x^2 - 2x + 3$; remainder $= 0$. Since the remainder is zero, then the binomial, $x + 2$, is a factor of the dividend.

Note. Synthetic division can be used only when the divisor is a binomial of the form, $x - c$, in which c is a constant and the coefficient of x is 1. If the coefficient of x is other than 1, such as in $3x - 4$, synthetic division may be used by making a certain adjustment in the divisor. We may first divide the divisor by the coefficient of x, in this case, 3, making it $x - \frac{4}{3}$. If we make no other adjustment, the quotient, of course, will be 3 times as large as it should be. However, to get the correct answer, we may also divide the dividend by 3. If we leave the dividend as it is, we finally divide the quotient by 3.

Example 3

Divide by synthetic division: $(12x^4 + x^3 - 12x^2 + x + 2) \div (3x - 2)$.

Solution

We change the divisor to $(x - \frac{2}{3})$. If we also divide the dividend by 3, we have

$$\begin{array}{r|rrrrr} \frac{2}{3} & 4 & \frac{1}{3} & -4 & \frac{1}{3} & \frac{2}{3} \\ & & \frac{8}{3} & 2 & -\frac{4}{3} & -\frac{2}{3} \\ \hline & 4 & 3 & -2 & -1 & 0 \end{array}$$

Quotient: $4x^3 + 3x^2 - 2x - 1$; remainder $= 0$. Then $3x - 2$ is a factor of the polynomial, $12x^4 + x^3 - 12x^2 + x + 2$.

Exercise 8-4

Use synthetic division to divide each of the following polynomials by the binomials shown. In each case, write the complete quotient and the remainder. Tell in which cases the polynomial can be factored, and state the binomial factor, or factors.

1. $2x^3 - 6x^2 + 3x + 2$ divided by $(x - 1)$; $x + 1$; $x - 2$; $x - 3$; $x + 2$.

2. $3x^4 - 6x^2 + 2x + 4x^3 - 5$ divided by $x - 1$; $x + 1$; $x - 2$; $x + 2$; $x - 3$.

3. $24x - 8 + 8x^2 - 21x^3 + 6x^4$ divided by $x - 2$; $x + 1$; $x + 2$; $x - 3$; $x - 4$.

4. $10x - 9x^3 + 2x^4 - 12 + 7x^2$ divided by $x - 1$; $x + 1$; $x - 2$; $x + 2$; $x - 3$.

5. $10x^2 - 7x^3 - 12 - 8x + 3x^4$ divided by $x - 1$; $x + 1$; $x - 2$; $x - 3$; $x - 4$.

6. $5x^3 - 18x - 4$ divided by $x - 1$; $x + 1$; $x - 2$; $x + 2$; $x - 3$; $x - 5$.

7. $6 - 5x^3 + 4x^4 - 15x$ divided by $x - 1$; $x + 1$; $x - 2$; $x + 2$; $x - 3$; $x + 3$.

8. $3x^4 - 12 + 40x - 10x^3 - 9x^2$ divided by $x - 1$; $x + 1$; $x - 2$; $x + 2$; $x - 3$.

9. $3x^4 - 2x^3 + x^2 - x - 7$ divided by $x - 1$; $x + 1$; $x - 2$; $x + 2$; $x - 4$.

10. $3x^4 + 16 - 7x^2 - 12x$ divided by $x - 1$; $x + 1$; $x - 2$; $x + 2$; $x - 4$.

11. $3x^4 - 5x^3 - 9x^2 + 20x - 12$; try to find one binomial factor.

12. $2x^5 + x^4 - 31x^3 - 36x^2 + 52x + 48$; find three binomial factors.

13. $x^4 - 15x^2 + 10x + 24$; find four binomial factors.

14. $2x^4 - 11x^3 - 13x^2 + 99x - 45$; find three binomial factors.

15. $3x^4 + 4x^3 - 41x^2 - 86x - 24$; find three binomial factors.

16. Divide $x^5 - 32$ by $x - 2$. **17.** Divide $x^7 + 128$ by $x + 2$.

18. Divide $10x^4 - 7x^3 - 18x^2 + 5x + 6$ by $2x - 3$.

QUIZ ON CHAPTER 8, MULTIPLICATION AND DIVISION. FORM A.

1. Multiply as indicated:

$$(-3x^2y)(+5x^3yz^2)(-4axz^3)(-x)$$

2. Multiply as indicated:

$$-2xy^2(4x^2 - 5xy + 3y^2 + 1)$$

3. Multiply:

$(2x - 3)(6x^2 - 5x + 4x^3 - 6)$; (first rearrange terms)

4. Multiply:

$$(3x - 5)(9x^2 + 15x + 25)$$

5. Divide as indicated:

$$(24ab^3c^4) \div (-3ab^2c)$$

6. Divide as indicated:

$$(12x^3y^2 - 8x^2yz + 2xy) \div (2xy)$$

7. Divide as indicated:

$(22x + 6x^4 - 6 - 17x^3) \div (2x - 3)$; (rearrange terms)

8. Divide as indicated:

$(8x^4 + x - 24x^2 + 10) \div (2x + 3)$; (rearrange terms)

9. Divide by synthetic division:

$$(3x^4 - 12x - 7x^2 + 4) \div (x - 2)$$

10. Divide by synthetic division:

$$(2x^4 - 3x^3 + 7 - 8x^2) \div (x + 1)$$

CHAPTER NINE The Equation

9-1 THE EQUATION DEFINITION

The equation is probably the most important idea in all mathematics. In fact, *algebra can be called the study of the equation.* The equation enables us to solve practical problems. All other work in algebra is pointed toward the solving of equations. It is the equation that makes algebra useful in finding answers to problems.

What, then, is an equation?

An equation is a statement of equality between two equal quantities. There are three requirements for an equation:

1. *We must have two quantities.*
2. *The quantities must be equal.*
3. *We must say they are equal.*

Let us consider the following statement: $3 + 5 = 8$. Here we have two separate quantities. One quantity is $3 + 5$. The other quantity is 8. The two quantities are equal. The equal sign (=) makes the statement that they are equal.

The equal sign (=) is one of the most important symbols in mathematics. It is the symbol that says something. It is the "verb" in grammar. Without the equal sign, nothing is stated. The equal sign is necessary in making the statement. However, we must also have something after the equal sign, otherwise the statement is not complete.

9-2 IDENTITIES AND CONDITIONAL EQUATIONS

There are two kinds of equations, *identities* and *conditional equations.* An *identity* is an equation that is true for *all values* of the letter used. A *conditional equation* is an equation that is true for *only some particular value or values* of the unknown or letter used.

Let us see whether the statement $8x - 5x = 3x$ is an identity or a conditional equation. We try some values of x to determine whether or not the statement is true for any particular value of x.

If x is equal to 7, we have $56 - 35 = 21$. This statement is true.
If x is equal to 11, we have $88 - 55 = 33$. This also is true.

If we try several other values for x, such as $x = 9$, $x = 12$, $x = 25$, or any other value, we find the statement $8x - 5x = 3x$ is true, no matter what value we use for x. This equation is therefore an *identity* because it is true for *all* values of x. For an identity we often use an equality sign consisting of three parallel lines ($\equiv$). Thus, $8x - 5x \equiv 3x$.

As another example, suppose we consider the statement, $3x + 2 = 17$. By inspection, we see that the statement is true *only* on the *condition* that x is equal to 5. We say that the number 5 *satisfies* the equation; that is, if we substitute 5 for x in the equation, the statement is true. If we try any other value for x, say $x = 4$, we shall find that the statement is not true. If $x = 4$, we have $12 + 2 = 17$, which is not true. Since the equation $3x + 2 = 17$ is true for some but not for all values of x, the equation is a *conditional equation.*

An equation may be true for more than one value of the letter and yet not be an identity. Consider the equation

$$x^2 - 5x + 6 = 0$$

Let us try some values for x in this equation. If we try $x = 1$, we find that the equation is not true. If we try the value $x = 2$, we get $4 - 10 + 6 = 0$, which is true. If we try the value $x = 3$, we get $9 - 15 + 6 = 0$, which is also true. In this equation we find that x may have two different values that make the equation true. Yet it is not an identity, since it is not true for *all* values of x.

9-3 ROOT OF AN EQUATION

The *root* of a conditional equation is any value for the letter that makes the equation true. In the equation $3x + 2 = 17$, the only value for x that makes the equation true is $x = 5$. Then we say that 5 is the only root of this equation. The root is also called the *solution* of the equation.

In some instances it may happen that an equation as written has no root; that is, there is no number that will make the equation true. Then we say the equation has no solution. Actually, in that case, it should not be called an equation because it is not true.

As an example, consider the following statement:

$$3x - 3x = 5$$

If we combine the terms on the left, we get the statement $0 = 5$. This result means that the original statement is impossible for any value of x, and the statement has no solution.

Exercise 9-1

Tell whether each of the following equations is an identity or a conditional equation. Then find, by inspection, the root of each conditional equation. Also tell whether the equation has no solution.

1. $x + 4 = 13$	**2.** $x + 2x = 3x$	**3.** $x + 3 = x$
4. $x + 7 = 7$	**5.** $x - 6 = 0$	**6.** $7 - x = 1$
7. $10 = 8 + x$	**8.** $5 + 2x = 2x$	**9.** $2x = 16$
10. $30 = 3x$	**11.** $4x = 48$	**12.** $24 = 6x$
13. $3x + 6 = 6$	**14.** $x = 8x - 7x$	**15.** $2x + 7 = 15$
16. $x = 3x - 2x$	**17.** $3x = -21$	**18.** $-10 = 5x$
19. $2x = 7$	**20.** $2x - 3 = 19$	**21.** $3x = 14 - 4x$
22. $3x + 4x = 7x$	**23.** $4x = 18$	**24.** $x + 8 = 3x$

9-4 SOLVING AN EQUATION

To *solve* an equation means to find the root. In the equation $3x + 2 = 17$ we know that the root is 5, since that is the only value for x that will make the statement a true equation.

In the foregoing equation we can easily determine the root by inspection. However, the root of an equation is not always easy to find. We need a more systematic method for solving an equation. For instance, in the equation $5x + 4 = 2x - 7$, we may have to guess a long while before finding the root.

9-5 THE EQUATION AS A BALANCED SCALE

Before beginning to solve equations, let us see how an equation resembles a balanced scale, as shown in Fig. 9-1. On the left pan of the scale is a bag of sand of an unknown weight, x. On the right pan is an 8-ounce weight. Let us suppose the scale is perfectly balanced. This means that we have exactly the same quantity on both sides of the scale.

An equation is like a balanced scale. The scale in Fig. 9-1 may be represented by the equation

$$x = 8$$

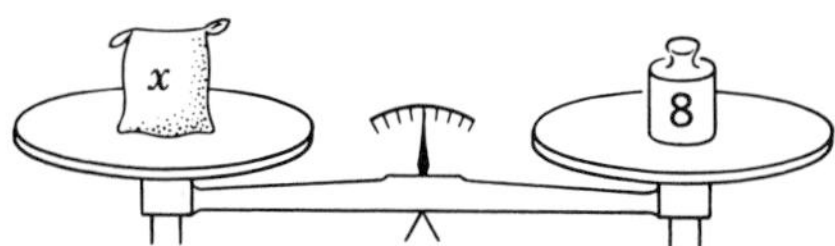

Figure 9-1

The quantity on the left side of the scale, or equation, is x. The quantity on the right of the scale, or equation, is 8. The equal sign (=) says that the quantities are equal. This is the same as saying that the scale balances.

If you think of an equation at all times as a balanced scale, you will find many of the algebraic processes very simple.

9-6 THE AXIOMS

To solve equations we make use of several important principles called *axioms. An axiom is a general statement that is accepted as true without proof.* In order to solve equations, we need to know what the axioms mean and how they are used. The axioms are best understood if they are applied to a balanced scale.

Consider the scale in Fig. 9-1. Suppose we add a 3-ounce weight to the left side and a 3-ounce weight to the right side of the scale. (Fig. 9-2.) We feel sure that the scale will still balance. We seem to know that the following statement is true:

If we add the same quantity to both sides of a balanced scale, the scale will still balance.

The foregoing statement may be tested in a laboratory by adding various weights to one side and then always adding an equal weight to the other side. We might verify the statement with many examples. For instance, we may add 4 ounces to both sides, then 5 ounces, then 6 ounces, and so on. In each case the scale would still balance.

However, notice that the statement says that, *no matter how much is added to one side, if the same amount is added to the other side, then the scale will still balance.* To verify the statement for *all* cases is impossible. *All* cases would also have to include fractions. Yet we feel that the general statement is true for all cases. *The truth of the statement seems so obvious that it does not require proof.*

The axiom we have just illustrated with a balanced scale is called the *addition axiom.* It is often referred to as the first axiom. It is stated as follows:

Axiom 1. *If the same quantity is added to both sides of an equation, the new equation will still be true.*

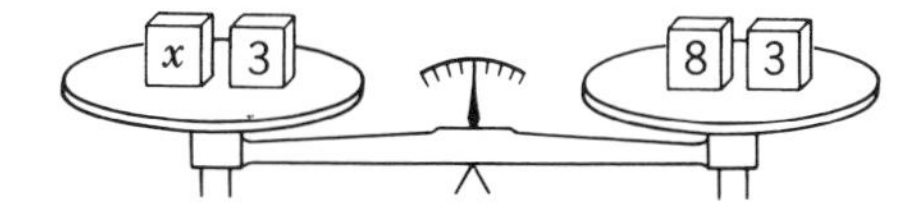

Figure 9-2

Let us see how this axiom applies to an equation. Consider again the equation

$$x = 8$$

Now, if we add some quantity, such as 3, to both sides of the equation, we get a new equation:

$$x + 3 = 8 + 3$$

or

$$x + 3 = 11$$

Then, by Axiom 1, the new equation is also true because we have added the same quantity to both sides of the original equation. We can easily see that this is so. If we substitute 8 for x in the new equation, we get

$$8 + 3 = 11$$

which is true.

Now, consider a second possibility. Suppose we start with the balanced scale in Fig. 9-1. If we subtract the same quantity from both sides of a balanced scale, the scale will still balance. This is the basis of the *subtraction axiom:*

Axiom 2. *If the same quantity is subtracted from both sides of an equation, the new equation will still be true.*

Let us see how this applies to the equation

$$x = 8$$

If we subtract some quantity, say 5, from both sides of the equation, we get

$$x - 5 = 8 - 5$$

or

$$x - 5 = 3$$

Then, by Axiom 2, the new equation is also true because we have subtracted the same quantity from both sides of the original equation. We can verify the new equation, for, if we substitute 8 for x, we get

$$8 - 5 = 3$$

which is true.

As a third possibility, if we start with the balanced scale in Fig. 9-1 and multiply both sides of the scale by some quantity, say 7, then each side of the scale will contain seven times as much as before, but the scale will still balance. Thus we have the *multiplication axiom:*

Axiom 3. *If both sides of an equation are multiplied by the same quantity, the new equation will still be true.*

Let us see how this applies to the equation

$$x = 8$$

If we multiply both sides of the equation by 7, we get

$$7x = 56$$

Then, by Axiom 3, the new equation is also true because we have multiplied both sides of the original equation by the same quantity. We can verify this equation by substituting 8 for x. We get

$$7 \cdot 8 = 56$$

which is true.

As a fourth possibility, we may divide both sides of the scale by some number, say 4, and the scale will still balance. Each side of the scale will contain only one-fourth as much as the original scale, but it will still balance. This principle is known as the *division axiom. The one exception is that we cannot divide by zero.*

Axiom 4. *If both sides of an equation are divided by the same quantity (not zero), the new equation will still be true.*

Let us apply this axiom to the equation

$$x = 8$$

If we divide both sides of the equation by 4, we get the new equation

$$\frac{x}{4} = 2$$

By Axiom 4, this equation is still true because we have divided both sides of the original equation by 4.

We can verify the new equation by substituting 8 for x. We get

$$\frac{8}{4} = 2$$

which is true.

The four axioms stated here are used in solving equations systematically. Other axioms will be needed later. All axioms may be summarized in one *general principle;*

Whatever operation is performed on one side of an equation, the same operation must be performed on the other side of the equation if the equation is to be true.

If a scale is to balance, both sides of the scale must be kept equal in value. The same is true with the two sides of an equation. We can make changes in the amount on each side of the scale, but *whatever change is made on one side of the scale, the same change must be made on the other side. The same is true regarding the equation.* This principle is the basis of solving equations.

9-7 USE OF AXIOMS IN SOLVING EQUATIONS

The axioms are used to solve equations. To show how this is done, we begin with easy examples. Remember, guessing at the answer in easy examples will be of no help when you come to more difficult equations. In learning how to solve equations, the procedure is more important than the answer itself.

Example 1.

Suppose we have the following equation and we wish to solve for x; that is, find the value of x that will make the equation true:

$$x + 3 = 15$$

By use of Axiom 2, we can subtract 3 from both sides of the equation,

$$\begin{array}{r} x + 3 = 15 \\ \underline{\quad 3 \quad\;\; 3} \end{array}$$

Subtracting 3 from both sides, $x = 15 - 3$

or $x = 12$

To check this answer, we substitute 12 for x in the original equation and ask:

$$\text{does} \quad 12 + 3 = 15? \quad \text{Yes.}$$

The statement is true, and therefore 12 is the root.

Example 2.

Suppose we have the equation $x - 5 = 14$.
By use of Axiom 1, we add 5 to both sides of the equation.

$$x - 5 = 14$$

Adding 5 to both sides, $\underline{\quad 5 \quad\;\; 5}$

$$x = 14 + 5$$

or $x = 19$

The root $x = 19$ can be checked in the same way as in the preceding example.

9-8 TRANSPOSING

In applying Axioms 1 and 2 in solving equations, we often use a "trick" called *transposing*. Although transposing is not a mathematical process, it is a shortcut that is simple, quick, and convenient if used properly. Remember, any "trick" in mathematics must be understood and used correctly. This is true even for such a trick as inverting the divisor when dividing fractions. Moreover, such "tricks" must be based on sound mathematical principles.

In the two foregoing examples notice that when a term is added to or subtracted from both sides of an equation, one term disappears from one side and reappears on the *other* side with its *sign changed.* For instance,

the equation $x + 3 = 15$
becomes $x = 15 - 3.$

When the quantity 3 is subtracted from both sides of the equation, the $+3$ disappears from the left side and reappears on the right side with its sign changed from $+$ to $-$. As another example,

the equation $x - 5 = 14$
becomes $x = 14 + 5.$

When 5 is added to both sides of the equation, the -5 disappears from the left side and reappears on the right side with its sign changed from $-$ to $+$.

A simple way to apply Axioms 1 and 2 is to *transfer any term we wish from one side of an equation to the other and then change its sign.* This process is called *transposing.* By *transposition* we mean that any term may be moved from one side of an equation to the other side, provided its *sign is changed.* An *entire term* must be transposed, not simply a factor.

Remember, *transposing* is only a "trick," but it is useful because it is simple, easy, and fast. It can always be used because it is based on the sound mathematical principles of Axioms 1 and 2.

Example 1.

Now let us solve the following equation by use of the axioms:

$$5x - 4 = 2x + 17$$

In solving any equation, one of our main objectives is to get all of the x terms on one side of the equation and all the other terms on the other side; that is, we isolate the terms containing x. This we do by "transposing." We transpose the -4 from the left side to the right side of the equation and change its sign to plus $(+)$. The $+2x$ is transposed from the right side to the left side and its sign is changed to minus $(-)$. The original equation is

$$5x - 4 = 2x + 17$$

Transposing, $5x - 2x = 17 + 4$

Combining like terms, $3x = 21$

Now we make use of Axiom 4 and divide *both* sides of the equation by 3.

We get

$$x = 7$$

The equation is solved when we have one single x on one side of the equation. Then the root appears on the other side of the equation.

The root of the foregoing equation is 7, provided there has been no error in the solution. This means that if 7 is substituted for x in the original equation, the statement should be true.

To check the answer, we substitute the 7 in place of x in the equation, and ask

does $(5)(7) - 4 = (2)(7) + 17$?
does $35 - 4 = 14 + 17$?
does $31 = 31$? Yes

As soon as we see that one side of the equation is equal to the other side, we know we have the root of the equation. The check must result in a statement that is true.

Example 2.

Solve the equation $3x + 2 - x + 9 = 5x - 12 + x + 3$.

Solution.

Transposing, $3x - x - 5x - x = -12 + 3 - 2 - 9$
Combining, $-4x = -20$
Dividing both sides by -4, $x = 5$

The check is left for the student.

Although the terms containing the unknown x are usually isolated on the left side of the equal sign, this is not essential. The x terms can just as well be transposed to the right side, and the other terms to the left side. Sometimes this may reduce the chances of error.

Example 3.

Solve $4x - 5 - 5x - 7 = x - 5 + 3x + 8$.

Solution.

Transposing, we get $-5 - 7 + 5 - 8 = x + 3x - 4x + 5x$
Combining like terms, $-15 = 5x$
Dividing both sides of the equation by 5, the coefficient of x, $-3 = x$

Therefore the root of the equation is -3. This can be checked by substituting -3 for x in the original equation.

Note. If a term on one side of the equation is the same as a term on the other side, these terms may be dropped before transposing; that is, we subtract the term from both sides by use of Axiom 2. In the foregoing equation the -5 may be dropped, since it can be subtracted from both sides of the equation.

Exercise 9-2

Solve the following equations and check the root of the even numbers.

1. $3x + 7 = 19$ **2.** $4y - 7 = 13$

3. $4 - 3y = 10$ **4.** $3 - 4x = 27$

5. $5x = 7 + 2x$ **6.** $9 + 7x = 3x$

7. $2x = 7x + 3 - 5x$ **8.** $6x - 2x = 5 + 4x$

9. $4a + 9 = a - 6$ **10** $5t + 8 = t - 9$

11. $10 - 3x = 4x - 18$ **12.** $2R - 9 = 7R + 21$

13. $4x + 3 = 7 + 4x$ **14.** $8 - x = 3x - 7$

15. $12 - 2R = 5R + 21$ **16.** $x - 9 = 5x + 8$

17. $5 - 2x = 3x + 5$ **18.** $y - 23 = 6y - 3$

19. $6 - 7x = 2x + 3$ **20.** $7 - 5x = x + 4$

21. $7t - 11 = 5t - 11$ **22.** $5h + 7 = 2h + 7$

23. $4x - 3 = 5 + 4x$ **24.** $2 - 3x = 8 + x$

25. $3x + 5 = 4x - 3$ **26.** $5x - 3 = 5 + 4x$

27. $6x + 5 - x = 2x - 7$ **28.** $7x + 5 = 2x + 5 - x$

29. $7x + 5 - x + 3 = 9x - 15 + 3x + 5$

30. $6 + 2x - 3 - 5x = x - 4 - 2x + 9$

9-9 PARENTHESIS IN EQUATIONS

Some equations contain terms or polynomials enclosed within parentheses, such as the equation $5x - 2(4x + 3) = 7 + (3x - 2)$. In such equations the first step is to remove the parentheses. In this step we observe the following rule:

Rule. *If a quantity within parentheses is preceded by a plus sign (+), this plus sign and the parentheses may be omitted without changing the sign of any term within the parentheses.*

If a quantity within parentheses is preceded by a minus sign (−), this minus sign and the parentheses may be omitted, provided that the sign of each term within the parenthesis is changed.

This rule is easily understood if we remember that a minus sign before parentheses is really a sign of the operation (subtraction).

Remember, if a coefficient appears before the parentheses, this coefficient is multiplied by each term within the parentheses. If no coefficient appears before the parentheses, it should be understood that the numerical coefficient is 1 (or -1 if the parentheses is preceded by a minus sign).

The foregoing rule holds true for any sign or symbol indicating a quantity, such as brackets, braces, vinculum, or for any other symbol of aggregation.

Example 1.

Solve $4x - 3(2x - 5) - 9 + (3x + 4) = 12 - (x - 5) - 2(7 - x) - 3x$.

Solution.

Removing parentheses,

$$4x - 6x + 15 - 9 + 3x + 4 = 12 - x + 5 - 14 + 2x - 3x$$

Transposing,

$$4x - 6x + 3x + x - 2x + 3x = 12 + 5 - 14 + 9 - 15 - 4$$

Combining,

$$3x = -7$$

Dividing both sides of the equation by 3, the coefficient of x,

$$x = \frac{-7}{3}$$

Note. After removing the parentheses, some like terms may be combined before transposing. However, to do so would mean an extra step, since like terms must later be combined again.

Some equations appear at first sight to be quadratics, they contain second-degree terms; that is, terms in x^2. The study of quadratic equations will be taken up later. At this point, we simply mention that in many cases the terms in x^2 disappear in the process of solving x.

Example 2.

Solve $(x - 3)(x + 5) - (x + 2)(x - 4) = 2x - 5(x + 4)$.

Solution.

In removing the parentheses, we must perform the indicated multiplication. This multiplication step may be called *expanding the products*. The minus sign between the sets of parentheses is a sign of subtraction, so that the signs of the terms following it are changed after the products are expanded.

Expanding the products, $(x^2 + 2x - 15) - (x^2 - 2x - 8) = 2x - 5x - 20$
Removing parentheses, $x^2 + 2x - 15 - x^2 + 2x + 8 = 2x - 5x - 20.$
Transposing and combining (x^2 disappears), $7x = -13$

Dividing both sides of the equation by 7, the coefficient of x,

$$x = \frac{-13}{7}$$

Check.

If checking is done by substituting the value of x as found, the substitution should always be done in the *original* equation. However, *the best check is to be sure that you are using the correct method and then to be sure you do not make mistakes in the process of solving. A second check is to go over your work a second time.*

Exercise 9-3

Solve the following equations:

1. $2x - 5 = 15 - 7x$ **2.** $7n + 10 = 2n - 7$

3. $4x + 3(2x - 5) = 7 + 5(3x - 7)$

4. $6y - 2(y - 8) = 9 + 4(3y - 2)$

5. $5c - 4(c + 3) = 14 - 2(c - 5)$

6. $2x + 5(x + 2) = 3x - (8 - x)$

7. $10 - 3(2 - 4t) = 9t - 2(5 + 3)$

8. $8 - (3x + 4) = x - (2x - 5)$

9. $3x - 4(2x + 3) - 5 = 7 - 3(4 - 2x) - 5x$

10. $4 - (x + 3) - 3x = 2(x - 4) + 5(8 - x)$

11. $5n - 3(n + 4) + 11 = 3 - 4(2n - 1) - 6n$

12. $2(7 - n) + 3(n + 2) = 4 + 3(4n - 3) - n$

13. $3(y - 1) - (4y - 5) + 4 = 5 - 2(1 - y) - 5y$

14. $4(7v - 3) - 2(6 + 5v) + 2 = 3v - 5(v - 1) - 1$

15. $(x + 2)(x - 5) - x(3x + 2) = 2 - 2x(3 + x) + 5x$

16. $5x - (x + 3)(x - 4) + 3 = 7 - x(x - 7)$

17. $3y - y(2y - 5) + 5 = 13 + 2y(5 - y)$
18. $4 + 3c(5 - c) - 2c(3 - c) = 7 - c(c - 3)$
19. $9 - 2(3h + 1) + 2h = 3 - 4(h - 1)$
20. $7k + 4(2 - k) + 5 = 4 + 3(k + 4)$
21. $4 - (3t - 2)(t - 3) - 5t = 3t - (t - 3)(3t - 4) - 5$
22. $5y - y(2y + 3) - 7 = 6 - (5 + y)(2y - 3) - 2y$
23. $7n - 2n(2n - 3) - 2 = n - (n - 2)(4n + 3) - 8$
24. $3c - (2c - 1)(3c + 2) + 5 = 2c - 2(c - 3)(3c + 1) + 5$
25. $(2b - 1)(2b + 3) - b(b - 2) = 5b - (4 + 3b)(1 - b)$
26. $6 - 2k(4 - 3k) - 3(k - 2)(k - 3) = 5 - 3(2 - k)(2 + k)$
27. $10x - 3(x + 3)(2x - 5) = 3 - 2(4x - 7)(x + 1) - x(3 - 2x)$
28. $8x - 3(2x - 1)(3 - x) - 2x(x - 3) = 5 - 4(x + 2)(5 - x)$
29. $3 + (3x - 2)(4x + 3) - 3(3x - 1)(2x - 3)$
    $= 7x - (2x - 1)(3x + 4)$
30. $5x - (2x - 3)(x + 2) - (4x + 3)(x - 1) = 3 - (3x + 2)(2x - 3)$
31. $4x - (2x - 1)(x + 2) - x(x - 2) = 1 - 3(x - 2)(x + 1)$

QUIZ ON CHAPTER 9, THE EQUATION, FORM A.

Solve the following equations, or tell whether the equation is an identity or has no solution.

1. $3x - 4 = 16 - 2x$
2. $2x + 3 = 5x + 18$
3. $5x - 2 = 3x + 9$
4. $5 + 3x = 4x + 17$
5. $3 + 4x = 7x + 3$
6. $7x + 4 = 5x + 2x$
7. $6 + 7x - 2 = 5x + 4 + 2x$
8. $2(3x - 4) - 6 = 4(2x - 1) + x$
9. $6x - 2(x - 3) - 5 = 7 - (x - 1)$
10. $13 - 3(3x - 1) - 2x = 8 - (3x - 2)$
11. $7 - 3(x - 2) + 2x = 9 - 2(x - 5) - 4x$
12. $8x(x + 3) - 3x(2x + 3) = 7 + (2x - 1)(x + 3) - 6x$
13. $3x - (2x + 1)(4x - 3) - 5 = 7 - x(8x + 1)$
14. $5x - (x + 3)(x - 2) - 1 = 5 - x(x - 3) + x$
15. $3x - (2x - 1)(2x + 1) + 2 = 7 - x(4x + 1) + 4x$
16. $2x - 2(x - 1)(2x + 3) + 1 = 7 - x(4x + 5) + x$

CHAPTER TEN Solving Stated Word Problems

10-1 THE IMPORTANCE OF STATED PROBLEMS

In this chapter we shall show how equations can be used to solve the so-called "stated problems" or "word problems." By this, we mean the problems that arise in connection with our work in everyday life.

Equations in algebra are used to find answers to questions that arise in connection with many kinds of work. We use algebra to solve problems concerning electric circuits, laws of motion, pressures of gases, speeds of electrons and satellites, equilibrium of forces, and many other scientific ideas. Algebra is also used to solve practical problems in business, in industry, and in the home. Solving equations enables us to find answers to practical problems, not only in scientific fields but in all kinds of activity.

We do not begin with problems about unfamiliar things such as the laws of electric circuits, laws of motion, angular velocity, or efficient power supply simply because these topics are unfamiliar to most people beginning the study of algebra. Instead, we shall first solve problems dealing with familiar subject matter. We shall work problems about ages of people, speeds of cars, costs of articles, scores in games, measurements, and other everyday matters.

Remember, your purpose here is to learn how to apply the principles of algebra to the solution of problems. The kind of thinking involved in easy problems that we shall solve is exactly the same kind of thinking that must be done in solving more difficult problems in science. The approach to all problems is the same. Your purpose here should be as follows:

1. To learn the correct approach to the solution of a problem.
2. To learn to analyze a problem carefully; to look at it from every angle.
3. To learn how to write the correct equation from the given information.

If you master the correct method for an easy problem, you can use the same approach to the more difficult problems you will encounter later.

The chief difficulty in solving word problems is that we must make up the equations from the information that we find in the particular problem we are facing. In word problems, we have two chief hurdles to overcome:

1. *First, we must make up an equation from the given information.*
2. *Then we must solve the equation.*

In working word problems, students usually have most trouble making up the correct equation from the given information. After the equation is correctly set up, it can usually be solved without too much difficulty.

10-2 EXPRESSING ALGEBRAIC QUANTITIES

In solving problems, it is necessary to express various quantities in algebraic symbols. In Chapter 7 we saw how to express the sum, the difference, the product, and the quotient of literal quantities. For instance, the sum of x and y is expressed as $x + y$.

The sum of two quantities is expressed by means of the plus sign (+). The difference is expressed by the minus sign (−). The product in many cases is expressed by writing one factor next to another if there is no confusion in meaning. The product is sometimes indicated by placing a raised dot between the factors. Division is expressed by the division sign (÷). Division may also be expressed in fractional form; thus $12x \div 5y$ can be written $\frac{12x}{5y}$.

If the sum or difference of two terms is to be considered one single quantity, it should be enclosed in parentheses. For instance, if we wish to indicate three times the sum of a and b, we should write it $3(a + b)$.

Exercise 10-1

Express the following:

1. The sum of $2x$ and $3y$.
2. The difference between m and n if m is greater.
3. The product of $5x$ and $7y$.
4. The quantity x plus 6 divided by 3.
5. A number 13 more than $2x$.
6. A number 5 less than $3a$.
7. Twice the sum of x and $5x$.

8. Three times the difference between $7x$ and $3y$ if x is greater than y.

9. The sum of x and y divided by their product.

10. The difference between x and y divided by the sum of a and b.

11. If A's age is x years and B is 5 years older, express B's age. Then express the sum of their ages.

12. If C is x years old and D is three years older than twice C's age, express D's age in years. Then express the sum of their ages.

13. If a boy is x years old and his father is four times as old, express the father's age in years. Express the age of each 5 years ago.

14. How many months are there in x years and 5 months?

15. If a radio costs x dollars and a TV set costs $20 more than 6 times the cost of the radio, express the cost of both.

16. Express the number of cents in x nickels; in $(32-x)$ dimes; then express the total number of cents in both.

17. Express the cost of x pounds of coffee at $2.80 a pound.

18. If x is an integer (whole number), express the next consecutive integer.

19. If n is an even number, express the next consecutive even number.

20. If c is an odd number, express the next two consecutive odd numbers.

21. If x is an even number and y is even, is $(x + y)$ odd or even?

22. If c is an odd number and d is odd, is $(c + d)$ odd or even?

23. At 50 miles per hour, find the distance driven in $(8 - x)$ hours.

24. Express the number of minutes in x hours and 15 minutes.

25. If x represents the number of inches in the width of a rectangle, express the length and the perimeter if the length is 5 in. less than 3 times the width.

26. A man gets a salary increase of $1200 per year. If x represents his salary for the first year, express his salary for each of the next 4 years.

27. On a trip by car, you decrease your mileage by 50 miles per day each day. If you drive x miles the first day, express the distance for each of the next 3 days.

10-3 THE APPROACH TO PROBLEM SOLVING

The first thing to remember in beginning to solve a problem is this: *never think about the answer.* If you start out trying to guess the answer, you are starting out the wrong way. Your first concern should always be with the proper approach.

Remember, at this time you are not simply looking for answers. If your only concern is the answer to the problem you can usually find it in the back of the book or you can hire someone to work the problem for you. Instead, you are now trying to learn how to use algebra to solve problems.

Some students say: "Oh, I can work this problem by arithmetic. I can find the answer without using algebra." That is probably true in easy problems. However, do not lose sight of your purpose here. Your purpose is not only to get an answer to a problem but to learn how to apply the principles of algebra. Working an easy problem without using algebra will not furnish practice in the use of algebra.

In learning how to use algebra in solving problems, we begin with simple examples. No matter how easy it may be to get the answer to a problem, you should use algebra to solve it. That is the only way to acquire the necessary practice so that you can solve more difficult problems later on.

In every problem there is some quantity that is already known, sometimes several. Also, in every problem there is at least one quantity that is not known. Such a quantity is called an *unknown.* Sometimes a problem contains several unknowns.

In solving a problem by algebra, we use some letter, say x to represent an unknown quantity. Then we work with the letter just as though we know its value. An example may help to make this procedure clear.

Let us suppose you are 23 years old. From that given fact we can tell many things:

1. Four years from now you will be $23 + 4$, or 27 years old.
2. Five years ago you were $23 - 5$, or 18 years old.
3. If your father is twice your age, he is $(2)(23)$, or 46 years old.
4. If your uncle is three years older than your father, he is $46 + 3$, or 49 years old.
5. If you have a sister two years younger than you, she is $23 - 2$, or 21 years old.
6. If your mother is one year younger than your father, she is $46 - 1$, or 45 years old.

Now, let us begin again and assume that we do not know your age. We shall say that the number of years in your age will be represented by the letter x. Then we can repeat the foregoing facts, using the letter x for your age in years.

1. Four years from now you will be $x + 4$ years old.
2. Your age (in years) five years ago can be represented by $x - 5$.
3. If your father is twice your age, his age (in years) can be represented by $2x$.
4. If your uncle is three years older than your father, his age (in years) can be represented by $2x + 3$.
5. If you have a sister two years younger than you, her age (in years) can be represented by $x - 2$.
6. If your mother is one year younger than your father, her age (in years) can be represented by $2x - 1$.

There is a definite procedure you can follow to get started on the solution of a problem. There are *five definite steps* you can follow in solving any problem. You may have some difficulty in following one or two of the steps, but you can always use these five as a guide.

Here are the "Five Golden Rules" for solving problems:

1. *Let some letter, such as x, represent one of the unknowns.* (This is usually though not necessarily the smallest.)
2. *Then, try to express the other unknowns by using the same letter.*
3. *Write a true equation from the information given in the problem. Make your equation say in symbols exactly what the problem says in words.*
4. *Solve the equation.*
5. *Check your answer to see whether it satisfies the conditions given in the problem.* Do *not* check in the equation you have made up.

If you can do the *first three* steps, the problem is practically done. Most people have trouble with the first three steps. In fact, many people have trouble with the very first step. *If you can let some letter represent one of the unknowns and if you state definitely what the letter is to represent, you have the right start.*

After the first two steps have been completed and the unknown quantities have been properly stated in terms of a letter, there is often some difficulty in analyzing the problem before the equation can be written. In this analysis between the second and third steps, try to discover the different relationships among the various quantities in the problem. Try to determine whether one quantity is equal to another or whether the sum or difference of two quantities is equal to a third quantity. Then try to make the equation state this relationship.

We shall now work out several problems, following the five steps.

10-4 GENERAL PROBLEMS

Example

A chair and a desk together cost \$68. The desk cost \$5 more than twice the cost of the chair. Find the cost of each.

Solution

Here we have at least two unknowns, the cost of the chair and the cost of the desk. Your thinking should be, "I am going to let the letter x represent the number of dollars the chair cost." Begin your statement with the word "Let."

Step 1. Let x = the number of dollars the chair cost.
Be sure you state definitely just what the letter x is to represent. Now we express the cost of the desk, using the same letter.
Step 2. Then $2x + 5$ = the number of dollars the desk cost.
At this point, we see that the problem says that the sum of the two costs was \$68. So we make that statement by means of an equation.
Step 3. Equation: $x + (2x + 5) = 68$
If you can get this far in the problem, the rest is usually easy.
Step 4. Solving the equation

$$x + 2x + 5 = 68$$
$$3x = 68 - 5$$
$$3x = 63$$
$$x = 21$$

Therefore, the cost of the chair was \$21 and the cost of the desk was \$47, which is \$5 more than twice as much as the cost of the chair.
Step 5. *Check*:

Cost of chair	\$21
Cost of desk	47
Total cost	\$68

Warning. Be sure the checking is done in the *original problem*, not in the equation that you have made up.

Note concerning the use of a letter x to represent a quantity. Be sure to let x (or the letter used) represent a *number.* A letter such as x cannot represent the length of a rectangle. Instead, it may represent the *number of inches* or the *number of feet* in the length. The letter x cannot represent a person's age. Instead, it may represent the *number of years* or the *number of months* in his age. A letter, such as x, cannot represent an amount of money. It may represent the *number of dollars,* the *number of cents,* or the *number of dimes.* It cannot represent the weight of an object. Instead, it may represent the *number of pounds*, the *number of ounces*, the *number of tons* in the weight.

Any letter may be used to represent an unknown number. We may use x, y, z, n, t, or any other letter. Some students often use the first letter of a word to which it refers, such as d for the number of dimes, n for the number of nickels, and so on. This is convenient provided that the letter is understood to represent a *number*. When this is done, however, students sometimes take the letter to mean something other than a number. For example, using d for the number of dimes, they mistake d for the amount of money rather than the number of coins. Some even go so far as to say that d must be 10 because there are 10 cents in one dime. Remember, whatever letter you use, it always represents a *number*. If you use d for the number of dimes, then it represents just so many pieces of money.

A good way to be sure of a correct statement is to start the problem with the form shown here:

"Let x = the *number* of . . ."

Another correct way is to say:

"Let x = the width of the rectangle (in inches)"

Exercise 10-2

1. One number is three times another number, and their sum is 68. What are the two numbers?

2. One number is two less than five times another and their sum is 76. What are the numbers?

3. One number is four times another and their difference is 69. What are the two numbers?

4. Divide the number 40 into two parts so that three times the smaller part is equal to twice the larger part.

5. A student bought a slide rule and a drawing set for a total of \$39. The drawing set cost \$8 more than the slide rule. Find the cost of each.

6. A lady bought a purse and a pair of shoes for a total cost of \$45. If the shoes cost \$2.50 more than the purse, what was the cost of each?

7. A fishing rod and reel together cost \$40. If the reel cost \$7 less than the rod, what was the cost of each?

8. A lady bought a dress, a hat, and a pair of gloves for a total of \$63. The hat cost \$3 less than twice the cost of the gloves, and the dress cost $12 more than the combined cost of the gloves and the hat. Find the cost of each.

9. In a football game the home team made five points less than twice the score of the visiting team. If the total score made by both teams was 52, how many points were made by each team?

10. In a certain baseball game one team made a total score of one more than four times the score of the other team. If the total score made by both teams was 11, what score was made by each team?

11. In a certain basketball game one team made a total of 81 points. The number of field goals was five less than three times the number of free throws. How many field goals and how many free throws did the team make? (A free throw counts one point, a field goal counts two points.)

12. In a duckpin bowling match one man averaged a score of 13 points more *per game* than the other. If the total combined score of both men for three games was 611 points, what was the average score of each man *per game*?

13. A family on a tour of 820 miles in three days drove twice as many miles the second day as the first. The third day they drove 60 miles less than on the second day. Find the distance traveled each day.

14. The length of a rectangle is 6 in. greater than the width. If the perimeter of the rectangle is 50 in., how wide and how long is the rectangle? (Remember, a rectangle has four sides. The perimeter is the distance around the rectangle.)

15. It requires 62 ft of picture molding to reach around the four walls of a room. The length of the room is 5 ft less than twice the width. What is its length and its width?

16. A tract of land along a river is to be fenced along one side and both ends. The length of the tract is 20 rd less than three times the width. If the total length of fence required is 120 rd, find the width and the length of the tract.

17. In a certain triangle one side is $4\frac{1}{2}$ in. longer than another side, and the third side is 3 in. shorter than the sum of the other two. If the perimeter of the triangle is 34 in., what is the length of each side?

18. Smith, Jones, and Brown invest a total of $48,000 in a business. Smith invests one and one half times as much as Jones, and Brown invests $2400 less than Smith. How much did each invest in the business?

10-5 AGE PROBLEMS

Example

A man is three years older than four times his son's age. Five years from now, the sum of their ages will be 48 years. What are their ages now?

Solution

Let x = the number of years in the son's age *now*.
Then $4x + 3$ = the number of years in the father's age *now*.

In this problem we must analyze carefully the information given. The problem does not say the sum of their ages is 48 years *now*. Instead, the sum of their ages will be 48 years *five years from now*. Let us look at their ages five years from now. Remember, both people will be five years older then.

Five years from now the son will be $x + 5$ years, and the father will be $4x + 3 + 5$, or $4x + 8$. These are the two quantities whose sum is 48. Therefore, we write the equation which makes this statement:

Equation: $(x + 5) + (4x + 8) = 48$

When you have once written the equation correctly, the rest of the problem is usually easy. If you solve this equation, the value you get for x will be exactly what x represents as indicated in the first statement in the solution. It will be the number of years in the son's age *now*.

Exercise 10-3

1. A man is 27 years older than his son. The sum of their ages is 45 years. How old is each?

2. A mother is three times as old as her daughter. If the sum of their ages is 52 years, how old is each?

3. Find the age of a man and the age of his son if the father is seven years older than three times the son's age and the sum of their ages is 41 years.

4. A man is now five times as old as his son. Four years hence the sum of their ages will be 47 years. How old is each one *now?*

5. Vernon is eight years older than Donald. Two years hence Vernon will be three times as old as Donald. How old is each one *now*?

6. A man is now six times as old as his son. In three years he will be four times as old as his son. How old is each one *now?*

7. A man is six years older than three times his son's age. Five years hence the sum of their ages will be 56 years. How old is each one now?

8. John is six times as old as Gregory. Four years hence John will be twice as old as Gregory. How old is each one now?

9. A man is now twice as old as his son. Eighteen years ago he was 5 times as old as his son. Find the age of each.

10. A man's age is 7 years more than 3 times his son's age. Eight years hence he will be 6 times as old as his son was 2 years ago. How old is each now?

10-6 CONSECUTIVE NUMBER PROBLEMS

An integer is a whole number, such as 3, 8, 17, 546. *Consecutive integers* are integers that have no other integer between them, such as 6, 7, 8, 9. If we begin with the integer 15, then the next consecutive integer is $15 + 1$, or 16.

If we say that the letter x represents the first of a series of four consecutive integers, then the second is $x + 1$, the third is $x + 2$, and the fourth is $x + 3$.

Consecutive even integers are even integers that have no other even integer between them, such as 6, 8, 10. Notice that two consecutive even integers have a difference of 2. In a series of three consecutive *even* integers, if we let x represent the first, then $x + 2$ will represent the second, and $x + 4$ will represent the third.

Consecutive odd integers are odd integers with no other odd integer between them, such as 7, 9, 11. Notice that two consecutive odd integers have a difference of 2. In a series of three consecutive *odd* integers, if we let x represent the first, then $x + 2$ will represent the second, and $x + 4$ will represent the third.

Example

Find four consecutive odd integers whose sum is 216.

Solution

Let x = the first of the four consecutive odd integers.
Then $x + 2$ = the second of these odd integers.
and $x + 4$ = the third of these odd integers.
and $x + 6$ = the fourth of these odd integers.
Equation: $x + (x + 2) + (x + 4) + (x+6) = 216$
The equation simply states that the sum of the four numbers is 216, as the problem says. When we solve the equation, the value we get for x will be the number represented by x in the first statement.

Solving the equation,

$$\begin{aligned} x + x + 2 + x + 4 + x + 6 &= 216 \\ 4x &= 216 - 2 - 4 - 6 \\ 4x &= 204 \\ x &= 51 \end{aligned}$$

Since we let x represent the first of the four integers, this number is 51. The four consecutive odd integers are 51, 53, 55, and 57.

Check:

$$51 + 53 + 55 + 57 = 216.$$

Exercise 10-4

1. Find four consecutive integers whose sum is 270.
2. Find five consecutive even integers whose sum is −240.
3. Find six consecutive odd integers whose sum is 252.
4. A mechanic wishes to make a trapezoidal tray that requires five strips, each one after the first to be $\frac{3}{4}$ in. longer than the previous strip. If he has a piece of metal 40 in. long to be used for the five strips, how long should each strip be?
5. On a vacation trip of 1400 miles to be covered in 4 days, a family wishes to arrange their driving so that each day after the first they may drive 60 miles less than on the previous day. How many miles do they travel each day?
6. Each year a woman received a salary increase of $600 over the previous year. Her total salary for a five-year period was $48,000. Find her salary each year.
7. A ladder is to have 6 steps, the longest at the bottom, and each step above the bottom is to be $1\frac{3}{4}$ in. shorter than the one immediately below it. If the steps are to be cut from a 7-ft board without any part left over, find the length of each step.
8. Four resistors in series have a total resistance of 3900 ohms. If each one after the first is 250 ohms more than the previous one, find each resistor.
9. Three voltages in series have a total voltage of 1290 volts. The second is 80 volts less than the first, and the third is 80 volts less than the second. Find each voltage.
10. Three capacitors in parallel have a total capacitance of 86.4 microfarads (mf). The first is 6 mf more than the second, and the second is 6 more than the third. Find the size of each.

10-7 COIN AND OTHER PROBLEMS ABOUT MONEY

Example

A collection of 71 coins, consisting of nickels and quarters, is worth \$8.15. Find the number of coins of each kind.

Solution

In this problem there are two unknowns: the number of nickels and the number of quarters. Follow Steps 1 and 2:

Let x = the number of nickels in this collection
Then $71 - x$ = the number of quarters in the collection

Now, if we study the problem carefully, we see that we have used all the information except the \$8.15. This is the total value. It is not the total number of coins.
To write the equation, we must remember that the total *value* of the coins is \$8.15, or 815 cents. If we express the number of cents in all the coins, we can say that the total number of cents is equal to 815.
The value of x nickels is $5x$ cents.
The value of $(71 - x)$ quarters is $25(71 - x)$ cents.
The equation will simply state that the total number of cents = 815.
Equation: $5x + 25(71 - x) = 815$

When we solve this equation, the number we shall get for x will be the number of nickels in the collection, since that is the number that was represented by x in the first statement.

Exercise 10-5

1. A collection of 89 coins, consisting of nickels and dimes, is worth \$5.75. Find the number of coins of each kind in the collection.

2. A collection of 77 coins, consisting of dimes and quarters, is worth \$10.85. Find the number of coins of each kind in the collection.

3. A collection of nickels, dimes, and quarters is worth \$9.75. The number of nickels is 3 times the number of dimes. The number of quarters is 5 less than the number of nickels. Find the number of coins of each kind in the collection.

4. A man buys some 3-cent stamps, some 9-cent stamps, and some 15-cent stamps. The number of 9-cent stamps is twice the number of 3-cent stamps. If the total cost is \$5.88 for a total of 52 stamps, find the number of stamps of each denomination.

5. At an entertainment, the price of admission was 65 cents for children and $1.50 for adults. If the total amount received was $95 for a total of 103 tickets, find the number of admissions of each kind.

6. At an entertainment, the price of admission was 60 cents for children and $1.25 for adults. The total paid attendance was 222. If the total amount received for children's tickets was the same as the total amount received for the adults' tickets, how many tickets of each were sold?

7. A merchant has one kind of nuts selling for 95 cents a pound and another kind selling at 50 cents a pound. He wishes to make a mixture of the two kinds to sell at 65 cents a pound. How many pounds of each kind should he use for a mixture of 100 pounds so that he will receive the same amount as if they were sold separately?

8. At a ball game the price of admission was $1.50 for general admission, $2.50 for grandstand seats, and $3.25 for box seats. If 19,480 tickets were sold and the total receipts were $34,108, find the number of tickets sold at each rate if there were 4 times as many general admission tickets sold as the number of grandstand tickets.

9. A 1-dollar bill has a portrait of Washington, a 2-dollar bill has a portrait of Jefferson, and a 5-dollar bill has a portrait of Lincoln. A man has 48 bills consisting of Washingtons, Jeffersons, and Lincolns, with a total value of $175. If the number of Lincolns is 3 more than twice the number of Washingtons, find the number of bills of each kind.

10. On a trip to Latin America a traveler picked up some *pesos* in Nicaragua worth about 11.6 cents each, some *lempira* in Honduras worth 50 cents each, and some *sucres* in Ecuador worth 7.4 cents each. On his return he found he had a total number of 425 pieces of money worth $95.20. If he had twice as many pesos as lempira, find the number of pieces of each kind of money he had.

11. Later the traveler took a trip to northern Europe and collected some *marks* from West Germany worth about 23.8 cents each, some *kroner* from Sweden worth about 19.3 cents each, and some *gulden* from Netherlands worth about 26.3 cents each. On his return he found he had 500 pieces of money worth $110.20 and that he had 60 more kroner than marks. How many pieces of money of each kind did he have?

12. A sand-and-gravel dealer sells sand at \$1.70 a ton, fine gravel at \$2.10 a ton, and coarse gravel at \$2.30 a ton. For a concrete preparation, he wishes to mix the three kinds to sell at \$2.00 a ton, and he wishes to use twice as much fine gravel as sand. How many tons of each kind should he use to prepare a mixture of 100 tons?

10-8 PROBLEMS IN UNIFORM MOTION

If we drive a car for a distance of 120 miles in exactly 4 hours we call the 120 miles the *distance* and the 4 hours the *time*. We can find the average *rate* of speed by dividing the distance traveled by the time of traveling. If the distance is 120 miles and the time required is exactly 4 hours, the average rate of speed is

$$120 \div 4 = 30$$

Our average rate of speed is 30 miles per hour (mph).

When we say *average speed*, we do not mean that the speed was exactly the same rate, 30 mph, during every minute of the time. It is impossible to drive at exactly 30 mph every minute for 4 hours.

In working problems in uniform motion, we assume that the speed does not change. If there is any change in the speed, we must make some adjustment for the change. We assume also that the speed was uniform even at the start. It is incorrect to say, "A car starts out from a certain town at 30 mph." A car cannot start at 30 miles per hour.

In problems in motion we assume that average speed indicated will continue at the same average. For instance, if a car travels at an average rate of speed of 45 miles per hour for 3 hours, it will travel 135 miles in all. In this example

45 mph is called the *rate* (r)
3 hr is called the *time* (t)
135 miles is called the *distance* (d)

To find the total distance traveled in a given time at a given average rate of speed, we multiply the *rate* by the *time*. Of course, the factors must be stated in the proper units of measurements. The formula for the distance is

$$d = rt$$

Example

A car starts out from a town at 7 A.M. traveling at an average rate of 35 mph. At 10 A.M. a second car starts out from the same town traveling along the same road at an average rate of 50 mph. At what time of the day will the second car overtake the first?

Solution

In this problem several facts are unknown: the number of hours each car travels and the distance each car travels. We could let some letter such as x represent any one of these unknown quantities. However, it is probably best to let x represent the number of hours for one of the cars.

Let x = the number of hours the first car travels
Then $x - 3$ = the number of hours the second car travels

Let us set up a table of *values*:

	Rate (mph)	Time (hrs)	Distance (miles)
First car	35	x	$35x$
Second car	50	$x - 3$	$50(x - 3)$

We then fill in the given values. When we have expressed the number of hours each car travels, we fill in the remaining blanks. In the "distance" column we fill in the expression for the distance each car travels. This distance is found simply by multiplying the time by the rate for each car.

The first car travels $35x$ miles
The second car travels $50(x - 3)$ miles

Now, we suddenly realize that the two cars travel exactly the same distance. This fact is stated in equation form:

$$35x = 50(x - 3)$$

The equation says exactly what is implied in the problem. An equation such as this shows the power of the symbol of equality. This symbol, the equal sign (=), is probably the most powerful symbol in all mathematics.

When the equation is solved for x, this value will show the number of hours traveled by the first car. However, we must be careful to answer the question asked in the problem. The question was: "At what *time of the day* will the second car overtake the first?" The answer is 5 P.M.

Exercise 10-6

1. One automobile starts out from a town at 8 A.M. and travels at an average rate of 40 mph. Two hours later a second automobile starts out to overtake the first. If the second automobile travels at an average rate of 55 mph, how long will it take the second to overtake the first?
2. Two cars travel in opposite directions, the first at an average rate of 60 mph and the second at an average rate of 45 mph. How long will it take until they are 400 miles apart?
3. Two cars start out at the same time, one from town A, traveling

toward town B at an average rate of 50 mph, the other from town B, traveling toward town A at an average rate of 35 mph. How long will it be until they meet if the towns are 500 miles apart?

4. One car starts out from town C at 8 A.M., traveling toward town D at an average rate of 40 mph. A second car starts out from town D at 11 A.M., traveling toward town C at an average rate of 45 mph. At what time of day will they meet if towns C and D are 460 miles apart?

5. One car starts out at 10 A.M., traveling at an average rate of 42 mph. At 12 noon a second car starts out to overtake the first. How fast must the second car travel to overtake the first by 7 P.M.?

6. A boy returns a bicycle to his friend. He cycles at 8 mph and walks back at 3 mph, after spending an hour with his friend. If he arrives home 6 hr after he left, how far has he traveled *one way?*

7. A man takes a trip of 675 miles, part way by train at 60 mph and the rest of the way by car at 50 mph. If the entire trip takes 12 hr, how far has he traveled by each mode of transportation?

8. A bus driver makes a regular run at an average rate of 50 mph. On one particular run he finds it necessary to reduce his average speed by 10 mph, and, as a result, he is 2 hr late. How long is the trip?

9. A man starts out at 10 A.M. on a 290-mile trip and drives part of the way at an average of 40 mph. He stops $\frac{1}{2}$ hr for lunch and then finds that he must drive the rest of the way at 50 mph to finish his trip by 5 P.M. When does he stop for lunch?

10. A man makes a trip of 360 miles in a total of 12 hr. He travels the first part of the trip by motor boat at 18 mph, then transfers to a car and travels at 40 mph. The last part of the trip is by train at 55 mph. If the trip by motor boat is 2 hr longer than the trip by car, how far does he travel by each mode of transportation?

11. A jet plane flying at a speed of 525 mph makes a trip in 2 hr less time than another plane flying at a speed of 350 mph. What is the length of the trip?

Exercise 10-7 (Review)

1. The sum of two numbers is 71. The larger number is five less than three times the smaller. What are the two numbers?

2. The sum of three numbers is 89. The second number is three times

the first and the third is four more than the first. What are the three numbers?

3. In a certain basketball game the total score made by both teams was 108. The home team made 21 points less than twice the score of the visiting team. Find the score made by each team.

4. In a certain football game team A made 31 points more than the team B. The score of team A was 5 points less than four times the score of team B. Find the score made by each team.

5. In a certain basketball game a team made a score of 93 points. The number of field goals was three less than four times the number of free throws. How many field goals and how many free throws did the team make? (A field goal counts 2 points, a free throw 1 point.)

6. The total cost of a suit and a hat was \$106. The suit cost \$7 more than five times the cost of the hat. Find the cost of each.

7. The total cost of a house and lot together was \$37,800. If the house cost \$360 more than eight times the cost of the lot, find the cost of each.

8. Three assignments together contain 63 problems. The second assignment contains four problems more than the first and the third contains seven less than the first. How many problems are there in each assignment?

9. A student wishes to arrange his reading assignment so that he may read five pages less each day than the day before. He has 350 pages to read in five days. How many pages should he read the first day and how many each day thereafter?

10. A strip of aluminum 30 in. long is to be cut into six pieces. It is necessary that each piece cut off be $\frac{1}{2}$ in. longer than the piece before it. What should the length of each piece be?

11. The current through one branch of an electric circuit is 1.46 amperes more than the current through another branch. When the branches join, the total current is 3.70 amperes. Find the current in each branch.

12. Two resistors connected in series in a circuit have a total resistance of 15.7 ohms. The first has a resistance of 4.3 ohms more than the second. Find the resistance of each.

13. Suppose it is necessary that two resistors connected in series have a total resistance of 50,000 ohms. When each is used alone, one must have a resistance of 45,200 ohms more than the other. What must be the resistance of each?

14. Three resistors connected in series have a total resistance of 340 ohms. The first has a resistance of 6 ohms more than the second, and the third has a resistance three times as great as the first. Find resistance of each.

15. An electric current of 0.195 amperes is branched off into two circuits so that one branch carries a current of 0.03 amperes less than twice the other. Find the current in each branch.

16. The flow capacity of a water main is 600 gal per min. The main separates into three branches. It is necessary that the second branch have three times the capacity of the first, and the third must have a capacity of 50 gal per min more than the first. Find the capacity of each branch.

17. A man is now four times as old as his son. Three years ago the father was six times as old as his son. Find the present age of each.

18. The length of a room is 5 ft less than twice the width. Its perimeter is 71 ft. Find its dimensions.

19. A wire 58 in. long is bent to form a rectangle such that the length of the rectangle is 3 in. longer than twice the width. Find the dimensions of the rectangle.

20. The sum of the three angles of any triangle is always 180°, whatever the shape or size of the triangle. In a certain triangle angle A is twice as large as angle B and angle C is 8° more than angle A. Which angle is the smallest? Find the number of degrees in each angle. Make a sketch of the triangle.

21. In a certain triangle the first angle is 10° less than the second and the third angle is twice the first. Find the size of each angle.

22. In a certain right triangle one acute angle is twice the other. Find the size of each angle.

23. In an isosceles triangle two angles are always equal. In a particular triangle, ABC, angle A and angle B are equal. Angle C contains 12.6° more than angle A. Find the number of degrees in each angle of the triangle.

24. In a certain triangle the first side is 4.5 in. longer than the second and the third side is 10 in. shorter than twice the first. The perimeter of the triangle is 64.3 in. Find the length of each side.

25. The sum of three consecutive even integers is 258. What are the numbers?

26. Find five consecutive odd integers whose sum is 455.

27. A collection of nickels, dimes, and quarters is worth \$7.05. There are six more dimes than nickels, and the total number of coins is 50. Find the number of each.

28. A collection of nickels, dimes, and quarters is worth \$7.20 The number of dimes is three less than the number of quarters, and the number of nickels is equal to the number of dimes and quarters together. Find the number of coins of each kind.

29. At an entertainment a total of 260 tickets were sold, children's tickets at 30 cents each and adults' tickets at 75 cents each. The total receipts were \$109.05. How many tickets of each kind were sold?

30. At a football game 750 tickets were sold. General admission cost \$1.60 per ticket, and reserved seats were \$2.40 each. Total receipts were \$1459.20. How many tickets of each kind were sold?

31. How many pounds each of 70-cent coffee and \$1.20-coffee should be mixed together for 50 lb of the mixture to sell at 85 cents a pound?

32. A car starts out on a trip and travels at an average rate of 35 mph. Two hours later a second car starts out from the same point, travels the same route, and averages 50 mph. How long will it take the second car to overtake the first?

33. A man takes a trip of 960 miles, part of the way by train at 160 mph and the rest of the way by car at 45 mph. The total travel time was 18 hr. How far did he travel at each rate?

34. A messenger starts out at 7 A.M. and travels at an average speed of 35 mph. Three hours later it is found that the message must be changed, so a second messenger starts out to overtake the first. How fast must the second one travel in order to overtake the first by 5 P.M.?

35. At a certain banquet there were 76 people seated at five different tables. The tables were lettered A, B, C, D, and E. The number seated at table A was one more than the number at table D. The number at table B was two more than the number at table A. The number at table C was three less than twice the number at table A. The number at table E was four less than twice the number at table D. Find the number seated at each table.

QUIZ ON CHAPTER 10, SOLVING STATED WORD PROBLEMS. FORM A.

1. The length of a rectangle is 4 in. less than 3 times the width. The perimeter of the rectangle (the distance around it) is 44 in. Find the width and the length of the rectangle.
2. A man's age is 7 years more than 3 times his son's age. Two years ago, the sum of their ages was 45 years. How old is each one now?
3. Find 4 consecutive odd numbers whose sum is 120.
4. A collection of 62 coins, consisting of nickels and quarters, is worth $7.70. Find the number of coins of each kind in the collection.
5. At an entertainment, children's tickets cost 50 cents each, and adults' tickets cost $1.25 each. The total paid attendance was 91. If the same amount was received from all the children's tickets as from all the adults' tickets, how many tickets of each kind were sold?
6. At an entertainment, children's tickets cost 55 cents each, and adults' tickets cost $1.30 each. If the total amount received was $89 from a total of 110 tickets, how many tickets of each kind were sold?
7. One car starts on a trip at 8 A.M., and averages 40 miles per hour. At 10 A.M. the same day, a second car starts on the same trip to overtake the first car and averages 55 miles per hour. At what time of day will the second car overtake the first?
8. A man takes a trip of 402 miles, part of the way by car at an average of 48 miles per hour, and the rest of the way by train at an average of 60 miles per hour. If the total travel time is 7 hours, how many hours and how far does he travel by each mode of travel?
9. Three resistors in series have a total resistance of 1600 ohms. If the second resistor has a resistance twice that of the first and the third has a resistance 200 ohms less than the second, find the resistance of each resistor.
10. A strip of brass 25 in. long is to be cut into 4 pieces. Each piece after the first is to be $\frac{1}{2}$ in. longer than the previous piece. Find the length of each piece.

CHAPTER ELEVEN Special Products and Factoring

11-1 DEFINITION

Multiplication occurs so often in algebra that we should be able to do it quickly. In many cases multiplication can be done by *inspection;* that is, it can be done mentally and the product written down at once.

A product that can be found by inspection may be called a *special product.* We shall study several of these products. A special product does not give us a different answer. Any such product may be found by the usual long method of multiplication. The only thing unusual about a special product is that we find the product quickly. It may be called a "shortcut."

11-2 MULTIPLICATION OF MONOMIALS BY INSPECTION

In multiplying monomials we take note of three things:

1. *The sign of the product, observing the rule for signs in multiplication.*
2. *The product of the numerical coefficients.*
3. *The literal numbers (letters), adding exponents of the same letter.*

Example

Multiply $(-5x^3y^2z)(-4x^2y)(-3yz^3)$.

For the sign of the product we have the product of three negative numbers, which is negative. For the coefficients we have: $(5)(4)(3) = 60$. Adding the exponents of like letters we get the product

$$(-5x^3y^2z)(-4x^2y)(-3yz^3) = -60x^5y^4z^4$$

Exercise 11-1

Multiply the following:

1. $(-2x)(4x^2y)(-3xy^2z)$

2. $(-4a)(-2a^2b^3c)(-ac)$

3. $(5mn)(-3m^2n^3)(4mn^2)$

4. $(3x)(-x^2)(-x^3y)(4y^3)$

5. $(-xy)(-3y^2)(-2x)(-x^3y)$

6. $(xy)^2(-x)^2(-x^2)(-x^2y^3)$

7. $(-7ab^2)(-3a^2c)(-8a^4bc^5)$

8. $(abc)^2(a^2c)^2(-bc^2)(-3ac^2d)$

9. $(-3x^2y)^3(-4xy^3)(x^3y)(-5)$
10. $(a^2b^3c)(-2a^3bc^2)(-4ac)(-b)$
11. $(-9x)(-3x^2yz)(-4a^3b)^3$
12. $(-3xy)(-6abc)(-4mn)(-yz)$
13. $(-2r^2s)(-3xy^2)^3(-r^2st)^2$
14. $(rt)(-st)^2(-rs^3)(3r^2st^3)$

11-3 FACTORING MONOMIALS

Factoring is the reverse of multiplication. If two or more quantities are multiplied together, the answer is called a *product.* Each of the quantities multiplied is called a *factor* of the product. For instance, if we multiply the two numbers 5 and 7, we get the product 35; 5 and 7 are the factors of 35.

It is often necessary to find the factors of a given product. When we say, "Find the factors of 323," we mean to find two numbers that can be multiplied to produce 323. In arithmetic we can often tell the factors at sight. For instance, the factors of 21 are 3 and 7. If we are told one of the factors of a product, we can find the other factor by division. If we know that one of the factors of 323 is 17, we can find the other factor by division.

It must be remembered that some quantities cannot be factored. The number 37 is *prime* and cannot be factored. Just so, the expression $3x^2 - 7xy + 3y^2$ cannot be factored in terms of real rational factors. We exclude the factor 1 because 1 is a factor of any number.

11-4 PRIME FACTORS

A *prime factor* is a factor that cannot be further separated into any factors except itself and 1. The factors of 120 can be considered to be 10 and 12. However, 10 and 12 can be further separated into factors. The prime factors of 120 are 2, 2, 2, 3, and 5.

In algebra a monomial can easily be separated into its prime factors simply by writing all factors separately. As an example,

$$30x^3y^4z = 2 \cdot 3 \cdot 5 \cdot x \cdot x \cdot x \cdot y \cdot y \cdot y \cdot y \cdot z$$

However, in most work in algebra, a monomial is not usually separated into its prime factors in this manner.

11-5 MULTIPLICATION OF A POLYNOMIAL BY A MONOMIAL

We have already mentioned (Chapter 8) that multiplication can be considered as a shortened form of addition. For instance, if we wish to find three times the quantity $7x^2 + 5x - 4$, we could write down the quantity three times and then add, as shown here:

$$\begin{array}{r} 7x^2 + 5x - 4 \\ 7x^2 + 5x - 4 \\ 7x^2 + 5x - 4 \\ \hline 21x^2 + 15x - 12 \end{array}$$

Here we see that each term of the polynomial $7x^2 + 5x - 4$ is multiplied by 3. If we wish to indicate three times the polynomial, we can write 3 next to the polynomial. However, we must then enclose the polynomial in parentheses, to indicate that it is to be taken as a quantity; thus

$$3(7x^2 + 5x - 4)$$

If parentheses are omitted and the 3 is placed next to the first term of the polynomial, the meaning will be changed. The expression $3 \cdot 7x^2 + 5x - 4$ means that only the first term, $7x^2$, is to be multiplied by 3.

Compare the following two expressions and you will see the extreme importance of parentheses to indicate the intended meaning.

$$3 \cdot 7x^2 + 5x - 4 = 21x^2 + 5x - 4$$
$$3(7x^2 + 5x - 4) = 21x^2 + 15x - 12$$

For this kind of product we have the following rule:

Rule. *To find the product of a monomial and a polynomial, multiply each term of the polynomial by the monomial. Such a product may be written down at sight.*

Example 1

Multiply $3(4x + 5y) = 12x + 15y$.

Example 2

Multiply $3xy(4x - 5xy + 2y) = 12x^2y - 15x^2y^2 + 6xy^2$.

Example 3

Multiply $2x^2(3x^3 - 4xy + 5y^2)yz =$

In Example 3 part of the monomial precedes the polynomial and part follows it. However, the entire monomial multiplier is the quantity $2x^2yz$. The product is $6x^5yz - 8x^3y^2z + 10x^2y^3z$.

Exercise 11-2

Multiply the following by inspection:

1. $2(3x - 5y + 2z)$

2. $-7x(2 - 3x - x^2)$

3. $4(5x - 7y + 3)$

4. $2x(3 - 7x - xy)$

5. $-3n(5n^2 - 9)2$ **6.** $5x^2(6x^2 - 3xy - 1)$

7. $-2xy^2(3x - 4y)$ **8.** $4x(2x - 5y + 1)$

9. $-5x^2y(x^3 - y^2)$ **10.** $-4x(5x^2 - 3x + 1)$

11. $4x^2y(1 + 2zy^3 - 3x^3y^2)$ **12.** $(1 - 3x^2 + 4y^2)x^2y$

13. $(2xy + 3x^2 - 5y^3 - 1)(-4x^2y)$

14. $5x(4x^3y^2 - x^2 + 2y^3 - x)y^3$

15. $x^2(5x^3y + 4x^2y^2 - z^2 + 1)yz^3$

16. $3m(7m^2n - mn^2 + 2n - 1)n$

17. $h^2(5h - 3hk - 4k)3k$ **18.** $rs^2(3r^2s - rt^2 + 5s^2t - 5t)t$

19. $-a(2ab^2c^3 - 9ab^2 + 4b^3x - 2)bc$

20. $(x^2y - 3xz^2 + 5y^2z + xy)(-2xyz^2)$

11-6 FACTORING BY TAKING OUT A COMMON FACTOR

If we multiply the polynomial $5x + 4y$ by the monomial $3x^2$, we get the product $15x^3 + 12x^2y$. Therefore, we know the quantity $15x^3 + 12x^2y$ is a product and can be separated into factors. One factor is the monomial $3x^2$, and the other factor is the binomial $5x + 4y$.

It is often necessary to factor an expression such as $20x^4 - 15x^2y$. If we are told that one factor is the monomial $5x^2$, we can find the other factor by division. It is $4x^2 - 3y$.

When we divide the factor $5x^2$ into the binomial. $20x^4 - 15x^2y$, we must divide the monomial into *each* term of the polynomial. We know that $5x^2$ and the binomial $4x^2 - 3y$ are the factors of the expression $20x^4 - 15x^2y$ because they can be multiplied together to produce that expression. That is the real test for factors.

The factor $5x^2$ is called a *common factor* because it can be divided into each term of the polynomial. To factor expressions of this kind, our problem is, first, to find a common factor that can be divided into each term of the polynomial. For this kind of factoring, we have the following rule:

Rule. *Inspect the terms of the polynomial to determine the greatest common factor that is contained in each term. This common factor is one factor of the given expression.*

Divide this common factor into each term of the polynomial. The quotient is the other factor.

Example 1

Factor $6x^3y^2 + 9x^2y^3 - 12xy^4$.

Solution

The greatest common factor contained in each of the terms is $3xy^2$. This is one factor of the expression. Dividing this factor into each term of the polynomial, we get the quotient $2x^2 + 3xy - 4y^2$. This is the other factor. The factors are usually written to indicate multiplication; thus

$$6x^3y^2 + 9x^2y^3 - 12xy^4 = 3xy^2(2x^2 + 3xy - 4y^2).$$

Example 2

Factor $5x^2 + x$.

Solution

The common factor is x. For the other factor, we divide each term by x.

$$5x^2 + x = x(5x + 1)$$

Exercise 11-3

Find the factors of the following expressions:

1. $4x + 20$ **2.** $18 - 6y$ **3.** $24n - 8$

4. $15y - 40ay$ **5.** $6n^2 - 32mn$ **6.** $12x + 12x^3$

7. $30x^4 - 20x^3$ **8.** $33x^2y^3 - 44xy^2$ **9.** $12x^5y - 8x^3y^3$

10. $16R^2h - 9r^2h$ **11.** $\pi R^2h - \pi r^2h$ **12.** $40rs^2t^3 - 20rst$

13. $12x^3 + 9x^2 - 3x$ **14.** $6y^4 + 8y^3 - 10y^2 + 2y$

15. $20y^5 - 15y^4 - 10y^3 - 5y^2$ **16.** $6x^4y + 9x^3y + 3x^2y - 3xy$

17. $x^5 - x^4 + x^3 - x^2 - x$ **18.** $4x^2z^2 - 4x^2z^3 + x^2z^4$

19. $6n^2x + 12n^3x^2 - 9n^4x^3 - 3nx$

20. $8h^5k^2 - 16h^3k^3 + 12h^2k^4 - 4hk$

21. $(m + n)^2 + (m + n)$ **22.** $4(x - y) - (x - y)^2 - (x - y)^3$

23. $(x - y)^3 + (x - y)^2$ **24.** $(c + d)^3 - 2(c + d)^2 - 3(c + d)$

11-7 A SPECIAL PRODUCT: THE SUM OF TWO NUMBERS TIMES THEIR DIFFERENCE

Suppose we multiply the two binomials: $3x + 5$ and $3x - 5$.

$$\begin{array}{lll} 3x & + 5 & \\ 3x & - 5 & \\ \hline 9x^2 & + 15x & \\ & - 15x & - 25 \\ \hline 9x^2 & & - 25 \end{array}$$

Notice that the two binomials are exactly alike except for one sign. The first binomial may be called the *sum of two numbers;* the second binomial may be called their *difference.*

The product contains only two terms. These two terms are the squares of the first term, $3x$, and the second term, 5. The two squares are separated by a minus sign. The product is the *difference between the squares.*

If we multiply the *sum* of any two quantities, such as $a + b$, by their *difference,* $a - b$, we always get a product that is the difference between the squares. The a and b represent any two quantities.

For this kind of special product we have the following rule:

Rule. *The product of the sum of two numbers times their difference is equal to the difference between their squares.*

Stated as a formula, this rule is

$$(a + b)(a - b) = a^2 - b^2$$

Since this rule is true for any and all values of a and b, it is an identity. All special product rules are identities.

As an example of this rule, $(7x - 6)(7x + 6) = 49x^2 - 36$.

Exercise 11-4

Multiply the following by inspection:

1. $(x - 7)(x + 7)$
2. $(9 - n)(n + 9)$
3. $(4x + 5y)(4x - 5y)$
4. $(3r - 10s)(3r + 10s)$
5. $(5n - 8ab)(5n + 8ab)$
6. $(4y + 9x)(9x - 4y)$
7. $(x^2 - 1)(x^2 + 1)$
8. $(n^3 - 6)(n^3 + 6)$
9. $(6y^4 - \frac{1}{3})(6y^4 + \frac{1}{3})$
10. $(10n + \frac{1}{4})(10n - \frac{1}{4})$
11. $(7a + 9xy)(7a - 9xy)$
12. $(5x - 7abc)(5x + 7abc)$
13. $(20x - 13y)(13y + 20x)$
14. $(30n + 17)(17 - 30n)$
15. $(8x^2y^3 - z)(z + 8x^2y^3)$
16. $(20t + 1)(20t - 1)$

17. $(50 + 1)(50 - 1)$

18. $(40 + 3)(40 - 3)$

19. $(82)(78)$

20. $(300 + 1)(300 - 1)$

11-8 FACTORING THE DIFFERENCE BETWEEN TWO SQUARES

From the multiplication we have just seen, we know that any expression that represents the difference between two squares can be factored; that is, the expression must have two quantities with a minus sign $(-)$ between them, and each quantity must represent the square of some number.

For example, suppose we have the expression

$$9x^2 - 16y^2$$

This binomial represents the difference between two squares. Then we know that it can be factored into two binomial factors. We indicate the two factors by first setting down the parentheses:

$$(\quad)(\quad)$$

Now we take the square root of each term of the given binomial. The square root of $9x^2$ is $3x$, which becomes the first term of each factor:

$$(3x \quad)(3x \quad)$$

The square root of the second term, $16y^2$, is $4y$, which becomes the second term of each factor:

$$(3x \quad 4y)(3x \quad 4y)$$

We connect the square roots with a plus sign $(+)$ for one factor, and with a minus sign $(-)$ for the other factor. Then we have the two factors:

$$9x^2 - 16y^2 = (3x + 4y)(3x - 4y)$$

We can check the factoring by multiplication.

Note. *It is immaterial which factor is written first.*

To factor an expression which represents the difference between two squares, we have this rule:

Rule. *Find the square root of each quantity. Connect the square roots with a plus sign for one factor and with a minus sign for the other factor.*

Factoring the difference between two squares is one of the most important and useful types of factoring. It should be thoroughly understood and remembered. The type might be represented in this way:

$$(\text{Quantity})^2 - (\text{quantity})^2$$

or in symbols:

$$(Q)^2 - (q)^2 = [Q + q][Q - q]$$

The rule applies not only when we have the squares of single terms, but also when we have the squares of entire quantities.

Example 1

$4x^2 - 25 = (2x - 5)(2x + 5)$.

Example 2

Factor the expression $(x - 3)^2 - (y + 2)^2$.

Solution

Here we have the squares of entire quantities. Following the rule, we take the square root of each quantity and connect the square roots as directed. It is well to keep each square root in parentheses the first time around as shown here. We get

$$\begin{aligned} &[(x - 3) + (y + 2)][(x - 3) - (y + 2)] \\ &= [x - 3 + y + 2][x - 3 - y - 2] \\ &= [x + y - 1][x - y - 5] \end{aligned}$$

Example 3

Factor $121x^6 - 169y^2$.

Solution

Note that the square root of x^6 is x^3. Then we get

$$121x^6 - 169y^2 = (11x^3 - 13y)(11x^3 + 13y)$$

Exercise 11-5

Factor the following expressions, if possible:

1. $x^2 - 36$ **2.** $25 - y^2$ **3.** $h^2 - 49$

4. $R^2 - r^2$ **5.** $4x^2 - 49y^2$ **6.** $25n^2 - 144$

7. $16t^2 - 1$ **8.** $1 - 100x^2$ **9.** $81y^2 - 1$

10. $81x^4 - 16y^2$ **11.** $196a^6 - 25x^2$ **12.** $100s^2 - 289t^2$

13. $a^2b^4 - 9c^6$ **14.** $x^2 - 5$ **15.** $x^2 - 2$

16. $49x^2 - \frac{1}{4}$ **17.** $225n^2 - \frac{1}{9}$ **18.** $64t^2 - \frac{1}{16}$

19. $25n^2 - 0.16$ **20.** $0.36 - 49y^2$ **21.** $9x^4 - 0.04$

22. $9n^2 - 36$ **23.** $4x^2 - 64$ **24.** $18y^2 - 50$

25. $(x - y)^2 - z^2$ **26.** $(a - b)^2 - 36$ **27.** $x^2 - (y + 4)^2$

28. $(a - b)^2 - (c + d)^2$

29. $(x - 5)^2 - (y - 2)^2$ **30.** $(2x + 3)^2 - (3y - 4)^2$

11-9 THE SQUARE OF A BINOMIAL: A SPECIAL PRODUCT

If we multiply two binomials that are exactly alike, we call the product the *square of a binomial.* Usually we indicate such multiplication by only one factor and the exponent 2. For instance,

$$(3x + 5)(3x + 5) \quad \text{is usually written } (3x + 5)^2$$

This expression indicates a perfect square. By a perfect square, we mean a product of two factors that are exactly alike. The number 36 is a perfect square, since it is the product of the two identical factors, 6 and 6.

The expression $(3x + 5)^2$ indicates a perfect square, since we have two identical factors. If we multiply the two factors by the usual long method of multiplication, we get the product

$$(3x + 5)^2 = 9x^2 + 30x + 25$$

The product, $9x^2 + 30x + 25$, is called the *square of a binomial.*

Let us see how we may write out the square of a binomial by inspection. You will notice that the product has *three* terms. The *first* and *third* terms of the product are simply the *squares* of the two terms of the binomial. The middle term of the product is *twice the product of the two terms of the binomial.*

If the binomial indicates the difference between two numbers, such as $3x - 5$, then, by long multiplication, we find that

$$(3x - 5)^2 = 9x^2 - 30x + 25$$

If we square the binomial $3x - 5$, just as we squared the binomial $3x + 5$, we find the product is the same except for the sign of the middle term. The expression $(3x + 5)^2$ is called the square of the *sum*, and the expression $(3x - 5)^2$ is called the square of the *difference* between two numbers. The products are exactly alike except for one sign, the sign of the middle term.

For the squares of binomials, we have these rules:

Rule 1. *The square of the sum of two quantities, such as $(a + b)^2$, is equal to the square of the first term, plus twice the product of the two terms, plus the square of the second term. Stated as a formula, this rule is*

$$(a + b)^2 = a^2 + 2ab + b^2$$

Rule 2. *The square of the difference between two quantities, such as $(a - b)^2$, is equal to the square of the first term, minus twice the product of the two terms, plus the square of the second term. Stated as a formula, this rule is*

$$(a - b)^2 = a^2 - 2ab + b^2$$

Exercise 11-6

Multiply the following by inspection:

1. $(x + 8)(x + 8)$ **2.** $(x - 10)^2$ **3.** $(5x + 7)^2$

4. $(3xy - 5)^2$ **5.** $(4r + 9s)^2$ **6.** $(7ab - 8c)^2$

7. $(1 + 11x)^2$ **8.** $(9c - 1)^2$ **9.** $(10x + 3)^2$

10. $(6x^2 - 1)^2$ **11.** $(2x^3 + 5)^2$ **12.** $(4xy - 7z)^2$

13. $(8y + \frac{1}{2})^2$ **14.** $(9x - \frac{1}{3})^2$ **15.** $(2x + \frac{1}{4})^2$

16. $(R - r)^2$ **17.** $(5x + 0.2)^2$ **18.** $(12x - 0.25)^2$

19. $(70 + 1)^2$ **20.** $(69)^2$ **21.** $(299)^2$

22. $[(x - 3) - y]^2$ **23.** $[(a + b) + c]^2$ **24.** $[x - (y - 5)]^2$

Write the square of each of the following binomials:

25. $5x + 8$ **26.** $3n - 7$ **27.** $9x + 4$ **28.** $6 - 5y$

29. $4n + \frac{1}{4}$ **30.** $3x - \frac{1}{6}$ **31.** $b^2 - 3x^3$ **32.** $1 + 10xy$

Square both sides of each of the following equations:

33. $x = 3a + 8$ **34.** $6x - 5 = y$ **35.** $\sqrt{x - 3} = 6$

11-10 FACTORING TRINOMIALS THAT ARE PERFECT SQUARES

We have seen that the square of the binomial $3x + 5$ is $9x^2 + 30x + 25$. Therefore, we know that the trinomial $9x^2 + 30x + 25$ can be factored. Moreover, the two factors will be exactly alike, and the trinomial therefore is a perfect square.

Our problem is this: how can we recognize that a certain given trinomial is a perfect square? Suppose we have the trinomial $16x^2 + 40x + 25$. Our questions are, first, can the expression be factored and, second, if so, is the expression a perfect square? That is, will the two factors be exactly alike?

Perfect squares can often be recognized by inspection, and the factors can be written down at once. First of all, we know the expression must be a trinomial; that is, it must contain three terms. Moreover, the first and the third terms must be perfect square terms.

Look again at the trinomial $16x^2 + 40x + 25$. The first and third terms are perfect squares; they are the squares of the quantities $4x$ and 5.

However, before we can be sure the entire expression is the complete square of a binomial, we must be sure the middle term is the proper term for such a square. The middle term must be *twice* the product of the *square roots* of the other terms. In this case we have twice the product of $4x$ and 5, which makes $40x$. This means that the given expression is a perfect square of the binomial $(4x + 5)$. Therefore, we can say

$$16x^2 + 40x + 25 = (4x + 5)(4x + 5) \quad \text{or} \quad (4x + 5)^2$$

In some perfect squares the middle term is negative. You will recall that the square of the binomial $3x - 5$ is the trinomial $9x^2 - 30x + 25$. This is exactly like the square of $3x + 5$ except for the sign of the middle term.

In the trinomial $9x^2 - 30x + 25$ we notice that the first and third terms are perfect squares of the quantities $3x$ and 5, respectively. However, we recall that the square root of 25 is either a $+5$ or a -5, and twice the product of the two square roots is $-30x$ if we choose the -5. The expression is therefore a perfect square of a binomial:

$$9x^2 - 30x + 25 = (3x - 5)(3x - 5) \quad \text{or} \quad (3x - 5)^2$$

The factors of $9x^2 - 30x + 25$ can also be $(5 - 3x)^2$.

Example

$x^2 - 8x + 16 = (x - 4)(x - 4)$ or $(x - 4)^2$. The factors of $x^2 - 8x + 16$ can also be $(4 - x)^2$.

Note. Although the first and third terms are positive and perfect squares, the entire expression is not always a perfect square. Before we can be sure of having a perfect square, the middle term must be checked. The following expression is not a perfect square: $x^2 + 9x + 16$.

Exercise 11-7

Determine which of the following expressions are perfect squares. Then factor the perfect squares and write the two factors as the square of a binomial.

1. $x^2 - 10x + 16$

2. $y^2 + 13y + 36$

3. $y^2 + 10y + 25$

4. $x^2 - 16x + 64$

5. $9x^2 - 48x + 64$

6. $4y^2 + 12y + 9$

7. $49t^2 + 26t + 4$

8. $16t^2 - 50t + 25$

9. $36 - 60n^2 + 25n^4$

10. $4 + 28x^3 + 49x^6$

11. $y^2 + 20y + 81$

12. $t^2 - 25t + 100$

13 $9t^4 - 24t + 16$

14. $4n^6 - 20n^2 + 25$

15. $4x^2 - 2x + \frac{1}{4}$

16. $t^2 + t + \frac{1}{4}$

17. $225n^2 - 30n + 1$

18. $81n^2 + 18n + 1$

19. $x^2y^2 - 3axy + (\frac{9}{4})a^2$

20. $144t^2 + 8t + \frac{1}{9}$

11-11 FINDING A MISSING MIDDLE TERM OF A PERFECT SQUARE

Suppose we have given the two end terms of a perfect square, such as $25x^2 \cdots + 16y^2$. Our problem may be to determine the proper middle term for a perfect square trinomial. To find the necessary term, we first take the square roots, $5x$ and $4y$, of the two given terms. Then, the middle term must be twice the product of these two terms: $(2)(5x)(4y)$, or $40xy$. This middle term may be either positive or negative. If we call the middle term positive, we have

$$25x^2 + 40xy + 16y^2 = (5x + 4y)(5x + 4y) \quad \text{or} \quad (5x + 4y)^2$$

If we call the middle term negative, we have

$$25x^2 - 40xy + 16y^2 = (5x - 4y)(5x - 4y) \quad \text{or} \quad (5x - 4y)^2$$

Exercise 11-8

Supply a proper middle term to make each of the following expressions a perfect square. Then factor each expression and write as the square of a binomial:

1. $x^2 + \cdots + 9$

2. $y^2 + \cdots + 36$

3. $n^2 - \cdots + 64$

4. $a^2 - \cdots + 64$

5. $4x^2 + \cdots + 9$

6. $9n^2 - \cdots + 16$

7. $25x^2 + \cdots + 1$

8. $49y^2 - \cdots + 4$

9. $16x^2 + \cdots + 81$

10. $x^2y^2 - \cdots + \frac{1}{9}$	**11.** $36x^2 + \cdots + 25$	**12.** $121 - \cdots + 36x^2$
13. $9x^2 - \cdots + \frac{1}{9}$	**14.** $4r^2 + \cdots + \frac{1}{16}$	**15.** $a^2b^4 - \cdots + 9$
16. $16 + \cdots + 49t^2$	**17.** $x^2 - \cdots + \frac{1}{4}$	**18.** $n^2 + \cdots + 324$
19. $9x^2 - \cdots + \frac{1}{4}$	**20.** $\frac{4}{9} - \cdots + 25x^2$	**21.** $144 + \cdots + 4x^2$
22. $25 - \cdots + 9i^2$	**23.** $F^2 + \cdots + 400$	**24.** $I^2 - \cdots + 361$

11-12 COMPLETING A SQUARE BY ADDITION OF A THIRD TERM

In some expressions it happens that we have the first two terms of a perfect square. As an example, suppose we have the expression $x^2 - 6x$. Our problem in many cases is to determine what number should be added to the expression to produce a perfect square. This process is called *completing the square.* The device of completing a square is very useful in much work in mathematics.

If we remember that the middle term is *twice* the product of the two square roots of the end terms, respectively, we see that this product is $-6x$. Therefore, the product itself of the two square roots is $-3x$. Since the square root of the first term is x, the square root of the last term must be -3. Therefore, the last term is 9.

For completing the square, in some cases it may be a little more difficult to determine the number to be added. Take the example $9x^2 + 30x \ldots$. Here, the $30x$ is twice the product of the square roots, so that the product itself is $15x$. Now, the square root of the first term is $3x$. Therefore, the square root of the last term must be 5. The number to be added is the square of 5, which is 25.

For completing a square, we have the following rule:

Rule. *If the coefficient of x^2 is 1, take one-half the coefficient of x and square this number. The result is the number to be added.*

If the coefficient of x^2 is something other than 1, take one-half the coefficient of x, divide this by the square root of the coefficient of x^2, and square the result. This is the number to be added.

Exercise 11-9

Supply the missing term that will make each of the following expressions a perfect square. Then express each as the square of a binomial:

1. $x^2 + 8x \ldots$	**2.** $y^2 - 14y \ldots$	**3.** $x^2 + 16x \ldots$
4. $n^2 - 20n \ldots$	**5.** $x^2 + 30x \ldots$	**6.** $y^2 - 24y \ldots$
7. $y^2 + y \ldots$	**8.** $t^2 - 3t \ldots$	**9.** $n^2 + 5n \ldots$

10. $x^2 - \frac{x}{2} \ldots$ **11.** $y^2 + 0.8y \ldots$ **12.** $x^2 - \frac{3}{4}x \ldots$

13. $4n^2 + 28n \ldots$ **14.** $9x^2 - 30x \ldots$ **15.** $16y^2 + 40y \ldots$

16. $25y^2 - 60y \ldots$ **17.** $36t^2 + 84t \ldots$ **18.** $81n^2 - 72n \ldots$

19. $4x^2 + 5x \ldots$ **20.** $16n^2 - 4n \ldots$ **21.** $9x^2 + 7x \ldots$

22. $16n^2 - 3n \ldots$ **23.** $9x^2 + 5x \ldots$ **24.** $4x^2 - 9x \ldots$

11-13 A SPECIAL PRODUCT: $(x + a)(x + b)$

If we have the two factors $(x + 3)(x + 4)$ to be multiplied, we notice that this set of factors is different from the kinds of sets we have already considered. If we multiply these two factors by the usual long method, we get

$$(x + 3)(x + 4) = x^2 + 7x + 12$$

Note that the first term of the product is simply x^2. The third term is the product of the last two terms of the binomials; that is, (3)(4). The coefficient of x is equal to the *sum* of the two terms, +3 and +4.

Such a product may be quickly written from inspection. As an example, suppose we have the two factors

$$(x + 7)(x + 5)$$

The first term of the product is, of course, x^2. The last term of the product is (+7)(+5), or 35. The middle term has the sum of +7 and +5 as the coefficient of x. The product is therefore

$$(x + 7)(x + 5) = x^2 + 12x + 35$$

As a problem of this kind, we can say that

$$(x + a)(x + b) = x^2 + (a + b)x + ab$$

The signs, plus and minus, must be carefully observed in multiplication by inspection. Consider the problem

$$(x - 7)(x + 4)$$

In the product the first term is x^2. The last term is (−7)(+4), or −28. To get the middle term, we take the sum of (−7) and (+4), which is −3, as the coefficient of x. The complete product is: $x^2 - 3x - 28$.

Rule. *The first term of the product is the product $(x)(x)$, which is x^2.*

The second term is a term in x. The coefficient of x is the algebraic sum of the two last terms of the binomials.

The third term is the algebraic product of the last two terms of the binomials.

Exercise 11-10

Multiply by inspection:

1. $(x - 5)(x - 4)$	**2.** $(x + 3)(x + 7)$	**3.** $(x - 6)(x - 5)$
4. $(x + 7)(x + 5)$	**5.** $(x - 6)(x - 2)$	**6.** $(x + 3)(x + 8)$
7. $(x + 3)(x - 4)$	**8.** $(x - 7)(x + 6)$	**9.** $(x - 7)(x + 8)$
10. $(x - 7)(x + 4)$	**11.** $(x - 3)(x + 8)$	**12.** $(x - 9)(x + 5)$
13. $(x + 12)(x - 1)$	**14.** $(x + 16)(x - 1)$	**15.** $(x + 10)(x - 1)$
16. $(x + 4)(x - 6)$	**17.** $(x + 5)(x - 10)$	**18.** $(x - 7)(x + 9)$
19. $(y - 12)(y + 3)$	**20.** $(y + 11)(y - 4)$	**21.** $(y + 5)(y - 13)$
22. $(t - 8)(t + 8)$	**23.** $(t + 6)(t - 6)$	**24.** $(t - 9)(t + 9)$
25. $(n - 18)(n + 5)$	**26.** $(n + 30)(n - 2)$	**27.** $(n + 48)(n - 3)$
28. $(x^2 - 3)(x^2 + 7)$	**29.** $(x^3 + 9)(x^3 - 4)$	**30.** $(x^4 - 8)(x^4 + 2)$

11-14 FACTORING EXPRESSIONS OF THE TYPE $x^2 + px + q$

To find the factors of trinomials of this type, we write x as the first term of each factor: $(x \quad)(x \quad)$. Then we find two factors of the q term whose algebraic sum is p.

Example 1

Factor the trinomial $x^2 - 4x - 12$.

Solution

We set down x as the first term of each factor:

$$(x \quad)(x \quad)$$

Now we must find two factors of -12 whose algebraic sum is -4. By inspection, we see that these factors are -6 and $+2$. These numbers are the second terms of the binomials. The factors of the trinomial are

$$x^2 - 4x - 12 = (x - 6)(x + 2)$$

If the third term in a trinomial of this kind is positive, the factors of this term must have the same sign; *both* may be positive *or both* negative. Consider the example

$$x^2 - 5x + 6 = (\quad)(\quad)$$

The factors of +6 must have the same sign. Since their algebraic sum is −5, the factors must be −3 and −2. Then the factors of the trinomial are

$$x^2 - 5x + 6 = (x - 3)(x - 2)$$

Note. If there are no factors of the third term whose sum is equal to the coefficient in the middle term, then the trinomial cannot be factored in rational terms. If the two required factors of the third term are exactly alike, the two factors are identical, and the trinomial is a perfect square.

Example 2

$x^2 + 9x + 16$. This trinomial cannot be factored.

Example 3

$x^2 + 10x + 25 = (x + 5)(x + 5)$, a perfect square.

Exercise 11-11

Factor the following, if possible. Tell which are perfect squares.

1. $x^2 + 9x + 20$	**2.** $x^2 - 11x + 30$	**3.** $x^2 + 11x + 18$
4. $x^2 - 11x + 28$	**5.** $x^2 + 12x + 35$	**6.** $x^2 - 13x + 36$
7. $t^2 + 2t - 15$	**8.** $t^2 - 5t - 24$	**9.** $t^2 + 4t - 12$
10. $y^2 + 8y - 9$	**11.** $y^2 - 10y - 11$	**12.** $y^2 + 12y - 13$
13. $n^2 - n - 30$	**14.** $n^2 + n - 72$	**15.** $n^2 + n - 90$
16. $x^2 - 16x + 64$	**17.** $x^2 + 12x + 36$	**18.** $x^2 - 20x - 100$
19. $y^2 + 11y - 60$	**20.** $y^2 - 38y - 80$	**21.** $y^2 + 13y - 48$
22. $x^2 - 20x + 36$	**23.** $x^2 + 20x + 64$	**24.** $x^2 - 30x + 81$
25. $y^2 - 14y - 72$	**26.** $y^2 + 21y - 72$	**27.** $y^2 - 34y - 72$
28. $x^2 + 8x - 84$	**29.** $x^2 - 17x - 84$	**30.** $x^2 + 40x - 84$

11-15 A MORE DIFFICULT SPECIAL PRODUCT: $(ax + b)(cx + d)$

We now come to a special product of two binomials that is more difficult than those heretofore mentioned. Suppose we have the two factors $(3x - 4)(5x + 2)$. If we perform the multiplication by the usual long method, we find that the product is $15x^2 - 14x - 8$.

$$
\begin{array}{r}
3x - 4 \phantom{{}-8} \\
5x + 2 \phantom{{}-8} \\
\hline
15x^2 - 20x \phantom{{}-8} \\
+ 6x - 8 \\
\hline
15x^2 - 14x - 8
\end{array}
$$

Our problem now is to see how this product may be found quickly by inspection. Notice that the *first* term of the product, $15x^2$, is obtained by multiplying the first terms of the two binomials: $(3x)(5x)$.

The *third* term of the product is found by multiplying the second terms of the binomials: $(-4)(+2)$, which is -8.

The *middle* term of the product, $-14x$, is the algebraic sum of the *cross products:* $(5x)(-4)$ and $(3x)(2)$. If we write the two factors in horizontal form, we can show these cross products by arrows in this way:

$$(3x - 4)(5x + 2)$$

The middle term of the product is the sum of these *cross products:*

$$
\begin{array}{r}
-20x \\
+6x \\
\hline
-14x
\end{array}
$$

For the multiplication of two binomials of this form, we have the following rule:

Rule. *The first term of the answer is the product of the first terms of the binomials.*

The middle term of the answer is the algebraic sum of the cross products.

The third term of the answer is the product of the second terms of the binomials.

Exercise 11-12

Multiply the following by inspection:

1. $(2x + 3)(3x + 5)$ **2.** $(3x - 2)(2x - 5)$

3. $(4x + 5)(3x - 4)$ **4.** $(5x - 6)(4x + 5)$

5. $(3x - 4)(5x + 2)$ **6.** $(6x - 5)(5x + 3)$

7. $(4x - 5)(5x + 4)$

8. $(5x - 3)(3x + 5)$

9. $(6y - 5)(2y - 3)$

10. $(3y - 2)(4y - 7)$

11. $(7t - 4)(5t + 3)$

12. $(4t + 3)(5t - 4)$

13. $(4n - 1)(n - 6)$

14. $(n + 3)(8n + 1)$

15. $(10a + 3)(3a - 2)$

16. $(9c - 5)(2c + 3)$

17. $(6 - 5xy)(4 + 3xy)$

18. $(5ab - 7)(4ab + 5)$

19. $(3x^2 + 4y)(4x^2 - 3y)$

20. $(2a^2 - 3b)(8a^2 + 5b)$

21. $(4n - 1)(n - 16)$

22. $(9n + 1)(n + 4)$

23. $(6x + 1)(x + 8)$

24. $(8x + 1)(x - 10)$

11-16 FACTORING A TRINOMIAL OF THE TYPE $ax^2 + bx + c$

This kind of trinomial is one of the most difficult to factor. The method is shown by an example.

Example 1

Factor the expression $3x^2 - 7x + 2$.

Solution

Factoring a trinomial of this type is largely a matter of trial and error. At first we may have some difficulty in finding the correct set of factors. However, there is a definite approach that we can follow.

We first write down the form of the binomial factors by the use of parentheses:

$$3x^2 - 7x + 2 = (\quad)(\quad)$$

The first terms of the binomials must be factors of the first term $3x^2$. The only possible factors are $3x$ and x. So we write them as the first terms of the binomial factors:

$$3x^2 - 7x + 2 = (3x \quad)(x \quad)$$

The only possible factors of the last term 2 are 2 and 1. Both factors must be plus (+) or both minus (−), in order to make the product +2, as the last term of the trinomial.

To determine the proper factors of the trinomial, we must match up the factors of the first term, $3x$ and x, with the factors of the last term in such a way that the sum of the cross products will be equal to the middle term $-7x$. After some trial, we find that the factor -2 must be placed with the x

term and the factor -1 with the $3x$ term:

$$3x^2 - 7x + 2 = (3x - 1)(x - 2)$$

Matching up the terms in any other manner would not produce the middle term $-7x$. Remember, the middle term must be the sum of the *cross products.*

Example 2

Factor $4x^2 - 5x - 6 = (\quad)(\quad)$.

Solution

We know that the first terms of the factors must be factors of $4x^2$. One pair of such factors is the set $2x$ and $2x$. Another set is $4x$ and x.

To determine the proper factors of the trinomial, we must match up the factors of the first term, $4x^2$, with the proper set of factors of the last term, -6, in such a way that the sum of the cross products will be $-5x$. After some practice, we are usually able to select the proper sets of factors without too much difficulty. However, in a problem of this kind, it is well at first to write down all possible combinations.

In each of the following sets of factors the first terms are factors of $4x^2$ and the second terms are factors of -6.

$(4x - 6)(x + 1)$	$(4x - 2)(x + 3)$	$(2x - 3)(2x + 2)$
$(4x - 1)(x + 6)$	$(4x + 2)(x - 3)$	$(2x + 3)(2x - 2)$
$(4x + 6)(x - 1)$	$(4x - 3)(x + 2)$	$(2x - 6)(2x + 1)$
$(4x + 1)(x - 6)$	$(4x + 3)(x - 2)$	$(2x + 6)(2x - 1)$

In all of these sets of factors only one has the correct arrangement to make the middle term $-5x$. This is the set $(4x + 3)(x - 2)$.

At first you may have difficulty in selecting the correct set of factors in some examples of this type. With practice you will be able to find the correct set rather quickly.

As a help in factoring, you might remember one important point. If the original expression has no common factor, then there cannot be a common factor in any one of the separate factors. This fact enables us to omit many possible pairs immediately. Note the common factor 2 in eight of the pairs of factors we have set down. These pairs can be omitted at once because the original expression has no such common factor.

Exercise 11-13

Factor the following. Identify perfect squares.

1. $3x^2 - 13x - 10$ **2.** $2x^2 + 5x - 12$ **3.** $3n^2 - 11n - 4$

4. $3r^2 + 10r - 8$ **5.** $6c^2 - 5c - 6$ **6.** $4t^2 + 7t + 3$

7. $5y^2 - 11y + 6$ **8.** $15n^2 + n - 40$ **9.** $30a^2 - a - 42$

10. $6y^2 + y - 12$ **11.** $10n^2 - n - 21$ **12.** $4x^2 + 20x + 25$

13. $9t^2 - 24t + 16$ **14.** $18t^2 + 3t - 10$ **15.** $24t^2 - 23t - 20$

16. $16y^2 + 30y + 9$ **17.** $9y^2 - 48y + 64$ **18.** $4y^2 + 65y + 16$

19. $9x^2 - 50y + 36$

20. $6a^2 - 23ab - 20b^2$

21. $6n^2 + 7nx - 24x^2$

22. $16y^2 + 26y + 9$ **23.** $16k^2 + 56k + 49$ **24.** $25t^2 - 61t + 36$

25. $36n^2 - 75n + 25$ **26.** $9n^2 - 50n + 25$ **27.** $4n^2 + 35n + 49$

28. $4h^2 + 45h + 81$ **29.** $25r^2 + 50r + 49$ **30.** $21x^2 - 40x - 21$

11-17 FACTORING THE SUM AND THE DIFFERENCE OF TWO CUBES

If we divide the expression $a^3 + b^3$ by the quantity $a + b$, we get the quotient $a^2 - ab + b^2$. Therefore, we know that

$$a^3 + b^3 = (a + b)(a^2 - ab + b^2)$$

Also, if we divide $a^3 - b^3$ by $a - b$, we get the quotient $a^2 + ab + b^2$. Therefore,

$$a^3 - b^3 = (a - b)(a^2 + ab + b^2)$$

The best way to learn to factor the sum or difference of two cubes is to become thoroughly familiar with the pattern of the factors. In either case one factor contains *two* terms, the other factor contains *three* terms. Notice that the binomial factor is made up of the *cube roots* of the two cubes. The sign between them is the same as the sign between the cubes.

The second factor, the trinomial, in each case is made up of the squares of the cube roots, with the product of the cube roots as the middle term. In either case the two factors have only one minus $(-)$ sign. If you forget the second factor, you can always find it by long division.

Exercise 11-14

Factor the following:

1. $x^3 - 8$ **2.** $125 - n^3$ **3.** $T^3 + 1000$

4. $216x^3 + 1$ **5.** $x^6 - 27$ **6.** $x^3 + \frac{1}{8}$

7. $343t^3 - 64$ **8.** $X^{15} + 216$ **9.** $x^9 - 1$

10. $R^3 - r^3$ **11.** $8y^3 + 125$ **12.** $27x^3 - \frac{1}{27}$

13. $64n^3 - 729$ **14.** $(a + b)^3 + 8$ **15.** $(x + y)^3 - (a - b)^3$

11-18 FACTORING THE SUM OR DIFFERENCE OF TWO EQUAL POWERS

The difference between equal powers (odd or even) of two numbers, such as $a^5 - b^5$ or $a^6 - b^6$, is always divisible by the difference between the numbers $a - b$. Therefore, in factoring such expressions we may take the difference between the numbers as one of the factors. The best way to get the other factor is by long division:

$$x^5 - y^5 = (x - y)(x^4 + x^3y + x^2y^2 + xy^3 + y^4)$$

However, if the powers are even powers, such as $x^8 - y^8$, we should first factor the expression as the difference between two squares.

The *sum* of equal powers of two numbers, such as $a^n + b^n$, is divisible by the sum of the numbers, *provided* the powers are *odd* powers. Thus

$$x^5 + y^5 = (x + y)(x^4 - x^3y + x^2y^2 - xy^3 + y^4)$$

The second factor is obtained by long division.

The *sum* of two *even* powers cannot be factored unless the powers can also be shown to be odd powers of some quantities. Thus, $x^4 + y^4$ is not factorable; $x^6 + y^6$ can be called the sum of two cubes $(x^2)^3 + (y^2)^3$, and therefore the expression can be factored as $(x^2 + y^2) \times (x^4 - x^2y^2 + y^4)$.

Exercise 11-15

Factor if possible:

1. $x^4 - y^4$ **2.** $x^5 + 32$ **3.** $x^7 - y^7$

4. $n^{12} + 1$ **5.** $y^4 + 16$ **6.** $x^8 - 1$

7. $x^2 + y^2$ **8.** $x^{10} - 1$ **9.** $x^6 - 64$

11-19 FACTORING BY VARIOUS METHODS OF GROUPING

Factoring by grouping of terms is shown here by examples.

Example 1

Factor $2x + ax - 2y - ay$.

Solution

Grouping terms:	$(2x + ax) - (2y + ay)$.
Common factors:	$x(2 + a) - y(2 + a)$.
Now, we take out the common factor	$(2 + a)$.
The factors are	$(2 + a)(x - y)$.

Example 2

Factor $x^2 + 2xy + y^2 - x - y$.

Solution

We write the first three terms as a perfect square and enclose the remaining terms in parentheses. We get

$$(x + y)^2 - (x + y)$$

We now take out the common factor $x + y$:

$$(x + y)(x + y - 1)$$

Example 3

Factor $x^2 + 6x + 9 - c^2$.

Solution

We write the first three terms as a perfect square:

$$(x + 3)^2 - c^2$$

We now factor the entire expression as the difference between two squares:

$$(x + 3 - c)(x + 3 + c)$$

Example 4

Factor $x^4 + 5x^2 + 9$.

Solution

If we add and then subtract the same quantity from the expression, the value will not be changed. Let us add and subtract the quantity x^2. Then we get

$$x^4 + 5x^2 + 9 + x^2 - x^2$$

The first four terms taken together form a perfect square:

$$(x^2 + 3)^2 - x^2$$

The expression can now be factored as the difference between two squares:

$$(x^2 + 3 - x)(x^2 + 3 + x)$$

Exercise 11-16

Find the prime factors of the following. Always take out a common factor first if possible. If any factor is itself a product, separate it into prime factors.

1. $16x^2 - 36$ **2.** $48n^2 - 75$ **3.** $2a^2 - 72b^2$

4. $9t^3 - 0.25t$ **5.** $9x^4 + 81x^2y^2$ **6.** $16x^3y - 100xy^3$

7. $x^4 - 16$ **8.** $x^8 - y^8$ **9.** $x^{10} - x^6$

10. $x^5y - 169x^3y^3$ **11.** $a^2h^2 - a^2k^2$ **12.** $64y^3 - 4y^7$

13. $x^4 - 81x$ **14.** $8x^2 + x^5$ **15.** $5xy^4 + 320xy$

16. $27n^8 + n^5$ **17.** $250n^3 - 2n^6$ **18.** $\pi R^3 - \pi r^3$

19. $x^5 - x^4 - 42x^3$ **20.** $3x^3 + 3x^2 - 60x$ **21.** $4y^4 - 4y^3 - 48y^2$

22. $9y^3 + 63y^2 + 90y$ **23.** $6n^4 - 24n^3 - 72n^2$

24. $4t^5 + 36t^4 + 72t^3$ **25.** $18x^4 - 33x^3y - 30x^2y^2$

26. $8ay^3 + 10a^2y^2 - 12a^3y$ **27.** $36n^3 + 96n^2 + 64n$

28. $100x^2y - 80xy + 16y$ **29.** $a^2 - 6ab + 9b^2 - x^2$

30. $16y^2 + x^2 - 16 - 8xy$ **31.** $x^5 - x^3 + 8x^2 + 8$

32. $x^3 + 3x^2 - 9x - 27$ **33.** $16x^4 + 7x^2 + 1$

34. $9x^4 + 3x^2 + 4$ **35.** $6x + 4y^2 - 9 - x^2$

36. $a^2 - 36 - 12b - b^2$ **37.** $x^2 - y^2 + 2x + 6y - 8$

38. $x^2 - y^2 - 10x + 4y + 21$ **39.** $x^2 - 9y^2 + 4x + 12y$

40. $4x^2 - 9y^2 - 8x + 18y - 5$

QUIZ ON CHAPTER 11, FACTORING. FORM A.

Factor each expression into prime factors.

1. $b^2 - 144$ **2.** $8n^4 - 50n^2$ **3.** $2x^7 - 2x^3$

4. $y^2 + y - 56$ **5.** $x^3 - x^2 - 20x$ **6.** $2x^2 - 6x - 56$

7. $x^2 + 9x - 52$ **8.** $y^2 - 16y - 80$ **9.** $x^2 + 13x - 30$

10. $n^2 - 9n - 90$ **11.** $x^2 + 10x - 96$ **12.** $64x^3 + 27t^3$

13. $8a^3b^3 - 1$ **14.** $3x^2 - 7x + 2$ **15.** $16x^2 + 56x + 49$

16. $24x^2 - 25x - 25$ **17.** $9x^2 + 26x + 16$ **18.** $25x^2 - 30x - 16$

19. $a^6 - b^6$ **20.** $x^8 - 1$ **21.** $x^2 - y^2 + 6x + 9$

22. Tell the difference between these two expressions and state the factors of each: $m^3 + n^3$ and $(m + n)^3$.

Add the proper number to each of the following expressions to form a perfect square and then factor the result and express it as a square:

23. $x^2 + 16x + \cdots$ **24.** $4x^2 - 28x + \cdots$ **25.** $25x^2 + 3x \cdots$

Factor each of the following:

26. $x^3 + 2x^2 - 9x - 18$ **27.** $x^4 - 5x^2 + 9$

28. $x^2 - y^2 - 8x - 2y + 15$

CHAPTER TWELVE Fractions

12-1 DEFINITION

In algebra, as in arithmetic, it often happens that we must operate with fractions. In order to understand clearly how to perform the operations, let us recall some of the facts about fractions in arithmetic.

From arithmetic we know that a common fraction consists of two numbers, one written above the other, with a horizontal line between them: thus $\frac{3}{4}$. The number above the line is called the *numerator.* The number below the line is called the *denominator. The numerator and the denominator are called the terms of a fraction.*

The denominator of a fraction indicates the number of parts into which a whole quantity is divided. The numerator indicates the number of these parts to be taken.

A fraction may also be considered as an indicated division. The line separating the numerator and the denominator can be called a sign of division. Then, a fraction, such as $\frac{13}{5}$, can be taken to mean: $13 \div 5$. Any division can be written as a fraction. For example, the division: $2x \div 3y$, can be written: $\dfrac{2x}{3y}$.

The fraction $\dfrac{x^2 - 5x + 6}{x - 2}$ can be written as division:

$$(x^2 - 5x + 6) \div (x - 2)$$

12-2 FUNDAMENTAL PRINCIPLE OF FRACTIONS

Before we consider the different operations with fractions, it is important that we thoroughly understand the following principle:

Fundamental Principle of Fractions. *If the numerator and the denominator of any fraction are multiplied or divided by the same quantity (other than zero), the value of the fraction will not be changed.*

We have seen how this principle applies to a fraction in arithmetic. Let us now consider its application to an algebraic fraction:

$$\frac{3x - 4}{5y + 2}$$

The fundamental principle says that if the numerator and the denominator are multiplied by the same quantity, say, $7x$, the value of the fraction will not be changed:

$$\frac{7x(3x-4)}{7x(5y+2)} = \frac{21x^2-28x}{35xy+14x}$$

When we multiply numerator and denominator by the same quantity, we are only multiplying the fraction by 1. We change the form of the fraction but not its value.

The fundamental principle also says that if the numerator and denominator of any fraction are divided by the same quantity, the value of the fraction will not be changed. Consider the fraction

$$\frac{15ax^2y}{20ax^3}$$

If we divide numerator and denominator by the quantity $5ax^2$, we get

$$\frac{15ax^2y \div 5ax^2}{20ax^3 \div 5ax^2} = \frac{3y}{4x}$$

Here again we change the form of the fraction but not its value. This is the procedure we follow when we reduce a fraction to lower terms.

Note. Remember that multiplying both numerator and denominator of a fraction by some quantity, for instance, 7, is *not* the same as multiplying the fraction itself by that number. For example, consider the fraction $\frac{2}{3}$:

$$\frac{(7)(2)}{(7)(3)} = \frac{14}{21} \quad \text{but} \quad (7)\left(\frac{2}{3}\right) = \frac{14}{3}$$

It might also be noted that if the same number (not zero) is *added* to, or *subtracted* from, both terms of a fraction, the value of the fraction *will be changed.* As an example, the fraction $\frac{2}{3}$ will not have the same value if some number, such as 4, is added to both numerator and denominator. The new fraction in that case will be $\frac{6}{7}$, which does *not* have the same value as $\frac{2}{3}$.

12-3 REDUCING FRACTIONS TO LOWEST TERMS

A fraction is said to be reduced to *lowest terms* when the numerator and the denominator have no common factor other than 1.

To reduce fractions to lowest terms, we make use of the fundamental principle by dividing the numerator and denominator by some common factor. In some fractions we can tell at a glance what number is a divisor of both numerator and denominator. In the fraction $\frac{21}{35}$ we see at once that the numerator and denominator are divisible by 7. One way to show this is to factor numerator and denominator:

$$\frac{21}{35} = \frac{3 \cdot \not{7}}{\not{7} \cdot 5} = \frac{3}{5}$$

We cross out the "7" in numerator and denominator to show that both have been divided by 7.

In reducing fractions, some students use the word "cancel." In the foregoing example we cross out the "7" to show that numerator and denominator have been divided by 7. *Cancel* is not a proper word to use in this connection. However, the word does have one correct use in mathematics. *Cancel* means to nullify, to neutralize, or to make void, as when we say we *cancel* an order. A payment of \$10 will cancel a debt of \$10. In the form of an equation, $-10 + 10 = 0$. In an equation such as this, $x = 12 + 5 - 5$, the $+5$ and -5 cancel each other. This is the correct use of the word.

The word *cancel* often causes students much trouble. In connection with reducing fractions, if we say "cancel" a factor in numerator and denominator, we really mean that we *divide numerator and denominator by the same factor.* Students of algebra can save themselves a lot of trouble and mistakes if they will consider carefully the actual procedure involved in the so-called "canceling" in reducing fractions.

In some fractions it may not be easy to determine the common divisor of numerator and denominator. Consider the following fraction:

$$\frac{119}{323}$$

We can easily reduce the fraction if we first factor the numerator and the denominator as shown here:

$$\frac{119}{323} = \frac{7 \cdot 17}{17 \cdot 19}$$

Now we see at once that numerator and denominator can be divided by 17, and the reduced fraction becomes $\frac{7}{19}$.

A common mistake in working with fractions is to cross out any *term* that appears in both numerator and denominator. For example, take the fraction

$$\frac{x^2 + 5x + 6}{x^2 - 9}$$

A mistake sometimes made is to cross out the "x^2" in numerator and denominator. This error happens often when we use the word *cancel.*

In fact, some people cross out any two things that look alike wherever they are found. Crossing out the "x^2's" in this fraction would be somewhat like crossing the "3's" in a fraction such as $\frac{34}{37}$, which would, of course, be incorrect.

Remember, any quantity in numerator or denominator *cannot* be split up by crossing out a separate term if the term is connected to another term by a plus (+) or a minus (−). Only *factors* can be divided into numerator and denominator. Do not cross out separate terms unless the entire term or quantity is a *factor* of the numerator and the denominator.

However, the algebraic fraction shown may be reduced to lower terms if the numerator and denominator are factored:

$$\frac{x^2 + 5x + 6}{x^2 - 9} = \frac{\cancel{(x + 3)}(x + 2)}{(x - 3)\cancel{(x + 3)}} = \frac{x + 2}{x - 3}$$

We see that numerator and denominator are each made up as a product of factors. The binomial $x + 3$ is a factor of both numerator and denominator. The factor $x + 3$ can be crossed out by a slanted line to show that numerator and denominator are both divided by the quantity $x + 3$.

To avoid mistakes, follow these steps.

Steps in reducing fractions:

1. *Factor numerator and denominator into prime factors.*
2. *Divide numerator and denominator by any factor or factors found in both.*

To show how algebraic fractions may be reduced, we shall work out a few examples. If the numerator and denominator of a fraction consist of only monomial factors, the prime factors need not be written out separately, since no additions or subtractions are involved.

Example 1

Reduce the fraction

$$\frac{18x^3y^3z}{24ax^4y}$$

We shall first write out all the prime factors separately and then later show that this need not be done in such fractions. Writing the prime factors of numerator and denominator, we have

$$\frac{18x^3y^2z}{24ax^4y} = \frac{\cancel{2} \cdot \cancel{3} \cdot 3 \cdot \cancel{x} \cdot \cancel{x} \cdot \cancel{x} \cdot \cancel{y} \cdot y \cdot z}{\cancel{2} \cdot 2 \cdot 2 \cdot \cancel{3} \cdot a \cdot \cancel{x} \cdot \cancel{x} \cdot \cancel{x} \cdot x \cdot \cancel{y}} = \frac{3yz}{4ax}$$

The foregoing fraction might have been reduced simply by dividing numerator and denominator by common factors of each: thus

$$\frac{\overset{3}{\cancel{18}}\,\cancel{x^3}\,\overset{y}{\cancel{y^2}}\,z}{\underset{4}{\cancel{24}}\,a\,\underset{x}{\cancel{x^4}}\,\cancel{y}} = \frac{3yz}{4ax}$$

Example 2

Reduce the fraction

$$\frac{3x^2 + 7x - 6}{3x^2 + 5x - 12}$$

In this fraction we first factor the numerator and denominator. Then we divide the numerator and denominator by any factor found in both:

$$\frac{3x^2 + 7x - 6}{3x^2 + 5x - 12} = \frac{(3x - 2)\cancel{(x + 3)}}{(3x - 4)\cancel{(x + 3)}} = \frac{3x - 2}{3x - 4}$$

We see that the factor $x + 3$ is found in the numerator and denominator. Therefore, each may be divided by this factor. The final reduced fraction, as shown, *cannot* be further reduced.

Note. As a student of mathematics, you should always remember one important fact: the correct procedure is more important than getting the right answer. This is true in all mathematics. Consider the fraction

$$\frac{x^2 - 9}{x - 3}$$

The correct way to reduce this fraction is to factor the numerator and denominator and then to divide each by any common factor: thus

$$\frac{x^2 - 9}{x - 3} = \frac{\cancel{(x - 3)}(x + 3)}{\cancel{x - 3}} = x + 3$$

The factor $x - 3$ is divided into both numerator and denominator and we get the correct answer.

One student reduced the foregoing fraction in this way:

$$\frac{\overset{x}{\cancel{x^2}}\ \overset{+}{\cancel{-}}\ \overset{3}{\cancel{9}}}{\cancel{x}\ \cancel{-}\ \cancel{3}} = x + 3$$

His "cancellation" method was as follows: x into x^2 is x, 3 into 9 is 3, and minus $(-)$ into a minus $(-)$ is plus $(+)$. When he got the correct answer, he wondered why he should not be given credit for working the problem. He was told that the work was no more correct than to cancel the 6's in the fraction $\frac{16}{64} = \frac{1}{4}$. Although the answer is correct, the work is entirely wrong.

The numerator and denominator of a fraction may contain monomial factors as well as others. The first step in factoring is to take out any common factor, as shown in this example.

Example 3

$$\frac{9x^4 - 36x^2}{6x^3 - 30x^2 + 36x} = \frac{9x^2(x^2 - 4)}{6x(x^2 - 5x + 6)}$$
$$= \frac{9x^2(x - 2)(x + 2)}{6x(x - 3)(x - 2)} = \frac{3x(x + 2)}{2(x - 3)}$$

Exercise 12-1

Reduce the following fractions to lowest terms:

1. $\frac{15x^2y^3}{25xy^5}$
2. $\frac{35an^2y^3}{42bn^2y^3}$
3. $\frac{16c^2n^3x^5y}{24cn^5xyz}$
4. $\frac{36a^3bc^2}{45a^2c^3d}$
5. $\frac{60x^5y^3z}{80x^4y^4ab}$
6. $\frac{14R^2h}{28r^2h}$
7. $\frac{26x^5y^3z}{13x^3y^2z}$
8. $\frac{15a^2bc^3}{15a^2bc^3}$
9. $\frac{4ab}{12a^2b^2x}$
10. $\frac{20x^3yz^4}{-4x^3z^4}$
11. $\frac{-7a^2m^3n}{7a^2m^3n}$
12. $\frac{12x^3(x - 5)^2}{8x^2(x - 5)}$
13. $\frac{x^2 - 16}{x - 4}$
14. $\frac{c^2 - 25}{5 - c}$
15. $\frac{(x - 1)^4}{(x - 1)^3}$
16. $\frac{x^3 - 1}{x - 1}$
17. $\frac{(x - 1)^3}{x^3 - 1}$
18. $\frac{n^2 - 4}{n^3 - 8}$
19. $\frac{a + b}{a^2 + 2ab + b^2}$
20. $\frac{x^2 - 9}{x^2 - 6x + 9}$
21. $\frac{R^2 - r^2}{R^2 - 2Rr + r^2}$
22. $\frac{x^2 - 2xy + y^2}{y^2 - x^2}$
23. $\frac{x^2 + 5x + 6}{12 - 4x - x^2}$
24. $\frac{2x^2 - 50}{3x^2 - 3x - 60}$
25. $\frac{2x^3 + 6x^2 - 8x}{4x^3 + 4x^2 - 8x}$
26. $\frac{3c^4 - 12c^3 + 12c^2}{3c^2 - 18c + 24}$
27. $\frac{3y^3 + 3y^2 - 18y}{6y^4 + 24y^3 - 72y^2}$
28. $\frac{4a^3 + 12a^2 + 9a}{18a^2 + 12a^3}$
29. $\frac{3n^3x - 108x}{2n^2y - 6ny - 36y}$
30. $\frac{n^5 - 9n^3}{n^4 - 2n^3 - 3n^2}$

31. $\dfrac{4x^3y - 4x^2y - 48xy}{6x^3 + 6x^2 - 36x}$

32. $\dfrac{3x^3 + 15x^2 - 18x}{6x^2y + 42xy + 36y}$

33. $\dfrac{36n^2 + 24n - 45}{30n^2 + 25n - 30}$

34. $\dfrac{20nx - 20nx^2 - 75nx^3}{90x^4 - 40x^2}$

35. $\dfrac{9c^3 - 30c^2 + 25c}{6c^4 - 22c^3 + 20c^2}$

36. $\dfrac{10n^5 + 5n^4 - 30n^3}{4n^3 - 4n^2 - 24n}$

37. $\dfrac{8ax^2 - 10ax - 12a}{72bx^2 - 42bx - 72b}$

38. $\dfrac{12abx^2 - 28abx + 8ab}{54ax^3 - 6ax}$

39. $\dfrac{u^4 - uv^3}{2u^4 - u^3v - u^2v^2}$

40. $\dfrac{(a + b)^2 + a + b}{(a + b)^2}$

41. $\dfrac{a^2 - b^2 - a - b}{a^2 - 2ab + b^2 - 1}$

42. $\dfrac{ax + 3x - ay - 3y}{a^2 + a - 12}$

43. $\dfrac{x^2 - a^2 - 6x + 9}{3x^2 - 15x^2 + 18x}$

44. $\dfrac{n^4 + 5n^2 + 9}{3n^3 - 3n^2 + 9n}$

12-4 MULTIPLICATION OF FRACTIONS

In multiplication and division of fractions we also often use the word *cancel.* Actually, here, again, we mean "divide" the numerator and denominator of the fraction by the same quantity. Let us take an example from arithmetic:

$$\frac{12}{35} \cdot \frac{25}{8} \cdot \frac{21}{4} =$$

We know from arithmetic that we can get the product of these fractions by multiplying all the numerators together for the numerator of the product.

However, before we do the multiplication, we usually look for common factors in numerator and denominator. We see that the factor 4 can be divided into a numerator and a denominator. We can also divide numerator and denominator by the numbers 5 and 7.

$$\frac{\overset{3}{\cancel{12}}}{\underset{7}{\cancel{35}}} \cdot \frac{\overset{5}{\cancel{25}}}{8} \cdot \frac{\overset{3}{\cancel{21}}}{\underset{1}{\cancel{4}}} = \frac{45}{8}$$

Example 1

Multiply

$$\frac{8x^2y^3}{15a^2c} \cdot \frac{9a^3b}{16cd^2} \cdot \frac{4ac^3d}{21bx^3y^2} \cdot (5x)$$

Solution

$$\frac{\overset{y}{\cancel{8}\cancel{x^2}\cancel{y^3}}}{\underset{5}{\cancel{15}\cancel{a^2}\cancel{c}}} \cdot \frac{\overset{3\,a}{\cancel{9}\cancel{a^3}\cancel{b}}}{\underset{2\,d}{\cancel{16}\cancel{c}\cancel{d^2}}} \cdot \frac{\overset{2\,c}{\cancel{4}\,a\cancel{c^3}\cancel{d}}}{\underset{7\,x}{\cancel{21}\cancel{b}\cancel{x^3}\cancel{y^2}}} \cdot \frac{\cancel{3}\,x}{1} = \frac{2a^2cy}{7d}$$

Example 2

Multiply

$$\frac{3x^2 - 12}{2x^3 + 2x^2 - 24x} \cdot \frac{2x^5 - 6x^4}{x^2 + 2x - 8} \cdot \frac{x^2 + 8x + 16}{9x^3 + 54x^2}$$

Solution

First find prime factors:

$$\frac{\cancel{3}\cancel{(x-2)}(x+2)}{\cancel{2x}\cancel{(x-3)}\cancel{(x+4)}} \cdot \frac{\overset{x}{\cancel{2x^4}}\cancel{(x-3)}}{\cancel{(x-2)}\cancel{(x+4)}} \cdot \frac{\cancel{(x+4)}\cancel{(x+4)}}{\underset{3}{\cancel{9x^2}}(x+6)} = \frac{x(x+2)}{3(x+6)}$$

This answer cannot be further reduced.

12-5 DIVISION OF FRACTIONS

In division involving algebraic fractions, we follow the same rule as in division in arithmetic. That is, we multiply the dividend by the *reciprocal* of the divisor. Remember, the divisor is the quantity that *follows* the division sign. That is, we invert the divisor and then multiply.

When multiplications and divisions of fractions occur in the same problem, these operations are performed *in the order in which they occur* unless otherwise indicated by parentheses (including brackets, braces, or other forms indicating a quantity).

Example 1

Simplify

$$\frac{15}{28} \div \frac{5}{21}.$$

Solution

$$\frac{15}{28} \div \frac{5}{21} = \frac{15}{28} \cdot \frac{21}{5} = \frac{9}{4}$$

Example 2

Simplify:

$$\frac{9a^2x}{20by^2} \div \frac{6ax^2}{25by^3}$$

Solution

Inverting the divisor,

$$\frac{9a^2x}{20by^2} \cdot \frac{25by^3}{6ax^2} = \frac{15ay}{8x}$$

Example 3

Simplify:

$$\frac{8x^3}{15y^2} \cdot \frac{3ay}{4bx} \div \frac{12bx}{5cy}$$

Solution

We write:

$$\frac{8x^3}{15y^2} \cdot \frac{3ay}{4bx} \cdot \frac{5cy}{12bx} = \frac{acx}{6b^2}$$

Example 4

Simplify:

$$\frac{16xy}{15a^2b} \div \frac{12x^2y}{5abc} \cdot \frac{9ay}{10bc}$$

Solution

Inverting the second fraction only, we get:

$$\frac{16xy}{15a^2b} \cdot \frac{5abc}{12x^2y} \cdot \frac{9ay}{10bc} = \frac{2y}{5bx}$$

Example 5

Simplify:

$$\frac{4ab^3}{3x^2y^2} \div \frac{8bc}{9xy} \div \frac{12a^2b}{7x^2y}$$

Solution

Inverting both second and third fractions, we get:

$$\frac{4ab^3}{3x^2y^2} \cdot \frac{9xy}{8bc} \cdot \frac{7x^2y}{12a^2b} = \frac{7bx}{8ac}$$

Example 6

Simplify:

$$\frac{3a^2b}{4x^2y} \div \left(\frac{9a^3b}{8xy^2} \div \frac{6a^2b}{5xy}\right)$$

Solution

Simplifying the portion within parentheses first, we get:

$$\frac{3a^2b}{4x^2y} \div \left(\frac{9a^3b}{8xy^2} \cdot \frac{5xy}{6a^2b}\right) = \frac{3a^2b}{4x^2y} \div \left(\frac{15a}{16y}\right)$$

$$= \frac{3a^2b}{4x^2y} \cdot \frac{16y}{15a} = \frac{4ab}{5x^2}$$

Exercise 12-2

Perform the indicated operations:

1. $\frac{4x^2y^3}{3a^3c} \cdot \frac{7ax^2}{8b} \cdot \frac{9ac^2}{14x^3y}$
2. $\frac{5xy^2}{6ac^2} \cdot \frac{9c^3}{10xy} \div \frac{3cy}{4a}$
3. $\frac{22xy}{21abc^2} \div \frac{8y}{9b^2c^2} \cdot \frac{14a^2bc}{33x^2}$
4. $\frac{14by^2}{15a^2x} \div \frac{21by}{10a} \cdot \frac{9ax^2}{4y}$
5. $\frac{16a^2}{15x^2} \cdot \frac{3cx}{2ab} \div 4ac$
6. $\frac{8x^2}{15a^3} \div \frac{6c}{5b} \div \frac{3a^2}{4bx}$
7. $\frac{5ac^2}{8xy^3} \div \left(\frac{3bc}{4x^2y} \cdot \frac{25ac}{6by}\right)$
8. $\frac{3ab}{4xy} \div \left(\frac{15x^2}{16y^2} \div \frac{5ax^2}{12by}\right)$
9. $\frac{2x^2 - 18}{x^2 - 25} \cdot \frac{5x - 25}{2x^2 - 6x}$
10. $\frac{2x^2 - 72}{x^4 - 4x^2} \cdot \frac{x^3 - 2x^2}{6x - 36}$
11. $\frac{x^2 - 6x - 16}{x^2 - 8x + 16} \div \frac{x + 2}{x - 4}$
12. $\frac{x^2 - 3x - 10}{x^2 - 10x + 25} \div \frac{x + 6}{x - 5}$
13. $\frac{8x^4 - 32x^2}{3x^3 - 21x^2 + 18x} \div \frac{4x^3 + 8x^2 - 32x}{x^2 + x - 2}$
14. $\frac{3x^3 - 75x}{2x^2 - 2x - 24} \div \frac{x^4 - 8x^3 + 15x^2}{4x^2 + 4x - 80}$
15. $\frac{a + 3}{a^2 + 8a - 9} \cdot (3a - 3)$
16. $(4x - 8) \cdot \frac{x + 4}{x^2 + 6x - 16}$
17. $(4x^2 - 1) \cdot \frac{4x}{4x^2 - 4x + 1}$
18. $(9x^2 - 1) \cdot \frac{3x}{9x^2 - 6x + 1}$

19. $(9c - 12) \div \dfrac{9c^2 - 16}{c + 2}$ **20.** $\dfrac{4x^2 - 25}{x + 2} \div (8x - 20)$

21. $\dfrac{x^3 - 6x^2 + 9x}{3x^2 + 6x - 9} \cdot \dfrac{6x^2 - 12x + 6}{x^3 - 2x^2 - 3x} \div \dfrac{2x^2 - 18}{x^2 - 1}$

22. $\dfrac{x^3 - 5x^2 + 4x}{6x^2 + 30x + 36} \div \dfrac{x^4 - 16x^2}{3x^2 - 6x - 45} \cdot \dfrac{x^3 + 8x^2 + 16x}{x^2 - 6x + 5}$

23. $\dfrac{4r^3 - 24r^2 + 20r}{4r^2 - 9} \div \dfrac{2r^4 - 6r^3 - 20r^2}{r^2 - r - 12} \cdot \dfrac{2r^3 + r^2 - 6r}{r^2 - 5r + 4}$

24. $\dfrac{n^4 + 5n^3 - 14n^2}{6n^2 - 6n - 36} \div \dfrac{n^3 + 8n^2 + 7n}{2n^2 - 18n + 36} \cdot \dfrac{3n^2 + 7n + 2}{n^3 - 8n^2 + 12n}$

25. $\dfrac{y^3 + 9y^2 + 8y}{y^2 - y - 20} \cdot \dfrac{y^3 - 7y^2 + 10y}{6y^2 - 18y - 24} \div \dfrac{2y^4 - 8y^2}{3y^2 - 48}$

26. $\dfrac{2t^2 - 20t + 18}{t^3 + t^2 - 20t} \cdot \dfrac{2t^2 - 2t - 24}{3t^3 - 18t^2 + 15t} \div \dfrac{4t^2 - 36}{t^4 - 25t^2}$

27. $\dfrac{a^2 - b^2 + 2a + 1}{a^3 - b^3} \div \dfrac{3a + 3b + 3}{a^2 - b^2}$

28. $\dfrac{c^2 - d^2 - 4c + 4}{c^3 + d^3} \div \dfrac{4c - 4d - 8}{c^2 - d^2}$

12-6 SIGNS IN A FRACTION

Before beginning the study of adding and subtracting fractions, it is necessary to consider the signs in a fraction. In connection with any fraction, there are three signs that must be considered.

1. The sign of the numerator
2. The sign of the denominator.
3. The sign of the fraction itself.

If any *two* of the three signs mentioned are changed, the value of the fraction will not be changed. Therefore, it is permissible to change any two of these signs without changing in any way the value of the fraction. We show this by an example:

$$+\frac{+8}{+2} = +\frac{-8}{-2} = -\frac{+8}{-2} = -\frac{-8}{+2} = +4$$

If the numerator and/or denominator are polynomials, a change in sign means a change in *all* the signs of the polynomial. Changing the sign of a polynomial is equivalent to multiplying or dividing it by -1.

Consider this fraction:

$$-\frac{x^2 - 4x - 7}{3x - 5 - x^2}$$

This fraction may be written in any of the following ways:

$$-\frac{7 + 4x - x^2}{x^2 - 3x + 5} = +\frac{x^2 - 4x - 7}{x^2 - 3x + 5} = +\frac{7 + 4x - x^2}{3x - 5 - x^2}$$

In the first fraction the signs of numerator and denominator have been changed. In the second the signs of the fraction and the denominator have been changed. In the third the signs of the fraction and the numerator have been changed.

However, suppose a polynomial is stated in factored form. Then, if we change the signs of two factors, we are, in effect, multiplying by $(-1) \cdot (-1)$, which is equivalent to $+1$. Therefore, changing the signs of *two factors* does not change the sign of their product. If we wish to change the signs of a polynomial in factored form, we must change the signs of an *odd* number of factors.

12-7 ADDITION AND SUBTRACTION OF FRACTIONS

You will recall from arithmetic that in order that fractions may be added or subtracted they must have the *same* denominator. If two or more fractions have the *same* denominator, they are added or subtracted simply by combining the numerators and then placing the combined numerator over the common denominator: thus

$$\frac{3}{7} + \frac{5}{7} - \frac{2}{7} = \frac{3 + 5 - 2}{7} = \frac{6}{7}$$

As another example, the following fractions can be combined into one fraction simply by combining the numerators and placing the result over the denominator:

$$\frac{3x + 1}{5x} + \frac{x - 3}{5x} - \frac{2x - 7}{5x} = \frac{(3x + 1) + (x - 3) - (2x - 7)}{5x}$$

$$= \frac{3x + 1 + x - 3 - 2x + 7}{5x} = \frac{2x + 5}{5x}$$

Notice that the minus sign $(-)$ before the third fraction brings about a change in the signs of the numerator. Each fraction must be considered as one quantity, and the effect is the same as though the fraction were enclosed in parentheses.

If we wish to add or subtract fractions with denominators, we must first state all the fractions with the same denominator. To do so, we find the lowest common denominator (LCD). Then, some or all fractions must be changed to higher terms by multiplying the numerator and the denominator of each fraction by some quantity so that all denominators will be the same.

It should be carefully noted that when a fraction is changed to higher terms, the *value* of the fraction is *not* changed. For example, if the fraction $\frac{3}{5}$ is changed to $\frac{12}{20}$, the value of the fraction is not changed.

After the fractions are all stated with the same denominator, they are added or subtracted by combining the numerators and placing the result over the common denominator. In adding or subtracting algebraic fractions, we follow the same steps as in arithmetic:

1. *Find the lowest common denominator (LCD).*
2. *Rewrite each fraction with the LCD as its denominator.*
3. *Combine the numerators and place the result over the LCD.*

Finding the lowest common denominator. To find the LCD, first factor all denominators into prime factors. The LCD must contain all the *different* factors found in all the denominators. If a certain factor is found *twice* in any *one* denominator, it must be used twice in the LCD. If a certain factor is found only *once* in any *one* denominator, it is used only *once* in the LCD. Do not use any factor more times than it is found in any *one* denominator.

Briefly, the LCD is the product of all the *different* factors found in all the denominators, and each factor is used only as many times as it is found in any *one* denominator. Remember, the LCD must be such that it is divisible by each one of the given denominators.

Example 1

Combine

$$\frac{7x}{10} + \frac{3x + 1}{12} + \frac{5x}{18}.$$

Solution

To find the LCD, we first separate each denominator into prime factors.

$$10 = 2 \cdot 5$$
$$12 = 2 \cdot 2 \cdot 3$$
$$18 = 2 \cdot 3 \cdot 3$$

The LCD must contain all the different factors, 2, 3, and 5. However, we must use each factor as many times as it is found in any one denominator. The factor 2 is found twice in 12 and the factor 3 is found twice in 18. Therefore, for the LCD, we have

$$2 \cdot 2 \cdot 3 \cdot 3 \cdot 5$$

which is equal to 180. Now, we change each fraction to a new fraction having a denominator of 180. The original expression then becomes

$$\frac{126x}{180} + \frac{15(3x + 1)}{180} + \frac{50x}{180}$$

At this point it should be carefully noted that the value of each fraction has *not* been changed. Each fraction has simply been changed to a new form having exactly the *same value* as before.

We now write all the numerators over the LCD. Then we remove the parentheses and combine like terms.

$$\frac{126x + 15(3x + 1) + 50x}{180} = \frac{126x + 45x + 15 + 50x}{180} = \frac{221x + 15}{180}$$

Example 2

Combine

$$\frac{3x + 2}{4} - \frac{x - 3}{5} - \frac{4x - 3}{20}.$$

Solution

Notice the minus $(-)$ signs before some fractions. This means that when the numerators of such fractions are placed over a common denominator, the numerators must be considered as quantities.

In this example the LCD is 20. All the fractions are changed to new fractions having the denominator 20. They are

$$\frac{5(3x + 2)}{20} - \frac{4(x - 3)}{20} - \frac{4x - 3}{20}$$

$$= \frac{5(3x + 2) - 4(x - 3) - (4x - 3)}{20}$$

$$= \frac{15x + 10 - 4x + 12 - 4x + 3}{20} = \frac{7x + 25}{20}$$

Example 3

Combine into a single fraction: $\frac{4}{a} + \frac{3}{b}$

Solution

The common denominator is the product, ab. We change each fraction so that its denominator is ab. Then we get

$$\frac{4b}{ab} + \frac{3a}{ab} = \frac{4b + 3a}{ab}$$

Example 4

Combine into a single fraction: $\dfrac{3a}{2x} + \dfrac{2}{y} - 4$

Solution

The lowest common denominator is $2xy$. If the 4 is to be placed over the LCD, then the 4 must be multiplied by the LCD, $2xy$. We might think of the denominator 1 for the number 4. Then we get

$$\frac{3a}{2x} + \frac{2}{y} - 4 = \frac{3ay + 4x - 8xy}{2xy}$$

Example 5

Combine into a single fraction: $\dfrac{5x + 4}{3x} - \dfrac{y - 2}{5y} - 2$

Solution

Here the LCD is $15xy$. Then we get

$$\frac{5x + 4}{3x} - \frac{y - 2}{5y} - 2 = \frac{25xy + 20y - 3xy + 6x - 30xy}{15xy} = \frac{20y + 6x - 8xy}{15xy}$$

Example 6

Combine into a single fraction: $2a - \dfrac{4 - a}{2a - 1} + \dfrac{a}{3}$

Solution

The LCD is $3(2a - 1)$. If we consider the first term as $\dfrac{2a}{1}$, we have:

$$\frac{2a(3)(2a - 1) - 3(4 - a) + a(2a - 1)}{3(2a - 1)}$$

$$= \frac{12a^2 - 6a - 12 + 3a + 2a^2 - a}{3(2a - 1)} = \frac{14a^2 - 4a - 12}{3(2a - 1)}$$

Example 7

Combine

$$\frac{5}{3x} + \frac{3x - 1}{2x^2} - \frac{x + 2}{x^3}$$

Solution

The LCD is $6x^3$, since this denominator is great enough to contain all of the given denominators by division. The fractions are changed to new forms having the denominator $6x^3$:

$$\frac{10x^2}{6x^3} + \frac{3x(3x - 1)}{6x^3} - \frac{6(x + 2)}{6x^3} = \frac{10x^2 + 3x(3x - 1) - 6(x + 2)}{6x^3}$$

$$= \frac{10x^2 + 9x^2 - 3x - 6x - 12}{6x^3} = \frac{19x^2 - 9x - 12}{6x^3}$$

Example 8

Combine

$$\frac{4}{x - 3} + \frac{x + 2}{x^2 - 9} - \frac{5x - 2}{x^2 - 6x + 9}$$

Solution

In this example the LCD is $(x + 3)(x - 3)(x - 3)$, which may be written $(x + 3)(x - 3)^2$. We must use the factor $(x - 3)$ *twice* because it is contained *twice* in one of the given denominators. The fractions are changed to new forms having the LCD as the denominators:

$$\frac{4(x - 3)(x + 3)}{(x + 3)(x - 3)^2} + \frac{(x + 2)(x - 3)}{(x + 3)(x - 3)^2} - \frac{(5x - 2)(x + 3)}{(x + 3)(x - 3)^2}$$

$$= \frac{4(x - 3)(x + 3) + (x + 2)(x - 3) - (5x - 2)(x + 3)}{(x + 3)(x - 3)^2}$$

$$= \frac{4x^2 - 36 + x^2 - x - 6 - 5x^2 - 13x + 6}{(x + 3)(x - 3)^2} = \frac{-14x - 36}{(x + 3)(x - 3)^2}$$

The denominator is usually left in factored form. The numerator should also be factored, if possible, to determine whether the fraction can be reduced.

Example 9

Combine

$$\frac{7}{x - 5} - \frac{x - 3}{5 - x}$$

Solution

These fractions cannot be combined as they stand, since the denominators are not identical. However, they may be made identical by changing the sign of the second fraction and its denominator:

$$\frac{7}{x-5} + \frac{x-3}{x-5}$$

Now the numerators are combined and the result is placed over the common denominator. The sum is $\frac{7+x-3}{x-5}$, or $\frac{x+4}{x-5}$.

Example 10

Combine

$$5x - 2 - \frac{2x-1}{3} - \frac{3x+2}{2}$$

Solution

The LCD is 6. If the terms $5x$ and -2 are to be placed over the LCD, 6, they must be multiplied by 6 for the numerator. We get

$$\frac{30x}{6} - \frac{12}{6} - \frac{4x-2}{6} - \frac{9x+6}{6} = \frac{17x-16}{6}$$

Example 11

Combine

$$\frac{3x-4}{x-3} - \frac{5}{x} - 2x - 3$$

Solution

Here the LCD is $x(x-3)$. When each term is placed over this denominator, we get

$$\frac{x(3x-4)}{x(x-3)} - \frac{5(x-3)}{x(x-3)} - \frac{2x^2(x-3)}{x(x-3)} - \frac{3x(x-3)}{x(x-3)}$$

$$= \frac{3x^2 - 4x - 5x + 15 - 2x^3 + 6x^2 - 3x^2 + 9x}{x(x-3)} = \frac{6x^2 - 2x^3 + 15}{x(x-3)}$$

Exercise 12-3

Combine the terms into a single fraction as indicated, and simplify:

1. $\frac{1}{5} + \frac{1}{9}$ **2.** $\frac{1}{6} - \frac{1}{7}$ **3.** $\frac{1}{x} + \frac{1}{y}$

4. $\frac{1}{a} - \frac{1}{b}$

5. $\frac{2x}{3} - \frac{4x}{15}$

6. $\frac{7x}{4} - \frac{5x}{12}$

7. $\frac{x}{3} - \frac{x}{7} + \frac{2x}{21}$

8. $\frac{3x}{4} + \frac{5x}{8} - \frac{x}{2}$

9. $\frac{y + 2}{3} + \frac{y - 3}{4} + 3$

10. $\frac{n - 3}{5} + \frac{n + 5}{4} - 2$

11. $\frac{n + 3}{4} + \frac{n - 2}{2} - \frac{3n}{8}$

12. $\frac{y - 2}{3} + \frac{y - 1}{4} - \frac{5y}{12}$

13. $\frac{2r + 5}{3} - \frac{5r}{6} - \frac{r - 3}{12}$

14. $\frac{3t - 4}{5} - \frac{2t}{3} - \frac{t + 4}{15}$

15. $\frac{4x - 5}{3} - \frac{2x + 3}{15} - \frac{4 - 3x}{6}$

16. $\frac{2x - 5}{3} - \frac{3x - 7}{18} - \frac{6 + 5x}{12}$

17. $\frac{2x}{3y} + \frac{3}{x} + \frac{4}{y} - 2$

18. $\frac{n}{2x} + \frac{3}{nx} + \frac{x}{n} - 3$

19. $\frac{3}{2x^2} + \frac{4}{3x} - \frac{1}{x^3} - 3$

20. $\frac{5}{3a^2} + \frac{4}{a^3} - \frac{x}{4a} - 2$

21. $\frac{x + 3}{2x} - \frac{y - 2}{3y} - 5$

22. $\frac{y + 2}{3y} - \frac{x - 3}{4x} - 2$

23. $\frac{1}{r_1} - \frac{1}{r_2} + \frac{1}{r_3}$

24. $\frac{1}{c_1} + \frac{1}{c_2} + \frac{1}{c_3}$

25. $\frac{5}{x - 4} - \frac{3}{4 - x}$

26. $\frac{7}{x - 5} - \frac{2}{5 - x}$

27. $\frac{3n}{n + 4} - \frac{2n}{3 - n}$

28. $\frac{5}{x - 4y} - \frac{3}{x + 2y}$

29. $\frac{x + 4}{x^2 - 9} - \frac{x - 2}{3 - x}$

30. $\frac{x + 5}{x^2 - 16} + \frac{x - 3}{4 - x}$

31. $\frac{3x + 4}{x^2 - 6x + 8} - \frac{x + 3}{2 - x}$

32. $\frac{3x + 2}{x^2 - 5x + 6} - \frac{x + 2}{3 - x}$

33. $\frac{4}{x + 3} - \frac{2x - 3}{x^2 + 3x} - \frac{3}{x}$

34. $\frac{5}{x - 4} - \frac{3x + 4}{x^2 - 4x} - \frac{2}{x}$

35. $\frac{2x - 7}{x - 5} - \frac{3}{x} - 4x - 2$

36. $\frac{2x - 3}{x - 2} - \frac{4}{x} - 3x - 4$

37. $\dfrac{3x-5}{x+3} - \dfrac{4}{x} - 2x - 3$ **38.** $\dfrac{4-3x}{x-3} - \dfrac{5}{x} - 3x - 2$

39. $\dfrac{3x^2-2x+1}{x^2+3x-10} - \dfrac{2x}{x+5} - \dfrac{1}{x-2}$

40. $\dfrac{2x^2+x+9}{x^2-2x-15} - \dfrac{2}{x-5} - \dfrac{x}{x+3}$

12-8 COMPLEX FRACTIONS

A *complex fraction* is a fraction whose numerator or denominator, or both, also contain fractions. In fact, the division of two fractions may be shown as a complex fraction.

Example 1

The indicated division, $\frac{3}{4} \div \frac{5}{7}$, may be written as the complex fraction:

$$\frac{\frac{3}{4}}{\frac{5}{7}}$$

The heavy line separating the two fractions indicates division. Of course, in working out the division, we usually use the regular division form:

$$\frac{3}{4} \div \frac{5}{7} = \frac{3}{4} \cdot \frac{7}{5} = \frac{21}{20}$$

Example 2

$$\frac{\frac{4}{9}}{\frac{5}{6}} = \frac{4}{9} \div \frac{5}{6} = \frac{4}{9} \cdot \frac{6}{5} = \frac{8}{15}$$

Sometimes a numerator or denominator, or both, contain several fractions.

Example 3

$$\frac{\frac{3}{4} + \frac{7}{8} - \frac{1}{2}}{\frac{2}{3} - \frac{5}{6} + \frac{1}{2}}$$

To work out the expression, we may first combine all the fractions of the numerator into a single fraction; then we combine all the fractions of the denominator into a single fraction and divide the numerator by the denominator.

$$\frac{\frac{3}{4}+\frac{7}{8}-\frac{1}{2}}{\frac{2}{3}-\frac{5}{6}+\frac{1}{2}}=\frac{\frac{6+7-4}{8}}{\frac{4-5+3}{6}}=\frac{\frac{9}{8}}{\frac{2}{6}}$$

The answer is found by

$$\frac{9}{8}\div\frac{2}{6}=\frac{9}{8}\cdot\frac{6}{2}=\frac{27}{8}$$

Another method that is often simpler is to make use of the fundamental principle of fractions. In the fraction shown in Example 3, we may multiply the *entire numerator* and the *entire denominator* of the original fraction by 24, which is the LCD of all the denominators that appear anywhere in the fraction. Multiplying entire numerator and denominator by 24, we get

$$\frac{24)\ \frac{3}{4}+\frac{7}{8}-\frac{1}{2}}{24)\ \frac{2}{3}-\frac{5}{6}+\frac{1}{2}}=\frac{18+21-12}{16-20+12}=\frac{27}{8}$$

Example 4

Simplify the fraction:

$$\frac{2-\frac{7x-6}{x^2}}{4x-\frac{9}{x}}$$

Solution

We multiply the entire numerator and denominator by x^2, and get

$$\frac{2x^2-7x+6}{4x^3-9x}=\frac{(2x-3)(x-2)}{x(2x-3)(2x+3)}=\frac{x-2}{x(2x+3)}$$

Exercise 12-4

1. $\dfrac{3+\frac{1}{2}}{5-\frac{5}{8}}$ **2.** $\dfrac{\frac{2}{3}-4}{\frac{4}{9}-6}$ **3.** $\dfrac{9x-\frac{4}{x}}{3+\frac{2}{x}}$

4. $\dfrac{x - \dfrac{1}{4}}{4 - \dfrac{1}{x}}$ 5. $\dfrac{16x - \dfrac{9}{x}}{1 + \dfrac{3}{4x}}$ 6. $\dfrac{\dfrac{3x}{2} - 1}{6x - \dfrac{8}{3x}}$

7. $\dfrac{\dfrac{x}{3} + \dfrac{x}{4}}{\dfrac{5x}{12} + 1}$ 8. $\dfrac{\dfrac{1}{9} - \dfrac{x}{3}}{\dfrac{x}{6} - \dfrac{1}{18}}$ 9. $\dfrac{\dfrac{x}{a} + \dfrac{x}{b}}{\dfrac{5x}{ab}}$

10. $\dfrac{\dfrac{x}{3} - \dfrac{y}{5}}{\dfrac{4x}{15} + y}$ 11. $\dfrac{2 - \dfrac{5x - 3}{x^2}}{x - \dfrac{9}{4x}}$ 12. $\dfrac{x - \dfrac{2x + 8}{3x}}{2 - \dfrac{8}{x^2}}$

13. $\dfrac{\dfrac{9}{x} - 4x}{4x + 4 - \dfrac{3}{x}}$ 14. $\dfrac{2 - \dfrac{3x + 4}{5}}{3x - \dfrac{2 + 5x}{2}}$ 15. $\dfrac{1 - \dfrac{5x - 4}{3}}{2 - \dfrac{x + 3}{4}}$

16. $\dfrac{\dfrac{1}{9} - \dfrac{2}{x} + \dfrac{5}{x^2}}{3 - \dfrac{x^2}{3}}$ 17. $\dfrac{3x + 4}{2 - \dfrac{5x}{x - 2}}$ 18. $\dfrac{1 - \dfrac{3x}{x - 2}}{\dfrac{3}{x^2 - 4} + 1}$

19. $\dfrac{\dfrac{x}{x - 2y} - 1}{1 - \dfrac{x}{x + 2y}}$ 20. $\dfrac{\dfrac{x}{x - y} - 1}{\dfrac{y}{x - y} + 1}$ 21. $\dfrac{\dfrac{1}{2x - 3} + 3}{\dfrac{10}{3x + 1} - 2}$

QUIZ ON CHAPTER 12, FRACTIONS. FORM A.

1. Reduce to lowest terms:

$$\frac{x^2 - 25}{x^2 - 10x + 25}$$

2. Reduce to lowest terms:

$$\frac{9x^3(x - 4)^2}{12x^2 - 3x^3}$$

In Nos. 3, 4, and 5, multiply and divide as indicated:

3. $\dfrac{5xy^2}{9ac^2} \cdot \dfrac{3c^3}{10xy} \div \dfrac{4cy}{3ab}$ 4. $\dfrac{16x^2y}{15ab^2} \div \dfrac{9cx}{10ab} \cdot \dfrac{21ac}{8y^2}$

5. $\dfrac{12x^2 - 27}{x - 3} \div (8x - 12)$

6. Simplify the following expression by first factoring numerators and denominators into prime factors. Then divide numerators and denominators by all factors found in numerator and denominator.

$$\frac{x^2 - 2x - 3}{2x^4 - 72x^2} \cdot \frac{x^3 - 4x^2 - 12x}{x^2 - 4x + 3} \div \frac{2x^2 + 5x + 2}{4x^3 + 20x^2 - 24x}$$

7. Combine the fractions at the right into a single fraction reduced to lowest terms:

$$\frac{3x - 5}{4} - \frac{4x - 3}{7} - \frac{2x - 3}{28}$$

8. Combine into a single fraction reduced to lowest terms:

$$\frac{5x - 4}{x - 2} - \frac{2x - 5}{x + 3} - 3$$

9. Combine into a single fraction

$$\frac{3x - 4}{x - 5} - \frac{6}{x} - 3x - 2$$

10. Simplify this complex fraction:

$$\frac{1 - \dfrac{3x + 2}{2x^2}}{x - \dfrac{4}{x}}$$

CHAPTER THIRTEEN Fractional Equations

13-1 DEFINITION

A *fractional equation* is an equation containing a fraction, either a common fraction or a decimal. The denominator may contain the variable x.

Let us recall the definition of an equation. An equation is a statement of equality between two equal quantities. The expression must contain the equal sign (=) with some quantity on each side. One side may be only zero.

The following expression is not an equation. It simply indicates the sum of fractions:

$$\frac{3x+2}{4}+\frac{x-3}{5}-\frac{3x}{2}$$

All we can do with these fractions is to combine them and simplify the sum. We cannot solve for x because it is not an equation. On the other hand, the following expression is a fractional equation and it can be solved:

$$\frac{3x+2}{4}+\frac{x-3}{5}-\frac{3x}{2}=0$$

Here are some examples of fractional equations:

(a) $\dfrac{2x}{3}=8$ (b) $\dfrac{2x}{5}+6=4x-\dfrac{3}{4}$ (c) $0.23x-15=1.3x+2.4$

(d) $\dfrac{3}{x}+\dfrac{2}{x-1}=\dfrac{12}{x^2-x}$ (e) $\dfrac{3}{x+3}-\dfrac{5x+2}{x-4}=\dfrac{x-3}{x^2-x-12}$

The expression "fractional equation" is sometimes restricted to refer only to equations in which the unknown x appears in the denominator, as in examples (d) and (e). Then the expression is not used to refer to equations having only constant denominators. However, the same procedure is used in solving all types.

13-2 USE OF MULTIPLICATION AXIOM

As a general rule, the best procedure to follow in solving a fractional equation is to get rid of the denominators first; that is, first clear the equation of fractions. As a first step, therefore, we multiply both sides of the equation by the lowest common denominator (LCD). The LCD may be called a "multiplying operator." This first step is simply the application of the multiplication axiom. This axiom, or rule, says:

If both sides of an equation are multiplied by the same quantity, the equation is still true.

In some instances it may be possible to solve simple fractional equations without first eliminating denominators. However, as a general rule, the *best first step* is to clear the equation of fractions.

It is important to see exactly what happens when both sides of an equation are multiplied by some quantity, as shown in the following examples.

Example 1

Solve for x:

$$\frac{2x}{3} = 8$$

Solution

To eliminate the fraction, we multiply both sides of the equation by 3, as shown here:

$$(3)\left(\frac{2x}{3}\right) = (3)(8)$$

The equation reduces to

$$2x = 24$$

The new equation is not exactly the same as the original. Each side has been made 3 times as much. However, we know that if the original equation is true, then the new equation is true. This is the meaning of the multiplication axiom. The new equation can be solved in the usual way. Dividing both sides by 2, we get

$$x = 12$$

As a check, we substitute 12 for x in the *original* equation and ask:

$$\text{does } \frac{2(12)}{3} = 8? \quad \text{Yes.}$$

The number 12 satisfies the equation. It is therefore the root or solution.

Example 2

Solve for x:

$$\frac{2x}{5} + 6 = 4x - \frac{3}{4}$$

Solution

Multiplying both sides of the equation by the LCD, 20, we get

$$8x + 120 = 80x - 15$$

Transposing,

$$15 + 120 = 80x - 8x$$

Combining and solving for x,

$$\frac{15}{8} = x$$

To check, we substitute $\frac{15}{8}$ for x in the *original* equation, and ask if

$$\frac{(2)(15/8)}{5} + 6 = 4\left(\frac{15}{8}\right) - \frac{3}{4}$$

Working out both sides, we get,

$$\frac{27}{4} = \frac{27}{4}$$

For solving a fractional equation we have the following steps:

1. *Multiply both sides of the equation by the lowest common denominator. This will clear the equation of fractions.*
2. *Solve the resulting equation as usual for a simple equation.*

Exercise 13-1

Solve each equation for the unknown and check in the original equation.

1. $\frac{5x}{3} = 20$ **2.** $18 = \frac{3y}{4}$ **3.** $\frac{2n}{5} = n - 12$

4. $b + 4 = \frac{3b}{8}$ **5.** $\frac{3c}{7} = c + 5$ **6.** $\frac{2d}{9} = d + 4$

7. $n - 3 = \frac{n + 6}{7}$ **8.** $4 - k = \frac{2 + k}{8}$ **9.** $\frac{n - 8}{6} = 4 - n$

10. $\frac{x}{2} - \frac{x}{3} = x - 15$ **11.** $\frac{4x}{3} - \frac{3x}{5} = x - 7$ **12.** $\frac{5x}{4} - \frac{2x}{5} = x - 9$

13. $\frac{3x}{8} - x = 2 + \frac{x}{12}$ **14.** $\frac{8x}{15} - x = \frac{7x}{10} + 4$ **15.** $\frac{3x}{4} - x = 2 + \frac{x}{6}$

16. $x - \frac{5x}{6} = 3 + \frac{4x}{9}$ **17.** $\frac{3x}{8} + 2 = x - \frac{5x}{6}$ **18.** $\frac{4x}{9} - x = 1 - \frac{5x}{12}$

19. $\frac{2n}{3} - \frac{3n}{4} = 2 - \frac{5n}{6}$ **20.** $\frac{5y}{6} - \frac{7y}{18} = \frac{2y}{3} + 2$

21. $\frac{2t}{5} - 2 = \frac{4t}{15} + \frac{3t}{10}$ **22.** $\frac{2x + 3}{5} = \frac{3x - 1}{4} + 3$

23. $\frac{9x - 5}{7} = \frac{5x - 4}{3} + 2$ **24.** $\frac{2x + 5}{4} = \frac{4x - 3}{5} + 4$

25. A man paid $\frac{5}{8}$ of his money as a half-payment on the cost of a house and $\frac{1}{10}$ of his money for a car. If he had $9900 left, how much had he at first? What was the total cost of the house?

26. One-eighth of a street was paved the first week and $\frac{3}{16}$ the following week. If 2970 ft of paving was left to be done, what was the total length of the street?

27. The current in one branch of an electric circuit is $\frac{3}{5}$ as much as the current in a second branch. When the currents unite, the total current is 20 amperes. Find the current in each branch.

28. When 3 voltage sources are connected in series (same polarity), the total voltage is 238 volts. The second voltage is $\frac{2}{3}$ of the first, and the third is $\frac{3}{5}$ of the first. Find each voltage.

29. A woman spent $\frac{1}{10}$ of her money for a hat. She bought a dress that cost 6 times as much as the hat, and then had $22.50 left. How much did she have at first, and how much was spent for each item?

30. One fraction is $\frac{3}{4}$ as much as another and their sum is $\frac{7}{5}$. What are the fractions?

31. The width of a rectangle is 3 in. more than $\frac{3}{4}$ of the length. Find the length and the width if the perimeter is 111 in.

13-3 SOLVING FRACTIONAL EQUATIONS VERSUS COMBINING FRACTIONS

It is important to see the difference between solving a fractional equation and simply combining fractions by addition or subtraction. Let us take the following two expressions and note the similarity and the difference between the two.

Example 1

(a) $\frac{5x + 7}{2} + \frac{4x - 1}{3}$ (b) $\frac{5x + 7}{2} + \frac{4x - 1}{3} = 2$

The first expression is *not* an equation. An equation must have something on each side of the equal sign, even though one side may be only a zero. The expression *cannot* be solved for x. All we can do is to combine the fractions.

The second expression is an equation. It has something on each side of the equal sign. The equation *can* be solved for x.

Now notice what happens in the addition of the fractions in (a). To combine the fractions we must have a common denominator. We change the fractions into new fractions having 6 as the lowest common denominator. The new fractions are

$$\frac{3(5x + 7)}{6} + \frac{2(4x - 1)}{6}$$

Note especially that each fraction has exactly the *same value* as in the original. We have *not* changed the *value* of either fraction; we have changed only its *form*. Now we can add the fractions, and get

$$\frac{3(5x + 7) + 2(4x - 1)}{6} = \frac{15x + 21 + 8x - 2}{6} = \frac{23x + 19}{6}$$

Note that we do not lose the denominator.

Now let us consider the equation in (b):

$$\frac{5x + 7}{2} + \frac{4x - 1}{3} = 2$$

Here the multiplying operator is 6, the LCD. When we multiply both sides of the equation by 6, the quantity on each side becomes 6 times as much as in the original equation. Then we get

$$3(5x + 7) + 2(4x - 1) = 12$$

The denominator has disappeared. Solving, we get

$$x = -\frac{7}{23}$$

Consider a second example. Compare the following expressions.

Example 2

(a) $\dfrac{5x - 2}{3} - \dfrac{2x - 9}{4}$ (b) $\dfrac{5x - 2}{3} - \dfrac{2x - 9}{4} = 0$

First, let us summarize the situation with respect to each expression. For the first expression (a), we can say

1. It is *not* an equation.
2. It *cannot* be solved for x.
3. When we combine the fractions, the denominator (LCD) remains.

For the second expression (b), we can say

1. It *is* an equation.
2. It *can* be solved for x.
3. When we solve the equation, the denominators *disappear.*

To combine the fractions in (a), we change the fractions to new fractions with 12 as a common denominator. Combining the fractions, we get the new fraction

$$\frac{14x + 19}{12}$$

This is the answer to the problem. It cannot be solved for x.

To solve the equation in (b) we multiply both sides by 12 and get

$$4(5x - 2) - 3(2x - 9) = 0$$

Solving the equation, we get

$$x = -19/14$$

Example 3

Solve for t:

$$0.36t - 11 = 1.2t - 5.6$$

Solution

If we write the decimals as common fractions, we see that the LCD is 100. Then we multiply both sides by 100 and get

$$36t - 1100 = 120t - 560$$

Transposing and combining,

$$-84t = 540$$

Dividing both sides by -84,

$$t = -\tfrac{540}{84} = -\tfrac{45}{7}$$

$$t = -6.4 \text{ (approx.)}$$

Exercise 13-2

Solve each equation for the unknown. Check as directed by the instructor.

1. $\dfrac{3x + 2}{4} + 5 = \dfrac{x - 9}{3} - x$

2. $\dfrac{2x + 1}{5} + 2 = \dfrac{4 - 3x}{4} + x$

3. $\dfrac{4x + 1}{3} - x = 1 + \dfrac{3x - 8}{5}$

4. $\dfrac{2x - 5}{7} + 1 = \dfrac{3x - 1}{2} - x$

5. $x + \frac{6x - 5}{10} = \frac{2x - 7}{5} - 3$ **6.** $2 + \frac{7x - 5}{12} = \frac{x - 5}{4} + x$

7. $\frac{5a - 1}{2} = \frac{4a - 5}{6} + \frac{a - 2}{3} - 3$ **8.** $\frac{4n - 3}{5} = \frac{5 - 3n}{10} + \frac{n - 1}{2} - 2$

9. $\frac{3x - 4}{9} + \frac{2x - 1}{6} = \frac{x - 3}{4} - 1$ **10.** $\frac{8x - 3}{6} + \frac{2 - 3x}{8} = \frac{x - 5}{3} - 1$

11. $0.3x + 20 = 12 + 0.5x$ **12.** $36 - 0.15n = 45 - 0.6n$

13. $7 - 0.32y = 2.6 + 0.2y$ **14.** $0.4c + 2.2 = 0.48c - 5$

15. $0.375x - 3.4 = 4.35 + 0.25x$ **16.** $3 - 1.25x = x - 0.375$

17. $3 - 0.2y = 0.05y - 1.25$ **18.** $15.6 + 0.4t = 1.15t - 8.4$

19. $0.6n - 4 = 0.875 - 0.15n$ **20.** $0.325y - 0.05 = 1.2 + 0.3y$

13-4 A FRACTION AS AN INDICATED QUANTITY

If a fraction is preceded by a minus (−), we must remember that this sign makes the *entire fraction negative.* The fraction must always be considered as a *quantity*. The fraction line separating the numerator from the denominator can be considered a vinculum, or the sign of grouping. If you are in doubt about the meaning in any expression, you may enclose a fraction in a set of parentheses.

As an example, consider the following equation:

Example 1

Solve for x:

$$\frac{3x - 2}{5} - \frac{2x - 7}{4} = 2$$

Solution

Multiplying both sides of the equation by 20, we get

$$4(3x - 2) - 5(2x - 7) = 40$$

Removing parentheses, we get

$$12x - 8 - 10x + 35 = 40$$

Solving, $x = 6.5$

Example 2

Solve for x:

$$\frac{4x + 3}{5} - \frac{x + 4}{3} - \frac{2x - 1}{15} = 2$$

Solution

This example shows one of the great danger spots in working with fractions, whether we are simply combining fractions or solving fractional equations. As a first step, we multiply both sides of the equation by the LCD, 15. We get $3(4x + 3) - 5(x + 4) - (2x - 1) = 30$. Note the minus signs before the second and the third fractions. The entire fraction after each sign is to be considered as a quantity, and the minus signs can be considered as signs of operations (subtraction). The net result is that the signs of the terms in their numerators will eventually be changed. Continuing the solution,

$$3(4x + 3) - 5(x + 4) - (2x - 1) = 30$$

$$12x + 9 - 5x - 20 - 2x + 1 = 30$$

Transposing, $\quad 12x - 5x - 2x = 30 - 9 + 20 - 1$

Combining, $\quad 5x = 40$

Dividing both sides by 5, $\quad x = 8$

The check is left for the student. Remember, it must be started in the *original equation* and both sides worked out in fraction form.

Exercise 13-3

Solve the following equations. Check any five.

1. $\dfrac{3x + 1}{4} - \dfrac{x - 3}{8} = \dfrac{2x - 5}{2} - 1$ **2.** $\dfrac{5 - 2x}{2} - \dfrac{1 - 3x}{3} - \dfrac{2x - 3}{6} = 4$

3. $\dfrac{2x}{3} - \dfrac{x - 5}{5} = 2 - \dfrac{5 + 3x}{15}$ **4.** $\dfrac{4x - 5}{10} - \dfrac{x - 2}{4} = \dfrac{x}{20} - 1$

5. $\dfrac{2x - 7}{3} - \dfrac{3x + 5}{12} + \dfrac{x - 3}{4} = 0$ **6.** $\dfrac{5x}{4} + \dfrac{x - 2}{6} - \dfrac{5 - x}{12} = 0$

7. $\dfrac{x - 3}{4} + \dfrac{2x - 1}{5} - \dfrac{5x - 3}{20} = 1$

8. $\dfrac{4 - x}{2} - \dfrac{2(7 - 2x)}{5} - \dfrac{3(x - 2)}{4} = 1$

9. $\dfrac{5x - 1}{3} - \dfrac{x - 2}{21} - \dfrac{2x + 3}{7} = 0$

10. $\dfrac{2x}{15} - \dfrac{x - 4}{9} - \dfrac{3x + 1}{45} = 0$

11. $\dfrac{2(x + 5)}{15} - \dfrac{3(x + 4)}{2} - \dfrac{4(3 - x)}{3} = 0$

12. $\dfrac{3(2x + 1)}{4} - \dfrac{2(3x - 2)}{3} - \dfrac{3 - 2x}{12} = 0$

13. $3(0.06x + 4) = 0.05(1800 + x)$

14. $0.04n - 0.08(3000 - n) = 15$

15. $0.075y - 0.05(9000 - y) = 24$

16. $0.0425t + 0.06(3500 - t) = 168$

17. $0.06x + 0.03(1800 - x) = 67.5$

18. $0.04x - 0.07(8400 - x) = 83$

19. $0.15S - 0.35(400 - S) = 1.5$

20. $0.045(6000 - N) - 0.06N = 133.5$

13-5 FRACTIONS CONTAINING VARIABLE DENOMINATORS

As stated at the beginning of this chapter, some fractions have variable denominators. This often happens in fractional equations. To solve such equations we proceed in the same way as we do with other fractional equations. We first multiply both sides of the equation by the LCD.

Example 1

Solve for x:

$$\frac{3}{x} + \frac{2}{x - 1} = \frac{12}{x^2 - x}$$

Solution

If we factor the third denominator, we see that it consists of the two other denominators as factors. Therefore, $x(x - 1)$ is the lowest common denominator. We multiply both sides of the equation by the multiplying operator $x(x - 1)$. We get

$$3(x - 1) + 2x = 12$$

Solving,

$$3x - 3 + 2x = 12$$

$$5x = 15$$

$$x = 3$$

Example 2

Solve for x:

$$\frac{5x}{x + 3} - \frac{3x + 2}{x - 4} - \frac{x - 3}{x^2 - x - 12} - 2 = 0$$

Solution

The trinomial denominator can be factored into the factors $(x + 3)(x - 4)$. This denominator is the LCD, since it is divisible by all the denominators.

Therefore, we use $(x + 3)(x - 4)$ as a multiplying operator. Multiplying both sides of the equation by $(x + 3)(x - 4)$, we get

$$5x(x - 4) - (3x + 2)(x + 3) - (x - 3) - 2(x + 3)(x - 4) = 0$$
$$5x^2 - 20x - (3x^2 + 11x + 6) - x + 3 - 2(x^2 - x - 12) = 0$$
$$5x^2 - 20x - 3x^2 - 11x - 6 - x + 3 - 2x^2 + 2x + 24 = 0$$
$$21 = 30x$$
$$\frac{7}{10} = x$$

Example 3

Solve for x:

$$\frac{3}{x + 2} = \frac{1}{x - 2} - \frac{4}{x^2 - 4}$$

Solution

Multiplying both sides of the equation by the lowest common denominator, $x^2 - 4$, we get

$$3(x - 2) = x + 2 - 4$$

Removing parentheses, $3x - 6 = x + 2 - 4$

Transposing, $3x - x = 6 + 2 - 4$

Combining, $2x = 4$

Dividing both sides by 2, $x = 2$

The answer, $x = 2$, seems like a very logical one to this problem. The procedure is entirely correct. Yet, when we check the answer in the original equation, we get zero in a denominator. In fact, in this example the check produces a zero for each of two denominators.

Whenever the check of a solution produces a denominator of zero, that solution must be discarded. The foregoing equation, therefore, has *no solution.* If we carelessly say that a number divided by zero is equal to infinity, we are no better off. We might be inclined to write the check as follows:

$$\frac{3}{2 + 2} = \frac{1}{2 - 2} - \frac{4}{4 - 4}$$
$$\frac{3}{4} = \frac{1}{0} - \frac{4}{0}$$
$$\frac{3}{4} = \infty - \infty \qquad \text{(wrong)}$$

The result proves that we cannot operate in this manner with infinity. The important fact here is that we cannot divide any number by zero and that the expression $1 \div 0$ is *not* equal to infinity.

Exercise 13-4

Solve the following equations.

1. $\dfrac{x-3}{x+3}=7$

2. $\dfrac{y+4}{y-8}=5$

3. $\dfrac{n-2}{n+8}=6$

4. $\dfrac{2n+1}{3n-2}=\dfrac{4}{5}$

5. $\dfrac{3y-4}{5y+1}=\dfrac{2}{5}$

6. $\dfrac{4t+5}{3t-1}=\dfrac{6}{7}$

7. $\dfrac{5}{2x}-\dfrac{2}{3x^2}=\dfrac{9}{4x}$

8. $\dfrac{4}{3x}-\dfrac{3}{x}=\dfrac{6}{5x^2}$

9. $\dfrac{5}{4x}-\dfrac{3}{2x^2}=\dfrac{4}{5x}$

10. $\dfrac{3}{x-2}-\dfrac{7}{x-8}=0$

11. $\dfrac{5}{x+2}-\dfrac{9}{x-3}=0$

12. $\dfrac{7}{3x-2}-\dfrac{6}{2x-1}=0$

13. $\dfrac{3n}{n+4}-\dfrac{n}{n-3}=2$

14. $\dfrac{3t}{t-3}-\dfrac{4t}{2t-5}=1$

15. $\dfrac{5y}{y-2}-\dfrac{4y}{2y-7}=3$

16. $\dfrac{4x-1}{x-3}-2=\dfrac{2x}{x-4}$

17. $\dfrac{3x}{x+2}-\dfrac{2x-3}{2x-1}=2$

18. $\dfrac{x}{x+1}=3-\dfrac{4x-1}{2x-3}$

19. $\dfrac{4n-1}{2n+3}=3-\dfrac{n-4}{n-3}$

20. $\dfrac{t+2}{t-3}=3-\dfrac{4t+5}{2t-3}$

21. $\dfrac{6n-1}{2n-3}-2=\dfrac{n+3}{n+2}$

22. $\dfrac{8x-5}{2x-3}=\dfrac{x-2}{x+6}+3$

23. $\dfrac{x+4}{x+1}=2-\dfrac{4x-5}{2x-3}$

24. $\dfrac{3x+2}{x-1}=5-\dfrac{4x+5}{2x-3}$

25. $\dfrac{n-2}{n+3}-\dfrac{n-4}{n-3}=\dfrac{n+4}{n^2-9}$

26. $\dfrac{2t-3}{t-4}-\dfrac{2t+1}{t+4}=\dfrac{6t+7}{t^2-16}$

27. $\dfrac{y-3}{y-2}-\dfrac{y+3}{y+2}=\dfrac{4y+3}{y^2-4}$

28. $\dfrac{3n-2}{n-3}-\dfrac{3n+1}{n+3}=\dfrac{5n-8}{n^2-9}$

29. $\dfrac{6}{x^2-9}-2=\dfrac{2}{x+3}-\dfrac{2x-1}{x-3}$

30. $\dfrac{6}{x^2-4}-1=\dfrac{4}{x+2}-\dfrac{x-3}{x-2}$

31. $\dfrac{10}{h^2-25}-2=\dfrac{h-1}{h-5}-\dfrac{3h-1}{h+5}$

32. $\dfrac{t-3}{t^2-16}-3=\dfrac{6}{t+4}-\dfrac{3t-1}{t-4}$

33. $\dfrac{x-1}{x-2} - \dfrac{x+2}{x-3} - \dfrac{2x-5}{x^2-5x+6} = 0$

34. $\dfrac{x+1}{x+3} - \dfrac{2x-5}{x^2-x-12} - \dfrac{x+2}{x-4} = 0$

35. $\dfrac{x+3}{2x+3} - \dfrac{x+1}{2x-5} = \dfrac{2x-3}{2x^2-4x-15}$

36. $\dfrac{x-1}{2x-3} - \dfrac{x+2}{2x+1} = \dfrac{4x-5}{4x^2-4x-3}$

13-6 LITERAL EQUATIONS

A *literal equation* is an equation in which constants are represented by letters of the alphabet; that is, letters appear where we would expect arithmetic numbers. This is a literal equation: $ax = b$.

In general, a literal equation is solved in the same way as any other equation. Let us first look at a simple equation such as those we have already solved.

Example 1

Solve for x:

$$3x = 19$$

Solution

To find the value of x, we divide both sides of the equation by 3, which is the coefficient of x. We get

$$x = \frac{19}{3}$$

Now, suppose we have a literal equation in which we find other letters in place of 3 and 19.

Example 2

Solve for x:

$$ax = b$$

Solution

Our problem is to solve for the value of x. Again we divide both sides of the equation by the coefficient of x. Dividing both sides by a,

$$x = \frac{b}{a}$$

The equation is now solved for x. The quantity b/a is the root of the equation.

If an equation has several terms containing the unknown, x or whatever unknown we wish to find, then all the terms containing the unknown are first isolated, just as in any other equation. To see the similarity to equations we have already solved, let us first solve an equation with arithmetic constants.

Example 3

Solve the equation for x:

$$7x - 4 = 21 - 5x$$

Solution

We first transpose to isolate the terms in x.

$$7x + 5x = 21 + 4$$

Combining like terms, $$12x = 25$$

Dividing both sides by 12, $$x = \frac{25}{12}, \quad \text{the root}$$

Example 4

Solve this literal equation for x:

$$ax - 4 = b - 8x$$

Solution

Transposing to isolate x terms,

$$ax + 8x = b + 4$$

We cannot combine the x terms into a single term. However, by factoring the expression $ax + 8x$ into two factors, $x(a + 8)$, we see that the coefficient of x is the binomial $a + 8$. Therefore, we divide both sides of the equation by $(a + 8)$:

$$x = \frac{b + 4}{a + 8}$$

This answer cannot be reduced to lower terms.

In some literal equations it is necessary to clear the equation of fractions by multiplying by the LCD, as shown in the following examples.

Example 5

Solve this equation for R:

$$I = \frac{E - e}{R}$$

Solution

To solve for R, we must get R alone on one side of the equation. Multiplying by R, we get

$$RI = E - e$$

Dividing both sides by I, the coefficient of R, we get

$$R = \frac{E - e}{I}$$

Example 6

Solve this equation for A:

$$A = \frac{B + A}{C + 2}$$

Solution

Multiplying by $(C + 2)$, $\quad A(C + 2) = B + A$

or, $\quad AC + 2A = B + A$

Transposing, we get

$$AC + A = B; \qquad \text{then} \quad A = \frac{B}{C + 1}$$

Example 7

Solve for b:

$$C = \frac{Kab}{b - a}$$

Solution

Clearing of fractions, $\quad C(b - a) = Kab$

or $\quad bC - aC = Kab$

Isolating the b terms, $\quad bC - Kab = aC$

Dividing by the coefficient of b, $\quad b = \dfrac{aC}{C - Ka}$

Example 8

Solve for t:

$$A + 2 = \frac{B - 3}{t - 1}$$

Solution

Multiplying by $(t - 1)$, we get

$$(A + 2)(t - 1) = B - 3$$

Expanding the left side, we get $At - A + 2t - 2 = B - 3$

Transposing to isolate t terms, $At + 2t = A + B - 1$

Dividing by the coefficient of t, $t = \dfrac{A + B - 1}{A + 2}$

Example 9

Solve for c:

$$\frac{a}{b + c} = \frac{d}{c - 2}$$

Solution

Multiplying by the LCD, $a(c - 2) = d(b + c)$

Expanding the products, $ac - 2a = bd + cd$

Transposing to isolate the c terms, $ac - cd = bd + 2a$

Dividing by the coefficient of c, $c = \dfrac{bd + 2a}{a - d}$

Example 10

Solve the following formula for n:

$$I = \frac{nE}{R + nr}$$

Solution

Multiplying both sides of the equation by $R + nr$,

$$I(R + nr) = nE$$

Removing parentheses, $IR + Inr = nE$

Isolating terms in n, $IR = nE - Inr$

Factoring to show coefficient of n, $IR = n(E - Ir)$

Dividing both sides by $E - Ir$, $\dfrac{IR}{E - Ir} = n$

Note. The equation may be solved for any letter we wish by isolating all the terms containing that particular letter and then dividing both sides of the equation by its coefficient. This is called "changing the subject of the formula" and is important in much work in connection with formulas.

Exercise 13-5

Solve each formula for the letter indicated at the right of each.

1. $V = LWH$ (W) **2.** $P = 2L + 2W$ (L)

3. $t = \frac{d}{r}$ (r) **4.** $A = \frac{ab}{2}$ (b)

5. $V = \pi r^2 h$ (h) **6.** $A = 2\pi rh$ (r)

7. $I = \frac{E}{R + r}$ (R) **8.** $F = \frac{9}{5}C + 32$ (C)

9. $\frac{1}{f} = \frac{1}{a} + \frac{1}{b}$ (a) **10.** $\frac{W_1}{W_2} = \frac{L_2}{L_1}$ (W_2)

11. $F = K\frac{M_1 M_2}{d^2}$ (M_1) **12.** $C = \frac{5}{9}(F - 32)$ (F)

13. $S = N\frac{(A + L)}{2}$ (A) **14.** $A = \frac{h}{2}(B + b)$ (B)

15. $l = a + (n - 1)d$ (n) **16.** $A = 2\pi r(h + r)$ (h)

17. $S = \frac{a - rl}{1 - r}$ (r) **18.** $\frac{1}{T} = \frac{1}{a} + \frac{1}{b} + \frac{1}{c}$ (c)

19. $V = \frac{1}{3}\pi r^2 h$ (h) **20.** $\frac{E}{e} = \frac{R + r}{r}$ (r)

21. $A = 2\pi rh + 2\pi r^2$ (h) **22.** $V = \pi R^2 h - \pi r^2 h$ (h)

23. $Z_t = \frac{Z_1 Z_2}{Z_1 + Z_2}$ (Z_2) **24.** $S = Vt - \frac{1}{2}gt^2$ (g)

25. Formula 1 (above) is used to find the volume V of a rectangular solid when the length L, the width W, and the height H are known. After solving the formula for W, find W when $H = 15$ in., $L = 24$ in., and $V = 4500$ cu in.

26. Formula 2 is used to find the perimeter P of a rectangle when the length L and the width W are known. After solving the formula for L, find L when $W = 15$ in. and the perimeter P equals 75 in.

27. Formula 3 is used to find the time, t, it takes to drive a distance d at a rate r. After solving for r, find r when $t = 7.2$ hours and $d = 378$ miles.

28. Formula 4 is used to find the area A of a triangle when the altitude a and the base b are known. Solve for a, and then find a when $b = 24$ in. and $A = 210$ sq in.

29. Formula 5 is used to find the volume V of a circular cylinder when the radius r and the height h are known. After solving for h, find h when $r = 8$ in. and $V = 480\pi$ cu in.

30. Formula 6 is used to find the lateral area A of a right circular cylinder when the radius r and the altitude h are known. After solving for r, find r when $h = 14$ in., and $A = 182\pi$ sq in.

31. Formula 7 is used to find the current I in an electric circuit when voltage E, external resistance R, and internal resistance r are known. Solve the formula for r and then find r when $E = 33$ volts, $R = 27$ ohms, and $I = 1.2$ amperes.

32. Formula 8 expresses the relation between Fahrenheit temperature reading F and Celsius reading C. After solving for C, find C when $F = 68°$.

33. Formula 9 is used to express the relation among the focal length l of a lens, the distance a to an object, and the distance b to the image. Solve for b and then find b when $a = 300$ cm and $f = 50$ cm.

34. Formula 10 expresses the relation between two weights, W_1 and W_a, and two arm lengths, L_1 and L_2, of a balanced lever. After solving for W_2, find W_2 when $L_1 = 15$ in., $L_2 = 12$ in., and $W_1 = 40$ lb.

35. Formula 11 is used to find the force F between two objects, M_1 and M_2, at a distance d between them, where K is some constant. After solving for M_1, find M_1 when $M_2 = 30$, $K = 10$, $d = 5$, and $F = 240$.

36. Formula 14 is used to find the area A of a trapezoid when the altitude h and the bases B and b are known. Solve for b and then find b when $h = 9$, $B = 14.5$, and $A = 108$.

37. Formula 18 is used to refer to the total resistance T in an electric circuit, when three resistors having resistances of a, b, and c are arranged in parallel. Solve for T, and then find T when $a = 30$ ohms, $b = 40$ ohms, and $c = 60$ ohms.

38. Formula 24 is used to find the height s reached by an object when projected directly upward from the earth's surface, where t = time in seconds, V = initial velocity, and g = acceleration due to gravity. Solve for V, and then find V when $s = 84$ ft, $t = 3.5$ sec, and $g = 32\ \text{ft/sec}^2$.

Exercise 13-6

1. The difference between two numbers is 3. If five times the smaller number is divided by the larger, the quotient is 4. What are the numbers?

2. The numerator of a fraction is 5 less than the denominator. If the numerator is decreased by 2 and the denominator is increased by 3, the value of the new fraction is $\frac{1}{3}$. Find the original fraction.

3. The denominator of a fraction is 3 more than the numerator. If the denominator is increased by 2 and the numerator is decreased by 2, the value of the new fraction is $\frac{7}{8}$. Find the original fraction.

4. Find two consecutive integers such that $\frac{2}{3}$ of the first added to $\frac{3}{5}$ of the second becomes 7 more than the first.

5. Frank can mow a lawn alone in 6 hr and Joe can mow it alone in 8 hr. If each gets a lawn mower and both work at the same time, how long should it take them?

Solution

Before we can work the problem, we must assume that they work at the same rate as when each is working alone. We cannot make allowance for their getting into each other's way, which actually might happen. Assuming that the work goes along steadily,

let x = the number of hours it will take them working together

Since Frank can mow the lawn alone in 6 hr, he can mow $\frac{1}{6}$ of it in 1 hr. Since Joe can mow the lawn alone in 8 hr, he can mow $\frac{1}{8}$ of it in 1 hr. Note that the part done in 1 hr is simply the *reciprocal* of the number of hours. Then $1/x$ is part both can do together in 1 hr. Therefore, we can write the equation to show the total part of the work done in 1 hr:

$$\frac{1}{6} + \frac{1}{8} = \frac{1}{x}$$

The equation can be solved for x, and this value will be the number of hours it will take both working together.

6. A water tank is being filled through two pipes. One pipe alone can fill the tank in 8 hr, the other can fill the tank alone in 12 hr. How long will it take to fill the tank when both pipes are used?

7. A certain tank can be filled with water through one pipe in 6 hr. It can be filled through a second pipe alone in 4 hr. How long will it take to fill the tank if both pipes are used?

8. A swimming pool can be filled in 15 hr through one opening near the top and through a second opening in 12 hr. How long will it take to fill the pool if both openings are used?

9. In Problem 8 a third opening which can fill the pool alone in 20 hr is sometimes used. How long will it take to fill the pool if all three openings are used?

10. The pool in Problem 8 can be drained in 18 hr through a hole in the bottom of the pool. One day when the pool was empty it was decided to fill the pool by using the same two openings mentioned in Problem 8. However, someone had left the drain hole open. How long would it take to fill the pool under these conditions?

11. A businessman finds it necessary to get out a number of form letters. He has three typists. Doris can type them all alone in 6 hr, Emily can type them alone in $7\frac{1}{2}$ hr, and Lilas can type them alone in 8 hr. How long should it take them all working at the same time?

12. The width of a certain rectangle is $\frac{2}{3}$ of the length. if the width is decreased by 3 and the length increased by 3, the area is 36 sq in. less than the original rectangle. Find the dimensions of the original rectangle. What change takes place in the perimeter?

13. A rectangle is $\frac{3}{5}$ as wide as it is long. If the width is increased by 2 and the length decreased by 3, the area remains unchanged. Find the dimensions of the original rectangle. What change takes place in the perimeter?

14. If 5 is added to a certain number, $\frac{3}{4}$ of the result is 1 less than the number. What is the number?

15. If a certain number is subtracted from the numerator and twice the number is subtracted from the denominator of the fraction $\frac{12}{19}$, the resulting fraction is equal to $\frac{2}{3}$. What is the number?

16. What number must be subtracted from the numerator and from the denominator of the fraction $\frac{17}{19}$ so that the resulting fraction will be equal to the value $\frac{3}{4}$?

17. What number must be subtracted from the numerator and the denominator of the fraction $\frac{15}{4}$ so that the value will be equal to $\frac{1}{2}$?

18. What number must be added to the numerator and the denominator of the fraction $\frac{3}{8}$ to make the new fraction equal in value to $\frac{4}{7}$?

19. A woman has \$4000 invested, part of which brings her a return of 3% a year and the remainder a return of 7% a year. If her total annual return on these two investments is \$180, what is the amount invested at each rate?

Solution

Let x = the number of dollars invested at 3%

Then $4000 - x$ = the number of dollars invested at 7%

The income in each case equals the amount of money invested multiplied by the rate:

$$\text{income} = \text{rate} \times \text{amount invested}$$

The income from the 3% investment is $0.03x$ dollars per year. The income from the 7% investment is $0.07(4000 - x)$ dollars per year. Since the total income from both investments is \$180, we write

$$0.03x + 0.07(4000 - x) = 180$$

To eliminate the fractions, both sides of the equation are multiplied by 100. The equation is then easily solved.

20. A man has \$7800 invested, part of which brings him an income of 5% and the other an income of 7% per year. If the two incomes are equal, find the amount of money invested at each rate.

21. In Problem 20 find the amount of each investment if the 5% investment brings a return of \$30 more per year than the 7% investment.

22. A solution of 24 gal of alcohol and water contains 15% alcohol. How much pure alcohol should be added to the 24 gal to produce a new solution testing 25% alcohol? (Disregard the very slight solubility of alcohol in water.)

Solution

In a problem of this kind keep one important principle in mind: the total amount of alcohol in the final solution will be equal to the amount of alcohol in the original solution, plus the actual amount of alcohol added; that is,

$$\left\{\begin{matrix}\text{alcohol in}\\ \text{original solution}\end{matrix}\right\} + \left\{\begin{matrix}\text{amount of}\\ \text{alcohol added}\end{matrix}\right\} = \left\{\begin{matrix}\text{total alcohol in}\\ \text{final solution}\end{matrix}\right\}$$

Let x = the number of gallons of pure alcohol to be added

Then $24 + x$ = the total amount of the new solution

Now we investigate the amount of alcohol. The original solution already contains some alcohol: 15% of 24 gal = 3.6 gal. Since x gal of alcohol are to be added, the total amount of alcohol in the final solution will be $3.6 + x$ gal. We simply state that this amount is 25% of the entire final solution:

$$3.6 + x = 0.25(24 + x)$$

The value of x obtained in solving the equation is the amount of pure alcohol that must be added.

23. How much pure alcohol must be added to 50 gal of a 10% solution to make a new solution testing 24% alcohol?

24. How many quarts of an 80% solution must be added to 15 qt of a 12% solution to make a new solution testing 30%?

25. How much water should be added to 15 qt of a 30% solution to produce a new solution testing 20%?

26. How much pure silver must be added to 20 oz of an alloy now testing 15% silver to produce an alloy testing 35%?

27. How many cubic centimeters of water must be added to 10 cc of a solution of 80% carbolic acid to make a 1.5% solution?

28. A creamery has a vat containing 500 gal of milk testing 2.5% butterfat. How many gallons of 30% cream must be added to the milk to make the milk test 3.2% butterfat?

29. An automobile radiator contains 18 qt of a 10% solution of alcohol. How many quarts should be drained out and replaced with pure alcohol to make the solution test 25%?

30. After a 20% reduction in price, a coat was sold for $68. What was the price before the reduction?

31. One year the population census of a city was 162,000. If this was an increase of 8% over the population of the preceding year, what was the population of the preceding year and what was the increase?

32. After a lady's weight decreased 6%, it was 117.5 lb. What did she weigh before the decrease?

33. On a particular trip a man averages a speed of 45 mph. On the return trip his average speed is 36 mph and the trip requires 1 hr longer. How long was his trip (one way)?

34. A man has a motor boat that travels 12 mph. The current of a particular river has a rate of 4 mph. How far down the stream can he travel so that he can be back at his starting point in 6 hr?

35. A man and his wife walk up a moving escalator. The man walks twice as fast as his wife. When he arrives at the top, he has taken

28 steps. When she arrives at the top, she has taken 21 steps. How many steps are visible in the escalator at any one time?

Exercise 13-7

The principles used in solving fractional equations should be thoroughly understood and the procedure should be followed correctly. If you feel that you need more practice in applying the rules for solving fractional equations, you might solve some or all of the following exercises:

1. $\frac{3x}{5} = 6$

2. $10 - \frac{5x}{3} = 0$

3. $\frac{3x}{4} = 5 - \frac{x}{2}$

4. $\frac{x}{2} - 4 - \frac{x}{3} = 0$

5. $\frac{5x}{3} = \frac{3x}{2} - 2$

6. $5 = \frac{3x}{4} - \frac{2x}{3}$

7. $\frac{4}{x} - \frac{3}{2x} - \frac{5}{x^2} = 0$

8. $\frac{5}{x} - \frac{7}{3x} = \frac{2}{x^2}$

9. $\frac{4}{x^2} = \frac{7}{5x} - \frac{5}{3x}$

10. $\frac{2x}{3} - \frac{4x}{5} + \frac{3}{5} = 1$

11. $\frac{n}{3} - \frac{3}{2} = \frac{2n}{5} - \frac{11}{6}$

12. $\frac{15}{4} - \frac{y}{2} = \frac{3}{5} - \frac{3y}{20}$

13. $\frac{5}{2} - \frac{2v}{3} + \frac{v}{6} = 2$

14. $\frac{3a}{4} - \frac{2a}{5} = \frac{a}{3} + 2$

15. $\frac{x+2}{4} - \frac{3-x}{6} = 2 + \frac{x}{3}$

16. $\frac{2x-3}{5} + x - \frac{5x-2}{4} - 2 = 0$

17. $\frac{x}{3} - \frac{x+4}{4} = 4 - \frac{x-4}{2}$

18. $3 - \frac{n-4}{10} + \frac{n-3}{5} = \frac{n}{2}$

19. $\frac{3x-4}{6} + 2 - \frac{2x+6}{9} = -\frac{3}{2}$

20. $\frac{c+3}{5} - \frac{3c-2}{4} + \frac{c}{2} = 1$

21. $\frac{3d+2}{4} - \frac{d-3}{5} - \frac{2d-1}{3} = 2$

22. $\frac{e+3}{5} - \frac{2e+1}{7} = 1 - \frac{e-3}{7}$

23. $\frac{2t-3}{5} + 1 = \frac{3t+5}{4} - \frac{t-1}{2}$

24. $\frac{3}{z} + \frac{4}{5} - \frac{z+1}{z} = 0$

25. $\dfrac{2r + 3}{2r} + \dfrac{r + 8}{r} = 7$

26. $\dfrac{2u - 3}{4} - \dfrac{u + 2}{3u} - \dfrac{u - 1}{2} = 0$

27. $\dfrac{5w - 3}{w} - \dfrac{w - 1}{w + 3} - 4 = 0$

28. $\dfrac{3x + 4}{x + 3} - 2 - \dfrac{2x - 5}{2x} = 0$

29. $\dfrac{2x + 6}{x} - \dfrac{x}{x + 3} = 1$

30. $\dfrac{x + 5}{x + 6} - \dfrac{2}{3} - \dfrac{x - 1}{3x} = 0$

31. $\dfrac{x}{x - 1} + \dfrac{2x}{x + 6} - 3 = 0$

32. $\dfrac{3x - 4}{x - 2} = 5 - \dfrac{2x + 7}{x - 3}$

QUIZ ON CHAPTER 13, FRACTIONAL EQUATIONS. FORM A.

1. Solve this equation for x:

$$\frac{2x + 3}{5} - \frac{4x - 1}{6} - \frac{5x - 2}{30} = 2$$

2. Solve:

$$\frac{x - 2}{4} - \frac{2x + 5}{6} - \frac{3x - 2}{12} = 0$$

3. Solve:

$$0.2x - 6 = 0.3 - 0.05x$$

4. Solve the equation:

$$\frac{4x - 3}{x - 3} - \frac{2x + 3}{x + 4} - 2 = 0$$

5. Solve this literal equation for t:

$$A - 2 = \frac{BC - t}{t + 3}$$

6. A man has \$8400 invested, part of it at 5% and the remainder at 7% income per year. If his yearly income from both investments is the same, find the amount invested at each rate.

7. A man has \$9000 invested, part of it at 5% and the remainder at 8%. If his total income from these two investments is \$546 per year, how much has he invested at each rate?

8. How much pure alcohol should be added to 20 gal of a solution testing 16% so that the new solution will test 36%?

9. How much of an 8% solution and how much of a 60% solution should be mixed together for 50 qt of a solution testing 30%?
10. One boy can mow a lawn in 8 hours, and a second boy can mow the same lawn in 12 hours working alone. How long should it take them both working together to mow the same lawn working at the same rate as when working alone?

CHAPTER FOURTEEN Systems of Equations

14-1 INDETERMINATE EQUATIONS

We have solved equations such as $3x + 2 = 17$. This equation is true only on the condition that $x = 5$. Therefore, the value 5 is the root or solution of the equation. Remember that a solution is any number that makes an equation true.

Let us now consider a different kind of equation in which there are two unknowns. Suppose we have the equation

$$x + y = 11$$

This equation is true if $x = 8$ and $y = 3$. The pair of numbers may be written in parentheses as (8, 3), if we always understand that the first number, 8, is the x-value, and the second number, 3, is the y-value. However, x and y may have other values. The equation is also true if $x = 10$ and $y = 1$; that is, for the pair (10, 1). There are also other pairs of values that make the equation true. We list several such pairs in the table at the right. Any one of these pairs of numbers will satisfy the equation. If we accept fractions and negatives, then there is an infinite number of pairs of values for x and y that will satisfy this equation.

If $x =$	$y =$
8	3
10	1
4	7
11	0
14	−3
3.5	7.5
−2	13

To get a pair of values for this equation, we first give one letter any value we choose. Then we compute the value of the other letter by using the equation. For example, if we first say x shall be 4, then from the equation we find that y must be 7. Of course, if we wish, we may first assign a value to y and then compute the corresponding value of x.

In such an equation as $x + y = 11$, the numerical value of one letter cannot be determined without affecting or being affected by the other letter value. Such an equation is called an *indeterminate equation* (sometimes called a *Diophantine* equation). In an indeterminate equation, all we can do is to find pairs of values that make the equation true.

In an equation of this kind, the algebraic numbers, x and y, are called *variables* because they may take on more than one value. The values of x and y are dependent upon each other. It is customary to call one the *independent* variable and the other the *dependant* variable.

The independent variable is one that is first assigned a specific value. This is usually taken to be x. Then y becomes the dependent variable.

Let us consider another indeterminate equation:

$$x - y = 3$$

Here, again, we can find pairs of numbers that make the equation true. For example, if $x = 10$, $y = 7$. If $x = 15$, $y = 12$. Again, x and y are variables. The value of either is dependent upon the value of the other. The numerical value of one cannot be given without affecting or being affected by the value of the other. All we can do is to find pairs of values for x and y that constitute a solution of the equation. Again, this equation has an infinite number of solutions. Some of the pairs of values are listed in the table at the right.

If $x =$	$y =$
10	7
4	1
14	11
5.4	2.4
−1	−4
28	25
9.15	6.15

14-2 SYSTEMS OF EQUATIONS

Let us consider these two equations together:

$$x + y = 11 \qquad \text{and} \qquad x - y = 3$$

Suppose it is required that both equations be true for the same pair of values for x and y. Then we call these equations a *system* of equations.

If we try a pair of values for the first equation, such as $x = 9$ and $y = 2$, we find that this pair does not satisfy the second equation. On the other hand, if we take a pair that satisfies the second equation, such as $x = 10$ and $y = 7$, we find that this pair does not satisfy the first equation. After some trial and error, we find that one particular pair will make both equations true. This is the pair $x = 7$ and $y = 4$. The pair of numbers that satisfies both equations can be written in parentheses as $(7, 4)$, *in that order.* For this system of equations the order cannot be reversed. It is an *ordered pair.*

Our problem in solving a system of equations is to find a pair of values for the two unknowns that will make both equations true. Such a pair of numbers is called *a solution* of the system.

In this chapter we consider equations containing terms in only the first power of x and y. Such equations are called *first degree* equations or *linear* equations. At this time we shall not be concerned with equations containing terms in x^2 or y^2 or terms with higher powers of the variables.

Note. A system of equations is sometimes called a set of "simultaneous" equations. The name *simultaneous* is not a good term to use in this connection since the word *simultaneous* means "being at the same time." It is *not* the idea of *time* that is involved here. We do not mean that the equations must necessarily be true only at some particular time, such as today, tomorrow, or at 10.30 A.M. What we mean is that the *same pair of numbers* for x and y will make the equation true *at any time*. It is therefore better to call the two equations a *system*.

14-3 SOLVING A SYSTEM OF EQUATIONS BY ADDITION OR SUBTRACTION

In the foregoing system we found a solution by inspection: $x = 7$, $y = 4$. However, we need a systematic method for solving systems of equations. There are several methods that can be used for this purpose. One of the most convenient is the method called *addition or subtraction.*

Let us consider again the two equations $x + y = 11$ and $x - y = 3$. If we consider an equation to be a balanced scale, we can show a scale for each equation. For the equation $x + y = 11$, we have the scale in Fig. 14-1*a*. For the equation $x - y = 3$, we have the second scale (Fig. 14-1*b*).

If we add the two left-hand pans of the scales and also add the two right-hand pans, we get a new scale that may be represented by the third diagram (Fig. 14-1*c*). This scale will also balance because of the addition axiom.

You will note on the combined scale that the entire quantity on the left results in $2x$, whereas the total quantity on the right is 14. So we see that x must have a value of 7.

These scales may be shown in equation form. When the two equations are added, the y terms drop out.

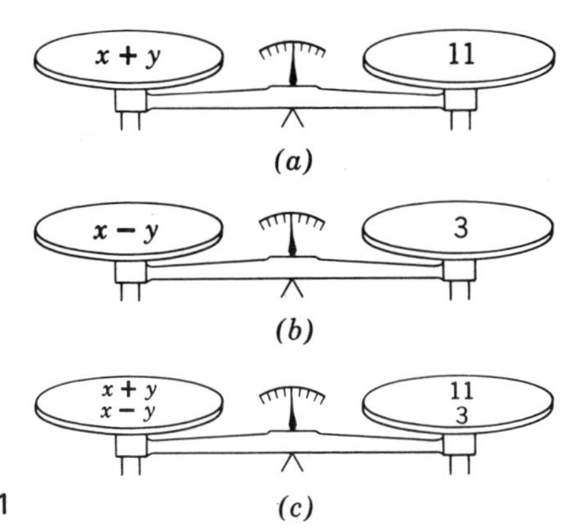

Figure 14-1

First scale:	$x + y = 11$
Second scale:	$x - y = 3$
Adding the two equations, we get	$2x = 14$
Solving for x (division axiom),	$x = 7$
Now we find the value of y by substituting 7 for x in either equation:	$y = 4$

In this example we might have subtracted the two equations instead of adding them. If we place one equation below the other, as in addition, and then subtract one equation from the other, we get $2y = 8$.

	$x + y = 11$
	$x - y = 3$
Subtracting,	$2y = 8$
Solving for y,	$y = 4$
Then	$x = 7$

The answers are the same whether we add or subtract the two equations. Notice that in either case one letter is *eliminated* in the process. The purpose of adding or subtracting the equations is to *eliminate* one letter so that the resulting equation has only one unknown.

In solving a system of equations by the method of addition or subtraction, we sometimes run into some complications. Consider these two equations:

	$5x - 3y = 7$
	$4x + y = 9$
Adding,	$9x - 2y = 16$

If we add the two equations, we get $9x - 2y = 16$. This is a perfectly legal operation in mathematics, but it does not lead to a solution. If we subtract the two equations, we are no better off. The new equation still contains both x and y terms. By adding or subtracting the equations, we hope to eliminate one of the letters. This can happen only if the two *coefficients of one letter have the same numerical value.*

In the foregoing set of equations we can make the two coefficients of y numerically equal by multiplying the second equation by 3 (Axiom 3). The second equation then becomes $12x + 3y = 27$. This result is written below or above the first equation. The two equations are then added.

First equation:	$5x - 3y = 7$
Multiplying the second equation by 3,	$12x + 3y = 27$
Adding the two equations,	$17x = 34$
Solving for x,	$x = 2$
Then	$y = 1$

After the value of x has been found, this value is substituted for x in any one of the equations. From the resulting equation, we compute the value of y.

To check the set of values for x and y, the values must be checked in *both* of the original equations. It is entirely possible to get some pair of values for x and y that will satisfy one equation but not the other. To be a correct solution, the set of answers must satisfy *both* equations.

It should be remembered that checking a pair of values must *always* be done in the *original* equations.

In solving a system of equations by the method of addition or subtraction, it is often necessary to multiply *each* of the given equations by some factor, as in the example shown here.

$$3x + 2y = 7$$
$$4x - 5y = 13$$

Multiplying the first equation by 5, $\quad 15x + 10y = 35$

Multiplying the second equation by 2, $\quad 8x - 10y = 26$

Adding the two equations, $\quad 23x = 61$

$$x = \tfrac{61}{23}$$

To find the value of y in the foregoing equations, we may substitute the known value of x in any one of the equations. However, when such substitution is difficult because of fractional values, we may begin again with the original equations and equalize the coefficients of a different variable.

Multiplying the first equation by 4, $\quad 12x + 8y = 28$

Multiplying the second equation by 3, $\quad 12x - 15y = 39$

Subtracting the equations, $\quad +23y = -11$

Solving for y, $\quad y = -\tfrac{11}{23}$

Steps in solving systems of equations by addition or subtraction.

1. *Multiply one or both of the given equations, if necessary, by some factor, or factors, that will make the coefficients of one variable numerically equal.*
2. *Eliminate this variable by addition or subtraction.*
3. *Solve the resulting equation for one variable.*
4. *Find the value of the other variable by substituting the known value of one variable in one of the equations. (The second variable may be found in the same way as the first.)*

Exercise 14-1

Solve by addition or subtraction:

1. $3x + 2y = 4$, $x - 4y = 6$

2. $2x - 3y = 12$, $5x + y = 13$

3. $5x - 6y = 7$, $4x + 3y = -10$

4. $5x + 3y = 1$, $4y + 3x = -6$

5. $4x + 5y = -8$, $2y + 3x = 1$

6. $7x - 4y = -19$, $5y + 3x = 12$

7. $3r = 10 + 5s$, $3s + 7r = 14$

8. $3c = 6 + 5d$, $8d + 5c = 4$

9. $4n + 8 = 5m$, $4m + 7n = 3$

10. $3c + 4d = 5$
$5d = 4 - 4c$

11. $3h = 2 + 4k$
$7k = 5h + 5$

12. $7r = 2 + 6s$
$7s = 8r - 3$

13. $3I_1 - 10I_2 = 15$
$7I_1 - 15I_2 = 35$

14. $6E_1 - 5E_2 = 3$
$9E_2 - 8E_1 = -11$

15. $7R_1 - 12R_2 = 8$
$16R_2 = 5R_1 - 28$

16. $5x + 8y = 10$
$24x = 20 - 15y$

17. $20x = 20 - 3y$
$12x = 30 + 5y$

18. $5y = 10 - 6x$
$15x - 18y = 40$

19. $2.5I + 1.8i = 7$
$3.5I - 2.7i = 3$

20. $3.2R + 1.5r = 9$
$2.4R - 2.5r = 6$

21. $5.6E + 2.7e = 260$
$3.2E + 4.5e = 240$

22. $bx - ky = m$
$cx + ky = n$

23. $mx + ny = b$
$nx - my = c$

24. $ax + by = e$
$cx + dy = f$

14-4 SOLVING SYSTEMS OF EQUATIONS BY SUBSTITUTION

Substitution is another method for solving systems of equations. This method can sometimes be used to advantage, although it is perhaps not so practical as addition or subtraction. One advantage of this method is that it can often be used to solve equations of *higher degree,* as we shall see later.

Steps in solving by the substitution method.

1. *Solve one of the equations immediately for one letter in terms of the other letter.*
2. *Substitute this value for its equal in the other equation.*
3. *Solve the resulting equation for the value of one variable.*
4. *Find the value of the second variable as usual.*

Example 1

Solve by substitution:

$$\begin{aligned} 3x - 4y &= 5 \\ 5x + y &= 16 \end{aligned}$$

Solution

Solving the second equation for y, $y = 16 - 5x$
Substitute $(16 - 5x)$ for y in the *other* equation $3x - 4(16 - 5x) = 5$
Solving the resulting equation for x: $x = 3$
Then $y = 1$
To find the value of y, we may use the equation, $y = 16 - 5x$.
Note that in the second step, one variable is *eliminated.*

Example 2

Solve by substitution:

$$\begin{aligned} 4x - 3y &= 11 \\ 5x + 4y &= 6 \end{aligned}$$

Solution

Solve one of the equations for one letter in terms of the other. Suppose we solve the first for x:

$$4x = 11 + 3y$$

$$x = \frac{11 + 3y}{4}$$

Here, x is equal to a fraction:

Substitute this fraction for x in the *other* equation:

$$5\left(\frac{11 + 3y}{4}\right) + 4y = 6$$

The resulting equation is now solved for y. Then the value of x is found.

$$y = -1$$
$$x = 2$$

Exercise 14-2

Solve by substitution:

1. $5x + 3y = 9$, $4x + y = 10$ **2.** $4x + 7y = 18$, $3x + y = 5$ **3.** $5x + y = 9$, $4x + 5y = 3$

4. $4a - 5b = 22$, $3a + b = 7$ **5.** $5c - 3d = -23$, $4c + d = 2$ **6.** $3h - 5k = 21$, $5h + k = 7$

7. $4x - 3y = 19$, $5x - y = 4$ **8.** $5x - 4y = 31$, $3x - y = 5$ **9.** $7x - 3y = 26$, $4x - y = 6$

10. $7x - 4y = 6$, $x + 3y = -4$ **11.** $5x - 8y = 6$, $x + 5y = -3$ **12.** $8x - 3y = 40$, $x + 4y = -5$

13. $4x + 5y = 17$, $3x + 2y = 4$ **14.** $5x - 3y = 21$, $6x + 5y = 8$ **15.** $7x - 4y = 9$, $3x - 5y = 17$

16. $5c - 9d = 12$, $7c + 5d = 8$ **17.** $7m - 8n = 9$, $4m - 5n = 7$ **18.** $3h - 5k = 7$, $5h - 7k = 8$

19. $7x - 8y = 9$, $6x - 5y = 7$ **20.** $4x - 3y = 11$, $5x - 2y = 9$ **21.** $7x + 6y = 10$, $4x + 5y = 2$

14-5 SOLVING SYSTEMS OF EQUATIONS BY COMPARISON

In the method of *comparison* we solve both equations for the same variable and then equate the resulting expressions.

Example

Solve by comparison:

$$4x - 3y = 11$$
$$5x + 4y = 6$$

Solution

Solve both equations for one letter. Suppose we solve both for x.

From the first equation, we get

$$x = \frac{11 + 3y}{4}$$

From the second equation, we get $x = \frac{6-4y}{5}$

Since x must have the same value in both equations, we set one fraction equal to the other: $\frac{11 + 3y}{4} = \frac{6 - 4y}{5}$

Solving the resulting *fractional* equation for y, we get $y = -1$
Then $x = 2$

Steps in solving by comparison method.

1. *Solve each of the equations for the same letter.*
2. *Set the two values obtained equal to each other.*
3. *Solve the resulting equation for one unknown.*
4. *Find the remaining unknown by the usual method.*

Exercise 14-3

Solve the first six problems by the comparison method:

1. $3x - 2y = 4$, $4x - 5y = 7$
2. $4x - 3y = 8$, $3x - 5y = 8$
3. $5x - 6y = 2$, $3x - 7y = 9$
4. $7x - 2y = 6$, $4x + 3y = 2$
5. $2x + 3y = -3$, $3x - 7y = 10$
6. $4x + 5y = -2$, $5x - 2y = 3$

Solve the following equations by any convenient method:

7. $\frac{x-3}{2} - \frac{y-2}{3} = 7$, $\frac{x-5}{5} - \frac{y+3}{4} = 0$

8. $\frac{x+2}{3} + \frac{y-4}{5} = -1$, $\frac{5-x}{2} + \frac{2-y}{7} = 6$

9. $\frac{2x-3}{4} - \frac{y-4}{5} = 1$, $\frac{x+2}{3} - \frac{2y-5}{7} = 3$

10. $1 - \frac{x+3}{3} - \frac{2y-1}{5} = 0$, $3 - \frac{5-2x}{4} - \frac{y+7}{6} = 0$

11. $4 - \frac{5-x}{2} - \frac{3+2y}{3} = 0$, $1 - \frac{3x+7}{5} - \frac{2-y}{4} = 0$

12. $\frac{2x-1}{4} - \frac{5-2y}{3} = 0$, $2 - \frac{7-2x}{6} - \frac{2y+5}{9} = 0$

14-6 SOLVING STATED PROBLEMS BY SYSTEMS

In solving stated problems by the use of two unknowns, we use a different letter for each unknown. For example, we may use x to represent one unknown and y to represent a second unknown.

From the information given in the problem, we write equations relating the unknowns. If we use two unknowns, we must write two equations. To be certain that the equations are independent, we make use of different information for each equation.

Example 1

A collection of 42 coins consisting of nickels and quarters is worth \$4.70. Find the number of each kind of coins in the collection.

Solution

We have already worked such a problem using only one letter. However, now let us use the following:

let x = the number of nickels

let y = the number of quarters

Then, for the number of coins, we have the equation:

$$x + y = 42$$

Now we state the value of each in cents. The x nickels are worth $5x$ cents. The y quarters are worth $25y$ cents. The total value of the coins is 470 cents. Then, for the total value, we have the equation

$$5x + 25y = 470$$

Now we solve the two equations as a system:

$$\begin{cases} x + y = 42 \\ 5x + 25y = 470 \end{cases}$$

Multiplying the first equation by 5, we get $5x + 5y = 210$

By subtracting one equation from the other, we get $20y = 260$

Solving for y, $y = 13$, number of quarters

Then we have the solution: 29 nickels; 13 quarters.
By checking, we shall find that the total value is \$4.70.

Example 2

A man has \$6000 invested, part of it at a yearly rate of 5% and the rest at 7%. If the total yearly income from these two investments is \$344, find the amount invested at each rate.

Solution

Let x = the number of dollars invested at 5%
let y = the number of dollars at 7%
The yearly income on the 5% investment is $0.05x$. The yearly income on the 7% investment is $0.07y$. Then we write the two equations:

For the total investment: $x + y = 6000$
For the total yearly income: $0.05x + 0.07y = 344$

To solve the system, we multiply the second equation by 100, and the first equation by 5, and have

$$\begin{aligned} 5x + 7y &= 34400 \\ 5x + 5y &= 30000 \end{aligned}$$

Subtracting, $2y = 4400$

Solving for y, $y = 2200$; then $x = 3800$.

Now we have the solution: \$3800 at 5%; \$2200 at 7%.

Example 3

In Example 2, if the two incomes are equal, find the amount invested at each rate.

Solution

For the two incomes, we have the equation: $0.05x = 0.07y$.
This equation can be written: $0.05x - 0.07y = 0$
Multiplying by 100, we get $5x - 7y = 0$
For the total investment, we have the equation, $x + y = 6000$
Multiplying this equation by 5, we get $5x + 5y = 30000$
Now we subtract the first equation, $5x - 7y = 0$
Subtracting, we get $12y = 30000$
Solving for y, $y = 2500$, then $x = 3500$
Now we have the solution: \$3500 at 5%; \$2500 at 7%.
Checking, we shall find that the two incomes are equal.

Example 4

How much of a 15% solution and how much of a 65% solution of alcohol should be mixed together for 20 quarts of a solution testing 35%?

Solution

Let x = number of quarts of the 15% solution
let y = number of quarts of the 65% solution

First, we note that the amount of alcohol in the new solution must be 35% of 20 quarts, or 7 quarts, (alcohol in new solution)

Now we write the two equations:

For the total amount of the new solution: $x + y = 20$
For the amount of alcohol in the new solution: $0.15x + 0.65y = 7$
Multiplying the second equation by 100, $15x + 65y = 700$
Multiplying the first equation by 15, $15x + 15y = 300$
Subtracting, $50y = 400$
Solving for y, $y = 8$; then $x = 12$
Now we have the solution: 12 qt of the 15% solution: 8 qt of the 65% solution.

Exercise 14-4

Solve the following problems using two letters.

1. A collection of 73 coins, consisting of nickels and quarters, is worth $12.85. Find the number of coins of each kind in the collection.

2. At an entertainment, children's tickets cost 50 cents each, and adults' tickets cost $1.25 each. If the total amount received was $112 for a total of 173 tickets, find the number of tickets of each kind sold.

3. A man has $8400 invested, part of it at 5% and the remainder at 7%. If his total annual income from these two investments was $476, find the amount invested at each rate.

4. A woman has $9100 invested, part at 5% and the rest at 8%. Find the amount invested at each rate if the return on the two investments is the same.

5. A man took a trip of 730 miles, part of the way by car at an average of 45 mph, and the rest of the way by train at an average of 65 mph. If his total travel time was 12 hours, how many hours and how many miles did he travel by each method of transportation?

6. A motorboat travels downstream for a distance of 45 miles in 3 hours and makes the return trip in 4.5 hours. Find the rate of the current in the river and the rate of motorboat in still water. (*Hint*: For the rate downstream the two rates are added; for the rate upstream the rate of the river current is subtracted from the rate of the motorboat in still water.)

7. A plane makes a trip of 640 miles in 2.5 hours with the advantage of a tail wind. On the return trip the plane requires 3 hours flying time against the wind. Find the speed of the wind and the speed of the plane in still air.

8. How much of a 15% solution and how much of a 40% solution of alcohol should be mixed together for 50 gal of a solution testing 25%?

9. A chemist has one solution testing 7.5% and another solution testing 82.5% of a certain chemical. How much of each should be mixed together for 24 qt of a new solution testing 15%?

10. In a certain electric circuit, two resistors, R_1 and R_2, have a total resistance of 71.5 ohms, when connected in series. Find the resistance of the two resistors if R_1 has a resistance of 25 ohms less than R_2.

11. Two voltages, E_1 and E_2, have a total voltage of 85 volts when connected in series with the same polarity. If the polarity of E_2 is reversed, the total voltage is 20 volts. Find E_1 and E_2. (If one polarity is reversed, the voltages are subtracted.)

14-7 SYSTEMS OF EQUATIONS IN THREE OR MORE UNKNOWNS

Three equations in three unknowns may be solved by methods similar to those used for two equations. This is also true for equations containing more than three unknowns. One of the best methods to use for such equations is addition or subtraction.

If we have three unknowns (that is, three different letters) in a set of equations, we must have three independent equations. In general, the number of independent equations must be the same as the number of unknowns.

Note. Checking answers by substitution of values in the original equations becomes difficult when answers are complicated fractions or decimals. If unending decimals are involved, checking will be only approximate. The best check in such problems is to go over each step carefully to see that no error has been made in working out the solution.

In most actual work in stated problems, such as those encountered in technology, the answers turn out to be fractions instead of integers. In such problems it is best to leave the answers in the form of common fractions instead of decimals until all answers have been found. Then they may be changed to decimal form if it is so desired. Then unending decimals are usually rounded off to three or four significant digits.

In solving a system of equations containing several unknowns, we try to eliminate one unknown by addition or subtraction. If the coefficients of one letter are *numerically* equal in two equations, then that letter may be eliminated by adding or subtracting the two equations.

To solve a system of three equations in three unknowns, we have the following steps:

1. *Eliminate* one *letter using a set of two equations. The result is a new equation containing, at most, two letters.*
2. *Eliminate the* same *letter using another set or combination of the given equations. The result is another new equation containing, at most, two letters.*
3. *Solve the two new equations obtained in Steps* 1 *and* 2.
4. *When the value of one letter has been found, work backward through the equations to find the other values.*

Example 1

Solve the following system of three equations. Call the equations A, B, and C, for convenience.

$$\begin{aligned} &(A)\ 2x + 5y + 3z = 7 \\ &(B)\ \ x - 2y + 5z = 15 \\ &(C)\ 3x - y - z = 8 \end{aligned}$$

Solution Suppose we decide to eliminate z first.

Then we write equation A: (A) $2x + 5y + 3z = 7$

Multiply equation C by 3 to make the z coefficients numerically equal: $9x - 3y - 3z = 24$

Adding, we get a new equation we might call D. (D) $11x + 2y = 31$

This new equation contains only two letters. The z term has been eliminated.

Now, we eliminate the same letter z, using another combination of equations. Suppose we use equations B and C.

We write equation B as it is: $x - 2y + 5z = 15$

Next, we multiply equation C by 5: $15x - 5y - 5z = 40$

Adding these two equations, we get another equation we may call E: (E) $16x - 7y = 55$

We now have two new equations, D and E, from which the z has been eliminated.

$$\begin{aligned} &(D)\ 11x + 2y = 31 \\ &(E)\ 16x - 7y = 55 \end{aligned}$$

Solving these two equations just as we solve any system of two equations, we get values for x and y:

$$x = 3 \quad \text{and} \quad y = -1$$

After we have found two of the unknowns, we go back to any previous equations in three unknowns and substitute the known values. Let us substitute the known values in equation A to find z.

$$6 - 5 + 3z = 7$$

Solving for z, $z = 2$

We now have all the values of the unknowns. Checking should always be done in the original equations.

Example 2

Solve the following system of equations:

$$\begin{aligned} &(A)\ 2x + 3y - 4z = 15 \\ &(B)\ 3x \qquad + 7z = -6 \\ &(C)\ 5x - y + 3z = 1 \end{aligned}$$

Solution

Since the second equation has no y term, we eliminate y, using A and C:

We write equation A:	$2x + 3y - 4z = 15$
Multiplying equation C by 3,	$15x - 3y + 9z = 3$
Adding, we get a new equation:	(D) $17x + 5z = 18$
We now solve this equation with B:	(B) $3x + 7z = -6$

The complete solution is left as an exercise for the student. The answers to the problem are 2, $-\frac{3}{2}$, and $\frac{3}{2}$, although these answers are not necessarily in the same order as the unknowns.

Exercise 14-5 Solve the following systems of equations. The answers are given for some, although not necessarily in the same order as the unknowns.

1.
$$\begin{aligned} 2x + 3y - z &= 11 \\ 3x - y + 2z &= 4 \\ 4x + 2y + 3z &= 8 \end{aligned}$$
(Answers: 1, 3 −2)

2.
$$\begin{aligned} x - 2y + 4z &= -4 \\ 3x + 4y - 5z &= 25 \\ 5x - 3y + 2z &= 12 \end{aligned}$$
(Answers: −1, 2, 4)

3.
$$\begin{aligned} 3x + 2y + 4z &= 9 \\ 4x + 3y - 2z &= 6 \\ 5x + 4y - 3z &= 8 \end{aligned}$$
(Answers: 4, 1 −1)

4.
$$\begin{aligned} 3r + 4s - 2t &= -2 \\ 5r + 7t &= -1 \\ 2r - 3s &= 23 \end{aligned}$$
(Answers: 4, −5, −3)

5.
$$\begin{aligned} 3a - 3b + 4c &= 0 \\ 3b - 5c + 2a &= 13 \\ 5a - 3c - 4b &= 27 \end{aligned}$$
(Answers: 2, −3, −2)

6.
$$\begin{aligned} 3x - 2y + 4z &= 8 \\ y - z + 5x &= 5 \\ 2z + 2x - 3y &= 4 \end{aligned}$$
(Answers: fractions)

7.
$$\begin{aligned} 3x + 4y - 5z &= 16 \\ 2x + 4z + 3y &= 8 \\ x - 2y - 3z &= -4 \end{aligned}$$
(Some halves)

8.
$$\begin{aligned} 5x - 2y + 3z &= 12 \\ 2z - 4x - 3y &= -5 \\ 3y - 4z - 2x &= -7 \end{aligned}$$
(Two answers: −1, 2)

9.
$$\begin{aligned} c + 3d - 4e &= 6 \\ 4e - 3c - 3d &= -8 \\ 2d - e - 5c &= -9 \end{aligned}$$
(Fractions)

10.
$$\begin{aligned} 5x - 3y + 4z &= 10 \\ 2y - 4x - 5z &= 10 \\ 3z - 2x - 4y &= 0 \end{aligned}$$

11.
$$\begin{aligned} 2w - 3x + 4y - 2z &= 5 \\ 3w + x - 3y + 4z &= 8 \\ w - 2x + y - 3z &= 7 \\ 4w + 3x - 2y + 5z &= 11 \end{aligned}$$
(Answers: 1, −2, 4, −1)

12.
$$\begin{aligned} 3r + 2s + 4t - 5u &= 8 \\ 2r - 3s - 2t - 3u &= 4 \\ 4r - 5s + 3t - 2u &= 22 \\ 5r + 4s + 5t + 4u &= -6 \end{aligned}$$
(Three answers: 3, −3, −1)

13.
$$\begin{aligned} 3a + 2b - 3c + d &= 14 \\ 2a - 3b + 4c &= -3 \\ 3d - 2c + 4a &= 12 \\ 5b - 5c + 4d &= 2 \end{aligned}$$

14.
$$\begin{aligned} 3w - 2x - z &= 4 \\ 2w - 3y + 3z &= 5 \\ x + y - 2z &= 0 \\ 4w + 3x + 2y &= 6 \end{aligned}$$

15.
$$\begin{aligned} 3w - 4z - 2x &= 4 \\ 3z - 3y + 2w &= 6 \\ 2z - 4y - 5x &= -9 \\ 5y - 3x + 4w &= 11 \end{aligned}$$

16.
$$\begin{aligned} 2a - 4c + d + 3b &= 9 \\ c - 3d + 3a - 2b &= 7 \\ 4b + 5a - 4d - 3c &= 8 \\ 2d - c + 3b - a &= -6 \end{aligned}$$

17.
$$\begin{aligned} x - 3y + 4z - 2t &= 7 \\ 2x + 5y - 3z - 5t &= -1 \\ 3x - y + z + t &= 9 \\ x + y - 2z - 3t &= 0 \end{aligned}$$

18.
$$\begin{aligned} 3A + 2B - C - 3D &= 5 \\ A - 3B + 2C - 5D &= 7 \\ 2A - B + 3C + D &= 6 \\ A + 3B - 2C - D &= -2 \end{aligned}$$

19.
$$\begin{aligned} a - 3b + c + 2d - e &= 5 \\ a + b - 2c - d - 3e &= 4 \\ 2a - b - 3c + 2d - 2e &= 3 \\ 3a + 2b - c + 3d + e &= 9 \\ 2a + 4b + c - d + e &= 0 \end{aligned}$$

20.
$$\begin{aligned} u + v + 2x + 3y + 2z &= 5 \\ u - v + 3x - 2y + z &= 16 \\ 2u - 3v - x + y - 2z &= 4 \\ 2u + 3v - 2x - y + z &= -16 \\ 3u - 2v + x - 3y - 3z &= 13 \end{aligned}$$
(Answers are integers)

21.
$$\begin{aligned} 3R_1 - 5R_2 + 7R_3 &= 20 \\ 8R_2 - 3R_3 &= 17 \\ 6R_3 - 5R_1 &= 13 \end{aligned}$$

22.
$$\begin{aligned} 21.5I_1 + 30I_2 - 15.4I_3 &= 40 \\ 13.4I_1 - 22I_2 + 31.2I_3 &= 40 \\ 15.2I_1 - 31.5I_2 - 24.21I_3 &= 0 \end{aligned}$$

14-8 PROBLEM SOLVING BY USING THREE OR MORE LETTERS

In solving verbal problems involving three or more unknowns, we can use a different letter to represent each unknown. Next, from the information given in the problem, we write equations relating the unknowns. For three unknowns we must have three independent equations. The number of independent equations must be as many as the number of letters used.

Example 1

A collection of 46 coins, consisting of nickels, dimes, and quarters, is worth \$6.20. If the number of nickels is 4 more than 3 times the number of dimes, find the number of coins of each kind in the collection.

Solution

Let x = number of nickels
let y = number of dimes
let z = number of quarters

For the value of the coins of each kind, we have:
the value of x nickels is $5x$ cents; of the y dimes, $10y$ cents; and of the z quarters, $25z$ cents. Since the number of nickels is 4 more than 3 times the number of dimes, we can say:

$$x = 3y + 4$$

Now we write three equations:

For the number of coins: $x + y + z = 46$

For the value, $5x + 10y + 25z = 620$

We also write the equation: $x - 3y = 4$

Solving, we get the answer: 22 nickels; 6 dimes; 18 quarters.

Exercise 14-6

Solve the following problems by the use of three letters in each.

1. A collection of 93 coins consisting of nickels, dimes, and quarters is worth \$9.15. The number of dimes is 4 more than the number of nickels. Find the number of coins of each denomination in the collection.

2. A woman buys a total of 44 stamps, some 3-cent, some 9-cent, and the rest 15-cent stamps. The total cost is \$4.92. If the number of 9-cent stamps is twice the number of 3-cent stamps, how many of each denomination does she buy?

3. A man has \$12,000 invested in three investments. On the first, he receives 4% a year, on the second, 7% a year, and on the third he receives 9%. The total yearly income from these investments is \$687. If the first two investments had been reversed, he would have received \$780 a year. Find the amount invested at each rate.

4. At a ball game, 1545 tickets were sold, some at \$1.20 each, others at \$1.80 each, and reserved seats at \$2.50 each. The total amount received for all the tickets was \$2715. If the number of \$1.80 tickets was 125 more than the number of \$1.20 tickets, how many tickets of each kind were sold?

5. Three solutions, testing 4%, 28%, and 60%, respectively, of a certain chemical, are to be mixed for 1000 cc of a new solution to test 28%. If the new solution is to contain twice as much of the 4% solution as of the 28% solution, how much of each should be mixed together?

6. Three resistors, R_1, R_2, and R_3, in series have a total resistance of 2600 ohms. If the resistance of R_1 is 420 ohms more than 3 times R_2, and the resistance of R_3 is 480 ohms more than R_2, find the rating of each resistor.

7. A man takes a trip of 400 miles in a total travel time of 12 hours. He travels the first part of the trip by motorboat at 18 mph, then transfers to a car and travels at 40 mph. The last part of the trip is by train at an average of 55 mph. If the trip by motorboat is 2 hours longer than the trip by car, how far does he travel by each mode of travel?

14-9 EQUATIONS DEPENDENT OR DERIVED; INDEPENDENT

Let us consider the following equation in two unknowns:

$$3x - 5y = 11$$

If we multiply both sides of this equation by some quantity, say 4, we get a new equation:

$$12x - 20y = 44$$

The new equation is called a *derived equation* because it is derived from the first by the use of one of the axioms. If the second equation is given, we can obtain the first by dividing both sides of the equation by 4. In either case, we can say that one of these two equations is derived from the other. The two equations may be called *dependent* equations.

If we have two given equations and if it is possible to obtain one of them from the other simply by use of one or more of the four axioms, then we can say the equations are *dependent.*

Two dependent equations may be reduced to one. Therefore, any set of values for x and y that satisfies one equation will also satisfy the other. For instance, the values $x = 7$ and $y = 2$ will satisfy both of the foregoing equations.

Let us see what happens when we attempt to solve a set of dependent equations. Try to solve the following:

	$3x - 2y = 7$
	$7x - 21 = 6y - 2x$
Transposing in the second equation,	$7x + 2x - 6y = 21$
Combining,	$9x - 6y = 21$
Multiplying the first equation by 3,	$9x - 6y = 21$
Subtracting,	$0 = 0$

Such an answer indicates a pair of dependent or derived equations. A set of dependent equations has an infinite number of solutions, since any set of values that satisfies one equation will also satisfy the other. The result we get, $0 = 0$, is true, but it offers no help in finding the answer to a practical problem.

Now, let us consider another set of equations. Let us see what happens when we attempt to solve this system:

$$2x + 5y = 21$$
$$4x + 10y = 37$$

Multiplying the first equation by 2, $\quad 4x + 10y = 42$

Subtracting from the second equation, $\quad 0 = -5$

Since the answer is impossible, the set of equations is called a pair of *inconsistent* equations. A set of inconsistent equations has *no* common solution. There is no set of values for the letters that will satisfy both equations.

In order to solve a system of equations, the equations must be *independent* and they must *not be inconsistent.* Almost all equations encountered in practical work are independent and can be solved. If two dependent or derived equations appear in practical work, then the problem must be analyzed more carefully to determine another *independent* equation or the problem cannot be solved. If two inconsistent equations appear in the course of the solution of a practical problem, the two describe an impossible condition, and the problem must be further analyzed for errors in the setting up of the equations.

To summarize:

1. Dependent equations have an infinite number of pairs of values for x and y that will satisfy both equations.
2. Inconsistent equations have no common solution.
3. Independent equations have a particular pair, or pairs, of values that satisfy the equations. In the case of linear equations there is one and only one pair of values for the unknowns that will satisfy the independent equations.

Exercise 14-7

Try to solve the following sets of equations. Tell whether each set is inconsistent, whether the equations are dependent, or whether they are independent. Solve the systems containing independent equations.

1. $3x - 5y = 7$
$6x - 4y = 13$

2. $4x - 5y = 8$
$10y + 15 = 8x$

3. $2x + 7y = 6$
$3x - 5y = 9$

4. $3x + 4y = 13$
$6x = 26 - 8y$

5. $x - 3y = 5$
$6y - 2x = 7$

6. $10x - 3y = 24$
$40 + 5y = 7x$

7. $3a - 6b = 21$
$2a - 14 = 4b$

8. $4y + 3x = -3$
$5x - 3y = 14$

9. $4y = 12 - 7x$
$3x = 6 - 2y$

10. $3x = 16 + 2y$
$4y = 6x - 32$

11. $5x - y = 13$
$2y + 18 = 10x$

12. $2x + 15 = 4y$
$6x + 8y = 30$

13.
$$\begin{aligned} 2x - 5y + 3z &= 13 \\ 5z + 4x - 10y &= 17 \\ 2y - 4z - 3x &= -4 \end{aligned}$$

14.
$$\begin{aligned} x - 4z + 3y &= 11 \\ 6y - 5z + 2x &= 7 \\ 9y + 6 + z &= -3x \end{aligned}$$

15.
$$\begin{aligned} 4x - 3z + 5 &= 2y \\ 2z - 5y - 2x &= 7 \\ y + 2x - 3z &= 0 \end{aligned}$$

16.
$$\begin{aligned} x + 3y - 2z &= 3 \\ y - 4z + 2x &= 7 \\ 3x + 4y - z &= 8 \end{aligned}$$

14-10 DETERMINANTS

We have already solved systems of equations such as the following:

$$ax + by = e \qquad (1)$$

$$cx + dy = f \qquad (2)$$

This system can be solved by any one of several methods, one of which is subtraction. We shall see that the system can also be solved conveniently by the use of *determinants.*

Before considering the method of determinants, let us first see just what is done in solving the system by subtraction. In these equations the letters a, b, c, d, e, and f are constants. To solve for x, we first equalize the y-coefficients and then subtract one equation from the other. We multiply equation (1) by d, and equation (2) by b, so that the coefficients of y become equal. Then we subtract and get

$$\begin{aligned} adx + bdy &= de \\ bcx + bdy &= bf \end{aligned}$$

Subtracting, $$adx - bcx = de - bf$$

Solving for x,

$$x = \frac{de - bf}{ad - bc}$$

In a similar manner, equalizing x-coefficients, we get

$$y = \frac{af - ce}{ad - bc}$$

Note that the denominators are the same for both values.

Let us see now what is meant by a determinant. Let us begin with an array of numbers called a *matrix.* A matrix is often denoted by enclosing the numbers, called elements, in a pair of braces or parentheses:

$$\begin{Bmatrix} 5 & 4 \\ 2 & 3 \end{Bmatrix} \qquad \begin{pmatrix} 5 & 4 \\ 2 & 3 \end{pmatrix}$$

This symbol for a matrix does *not* indicate any operation upon the elements.

Now suppose we wish to indicate a particular operation upon the numbers. We use a pair of vertical lines, one on each side of the array. The arrangement then denotes a *determinant:*

$$\begin{vmatrix} 5 & 4 \\ 2 & 3 \end{vmatrix}$$

The *determinant* is a symbol that indicates a specific rule for operating upon the numbers in the array. Now let us see how this operation must be done.

Note that this determinant has two *rows,* the first row consisting of the elements 5 and 4; and the second row, of the elements 2 and 3. It has two *columns,* the first consisting of the numbers 5 and 2, and the second column, of 4 and 3. The *principal diagonal* is the diagonal extending from *upper left* to *lower right.* In this example the principal diagonal consists of the numbers 5 and 3.

A determinant is expanded according to the following rule, as shown by the arrows in this array:

$$\begin{vmatrix} 5 & 4 \\ 2 & 3 \end{vmatrix}$$

First, we find the product of the elements along the *principal* diagonal. In this example, this product is (5)(3) = 15. Next, we find the product of the elements along the other diagonal extending from *lower left* to *upper right.* In this case, this product is (2)(4) = 8. This second product must be *subtracted* from the first product. This is the meaning of the symbol of the determinant. The result in this example is 15 − 8 = 7. The following expression shows how the determinant is expanded:

$$\begin{vmatrix} 5 & 4 \\ 2 & 3 \end{vmatrix} = (5)(3) - (2)(4) = 15 - 8 = 7$$

The value of this determinant is 7.

In a determinant the products must be taken *in exactly the order shown.* If some of the elements are negative, we must be especially careful to observe algebraic signs of the products. As an example,

$$\begin{vmatrix} 2 & -3 \\ -4 & -1 \end{vmatrix} = (2)(-1) - (-4)(-3) = -2 - (+12) = -2 - 12 = -14$$

In general terms, let us expand the following determinant:

$$\begin{vmatrix} a & b \\ c & d \end{vmatrix} = ad - bc$$

Note that the expanded determinant is exactly the *denominator* we obtained in solving the following system of equations:

$$ax + by = e$$

$$cx + dy = f$$

Recall that the solution of this system is

$$x = \frac{de - bf}{ad - bc} \quad \text{and} \quad \frac{af - ce}{ad - bc}$$

Each answer in the solution can be written in the form of a fraction having one determinant for the numerator and another for the denominator:

$$x = \frac{\begin{vmatrix} e & b \\ f & d \end{vmatrix}}{\begin{vmatrix} a & b \\ c & d \end{vmatrix}} \qquad y = \frac{\begin{vmatrix} a & e \\ c & f \end{vmatrix}}{\begin{vmatrix} a & b \\ c & d \end{vmatrix}}$$

Our problem now is this: How can the constants, a, b, c, d, e, and f, be set up in the proper array for the determinants so that we shall have immediately the correct expressions for the solution? Note that in each case the *denominator* consists of the coefficients of x and y; that is, a, b, c, and d, in the same order as they appear in the equations:

$$\begin{vmatrix} a & b \\ c & d \end{vmatrix}$$

This is called the *determinant of the system.*

The *numerators* are the same as the denominator *except* for one difference. In the solution for x, the numerator is the same as the denominator *except* that the x-coefficients are replaced with the constant terms, e and f. In like manner, in the solution for y, the numerator is formed by replacing the y-coefficients with the constant terms, e and f.

When the coefficients and constants are set up in this proper form in the numerators and in the denominator, the expansion of the determinants gives the solution of the system. This procedure is known as *Cramer's Rule.*

Example 1

Solve by determinants:

$$5x + 4y = 9$$

$$2x + 3y = 6$$

Solution

The denominator for both values is formed by writing the coefficients of x and y as they appear in the equations. This becomes the *determinant of the system:*

$$\begin{vmatrix} 5 & 4 \\ 2 & 3 \end{vmatrix}$$

The numerator determinants are exactly the same as the denominator *except* that the constant terms, 9 and 6, replace a column of coefficients: in each case the coefficients of the letter whose value we wish to find.

$$x = \frac{\begin{vmatrix} 9 & 4 \\ 6 & 3 \end{vmatrix}}{\begin{vmatrix} 5 & 4 \\ 2 & 3 \end{vmatrix}} = \frac{27-24}{15-8} = \frac{3}{7}; \qquad y = \frac{\begin{vmatrix} 5 & 9 \\ 2 & 6 \end{vmatrix}}{\begin{vmatrix} 5 & 4 \\ 2 & 3 \end{vmatrix}} = \frac{30-18}{15-8} = \frac{12}{7}$$

Example 2

Solve by determinants:

$$3x - 5y + 4 = 0$$

$$4y - 7x + 2 = 0$$

Solution

First, we arrange the terms so that the letters appear on the left side in the same order in both equations, and the constant terms on the right side:

$$3x - 5y = -4$$

$$-7x + 4y = -2$$

Then

$$x = \frac{\begin{vmatrix} -4 & -5 \\ -2 & 4 \end{vmatrix}}{\begin{vmatrix} 3 & -5 \\ -7 & 4 \end{vmatrix}} = \frac{-16 - 10}{12 - 35} = \frac{26}{23} \qquad y = \frac{\begin{vmatrix} 3 & -4 \\ -7 & -2 \end{vmatrix}}{-23} = \frac{-6 - 28}{-23} = \frac{34}{23}$$

Note that when the denominator determinant is once expanded, its value can be written immediately as the denominator for the value of y.

Two special cases should be mentioned here. If the numerator and the denominator in the solution are both zero, the system of equations has an infinite number of solutions, and the two equations are dependent. The following system is an example of this case:

$$2x + 3y = 5$$

$$4x + 6y = 10$$

Another special case is that in which the denominator is equal to zero but the numerator is not equal to zero. The equations are then inconsistent and they have no common solution. The following is an example of this case:

$$2x + 3y = 5$$

$$4x + 6y = 8$$

Exercise 14-8

Solve the systems in Exercises 14-1 and 14-2 by determinants.

14-11 THIRD-ORDER DETERMINANT

A system of three equations in three unknowns can also be solved by determinants. First we must know how to expand a third-order determinant.

A *third-order determinant* is one that contains three rows and three columns. The expansion of such a determinant is considerably more complicated than that of a second-order determinant. You will recall that the expansion of a determinant with two rows and two columns involves the multiplication along *two* diagonals. There are just two separate multiplications. On the other hand, the expansion of a determinant with three rows and three columns involves multiplication along *six* diagonals. There are six separate multiplications. The proper expansion is shown by an example.

Suppose we have the following third-order determinant:

$$\begin{vmatrix} 8 & 5 & 9 \\ 2 & 1 & 4 \\ 6 & 3 & 7 \end{vmatrix}$$

To expand this determinant, we first rewrite the first and the second column at the right of the given determinant as shown here:

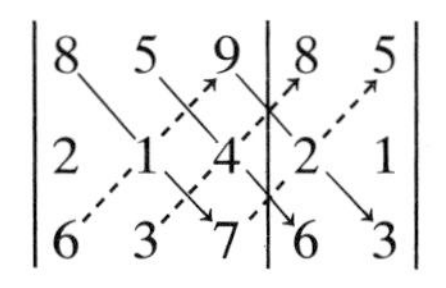

Now, starting with the three elements, 8, 5, and 9, in the top row of the determinant, we have three diagonals extending *downward to the right*, as shown by arrows. Each diagonal includes three elements. We call these the *downward diagonals*. The elements in each of these diagonals are multiplied together and their products are then added. We get

$$\begin{aligned} &(8)(1)(7) + (5)(4)(6) + (9)(2)(3) \\ &= \quad 56 \quad + \quad 120 \quad + \quad 54 \quad = 230 \end{aligned}$$

Next, starting with the three elements, 6, 3, and 7, in the bottom row, we have three diagonals extending *upward to the right*, shown by broken arrows. We call these the *upward diagonals*. The elements in each of these diagonals are multiplied together and their products are all *subtracted* from the products of the downward diagonals. For the upward diagonals we have the products:

$$(6)(1)(9); \qquad (3)(4)(8); \qquad (7)(2)(5)$$

The expansion of the determinant, in which the addition of the downward products and the subtraction of the upward products are indicated, can be shown as follows:

$$
\begin{aligned}
&(8)(1)(7) + (5)(4)(6) + (9)(2)(3) - (6)(1)(9) - (3)(4)(8) - (7)(2)(5) \\
&= 56 + 120 + 54 - 54 - 96 - 70 \\
&= 10
\end{aligned}
$$

In evaluating a third-order determinant, we need not rewrite the first two columns if we simply imagine them in place and then select the elements correctly according to the directions indicated by the arrows. Remember, each diagonal multiplication must involve *three* elements.

The elements of a determinant are often shown with subscripts. Then the expansion of a third-order determinant is shown in the following way:

$$
\begin{vmatrix} a_1 & b_1 & c_1 \\ a_2 & b_2 & c_2 \\ a_3 & b_3 & c_3 \end{vmatrix} = a_1b_2c_3 + b_1c_2a_3 + c_1a_2b_3 - a_3b_2c_1 - b_3c_2a_1 - c_3a_2b_1
$$

The solution of three equations by determinants follows the same general plan as that used for two equations. The solution for each of x, y, and z, will be the quotient of two determinants, the denominator in each case being formed by taking the coefficients of x, y, and z as they appear in the equations, in proper order. In each case, the numerator will be the same as the denominator *except* that one column will be replaced by the constant terms.

Example 3

Solve by determinants:

$$
\begin{aligned}
4x + 3y - z &= 7 \\
3x + 5y + z &= 6 \\
5x + 2y - 2z &= 9
\end{aligned}
$$

Solution

For each unknown, the denominator will be the *determinant of the system* formed by taking the coefficients as shown here:

$$
\begin{vmatrix} 4 & 3 & -1 \\ 3 & 5 & 1 \\ 5 & 2 & -2 \end{vmatrix}
$$

Now we make each numerator determinant exactly the same as the denominator with the following exception. In the numerator for the x-solution, we replace the x-coefficients with the constant terms, 7, 6, and 9. In the numerator for the y-solution, we replace the y-coefficients with the constant terms. We follow the same plan in solving for z. Then we have the forms:

$$x = \frac{\begin{vmatrix} 7 & 3 & -1 \\ 6 & 5 & 1 \\ 9 & 2 & -2 \end{vmatrix}}{\begin{vmatrix} 4 & 3 & -1 \\ 3 & 5 & 1 \\ 5 & 2 & -2 \end{vmatrix}} \qquad y = \frac{\begin{vmatrix} 4 & 7 & -1 \\ 3 & 6 & 1 \\ 5 & 9 & -2 \end{vmatrix}}{\begin{vmatrix} 4 & 3 & -1 \\ 3 & 5 & 1 \\ 5 & 2 & -2 \end{vmatrix}} \qquad z = \frac{\begin{vmatrix} 4 & 3 & 7 \\ 3 & 5 & 6 \\ 5 & 2 & 9 \end{vmatrix}}{\begin{vmatrix} 4 & 3 & -1 \\ 3 & 5 & 1 \\ 5 & 2 & -2 \end{vmatrix}}$$

Expanding the numerator and denominator for the value of x, we get

$$x = \frac{-70 + 27 - 12 + 45 - 14 + 36}{-40 + 15 - 6 + 25 - 8 + 18} = \frac{12}{4} = 3$$

Note that the denominator is the same for all the unknowns. Then, instead of writing the denominator determinant for each, its value, 4, can be written down at once for each of the other unknowns. Expanding the determinants, we find the values,

$$y = -1; \qquad z = 2$$

Exercise 14-9

Solve by determinants Nos. 1–10 in Exercise 14-5.

Note. Determinants of the fourth and higher orders can be used to solve equations of more than three unknowns. However, their expansion is much more complicated than that of third-order determinants. If you wish to learn the method, you will find it explained in almost any book on college algebra.

QUIZ ON CHAPTER 14, SYSTEMS OF EQUATIONS. FORM A.

Solve No. 1 and No. 2 by addition or subtraction: Solve No. 3 by substitution:

1. $\begin{aligned} 5x - 3y &= 4 \\ 3x - 4y &= 9 \end{aligned}$ **2.** $\begin{aligned} 3x + 2y &= 1 \\ 4x - 3y &= 5 \end{aligned}$ **3.** $\begin{aligned} 4x - 3y &= 5 \\ 3x + y &= 2 \end{aligned}$

4. Solve by comparison: $\begin{aligned} 4x + 3y &= 5 \\ 5x + 2y &= 6 \end{aligned}$

5. Solve by determinants: $\begin{aligned} 5x + y &= 6 \\ 4x - 3y &= 7 \end{aligned}$

6. Solve this system of three unknowns: $\begin{aligned} 3x + 2y - z &= 6 \\ 5x + 3y + 2z &= 7 \\ 4x - y + z &= 13 \end{aligned}$

7. Expand this determinant: $\begin{vmatrix} 3 & -1 & -2 \\ 2 & 2 & 3 \\ 3 & -1 & -4 \end{vmatrix}$

Solve each of the following problems by using two letters. Show all steps.

8. At the entertainment, children's tickets cost 35 cents each, and adults' tickets cost 80 cents each. If the total amount received was $53 from a total of 109 tickets, how many tickets of each kind were sold?

9. A man has $7500 invested, part of it at 5% interest per year, and the rest at 8% interest per year. If his total yearly income from these two investments is $414, find the amount invested at each rate.

10. A man took a trip of 510 miles, part of the way by car at an average of 48 mph, and the rest of the way by train at an average of 60 mph. If his total travel time was 9 hours, how many hours and how many miles did he travel by each mode of travel?

11. How many gallons of a 15% solution and how many gallons of a 75% solution of a certain chemical should be mixed together for 40 gal of a new solution that is to test 35%?

12. In an electric circuit, three resistors, R_1, R_2, and R_3, have a total resistance of 360 ohms when connected in series. The resistor R_2 has a resistance that is 15 ohms more than twice the resistance of R_1. The resistance of R_3 is equal to the sum of the resistances of R_1 and R_2. Find the resistance of each resistor.

CHAPTER FIFTEEN Graphing

15-1 INTRODUCTION

A day seldom goes by that we do not see a graph of some kind. A graph is a pictorial representation of certain factual data. We see graphs in newspapers, in magazines, in books, and in almost every technical article. We see graphs of temperature changes, business losses and gains, and economic trends. Many important facts in electrical engineering are represented by graphs of some kind, such as graphs of current, voltage, and power.

The advantage of a graph is that it shows important facts at a glance. Looking at a business graph, we can tell quickly the high points and the periods of rapid gains and losses. In electrical engineering a graph shows instantly the relation between current and power. Many ideas in mathematics are more easily understood when they are interpreted graphically. Some equations can be solved only by means of graphs. For these reasons graphs are extremely important in the study of mathematics and science.

15-2 THE RECTANGULAR COORDINATE SYSTEM

In most cities at least part of the street system is laid out so that some streets have an east-west direction and others have a north-south direction. Imagine a city laid out with two main streets perpendicular to each other and all other streets parallel to one or the other of these two main streets.

We have one main street extending east and west, which we might call street X and another main street extending north and south, which we might call street Y (Fig. 15-1). All points in the city can be located with reference to these two perpendicular streets. All distances are measured from the two main streets.

If we wish to direct someone to a particular house in the city, we might say, starting at the point of intersection of the main streets, "Go five blocks east and seven blocks north." This location is indicated by point A in Fig. 15-1. To another we might say, "Go eight blocks west and three blocks south." This location is shown by point B in the figure.

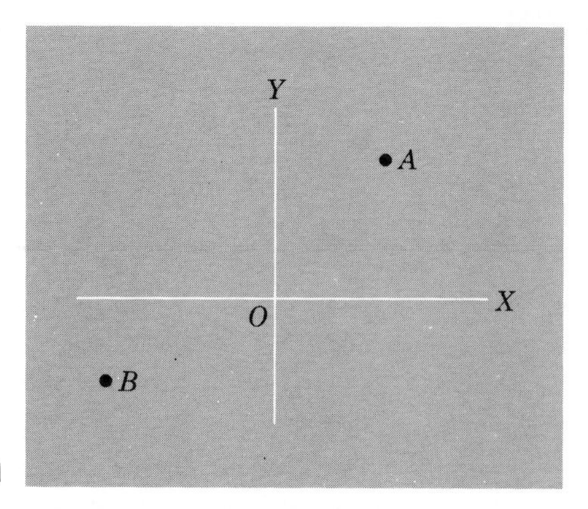

Figure 15-1

If we wish to locate a point in a plane, we use a system like that of our perpendicular streets. We set up two perpendicular lines for reference lines (Fig. 15-2), which correspond to the two main streets. Each of the reference lines is called an *axis.* We call the horizontal line the x-axis and the vertical line the y-axis. The point of intersection is called the *origin,* denoted by O.

By means of this arrangement, we can tell the location of any point with reference to the two axes. This system of locating points is called the *rectangular coordinate system.* The two axes divide the entire plane into four quarters called *quadrants.* The quadrants are numbered, usually with roman numerals, I, II, III, IV, starting in the upper right-hand quadrant and going around in a counterclockwise direction. To tell the location of a point, we measure its distance from each axis.

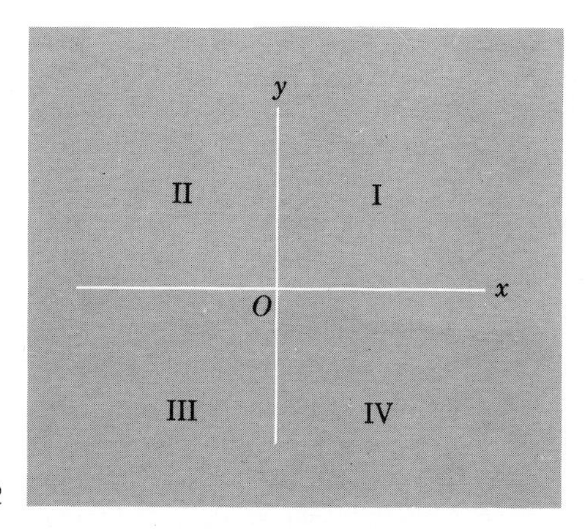

Figure 15-2

Suppose a point is located 8 units to the right of the y-axis and 5 units upward from the x-axis (Fig. 15-3). Then we can say that these two numbers, 8 and 5, *taken in that order*, will definitely tell the location of the point. We call the numbers 8 and 5 the *coordinates* of the point. They are like a house number in a city.

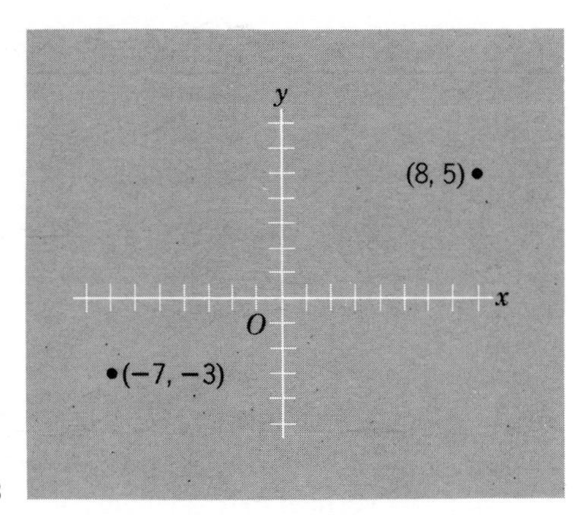

Figure 15-3

The distance of the point in the x-direction is called the *abscissa* of the point. The distance in the y-direction is called the *ordinate* of the point. Note that the abscissa of a point is actually measured from the y-axis and the ordinate of the point is measured from the x-axis. Be careful *not* to call the x-axis the abscissa and the y-axis the ordinate. The abscissa and the ordinate are *distances*, not the lines of reference. For the point shown in Fig. 15-3, the abscissa is 8; the ordinate is 5.

In a city we speak of distances east or west and north or south. In the rectangular coordinate system distances to the right are called positive (+); distances to the left are called negative (−); distances upward are called positive; and distances downward are called negative.

If we wish to indicate a point 7 units to the left of the y-axis, we say that the abscissa, or x-distance, is -7. If the point is located 3 units downward from the x-axis, we say that the ordinate, or y-distance, is -3. The two numbers -7 and -3 are the coordinates of the point. The coordinates are enclosed in parentheses and separated by a comma: thus $(-7, -3)$. The abscissa of the point is always written first. With this understanding, the set of numbers $(-7, -3)$ will definitely determine one and only one point (Fig. 15-3). Since the two numbers must be taken in the proper order, they are called an *ordered pair.*

The following points are shown on the graph in Fig. 15-4: $A(8, 2)$, $B(2, 8)$, $C(6, 0)$, $D(0, 6)$, $E(-5, 7)$, $F(-7, 0)$, $G(-6, -5)$, $H(0, -4)$, $I(3, -6)$, $J(9, -1)$.

The location of points on a graph is most conveniently done on "rectangular coordinate" graph paper. On this kind of paper, lines are evenly spaced horizontally and vertically.

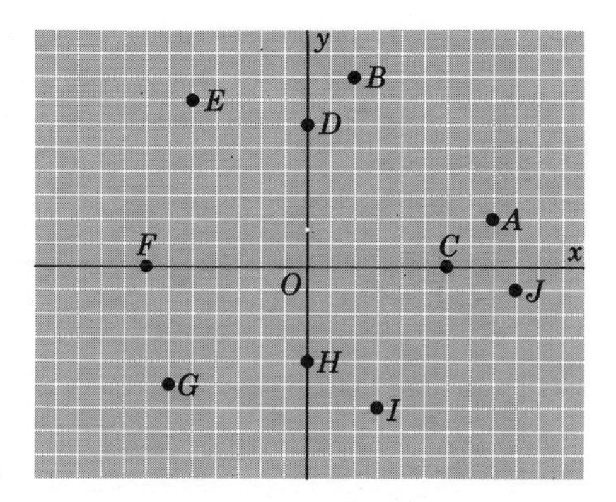

Figure 15-4

To locate points on a graph, we first draw two axes, x and y, at right angles so that the origin is approximately in the center of the portion of the paper to be used for the graph. The horizontal axis is labeled x and the vertical axis is labeled y. The exact position of the axes can be varied to suit each particular problem. In measuring the distances indicated by the coordinates of a point, the units may be of any convenient length, depending on the size of the paper and on the size of the numbers used. In most cases the units should have the same length on both axes.

The coordinates of each point are often written near the point on the graph. For practice, several points may be located on the same graph. To *plot* points means to locate the points on the graph.

Note. Notice especially that the rectangular system of coordinates and the plotting of points is *geometric*. It involves *geometry*. It has nothing to do with algebra.

Exercise 15-1

1. Plot these points on one graph: $A(7, 3)$, $B(1, 6)$, $C(0, 3)$, $D(-8, 2)$, $E(-5, -2)$, $F(0, 0)$, $G(3, -5)$, $H(0, -7)$, $I(6, -6)$, $J(9, 0)$.

2. Plot the following points on a graph and then connect the points in the order in which they are given with straight line segments: $A(8, 0)$, $B(0, 8)$, $C(-8, 0)$, $D(0, -8)$, $A(8, 0)$, $E(4, 0)$, $F(0, 4)$, $G(-4, 0)$, $H(0, -4)$, $E(4, 0)$.

Plot the points in each of the following sets on a graph and then connect the points in order by straight line segments to form polygons. Use a different portion of the graph paper for each set. End with A.

3. $A(8, 7)$, $B(-5, 3)$, $C(2, -6)$

4. $A(7, 1)$, $B(-2, 8)$, $C(4, -3)$

5. $A(3, 0)$, $B(-3, 7)$, $C(-8, -5)$

6. $A(-2, 8)$, $B(-6, 0)$, $C(7, -4)$

7. $A(5, -1)$, $B(0, 4)$, $C(-6, -5)$, $D(0, -1)$

8. $A(7, 5)$, $B(-3, 2)$, $C(-6, -5)$, $D(4, -2)$

9. $A(9, -2)$, $B(-3, 6)$, $C(-3, -1)$, $D(0, -3)$

10. $A(8, 0)$, $B(2, 6)$, $C(-5, 0)$, $D(0, -7)$, $E(6, -5)$

11. $A(5, 6)$, $B(-7, 0)$, $C(6, -2)$, $D(-3, 7)$, $E(-1, -6)$, $A(5, 6)$

12. $A(12, -3)$, $B(9, -1)$, $C(3, 3)$, $D(0, 5)$, $E(-3, 7)$

13. $A(9, 9)$, $B(5, 6)$, $C(1, 3)$, $D(-3, 0)$, $E(-7, -3)$, $F(-11, -6)$

14. $A(12, 8)$, $B(0, 4)$, $C(-3, 2)$, $D(-4, 0)$, $E(-3, -2)$, $F(0, -4)$, $G(12, -8)$
(Connect the points in Exercise 14 by a smooth continuous *curve*, instead of by straight line segments.)

15. Connect the following points by a smooth curve: $A(5, 0)$, $B(4, 3)$, $C(3, 4)$, $D(0, 5)$, $E(-3, 4)$, $F(-4, 3)$, $G(-5, 0)$, $H(-4, -3)$, $I(-3, -4)$, $J(0, -5)$, $K(3, -4)$, $L(4, -3)$, $A(5, 0)$.

15-3 GRAPHING AN EQUATION

Let us leave the problem of graphing for a moment and consider instead an algebraic equation, such as $x + y = 11$. A graph is a *geometric* concept, whereas an equation is *algebraic*.

x	y
8	3
10	1
2	9
−2	13
0	11
11	0

We have seen that an equation containing x and y may have many *solution sets* or pairs of numbers that make the equation true. For example, the equation $x + y = 11$ is satisfied by the pair of numbers $x = 8$ and $y = 3$. We can find many other solution sets, some of which are shown in the table at the right. Any one of these pairs of values will satisfy this equation.

Note especially that a pair of numbers satisfying the equation, such as $x = 8$ and $y = 3$, involves algebra. Finding a solution set for an equation is an algebraic process. It has nothing to do with geometry.

Now, here is one of the amazing stories in mathematics. For more than a thousand years the people of some nations had gone on happily finding numbers that would satisfy equations. They were chiefly interested in algebra. They had little or no interest in geometry. On the other hand, in other nations, many people were completely devoted to geometry. They cared little about algebra. They studied diagrams and geometric figures of all kinds and showed the location of points on a line and in a plane.

The algebraists were interested in finding numbers that would satisfy equations. The geometers were interested in points, lines, and planes. Yet, no one, for more than a thousand years, ever thought of connecting geometry and algebra.

Then along came Rene Descartes, born in 1596. While he was still a young man in his "teens," a brilliant idea flashed into his mind—an idea that marks one of the major advances in mathematics. This was his idea: if the numbers $x = 8$ and $y = 3$ satisfy the equation $x + y = 11$, let us call this pair of numbers the coordinates of a point on the rectangular system. In other words, we take a pair of numbers from an algebraic equation and then use this pair to represent a point in geometry.

The idea was revolutionary, and Descartes probably realized he was on the trail of something great. The amazing thing about the idea is that no one had thought of it before. Descartes spent many years developing the idea, which he later called *analytic geometry*. The result was a uniting of algebra and geometry into one of the most powerful tools in mathematical and scientific study.

Let us follow through on the equation $x + y = 11$. We have seen that several pairs of numbers satisfy the equation, and we call each pair of numbers the coordinates of a point. The result is several points; in fact, as many as we wish to find.

Here is Descartes' amazing discovery: if the points for this equation are connected by lines, the result is a straight line! If we find other pairs of numbers that satisfy the equation, each pair represents a point on the line (Fig. 15-5).

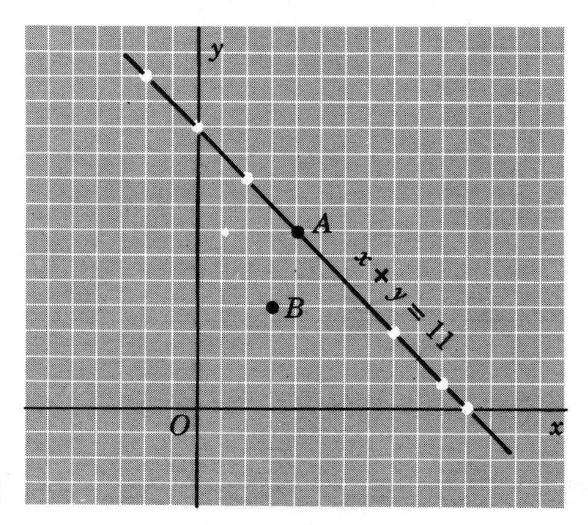

Figure 15-5

There are two things we should note here:

1. We have written down six pairs of numbers in tabular form that satisfy the equation. No matter how many pairs we might find for the equation, every pair would represent a point on the line.

2. On the other hand, if we take any other point on the line, such as point A, and then find its coordinates from the graph, this pair of numbers will satisfy the equation. The coordinates of point B will not satisfy the equation because the point does not lie on the line.

Therefore, the straight line (a geometric figure) shown in Fig. 15-5 is the graph or picture of the algebraic equation $x + y = 11$. Moreover, an equation such as $x + y = 11$ is called a *linear* equation because its graph is a straight line.

To summarize:

1. If we find any pair of numbers that satisfies the equation, this pair will be the coordinates of a point on the line.
2. If we take any point on the line and find its coordinates, this pair of numbers will satisfy the equation.

We have seen that the graph of the equation $x + y = 11$ is a straight line. If we graph any equation containing only first-degree terms, the result will be a straight line.

A *first-degree term* is any term that contains only one variable raised to the first power. A *second-degree term* is a term that is the product of two variables, either the same or different. For instance, the terms x^2, $5y^2$, and $7xy$ are second-degree terms, since each term contains a variable multiplied by a variable.

A *first-degree equation* is an equation containing only first-degree terms, such as $3x + 2y = 12$.

A *second-degree equation* is an equation containing terms of the second degree, but no higher, such as the following:

$$x^2 + y^2 = 25 \qquad y = x^2 - 3x - 4 \qquad xy = 12$$

$$x^2 - 3xy + 4y^2 - 5x - 3y + 7 = 0$$

A *third-degree equation* is one that contains at least one term of the third degree but no higher, such as the following:

$$y = x^3 \qquad x^3 + y^3 + xy - 4x = 0 \qquad x^2y + 5xy - 7x + 4y = 20$$

The following statement can be shown to be true: *The graph of every first-degree equation is a straight line.*

In finding values and constructing the graph for a linear equation, we should keep several facts in mind:

1. Since we know the graph is a straight line, it would be sufficient to find only two points because two points determine a straight line. However, one or two extra points should be plotted as a check on the accuracy of the work.
2. If only two points are used, they should not be too close together. If the two points are some distance apart, the straight line can usually be drawn more accurately.

3. The easiest points to find are those at which the line cuts the x- and the y-axis. These points are easily found by setting each variable equal to zero and solving for the other. First, we set $x = 0$ and solve for y; then we set $y = 0$ and solve for x. However, in some equations, such as in $3x = 2y$, we get $y = 0$ at the same time that $x = 0$. Therefore, we need at least one other point.
4. The straight line should be drawn *through* the points and should not end at any of the points plotted. It should always be understood that the line is infinite in extent and is not limited by the size of the paper.

Example

Graph the equation

$$2x - 5y = 20$$

Solution

First find the zero values. If $x = 0$, then $y = -4$. If $y = 0$, $x = 10$. Then other points may be found. For instance, if $x = -5$, $y = -6$. To find any pair of points, take any convenient value for one variable and then compute the value of the other variable. Some values may be fractional. The graph is shown in Fig. 15-6.

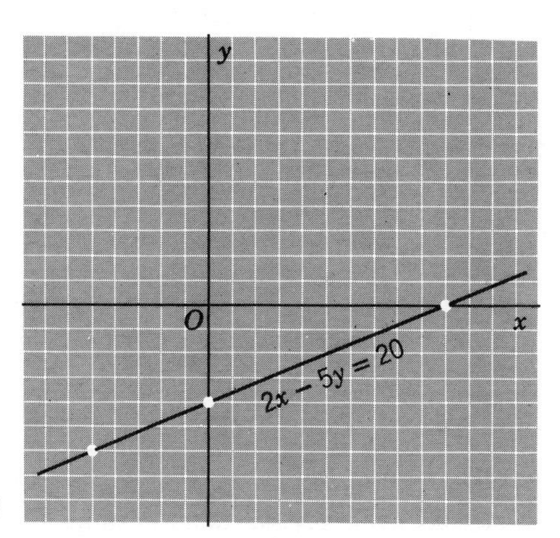

Figure 15-6

Exercise 15-2

Graph the following linear equations. The same graph paper may be used for three or four different equations.

1. $x + 2y = 6$ **2.** $2x - y = 8$ **3.** $2x + 3y = 12$

4. $3x - 2y = 18$	**5.** $4x - 5y = 40$	**6.** $2x + y = 9$
7. $x - 3y = 10$	**8.** $3x + 2y = 15$	**9.** $2x - 5y = 20$
10. $5x - 3y = 20$	**11.** $4x - 7y = 28$	**12.** $3x - 5y = 0$
13. $4x + 3y = 23$	**14.** $3x + 7y = 17$	**15.** $5x - 3y = 16$
16. $7x - 4y = 18$	**17.** $6x + 5y = 20$	**18.** $4x = 3y$
19. $2y = 5x$	**20.** $4x + 2y = 15$	**21.** $x - 3y = 7$
22. $5x - 7y = 21$	**23.** $6x + y = 12$	**24.** $x + 10y = 15$
25. $3x - 8y = 9$	**26.** $5x + 2y = 1$	**27.** $2x - 7y = 5$
28. $3x - 5y = 0$	**29.** $8x + 3y = 4$	**30.** $9x + 7y = 2$

15-4 SOLVING SYSTEMS OF EQUATIONS BY GRAPH

Graphing can be used to solve a system of equations. We have seen that the graph of the equation $x + y = 11$ is a straight line (Fig. 15-5). Let us graph the equation $x - y = 3$ on the same graph. We find several pairs of numbers that satisfy the equation. Some of the pairs are shown in the table at the right. Next, we assume that the two numbers in each ordered pair are the coordinates of a point. We plot the points on the graph and connect them with a line. Again we find that the graph is a straight line (Fig. 15-7.)

x	y
10	7
8	5
4	1
0	−3
−2	−5

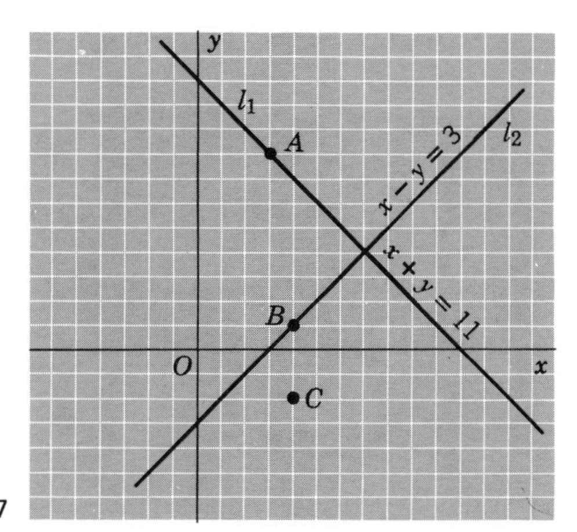

Figure 15-7

Figure 15-7 shows the lines for both of the equations:

$$x + y = 11 \qquad \text{and} \qquad x - y = 3$$

In the figure the line l_1 is the graph of the equation $x + y = 11$ and the line l_2 is the graph of the equation $x - y = 3$.

Since point A lies on line$_1$, its coordinates satisfy the equation $x + y = 11$. Since point B lies on line$_2$, its coordinates satisfy the equation $x - y = 3$. Point C does not lie on either line, and therefore its coordinates will not satisfy either equation.

Only one point has coordinates that will satisfy both equations. That is the point where the two lines intersect. From the graph it appears to be the point whose coordinates are approximately as follows: $x = 7$ and $y = 4$. If we check these values in the equations, we shall find that they satisfy both equations.

These equations, of course, could have been solved by algebraic methods. However, there is sometimes an advantage in the graphical method.

In solving a system of equations, our objective is to find a pair of numbers that will satisfy both equations. To do so by graphing, we draw the graph of each equation. Each graph is a straight line. We know that the straight lines can intersect at only one point, the point that lies on both lines. Therefore, its coordinates must be the pair of values for x and y that satisfy both equations. We illustrate the method by examples.

Example 1

Solve this system of equations by graphing:

$$x + 3y = 9 \qquad 2x - y = 11$$

Solution

We first find pairs of numbers that satisfy each equation. The table of values may be arranged in horizontal form:

$x + 3y = 9$

if $x =$	0	9	6	3
then $y =$	3	0	1	2

$2x - y = 11$

if $x =$	0	$5\frac{1}{2}$	4	2
then $y =$	-11	0	-3	-7

The graph is shown in Fig. 15-8.

The point of intersection appears to be approximately (6, 1). If we check these values in the equations, we find that the values satisfy both equations.

In this example the coordinates of the point of intersection are rather easily determined. However, in many cases it is not easy to read the exact coordinates from the graph. This is especially true when the values are fractions.

If the graphs are made carefully, it is possible to determine fairly accurately the coordinates of the point of intersection even when fractions are involved. However, it should always be remembered that any *answers read from a graph must be assumed to be only approximate.* Therefore, such values may not check exactly when substituted in the equations. The solution of a set of linear equations can be checked by solving by some algebraic method, such as addition, subtraction, substitution, or comparison. If the answers obtained by graphing are the same or nearly the same as those obtained by algebraic methods, then the solution can be considered correct.

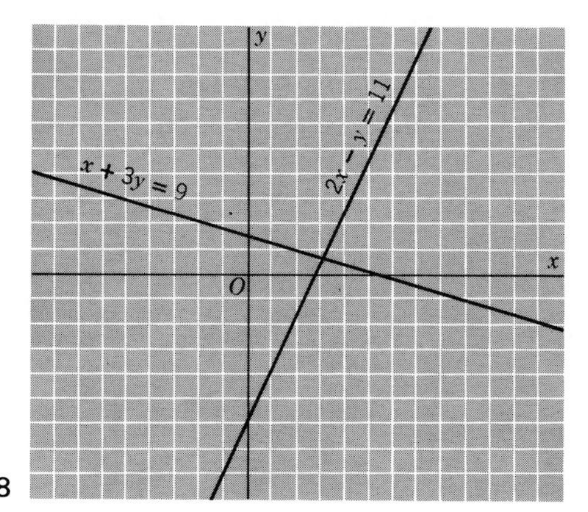

Figure 15-8

Example 2

Solve the set of equations: $3x + 2y = 12$

$4x - 5y = 20$

Solution

First, we find pairs of numbers that satisfy each equation:

$3x + 2y = 12$

if $x =$	0	4	6	−4
then $y =$	6	0	−3	12

$4x - 5y = 20$

if $x =$	0	5	$7\frac{1}{2}$	−5
then $y =$	−4	0	2	−8

The graph is shown in Fig. 15-9.

The only pair of values for x and y that will satisfy both equations is the pair for the point of intersection. From the graph, the abscissa of the point appears to be approximately $4\frac{1}{3}$ and the ordinate approximately $-\frac{1}{2}$. Therefore, we have the solution $x = 4\frac{1}{3}$ and $y = -\frac{1}{2}$.

If the equations are solved by algebraic methods, the answers are found to be exactly

$$x = 4\tfrac{8}{23} \qquad y = -\tfrac{12}{23}.$$

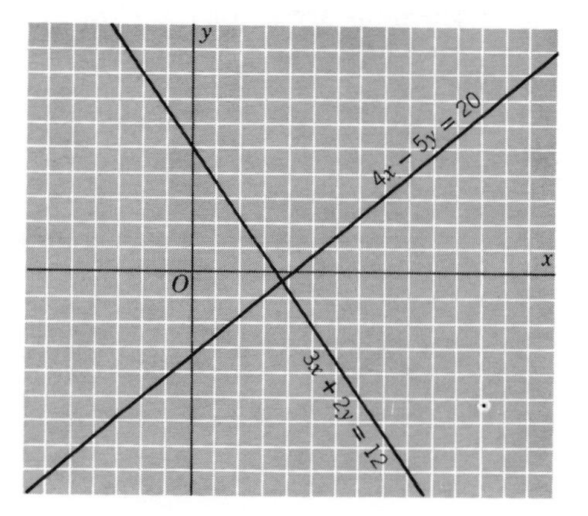

Figure 15-9

If the example calls for a graphical solution, the correct answer is the approximate solution, since it is almost impossible to read a value such as $4\frac{8}{23}$ on a graph. However, those answers may be estimated in tenths of a unit.

Exercise 15-3

Solve the following systems graphically:

1. $x + 2y = 10$, $3x - y = 9$

2. $3x - 5y = 25$, $4x + 3y = 14$

3. $5x - 2y = 13$, $3x + 5y = 14$

4. $x - 2y = -4$, $4x + 3y = 17$

5. $4x - 3y = -9$, $3x + 2y = -11$

6. $2x + 5y = 20$, $3x - 2y = 18$

7. $2x + y = -10$, $x - 3y = -19$

8. $2x + 3y = -5$, $4x - 5y = 23$

9. $2x + 5y = 24$, $3x - 2y = -21$

10. $3x - 5y = 30$, $2x + y = 10$

11. $5x - 6y = 30$, $3x + y = 12$

12. $2x - 5y = -20$, $5x + 4y = 2$

13. $4x - 5y = 16$, $7x - 3y = 5$

14. $4x + 5y = 1$, $5x + 7y = -1$

15. $6x + 5y = 8$, $7x + 6y = 10$

16. $5x - 2y = 0$, $3x + 5y = 0$

17. $4x + 5y = -14$, $8x - 7y = -28$

18. $3x - 5y = 25$, $5x - 7y = 35$

19. $4x - 5y = -20$, $3x + 4y = 0$

20. $5x - 6y = 30$, $6x - 5y = 15$

21. $8x - 3y = 24$, $7x - 2y = 14$

15-5 THE SLOPE OF A LINE

Besides the problem of drawing the graph of a given equation, we are often faced with the reverse type of problem. That is, if a particular line is given or indicated, we have the problem of writing the equation of the line. To do so, we must know what is meant by the *slope* of a line.

First, we must consider the various notations for points.

1. *Specific points* are denoted by specific numbers, such as (7, 3), (4, 0), and so on. These are the kinds of points we have used in graphing equations.
2. A *general* or variable point in any position on the graph is denoted by the general coordinates, as $P(x, y)$ in all quadrants.
3. In deriving formulas we often need to denote *specific points that may be anywhere on the graph.* For such points we often use subscripts, as $P_1(x_1, y_1)$, $P_2(x_2, y_2)$, and so on, or other letters, as $C(h, k)$, $Q(a, b)$, and so on. Points so denoted are considered as fixed and are not variable points.

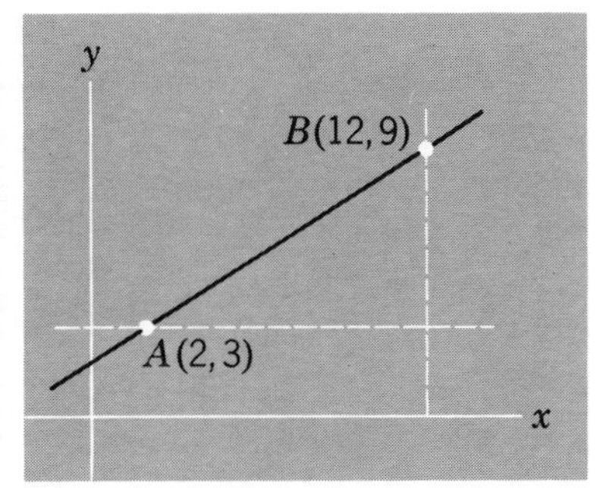

Figure 15-10

The *slope* of a line refers to its steepness. The slope is defined in terms of the *vertical* and *horizontal changes* as we move from one point to another on the line. Suppose a line passes through the two points, $A(2, 3)$ and $B(12, 9)$. (See Fig. 15-10.) To note the changes in x and in y, we draw a line through A parallel to the x-axis, and another line through B parallel to the y-axis. Then, as we move from A to B, y changes from 3 to 9, and x changes from 2 to 12. The change in y is called the *rise*. The change in x is called the *run*. Then we have

$$\text{rise} = 9 - 3 = 6 \text{ units}; \qquad \text{run} = 12 - 2 = 10 \text{ units}$$

Now, denoting the slope by m, we define the slope of the line by the ratio

$$\text{slope} = m = \frac{\text{rise}}{\text{run}} = \frac{9-3}{12-2} = \frac{6}{10} = \frac{3}{5}$$

Then the slope of this line is described by the fraction $\frac{3}{5}$.

When any two points on a line are known, we get the formula for the slope:

$$m = \frac{y_2 - y_1}{x_2 - x_1}$$

Either of the two points may be called P_1 and the other P_2.

If a line *rises* in moving from left to right, *the slope is positive.* If a line *falls* in moving from left to right, *the slope is negative.* A *horizontal* line has a *zero* slope since the rise is zero. A *vertical* line is said to have an *infinite* slope, since the denominator in the formula becomes zero. If two lines have the *same slope,* they are *parallel.* Conversely, if two lines are parallel, they have the same slope. If two lines are *perpendicular* to each other, the slope of one is the *negative reciprocal* of the other. If a line has a slope of $\frac{2}{5}$, the perpendicular to this line has a slope of $-\frac{5}{2}$.

Examples

Find the slope of the line through each of these pairs of points: Sketch the line on the graph. (a) (8, 6) and (−4, 2); (b) (−1, 6) and (3, −4); (c) (3, 5) and (−7, 5); (d) (2, 4) and (2, −3)

Solutions

For (a),

$$m = \frac{6 - 2}{8 + 4} = \frac{4}{12} = \frac{1}{3}$$

If we reverse the points, we get the same result. The positive slope indicates that the line slopes upward to the right, as the graph will show.

For (b),

$$m = \frac{6 + 4}{-1 - 3} = \frac{10}{-4} = -\frac{5}{2},$$ indicating a downward slope.

For (c),

$$m = \frac{5 - 5}{3 + 7} = \frac{0}{10} = 0,$$ indicating a horizontal line.

For (d),

$$m = \frac{4 + 3}{2 - 2} = \frac{7}{0},$$ indicating a vertical line.

15-6 EQUATION OF A STRAIGHT LINE

In analytic geometry it is shown that if a straight line has the equation

$$Ax + By = C$$

where A, B, and C are constants, then the slope of the line is given by the formula: $m = -A/B$. That is, the coefficients of x and y have the same ratio as the *negative of the slope.* As an example, the line, $3x + 4y = 20$, has a slope of $-\frac{3}{4}$. Therefore, if a line has a slope of $\frac{2}{5}$, the equations starts

$$2x - 5y = \_\_\_\_$$

or with other coefficients having the same ratio. Now, if we also know a point on the line, the coordinates of this point must satisfy the equation, and we can write the equation of the line.

Example 1

Write the equation of the line having a slope of $\frac{2}{5}$ and passing through the point (4, −1). Also write the equation of the perpendicular to this line through the same point. Sketch the line and its perpendicular.

Solution

The equation begins:

$$2x - 5y = \_\_\_\_$$

Since the given point $(4, -1)$ lies on the line, its coordinates satisfy the equation. Then we substitute 4 for x and -1 for y, and get the value 13. Then the equation of the line is: $2x - 5y = 13$. The graph is left for the student.

The perpendicular has a negative reciprocal slope, or $-\frac{5}{2}$. Since it passes through the same point, its equation is: $5x + 2y = 18$

Example 2

Write the equation of the line through the points $(3, -2)$ and $(-4, 1)$.

Solution

First we find the slope:

$$m = \frac{1 + 2}{-4 - 3} = -\frac{3}{7}$$

Then the equation begins:

$$3x + 7y = \_\_\_\_$$

Now, either of the two points determines the constant on the right side, and we get the equation:

$$3x + 7y = -5$$

Both given points will satisfy the equation.

The equation of a vertical or a horizontal line contains only one variable. If any point on a vertical line has the abscissa, $x = 3$, then every point on the line has $x = 3$. This is the equation of the line. If a is the abscissa of any point on a vertical line, the equation of the line is: $x = a$.

If any point on a horizontal line has the ordinate, $y = 5$, then every point on the line has $y = 5$. This is the equation of the line. If b is the ordinate of any point on a horizontal line, the equation of the line is: $y = b$.

Exercise 15-4

1. Find the slope of the line through each of the following pairs of points. Then write the equation of each line. Tell which are parallel and which are perpendicular.

(a) $(-4, -1)$; $(6, 5)$ (b) $(6, 1)$; $(-4, 5)$ (c) $(5, -1)$; $(-1, 9)$
(d) $(-7, 1)$; $(1, -3)$ (e) $(0, -6)$; $(10, 0)$ (f) $(-4, -1)$; $(0, 9)$
(g) $(4, 2)$; $(-2, -1)$ (h) $(-1, 3)$; $(5, 0)$ (i) $(2, -9)$; $(-4, 3)$

2. Same directions as for No. 1.
(a) $(4, -3); (-3, -5)$ (b) $(7, 0); (-1, 6)$ (c) $(6, -1); (-1, 4)$
(d) $(1, 7); (-7, 1)$ (e) $(-5, 3); (7, -1)$ (f) $(-5, 8); (-1, -6)$
(g) $(9, -4); (-3, 0)$ (h) $(0, 1); (-8, 7)$ (i) $(-1, -5); (2, 4)$

3. Write the equation of the line through each of these pairs of points:
(a) $(-3, -2); (5, -2)$ (b) $(4, -5); (4, 2)$ (c) $(6, 5); (-1, 5)$

4. Write the equation of the line through each of these pairs of points:
(a) $(-4, 5); (-4, -1)$ (b) $(3, -2); (-5, -2)$ (c) $(2, -4); (2, 5)$

5. A triangle has its vertices at the points: $A(3, 6)$, $B(-5, -4)$, $C(7, -2)$. Find the slope of each side of the triangle and write the equation of each side.

6. A triangle has its vertices at the points $A(1, 7)$, $B(5, 1)$, $C(-4, 5)$. Show by the slopes that this triangle is a right triangle.

7. A quadrilateral has its vertices at $A(7, 2)$, $B(1, 6)$, $C(-9, 0)$, $D(-3, -4)$. Show by the slopes of the sides that the figure is a parallelogram.

CHAPTER SIXTEEN Exponents, Powers, and Roots

16-1 MULTIPLICATION

In Chapter 8 we learned the meaning of an exponent. We also noted the rules to be observed in operations involving exponents. It is necessary at this time to review these rules carefully. Consider the example

$$(x^3)(x^5) =$$

We know that x^3 means $x \cdot x \cdot x$ and x^5 means $x \cdot x \cdot x \cdot x \cdot x$. Therefore,

$$(x^3)(x^5) \quad \text{means} \quad (x \cdot x \cdot x)(x \cdot x \cdot x \cdot x \cdot x).$$

Since multiplication can be done in any order,

$$(x^3)(x^5) \quad \text{means} \quad x \cdot x \cdot x \cdot x \cdot x \cdot x \cdot x \cdot x \quad \text{or} \quad x^8$$

This example is an illustration of the rule for exponents in multiplication.

Rule 1. *In multiplying quantities expressed with exponents, we add exponents of the same base.*

This rule may be stated as a formula: $x^m \cdot x^n = x^{m+n}$

Note carefully that the rule says, "When we *multiply quantities*, we *add exponents*." In the example

$$(x^3)(x^5) = x^8$$

we are not multiplying the exponents 3 and 5. Instead, we are multiplying the *quantity* x^3 by the *quantity* x^5.

To show that the rule holds true for positive integral exponents, we take the product $(x^m) \cdot (x^n)$ and analyze its meaning:

$$x^m = x \cdot x \cdots \text{to } m \text{ factors}$$

$$x^n = x \cdot x \cdots \text{to } n \text{ factors}$$

Therefore, $(x^m)(x^n) = (x \cdot x \cdots \text{to } m \text{ factors})(x \cdot x \cdots \text{to } n \text{ factors})$

$$= x \cdot x \cdot x \cdots \text{to } m + n \text{ factors}$$

$$= x^{m+n}$$

It can be shown that the rule holds true also for all types of exponents, such as negative, zero, fractional, and literal. If no exponent is expressed, the exponent is understood to be 1. The rule can be extended to cover the product of several factors expressed as powers on the same base:

$$10^5 \cdot 10^3 \cdot 10 = 10^9; \qquad x^4 \cdot x^0 \cdot x^{-7} = x^{-3}; \qquad x^n \cdot x^c \cdot x = x^{n+c+1}$$

The rule may be illustrated with an example involving only arithmetic numbers. Consider the example $(5^2)(5^4)$. By Rule 1, the product is $(5^2)(5^4) = 5^{2+4} = 5^6$. Note that the base does not change. If we expand each power before multiplying, we get the same product. The problem $(5^2)(5^4)$ becomes $(25)(625) = 15{,}625$. The product 15,625 can be expressed as 5^6.

Exponents on different bases cannot be added. For instance, in multiplying $(x^2) \cdot (y^4)$, we *cannot* add the exponents. The product can only be *expressed* as x^2y^4.

In an example involving only arithmetic numbers each factor can first be expanded and the results multiplied: thus

$$(3^2)(5^4) = (9)(625) = 5625$$

16-2 DIVISION

Division is the inverse of multiplication. Therefore, we can state the rule for division as the inverse of the rule for multiplication.

Rule 2. *When we divide two quantities expressed as powers of the same base, we subtract the exponent in the divisor from the exponent in the dividend.*

The rule may be stated as a formula:

$$x^m \div x^n = x^{m-n}$$

Let us see why this rule holds true for an example such as

$$x^7 \div x^4$$

If we write the division as a fraction and then factor each quantity, we have

$$\frac{x^7}{x^4} = \frac{x \cdot x \cdot x \cdot x \cdot x \cdot x \cdot x}{x \cdot x \cdot x \cdot x} = x^3$$

The numerator and the denominator can be divided by x four times, leaving three x's in the numerator to be multiplied together.

Note carefully that the rule says, "when we *divide two quantities,* we *subtract exponents.*" In the example shown, we are *not* dividing the exponents 7 and 4. Instead we are *dividing* the *quantity* x^7 by the *quantity* x^4.

To show that the rule holds true for positive integral exponents, we take the formula and analyze its meaning:

$$x^m \div x^n = x^{m-n}$$

$$\frac{x^m}{x^n} = \frac{x \cdot x \cdot x \cdots \text{to } m \text{ factors}}{x \cdot x \cdot x \cdots \text{to } n \text{ factors}} = x \cdot x \cdot x \cdots \text{to } (m - n) \text{ factors}$$

$$= x^{m-n}$$

Assuming that there are not more factors in the denominator than in the numerator, then the factor x can be divided into numerator and denominator n times, leaving only $(m - n)$ factors in the numerator.

It can be shown that the rule holds true also for all types of exponents, such as negative, zero, fractional, and literal. Thus

$$10^8 \div 10^2 = 10^6 \qquad x^n \div x = x^{n-1}$$

$$x^6 \div x^{-3} = x^9 \qquad x^{\frac{3}{4}} \div x^{\frac{1}{3}} = x^{\frac{5}{12}}$$

$$x^4 \div x^0 = x^4$$

The rule for division may be illustrated with an example involving only arithmetic numbers. Consider the example $3^6 \div 3^2$. By Rule 2, the quotient is $3^6 \div 3^2 = 3^{6-2} = 3^4$. Note that the base does not change. If we expand each power before dividing, we get the same quotient. The problem $3^6 \div 3^2$ becomes $729 \div 9 = 81$. The quotient, 81 can be expressed as 3^4.

Exponents on different bases cannot be subtracted. For instance, in the example $x^6 \div y^2$ we *cannot* subtract the exponents. The quotient can only be expressed as $x^6 \div y^2$ or x^6/y^2.

In an example involving only arithmetic numbers the dividend and the divisor can first be expanded. Then the division itself can be performed: thus

$$4^5 \div 3^2 = 1024 \div 9 = 113\tfrac{7}{9}$$

16-3 POWER OF A POWER

The example $(x^4)^3$ may be called a power of a power; that is, the third power of the fourth power of x. We can simplify this expression if we recall the meaning of parentheses. The expression $(x^4)^3$ means that the quantity x^4 is to be used three times as a factor: $(x^4) \cdot (x^4) \cdot (x^4)$. Now we apply the rule for multiplication:

$$(x^4)(x^4)(x^4) = x^{4+4+4} = x^{12}$$

Therefore, the expression $(x^4)^3$ can be simplified to x^{12}. The new exponent is the product of the two exponents 3 and 4.

Rule 3. *If finding a power of a power, we multiply the exponents to get the new exponent on the same base.*

Note that the base remains the same. The rule may be stated as a formula:

$$(x^m)^n = x^{mn}$$

To prove the formula for positive integral exponents, we analyze its meaning.

$$\begin{aligned}(x^m)^n &= x^m \cdot x^m \cdot x^m \cdots \text{to } n \text{ factors}\\ &= x^{m+m+m \cdots \text{to } n \text{ terms}}\\ &= x^{mn}\end{aligned}$$

The rule holds true for all kinds of exponents. The following examples show its applications:

$$(x^5)^3 = x^{15} \qquad (n^6)^{\frac{1}{2}} = n^3 \qquad (10^{20})^2 = 10^{40}$$
$$(x^4)^0 = x^0 \qquad (y^{-3})^4 = y^{-12} \qquad (x^{3.14})^2 = x^{6.28}$$

The rule may be illustrated with an example involving only arithmetic numbers. Consider the example $(3^2)^4$. By Rule 3,

$$(3^2)^4 = 3^{2 \cdot 4} = 3^8 = 6561$$

Now, if we first expand the quantity within parentheses, we get the same answer.

$$(3^2)^4 = (9)^4 = 6561$$

In the case of continued powers we multiply the several exponents together. As an example,

$$[(2^2)^3]^5 = 2^{(2)(3)(5)} = 2^{30}$$

16-4 POWER OF A PRODUCT

In the expression $(xy)^3$ the exponent 3 indicates that the entire quantity xy within the parentheses is to be raised to the third power; that is, $(xy)^3$ means $(xy)(xy)(xy)$. If we write the expression as separate prime factors, we have

$$\begin{aligned}(xy)^3 &= x \cdot y \cdot x \cdot y \cdot x \cdot y\\ &= x \cdot x \cdot x \cdot y \cdot y \cdot y\\ &= x^3y^3\end{aligned}$$

This example is an illustration of the rule for the power of a product.

Rule 4. *The power of a product of several factors is equivalent to the product of the same power of each of the separate factors.*

The rule may be stated as a formula:

$$(xyz)^n = x^n y^n z^n$$

It may be illustrated with an example involving arithmetic numbers:

$$(3 \cdot 5)^4 = 3^4 \cdot 5^4 = (81) \cdot (625) = 50{,}625.$$

If we expand the quantity within the parentheses, we get the same answer:

$$(3 \cdot 5)^4 = (15)^4 = 50{,}625.$$

Simplifying an expression often involves several rules in the same problem.

Example 1

Simplify

$$(5x^2yz^4)^3$$

Solution

The expression, first of all, is a power of a product. The exponent 3 is placed on each of the factors within the parentheses: thus

$$5^3(x^2)^3y^3(z^4)^3$$

Now we have examples of a power of a power. In such cases the exponents are multiplied together. The expression becomes

$$125x^6y^3z^{12}$$

Example 2

Simplify

$$(3x^{-4}y^5z^{-8})^{-\frac{1}{2}}$$

Solution

$$3^{-\frac{1}{2}}x^2y^{-\frac{5}{2}}z^4$$

If we have the product of several factors, each with separate powers, we must remember that *an exponent applies only to the factor on which it is placed*. For instance, in the expression xy^2 the exponent 2 applies only to the y.

Notice the difference in these values. If $x = 3$ and $y = 5$, then

$$xy^2 = 75; \qquad \text{that is, } 3 \cdot 25 = 75,$$

but

$$(xy)^2 = 225, \qquad \text{that is, } (15)^2 = 225$$

We must be especially careful when negative signs are involved. If $x = 7$, then

$$(-x)^2 = (-7)^2 = +49; \qquad \text{but } -x^2 = -7^2 = -49$$

16-5 POWER OF A FRACTION

In the power of a fraction, such as $(\frac{3}{5})^4$, the expression within the parentheses is to be treated as a single quantity. The exponent 4 applies to the entire fraction; that is,

$$\left(\frac{3}{5}\right)^4 \text{ means } \frac{3}{5} \cdot \frac{3}{5} \cdot \frac{3}{5} \cdot \frac{3}{5}$$

If we now apply the rule for multiplying fractions, we get

$$\frac{3}{5} \cdot \frac{3}{5} \cdot \frac{3}{5} \cdot \frac{3}{5} = \frac{3 \cdot 3 \cdot 3 \cdot 3}{5 \cdot 5 \cdot 5 \cdot 5}$$

The product may be indicated as

$$\frac{3^4}{5^4}$$

In more general terms

$$\left(\frac{x}{y}\right)^3 = \frac{x}{y} \cdot \frac{x}{y} \cdot \frac{x}{y} = \frac{x \cdot x \cdot x}{y \cdot y \cdot y} = \frac{x^3}{y^3}$$

For the power of a fraction we have this rule.

Rule 5. *An exponent placed on a fraction as a quantity indicates that both numerator and denominator are to be raised to the indicated power.*

The rule may be stated as a formula:

$$\left(\frac{x}{y}\right)^n = \frac{x^n}{y^n}$$

Example 1

Simplify

$$\left(\frac{4x^2y^4}{5z^5}\right)^3$$

Solution

By the fraction rule, the exponent 3 applies to numerator and denominator.

Therefore, we have

$$\left(\frac{4x^2y^4}{5z^5}\right)^3 = \frac{(4x^2y^4)^3}{(5z^5)^3}$$

Since the numerator and denominator consist of several factors, the exponent 3 applies to each factor by Rule 4, the product rule. Therefore, we get

$$\frac{(4)^3(x^2)^3(y^4)^3}{(5)^3(z^5)^3}$$

Now we have powers of other powers. Therefore, we apply Rule 3 for the power of a power and get

$$\frac{64x^6y^{12}}{125z^{15}}$$

The rules can be applied in a single step.

Example 2

Simplify

$$\left(\frac{2x^3y^0z^{\frac{1}{2}}}{3a^{-4}b^{-2}c^0}\right)^4$$

Solution

Applying several rules at once, we get

$$\left(\frac{2x^3y^0z^{\frac{1}{2}}}{3a^{-4}b^{-2}c^0}\right)^4 = \frac{16x^{12}y^0z^2}{81a^{-16}b^{-8}c^0}$$

Exercise 16-1

Multiply as indicated:

1. $(x^6)(x^3)$ **2.** $(y^3)(y)$ **3.** $(a^7)(a^{-2})$

4. $(x^{3n})(x^{2n})$ **5.** $(n^{-3})(n^{-4})$ **6.** $(b^{4n})(b^3)$

7. $(x^5)(x^0)$ **8.** $(x^{1.3})(x^{2.5})$ **9.** $(x^{\frac{2}{3}})(x^{\frac{5}{8}})$

10. $(x^n)(x^a)(x)$ **11.** $(10^4)(10^3)(10)$ **12.** $(2^{3n})(2)$

13. $(x^2)(x^8)(x^{-3})(x)$ **14.** $(y^{3.5})(y^{2.1})(y^{-1})$ **15.** $(y^{3a+2})(y^{2a-5})$

16. $(10^{4.153})(10^{1.572})$ **17.** $(10^{2.4})(10^{-3.25})$ **18.** $(3^2)(3^4)(2^3)$

Divide as indicated:

19. $x^8 \div x^2$ **20.** $x^2 \div x^{-5}$ **21.** $n^{-3} \div n^{-7}$

22. $x^n \div x$ **23.** $y^5 \div y^0$ **24.** $x^5 \div x^5$

25. $n^{25} \div n^5$ **26.** $x^{3n} \div x^n$ **27.** $y \div y^{4.326}$

28. $10^{2.9} \div 10^{1.53}$ **29.** $10^{1.42} \div 10^{3.71}$ **30.** $10^{2.4} \div 10^{-1.6}$

31. $x^{a^2} \div x^a$ **32.** $x^{n-1} \div x^{2n+3}$ **33.** $4^5 \div 3^2$

Find the power indicated:

34. $(-7)^3$ **35.** $-(-2)^4$ **36.** $(-\frac{1}{3})^4$ **37.** $(-0.1)^4$

38. $(x^5)^4$ **39.** $(-y^3)^5$ **40.** $(x^4)^0$ **41.** $(10^4)^{-3}$

42. $(y^{-3})^{-5}$ **43.** $(x^5)^{-\frac{1}{5}}$ **44.** $(a^{-6})^{-\frac{2}{3}}$ **45.** $(2^{2x})^x$

46. $(2^{-3})^{-6}$ **47.** $(10^{-4})^{1.5}$ **48.** $(10^{-2.7})^{-1.2}$ **49.** $(10^{3.56})^{\frac{1}{4}}$

50. $(-5x^3y^{-1})^4$ **51.** $x^3y^4)^4$ **52.** $(3x^3)^3$ **53.** $(4x^4)^4$

54. $-(a^0b^3)^5$ **55.** $(3x^2y^{-3})^3$

56. $(6a^{-3}b^4c^{-1})^{-2}$ **57.** $(2x^{-3})^2(x^{-2})^3$

58. $(a^{\frac{3}{2}}b^{\frac{2}{3}}c^{-\frac{1}{2}})^6$ **59.** $-(-4a^{-\frac{1}{6}}b^{\frac{1}{3}})^3$ **60.** $(3^{-1}x^{-\frac{1}{2}}y^{\frac{3}{4}})^{-6}$

61. $\left(\frac{4x^2}{5y^3}\right)^4$ **62.** $\left(\frac{3a^{-2}}{4b^5}\right)^3$ **63.** $\left(\frac{2n^{\frac{1}{2}}}{y^{-\frac{1}{3}}}\right)^6$

64. $\left(\frac{2x^{-2}y^3}{3a^4b^{-1}}\right)^5$ **65.** $\left(\frac{x^0y^4z^2}{a^6b^8c^4}\right)^{\frac{1}{2}}$. **66.** $\left(\frac{x^{-4}y^{-3}z^0}{x^0y^0z^{-3}}\right)^{-2}$

16-6 ZERO EXPONENT

In mathematics it often happens that we get a zero exponent, as in the terms x^0, $5x^0$, 7^0, and 10^0. Such expressions must be given some logical meaning.

Whenever an unusual expression forces itself into mathematics, we try to give it some reasonable and logical interpretation. At first sight some expressions appear to be illogical and meaningless. We have mentioned the fact that when negative numbers forced themselves into mathematics people discarded them as "fictitious." Now, when we try to explain the meaning of a zero exponent, as we explain other exponents, we run into trouble.

We have said that the expression 5^3 means that three 5's are to be multiplied together. In other words,

5^3 means (5)(5)(5) 5^1 means (5)

5^2 means (5)(5) 5^0 means ?

An exponent, such as 3, means that we write down three 5's and multiply them together. The same meaning is given to other positive integral exponents. If we try to explain that the zero (0) tells how many 5's to use as factors, the idea does not make sense. We might be inclined to say that the entire quantity is equal to zero. But we shall see that this is not correct.

The fact is that we cannot interpret a zero exponent in the same way we interpret other exponents. Yet we must give it some meaning. We *cannot* get at a correct meaning by the direct approach used with other exponents. We must get at the meaning in a roundabout way.

As a student, you should realize that in mathematics we do not deliberately set out to produce riddles or enigmas or concepts that are confusing. However, whenever a confusing or unusual idea forces itself into mathematics, such as a negative number or a zero exponent, the new idea must be given some logical interpretation. To get some meaning into the expression x^0, let us see how it comes about.

Suppose we have the division problem

$$x^5 \div x^5$$

Now, we know that the rule for division involving terms with exponents on the same base is shown by the formula $x^m \div x^n = x^{m-n}$; that is, when we divide quantities expressed as powers on the same base we subtract the exponents. Let us do the same with $x^5 \div x^5$. We get

$$x^5 \div x^5 = x^{5-5} = x^0$$

Whether or not we like it, the division results in x^0.

Now let us see what happens if we write the problem in a different form. The division can be written as a fraction:

$$\frac{x^5}{x^5}$$

This fraction reduces to 1. We know that any number divided by itself is 1. Yet, when we use the rule for exponents in division of quantities, we get $x^5 \div x^5 = x^0$. Therefore, we must conclude that the two answers are equal to each other, and

$$x^0 \text{ must be equal to } 1$$

Whatever example we use as an illustration, the answer always turns out to be 1. This means that if we have a zero exponent on any base (except 0) the *entire quantity* is equal to 1. Examples:

$$(3x)^0 = 1 \qquad (4x^2y^3)^0 = 1 \qquad (a^2 - 3ab - 5b^2)^0 = 1$$

However,

$$3x^0 = 3(1) = 3$$

Concerning a zero exponent, we have the following rule:

Rule 6. *Any quantity (except 0) expressed with a zero exponent is equal to 1.*

This meaning of a zero exponent is consistent with the rule for exponents in multiplication. For instance,

$$x^6 \cdot x^0 = x^{6+0} = x^6$$

By the rule in multiplication, the exponent in the product is $6 + 0 = 6$. If we remember that $x^0 = 1$, we see that $x^6 \cdot x^0 = x^6 \cdot 1 = x^6$.

16-7 NEGATIVE EXPONENT

In mathematics we sometimes get an expression with a negative exponent, such as 5^{-3}. Just as in the case of a zero exponent, a negative exponent must have some logical meaning. If we try to explain the meaning of a negative exponent as we explain the meaning of positive integral exponents, we again run into difficulty.

Yet, when an expression such as 5^{-3} does occur in mathematical computation, it must be given some logical meaning. Just as with a zero exponent, we can get some interpretation of its meaning by noting how a negative exponent comes about.

In the division of quantities expressed with exponents on the same base we have the rule that the exponents are subtracted; that is,

$$x^7 \div x^3 = x^4$$

By observing this rule carefully, we sometimes get negative exponents:

$$x^2 \div x^5 = x^{2-5} = x^{-3}$$

Now, if we write the division as a fraction and reduce, we get

$$\frac{x^2}{x^5} = \frac{x \cdot x}{x \cdot x \cdot x \cdot x \cdot x} = \frac{1}{x^3}$$

Therefore, we must conclude that

$$x^{-3} = \frac{1}{x^3}$$

This example is an illustration of the following rule.

Rule 7. *Any quantity with a negative exponent may be written as the reciprocal of the quantity with a positive exponent.*

The rule may be restated:

Any factor in the numerator of a fraction may be transferred to the denominator provided the sign of its exponent is changed; any factor in the denominator may be transferred to the numerator provided the sign of its exponent is changed.

Warning. Separate *terms* of polynomial numerators or denominators may *not* be shifted in this fashion. The rule says *factor*, not term.

As an example of the use of Rule 7, consider the following expression:

Example 1

$$\frac{a^{-1}b^2}{x^3y^{-4}}$$

In this expression the factor a^{-1} may be transferred to the denominator and the factor y^{-4} may be transferred to the numerator provided the signs of the exponents in these factors are changed: thus

$$\frac{a^{-1}b^2}{x^3y^{-4}} = \frac{b^2y^4}{ax^3}$$

The expression may also be written in a form in which the denominator is 1 by transferring all the terms to the numerator: thus

$$\frac{a^{-1}b^2}{x^3y^{-4}} = \frac{a^{-1}b^2x^{-3}y^4}{1} = a^{-1}b^2x^{-3}y^4$$

Example 2

Note especially the difference between the following two examples:

$$\text{(A)}\ \frac{a}{b^{-2}c^{-2}} \qquad \text{(B)}\ \frac{a}{b^{-2}+c^{-2}}$$

In Example A the two quantities, b^{-2} and c^{-2}, in the denominators are factors. Therefore, they can be transferred to the numerator if the signs of their exponents are changed. However, notice the difference in Example B, in which the denominator consists of two separate *terms*. These terms cannot be moved to the numerator because the denominator is an indicated *sum*. Example A can be changed to ab^2c^2. However, Example B is simplified as follows:

$$\frac{a}{b^{-2}+c^{-2}} = \frac{a}{\dfrac{1}{b^2}+\dfrac{1}{c^2}} = \frac{a}{\dfrac{c^2+b^2}{b^2c^2}} = \frac{ab^2c^2}{c^2+b^2}$$

A somewhat simpler method of simplifying Example B is to multiply the numerator and the denominator of the original fraction by the quantity b^2c^2. We get the simplified form in one step:

$$\frac{b^2c^2)}{b^2c^2)} \cdot \frac{a}{b^{-2}+c^{-2}} = \frac{ab^2c^2}{c^2+b^2}$$

Exercise 16-2

Simplify the following expressions as much as possible:

1. 3^0 **2.** $-x^0$ **3.** $(5y)^0$ **4.** $5x^0$

5. $-(6x)^0$ **6.** $-3a^0b^0$ **7.** $-7^0x^0y^0$ **8.** $(8x^2y^3)^0$

9. $8(x^2y^3)^0$

10. $-3(a^0b^3)^0$

11. $-6(x^2 - 5^0)^0$

12. $(4x)^0 + 7x^0 - 3y^0$

13. $\frac{(8x^0)(5y^0)}{4(2xy)^0}$

14. $\frac{3x^0 + 5x^0}{2y^0}$

15. $\frac{(3x)^0 - 7x^0}{4x^0 - (6x)^0}$

16. $\frac{9n^0 - (6n)^0}{(7n^0) - 4n^0}$

Express each of the following without negative exponents:

17. $-x^{-3}y^2z^{-1}$

18. $a^{-3}b^4c^{-2}$

19. $a^5b^{-4}c^{-1}d$

20. $2^{-1}a^2b^{-5}c^3$

21. $\frac{1}{-3^2ab^3c^{-1}}$

22. $\frac{-3^{-2}x^{-3}y^{-1}}{4^{-1}a^{-2}bx}$

23. $\frac{5x^{-3}y^2z^{-1}}{2ab^{-4}c^3}$

24. $\frac{1}{a^{-1} + b^{-1}}$

25. $\frac{a - b}{a^{-1} - b^{-1}}$

26. $\frac{x^{-2} - y^{-2}}{x^{-1}y^{-1}}$

27. $\frac{x^{-2} - y^{-2}}{x^{-1} + y^{-1}}$

28. $\frac{a^{-3} + b^{-3}}{a^{-1} - b^{-1}}$

Evaluate each of the following expressions:

29. $4^{-2} + 3^{-2}$

30. $5^{-2} - 2^{-3}$

31. $(5^{-2})(0.4)^{-1}$

32. $(3^{-1})(4^{-2})(6^0)$

33. $50^{-1} + 40^{-1}$

34. $2^{-3} + 3^{-2} - 6^{-1}$

35. $(4^0) + 3x^0 - 9^{-\frac{1}{2}}$

36. $\frac{1}{4^{-2} + 2^{-3}}$

37. $\frac{3^{-2} + 2^0}{2^{-3} + 3^0}$

38. $\frac{3^{-2} - 5^{-2}}{3^{-1} + 5^{-1}}$

Find the numerical value of each of the following, if $a = 3$, $b = 2$, $x = -2$.

39. $-a^2$

40. $(-a)^2$

41. $-x^2$

42. $-ax^2$

43. $(-ab)^2$

44. $(ax)^3$

45. $-x^4$

46. a^2bx^3

47. $-a^2bx$

48. $-(ax)^2b$

49. $a^{-2}x^{-1}$

50. $a^{-1} + b^{-1}$

51. $\frac{3a^0x^2}{10a^3x^0}$

52. $\frac{4a^2x}{9a^0b^{-1}}$

53. $\frac{ax^3b}{a^{-1}x^0b^{-1}}$

54. $\frac{1}{a^{-1} + b^{-1}}$

16-8 ROOTS OF NUMBERS

Before explaining a fractional exponent, we must mention briefly the meaning of a root of a number. A root of a number may be defined as one of its equal factors. The *square root* of a number is one of the *two* equal factors of the number. For example, the square root of 169 is 13 because 169 is equal to the product of the two factors, 13 and 13.

The *cube root* of a number is one of the *three* equal factors of the number. The cube root of 512 is 8 simply because 512 = (8)(8)(8). The *fourth root* of a number is one of the *four* equal factors of the number. Since 81 = (3)(3)(3)(3), then the fourth root of 81 is 3.

The symbol for a root is the *radical sign* ($\sqrt{\ }$). This symbol indicates that a root is to be found. The number under the radical sign is the number of which the root is to be found and is called the *radicand*. The particular root to be found is indicated by a small number placed in the notch of the radical sign. This number is called the *index* of the root.

(It is sometimes called the *index of the radical.*) Thus, the expression

$$\sqrt[3]{64}$$

means that the cube root of 64 is to be found. The entire expression is called a *radical.* In this example the radicand is 64 and the index is 3. The cube root of 64 is 4 because 4 is one of the three equal factors of 64. Therefore, we can say

$$\sqrt[3]{64} = 4$$

If the index of the root is not shown, then the square root is meant. Thus,

$$\sqrt{81} = 9$$

The following examples are illustrations of roots of numbers:

$$\sqrt{361} = 19 \qquad \sqrt[5]{32} = 2 \qquad \sqrt{9x^2} = 3x$$
$$\sqrt[3]{125} = 5 \qquad \sqrt[4]{10{,}000} = 10 \qquad \sqrt[3]{8x^6} = 2x^2$$

16-9 FRACTIONAL EXPONENTS

Whether or not we like it, we sometimes get a fractional exponent such as $5^{\frac{1}{2}}$. When this happens the expression must be given some meaning. The expression does *not* mean $\frac{1}{2}$ of 5, which is $2\frac{1}{2}$. If we take $\frac{1}{2}$ of 5, then the "$\frac{1}{2}$" is a factor, not a power. Our problem here is to determine the meaning when the "$\frac{1}{2}$" is the exponent of a power, as in $5^{\frac{1}{2}}$.

If we say that an exponent shows the number of times the base is used as a factor, then the expression has no meaning. Again, we must get at the meaning in another way. Let us see what meaning can be given to this kind of exponent. Consider the following example. Since

$$(17)(17) = 289$$

then

$$17 = \sqrt{289}$$

That is, if two equal factors are multiplied to form a product, one of these factors is the square root of the product. Now, we recall that when we multiply powers of the same base we add the exponents. Then

$$(x^{\frac{1}{2}})(x^{\frac{1}{2}}) = x^{\frac{1}{2}+\frac{1}{2}} = x$$

Here we have two equal factors whose product is x. One of the factors, $x^{\frac{1}{2}}$, must be the square root of x; that is,

$$x^{\frac{1}{2}} = \sqrt{x}$$

Therefore, we can say that *the fractional exponent,* $\frac{1}{2}$, means the same as the *square root of the base.* Here are two examples:

$$64^{\frac{1}{2}} = \sqrt{64} = 8; \qquad 5^{\frac{1}{2}} = \sqrt{5} = 2.236 \text{ (approx.)}$$

For the exponent $\frac{1}{3}$ we can get a meaning in the same way. Since

$$(7)(7)(7) = 343$$

then

$$7 = \sqrt[3]{343}$$

That is, if three equal factors are multiplied to form a product, one of these factors is the cube root of the product. By the rule for the multiplication of powers of the same base, we have

$$(x^{\frac{1}{3}})(x^{\frac{1}{3}})(x^{\frac{1}{3}}) = x^{\frac{1}{3}+\frac{1}{3}+\frac{1}{3}} = x$$

Here we have three equal factors whose product is x. One of the factors $x^{\frac{1}{3}}$, must be the cube root of x; that is,

$$x^{\frac{1}{3}} = \sqrt[3]{x}$$

Therefore, we can say that *the fractional exponent* $\frac{1}{3}$ means the same as *the cube root of the base.* The fractional exponents, $\frac{1}{4}$, $\frac{1}{5}$, etc., have similar meanings. Here are some examples showing the meaning of such exponents:

$$8^{\frac{1}{3}} = \sqrt[3]{8} = 2 \qquad 64^{\frac{1}{3}} = \sqrt[3]{64} = 4$$

$$16^{\frac{1}{4}} = \sqrt[4]{16} = 2 \qquad (100{,}000)^{\frac{1}{5}} = \sqrt[5]{100{,}000} = 10$$

Finally, let us consider the meaning of the exponent $\frac{2}{3}$, in which the numerator is different from 1. The meaning of such an exponent can be explained in the same way as in the previous examples. By the rule for the multiplication of powers of the same base, we have

$$(x^{\frac{2}{3}})(x^{\frac{2}{3}})x^{\frac{2}{3}}) = x^{\frac{6}{3}} = x^2$$

Here, again, we have three equal factors whose product is x^2. Therefore, one of the factors, $x^{\frac{2}{3}}$, must be the cube root of x^2, that is,

$$x^{\frac{2}{3}} = \sqrt[3]{x^2}$$

This final example is an illustration of the general rule for the meaning of fractional exponents:

Rule 8. *In a fraction exponent, such as p/r, the numerator, p, of the exponent, indicates a power, and the denominator, r, indicates a root.*

The rule may be stated as a formula.

$$N^{p/r} = \sqrt[r]{N^p}, \qquad \text{or} \qquad (\sqrt[r]{N})^p$$

That is, a number, N, raised to the p/r power is equal to the rth root of the pth power of N, or to the pth power of the rth root of N.

Example 1

Find the value of $8^{\frac{2}{3}}$.

The expression $8^{\frac{2}{3}}$ means the cube root of 8^2, or $\sqrt[3]{64}$, which is 4. The expression can also be simplified as the square of the cube root of 8; that is, we can first find the cube root of 8, which is 2. This value can be squared. The result, 4, is the same as in the first method.
In an example of this kind either process may be performed first. We may start out with the power or the root. If the value is rational, then the root should always be found first.

Example 2

Find the value of $125^{\frac{2}{3}}$.

The expression $125^{\frac{2}{3}}$ is the same as $(125^{\frac{1}{3}})^2$ or $(125^2)^{\frac{1}{3}}$, that is, we may first find the cube root of 125 and then square this value; or we may first square 125 and then find the cube root of 125^2. If we find the root first, we have

$$125^{\frac{1}{3}} = \sqrt[3]{125} = 5$$

Then we square 5 and get $5^2 = 25$
If we find the power first, we have $125^2 = 15{,}625$
Then we find the cube root and get $(15{,}625)^{\frac{1}{3}} = \sqrt[3]{15{,}625} = 25$
The answer is the same in both cases. However, the procedure is often simpler when the root is found first.

Example 3

Find the value of $7^{\frac{3}{2}}$.

In this example, if we first find the square root of 7, we get the approximate value 2.646. This number must be raised to the third power. In this instance it is easier to find the third power first and then find the square root of the result. Thus

$$7^3 = 343$$

then

$$\sqrt{343} = 18.52 \text{ (approx.)}$$

Exercise 16-3

Express the following with fractional exponents:

1. $\sqrt{ab}$ **2.** $\sqrt{29}$ **3.** $\sqrt[3]{343}$ **4.** $\sqrt{x^2 + y^2}$

5. $\sqrt[4]{x^3 + y^2}$ **6.** $\sqrt[5]{x + 5}$ **7.** $\sqrt[3]{x^2y^2}$ **8.** $\sqrt{(x^3 - 8)^3}$

Evaluate each of the following expressions:

9. $8^{\frac{3}{4}}$ **10.** $64^{\frac{2}{3}}$ **11.** $(4/9)^{\frac{1}{2}}$ **12.** $(25/49)^{\frac{1}{2}}$

13. $(9/25)^{\frac{3}{2}}$ **14.** $(-8)^{\frac{1}{3}}$ **15.** $(0.01)^{\frac{1}{2}}$ **16.** $8^{\frac{5}{3}}$

17. $32^{\frac{3}{5}}$ **18.** $125^{-\frac{2}{3}}$ **19.** $64^{-\frac{1}{2}}$ **20.** $(-32)^{\frac{4}{5}}$

21. $-(32)^{\frac{4}{5}}$ **22.** $[(4^2)(3^2]^{\frac{1}{2}}$ **23.** $(4^2 + 3^2)^{\frac{1}{2}}$ **24.** $(6^2 + 8^2)^{\frac{1}{2}}$

25. $(15^2 + 8^2)^{\frac{1}{2}}$ **26.** $(3^2 - 2)^{\frac{3}{2}}$ **27.** $(11^2 + 4)^{\frac{2}{3}}$ **28.** $(12^2 + 16^2)^{\frac{3}{2}}$

Exercise 16-4 (Review)

1. $9^{-\frac{1}{2}}$ **2.** 7^{-2} **3.** $16^{\frac{3}{2}}$ **4.** -5^2

5. $x^n \div x^{3n}$ **6.** $y^n \div y$ **7.** $(h)(h^{2n-3})$ **8.** $8^{-\frac{7}{3}}$

9. $(3x)^{-4}$ **10.** $4^{-1}x^2y^{-3}$ **11.** $3^2x^{-3}y^4$ **12.** $5(2^{-1})(x^2)$

13. $\left(\frac{-2x^2}{3a^0}\right)^2$ **14.** $\left(\frac{x^3y^{-2}}{3x^2y^{-1}}\right)^2$ **15.** $\frac{125^{\frac{2}{3}}}{10^2}$ **16.** $\frac{32^{\frac{2}{5}}}{8^0}$

17. $\frac{64^{\frac{2}{3}}}{(9 + 16)^{\frac{1}{2}}}$ **18.** $\frac{3xa^{-2}b^3}{2^{-1}x^{-3}a}$ **14.** $\frac{2^{-3}a^{-2}x^{-1}}{5^2a^{-2}y^3x^4}$ **20.** $\left(\frac{4x^2y^3z^{\frac{4}{3}}}{9^{\frac{1}{2}}a^4b^2c^{\frac{2}{3}}}\right)^{\frac{3}{2}}$.

16-10 SCIENTIFIC NOTATION

In scientific work we often use numbers that are very large or very small. If such numbers are written in the usual way, they are sometimes awkward to use in computation. For instance, the number of electrons in one coulomb is approximately 6,280,000,000,000,000,000. The wavelength of yellow light is approximately 0.000023 in. Such numbers are often written in a form called *scientific notation.*

A number is expressed in scientific notation in the following manner. Consider the number 93,200,000. In order to express this number in scientific notation, we first place the decimal point just to the right of the first significant digit. The first significant digit in a number is the first digit that is not zero, starting at the left. In the number 93,200,000 the first significant digit is 9. Therefore, the decimal point is placed between the 9 and the 3:

$$9.32$$

This position is called the *standard position* of the decimal point.

Now this number, 9.32, is multiplied by a power of 10 that will make the value equal to the original value. The correct power of 10 to be used is determined by counting the number of places *from standard position* to the decimal point in the original number. In the number 93,200,000 the decimal point is understood to be at the right of the number, since it is a whole number. From the standard position, between 9 and 3, we count seven places to the *right* to the decimal point in the original number.

```
93,200,000.
 ↑
Standard
position
```

The power of 10 to be used in this example is therefore 7. Written in scientific notation,

$$93{,}200{,}000 = (9.32)(10^7)$$

In order to see that the value of the original number is not changed, let us perform the multiplication $(9.32)(10^7)$. The expression means (9.32)(10000000). Multiplying, we get 93,200,000.00. The value of the original number has not been changed.

In the case of a decimal fraction the number can be written in a similar manner. However, the power of 10 will be negative. For instance, consider the number 0.00000327. The first significant digit in this number is 3. Therefore, the standard position of the decimal point is between the 3 and the 2; thus

$$3.27$$

Now we count the number of places from the standard position to the decimal point in the original number. Notice that, in counting, we move to the *left* instead of the right. The decimal point is six places to the left from standard position

```
0.00000327
       ↑
   standard
   position
```

Therefore, we multiply the number 3.27 by 10^{-6}. Writing the number in scientific notation, we have

$$0.00000327 = (3.27)(10^{-6})$$

The advantage of scientific notation is most evident in connection with extremely large or extremely small numbers. However, any number, no matter what its size may be, can be written in this form.

The number 25 can be written $(2.5)(10^1)$ or simply (2.5)(10).

The number 7.2 can be written $(7.2)(10^0)$, which is (7.2)(1).

The number 0.38 can be written $(3.8)(10^{-1})$.

One advantage, aside from its use in any form of computation, is that extremely large or extremely small numbers may be more easily compared when written in this form. For example, it is difficult at first glance to tell which of the following numbers is the larger:

$$870000000000000000 \quad \text{or} \quad 63000000000000000000$$

However, if these numbers are written in scientific notation, the larger one can be identified at a glance:

$$870000000000000000 = (8.7)(10^{17})$$

$$63000000000000000000 = (6.3)(10^{19})$$

The following examples show numbers written in scientific notation:

$$5423000000 = (5.423)(10^9)$$
$$165.4 = (1.654)(10^2)$$
$$0.00000007328 = (7.328)(10^{-8})$$
$$1000000000000000 = (1)(10^{15}) = 10^{15}$$
$$0.000000000000000000000001 = (1)(10^{-24}) = 10^{-24}$$
$$0.000000000006 = (6)(10^{-12})$$

Scientific notation is convenient in computation involving very large or very small numbers. Numbers are first written in scientific notation. Then the powers of 10 are combined to determine the position of the decimal point in the answer.

Example 1

Evaluate

$$(324000)(2250).$$

Solution

In scientific notation, we have $(3.24)(10^5)(2.25)(10^3)$.
Rearranging factors, $(3.24)(2.25)(10^5)(10^3)$.
Multiplying the first two factors, $(3.24)(2.25) = 7.29$ (rounded off). Combining the powers of 10, we get 10^8. The power of 10 determines the position of the decimal point. Then we get $(7.29)(10^8) = 729{,}000{,}000$.

Example 2

Evaluate

$$(268)(0.0000425)(0.820)$$

Solution

In scientific notation: $(2.68)(10^2)(4.25)(10^{-5})(8.20)(10^{-1})$.
Rearranging and combining powers of 10, $(2.68)(4.25)(8.20)(10^{-4})$.
For the product of the first three factors, we get $(93.4)(10^{-4})$. In expanded form, we get 0.00934; in scientific notation, $(9.34)(10^{-3})$. To indicate three-place accuracy, we write the third factor: $(8.20)(10^{-1})$

Example 3

Evaluate

$$\frac{8640000}{27100}$$

Solution

In scientific notation:

$$\frac{(8.64)(10^6)}{(2.71)(10^4)}, \quad \text{or} \quad \left(\frac{8.64}{2.71}\right)\left(\frac{10^6}{10^4}\right)$$

For the first fraction we get $(8.64) \div (2.71) = 3.18$. Subtracting powers of 10, we get 10^2. For the answer we have $(3.18)(10^2)$, or, in expanded form, 318.

Example 4

Evaluate

$$\frac{1}{41800}$$

Solution

In scientific notation, we have $\frac{1}{(4.18)(10^4)}$, or, $\left(\frac{1}{4.18}\right)\left(\frac{1}{10^4}\right)$
Dividing, $1 \div 4.18 = 0.239$; for the power of 10, we have 10^{-4}. For the answer we have $(0.239)(10^{-4})$, or in scientific notation, $2.39(10^{-5})$. In expanded form, we have 0.0000239.

Example 5

Evaluate

$$\frac{(43000)(0.00835)}{(0.900)(5100000)}$$

Solution

In scientific notation:

$$\frac{(4.30)(10^4)(8.35)(10^{-3})}{(9.00)(10^{-1})(5.10)(10^6)}$$

Written as two fractions:

$$\left(\frac{(4.30)(8.35)}{(9.00)(5.10)}\right)\left(\frac{(10^4)(10^{-3})}{(10^{-1})(10^6)}\right)$$

The first fraction reduces to 0.782. The second fraction reduces to 10^{-4}. The answer is 0.0000782, which is written $(7.82)(10^{-5})$.

Exercise 16-5

Write the number in each of the following examples in scientific notation:

1. The sun is about 93,000,000 miles from the earth.

2. Light travels about 186,200 miles per second.

3. The wavelength of red light is approximately 0.00063 mm.

4. The length of a wave of yellow light is about 0.0000228 in.

5. One angstrom unit is a measurement equal to one ten-millionth of a millimeter.

6. An atom weighs about 0.00000000000000000000000166 gram.

7. An atom is approximately 0.000000005 in. in diameter.

8. A certain radio station broadcasts at a frequency of 1260000 cps.

9. Light travels about 5,872,000,000,000 miles in 1 year.

10. One coulomb is equal to about 6,280,000,000,000,000,000 electrons.

Evaluate the following, using scientific notation. Express the answer in scientific notation and also in expanded form. Use three-digit accuracy. If a number begins with 1, we take four significant digits.

11. (38400)(2150) **12.** (4950000)(0.00732)

13. (0.0000615)(8600) **14.** (1285)(0.0000524)(0.00490)

15. $(780000) \div (9400)$ **16.** $(13850) \div (25400000)$

17. $(0.00465) \div (0.00000630)$ **18.** $(0.000830) \div (2600000)$

19. $\frac{(3800)(0.0560)}{420000}$ **20.** $\frac{(0.00935)(0.0720)}{(0.00000231)}$

21. $\frac{1}{247000}$ **22.** $\frac{(0.000430)(8250)}{(0.0000920)(740)}$

QUIZ ON CHAPTER 16, EXPONENTS, POWERS, AND ROOTS. FORM A.

1. Multiply as indicated:

(a) $(x^6)(x^{-2})(x)(x^0)$; **(b)** $(x^{a+2})(x^{2a-3})(x)$;
(c) $(10^7)(10^{-3})(10^0)(10^{-2})(10)$; **(d)** $(10^{2.435})(10^{0.832})$;
(e) $(10^{-4})(10^{0.243})$; **(f)** $(n^4)(n^{-3})(n)(n^{-2})$

2. Divide as indicated:

(a) $x^4 \div x$; **(b)** $n^6 \div n^{-4}$; **(c)** $x^{-3} \div x^2$;
(d) $x^{4.13} \div x^{1.46}$; **(e)** $(10^{5.23}) \div 10^{1.86}$; **(f)** $10^{-4.35} \div 10^{-1.24}$

3. Find the following powers:
(a) $(x^3)^4$; **(b)** $(-x^2)^5$; **(c)** $(x^8)^{\frac{3}{2}}$
(d) $-(-2x^2)^4$; **(e)** $-(-4n^2)^3$

4. Find these powers:

(a) $(3x^2y^{-3}z^0a^{\frac{1}{2}})^4$; **(b)** $(4a^{-1}bc^{-2})^{-3}$; **(c)** $\left(\dfrac{x^2y^{-3}}{a^4b}\right)^4$

5. Evaluate:

(a) $5x^0 + (3x^0) - 2^0 + (3^2)^0$; **(b)** $\dfrac{9x^0 - (4x)^0}{6x^0 - (8x)^0}$

6. Express the following without negative exponents:

(a) $\dfrac{2x^3y^{-1}z^0}{3^{-1}a^{-2}b^{-3}c}$; **(b)** $\dfrac{1}{x^{-1} - y^{-1}}$ **(c)** $\dfrac{x^{-1} + y^{-1}}{x^{-2} - y^{-2}}$

7. Evaluate:

$64^{\frac{1}{2}}$; $8^{\frac{2}{3}}$; $25^{-\frac{1}{2}}$; $9^{-\frac{3}{2}}$; $16^{-\frac{3}{2}}$; $16^{\frac{3}{4}}$; $88^{-\frac{1}{3}}$; $4^{\frac{5}{2}}$

8. Write the following numbers in scientific notation:

(a) 63,200; **(b)** 0.0000032; **(c)** 23.4; **(d)** 0.72

CHAPTER SEVENTEEN Radicals

17-1 SQUARE ROOTS

In much work in algebra, as in arithmetic, we are faced with problems involving radicals, chiefly square roots. In our work in arithmetic we have already found the square roots of arithmetic numbers. For instance, if a square floor has an area of 169 sq ft, the length of one side is found by taking the square root of 169, which is 13. We have usually written the problem

$$\sqrt{169} = 13$$

Negative numbers also appear as the square roots of numbers. In order to be able to work with square roots correctly, it is necessary to understand exactly what is meant by the square root of any number.

By the *square root* of any number, we mean one of the two equal *factors* of that number. For instance, the square root of 1225 is 35 simply because 35 can be multiplied by *itself* to make 1225.

Now, if we multiply the number -35 by *itself*, we get $+1225$; that is, $(-35)(-35) = +1225$. Therefore, we must say that the square root of 1225 can also be -35. The test for the square root of any number is whether the square root can be multiplied by *itself* to produce the given number.

Consider another example. If 17 times 17 is 289, then 17 is the square root of 289. Moreover, since -17 can be multiplied by *itself* to produce $+289$, then -17 is also a square root of 289.

Every number has two square roots, one positive and the other negative. The root indicated by the positive radical is called the *principal square root.* The two square roots of 9 are $+3$ and -3.

$$(+3)(+3) = +9$$

$$(-3)(-3) = +9$$

The principal square root of 9 is $+3$.

When we use the symbol $\sqrt{}$ to indicate a root, we must be careful to use it correctly. The symbol means the *principal root.* The statement $\sqrt{25} = +5$ can be read, "the principal square root of 25 is +5." If we wish to indicate both square roots by use of the symbol, we must write $\pm\sqrt{25} = \pm 5$.

This statement is correct: $\sqrt{25} = +5$ (correct)

This is *not* correct: $\sqrt{25} = \pm 5$ (wrong)

17-2 ANY ROOT OF A NUMBER

If we wish to indicate the multiplication of three 7's, $(7 \cdot 7 \cdot 7)$, we can show it by the exponent 3: thus

$$7^3$$

The expression is read "7 cubed" or "7 raised to the third power." The third power of 7 is 343.

Now, suppose we wish to find the *cube root* of the number 343. By cube root of a given number, we mean one of the *three equal* factors that can be multiplied to form the given number. The problem may be indicated thus:

$$\sqrt[3]{343}$$

In this problem we use the same radical sign ($\sqrt{\ }$). The small 3, placed in the notch of the radical sign, is called the *index* (plural: indices) of the root. It indicates the root to be found; thus

$$\sqrt[3]{343} = 7$$

The square root of a number is indicated by the index 2. However, the square root is used so often that the index is usually omitted.

In the problem $\sqrt[3]{343}$ the number 343, under the radical sign, is called the *radicand.* The radicand is the number of which a particular root is to be found. The *radical* is the entire expression indicating a root.

We have said that the principal root is the positive value of the radical. In some cases the principal root may actually be negative. Suppose we have the problem

$$\sqrt[3]{-64}$$

The radical is understood to be preceded by a plus sign (+), since no sign is shown before it. In this case the principal cube root of -64 is -4, since $(-4)(-4)(-4) = -64$. In fact, the *principal odd root of a negative number is negative.* The *principal odd root of a positive number is positive.*

The following additional examples show principal roots of some numbers:

$\sqrt[4]{81} = 3$ because $3^4 = 81$

$\sqrt[5]{-32} = -2$ because $(-2)^5 = -32$

$\sqrt[7]{-1} = -1$ because $(-1)^7 = -1$

$\sqrt[6]{64} = 2$ because $2^6 = 64$

$\sqrt[3]{-8} = -2$ because $(-2)^3 = -8$

In algebra we use the radical sign with the same meaning as in arithmetic, as shown in the following examples:

$$\sqrt{x^6} = +x^3 \qquad \text{because} \qquad (x^3)(x^3) = x^6$$

$$\sqrt[3]{x^{12}} = +x^4 \qquad \text{because} \qquad (x^4)(x^4)(x^4) = x^{12}$$

$$\sqrt{x^2 - 6x + 9} = (x - 3) \qquad \text{because} \qquad (x - 3)(x - 3) = x^2 - 6x + 9$$

From the foregoing examples we can formulate the following rule:

Rule. *To find an indicated root of a given power expressed with exponents, divide the exponents by the index of the root. This rule holds true for all types of exponents.* The rule may be stated as a formula:

$$\sqrt[r]{x^n} = x^{n/r}$$

that is, the rth root of the nth power of x is equal to x raised to a power which is n/r.

The following examples show how this rule is applied:

$$\sqrt[3]{10^6} = 10^{\frac{6}{3}} = 10^2 \qquad \sqrt[3]{x^5} = x^{\frac{5}{3}}$$

$$\sqrt[4]{10^{3.532}} = 10^{0.883} \qquad \sqrt[2]{x^7} = x^{\frac{7}{2}}$$

$$\sqrt[3]{-64x^6} = -4x^2 \qquad \text{because} \qquad (-4x^2)^3 = -64x^6$$

$$\sqrt{25x^4y^6} = 5x^2y^3 \qquad \text{because} \qquad (5x^2y^3)^2 = 25x^4y^6$$

$$\sqrt[6]{64a^6b^{12}} = 2ab^2 \qquad \text{because} \qquad (2ab^2)^6 = 64a^6b^{12}$$

$$\sqrt[3]{125x^{12}} = 5x^4 \qquad \text{because} \qquad (5x^4)^3 = 125x^{12}$$

In algebra we should remember also that any expression has two square roots, one positive, the other negative. If no sign (+ or −) is shown before the radical sign, then the principal root is meant. For instance, the square roots of x^2 are $+x$ and $-x$. However,

$$\sqrt{x^2} = +x$$

Exercise 17-1

Find the indicated root in each radical.

1. $\sqrt{x^8}$ **2.** $\sqrt{10^4}$ **3.** $\sqrt{10^{5.6}}$

4. $\sqrt{10^{3.472}}$ **5.** $\sqrt{4^6}$ **6.** $\sqrt{10^{16}}$

7. $\sqrt{16x^6}$ **8.** $\sqrt[3]{125x^6}$ **9.** $\sqrt[3]{-64x^9}$

10. $\sqrt[4]{16x^8}$ **11.** $\sqrt[3]{8x^6}$ **12.** $\sqrt{16x^{16}}$

13. $\sqrt{9x^{20}}$ **14.** $\sqrt[3]{27n^{12}}$ **15.** $-\sqrt[3]{-125a^3}$

16. $\sqrt{\frac{4}{9}}$ 17. $\sqrt{\frac{25}{49}}$ 18. $\sqrt{\frac{100}{121}}$

19. $\sqrt[4]{16x^{16}}$ 20. $\sqrt[5]{-32a^{30}}$ 21. $\sqrt[6]{64x^6}$

22. $\sqrt{(x-4)^2}$ 23. $\sqrt[3]{(x+y)^3}$ 24. $-\sqrt[5]{1024x^{10}}$

25. $\sqrt[3]{n^4}$ 26. $-\sqrt[5]{-243x^5y^5}$ 27. $-\sqrt{81x^6y^8z^2}$

28. $\sqrt{121n^4t^3s^2}$ 29. $\sqrt{x^2-4x+4}$ 30. $\sqrt{4n^2+12n+9}$

17-3 IRRATIONAL NUMBERS

Such numbers as $\sqrt{2}$, $\sqrt{3}$, $\sqrt{5}$, and π, are called *irrational* numbers to distinguish them from rational numbers. A *rational* number is defined as a number that can be expressed as the quotient of two integers. In other words, a rational number can be written as a common fraction with numerator and denominator as whole numbers, either positive or negative.

1. All common fractions are rational.

$\frac{3}{4}$ is the quotient of $3 \div 4$; $\frac{15}{37}$ means $15 \div 37$

2. All whole numbers are rational.

3 means $\frac{3}{1}$; -5 means $\frac{-5}{1}$

3. All decimal fractions are rational.

0.3 can be written $\frac{3}{10}$; 17.65 can be written $\frac{1765}{100}$

The number $\sqrt{2}$ is *irrational* because the exact square root of 2 cannot be expressed as any common fraction. The same is true of the numbers $\sqrt{3}$, $\sqrt{5}$, $\sqrt{6}$ and many others. There are other irrational numbers besides some square roots. The number π, for instance, is irrational. It never comes out even as a decimal or a common fraction.

We often use rational numbers as approximations. We have said the number $\sqrt{2}$ is irrational. It is only approximately equal to 1.414. However, the number 1.414 is itself rational because it can be written $\frac{1414}{1000}$. The number π is irrational. It is only approximately equal to 3.14159. The number 3.14159, however, is itself rational because it can be written $\frac{314159}{100000}$.

Let us consider a little more carefully the irrational number, the square root of 2. Suppose we wish to know the *exact* square root of 2. It is close to 1.414. However, if we multiply 1.414 by *itself*, we get 1.999396, which is less than 2. If we multiply the number 1.4142 by itself, we still get a number slightly less than 2. The square root of 2 is irrational and will never come out even as a common fraction or as a decimal no matter how far we carry out the decimal part. We cannot write the square root of 2 as we write other numbers, such as whole numbers, fractions, or decimals. It is some elusive thing that we cannot get hold of. Yet we can write the square root of 2 as a *symbol:*

$$\sqrt{2}$$

Now, if we say that this symbol is to represent the *exact square root* of 2, then we must conclude that this symbol, multiplied by itself, must equal 2. In other words,

$$(\sqrt{2})(\sqrt{2}) = 2$$

Remember, by definition, the square root of any given quantity means some other quantity that can be multiplied by itself to produce the given quantity; that is,

$$\sqrt{\text{quantity}} \cdot \sqrt{\text{quantity}} = \text{quantity}$$

Here are some other examples that show the meaning of square root:

$$(\sqrt{361})(\sqrt{361}) = 361 \qquad (\sqrt{41.2})(\sqrt{41.2}) = 41.2$$

$$(\sqrt{3})(\sqrt{3}) = 3 \qquad (\sqrt{x})(\sqrt{x}) = x$$

$$(\sqrt{\text{book}})(\sqrt{\text{book}}) = \text{book} \qquad (\sqrt{\text{chair}})(\sqrt{\text{chair}}) = \text{chair}$$

Of course, the last two examples have no meaning in the ordinary sense of square roots. Yet they should help to show the meaning of the expression *square root.*

17-4 SIMPLIFYING RADICALS

Whenever a radicand represents a perfect power of some number of which the indicated root may be found, then the root should be stated as a rational number. For instance, the expression $\sqrt{36}$ should always be stated as 6. The following expression should be simplified as shown:

$$\sqrt[3]{125x^6} = 5x^2$$

However, it often happens that we have irrational numbers, $\sqrt{2}$, $\sqrt{3}$, and so on. Such irrational numbers are often changed to approximate rational decimal values.

$$\sqrt{2} = 1.414 \text{ (approx.)}; \qquad \sqrt{3} = 1.732 \text{ (approx.)}$$

It is possible to combine a rational and an irrational number. The result is another irrational number, as shown by the following examples:

$$7 + \sqrt{5} = 7 + 2.236\ (\text{approx.}) = 9.236\ (\text{approx.})$$

$$2 - \sqrt{10} = 2 - 3.162\ (\text{approx.}) = -1.162\ (\text{approx.})$$

In many instances a radical can be *simplified* to facilitate computation. *There are two immediate goals in simplifying radicals*:

1. *The radicand should be made as small as possible.*
2. *The radicand should not contain a fraction.*

To make a radicand as small as possible, we proceed as follows:

1. First we separate the radicand into two factors so that one factor is a perfect square or other power indicated by the index of the root. A perfect power is a number whose root, as indicated by the index, is rational.
2. Then we separate the radical into the product of two radicals.
3. Finally, we take the root of the perfect power and place this quantity as a coefficient of the second radical.

Example 1

Simplify the radical $\sqrt{18}$.

$$\sqrt{18} = \sqrt{(9)(2)} = \sqrt{9} \cdot \sqrt{2} = 3 \cdot \sqrt{2}$$

Let us see whether we have a right to separate a radical into the product of two separate radicals by factoring the radicand.

$$\text{Is } \sqrt{36} = \sqrt{9 \cdot 4} = \sqrt{9} \cdot \sqrt{4} = 3 \cdot 2 = 6?\ \text{Yes.}$$

$$\text{Is } \sqrt{400} = \sqrt{16 \cdot 25} = \sqrt{16} \cdot \sqrt{25} = 4 \cdot 5 = 20?\ \text{Yes.}$$

In the two examples shown, the answers are correct. Remember, however, that two examples do not necessarily establish a rule. Yet the following principle happens to be true:

A radicand may be factored and the radical may be written as the product of two radicals.

Example 2

$$\sqrt{75} = \sqrt{25 \cdot 3} = \sqrt{25} \cdot \sqrt{3} = 5\sqrt{3}$$

Example 3

$$7\sqrt{20} = 7\sqrt{4 \cdot 5} = 7\sqrt{4} \cdot \sqrt{5} = 14\sqrt{5}$$

Example 4

$$\sqrt{\frac{5}{9}} = \sqrt{\frac{1}{9} \cdot 5} = \frac{1}{3}\sqrt{5}$$

Example 5

$$5\sqrt[3]{32} = 5\sqrt[3]{8 \cdot 4} = 5 \cdot 2\sqrt[3]{4} = 10\sqrt[3]{4}$$

Example 6

$$\sqrt{12x^3y} = \sqrt{(4x^2)(3xy)} = 2x\sqrt{3xy}$$

Example 7

$$\sqrt{80x^3y^8z^7} = \sqrt{(16x^2y^8z^6)(5xz)} = 4xy^4z^3\sqrt{5xz}$$

If a radicand is a fraction, it should be changed into a whole number. In this case the first step is to multiply the numerator and the denominator of the fraction by some number that will make the *denominator* a perfect power, as indicated by the root. Then the radical may be written as the product of two radicals, one of which is rational.

Example 8

Simplify the radical $\sqrt{\frac{2}{3}}$.

The first step in this example is to multiply the numerator and the denominator by 3. Then the denominator becomes a perfect square.

$$\sqrt{\frac{2}{3}} = \sqrt{\frac{2 \cdot 3}{3 \cdot 3}} = \sqrt{\frac{6}{9}} = \sqrt{\frac{1}{9} \cdot 6} = \frac{1}{3}\sqrt{6}.$$

Example 9

Simplify $7\sqrt{\frac{5}{8}}$.

$$7\sqrt{\frac{5}{8}} = 7\sqrt{\frac{5 \cdot 2}{8 \cdot 2}} = 7\sqrt{\frac{10}{16}} = 7\sqrt{\frac{1}{16} \cdot 10} = \frac{7}{4}\sqrt{10}$$

Warning. You should be very careful about removing quantities from under a radical sign. In the following statements can you tell which equal signs are wrong? (Some equal signs are correct.)

$$\sqrt{13} = \sqrt{9 + 4} = \sqrt{9} + \sqrt{4} = 3 + 2 = 5 \qquad \text{(wrong)}$$

$$\sqrt{x^2 + y^2} = \sqrt{x^2} + \sqrt{y^2} = x + y \qquad \text{(wrong)}$$

$$\sqrt{24} = \sqrt{49 - 25} = \sqrt{49} - \sqrt{25} = 7 - 5 = 2 \qquad \text{(wrong)}$$

$$\sqrt{x^2 - y^2} = \sqrt{x^2} - \sqrt{y^2} = x - y \qquad \text{(wrong)}$$

You cannot take the root of a separate term of a polynomial radicand and place the root on the outside of the radical sign. Roots may be taken *only of factors*, not of separate terms of a radicand.

Exercise 17-2

Simplify each of the following expressions as much as possible.

1. $\sqrt{48}$ **2.** $2\sqrt{75}$ **3.** $3\sqrt{45}$ **4.** $4\sqrt{147}$

5. $\frac{1}{2}\sqrt{72}$ **6.** $\frac{1}{5}\sqrt{300}$ **7.** $\frac{1}{4}\sqrt{24}$ **8.** $\frac{1}{2}\sqrt{99}$

9. $6\sqrt{\frac{3}{4}}$ **10.** $5\sqrt{\frac{7}{9}}$ **11.** $7\sqrt{\frac{9}{16}}$ **12.** $2\sqrt{\frac{3}{8}}$

13. $3\sqrt{\frac{5}{32}}$ **14.** $6\sqrt{\frac{2}{3}}$ **15.** $4\sqrt{\frac{3}{5}}$ **16.** $12\sqrt{\frac{20}{27}}$

17. $\sqrt[3]{16}$ **18.** $\sqrt[3]{54}$ **19.** $\sqrt[4]{48}$ **20.** $\sqrt[5]{64}$

21. $\sqrt{18x^3}$ **22.** $\sqrt{25x^5}$ **23.** $\sqrt{12x^7y^6}$ **24.** $\sqrt{80x^9y^8}$

25. $\sqrt{\frac{7}{x^2}}$ **26.** $\sqrt{\frac{9}{x}}$ **27.** $\sqrt{\frac{3a}{x^3}}$ **28.** $\sqrt{\frac{8a^2}{3x^2y}}$

29. $\sqrt[3]{16x^7}$ **30.** $\sqrt[4]{16x^7}$ **31.** $\sqrt[3]{81x^4y^5}$ **32.** $\sqrt[5]{64x^7y^{10}}$

33. $\sqrt{64 \cdot 36}$ **34.** $\sqrt{64 + 36}$ **35.** $\sqrt{25 \cdot 144}$ **36.** $\sqrt{36 + 9 + 4}$

37. $\sqrt{x^6 + x^4}$ **38.** $\sqrt{x^4 - x^2}$ **39.** $\sqrt{25 + 144}$ **40.** $\sqrt{49 + 36 - 4}$

17-5 ADDITION AND SUBTRACTION OF RADICALS

Only *like* or *similar radicals* can be added or subtracted. Like radicals are those that have the same *radicand* and the same indicated *root:*

$$\sqrt{7} \qquad 3\sqrt{7} \qquad 15\sqrt{7} \qquad -8\sqrt{7}$$

$$4\sqrt[3]{5xy} \qquad -2\sqrt[3]{5xy} \qquad 7\sqrt[3]{5xy} \qquad -\sqrt[3]{5xy}$$

The coefficient of a radical is the number appearing as a factor outside the entire radical. If a radical has no expressed coefficient, the coefficient is understood to be 1.

Like radicals are added or subtracted by adding or subtracting their coefficients, just as in algebraic terms.

$$5\sqrt{2} + 7\sqrt{2} - 3\sqrt{2} = 9\sqrt{2}$$

$$7\sqrt[3]{4} - 2\sqrt[3]{4} + 9\sqrt[3]{4} = 14\sqrt[3]{4}$$

Example 1

Simplify $\sqrt{3} - 5\sqrt{2} - 5\sqrt{3} + \sqrt{2} + 6\sqrt{3} + 8$.

Combining like radicals, we get

$$2\sqrt{3} - 4\sqrt{2} + 8$$

Thus, the approximate rational value may be more easily found. The $\sqrt{3}$ equals approximately 1.732 and the $\sqrt{2}$ equals approximately 1.414. The approximate value of the expression is

$$2(1.732) - 4(1.414) + 8 = 3.464 - 5.656 + 8 = 5.808 \text{ (approx.)}$$

Sometimes unlike radicals may be transformed into like radicals and the results combined.

Example 2

Simplify and combine

$$5\sqrt{2} + \sqrt{12} + 6\sqrt{32} - 7\sqrt{18} - 8\sqrt{3} + \sqrt{\tfrac{1}{8}} =$$

If the radicals are simplified by reducing the radicands, the expression becomes

$$5\sqrt{2} + 2\sqrt{3} + 24\sqrt{2} - 21\sqrt{2} - 8\sqrt{3} + \tfrac{1}{4}\sqrt{2} = 8\tfrac{1}{4}\sqrt{2} - 6\sqrt{3}$$

The approximate value is found as follows:

$$8\tfrac{1}{4}\sqrt{2} - 6\sqrt{3} = \tfrac{33}{4}(1.414) - 6(1.732) = 1.274 \text{ (approx.)}$$

Exercise 17-3

Simplify the radicals in each of the following examples, and combine as much as possible. Finally, find the approximate value of each example.

1. $\sqrt{2} + 3\sqrt{72} + 4\sqrt{50} - \sqrt{128}$ **2.** $\sqrt{5} + 2\sqrt{45} - \sqrt{80} + \sqrt{180}$

3. $\sqrt{3} - \sqrt{12} + 2\sqrt{8} + 5\sqrt{200}$ **4.** $\sqrt{75} - \sqrt{125} - 3\sqrt{108} + \sqrt{320}$

5. $\sqrt{27} + \sqrt{288} - 3\sqrt{48} - \sqrt{162}$ **6.** $6\sqrt{32} - \sqrt{98} + 2\sqrt{96} - 3\sqrt{54}$

7. $3\sqrt{40} + 5\sqrt{90} - \sqrt{250} + \sqrt{60}$ **8.** $\sqrt{300} + \sqrt{800} - 4\sqrt{45} + \sqrt{450}$

9. $\sqrt{18} + \sqrt{12} + \sqrt{20} + \sqrt{24} + \sqrt{28}$

10. $\sqrt{10} + \sqrt{100} + \sqrt{1000} + \sqrt{10000}$

11. $\sqrt{\frac{1}{32}} + \sqrt{\frac{1}{8}} + 6\sqrt{\frac{1}{2}} - \sqrt{8}$ **12.** $10\sqrt{\frac{2}{5}} + \sqrt{\frac{1}{10}} - \sqrt{40} + 2\sqrt{\frac{5}{8}}$

13. $6\sqrt{\frac{1}{6}} - \sqrt{\frac{2}{3}} - 2\sqrt{54} + 3\sqrt{24}$ **14.** $5\sqrt{\frac{1}{3}} - \sqrt{\frac{1}{27}} + 2\sqrt{12} - \sqrt{\frac{4}{3}}$

15. $8\sqrt{\frac{3}{8}} + \sqrt{\frac{9}{5}} + 5\sqrt{\frac{6}{25}} - \sqrt{\frac{3}{32}}$ **16.** $\sqrt{\frac{5}{32}} + 4\sqrt{\frac{4}{5}} - \sqrt{\frac{3}{16}} + 6\sqrt{\frac{5}{9}}$

17. $\sqrt[3]{16} + \sqrt[3]{54} + \sqrt[3]{128} - \sqrt[3]{343}$ **18.** $16^{\frac{1}{4}} - 9^{-\frac{1}{2}} + 8^{\frac{2}{3}} - 3^0$

19. $18^{\frac{1}{2}} + 72^{\frac{1}{2}} + 125^{\frac{1}{2}} + 64^{\frac{1}{3}}$ **20.** $\sqrt{5} + \sqrt[3]{5} + \sqrt[4]{5} + 5 + \sqrt{9}$

17-6 MULTIPLICATION OF RADICALS

In the multiplication of radicals there are several things we must consider. Let us look again at the definitions of the numbers used in connection with radicals. Consider the radical

$$-4\sqrt[2]{15}$$

In this radical expression the *index* of the root is 2. The *radicand* is 15. The radical has the *coefficient* -4, which means that the value of the radical is to be multiplied by -4.

If two or more radicals are to be multiplied together, they must have the same index. If two radicals do not have the same index, their product can only be indicated or they may be reduced to the same index. We shall consider, first, the case in which the indices are equal.

If two or more radicals have the same index, they can be multiplied together. The radicands need not be the same.

Note. Although the index *2* is usually omitted, we show it sometimes for emphasis.

For the multiplication of radicals with the same index, we have the following steps:

1. The sign of the product will follow from the rule for the multiplication of signed numbers. This applies to the signs of the coefficients.

2. Multiply the coefficients for the coefficient in the product.
3. Multiply the radicands for the radicand in the product.
4. Simplify the resulting radical if possible.

Note. One other point should be mentioned here. For the multiplication of radicals, all rules *except one* for multiplying signed numbers hold true. The usual rule for the multiplication of signed numbers when both are negative may be stated symbolically, $(-)(-) = +$. However, if two radicals have negative radicands, these two negative radicands *cannot* be written as a positive product for the radicand of the new radical. For instance, $\sqrt{-9} \cdot \sqrt{-4}$ is not equal to $\sqrt{(-9)(-4)} = \sqrt{+36}$. The radicand of the product is *not* positive. *Neither is it negative.* In this chapter we do not include examples of this kind. The multiplication of such numbers is fully explained in Chapter 19.

Example 1

Multiply $(-7\sqrt{3})(-2\sqrt{5})$.

For the sign of the product, we have $(-)(-) = +$.

For the coefficient of the product, we have $(7)(2) = 14$.

For the radicand in the product, we have $(3)(5) = 15$.

The complete answer is $+14\sqrt{15}$.

The entire product may, of course, be expressed approximately by finding the approximate square root of 15, which is 3.873. Then we have

$$14\sqrt{15} = 14(3.873) = 54.222 \text{ (approx.)}$$

Example 2

Multiply $(+4\sqrt{7})(-5\sqrt{3})(-2\sqrt{5})$.

For the sign of the product, we have $(+)(-)(-) = +$.

For the coefficient, we have $(4)(5)(2) = 40$.

For the product of the radicands, we have $(7)(3)(5) = 105$.

For the entire product of the radicals, we have $+40\sqrt{105}$.

The answer cannot be reduced. It can, of course, be expressed as an approximate number, since $\sqrt{105}$ is approximately 10.247.

$$40\sqrt{105} = 40(10.247) = 409.88 \text{ (approx.)}$$

Example 3

Multiply $(5\sqrt{3})(-7\sqrt{3})$.

In this example notice that the radicands are equal. Whenever the two radicands are equal, they should not be multiplied together to form a product under the radical sign. In this example we do not multiply $3 \cdot 3 = 9$. Instead, the product of the radicals should be stated at once: $(\sqrt{3})(\sqrt{3}) = 3$;

that is, we recall that the square root of a number multiplied by the square root of the number equals the number itself. Thus

$$(\sqrt{x})(\sqrt{x}) = x$$

If we multiply the radicands and then find the square root of the product, the method is not only longer but it can easily lead to errors. The answer to Example 3 is

$$(5\sqrt{3})(-7\sqrt{3}) = (-35)(3) = -105$$

Example 4

Multiply: $(+\sqrt{2})(\sqrt{8})$

In most instances, if an irrational number is multiplied by another irrational number, the product is also irrational. However, it sometimes happens that the product of two irrational numbers is rational. In Example 4 each of the radicals is irrational. Yet when they are multiplied, the product is rational.

$$(\sqrt{2})(\sqrt{8}) = \sqrt{16} = 4, \quad \text{a rational number.}$$

In the multiplication of polynomials involving radicals, we follow the same general procedure that we use in the multiplication of polynomials involving literal numbers.

Consider the following example.

Example 5

Multiply $2\sqrt{3}(4\sqrt{3} + 2\sqrt{5} - 3\sqrt{7})$

This example is similar in form to an example containing rational literal numbers:

$$2x(4x + 2y - 3z)$$

In multiplying a polynomial by a monomial, we multiply each term of the polynomial by the monomial. Thus, $2x(4x + 2y - 3z) = 8x^2 + 4xy - 6xz$.

In a similar way we find the product in the example containing radicals.

For the first term, we have $(2\sqrt{3})(4\sqrt{3}) = (8)(3) = 24$.

For the second term, we have $(2\sqrt{3})(2\sqrt{5}) = 4\sqrt{15}$.

For the third term, we have $(2\sqrt{3})(-3\sqrt{7}) = -6\sqrt{21}$.

The complete product is $24 + 4\sqrt{15} - 6\sqrt{21}$.

In the multiplication of two polynomials involving radicals, we follow the same general procedure as in the usual multiplication of polynomials.

Example 6

Multiply $(5\sqrt{3} - 4\sqrt{2})(7\sqrt{3} + 3\sqrt{2})$

This expression is similar in form to the expression $(5x - 4y)(7x + 3y)$. Let us show all the steps by writing one binomial below the other and taking note of each separate step in the multiplication. Note especially the multiplications.

$$\sqrt{3} \cdot \sqrt{3} = 3 \quad \text{and} \quad \sqrt{2} \cdot \sqrt{2} = 2$$

$$\begin{array}{l} 5\sqrt{3} - 4\sqrt{2} \\ \underline{7\sqrt{3} + 3\sqrt{2}} \\ 35 \cdot 3 - 28\sqrt{6} \\ \underline{ + 15\sqrt{6} - 12 \cdot 2} \\ 105 - 13\sqrt{6} - 24 \quad = 81 - 13\sqrt{6} \end{array}$$

We can easily see the advantage of this kind of simplification if we wish to compute the approximate rational value of the expression. To find the value from the original expression involves much work. The $\sqrt{3}$ is approximately equal to 1.732, and the $\sqrt{2}$ equals approximately 1.414. Substituting these values in the binomials, we get

$$\begin{aligned} &[5(1.732) - 4(1.414)][7(1.732) + 3(1.414)] \\ &= (8.660 - 5.656)(12.124 + 4.242) \\ &= (3.004)(16.366) = 49.163 \text{ (rounded off)} \end{aligned}$$

Instead, if the answer, $81 - 13\sqrt{6}$, is used to find the approximate rational value, we have

$$81 - 13\sqrt{6} = 81 - 13(2.449) = 49.163 \text{ (approx.)}$$

The approximate values do not always check in the last digit for both methods, since the values used are only approximate.

If radicals do not have the same index, their product can only be indicated. For instance, the product of $\sqrt[2]{5}$ and $\sqrt[3]{4}$ can only be indicated in some way such as $(\sqrt[2]{5})(\sqrt[3]{4})$. Of course, the approximate product may be found by multiplying the approximate values of the given radicals. The product may also be simplified by reducing the radicals to the same order by the following steps:

1. *Change the radicals to exponential form.*
2. *Reduce the fractional exponents so that they have the same denominator.*
3. *Change back to the radical form.*
4. *Multiply the radicands.*

Example 7

Multiply $(\sqrt{5})(\sqrt[3]{4})$.

Solution

Here we first change each radical so that they have the same index. This is done by writing each radical in exponential form and then changing the fractional exponents to fractions with the same denominators.

$$\sqrt{5} = 5^{\frac{1}{2}} = 5^{\frac{3}{6}} = \sqrt[6]{5^3} = \sqrt[6]{125}; \quad \sqrt[3]{4} = 4^{\frac{1}{3}} = 4^{\frac{2}{6}} = \sqrt[6]{4^2} = \sqrt[6]{16}$$

Now we can multiply the radicals, since they have the same index. We get

$$(\sqrt[6]{125})(\sqrt[6]{16}) = \sqrt[6]{2000}$$

In many, if not most, problems that we encounter in scientific work involving multiplication of radicals the radicals are of the same order; that is, they have the same index. Then the multiplication is rather easy.

Exercise 17-4

Multiply and find approximate decimal values:

1. $(3\sqrt{2})(4\sqrt{5})$

2. $(-4\sqrt{3})(5\sqrt{2})$

3. $(-5\sqrt{2})(-3\sqrt{2})$

4. $(-3\sqrt{7})(8\sqrt{7})(\sqrt{2})$

5. $(9\sqrt{8})(4\sqrt{2})$

6. $(5\sqrt{10})(4\sqrt{13})(\sqrt{2})$

7. $(4\sqrt{5})(3\sqrt{5})(2\sqrt{5})$

8. $(-2\sqrt{3})(4\sqrt{3})(-5\sqrt{3})$

9. $(\sqrt{42})(\sqrt{42})(\sqrt{42})$

10. $(3\sqrt{6})(-2\sqrt{6})(-\sqrt{6})$

11. $(5\sqrt{6})(2\sqrt{3})(\sqrt{2})$

12. $(3\sqrt{7})(5\sqrt{14})(2\sqrt{3})$

13. $3\sqrt{2}(5\sqrt{3} - 4\sqrt{5} - 5\sqrt{2})$

14. $-4\sqrt{3}(2\sqrt{5} - 3\sqrt{3} + 7\sqrt{6})$

15. $5\sqrt{5}(\sqrt{3} - \sqrt{5} + 4\sqrt{15})$

16. $(\sqrt{3} + 2\sqrt{5})(2\sqrt{3} - 7\sqrt{5})$

17. $(3\sqrt{2} - 5\sqrt{7})(4\sqrt{2} + 2\sqrt{7})$

18. $(2\sqrt{5} - 3\sqrt{2})(2\sqrt{5} + 3\sqrt{2})$

19. $(3\sqrt{2} + \sqrt{5})(\sqrt{3} - 4)$

20. $(4\sqrt{3} - 3\sqrt{5})(5\sqrt{2} + 4\sqrt{7})$

21. $(4\sqrt{3} - 2\sqrt{7})^2$ **22.** $(2\sqrt{5} + \sqrt{2})^2$

23. $(\sqrt{7} - 3)^2$ **24.** $(3\sqrt{2} - 5)^2$

25. Write each of the following as a single radical and simplify:

(a) $(\sqrt[3]{x})(\sqrt{y})$ (b) $(\sqrt[3]{3})(\sqrt[4]{2})$

(c) $(\sqrt{8})(\sqrt[4]{8})$ (d) $(\sqrt{5})(\sqrt[5]{6})$

17-7 DIVISION OF RADICALS

If, in the division of radicals, having the same indices, we have a monomial radical as a divisor and *if the radicand of the divisor can be divided evenly into the radicand of the dividend*, then the division is easily done.

Example 1

Divide as indicated: $3\sqrt{55} \div 7\sqrt{5}$

In this example, we write the coefficient as $\frac{3}{7}$, which is the same as $3 \div 7$. Dividing the radicands, we get 11 for the radicand of the quotient. The rules for the sign of a quotient must be observed just as in the division of algebraic terms involving literal numbers. The answer to Example 1 is $\frac{3}{7}\sqrt{11}$.

If, in the division of radicals, *the divisor contains a radical*, the division involves one main objective: when the division is indicated as a fraction, *eliminate any radical in the denominator.*
Suppose we have fractions such as the following:

$$\frac{3}{\sqrt{2}}; \quad \frac{2\sqrt{5}}{5\sqrt{3}}; \quad \frac{5\sqrt{3}}{3 + \sqrt{5}}$$

You will notice that all the denominators are irrational. If we wish to compute the value of each fraction, we might first find the approximate values of the denominators. Then we should have to use long division. The divisors would be long decimal fractions. Instead, there is an easier way.

Our main objective is to eliminate any radical in a denominator. To do this, we multiply the numerator and the denominator of the fraction by *some quantity* that will make the denominator *rational.* In other words, we *rationalize* the denominator.

Take a simple example. Suppose we have the fraction

$$\frac{1}{\sqrt{2}}$$

If we wish to work out the *numerical value* of the fraction as it stands, we have the following problem in *very long* division:

$$\frac{1}{1.4142} \quad \text{or} \quad 1.4142\overline{)1.}$$

There is a better way to find the numerical value of this fraction.

Instead of doing the long division, suppose we *multiply the numerator and the denominator* of the original fraction by the quantity $\sqrt{2}$. Then we get

$$\frac{1}{\sqrt{2}} = \frac{1 \cdot \sqrt{2}}{\sqrt{2} \cdot \sqrt{2}} = \frac{\sqrt{2}}{2} = \frac{1.4142}{2} = ?$$

The result is *easy division.* Note that we first made the denominator *rational.* Then the division is much easier.

If the denominator is a *binomial,* the *multiplying factor* will have to be the *conjugate form* of the denominator. *The conjugate form* of a binomial is the same binomial, *except that the middle sign is changed.* So the *numerator* and the *denominator* are *both* multiplied by the *conjugate form of the denominator.* The conjugate form of $4 - \sqrt{5}$ is $4 + \sqrt{5}$.

Example 2

Simplify

$$\frac{3}{4 - \sqrt{5}}$$

We multiply numerator and denominator by the rationalizing factor, $4 + \sqrt{5}$.

$$\frac{3(4 + \sqrt{5})}{(4 - \sqrt{5})(4 + \sqrt{5})} = \frac{3(4 + \sqrt{5})}{16 - 5} = \frac{3(4 + \sqrt{5})}{11}$$

$$= \frac{3(4 + 2.236)}{11} = \frac{3(6.236)}{11} = 1.701 \text{ (approx.)}$$

The decimal value of this fraction is approximately 1.701. This value can be computed much more easily from the final form than from the original fraction. The final form involves easy division.

Exercise 17-5

Rationalize the denominator in each of the following fractions:

1. $\frac{12}{5\sqrt{2}}$ **2.** $\frac{5}{3\sqrt{5}}$ **3.** $\frac{1}{\sqrt{3}}$

4. $\dfrac{2}{3\sqrt{5}}$ 5. $\dfrac{3}{\sqrt{13}}$ 6. $\dfrac{\sqrt{3}}{\sqrt{2}}$

7. $\dfrac{2\sqrt{5}}{5\sqrt{3}}$ 8. $\dfrac{\sqrt{2}}{4\sqrt{5}}$ 9. $\dfrac{2}{\sqrt{7}}$

10. $\dfrac{5}{\sqrt{34}}$ 11. $\dfrac{3}{3 - \sqrt{5}}$ 12. $\dfrac{\sqrt{2}}{4 + \sqrt{5}}$

13. $\dfrac{8}{2 - \sqrt{3}}$ 14. $\dfrac{6}{\sqrt{5} + \sqrt{2}}$ 15. $\dfrac{\sqrt{3}}{2\sqrt{3} - \sqrt{5}}$

16. $\dfrac{2\sqrt{3}}{5 - 2\sqrt{3}}$ 17. $\dfrac{3 - \sqrt{3}}{3 + \sqrt{3}}$ 18. $\dfrac{3 + \sqrt{2}}{\sqrt{5} + \sqrt{2}}$

19. $\dfrac{2 + \sqrt{3}}{2 - \sqrt{3}}$ 20. $\dfrac{\sqrt{3} + \sqrt{5}}{4\sqrt{3} - 2\sqrt{5}}$ 21. $\dfrac{2 + \sqrt{3}}{3\sqrt{3} - \sqrt{5} + 2}$

22. $\dfrac{4 - \sqrt{2}}{2\sqrt{2} + 2}$ 23. $\dfrac{\sqrt{6} - 1}{\sqrt{2} + \sqrt{3}}$ 24. $\dfrac{\sqrt{2} + 3}{\sqrt{3} + \sqrt{2} - 1}$

25. $\dfrac{2\sqrt{3} - \sqrt{5}}{\sqrt{15} - 2}$ 26. $\dfrac{1}{4\sqrt{2} - 3\sqrt{3}}$ 27. $\dfrac{-3\sqrt{2} + 5\sqrt{10}}{\sqrt{2} - 3\sqrt{10}}$

28. $\dfrac{4\sqrt{3} - 5\sqrt{6}}{2\sqrt{3} + \sqrt{6}}$ 29. $\dfrac{3\sqrt{6} - 2\sqrt{2}}{3\sqrt{2} - 5\sqrt{6}}$ 30. $\dfrac{3\sqrt{5} - \sqrt{10}}{\sqrt{5} + 2\sqrt{10}}$

Exercises 17-6 Review of Radicals

Simplify the following radicals:

1. $2\sqrt{60}$ 2. $7\sqrt{40}$ 3. $5\sqrt{125}$ 4. $6\sqrt{700}$

5. $\frac{1}{2}\sqrt{800}$ 6. $\frac{1}{4}\sqrt{288}$ 7. $\frac{1}{3}\sqrt{450}$ 8. $\frac{1}{6}\sqrt{192}$

9. $3\sqrt{\frac{18}{25}}$ 10. $8\sqrt{\frac{25}{6}}$ 11. $14\sqrt{\frac{4}{7}}$ 12. $5\sqrt{\frac{8}{11}}$

13. $\sqrt[3]{\frac{3}{8}}$ 14. $\sqrt[3]{\frac{5}{54}}$ 15. $\sqrt[4]{\frac{3}{32}}$ 16. $\sqrt[5]{\frac{243}{16}}$

17. $\sqrt{16x^{16}}$ 18. $\sqrt{72x^9y}$ 19. $\sqrt{96x^7y^6}$ 20. $\sqrt{45a^3b^8c}$

21. $\sqrt{\dfrac{4}{x^3}}$ **22.** $\sqrt{\dfrac{3}{2x^2y}}$ **23.** $\sqrt{\dfrac{5}{3ab^3}}$ **24.** $\sqrt{\dfrac{2ab}{5x^2y^5}}$

Perform the indicated operations. State answers in decimal form.

25. $\sqrt{160} + \sqrt{\frac{5}{8}} - \sqrt{90} - \sqrt{\frac{2}{5}}$

26. $\sqrt{192} - \sqrt{108} + \sqrt{\frac{3}{25}} - \sqrt{\frac{1}{3}} + \sqrt{\frac{16}{27}}$

27. $\sqrt{\frac{81}{50}} - \sqrt{\frac{9}{75}} + \sqrt{243} + \sqrt{\frac{25}{32}}$ **28.** $\sqrt{48} + \sqrt{\frac{8}{27}} - \sqrt{\frac{3}{8}} - \sqrt{\frac{2}{3}} + \sqrt{\frac{3}{50}}$

29. $(-1 + \sqrt{3})^3$ **30.** $(\sqrt{5} - 3 + \sqrt{2})(\sqrt{5} - 3 - \sqrt{2})$

31. $(2 - 2\sqrt{3})^3$

32. $(\sqrt{7} + \sqrt{3} - \sqrt{5})(\sqrt{7} - \sqrt{3} + \sqrt{5})$

33. $(\sqrt{3} + 1)^4$

34. $(3\sqrt{3} + \sqrt{2})(2 + \sqrt{6})(3\sqrt{3} - \sqrt{2})$

35. $(3\sqrt{2} - 2\sqrt{6}) \div (\sqrt{2} + \sqrt{6})$ **36.** $(3\sqrt{2}) \div (2\sqrt{3} - \sqrt{6})$

37. Find reciprocal of $(3 + \sqrt{5})$ **38.** Find reciprocal of $(5 + \sqrt{3} - \sqrt{2})$

39. The period of a pendulum refers to the time, in seconds, it takes the pendulum to make one cycle. The period is found by the formula

$$T = 2\pi\sqrt{L/g}$$

Here, L is the length of the pendulum (in feet), and g is the acceleration due to gravity. Taking $g = 32$, find the period, T, in seconds when $L = 2$ ft. Also find the number of cycles per minute.

40. Work Problem 39 if the pendulum is 3 ft long.

41. Work Problem 39 for a pendulum whose length is 9 in.

42. Work the problem if the length is 30 in.

QUIZ ON CHAPTER 17, RADICALS. FORM A.

1. Find the indicated root in each of the following:

(a) $\sqrt[3]{-8x^6}$; **(b)** $-\sqrt[4]{16x^4y^8z^0}$; **(c)** $\sqrt{100a^6b^8}$;

(d) $\sqrt{10^{8.52}}$; **(e)** $\sqrt[3]{10^{8.52}}$

2. Simplify these radicals:

(**a**) $5\sqrt{72}$; (**b**) $3\sqrt{75}$; (**c**) $6\sqrt{2/5}$

(**d**) $24\sqrt{7/8}$; (**e**) $\sqrt{4x^2 - 16x}$; (**f**) $\sqrt{36 + 64}$;

(**g**) $\sqrt{17^2 - 8^2}$

3. Add and subtract as indicated:

(**a**) $5\sqrt{48} - 2\sqrt{147} + 4\sqrt{108} + \sqrt{16}$;

(**b**) $5\sqrt{32} + 3\sqrt{50} - 2\sqrt{98} + 4\sqrt{200}$

4. Multiply as indicated:

(**a**) $(4\sqrt{2})(-5\sqrt{7})(-3\sqrt{14})$;

(**b**) $5\sqrt{3}(4\sqrt{5} - 5\sqrt{3} + 6\sqrt{2})$;

(**c**) $(3\sqrt{2} - 4\sqrt{6})(5\sqrt{2} + 2\sqrt{6})$

5. In the following expressions, eliminate the radical in the denominator:

(**a**) $\dfrac{8}{3\sqrt{2}}$; (**b**) $\dfrac{\sqrt{3}}{4 - \sqrt{5}}$;

(**c**) $\dfrac{5 + \sqrt{3}}{5 - \sqrt{3}}$; (**d**) $\dfrac{3\sqrt{6} - 4\sqrt{2}}{\sqrt{6} - 2\sqrt{2}}$

CHAPTER EIGHTEEN Quadratic Equations

18-1 DEFINITION A *quadratic* equation is an equation containing the second power of an unknown such as x^2, but no higher power. An example of a quadratic equation is $x^2 - 5x + 6 = 0$.

In order to understand better the meaning of a quadratic equation, let us first take a look at some other equations that are not quadratics.

An equation containing only the first power of x or any unknown is called a *linear* equation.

Linear equation: $3x + 7 = 24$

An equation containing a term in the second power of the unknown, but no higher power, is called a *quadratic* equation.

Quadratic equations:

$$3x^2 - 5x + 2 = 0$$
$$5x^2 = 2x - 7$$

An equation containing the third power of the unknown, but no higher power, is called a *cubic* equation.

Cubic equations:

$$x^3 + 5x^2 - 7x + 4 = 0$$
$$2y^3 - 4y^2 + 5 = 0$$

An equation containing the fourth power of the unknown, but no higher power, is called a *quartic* equation.

Quartic equations:

$$x^4 + 3x^3 - 5x^2 + 2x - 8 = 0$$
$$3n^4 + 5n^2 - 4n = 0$$

Equations above the second degree are often named simply after the highest degree term.

Third degree equation: $x^3 + 5x^2 - 7x + 4 = 0$
Fourth degree equation: $y^4 - 7y^3 + 3y = 0$
Fifth degree equation: $n^5 - 6n^2 + 4n + 9 = 0$

A *root* of any equation is *any* number that satisfies the equation. The real test of any root is to see whether it makes the equation true. We shall find that a quadratic equation has two roots, a cubic has three roots, and a quartic has four roots. In general, the number of roots of any equation is equal to the degree of the equation. In this book we do not solve any equation above the quadratic unless they are very easy. We mention others only occasionally. Our concern here is with quadratic equations.

The general form of a quadratic equation is

$$ax^2 + bx + c = 0$$

This means

1. The x^2 term will have some coefficient we call a.
2. The x term will have some coefficient we call b.
3. The constant term, which we call c, is the term or part that does not contain x in any form.

One or both of the coefficients, b and c, may be zero. Here are some examples of quadratic equations:

$$3x^2 + 5x + 2 = 0 \qquad 4x^2 = 5x \qquad 2x^2 + 7 = 0$$

$$6x^2 = 7x + 4 \qquad 3x^2 = 0 \qquad 7x^2 + 3x = 0$$

If $b = 0$, the equation contains no term in x. Then the equation is called a *pure quadratic*. For example, $3x^2 - 5 = 0$ is a pure quadratic equation.

If the equation contains both the x^2 term and the x term, it is called a *complete quadratic*. Here are two complete quadratics:

$$x^2 = 5x + 6 \quad \text{and} \quad 3x^2 - 7x = 0$$

A *root* of a quadratic equation is any number that satisfies the equation. Consider the following quadratic equation:

$$x^2 - 5x + 6 = 0$$

Let us try to find the roots by trial and error.

We try $x = 1$?	Does $1^2 - 5(1) + 6 = 0$?	No	
We try $x = 2$?	Does $2^2 - 5(2) + 6 = 0$?	Yes.	"2" is a root.
We try $x = 3$?	Does $3^2 - 5(3) + 6 = 0$?	Yes.	"3" is a root.
We try $x = 4$?	Does $4^2 - 5(4) + 6 = 0$?	No.	

We find that the equation has two roots, 2 and 3. We usually call the roots r_1 and r_2.

The two roots of a quadratic equation are sometimes called a *solution set*. The set of numbers is often enclosed in braces: {2, 3}. The solution set is then the set of numbers, {2, 3}, which may also be written as the set {3, 2}. Notice that this set of numbers does *not* mean an ordered pair.

The numbers of a set may be written in any order. The meaning is *not* the same as the solution we get in solving a system of equations, in which we get one particular value for x and a particular value for y. In that case, we have seen that we enclose the numbers in parentheses and call the numbers an *ordered pair* because they must be taken in a certain order to show the values of x and y, respectively. However, in the case of the equation $x^2 - 5x + 6 = 0$, the solution set is the set $\{2, 3\}$, in which the numbers may be shown in any order. The set simply means that either of the two numbers will satisfy the equation.

To *solve* an equation means to find the roots. In the foregoing equation we found the roots by trial and error. We need a more systematic method. There are several methods for solving quadratic equations. These will now be explained.

18-2 SOLVING A PURE QUADRATIC EQUATION

The method of solving pure quadratics has an approach similar to that for solving linear equations. In general, we may use the following steps:

1. *Isolate the term containing x^2, transposing if necessary.*
2. *If x^2 has a coefficient other than "1," divide both sides of the equation by the coefficient.*
3. *Take the square roots of both sides of the equation.*

We illustrate the procedure with examples.

Example 1

Solve the pure quadratic $\quad 3x^2 - 75 = 0$

Solution

Transposing the constant term, $\quad 3x^2 = 75$

Dividing both sides by 3, $\quad x^2 = 25$

Taking the square root of both sides of the equation,

$$x = \begin{cases} +5 \\ -5 \end{cases}$$

$$x = \pm 5$$

The answer is often written as the solution set $\{5, -5\}$.

Note. Actually, it should be understood that the square roots of the left side of the equation are $+x$ and $-x$, just as the square roots of the right side are $+5$ and -5. This would lead to the following condition:

$$\left.\begin{matrix} +x \\ -x \end{matrix}\right\} = \begin{cases} +5 \\ -5 \end{cases}$$

The foregoing statement really means four equations:

$$+x = +5$$
$$+x = -5$$
$$-x = +5$$
$$-x = -5$$

We notice, however, that the first and the fourth equations have the same meaning and the second and the third equations have the same meaning. For this reason, in actual work the "+" and "−" are omitted before the x on one side of the equation.

Example 2

Solve the pure quadratic: $4x^2 - 7 = 0$

Solution

Transposing the constant term, $4x^2 = 7$
Dividing both sides by 4, $x^2 = \frac{7}{4}$

Taking the square root of both sides of the equation, $x = \dfrac{\pm\sqrt{7}}{2}$

In this example the roots are irrational. In decimal form they may be stated as the following approximate values:

$$r_1 = \frac{+2.646}{2} = +1.323 \qquad r_2 = \frac{-2.646}{2} = -1.323$$

Example 3

Solve the pure quadratic: $2x^2 + 24 = 0$

Solution

Transposing the constant term, $2x^2 = -24$
Dividing both sides by 2, $x^2 = -12$

Taking the square root of both sides of the equation, $x = \pm\sqrt{-12}$

At this point we simply mention the fact that the *square roots of negative numbers* are called *imaginary numbers.* Imaginary numbers are more fully discussed in Chapter 19. All we do at this point is to say that in equations such as Example 3 the roots are *imaginary.*

Exercise 18-1

Solve the following pure quadratics by the foregoing method:

1. $x^2 - 36 = 0$ **2.** $81 - t^2 = 0$ **3.** $y^2 - 144 = 0$

4. $4x^2 - 25 = 0$ **5.** $16x^2 - 49 = 0$ **6.** $9x^2 - 1 = 0$

7. $x^2 - 3 = 0$ **8.** $3y^2 - 15 = 0$ **9.** $5t^2 - 35 = 0$

10. $18x^2 - 5 = 0$ **11.** $8x^2 - 7 = 0$ **12.** $32x^2 - 3 = 0$

13. $2t^2 + 18 = 0$ **14.** $3n^2 + 75 = 0$ **15.** $5x^2 + 80 = 0$

16. $5x^2 - 64 = 0$ **17.** $6x^2 - 169 = 0$ **18.** $7x^2 - 50 = 0$

19. $9y^2 - 20 = 0$ **20.** $4x^2 - 27 = 0$ **21.** $16n^2 - 75 = 0$

22. $R^2 - 10^8 = 0$ **23.** $h^2 - k^2 = 0$ **24.** $s - 16t^2 = 0$

25. $A - \pi r^2 = 0$ **26.** $V - \pi r^2 h = 0$ **27.** $V = (\frac{1}{3})\pi r^2 h$

28. $F = \dfrac{Kab}{d^2}$ (for d) **29.** $K = \dfrac{Mv^2}{2}$ (for v) **30.** $F = \dfrac{Mv^2}{r}$ (for v)

18-3 SOLVING A QUADRATIC EQUATION BY FACTORING

A quadratic equation may sometimes be solved by *factoring*. Consider the equation $x^2 - 5x + 6 = 0$. Factoring the left side, we get

$$(x - 2)(x - 3) = 0$$

Here we have two factors whose product is zero (0). Let us see what this implies.

Now, we know that if we multiply two or more factors and if one of the factors is zero, then the product must be zero. Examples:

$$(4) \cdot (0) = 0 \qquad (0) \cdot (5) = 0 \qquad (6) \cdot (0) \cdot (3) = 0$$
$$(2847) \cdot (5963) \cdot (791) \cdot (0) \cdot (645) = 0$$

Moreover, we know that if the product of two or more factors is zero, one of the factors *must* be zero.

If $a \cdot b = 0$, then either a or b *must* be zero (0).

If $x \cdot y \cdot z = 0$, then x, y, or z must be zero.

If $a \cdot b \cdot c \cdot d = 0$, then at least *one* of the factors must be zero. They need not all be zero.

Suppose we have the following equation:

$$(x - 4)(x - 2)(x + 3)(x)(x - 1) = 0$$

Here we have five factors whose product is equal to zero. Therefore, we know that at least *one* of the factors must be equal to zero. The equation will be true if *any* one of the factors is equal to zero. Therefore, we may set each one of the factors equal to zero and solve the resulting equation.

If the first factor, $x - 4$, is equal to 0, then $x = 4$. The number 4 is a root of the equation because if 4 is put in place of x in the equation the statement is true. Does $(4 - 4)(4 - 2)(4 + 3)(4)(4 - 1) = 0$? Yes.

You will note that all of the factors do not have to be equal to zero at the same time.

The equation $(x - 4)(x - 2)(x + 3)(x)(x - 1) = 0$ has five roots. They are found by setting each factor equal to zero and solving the resulting equations. The work usually takes the following form:

$$\begin{array}{lll} \text{if } x - 4 = 0 & \text{if } x - 2 = 0 & \text{if } x + 3 = 0 \\ \text{then } \quad x = 4; & \text{then } \quad x = 2; & \text{then } \quad x = -3; \end{array}$$

$$\begin{array}{ll} \text{if } \quad x = 0 & \text{if } x - 1 = 0 \\ \text{then } x = 0; & \text{then } x = 1 \end{array}$$

The foregoing equation has five roots: 4, 2, −3, 0, and 1. Any one of these roots will make the equation true. The roots are usually called

$$r_1 \qquad r_2 \qquad r_3 \qquad r_4 \qquad \text{and} \qquad r_5$$

Now let us go back to the equation $x^2 - 5x + 6 = 0$

Factoring the left side, $(x - 2)(x - 3) = 0$

We set each factor equal to zero and solve:

$$\begin{array}{ll} \text{if } x - 2 = 0 & \text{if } x - 3 = 0 \\ \text{then } \quad x = 2, \quad r_1 & \text{then } \quad x = 3, \quad r_2 \end{array}$$

It is immaterial which root is called r_1 and which r_2.

The entire procedure of solving a quadratic equation by factoring depends on having *zero* as the product of two factors; that is, one side of the equation must be zero. If the product of two factors is not zero, we know nothing about either factor. For instance, consider the equation

$$x^2 - 5x + 6 = 12$$

It is possible to factor the left side: $(x - 2)(x - 3) = 12$

However, we cannot solve the equation in this form because we know nothing about either factor. This problem may possibly be solved if we transpose the 12, so that one side of the equation becomes zero.

$$x^2 - 5x + 6 = 12$$

Transposing, we get $\quad x^2 - 5x + 6 - 12 = 0$

or $\quad x^2 - 5x - 6 = 0$

Factoring the le*f*t side, $\quad (x - 6)(x + 1) = 0$

$$\begin{array}{ll} \text{if } x - 6 = 0 & \text{if } x + 1 = 0 \\ \text{then } \quad x = 6, \quad r_1 & \text{then } \quad x = -1, \quad r_2 \end{array}$$

Although factoring can sometimes be used to solve quadratics, it cannot be used when an expression cannot be factored. For this reason, this method is not very useful in most practical problems involving quadratics.

To show how quadratic equations may be solved by factoring, we shall work out several examples.

Example 1

Solve by factoring: $3x^2 + 7x + 2 = 0$

Solution

Factoring, $(3x + 1)(x + 2) = 0$

Set each factor equal to zero and solve:

if $3x + 1 = 0$ then $x = -\frac{1}{3}$, r_1

if $x + 2 = 0$ then $x = -2$, r_2

Example 1

Solve by factoring: $4x^2 = 5x + 6$

Solution

Transposing, $4x^2 - 5x - 6 = 0$
Factoring, $(4x + 3)(x - 2) = 0$

Set each factor equal to zero and solve:

if $4x + 3 = 0$ then $x = -\frac{3}{4}$, r_1

if $x - 2 = 0$ then $x = 2$, r_2

Example 3

Solve by factoring: $3x^2 = 5x$

Solution

Transposing, $3x^2 - 5x = 0$
Factoring, $x(3x - 5) = 0$

Set each factor equal to zero and solve:

if $x = 0$ then $x = 0$, r_1

if $3x - 5 = 0$ then $x = \frac{5}{3}$, r_2

Note. In solving the equation $3x^2 = 5x$, we must be careful to solve for both roots. A common error is to divide both sides of the equation by x. If this is done, we get the new equation $3x = 5$, from which we get $x = \frac{5}{3}$. By dividing both sides of the equation by x, we lose one root, $x = 0$.

Example 4

Solve by factoring: $x^2 - 8x + 16 = 0$

Solution

Factoring, $(x - 4)(x - 4) = 0$

Set each factor equal to zero and solve:

if $x - 4 = 0$	if $x - 4 = 0$
then $x = 4$, r_1	then $x = 4$, r_2

In this equation, we note that the two roots are equal. The roots are +4 and +4. We might be inclined to say that the equation has only one root. It is true that the only number we need to use for a check is 4. However, we have said that every quadratic equation has two roots. For this reason, it is better to think of the equation in Example 4 as having two roots, but they are equal. Such an equation is sometimes said to have a *double root.*

Example 5

Even a pure quadratic may be solved by factoring.

Solve by factoring: $x^2 - 9 = 0$

Solution

Factoring, $(x - 3)(x + 3) = 0$

By setting each factor equal to zero, we get $x = 3$ and $x = -3$.

Exercise 18-2

Solve the following by factoring:

1. $x^2 - x = 6$	**2.** $x^2 + x = 42$	**3.** $x^2 - x = 20$
4. $x^2 = 12 - 4x$	**5.** $x^2 = 2x + 15$	**6.** $x^2 = 24 - 5x$
7. $y^2 + 18 = 9y$	**8.** $y^2 + 5y = 14$	**9.** $y^2 + 12 = 7y$
10. $9x^2 + 16 = 24x$	**11.** $4x^2 + 25 = 20x$	**12.** $9x^2 + 25 = -30x$
13. $5r^2 = 3r$	**14.** $3t^2 = 7t$	**15.** $4x^2 = 9x$
16. $4y^2 = 25$	**17.** $9t^2 = 16$	**18.** $25n^2 = 9$
19. $2x^2 = 5x + 6$	**20.** $6x^2 = 4 - 5x$	**21.** $4x^2 - 15 = 4x$
22. $6x^2 = 12 - x$	**23.** $12x^2 = 7x + 12$	**24.** $6 - 5x = 6x^2$

18-4 SOLVING A QUADRATIC EQUATION BY COMPLETING A SQUARE

We illustrate this method by an example.

Example 1

Solve by completing a square: $x^2 - 6x - 7 = 0$.

Solution

The first step is to see that the coefficient of x^2 is 1.

Transpose the constant term, −7: $x^2 - 6x = 7$.

Now we add a quantity to both sides of the equation that will make the left side a perfect square in x. In this case we add the number 9 to both sides of the equation and get

$$x^2 - 6x + 9 = 7 + 9.$$

We write the left side as a square and combine the numbers on the right:

$$(x - 3)^2 = 16$$

Take the square root of both sides: $x - 3 = \begin{cases} +4 \\ -4 \end{cases}$

This is usually written $x - 3 = \pm 4$

Solve for x: $x = 3 + 4$ and $x = 3 - 4$

$x = 7$ $\qquad x = -1$

Of course, the foregoing equation could have been solved more easily by factoring.

The method of solving a quadratic equation by completing a square can be used for all types of quadratics. However, the method is a rather long, complicated process. For this reason it is seldom used in practical work. In much work in advanced mathematics, however, the technique of completing a square is a useful device. Its chief use for our purpose here is in deriving a general formula for solving all quadratics. For this reason we shall work out a few examples by this method.

For the method of solving a quadratic by completing a square, we have this general procedure:

1. *See that the coefficient of x^2 is 1. If it is some other number than 1, divide both sides by the coefficient of x^2.*
2. *Transpose the constant term to the right side of the equation.*
3. *Add some quantity to both sides that will make the left side a perfect square. The quantity to be added will always be the square of one half of the coefficient of x.*
4. *Write the left side as a square and at the same time combine the terms on the right side.*
5. *Take the square root of both sides, using both signs (+ and −) on the right side.*
6. *Solve the resulting equation for x.*

Example 2

Solve by completing a square: $3x^2 + 7x + 2 = 0$

Solution

Divide both sides of the equation by 3: $$x^2 + \frac{7}{3}x + \frac{2}{3} = 0$$

Transpose $\frac{2}{3}$ to the right side: $$x^2 + \frac{7}{3}x = -\frac{2}{3}$$

Add the square of $\frac{1}{2}$ of $\frac{7}{3}$ to both sides; in this case $\frac{49}{36}$: $$x^2 + \frac{7}{3}x + \frac{49}{36} = -\frac{2}{3} + \frac{49}{36}$$

Write the left side as a square, and combine the terms on the right: $$\left(x + \frac{7}{6}\right)^2 = \frac{25}{36}$$

Take the square root of both sides: $$x + \frac{7}{6} = \pm\frac{5}{6}$$

Solve for x: $$x = -\frac{7}{6} \pm \frac{5}{6} = -\frac{1}{3} \quad \text{and} \quad -2$$

Example 3

Solve by completing a square:

$$x^2 - x - 5 = 0$$

Solution

Transpose the -5 to the right side: $$x^2 - x = 5$$

Add the square of $\frac{1}{2}$ of the coefficient of x to both sides: $$x^2 - x + \frac{1}{4} = 5 + \frac{1}{4}$$

Write the left side as a square: $$\left(x - \frac{1}{2}\right)^2 = \frac{21}{4}$$

Take the square root of both sides: $$x - \frac{1}{2} = \frac{\pm\sqrt{21}}{2}$$

Solve for x: $$x = \frac{1 \pm \sqrt{21}}{2}$$

Note that the roots of this equation are irrational. In decimal form they may be stated approximately:

$$r_1 = 2.7913 \qquad r_2 = -1.7913$$

Example 4

Solve by completing a square: $$5x^2 - 2x + 3 = 0$$

Solution

Divide both sides of the equation by 5: $$x^2 - \frac{2}{5}x + \frac{3}{5} = 0$$

Transpose $\frac{3}{5}$: $$x^2 - \frac{2}{5}x = -\frac{3}{5}$$

Add $(\frac{1}{5})^2$ to both sides: $$x^2 - \frac{2}{5}x + \frac{1}{25} = -\frac{3}{5} + \frac{1}{25}$$

Write the left side as a square: $$\left(x - \frac{1}{5}\right)^2 = \frac{-14}{25}$$

Take the square root of both sides: $$x - \frac{1}{5} = \frac{\pm\sqrt{-14}}{5}$$

Solve for x: $$x = \frac{1 \pm \sqrt{-14}}{5}$$

Note that the foregoing equation has *imaginary* roots.

Exercise 18-3

Solve by completing a square:

1. $x^2 - 4x - 5 = 0$ **2.** $x^2 + 10x + 9 = 0$

3. $x^2 - 8x - 9 = 0$ **4.** $x^2 + 12x + 11 = 0$

5. $x^2 - 12x + 20 = 0$ **6.** $x^2 + 10x + 16 = 0$

7. $x^2 - 3x - 10 = 0$ **8.** $x^2 + 5x - 14 = 0$

9. $2x^2 - 7x + 3 = 0$ **10.** $2x^2 + 3x - 2 = 0$

11. $3x^2 - 4x - 2 = 0$ **12.** $3x^2 + 8x - 3 = 0$

13. $5x^2 - 6x - 2 = 0$ **14.** $7x^2 + 2x - 5 = 0$

15. $4x^2 - 5x - 3 = 0$ **16.** $4x^2 + 5x - 6 = 0$

17. $3x^2 - 5x = 0$ **18.** $5x^2 + 3x = 0$

19. $9x^2 - 12x + 9 = 0$ **20.** $4x^2 - 20x + 25 = 0$

21. $x^2 - 8x + 9 = 0$ **22.** $x^2 - 10x + 7 = 0$

23. $x^2 + 4x + 13 = 0$ **24.** $x^2 - 6x + 10 = 0$

18-5 SOLVING A QUADRATIC EQUATION BY FORMULA

In the quadratic equation, $3x^2 + 7x + 2 = 0$, the roots are determined by the constants 3, 7, and 2. If these numbers are placed into the proper formula, the roots will be obtained. Here is the formula:

$$x = \frac{-b \pm \sqrt{b^2 - 4ac}}{2a}$$

In the quadratic formula a represents the coefficient of x^2; b represents the coefficient of x; c represents the constant term.

In the quadratic equation $3x^2 + 7x + 2 = 0$

$$a = 3 \qquad b = 7 \qquad c = 2$$

In order to identify properly the three constants, a, b, and c, in any quadratic equation, all the terms must be on one side of the equation and a zero (0) on the other side.

If the a, b, and c are placed properly in the formula, the result will be the roots of the equation. Let us see where the formula comes from.

The formula is derived by starting with the general quadratic equation $ax^2 + bx + c = 0$. We solve this general equation by completing a square in x.

We start with $$ax^2 + bx + c = 0$$

Divide through by a: $$x^2 + \frac{b}{a}x + \frac{c}{a} = 0$$

Transpose: $$x^2 + \frac{b}{a}x = -\frac{c}{a}$$

Add $\left(\frac{b}{2a}\right)^2$ to both sides: $$x^2 + \frac{b}{a}x + \frac{b^2}{4a^2} = -\frac{c}{a} + \frac{b^2}{4a^2}$$

Write the left side as a square: $$\left(x + \frac{b}{2a}\right)^2 = \frac{b^2 - 4ac}{4a^2}$$

Take the square root of both sides $$x + \frac{b}{2a} = \frac{\pm\sqrt{b^2 - 4ac}}{2a}$$

Solve for x: $$x = \frac{-b \pm \sqrt{b^2 - 4ac}}{2a}$$

This is the famous quadratic formula. It is important that we identify these constants correctly, especially with regard to sign. The equation should always be written so that the a is positive.

The quadratic formula can be used for *all* types of quadratic equations. For this reason it is the most useful method for solving quadratics. The formula should be thoroughly memorized. To illustrate its use, we solve several examples.

Example 1

Solve by formula: $3x^2 + 7x + 2 = 0$.

Solution

In this example $a = 3$, $b = 7$, $c = 2$. One of the best ways to memorize the formula is to write it down whenever it is to be used.

$$x = \frac{-b \pm \sqrt{b^2 - 4ac}}{2a}$$

Inserting the constants, $$x = \frac{-7 \pm \sqrt{49 - (4)(3)(2)}}{6}$$

$$= \frac{-7 \pm \sqrt{49 - 24}}{6}$$

$$= \frac{-7 \pm \sqrt{25}}{6} = \frac{-7 \pm 5}{6}$$

$$x = \frac{-2}{6} = -\frac{1}{3}, \quad r_1 \qquad x = \frac{-12}{6} = -2, \quad r_2$$

Example 2

Solve by formula: $4x^2 = 5x + 6$

Solution

Rewrite the equation: $4x^2 - 5x - 6 = 0$

In this equation, $a = 4 \qquad b = -5 \qquad c = -6$

Inserting the constants, $$x = \frac{+5 \pm \sqrt{25 - (4)(4)(-6)}}{8}$$

$$= \frac{+5 \pm \sqrt{25 + 96}}{8} = \frac{+5 \pm \sqrt{121}}{8}$$

For one root, $$x = \frac{+5 + 11}{8} = \frac{16}{8} = +2$$

For the other root, $$x = \frac{+5 - 11}{8} = \frac{-6}{8} = -\frac{3}{4}$$

Example 3

Solve by formula: $x^2 - 6x + 9 = 0$

Solution

Inserting the constants, $$x = \frac{+6 \pm \sqrt{36 - 36}}{2}$$

$$x = \frac{+6 \pm 0}{2} = \begin{cases} +3 \\ +3 \end{cases}$$

In this equation we note that the roots are equal. This will always be the case when the quantity under the radical sign is equal to zero (0). In such an equation, although we get only one distinct value for x, we should remember that there are two roots but that they are equal.

Example 4

Solve by formula: $x^2 - 6x + 4 = 0$

Solution

Inserting the constants, $$x = \frac{+6 \pm \sqrt{36 - 16}}{2}$$

$$= \frac{+6 \pm \sqrt{20}}{2}$$

In this example we note that the roots are irrational. However, the radical may be simplified and the roots reduced to simpler terms.

$$x = \frac{+6 \pm 2\sqrt{5}}{2} = +3 \pm \sqrt{5} \quad \text{or} \quad x = \begin{cases} 5.236 \text{ (approx.)} \\ 0.764 \text{ (approx.)} \end{cases}$$

Example 5

Solve by formula: $3x^2 + 5x + 4 = 0$

Solution

Inserting the constants, $a = 3;\ b = 5;\ c = 4$

$$x = \frac{-5 \pm \sqrt{25 - 48}}{6}$$

$$= \frac{-5 \pm \sqrt{-23}}{6}$$

At this point we simply state that the roots are imaginary, since we have the square root of a negative number.

Example 6

Solve by formula: $3x^2 - 5x = 0$

Solution

Inserting the constants, $a = 3;\ b = -5;\ c = 0$

$$x = \frac{+5 \pm \sqrt{25 - (4)(3)(0)}}{6}$$

$$x = \frac{+5 \pm \sqrt{25 - 0}}{6} = \frac{+5 \pm 5}{6}$$

In this equation one root is zero. $x = \frac{+5}{3}$ and $x = 0$

Exercise 18-4

Solve by use of the quadratic formula:

1. $2x^2 - 9x + 7 = 0$ **2.** $3x^2 + 7x + 4 = 0$ **3.** $5x^2 - 3x - 2 = 0$

4. $t^2 - 6t = 7$ **5.** $2h^2 + 7 = 15h$ **6.** $3n^2 + 4n = 7$

7. $9x^2 + 4 = 12x$ **8.** $16x^2 + 9 = 24x$ **9.** $4x^2 + 20x = -25$

10. $3x^2 + 6x = -2$ **11.** $7x^2 = 2x + 1$ **12.** $2x^2 + 6x = 3$

13. $5x^2 - 4 = 8x$ **14.** $x^2 + 3x + 2 = 0$ **15.** $9x^2 + 1 = 6x$

16. $6x^2 - 5x = 0$ **17.** $5x^2 - 4 = 0$ **18.** $3x^2 + 2x = 0$

19. $4c^2 + 7c = 2$ **20.** $3n^2 + 5 = 8n$ **21.** $h^2 + 4h = 5$

22. $5x^2 - 2 = 6x$ **23.** $5x^2 + 6x = 1$ **24.** $2x^2 - 7 = 2x$

25. $3n^2 + 12n = -8$ **26.** $9c^2 + 6c = 4$ **27.** $4d^2 + 3 = 12d$

28. $2r^2 = 5r + 25$ **29.** $4s^2 = 5s + 6$ **30.** $3t^2 + 5t = 12$

31. $4x^2 + 3x = -2$ **32.** $3x^2 + 2x = -1$ **33.** $5x^2 = 4x - 3$

34. $12x^2 + 17x = 40$ **35.** $6x^2 + 72 = 43x$ **36.** $8x^2 + 54x = 45$

37. $\frac{5}{x + 3} + 1 + \frac{2}{x - 3} = 0$ **38.** $\frac{4}{x - 2} + 3 - \frac{3}{x + 3} = 0$

39. $\frac{x}{x - 2} - \frac{3x}{x + 1} + 2 = 0$ **40.** $\frac{3}{x - 1} + \frac{2}{x + 3} - 1 = 0$

18-6 SOLVING STATED PROBLEMS INVOLVING QUADRATIC EQUATIONS

Solving practical problems often involves the solution of a quadratic equation. In solving such a problem, we begin as usual. We let x (or any letter) represent one of the unknown quantities. Then we express the other unknowns in terms of the same letter. We write the equation, which in this case becomes a quadratic.

We have seen that a quadratic equation has two roots. In the solution of a practical problem involving a quadratic equation, it often happens that one of the roots has no meaning in the particular problem. For example, for the width of a rectangle, we may get one negative root. Since the width of a rectangle cannot be negative, that root is discarded. Sometimes, only one of two positive roots has meaning in the given problem.

The following examples illustrate the solving of problems in which quadratic equations appear.

Example 1

One square has a side 5 in. less than twice the side of a smaller square. The larger square has an area 88 sq in. more than the smaller square. Find the length of the side of each square.

Solution

Let x = the number of inches in the side of the smaller square.
Then, $2x - 5$ = the number of inches in the side of the larger square. We express the area (square inches) of each square. For the larger square: area = $(2x - 5)^2$; for the smaller square: area = x^2. Because the difference between the areas is 88 sq in., we have the equation:

$$(2x - 5)^2 - x^2 = 88$$

Expanding and simplifying, $\quad 3x^2 - 20x - 63 = 0$
Solving by factoring, $(3x + 7)(x - 9) = 0$; then $x = 9$; and $x = -7/3$. Here we must take the positive value. Then we have the side of each square: smaller square: 9 in.; larger square: 13 in. To check, we find the area of each square: larger square: 169 sq in.; smaller square: 81 sq in.; difference: $169 - 81 = 88$.

Example 2

A rectangular lawn is 80 ft long and 60 ft wide. Two boys agree to mow half of the lawn each. The first boy mows a strip of uniform width around the entire lawn (along two sides and two ends) until half the lawn is mowed. Find the width of the mowed strip and the size of the remaining portion.

Solution

Let x = width of the strip (in feet)
Then, for the remaining portion we have

$$80 - 2x = \text{length (feet)}; \quad 60 - 2x = \text{width (feet)}$$

Since the entire lawn has an area of $(80)(60)\,\text{ft}^2$, or $4800\,\text{ft}^2$, the area of the remaining portion is 2400 sq ft. For the area of the remaining portion, we have the equation

$$(80 - 2x)(60 - 2x) = 2400$$

Expanding, $\quad 4800 - 280x + 4x^2 = 2400$

Simplifying, $\quad x^2 - 70x + 600 = 0$

Solving by factoring, $(x - 60)(x - 10) = 0$; then $x = 60$; and $x = 10$. For the width of the strip, the first root, 60, is discarded, because the strip obviously could not be 60 ft wide. Then the answer is 10. The width of the strip is 10 ft; the remaining portion is 60 ft by 40 ft.

Example 3

A number of people chartered a bus for \$200 for a trip, the expense to be shared equally. At the last moment, eight of the people were unable to go. As a result, the share of each of those who went on the trip was increased by \$1.25. How many people went on the trip?

Solution

Let x = the number of people scheduled to go on the trip.
Then, $x - 8$ = the number who went
For the price for each person, we divide the total cost by the number of people. We write \$1.25, as $\frac{5}{4}$ dollars.

Then, $\dfrac{200}{x}$ = the original price per person

$\dfrac{200}{x-8}$ = the new price for each person who went

The difference in price for each person was $\frac{5}{4}$ dollars. Then we have the equation:

$$\frac{200}{x-8} = \frac{200}{x} + \frac{5}{4}$$

Multiplying by the LCD, we get

$$800x = 800(x - 8) + 5x(x - 8)$$

Expanding and transposing, $5x^2 - 40x - 6400 = 0$
Dividing by 5, $x^2 - 8x - 1280 = 0$
Factoring, $(x + 32)(x - 40) = 0$
Then we have the two answers: $x = -32$; and $x = 40$
Since x represents the number of people originally scheduled to go on the trip, the answer cannot be -32, a negative. Then we take the answer: $x = 40$. Then, the number of people who went on the trip was $40 - 8$, or 32.

Exercise 18-5

1. Divide 20 into two parts whose product is 84.
2. Divide 11 into two parts whose product is $29\frac{3}{4}$.
3. The sum of a number and its square is 72. Find the number.
4. Find two consecutive odd numbers whose product is 143.
5. The sum of the squares of two consecutive even numbers is 340. What are the numbers?
6. The reciprocal of a fraction is $\frac{16}{15}$ more than the fraction. What is the fraction itself?

7. A rectangle whose length is 4 in. more than its width has an area of 45 sq in. Find the width and the length of the rectangle.

8. Two square fields have a combined area of 5625 sq rd. The side of one field is 15 rd longer than the side of the other square. Find the length of a side of each square.

9. A lawn is 120 ft long and 90 ft wide. After a strip of uniform width has been mowed around the two sides and the two ends, half the total area has been mowed. Find the width and the length of the remainder.

10. A rectangular lawn is 4 ft longer than it is wide. It is surrounded by a sidewalk of uniform width. The walk contains 32 sq ft more than the lawn itself. Find the size of the lawn if the walk is 4 ft wide.

11. A rectangular flower garden whose width is 7 ft less than its length is surrounded by a walk 3 ft wide. The total area of the walk and the garden is twice the area of the garden alone. Find the dimensions of the garden.

12. A variable electric current is given by the formula, $i = t^2 - 7t + 12$. If t is in seconds, at what two times is current equal to 2 amp?

13. A variable voltage is given by the formula, $e = t^2 - 7t + 9$. At what two values of t is voltage e equal to 3?

14. A man makes a regular trip of 180 miles by car at a regular average speed. One day, because of bad weather, his average speed was reduced by 9 mph. As a result, his trip took 1 hr longer. What was his usual speed?

15. A man made a trip of 360 miles. On the return trip his average speed was reduced by 15 mph, and, as a result, the return trip required 4 hr longer. Find his average speed each way.

16. A bus driver on a regular trip of 72 miles found that on one particular run his speed was reduced by 12 mph. As a result, he was $\frac{1}{2}$ hr late. What was his regular speed?

17. A bus driver on a scheduled trip of 312 miles found that he had to reduce his average speed by 9 mph. As a result, the trip took $1\frac{1}{2}$ hr longer than the usual time. What was his usual speed?

18. A ball is thrown directly upward from ground level. If its distance s above ground is given by the formula, $s = 96t - 16t^2$, how long will it take to reach the following heights: (a) 128 ft; (b) 140 ft; (c) 144 ft (s in ft, t in sec)? What is the significance of two answers for each?

19. A particular rifle bullet is fired directly upward from the ground. Its distance from the ground is given by the formula $s = 1200t - 16t^2$, in which s represents the distance (in feet) and t represents the time (in seconds). How long will it take the bullet to reach a height of 5600 ft?

20. An iron bar of uniform cross-sectional area weighs 180 lb. If the length is increased 3 ft by rolling, the weight per foot is 2 lb less. Find the length of the original bar.

21. A number of people chartered a bus for a trip for $270, the expense to be shared equally. At the last moment six of the people were unable to go, and, as a result, the share of each of those who went on the trip was increased by $1.50. How many people went on the trip?

22. A square, 3 in. on a side, is cut out of each corner of a square sheet of aluminum, and the sides are then turned up to form a rectangular container. If the volume of the container is 192 cu in., what was the size of the original sheet of aluminum?

23. A rectangular sheet of metal has a length 6 in. longer than the width. A 4 in. square is cut out of each corner and the edges are turned up to form a rectangular container. If the volume of the container is 640 cu in., find the dimensions of the original sheet of metal.

18-6 IRRATIONAL EQUATIONS

Expressions containing radicals, such as $\sqrt{x + 5}$, are called *irrational* in x. The radical cannot be eliminated until some numerical value is assigned to x. If x has the value 11, then the expression is rational because it has the value $\sqrt{11 + 5}$, which is equal to 4. Until some numerical value is assigned to the variable, the radical is irrational.

If an equation contains an irrational expression, it is called an *irrational equation*. Solving irrational equations involves the same steps involved in solving other equations. However, one additional step is necessary. First, we eliminate the radical by raising both sides of the equation to a higher power. Then the new derived equation can usually be solved. However, when the equation has been solved, all roots must be checked in the *original* equation.

Example 1

Solve

$$\sqrt{x + 1} = 7$$

Solution

First we raise both sides of the equation to a power that will eliminate the radical. In this example, we square both sides. Note especially that when we have the square root of some quantity and then square the square root, we get back to the original quantity. Squaring both sides, we get

$$x + 1 = 49$$

Solving for x,

$$x = 48$$

Now we must check the answer in the *original* equation, *not* in the derived equation. The original equation becomes

$$\sqrt{48 + 1} = 7; \qquad \text{or} \qquad 7 = 7$$

Since the check shows both sides equal, the answer is correct.

Whenever both sides of an equation are raised to a higher power, new roots may be introduced. Then the derived equation may have more roots than the original. Therefore, all roots must be checked. Any roots obtained for the derived equation that do not check in the original equation are called *extraneous* roots. Such answers are not solutions of the original equation.

To see why new roots may be introduced when a quantity is raised to a higher power, consider the equation

$$x - 3 = 0$$

We know there is only one root of this equation. It is +3. However, let us perform some operations on the equation. Transposing, we get

$$x = 3$$

Suppose we square both sides $\qquad x^2 = 9$

Transposing again, $\qquad x^2 - 9 = 0$

Factoring, $\qquad (x - 3)(x + 3) = 0$

Solving, $\qquad x = 3, \quad \text{and} \quad x = -3$

Every step in the solution is entirely legitimate. After squaring both sides of the equation, we end up with a quadratic, which has two roots. Both roots satisfy the derived equation, $x^2 = 9$, but not the original equation, $x - 3 = 0$.

Steps in solving equations containing radicals:

1. *If necessary, isolate the radical by transposing. If the equation contains two radicals, isolate the more complicated radical.*
2. *Raise both sides of the equation to a power that will eliminate one radical. This will be the power indicated by the index of the radical.*
3. *Simplify the resulting equation. If another radical is still present, repeat the process for eliminating the radical.*

4. *Solve for the unknown.*
5. *Check all roots in the original equation to discover any extraneous roots.*

Example 2

Solve

$$\sqrt{x + 4} + 7 = 2x$$

Solution

Transposing, $\sqrt{x + 4} = 2x - 7$
In squaring both sides of the equation, note that the right side is a binomial.

Squaring both sides,	$x + 4 = 4x^2 - 28x + 49$
Transposing and combining,	$0 = 4x^2 - 29x + 45$
Solving this quadratic, we get	$x = 5$ and $x = \frac{9}{4}$

To check, we substitute each value in the original equation. For the root 5, we get $\sqrt{5 + 4} + 7 = 10$. Since the two sides of the equation are equal, 5 is a true root.

To check $\frac{9}{4}$, we get $\sqrt{9/4 + 4} + 7 = \frac{9}{2}$, which is not true. Therefore, the answer $\frac{9}{4}$ is an *extraneous* root.

Example 3

Solve

$$(3x - 2)^{\frac{2}{3}} = 4$$

Solution

Cubing both sides, we get	$(3x - 2)^2 = 64$
Expanding and simplifying,	$9x^2 - 12x - 60 = 0$
Solving,	$x = -2$ and $x = \frac{10}{3}$

Both roots check and are therefore true roots of the original equation.

Example 4

Solve

$$\sqrt{3x - 5} + x = 1$$

Solution

Isolating the radical,	$\sqrt{3x - 5} = 1 - x$
Squaring both sides,	$3x - 5 = 1 - 2x + x^2$
Transposing and combining,	$0 = 6 - 5x + x^2$
Solving,	$x = 2$ and $x = 3$

Checking $x = 2$, we get $\sqrt{6 - 5} + 2 = 1$, or $1 + 2 = 1$, not true. Therefore 2 is not a root of the original equation.

Checking $x = 3$, we get $\sqrt{9 - 5} + 3 = 1$, or $2 + 3 = 1$, not true. Therefore both roots are extraneous and the original equation has no solution.

Example 5

Solve

$$\sqrt{3x - 2} + \sqrt{x - 1} = 3.$$

Solution

Transposing,	$\sqrt{3x - 2} = 3 - \sqrt{x - 1}$
Squaring both sides,	$3x - 2 = 9 - 6\sqrt{x - 1} + x - 1$
Transposing and simplifying,	$3\sqrt{x - 1} = 5 - x$
Squaring both sides,	$9(x - 1) = 25 - 10x + x^2$
The equation reduces to	$0 = x^2 - 19x + 34$
Solving,	$x = 2$ and $x = 17$

The value $x = 2$ checks in the original equation, but 17 is an extraneous root.

Exercise 18-6

Solve the following equations:

1. $\sqrt{x + 3} = 4$
2. $\sqrt{2x + 3} - x = 0$
3. $\sqrt{x - 2} = x - 4$
4. $5 - \sqrt{x + 7} = x$
5. $\sqrt{3x + 1} + 3 = x$
6. $\sqrt{3 - 2x} - 2x = 9$
7. $2x = \sqrt{13 + 2x} - 7$
8. $7 = \sqrt{5 - 2x} - 2x$
9. $\sqrt{10 + 2x} - 8 = 2x$
10. $\sqrt{3x - 2} - \sqrt{2x + 4} = 0$
11. $\sqrt{x} + \sqrt{6x + 1} = 3$
12. $\sqrt{6x + 1} - \sqrt{x} = 3$
13. $\sqrt{5x - 1} - x = 1$
14. $\sqrt{x} + \sqrt{5 - x} = 3$
15. $\sqrt{2x} + \sqrt{11 - x} = 5$
16. $\sqrt{5 - x} - \sqrt{x - 1} = 2$
17. $\sqrt{2x - 5} + \sqrt{x - 3} = 5$
18. $\sqrt{7x - 5} + \sqrt{x + 1} = 6$
19. $\sqrt{x} + \sqrt{2x + 1} = 5$
20. $2 = \sqrt{x} - \sqrt{5x - 4}$
21. $\sqrt{3 - 2x} + \sqrt{x - 1} = 1$
22. $(1 - 4x)^{\frac{1}{2}} + (2 - x)^{\frac{1}{2}} = 1$
23. $\sqrt[3]{x + 1} = 2$
24. $(x^2 - 6x)^{\frac{1}{3}} = 3$
25. $(x^2 - 2x - 7)^{\frac{1}{3}} = 2$
26. $x^{\frac{3}{2}} = 8$
27. $x^{\frac{3}{2}} = -27$
28. $x^{\frac{3}{2}} = -4$
29. $x^{\frac{2}{3}} = 64$
30. $\sqrt{5 - 2x} + \sqrt{3x - 1} = \sqrt{x}$

QUIZ ON CHAPTER 18, QUADRATIC EQUATIONS. FORM A.

Solve equations 1–6 by the method of pure quadratics:

1. $16x^2 - 25 = 0$ **2.** $8x^2 - 3 = 0$ **3.** $x^2 + 16 = 0$

4. $x^2 - 50 = 0$ **5.** $x^2 - 7 = 0$ **6.** $2x^2 - 9 = 0$

Solve equations 7–12 by factoring:

7. $x^2 = 13x + 48$ **8.** $5x + 2 = 3x^2$ **9.** $3x^2 = 7x$

10. $9x^2 = 49$ **11.** $x^2 + 36 = 12x$ **12.** $4x^2 = 5x + 6$

Solve equations 13 and 14 by completing the square:

13. $x^2 = 8x + 9$ **14.** $2x^2 + 3 = 7x$

15. $3x^2 = 4x + 3$ **16.** $109 = 12x - 4x^2$ **17.** $9x^2 + 4 = 12x$

18. Solve this equation and check all roots: $\sqrt{6 - x} + x = 4$.

CHAPTER NINETEEN Imaginary and Complex Numbers

19-1 IMAGINARY NUMBERS

In this chapter we study a new kind of number, called an *imaginary* number. This number is entirely different from any kind we have used up to this time. Imaginary numbers come about through the solution of some quadratic and higher degree equations. They are very useful in the study of electricity, especially in the theory of alternating currents. Let us see, then, what is meant by an imaginary number and how it comes about.

In mathematics it often happens that we are faced with finding the square root of some negative number, as in the equation

$$x^2 + 9 = 0$$

Transposing, we get $$x^2 = -9$$

To solve the equation, we must find a value for x. Taking the square root of both sides, we see that x must be equal to $\pm\sqrt{-9}$. In other words, x must be equal to some number such that this number multiplied by itself will equal -9.

Now, we know that the square root of -9 is neither $+3$ nor -3. If we multiply $+3$ by itself, we get $+9$. If we multiply -3 by itself, we also get $+9$. Whenever we square any positive or negative number we always get a positive number as the product. If we are to have the square root of -9, we must have some number that can be multiplied by *itself* to produce -9. This is the meaning of square root. At first thought this may seem impossible, yet the problem may not be so difficult as it first appears.

In solving an equation, if we get some answer such as $\sqrt{-9}$, our first question may be: "What does it mean?" The first thing to do is to recognize that the square root of a negative number must be a new kind of number different from those we have already studied. In order to get a better understanding of its meaning, let us first recall exactly what we mean by the square root of any number.

You will recall that the $\sqrt{2}$, $\sqrt{3}$, etc., *cannot* be expressed as exact decimals or as exact common fractions. Such numbers are called irrational numbers. Although we cannot express the exact square root of 2 as a decimal, we can write it as a symbol: $\sqrt{2}$.

Now, we have seen that the square root of a -9 is neither $+3$ nor -3. We know that we *cannot* express the square root of -9 as we express the square root of 25, for example. Yet, we must also know that if we were able to find the *exact* square root of -9 and then multiply it by *itself*, we must get -9 as the product. This must be so from the definition of square root.

Although we cannot express the square root of -9 as an ordinary number, we can write it as a symbol:

$$\sqrt{-9}$$

If we use the symbol $\sqrt{-9}$ as the exact square root of -9, then this symbol multiplied by itself must equal -9, that is,

$$(\sqrt{-9})(\sqrt{-9}) = -9$$

The symbol $\sqrt{-9}$ is called an imaginary number. *The square root of any negative number is called an imaginary number.* This is not a good name for such numbers. Here the word *imaginary* does not mean what is ordinarily meant by that word. Anything imaginary is usually understood to be something we can only imagine. These numbers were *called* imaginary at a time when people did not fully understand them.

However, we can do much more than merely imagine such numbers. We can actually write the square root of -9: thus $\sqrt{-9}$. Moreover, we can add, subtract, multiply, and divide such numbers. Therefore, they cannot be something that exists only in our imagination. Still, they are called *imaginary* numbers. All other numbers are called *real numbers.*

The terms *real numbers* and *imaginary numbers*, are probably misleading, since we can really work with all of them. Some people have proposed calling the square roots of negative numbers by some other name than *imaginary*, but such proposals have not been generally accepted. Therefore, at the present time we have these definitions:

1. The even roots of negative numbers are called *imaginary numbers.*
2. All other numbers are called *real numbers.*

Examples

Imaginary numbers: $\sqrt{-4}$, $\sqrt{-7}$, $-\sqrt{-16}$, $\sqrt{-3/5}$, $-\sqrt{-20}$.
Real numbers: 3, -5, 2/3, $-\sqrt{7}$, π, $-\pi$, $-3/7$, -4.37.

Imaginary numbers are always confusing to a student when he or she is faced with them for the first time. You may wonder, "What good are such numbers anyway?" It is true, we cannot count with imaginary numbers, but neither can we count with negative numbers. We cannot measure the length of a room with imaginary numbers, but neither can we do so with negative numbers. You will recall that people once called negative numbers "fictitious" and discarded them as having no meaning. Yet, negative numbers kept forcing themselves into the solution of equations, and now they have very important meanings. It has been the same with imaginary numbers. Although imaginary numbers may at first be confusing, they have important uses in engineering and other mathematics.

19-2 SIMPLIFYING IMAGINARY NUMBERS

When we come to imaginary numbers, the first problem we face is to devise methods and rules for working with them. The first thing we do is to simplify the form of the number. Imaginary numbers can be simplified in a way that is similar to the method of simplifying radicals. We have seen that a radical, such as $\sqrt{32}$, can be simplified thus:

$$\sqrt{32} = \sqrt{16 \cdot 2} = \sqrt{16} \cdot \sqrt{2} = 4\sqrt{2}$$

Imaginary numbers may be simplified in a similar manner. The radicand is factored so that one of its *factors* is (-1). The remaining factor is simplified as any ordinary radical. The following examples show how this is done:

$$\sqrt{-9} = \sqrt{9(-1)} = \sqrt{9} \cdot \sqrt{-1} = 3\sqrt{-1} \qquad \text{or 3 of these: } \sqrt{-1}$$

$$\sqrt{-25} = \sqrt{25(-1)} = \sqrt{25} \cdot \sqrt{-1} = 5\sqrt{-1} \qquad \text{or 5 of these: } \sqrt{-1}$$

$$\sqrt{-64} = \sqrt{64(-1)} = \sqrt{64} \cdot \sqrt{-1} = 8\sqrt{-1} \qquad \text{or 8 of these: } \sqrt{-1}$$

19-3 THE IMAGINARY UNIT

You will notice that in each of the foregoing examples the $\sqrt{-1}$ can be isolated as a factor. This factor, $\sqrt{-1}$, is a sort of measuring *unit* for determining what might be called the amount of an imaginary number. It is called the *imaginary unit*; that is,

$$\sqrt{-9} \text{ is 3 of these units} \quad \text{or} \quad 3\sqrt{-1}$$

$$\sqrt{-25} \text{ is 5 of these units} \quad \text{or} \quad 5\sqrt{-1}$$

$$\sqrt{-64} \text{ is 8 of these units} \quad \text{or} \quad 8\sqrt{-1}$$

Let us say, then, that

$$\sqrt{-1} = (i)\text{imaginary unit}$$

This definition is often shortened to read $\sqrt{-1} = i$. In this case the i is used because it is the initial letter of the word *imaginary*. Then we can say

$$\sqrt{-9} = 3i; \qquad \sqrt{-25} = 5i; \qquad \sqrt{-64} = 8i$$

Remember, we do not get rid of the radical simply by calling it i. The letter i is used only for convenience to represent the number $\sqrt{-1}$. In electrical engineering the number $\sqrt{-1}$ is represented by the letter j, since it is common practice to let the letter i represent current. In this chapter we shall use both forms so that the student may become familiar with both.

One important fact to keep in mind is that such numbers as $\sqrt{-9}$ and $\sqrt{-25}$ are actual numbers. We may at first be inclined to feel that these numbers do not exist, since we cannot express them by the use of our common number expressions, such as whole numbers, fractions, and decimals. However, we do call them numbers because we can perform arithmetic operations with them. They can be added, subtracted, multiplied, and divided.

19-4 ADDITION AND SUBTRACTION OF IMAGINARY NUMBERS

Imaginary numbers may be added or subtracted. First, we simplify them so that each one shows the imaginary unit, $\sqrt{-1}$, or i. Then the numbers are added or subtracted in the same way as similar algebraic terms.

Example 1

Addition: $\sqrt{-9} + \sqrt{-25} + \sqrt{-49} = 3i + 5i + 7i = 15i$

Example 2

Subtraction: $\sqrt{-81} - \sqrt{-16} = 9i - 4i = 5i$

Example 3

$$\sqrt{-4} + \sqrt{-36} - \sqrt{-144} + \sqrt{-1} = 2i + 6i - 12i + i = -3i$$

19-5 MULTIPLICATION OF IMAGINARY NUMBERS

Imaginary numbers may be multiplied by real numbers or by other imaginary numbers. In multiplication, however, we must remember that

$$\sqrt{-1} \cdot \sqrt{-1} = -1; \qquad i \cdot i = -1; \qquad i^2 = -1; \qquad j^2 = -1$$

In the multiplication of imaginary numbers, whenever we get i^2, we immediately call it -1.

Example 1

Multiply $(9i)(5i)$

Solution

The product of the coefficients is 45; and $(i)(i) = i^2 = -1$. Therefore,

$$(9i)(5i) = 45i^2 = 45(-1) = -45$$

Example 2

Multiply $(-4)(7j)(3j)(2j)$.

The product of the coefficients is -168. In multiplying $(j)(j)(j)$, we first multiply the first two factors: $(j)(j) = -1$. This answer, -1 is then multiplied by the third factor: $(-)(j) = -j$. Therefore,

$$(-4)(7j)(3j)(2j) = (-168)(-j) = +168j$$

Example 3

Multiply $(\sqrt{-4})(\sqrt{-9})$.

In this example we first express each factor in terms of i.

The expression $(\sqrt{-4})(\sqrt{-9})$

becomes $(2i)(3i) = 6i^2 = 6(-1) = -6$

An example of this kind can easily lead to error in multiplication. For example, the following method is *incorrect:*

$$(\sqrt{-4})(\sqrt{-9}) = \sqrt{(-4)(-9)} = \sqrt{+36} = +6 \text{ (wrong)}$$

The answer is wrong because the two negative radicands cannot be written as the product for the radicand of a single radical. The rule for signs in multiplication does *not* hold true for the multiplication of two separate radicands *when both radicands are negative.**

Note. Division of imaginary numbers will be explained in connection with complex numbers.

Exercise 19-1

Simplify each of the following and express each in terms of i (or j):

1. $\sqrt{-4}$ **2.** $\sqrt{-25}$ **3.** $\sqrt{-49}$ **4.** $\sqrt{-100}$

* Imaginary numbers are usually considered neither positive nor negative, since they cannot be compared in size with real numbers. We cannot say that $8i$ is more or less than the real number 8. To try to compare $8i$ with the number 8 to see which is the larger is like trying to compare the two quantities 8 hours and 8 apples to see which is the greater. We might even go one step further and try to see which is larger, 8 inches or the number 8.

However, imaginary numbers do have positive and negative signs before them, as we have seen. Whether or not we call them positive and negative, a number such as $-5i$ has a direction and value opposite to that of the number $+5i$. We may choose to place the number $+5i$ in a *positive* direction on a scale and then call it a *positive imaginary number.* Then the number $-5i$ may be said to be an imaginary number in a *negative* direction and, therefore, *a negative imaginary number.* The terms *positive real number* and *negative real number* refer to *real* number directions; these directions have no reference to imaginary numbers. In the same way, we may say that the terms *positive imaginary number* and *negative imaginary number* refer only to imaginary numbers and have no reference to real numbers.

5. $\sqrt{-169}$ **6.** $\sqrt{-400}$ **7.** $\sqrt{-16x^2}$ **8.** $\sqrt{-36b^2}$

9. $\sqrt{-64y^6}$ **10.** $\sqrt{-81a^4c^8}$ **11.** $\sqrt{-8}$ **12.** $\sqrt{-75}$

13. $\sqrt{-27}$ **14.** $\sqrt{-45x^3}$ **15.** $\sqrt{-\frac{1}{4}}$ **16.** $\sqrt{-9/16}$

17. $\sqrt{-3/8}$ **18.** $\sqrt{-7/32}$ **19.** $\sqrt{-50n^3r^5}$ **20.** $\sqrt{-5a/8x}$

Combine each of the following sets of imaginary numbers into a single term containing i (or j):

21. $\sqrt{-16} + \sqrt{-25} + \sqrt{-1}$ **22.** $\sqrt{-36} + \sqrt{-49} - \sqrt{-9}$

23. $\sqrt{-144} + \sqrt{-81} - \sqrt{-64}$ **24.** $\sqrt{-121} + \sqrt{-100} - \sqrt{-169}$

25. $3\sqrt{-4} + 5\sqrt{-9} - 2\sqrt{-25}$ **26.** $4\sqrt{-16} - 5\sqrt{-1} + 6\sqrt{-49}$

Multiply as indicated: (simplify answers):

27. $\sqrt{-25} \cdot \sqrt{-4}$ **28.** $\sqrt{-2} \cdot \sqrt{-32}$ **29.** $\sqrt{-5} \cdot \sqrt{-5}$

30. $(-3\sqrt{-4})(\sqrt{-49})$ **31.** $(-4\sqrt{-2})(-5\sqrt{-2})$ **32.** $\sqrt{-16} \cdot \sqrt{-9}$

33. $\sqrt{-4} \cdot \sqrt{-81}$ **34.** $(5i)(4i)$ **35.** $(-3i)(-7i)$

36. $(-8j)(+3j)$ **37.** $(\sqrt{-7})^2$ **38.** $(4j)(-9j)$

39. $(j)(j)(j)$ **40.** i^4 **41.** i^5

42. i^8 **43.** j^{10} **44.** i^{12}

19-6 COMPLEX NUMBERS

A complex number is a number that is partly real and partly imaginary. Such a number appears in the solution of some equations. Suppose we solve the quadratic equation $x^2 - 4x + 13. = 0$. By using the quadratic formula, we find the two roots of the equation are $2 + 3i$ and $2 - 3i$.

We must remember that the entire expression, $2 + 3i$, is to be considered as one number. It is one of the roots. The number $2 + 3i$, consists of two parts, one real and the other imaginary. The "2" is the real part; "$3i$" is the imaginary part. The entire number, $2 + 3i$, is called a *complex number.*

A complex number is defined as a number consisting of two parts, one real and the other imaginary. It has the form $a + bi$. By this expression, we mean that a represents the real part, and bi represents the imaginary part. The real part is separated from the imaginary part by a plus or minus sign. The coefficient b itself is real. The real part should always be written first, the imaginary part second.

If the letter j is used instead of i for $\sqrt{-1}$, as is usually done in electrical engineering, then the general form of a complex number is $a + bj$. The j factor is often written before its coefficient: thus, $3 + 7i$ is often written $3 + j7$; $-5 + 4.32i$, is often written $-5 + j4.32$.

In the general complex number, $a + bi$, the real part a and the coefficient b may have any values. If the real part a is equal to zero, then the number is entirely imaginary and is called a *pure imaginary*. If b is equal to zero, then the number is entirely *real*. The complex number $0 + 5i$ is a pure imaginary number. The number $6 + 0i$ is a real number. Any number that is entirely real may be written as a complex number. For example, the number 6 may be written $6 + 0i$.

We often get complex numbers as roots of quadratic and higher degree equations. Here are some examples of complex numbers:

$$3 + 4i \qquad 5 - 2i \qquad -6 + 9j \qquad 8 - 3j \qquad 4 + j \qquad \tfrac{3}{4} - \tfrac{2}{3}i$$

19-7 CONJUGATE COMPLEX NUMBERS

Conjugate complex numbers are defined as any two complex numbers that differ only in the sign of the imaginary part, such as the two numbers $3 + 4i$ and $3 - 4i$. Each of the numbers is called the conjugate of the other. In two conjugate complex numbers the real parts are identically the same. The imaginary parts are exactly alike, except that they have opposite signs.

Exercise 19-2

Solve the following quadratic equations by formula. Notice that the two roots of each equation are conjugate complex numbers.

1. $x^2 - 6x + 13 = 0$ **2.** $y^2 - 10y + 34 = 0$

3. $2n^2 + n + 1 = 0$ **4.** $c^2 - 8c + 20 = 0$

5. $n^2 - 4n + 9 = 0$ **6.** $p^2 - 2p + 2 = 0$

7. $x^2 + 6x + 10 = 0$ **8.** $5r^2 + 2r + 1 = 0$

9. $v^2 + 4v + 16 = 0$ **10.** $t^2 - 2t + 3 = 0$

11. $q^2 - 5q + 52 = 0$ **12.** $3x^2 + x + 2 = 0$

13. $2h^2 - 8h + 5 = 0$ **14.** $t^2 + 12t + 100 = 0$

15. $x^2 - 20x + 500 = 0$ **16.** $2I^2 + 4I + 3 = 0$

17. $x^2 + 32 = 0$ **18.** $y^2 + 64 = 0$

19-8 ADDITION AND SUBTRACTION OF COMPLEX NUMBERS

The two parts of a complex number cannot be combined as we combine real numbers. We have seen that a rational number may be combined with an irrational number, if both are real. For instance, if we solve the quadratic equation $x^2 - 4x + 1 = 0$, we get the following roots:

$$x = 2 + \sqrt{3} \qquad \text{and} \qquad x = 2 - \sqrt{3}$$

In the first answer, $2 + \sqrt{3}$, the two parts can be combined to become a new irrational number approximately equal to 3.732. The number $2 - \sqrt{3}$ becomes approximately equal to 0.268. The two parts of such numbers can be combined because both parts are real.

However, in the case of a complex number such as $2 + 3i$, the two parts are entirely different kinds of numbers. To attempt to combine the real number 2 with the imaginary number $3i$ is like trying to combine 2 pounds and 3 hours. The only way is to indicate the sum, as $2 + 3i$.

For this reason, if two complex numbers are said to be equal, the two real parts must be equal to each other, and the imaginary parts must be equal to each other. As an example, suppose we have the equation

$$x \text{ pounds} + y \text{ hours} = 2 \text{ pounds} + 3 \text{ hours}$$

The only possible way the left side of the equation can be equal to the right side is on the condition that x pounds are equal to 2 pounds and y hours are equal to 3 hours. In other words, $x = 2$ and $y = 3$.

In general terms,

$$\text{if } a + bi = c + di$$

$$\text{then } a = c \qquad \text{and} \qquad b = d$$

Two or more complex numbers are added by adding separately their real parts and their imaginary parts. The same is true with regard to subtraction. The addition or subtraction of two complex numbers, such as $3 - 4i$ and $-6 + 3i$, is similar to the addition or subtraction of two binomials, as $3 - 4x$ and $-6 + 3x$. The following examples show addition and subtraction of complex numbers.

Example 1

(a) Add:

$$\begin{array}{r} 3 - 4i \\ -6 + 3i \\ \hline -3 - i \end{array}$$

(b) Subtract:

$$\begin{array}{r} 3 - 4i \\ -6 + 3i \\ \hline 9 - 7i \end{array}$$

Example 2

(a) Add:

$$\begin{array}{r} -3.27 - j5.21 \\ 6.14 - j2.48 \\ \hline 2.87 - j7.69 \end{array}$$

(b) Subtract:

$$\begin{array}{r} -3.27 - j5.21 \\ 6.14 - j2.48 \\ \hline -9.41 - j2.73 \end{array}$$

Example 3

Combine the following into a single complex number:

$$(3 - 2i) + (-5 + 6i) - (4 + i) - (2 - 7i)$$

Removing parentheses: $3 - 2i - 5 + 6i - 4 - i - 2 + 7i = -8 + 10i$

Exercise 19-3

Combine the following as indicated:

1. $(3 + 2i) + (5 - 3i)$ **2.** $(4 - 3j) - (3 - 2j)$

3. $(2 - 3i) + (-4 - 5i)$ **4.** $(5 + 7j) - (-2 - 3j) + (-3 -$

5. $(-5 + 2j) - (3 + 4j) - (-6 + 0j)$

6. $(2 + 3i) + (0 - 5i) - (5 - 2i)$

Add the following complex numbers:

No.					
7.	$3 + 2i$	$-4 + 4j$	$-5 - 6i$	$2 - 5j$	$5 - 3i$
	$4 - 4i$	$7 - 3j$	$-2 - 3i$	$-3 + 4j$	$5 + 3i$
8.	$-3 - j$	$7 + 4i$	$8 - j3$	$4 - 5i$	$-1 - j6$
	$-3 + j$	$-7 + 0i$	$2 + j4$	$1 + 3i$	$4 - j5$
9.	$4 + 3i$	$-5 + 4j$	$-3 - 2i$	$-2 + j7$	$7 - 9j$
	$-7 + 2i$	$3 - 5j$	$-1 + 5i$	$-8 + j4$	$8 + 4j$
	$5 - i$	$-7 - 3j$	$2 - 4i$	$-1 - j3$	$-9 - 7j$
	$-6 - 6i$	$6 + 2j$	$-5 + i$	$9 + j5$	$-1 + j$

10–11. In each example in No. 7 and 8, subtract the bottom number from the top.

To each of the following complex numbers add its conjugate.

12. $5 + 2i$ **13.** $-7 - 3j$ **14.** $-4 + 5j$ **15.** $3 - j7$

16. $-8 + j$ **17.** $6 - 6j$ **18.** $9 + 0j$ **19.** $0 - 4j$

20–27. From each complex number in Nos. 12–19, subtract its conjugate.

28. Show that the sum of any two conjugate numbers is a real number.

29. Show that the difference between any two conjugate complex numbers is an imaginary number.

30. If $3 + 5j = 4x + 3yj$, find x and y.

31. Solve for x and y: $2 + 7j = (x - 3) - (4 - y)j$

32. Solve for x and y: $2x - 5 - 3j = 4y - 3 + j(2 - y)$.

19-9 MULTIPLICATION OF COMPLEX NUMBERS

In multiplication involving imaginary numbers, the letter i (or j) is treated first as any other literal number, such as x. The multiplication of complex numbers is analogous to the multiplication of any two binomials; that is, the problem $(3 - 4i)(-2 - 5i)$ is analogous to the problem of multiplying the two binomials $(3 - 4x)$ and $(-2 - 5x)$.

However, as soon as the expression i^2 appears, it should be immediately stated in its equivalent form, -1.

Example 1

Multiply $(3 - 4i)(-2 - 5i)$.

$$\begin{aligned}(3 - 4i)(-2 - 5i) &= -6 - 7i + 20i^2 \\ &= -6 - 7i + 20(-1) \\ &= -6 - 7i - 20 \\ &= -26 - 7i\end{aligned}$$

Example 2

Multiply $(-7 + 2j)(3 - j)$.

$$\begin{aligned}(-7 + 2j)(3 - j) &= -21 + 13j - 2j^2 \\ &= -21 + 13j - 2(-1). \\ &= -21 + 13j + 2 \\ &= -19 + 13j\end{aligned}$$

If the multiplier is only a real number or a pure imaginary, the multiplication is similar to the multiplication of a polynomial by a monomial.

Example 3

$$-3(5 - 4i) = -15 + 12i$$

Example 4

$$-2i(-3 - 7i) = +6i + 14i^2 = -14 + 6i$$

Exercise 19-4

Multiply as indicated and write the product with the real part first:

1. $-3(4 - 5i)$ **2.** $5i(3 + 2i)$ **3.** $-4i(-2 + 5i)$

4. $3j(-5 - 7j)$ **5.** $2j(-8 + 3j)$ **6.** $-7j(-3 - 4j)$

7. $(5 + 2j)(1 + 3j)$ **8.** $(-5 + 4j)(2 + 5j)$ **9.** $(3 + 2j)(-5 + 3j)$

10. $(-4 - 3j)(-5 + 4j)$ **11.** $(5 - 8j)(3 + 5j)$ **12.** $(-7 + j)(7 - j)$

13. $(2 - 5i)(2 - 5i)$ **14.** $(4 + 5i)(4 - 5i)$ **15.** $(-5 + i)(-1 + 2i)$

16. $(-6 - 5j)(3 + 2j)$ **17.** $(3 + 4j)(3 + 4j)$ **18.** $(-5 - 2j)(-5 + 2j)$

19. $(-3 + i)(3 + i)$ **20.** $(4 + 3i)(4 - 3i)$ **21.** $(-3 + 2i)(-3 + 2i)$

22. $(3 - 5j)^2$ **23.** $(1 + j)^2$ **24.** $(-7 - 3j)^2$

25. $(8 + 2j)^2$ **26.** $(7 + j)^2$ **27.** $(-1 - 5j)^2$

28. Multiply the complex number, $5 + 2j$, by the multiplier j; the answer by j again; that answer by j, and so on until you have used the j multiplier four times. Explain the result.

29. Beginning with the complex number, $-1 - 3j$, follow the procedure indicated in No. 28, until you have used the j multiplier four times.

30. Simplify i^n, for all values of n (integers) from 1 through 20.

31. Multiply each of the following complex numbers by its conjugate:

(a) $5 + 2j$ (b) $-7 - 3i$ (c) $4 - 5i$ (d) $-3 + 4i$
(e) $-8 + j$ (f) $6 - 6j$ (g) $9 + 0j$ (h) $0 - 4j$

32. Show that the product of any two conjugate complex numbers is a real number.

19-10 DIVISION OF COMPLEX NUMBERS

When we add, subtract, or multiply complex numbers, we get an answer that is another complex number. The answer can always be reduced or combined so that it contains only two parts, real and imaginary.

In division our objective is the same: to reduce the answer, or quotient, to a complex number that contains the two parts, real and imaginary, in such a way that they can be separated by a plus or minus sign. If the divisor is a real number, we merely divide each part of the complex number by the real divisor.

Example 1

Divide $\dfrac{-3 + 8i}{-7} = +\dfrac{3}{7} - \dfrac{8}{7}i.$

Notice that the real part is separated from the imaginary part.

In the division of complex numbers, the chief difficulty occurs when the divisor is an imaginary or a complex number; that is, when i appears in the denominator. When this happens, our main objective is to *eliminate any imaginary appearing in the denominator.* The purpose is to make it possible to state the quotient in the general form of all complex numbers, $a + bi$, in which the real part is separated from the imaginary part by a plus or minus sign. The procedure is shown by examples.

Example 2

Divide $(-2 + 5i) \div (3 - 2i)$.

First we write the division as a fraction:

$$\frac{-2 + 5i}{3 - 2i}$$

This fraction, in its present form, cannot be written so that the real part is separated from the imaginary part. Our first objective is to eliminate the imaginary from the denominator.

In order to make the denominator a real number, we multiply the numerator and the denominator of the fraction by the *conjugate form of the denominator*, that is, by $3 + 2i$.

$$\frac{(-2 + 5i)(3 + 2i)}{(3 - 2i)(3 + 2i)} = \frac{-6 + 11i + 10i^2}{9 - 4i^2} = \frac{-16 + 11i}{13}$$

The answer can now be written in two parts: $-\frac{16}{13} + \frac{11}{13}i$

We can now recognize the general form of the complex number, in which the real part, a, is equal to $-\frac{16}{13}$, and b, or coefficient of i, is equal to $+\frac{11}{13}$.

Whenever a complex number appears as the denominator of a fraction, the denominator can be transformed into a real number by multiplying it by the conjugate form of complex number. For the division of complex numbers, we have this rule:

Rule. *Multiply the numerator and denominator by the conjugate form of the denominator. Then expand both numerator and denominator. Since the denominator will be real, the entire answer should then be written in a form in which the real part is separated from the imaginary part by a plus or minus sign.*

Note. If the denominator is a pure imaginary, the multiplier may simply be i.

Example 3

Divide $\frac{6}{2 - 3i}$.

Multiply numerator and denominator by the conjugate form of the *denominator.* In this example we multiply both by $2 + 3i$.

$$\frac{6(2 + 3i)}{(2 - 3i)(2 + 3i)} = \frac{6(2 + 3i)}{4 - 9i^2} = \frac{6(2 + 3i)}{4 + 9} = \frac{12 + 18i}{13}$$

The answer may now be written in the form

$$\frac{12}{13} + \frac{18}{13}i$$

Exercise 19-5

Divide as indicated. Write the answer in the form, $a + bi$, with the real part separated from the imaginary part.

1. $\frac{5 + 2i}{1 + 3i}$ **2.** $\frac{-5 + 4i}{2 + 5i}$ **3.** $\frac{3 - i}{-5 + 3i}$ **4.** $\frac{-4 - 3i}{-2 + 7i}$

5. $\frac{5 - 3j}{3 + j}$ 6. $\frac{5 - j}{-3 + 4j}$ 7. $\frac{3 + 7j}{5 - 3j}$ 8. $\frac{-1 - 3j}{2 + 4j}$

9. $\frac{-8 + j}{-3 + j}$ 10. $\frac{5 + 7j}{-1 - 3j}$ 11. $\frac{16 + 2j}{5 - j}$ 12. $\frac{-3 + 5j}{-2 - 3j}$

13. $\frac{7 - 11i}{-2 + i}$ 14. $\frac{5 - 7i}{3 - i}$ 15. $\frac{1 - 5i}{1 - 3i}$ 16. $\frac{9 + 3i}{1 + 3i}$

17. $\frac{3}{5 - 2j}$ 18. $\frac{4j}{3 - 5j}$ 19. $\frac{5}{-4 + 3j}$ 20. $\frac{8}{-5 - 7j}$

21. $\frac{6j}{1 - 3j}$ 22. $\frac{1}{-3 - 5j}$ 23. $\frac{1}{-2 + 4j}$ 24. $\frac{5}{3j}$

25. Find reciprocals of (a) $3 + 2j$; (b) $5 - j$; (c) i; (d) $3i$.

19-11 GRAPHING OF IMAGINARY AND COMPLEX NUMBERS

Imaginary and complex numbers had been studied for a long time before any attempt was made to represent them on a number scale. Since early times, real numbers had been represented on a horizontal line.

To represent the positive integers on a horizontal line, we begin at a point we call zero (0) (Fig. 19-1).

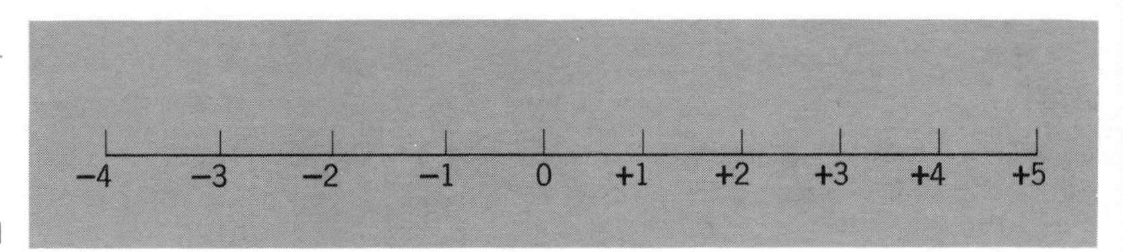

Figure 19-1

To show the points representing the integers, we start at 0 and then mark off equal distances to the right for each of the integers, 1, 2, 3, 4, and so on. We can say that each of these points represents an integer, or number.

Later, fractions and irrational numbers were shown on the same line. When negative numbers were to be shown on this line, they were laid off in a direction opposite to that used for positive numbers. On this line we can represent all of the real numbers: positive, negative, fractional, and irrational numbers.

When imaginary numbers first forced themselves upon the scene, there was no way to represent them on the common number line. Then, in 1797, Caspar Wessel, a Norwegian surveyor, demonstrated before a mathematical meeting in Europe how imaginary and complex numbers might be shown on a graph together with real numbers. His method was to represent imaginary numbers on another straight line drawn perpendicular to the line of real numbers. The arrangement was similar to the rectangular coordinate system consisting of the x-axis and the y-axis.

In the representation of real numbers and imaginary numbers, the horizontal axis is called the axis of *real* numbers. The vertical axis is the axis of *imaginaries.* Real numbers and pure imaginary numbers can be represented on these two axes as shown in Fig. 19-2.

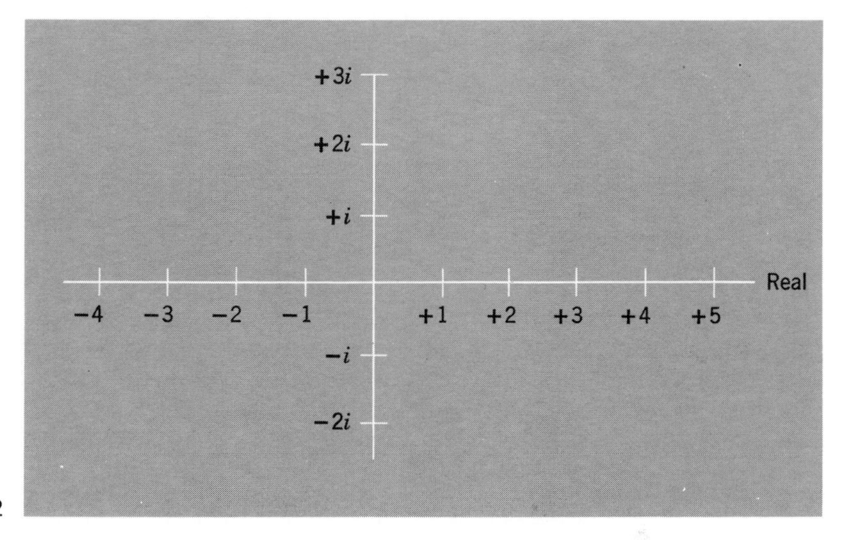

Figure 19-2

The graphing of imaginary and complex numbers on a pair of perpendicular axes in the same plane proved to be one of the most valuable inventions in mathematics. However, although the demonstration was made before 1800, almost one hundred years passed before its importance was fully realized. Then it was discovered that the graphing of complex numbers was a most useful idea in the study of electricity involving alternating currents.

Complex numbers that are partly real and partly imaginary, such as the number $5 + 2i$, can be represented as a point on the graph. The real part is laid off in the direction of the axis of reals, that is, to the right or left of zero. The imaginary part is laid off in the direction of the axis of imaginaries, that is, upward or downward. To locate the point representing the number $5 + 2i$, we count off five units to the *right of zero.* We then count two units in the *upward* direction.

In terms of the x- and y-axes coordinate system, the point representing $5 + 2i$ would be indicated by the notation $(5, 2)$. The two coordinates, 5 and 2, are written within parentheses and separated by a comma. However, when this point represents a complex number, we indicate the sum of the two parts, 5 and $2i$, and we do not enclose the complex number by parentheses. On the graph in Fig. 19-3 are shown several complex numbers.

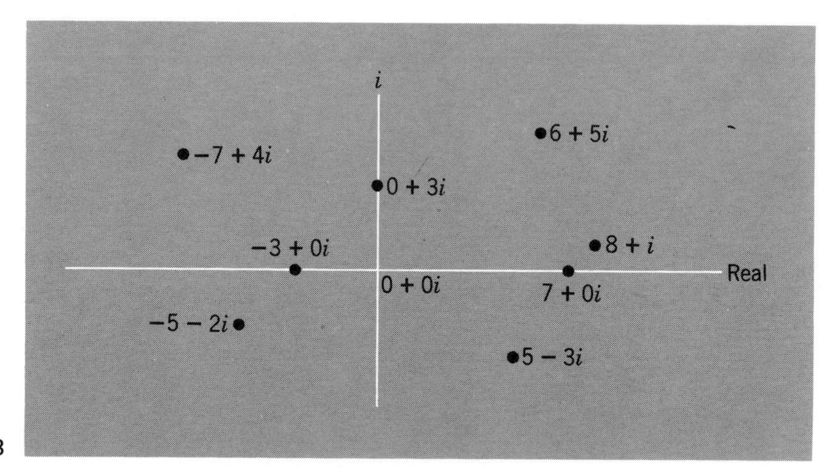

Figure 19-3

19-12 ADDITION OF COMPLEX NUMBERS BY GRAPHING

Complex numbers can be added graphically. The method is different from algebraic addition, but the result, the sum of the numbers, must be the same. Graphical addition is explained by an example.

Example

Add graphically $5 + 2i$ and $1 + 3i$.

Solution

The first step is to show each number as a point on the graph for complex numbers (Fig. 19-4).

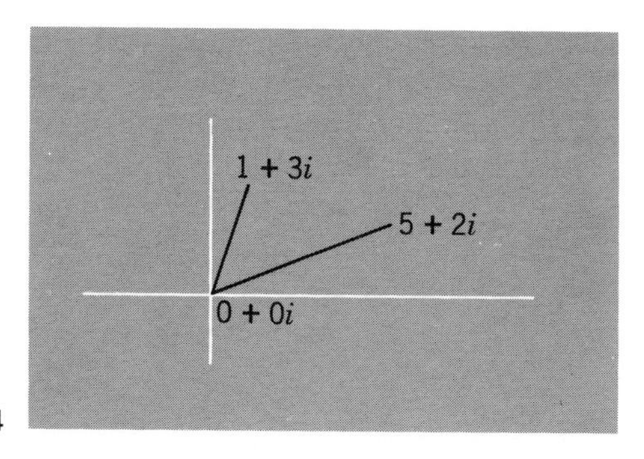

Figure 19-4

Next, we draw a line to each point from the point $0 + 0i$. The point $0 + 0i$ is the point of intersection of the two axes, real and imaginary, and corresponds to the origin in the x- and y-coordinate system. The two lines drawn are often called *vectors.*

On the two vectors we complete a parallelogram, using the vectors as two adjacent sides of the parallelogram (Fig. 19-5). The point at which the two other sides intersect represents the sum of the two complex numbers.

The sum of the two numbers can be further shown by drawing another vector from the point $0 + 0i$ to the point representing the sum.

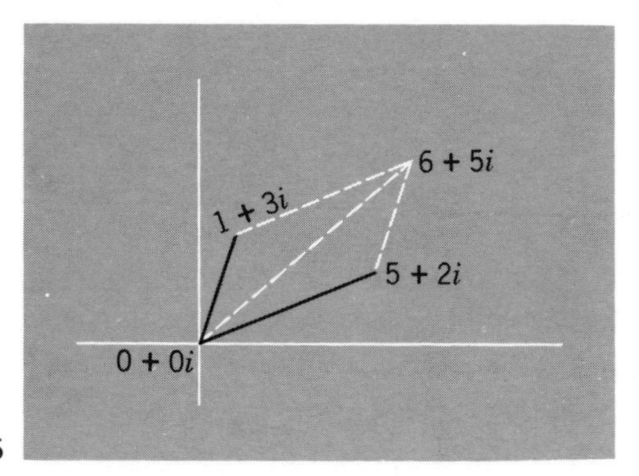

Figure 19-5

The sum of the two numbers can be read directly from the graph as a complex number. In order to determine the real part and the imaginary part, we estimate the distance in the direction of reals and the distance in the direction of imaginaries. If the construction is done carefully, the sum can be estimated fairly accurately. The sum of the two numbers, as indicated by graphical addition, can, of course, be checked by algebraic addition.

In the graphical addition of two complex numbers the necessary parallelogram can be constructed by different methods. One method makes use of the definition of a parallelogram: *a parallelogram is a quadrilateral whose opposite sides are parallel.* The two vectors representing the given complex numbers are drawn first. Let us assume that the two complex numbers are represented by the points A and B (Fig.

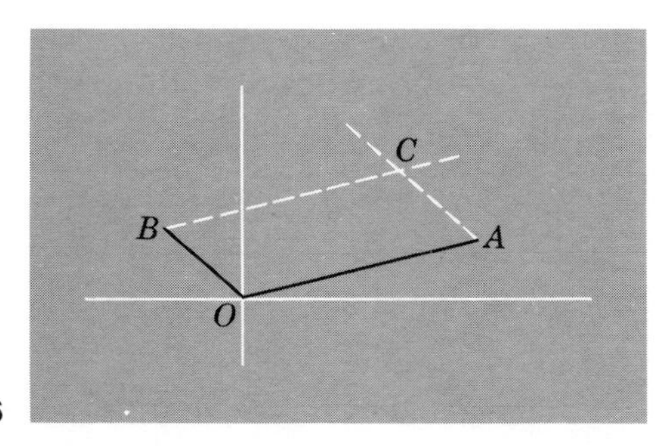

Figure 19-6

19-6) and that the two vectors, OA and OB, have been drawn. Then, through A, we draw a line parallel to OB, and, through B, we draw a line parallel to OA. These two lines intersect at point C. Point C represents the sum of the two numbers A and B. The parallels may be drawn by making equal angles with a horizontal line.

As another method, by use of a compass, we draw two intersecting arcs. For one arc we use A as a center and the vector OB as a radius. For the other arc we use B as a center and the vector OA as a radius. The point of intersection of the arcs is point C, which represents the sum of the two numbers A and B. If this point is connected with each of the points A and B, a parallelogram is formed. This method makes use of the theorem that *a quadrilateral is a parallelogram if its opposite sides are equal.*

If three complex numbers are to be added graphically, two of them are added first. The sum is then treated as a new single complex number, and this number is added to the third. Several complex numbers can be added in the same way.

19-13 SUBTRACTION OF COMPLEX NUMBERS BY GRAPH

Graphical subtraction follows the rule for algebraic subtraction of numbers. You will recall that when we wish to subtract two numbers in algebra we do not really subtract at all. Instead, we change the sign of the subtrahend, and then proceed as in algebraic addition. We do the same in graphical subtraction; that is, we change the sign of the subtrahend and then add.

Example

To subtract $\begin{array}{r} 6 + 3i \\ 5 - 2i \\ \hline \end{array}$ change to add $\begin{array}{r} 6 + 3i \\ -5 + 2i \\ \hline 1 + 5i \end{array}$

In algebraic subtraction we change the sign of the complex number to be subtracted. We do the same in graphical subtraction. In the foregoing example we first change the complex number $5 - 2i$ to its negative $-5 + 2i$. Then we add the two numbers $6 + 3i$ and $-5 + 2i$ graphically.

19-14 MULTIPLICATION AND DIVISION OF COMPLEX NUMBERS BY GRAPH

Graphical multiplication and division of complex numbers, although possible, is not practical. If we wish to represent a product graphically, the best procedure is to multiply the numbers algebraically, and then locate the point that represents the product. If we wish to represent a quotient on the graph, we divide the complex numbers algebraically and then locate on the graph the point that represents the quotient.

19-15 OPERATOR-*j* OR *j*-OPERATOR

In electrical work, as we have mentioned, the imaginary unit $\sqrt{-1}$ is represented by the letter j rather than by i, since the letter i is used to represent electric current. Whatever is said with reference to the j-operator also holds true if we use i to represent $\sqrt{-1}$.

Suppose we represent a pure imaginary or a complex number by a point on the graph. Then, multiplying that number by the multiplier j has the effect of rotating the number through 90° on the graph. (Counterclockwise rotation is called *positive* rotation.)

Suppose we begin with the real number 5. If we multiply 5 by j, we get a product of $5j$ (sometimes written $j5$). The rotation of 90° is shown in Fig. 19-7 by the curved arrow from 5 to $5j$. Now, if we multiply the product $5j$ by j, the number $5j$ rotates through another 90° and becomes $5j^2$, $5(-1)$, or -5. We have thus multiplied the original number 5 by j^2, or -1.

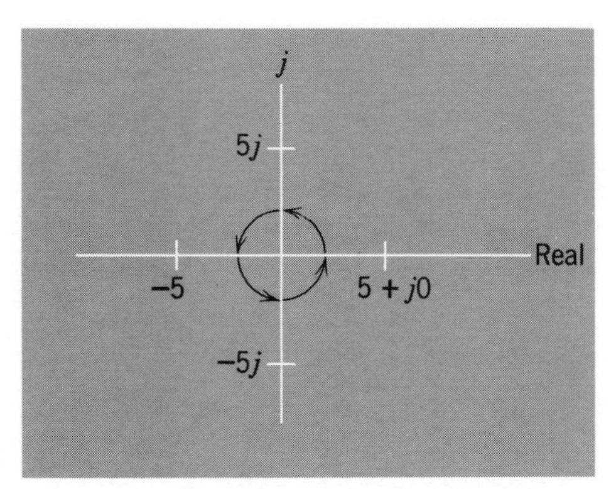

Figure 19-7

If we multiply this product, -5, again by j, we get $-5j$. The result is another 90° rotation. A fourth multiplication by j becomes $j(-5j) = -5j^2 = +5$. The result is still another 90° rotation.

The number 5 has now made one complete rotation of 360°, since it has been multiplied by j four times, which is j^4 or $+1$. A similar change takes place if we start with any real number, positive or negative.

Each time a number is multiplied by j the number is rotated through 90°. For this reason we refer to the multiplier j as "operator-j" or the "j-operator." (The same may be said with regard to the letter i, if that letter is used to represent $\sqrt{-1}$.)

In the same way we can show that multiplying any number by j^2 is the same as multiplying by the operator -1 and has the effect of rotating the number through an angle of 180°. This is a simple explanation of the fact that the product of two negative numbers is a positive number:

$$(-1)(-5) = +5, \text{ that is, } (j^2)(-5) = +5$$

In the foregoing example we began with the real number 5. The result after each multiplication by the j-operator was either another real number or a pure imaginary. Now, if we show a complex number on the graph and multiply the number by the j-operator several times, we still get a rotation of 90° for each multiplication by j.

Suppose we start with the number $5 + 2j$ (Fig. 19-8). If we multiply this number by the j-operator once, we get the first product, $-2 + 5j$.

$$j(5 + 2j) = 5j + 2j^2 = 5j + 2(-1) = -2 + 5j$$

The original number $5 + 2j$ has been rotated through 90° to the position $-2 + 5j$.

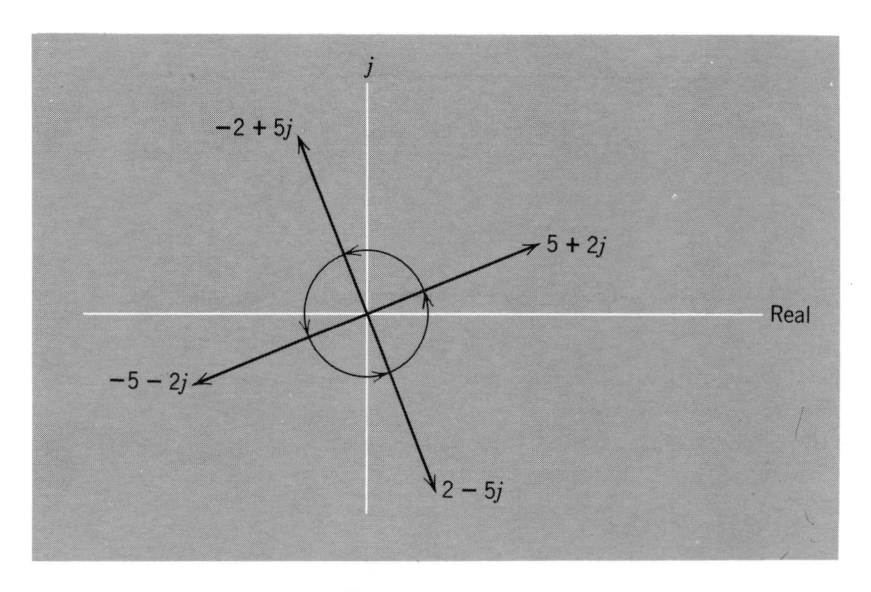

Figure 19-8

If this product, $-2 + 5j$, is then multiplied by j, the product is $-5 - 2j$. Now we have multiplied the original number $5 + 2j$ by j^2, or -1, and the rotation is another 90°, or a total rotation of 180°.

$$j^2(5 + 2j) = -1(5 + 2j) = -5 - 2j$$

Another multiplication by j makes a product of $2 - 5j$, and the result is another rotation of 90°. A fourth multiplication by j makes a product of $5 + 2j$. The number is back at its original position because it has been multiplied by j^4, or $+1$, and it has been rotated through 360°.

Exercise 19-6

Add the following graphically:

1. $5 + 2j$ $2 + 4i$	**2.** $6 + 2i$ $-2 + 4i$	**3.** $-5 + 4i$ $1 + 3i$	**4.** $-6 + 5j$ $-7 - 2j$
5. $4 + i$ $5 - 7i$	**6.** $-5 - 4i$ $1 - 3i$	**7.** $-3 + 3j$ $-3 - 5j$	**8.** $-5 + 3j$ $7 - 2j$
9. $3 + 4j$ $-7 - 3j$	**10.** $-5 + 2j$ $2 - 6j$	**11.** $7 + 3j$ $-4 - 6j$	**12.** $5 + 3j$ $5 - 3j$
13. $3 + 5i$ $-3 + 5i$	**14.** $-5 + j$ $-5 - j$	**15.** $5 + 2i$ $-6 - i$	**16.** $-9 - 8i$ $-8 + 7i$
17. $-6 - 3i$ $4 + 2i$ $-5 - 4i$	**18.** $-1 - 5j$ $4 - 3j$ $-4 + 2j$	**19.** $2 - 5i$ $-3 - 2i$ $-7 + 4i$	**20.** $4 + 3i$ $2 + 5i$ $-5 + i$

21–32. In the first twelve of the foregoing examples, subtract the bottom complex number from the top number graphically.

Graph the following complex numbers and then find the point that represents their product (33–38):

33. $(3 + 2i)(1 + 4i)$ **34.** $(4 + 2j)(1 + 5j)$ **35.** $(1 - 5j)(-2 + j)$

36. $(-2 - 4i)(3 - 2i)$ **37.** $(3 + i)(3 + 4i)$ **38.** $(1 + 4j)(-2 + j)$

39. (a) Graph the number $1 + 4j$.
(b) Multiply this number by j and show the product on the graph.
(c) Multiply the result in (b) by j again and show the product.
(d) Multiply by j again and show the product on the graph.
(e) Multiply by j a fourth time and show the product on the graph.

40. Graph the number $5 - 3j$. Then multiply by j^2 and show the result on the graph.

QUIZ ON A CHAPTER 19, IMAGINARY AND COMPLEX NUMBERS. FORM A.

1. Change to imaginary notation and then add or subtract as indicated:

$$4\sqrt{-9} + 3\sqrt{-49} - 3\sqrt{-4} + 2\sqrt{-121}$$

2. Multiply as indicated and simplify:

(a) $(6i)(-7i)(-2)$ **(b)** $(-3i)(-8i)(-5i)(-4)$; **(c)** $(\sqrt{-16})(\sqrt{-9})$

3. Solve the following equations and show that the roots of each are conjugate complex numbers:

(a) $x^2 - 4x + 20 = 0$; **(b)** $2x^2 - 4x + 5 = 0$;
(c) $5x^2 - 4x + 1 = 0$ **(d)** $2x^2 + 17 = 6x$

4. Add the complex numbers in each set:

(a)	(b)	(c)	(d)
$-3 - 6i$	$5 + 3i$	$1 - 3i$	$-4 - 3i$
$-3 + 2i$	$2 + 3i$	$-5 + 4i$	$4 - 7i$

(e)	(f)
$5 + 2i$	$-6 + 3i$
$-3 - 4i$	$4 + 0i$
$2 + 5i$	$-3 - 4i$
$-1 - 6i$	$2 + i$

5. In No. 4(abcd), subtract the bottom number from the top.

6. Multiply as indicated:

(a) $3i(4 - 5i)$; **(b)** $(5 - 3i)(-2 + 4i)$; **(c)** $(4 - 5i)(4 + 5i)$;
(d) $(-3 + 5i)^2$ **(e)** $(-4 + i)(3 - 6i)$

7. Divide as indicated:

(a) $4 \div 3i$; **(b)** $2 \div (3 - 5i)$; **(c)** $(5 - 2i) \div (2 - 3i)4$;

8. Find the reciprocal of this complex number: $-5 + 3i$

9. Add the following complex numbers graphically:
$(4+i)$ and $(-5+3i)$

CHAPTER TWENTY Roots of Equations

20-1 ROOTS OF A QUADRATIC EQUATION

We have seen that every quadratic equation has two roots. Of course, in some equations, the two roots are equal. In that case we might call the root a *double root.* An example is the equation

$$x^2 - 6x + 9 = 0$$

The roots of this equation are 3 and 3, or the double root 3.

We have found that some quadratic equations have imaginary roots, as distinguished from real roots. In some equations the roots are rational and in others they are irrational. In the equation

$$3x^2 - 7x + 2 = 0$$

the roots are $\frac{1}{3}$ and 2, which are real and rational. In the equation

$$x^2 - 4x + 13 = 0$$

the roots are $2 + 3i$ and $2 - 3i$, which are imaginary.

It is possible to tell something about the roots of a quadratic equation without knowing the roots themselves. By studying the equation, we can tell, without solving it, whether the roots are real or imaginary, whether they are rational, and whether they are equal. There is sometimes an advantage in knowing something about the roots even before solving the equation.

We shall work out several equations by formula to illustrate the different kinds of roots. Note especially the quantity $b^2 - 4ac$ under the radical sign. This quantity is called the *discriminant* because it enables us to discriminate between the different kinds of roots.

Example 1

First, consider the equation

$$3x^2 - 7x + 2 = 0$$

Solution

Let us solve this equation by the quadratic formula:

$$x = \frac{-b \pm \sqrt{b^2 - 4ac}}{2a}$$

Substituting numerical values, we get

$$x = \frac{7 \pm \sqrt{49 - 24}}{6} = \frac{7 \pm \sqrt{25}}{6}$$

At this point in the solution we see that the discriminant, 25, is a perfect square. Therefore, in the final solution the radical sign will disappear and we know at once that the roots are *rational.* Continuing the solution,

$$x = \frac{7 \pm 5}{6} = \begin{cases} \dfrac{7 + 5}{6} = 2, & r_1 \\ \dfrac{7 - 5}{6} = \dfrac{1}{3}, & r_2 \end{cases}$$

As we expected, the roots are *rational.* They are also *real.*

Example 2

Consider a second example:

$$x^2 - 6x + 7 = 0$$

Solution

Substituting numerical values in the quadratic formula, we have

$$x = \frac{6 \pm \sqrt{36 - 28}}{2} = \frac{6 \pm \sqrt{8}}{2}$$

At this point in the solution we see that the discriminant, 8, although positive, is not a perfect square. Therefore, in the final solution the radical sign will *not* disappear and we know at once that the roots are *irrational.* Continuing the solution,

$$x = \frac{6 \pm 2\sqrt{2}}{2} = \begin{cases} 3 + \sqrt{2} \\ 3 - \sqrt{2} \end{cases}$$

As we expected, the roots are *irrational.* They are also *real.*

Example 3

Consider a third example:

$$x^2 - 4x + 13 = 0$$

Solution

Substituting numerical values in the quadratic formula, we get

$$x = \frac{4 \pm \sqrt{16 - 52}}{2} = \frac{4 \pm \sqrt{-36}}{2}$$

At this point in the solution we see that the discriminant is negative, −36. Therefore, in the final solution a negative will appear under the radical sign and we know at once that the roots are *imaginary*. Continuing the solution,

$$x = \frac{4 \pm 6i}{2} = \begin{cases} 2 + 3i \\ 2 - 3i \end{cases}$$

The roots are *imaginary*, as we expected. We do not use the terms *rational* or *irrational* with reference to imaginary numbers.

Example 4

Consider the equation,

$$4x^2 - 12x + 9 = 0$$

Solution

Substituting numerical values in the quadratic formula, we get

$$x = \frac{12 \pm \sqrt{144 - 144}}{8} = \frac{12 \pm \sqrt{0}}{8}$$

At this point in the solution we see that the discriminant is zero (0). Now, if zero is added to, or subtracted from, the 12, the result is the same. Therefore, we know the roots are *equal*. This can happen only if the discriminant is *zero*.

Continuing the solution,

$$x = \frac{12}{8}; \quad x = \frac{3}{2} \quad \text{and} \quad \frac{3}{2}, \quad \text{a double root}$$

The two roots are *equal*, as we expected. They are also *real* and *rational*.

To summarize:

1. If the discriminant, $b^2 - 4ac$, is *negative*, the roots are *imaginary*; in all other cases the roots are real.
2. If the discriminant is *zero*, the roots are *equal*; in all other cases the roots are unequal.
3. If the discriminant is *zero* or a *positive perfect square*, the roots are *rational*; otherwise real roots are irrational.

Remember, the only condition that produces *imaginary roots* is a *negative discriminant*. The only condition that produces *equal roots* is a *zero discriminant*.

Exercise 20-1

Solve the following equations by use of the quadratic formula, and notice the relation between the discriminant and the kinds of roots in each:

1. $x^2 - 2x - 3 = 0$ **2.** $x^2 - 2x + 5 = 0$

3. $4x^2 - 20x + 25 = 0$

4. $2x^2 - 2x + 5 = 0$

5. $3x^2 - 7x + 2 = 0$

6. $2x^2 - 5x + 4 = 0$

7. $4x^2 + 9 = 12x$

8. $2x^2 - 4x = 7$

9. $x^2 + 6x = 7$

10. $3x^2 - 8x = 2$

11. $4x^2 + 2 = 5x$

12. $4x^2 + 5 = 9x$

13. $2x^2 + 4 = 7x$

14. $3x^2 - 7x = 4$

15. $3x^2 - x = 3$

16. $3x^2 - 5x + 2 = 0$

17. $9x^2 + 16 = 24x$

18. $5x^2 + 2 = 6x$

20-2 THE SIGNIFICANCE OF THE DISCRIMINANT: $b^2 - 4ac$

One value of the discriminant is its use in pointing out the presence or absence of imaginary roots in an equation. In practical problems it sometimes happens that imaginary roots can be discarded. Under such conditions, if we can determine by the discriminant that the roots of the equation are imaginary, we need not solve the equation, since it has no "practical" significance.

However, it should never be assumed that all imaginary roots must be discarded. In connection with electric circuits, for example, an imaginary or complex root of an equation is just as significant as a real root. It has often been said that "imaginary" volts are just as dangerous as "real" volts. An imaginary or complex root of an equation involving impedance to current flow simply indicates that the circuit contains some impedance besides pure resistance, such as inductive or capacitive reactance.

In the following equations compute the value of the discriminant and then tell the nature of the roots without solving the equations.

Example 1

$$4x^2 - 5x - 6 = 0.$$

The discriminant is given by the expression $b^2 - 4ac$.
In this equation $a = 4$, $b = -5$, $c = -6$.
Substituting numerical values, we find that the discriminant is

$$25 - 4 \cdot 4(-6) = +121$$

The discriminant, +121, is not negative. Therefore, the roots are not imaginary, but *real.* The discriminant is not zero. Therefore, the roots are *unequal.* The discriminant is a perfect square. Therefore, the roots are *rational,* In summary, the roots are *real, unequal,* and *rational.*

Remember, the discriminant does *not* show the roots themselves. It shows only some characteristics of the roots. If we wish to find the roots, we must solve the equation. As a check, the student should solve some of these examples for the roots themselves.

Example 2

$$9x^2 + 42x + 49 = 0.$$

In this equation $a = +9$, $b = +42$, $c = +49$.

Substituting numerical values in the expression $b^2 - 4ac$, we find that the discriminant is

$$(42)^2 - 4(9)(49) = 1764 - 1764 = 0$$

The discriminant 0 indicates that the roots of the equation are *real*, *equal*, and *rational*.

Example 3

$$3x^2 - 4x + 5 = 0.$$

In this equation $a = +3$, $b = -4$, $c = +5$.

Substituting numerical values, we find that the discriminant is

$$(-4)^2 - 4(3)(5) = 16 - 60 = -44$$

The negative discriminant, -44, shows that the roots are *imaginary*. Imaginary roots of quadratic equations are always unequal. The only condition that produces equal roots is a zero discriminant.

Example 4

$$2x^2 - x - 5 = 0.$$

In this equation $a = +2$, $b = -1$, $c = -5$.

Substituting numerical values, we find that the discriminant is

$$(-1)^2 - 4(2)(-5) = 1 + 40 = +41$$

The positive discriminant, $+41$, shows that the roots are *real*. They are also *unequal*. However, they are *irrational*, since 41 is not a perfect square.

Example 5

$$x^2 - 8x + 25 = 0.$$

In this equation $a = +1$, $b = -8$, $c = +25$.

Substituting numerical values, we find that the discriminant is

$$(-8)^2 - 4(1)(25) = 64 - 100 = -36$$

The negative discriminant indicates the roots are *imaginary* and *unequal*.

Exercise 20-2

Compute the value of the discriminant in each of these equations and, from its value, describe the roots of the equation.

1. $2x^2 + x - 3 = 0$

2. $5x^2 - 6x + 1 = 0$

3. $3x^2 - 10x + 8 = 0$

4. $5x^2 - 6x + 1 = 0$

5. $4x^2 + 7x - 2 = 0$
6. $2x^2 - 5x - 3 = 0$
7. $7x^2 + 3x = 4$
8. $3x^2 + 8x = 3$
9. $16x^2 + 40x + 25 = 0$
10. $9x^2 + 12x + 4 = 0$
11. $6x^2 + 5x = 6$
12. $8x^2 + 5 = 14x$
13. $3x^2 - 8x = 0$
14. $t^2 - t + \frac{1}{4} = 0$
15. $5x^2 + 3x = 0$
16. $4x^2 + 5 = 9x$
17. $5x^2 + 70 = 57x$
18. $3x^2 = 14x + 5$
19. $5x^2 + 2 = 10x$
20. $7x^2 + 4 = 12x$
21. $4x^2 + 1 = 6x$
22. $x^2 + 10x + 5 = 0$
23. $8x^2 - 4x + 5 = 0$
24. $9x^2 + 6x = 2$
25. $4x^2 + 5 = 4x$
26. $x^2 + 2x = 2$
27. $3x^2 = x - 3$
28. $2x^2 = 2x - 3$
29. $3x^2 = 4$
30. $2x^2 = 2x - 5$

20-3 WRITING AN EQUATION THAT SHALL HAVE EQUAL ROOTS

In some problems we may find it desirable to establish a specific condition so that the roots of a quadratic equation will be equal. Suppose we have a quadratic equation containing an undetermined constant such as k in the following example:

Example 1

$$3x^2 + 5x + k = 0$$

Solution

Now, we may wish to determine the proper value of k so that the two roots of the equation will be equal. We recall that the only condition that will produce equal roots is a *zero discriminant.* Therefore, we first set up the form of the discriminant $b^2 - 4ac$. In the foregoing example $a = +3$, $b = +5$, $c = k$. The discriminant is

$$25 - 4(3)(k)$$

If the roots are to be equal, the discriminant must equal zero. So we simply make the statement that

$$25 - 4(3)(k) = 0$$

Solving for k,

$$k = +\frac{25}{12}$$

If $\frac{25}{12}$ is substituted for k in the given equation, the resulting equation should have equal roots. Let us see if this is so. Substituting $\frac{25}{12}$ for k, we get

$$3x^2 + 5x + \frac{25}{12} = 0$$

Multiplying both sides of the equation by 12, $36x^2 + 60x + 25 = 0$

Factoring, $(6x + 5)(6x + 5) = 0$

Solving, $x = -\frac{5}{6}$ and $x = -\frac{5}{6}$

The roots are equal, as we intended they should be.

Example 2

In the equation

$$3kx^2 - 2x^2 + 5x - kx + 4 = 0$$

determine the value of the constant k so that the roots of the equation will be equal.

Solution

In this equation $a = 3k - 2$, $b = 5 - k$, $c = +4$. Therefore, the discriminant is

$$(5 - k)^2 - (4)(4)(3k - 2)$$

If the roots of the equation are to be equal, the discriminant must equal zero. So we simply state that

$$(5 - k)^2 - (4)(4)(3k - 2) = 0$$

Expanding, $25 - 10k + k^2 - 48k + 32 = 0$

Solving for k, $k = 1$, or $k = 57$

Now, if either 1 or 57 is substituted for k in the given equation, the resulting equation should have equal roots. Let us see if this is so. First, substituting 1 for k, we get

$$3x^2 - 2x^2 + 5x - x + 4 = 0$$

Combining, $x^2 + 4x + 4 = 0$

Solving for x, $x = -2$ and $x = -2$

The roots are equal, as we intended they should be.

Now, we try $k = 57$. Substituting 57 for k, we get

$$171x^2 - 2x^2 + 5x - 57x + 4 = 0$$

Combining, $169x^2 - 52x + 4 = 0$

Factoring, $(13x - 2)(13x - 2) = 0$

Solving for the roots, $r_1 = \frac{2}{13}$ $r_2 = \frac{2}{13}$

In this case, also, the roots are equal, as we intended them to be.

Exercise 20-3

In each of the following equations, find the value of k or m that will make the two roots of the equation equal:

1. $x^2 - 6x + k = 0$

2. $x^2 + 3x - k = 0$

3. $mx^2 - x - 2 = 0$

4. $3mx^2 + 4x + 1 = 0$

5. $2kx^2 - 3x + 4 = 0$

6. $x^2 - 3mx + 4 = 0$

7. $9x^2 + 5x + kx + 1 = 0$

8. $2mx^2 - 3x + 2m = 0$

9. $x^2 - 3mx - 1 = 0$

10. $3x^2 + 3kx + k + 1 = 0$

11. Suppose we wish to write the following equation, choosing a value for k that will make the roots of the equation real, rational, but unequal:

$$x^2 - 2x + k = 0$$

We write the discriminant and set it equal to some perfect square, such as 36. Now, find the value of k that will make the discriminant equal to 36 so that the roots will be unequal but rational.

12. In the following equation, find some value of k that will make the roots of the equation imaginary:

$$x^2 - 4x + k = 0$$

20-4 FORMING AN EQUATION FROM GIVEN ROOTS

It is occasionally desirable to write an equation so that it will have certain specified roots. For instance, we may wish to write an equation having the roots 4 and 6. To see how this is done, let us look carefully at the following example of a quadratic equation:

$$x^2 - 10x + 24 = 0$$

We solve the equation by the factoring method.

Factoring, $\quad (x - 4)(x - 6) = 0$

Solving, $\quad x = 4 \quad$ and $\quad x = 6$

that is, $\quad r_1 = 4 \quad$ and $\quad r_2 = 6$

The two roots are 4 and 6. Now, notice that in the factored form of the equation we have

$$(x - 4)(x - 6) = 0$$

We see that each factor consists of

x minus a root

If we indicate the two roots by r_1 and r_2, the factored form of the equation is

$$(x - r_1)(x - r_2) = 0$$

For this reason, if the roots of any quadratic equation are denoted by r_1 and r_2, then the equation can be formed by writing the two factors $(x - r_1)(x - r_2)$ and setting the product equal to zero. The same procedure can be followed in writing an equation of any degree if the roots are known.

Let us consider the cubic equation

$$x^3 - 3x^2 - 10x + 24 = 0$$

This equation can be solved by factoring. The quantity on the left side of the equation can be factored into three factors: thus

$$(x - 2)(x - 4)(x + 3) = 0$$

Now we set each factor equal to zero and solve. We find that the roots are 2, 4, and -3. Note that in the factored form of the equation, each factor consists of the quantity

$$x \text{ minus a root}$$

If we call the roots $2 = r_1$, $4 = r_2$, $-3 = r_3$, then the equation consists of

$$(x - r_1)(x - r_2)(x - r_3) = 0$$

Example 1

Write the quadratic equation that has the roots 7 and -4.

Solution

$$(x - 7)(x + 4) = 0$$

Expanding, $$x^2 - 3x - 28 = 0$$

Example 2

Write the quadratic equation that has the roots $-\frac{2}{3}$ and $+\frac{1}{4}$.

Solution

$$\left(x + \frac{2}{3}\right)\left(x - \frac{1}{4}\right) = 0$$

The equation can be expanded as it stands, or the fractions can first be eliminated in the following manner. Multiply both sides of the equation by

3 and by 4. The multiplier 3 will eliminate the denominator 3; the multiplier 4 will eliminate the denominator 4. The right side will still be zero. The result is

$$(3x + 2)(4x - 1) = 0$$

Expanding, $$12x^2 + 5x - 2 = 0$$

Example 3

Write the equation that has the roots

$$\frac{-3 + 5i}{2} \quad \text{and} \quad \frac{-3 - 5i}{2}$$

Solution

$$\left(x - \frac{-3 + 5i}{2}\right)\left(x - \frac{-3 - 5i}{2}\right) = 0$$

Multiplying both sides of the equation by (2)(2),

$$(2x + 3 - 5i)(2x + 3 + 5i) = 0$$

Expanding, $$4x^2 + 12x + 9 - 25i^2 = 0$$

Combining, $$4x^2 + 12x + 34 = 0$$

or

$$2x^2 + 6x + 17 = 0$$

The resulting equation can be checked by solving.

Example 4

Write the cubic equation that has the roots 2, −3, −3.

Solution

$$(x - 2)(x + 3)(x + 3) = 0$$

Expanding, $$x^3 + 4x^2 - 3x - 18 = 0$$

Exercise 20-4

Write equations having the following sets of roots:

1. $3, -5$ **2.** $-\frac{1}{2}, \frac{3}{4}$ **3.** $\frac{1}{3}, \frac{2}{3}$ **4.** $5, 5$

5. $-3, 0$ **6.** $\pm 3i$ **7.** $2 \pm \sqrt{3}$ **8.** $2 \pm 3j$

9. $-1 \pm \frac{\sqrt{3}}{2}$ **10.** $1 \pm \frac{2j}{3}$ **11.** $-2 \pm \frac{\sqrt{5}}{3}$ **12.** $3 \pm \frac{i\sqrt{2}}{2}$

13. $2, -3, 1$ (cubic equation) **14.** $-2, +3, -3$ (cubic equation)

15. $2, 3, -1, -\frac{1}{2}$ (quartic) **16.** $-2, 2, 0, 1 \pm 2j$ (5th degree)

20-5 ROOTS OF SIMPLE HIGHER DEGREE EQUATIONS

If an equation of any degree can be factored into linear and/or quadratic factors, then the roots can easily be found. Factoring a higher degree equation is often a somewhat difficult matter. It can sometimes be done by the "trial and error" method.

Example 1

Solve $x^3 - 3x^2 - 10x + 24 = 0$.

Solution

Suppose we try $x = 1$ to see if 1 is a root of the equation. Substituting 1 for x in the equation, we ask

does $\quad 1 - 3 - 10 + 24 = 0$? $\quad$ No

Therefore, 1 is not a root. Now we try 2. Substituting 2 for x in the equation, we ask

does $\quad 8 - 12 - 20 + 24 = 0$? $\quad$ Yes

Therefore, 2 is a root of the equation.

Since 2 is a root of the equation, we know $x - 2$ must be a factor of the expression $x^3 - 3x^2 - 10x + 24$. If we divide the expression by the factor $x - 2$, we get another factor, $x^2 - x - 12$. This factor can be further split up into two factors, $(x - 4)(x + 3)$. Therefore, the equation $x^3 - 3x^2 - 10x + 24 = 0$ can be written

$$(x - 2)(x - 4)(x + 3) = 0$$

Now, the three roots may be found by setting each factor equal to zero:

if	$x - 2 = 0$	if	$x - 4 = 0$	if	$x + 3 = 0$
then	$x = 2$	then	$x = 4$	then	$x = -3$

To check each root, it is substituted for x in the original equation.

Some higher degree equations contain quadratic factors that cannot be separated into linear factors. In such equations the roots may be found by setting each quadratic factor equal to zero and solving by formula.

Example 2

Solve $x^3 - 3x^2 - 16x + 6 = 0$.

Solution

After trying several small numbers as roots of the equation, we eventually discover that the value $x = -3$ will satisfy the equation. Dividing the expression $x^3 - 3x^2 - 16x + 6$ by $x + 3$, we get a quotient of $x^2 - 6x + 2$.

The equation $x^3 - 3x^2 - 16x + 6 = 0$ can be written $(x + 3)(x^2 - 6x + 2) = 0$.

Now, if we set each factor equal to zero, we shall get the roots. If $x + 3 = 0$, then $x = -3$. For the second factor, $x^2 - 6x + 2 = 0$, we use the quadratic formula.

$$x = \frac{6 \pm \sqrt{36 - 8}}{2} = \frac{6 \pm \sqrt{28}}{2} = \frac{6 \pm 2\sqrt{7}}{2} = 3 \pm \sqrt{7}$$

Therefore, the roots are $-3, 3 + \sqrt{7}$, and $3 - \sqrt{7}$.

Solving equations above the second degree can sometimes be done by "trial and error" as we did in Example 2. However, the technique of *synthetic division* can often be used to detect linear factors. If the equation can be reduced by synthetic division to a quadratic, then if the remaining quadratic is not factorable, it can be solved by the quadratic formula. This was done in Example 2. For a review of synthetic division, see Chapter 8. Its use is shown in the following example.

Example 3

Solve the equation $x^4 - 2x^3 - 7x^2 + 8x + 12 = 0$.

Solution

If we had the roots of this equation, then we could write it

$$(x - r_1)(x - r_2)(x - r_3)(x - r_4) = 0$$

Now, if any number r is a root of the equation, then $(x - r)$ is a factor of the expression. Then, dividing the polynomial by $(x - r)$ should give us a remainder of zero. To see if 1 is a root, we divide by $(x - 1)$. In synthetic division form, we have

$$\begin{array}{r|rrrrr} 1 & 1 & -2 & -7 & 8 & 12 \\ & & 1 & -1 & -8 & 0 \\ \hline & 1 & -1 & -8 & 0 & 12 \end{array}$$

Note that the remainder is 12. Since the remainder is not zero, then $(x - 1)$ is not a factor of the expression, and 1 is not a root of the equation.

To see whether (-1) is a root, we divide by $[x - (-1)]$, or $(x + 1)$. We have

$$\begin{array}{r|rrrrr} -1 & 1 & -2 & -7 & 8 & 12 \\ & & -1 & +3 & +4 & -12 \\ \hline & 1 & -3 & -4 & 12 & 0 \end{array}$$

Since the remainder is zero, then $(x + 1)$ is a factor of the polynomial, and (-1) is a root of the equation. Then one factor is $(x + 1)$ and the other factor is the quotient. Then the original equation can be written:

$$(x + 1)(x^3 - 3x^2 - 4x + 12) = 0$$

Now we set the second factor equal to zero and get the reduced equation,

$$x^3 - 3x^2 - 4x + 12 = 0$$

To find another root of the equation, we attempt to factor this expression, again by synthetic division. Let us try the factor $(x - 2)$:

$$\begin{array}{r|rrrr} 2 & 1 & -3 & -4 & 12 \\ & & 2 & -2 & -12 \\ \hline & 1 & -1 & -6 & 0 \end{array}$$

Since the remainder is zero, then $(x - 2)$ is a factor of the expression and therefore 2 is a root of the equation. Now we can write the second quotient as a further reduced equation:

$$x^2 - x - 6 = 0$$

Factoring, $\qquad (x + 2)(x - 3) = 0$

Solving for x, $\qquad x = -2 \quad \text{and} \quad +3$

If we wish to see the entire original equation written in factored form, we can write

$$(x + 1)(x - 2)(x + 2)(x - 3) = 0$$

Now we recognize all the roots: $x = -1, +2, -2, +3$. Note again, as we have stated, in the factored form of the equation, each factor consists of $x -$ (*a root*).

It might be pointed out that if the algebraic sum of all of the coefficients in the equation is equal to zero, then 1 is a root of the equation. This is equivalent to setting 1 for x in the equation.

If an equation cannot be reduced to a quadratic by factoring (so that the remaining reduced equation is a cubic or higher degree), then the equation can be solved only by complicated formulas. Moreover, such formulas can be used only for cubics or quartics. There are no formulas for solving equations of a degree higher than the fourth unless the polynomial can be factored. Fortunately, most equations encountered in practical work can be solved by the methods we have discussed.

Exercise 20-5

Solve the following equations:

1. $x^3 - 4x^2 - 12x = 0$

2. $x^3 + 3x^2 - 18x = 0$

3. $x^3 + 8 = 0$ (factor)

4. $x^3 - 125 = 0$ (factor)

5. $x^4 - 81 = 0$

6. $x^4 - 16 = 0$

7. $x^5 - 5x^3 + 4x = 0$

8. $x^5 - 10x^3 + 9x = 0$

9. $x^3 - x^2 - 4x + 4 = 0$

10. $x^3 - 6x^2 + 11x - 6 = 0$

11. $x^3 + 4x^2 + x - 6 = 0$

12. $x^3 + 2x^2 - 5x - 6 = 0$

13. $x^3 + x^2 + 3x - 5 = 0$

14. $x^3 + 5x^2 - 5x - 1 = 0$

15. $x^4 - 4x^3 + x^2 + 6x = 0$

16. $x^4 - 3x^3 - 4x^2 + 12x = 0$

17. $x^4 - x^3 + 9x + 18 = 11x^2$

18. $x^4 + x^3 - x + 6 = 7x^2$

19. $2x^4 - 5x^2 - 15x + 18 = 0$

20. $x^4 - 15x^2 + 10x + 24 = 0$

21. $2x^4 - 8x^3 + 19x^2 - 5x = 34$

22. $3x^4 - 11x^3 + 9x^2 + 5x = 6$

23. $3x^5 - 13x^4 - 8x^3 + 52x^2 = 16x$

24. $x^5 - 6x^4 + 5x^3 + 24x^2 = 36x$

*20-6 SPECIAL METHOD FOR SOLVING SOME HIGHER DEGREE EQUATIONS

We have seen that synthetic division can be used to discover the rational roots of some higher degree equations. Of course, in many equations, we may have to try several possible roots before the correct roots are discovered. If a number is a root of the equation, then, in synthetic division, the remainder must be zero.

There is one procedure by which it is possible to discover some roots rather easily. The method can be used only for integral real roots and where the coefficient of the highest degree term is 1.

First, we must understand what is meant by a *function.* In an equation such as $y = 3x + 4$, we say that y is a *function* of x, because the value of y depends on the value of x. For every value of x, there is a corresponding value of y. In the equation shown here, if $x = 2$, $y = 10$. If $x = 0$, $y = 4$.

If two variables are so related that for each value of the first there is a corresponding value of the second, we say the second variable is a function of the first. As an example, if $s = 5t + 2$, then s is a function of t. For every value of t, there is a corresponding value of s.

We can also say: any expression containing x is a function of x. When we say, $y = 3x + 4$, the function of x is the expression, $3x + 4$. We use y only to represent the function.

Definition. *A function of x is any expression containing x and whose value depends on the value of x.* The expression $x^2 + 3x - 2$ is a *function* of x.

We often use y to represent a function of x. Instead of using y, we can use the symbol $f(x)$ to represent the function. For example, for the function $x^2 + 3x - 4$, we can write

$$f(x) = x^2 + 3x - 4$$

This statement is read: "f of x equals $x^2 + 3x - 4$."

* Optional.

Now, if we wish to find the value of the function for some particular value of x, we substitute the value of x in the symbol as well as in the function. For example, if

$$f(x) = x^2 + 3x - 4$$

For the value $x = 5$ we write

$$f(5) = 25 + 15 - 4 = 36$$

The expression $f(5)$, by itself has no meaning unless we have the expression for the function $f(x)$.

As another example, if

$$f(x) = x^2 - 2x + 6$$

$$f(3) = 9; \qquad f(1) = 5; \qquad f(0) = 6; \qquad f(-1) = 9$$

Now we come to what is known as the *Remainder theorem.* As an example, if a function of x is divided by the quantity $x - 1$, the remainder will be equal to $f(1)$. Consider the example,

$$f(x) = x^3 + 5x^2 - 7x + 10$$

$$f(1) = 1 + 5 - 7 + 10 = 9$$

Now let us divide the function by $(x - 1)$ (by synthetic division):

$$\begin{array}{r|rrrr} 1 & 1 & 5 & -7 & 10 \\ & & 1 & 6 & -1 \\ \hline & 1 & 6 & -1 & 9 \end{array}$$

Notice that the remainder is 9, which is equal to $f(1)$. Also, if the function is divided by $(x + 1)$, the remainder will be $f(-1)$. If the function is divided by $(x - 2)$, the remainder will be $f(2)$. If the function is divided by $(x - r)$, the remainder will be $f(r)$. This principle is known as the *Remainder theorem.*

We now show by examples the special method of solving some higher degree equations.

Example 1

Solve the equation: $x^4 + 8x^3 - 4x^2 - 128x - 192 = 0$

Solution

If we depend on "trial and error" alone, we may have to try several possible roots before the correct ones are discovered. Of course, the roots must be factors of 192, the constant term.

Now, consider the function corresponding to the equation:

$$f(x) = x^4 + 8x^3 - 4x^2 - 128x - 192$$

If the function is divided by $(x - 1)$, the remainder will equal $f(1)$. If the function is divided by $(x + 1)$, the remainder will equal $f(-1)$.

Using synthetic division, we divide the function by $(x - 1)$ and by $(x + 1)$: Dividing by $(x - 1)$:

$$\begin{array}{r|rrrrr} 1 & 1 & 8 & -4 & -128 & -192 \\ & & 1 & 9 & 5 & -123 \\ \hline & 1 & 9 & 5 & -123 & -315 \end{array}$$

then $f(1) = -315$. Dividing by $(x + 1)$:

$$\begin{array}{r|rrrrr} -1 & 1 & 8 & -4 & -128 & -192 \\ & & -1 & -7 & 11 & 117 \\ \hline & 1 & 7 & -11 & -117 & -75 \end{array}$$

then $f(-1) = -75$. Now we write the factors of $f(1)$ and of $f(-1)$, disregarding negative signs:

$$\begin{array}{rrrrrrr} \text{factors of } f(1)\text{:} & 315\text{:} & 1 & 3 & 3 & 5 & 7 \\ \text{factors of } f(-1)\text{:} & 75\text{:} & 1 & 3 & 5 & 5 & \end{array}$$

We rearrange the factors in the order shown below:

$$\begin{array}{rrrrrrr} f(1)\text{:} & 315\text{:} & 1 & 3 & 3 & 5 & 7 \\ f(-1)\text{:} & 75\text{:} & & 1 & 5 & 3 & 5 \\ \hline \end{array}$$

In the second arrangement, note that the pair of factors in each column differs by 2. Such a pair of factors, one from $f(1)$ and the other from $f(-1)$, may be called enveloping factors, because *a possible root of the equation must lie between two such factors.*

In this equation, one root must lie between the 7 in $f(1)$ and the 5 in $f(-1)$. Then we guess that one root is -6. We choose the negative, -6, because the factor 7 in $f(1)$ is greater than the 5 in $f(-1)$.

Now, for a second root, we consider the pair of factors, the 5 in $f(1)$ and the 3 in $f(-1)$. Again, we guess the root is -4.

For a third root, we take the pair of factors, the 3 in $f(1)$ and the 5 in $f(-1)$, and we guess the root is $+4$, choosing the positive number because the greater factor is in $f(-1)$. For the fourth root, we consider the 3 in $f(1)$ and the 1 in $f(-1)$, and guess the root is -2. Then we have the four roots: -6; -4; $+4$; -2. These roots all check in the equation.

In considering the two factors, one from $f(1)$ and the other from $f(-1)$, the root is positive if the factor from $f(-1)$ is greater than the paired factor from $f(1)$.

Example 2

Solve the equation: $x^3 - 5x^2 - 18x + 72 = 0$.

Solution

We write the function: $f(x) = x^3 - 5x^2 - 18x + 72$.
Dividing the function by $(x - 1)$, we get the remainder: $f(1) = 50$.

Dividing the function by $(x + 1)$, we get the remainder: $f(-1) = 84$.
Now we write the factors of $f(1)$ and of $f(-1)$:

$$f(1){:}\ 50{:}\quad 1\ \ 2\ \ 5\ \ 5$$
$$f(-1){:}\ 84{:}\quad 1\ \ 2\ \ 2\ \ 3\ \ 7$$

The equation calls for three roots. Then, we rearrange the factors so that the two factors in each column, one from $f(1)$ and the other from $f(-1)$, form a pair of enveloping factors. Then, a possible root must lie between the two factors in each pair:

$f(1)$: 50:	1	2	5	5
$f(-1)$: 84:	1	4	3	7
Now we choose the possible roots:		3	-4	6

The 1's are not used. The chosen roots will all check in the equation.

Note. A factor **1** is not used except when associated with a 3.

Example 3

Solve the equation: $x^3 - 12x - 16 = 0$.

Solution

We write the function: $f(x) = x^3 - 12x - 16$.
Dividing the function by $(x - 1)$, we get the remainder: $f(1) = -27$.
Dividing the function by $(x + 1)$, we get the remainder: $f(-1) = -5$.
Now we write the factors of $f(1)$ and $f(-1)$, disregarding negative signs:

$$f(1){:}\ 27{:}\quad 1\ \ 3\ \ 3\ \ 3$$
$$f(-1){:}\ \ 5{:}\quad 1\ \ 5$$

The equation calls for three roots. In order to get three pairs of enveloping factors, we supply two extra 1's as factors, and write the factors as shown:

$f(1)$: 27:	1	3	3	3
$f(-1)$: 5:	1	1	1	5
Now we choose the three possible roots:		-2	-2	4

We shall find that these roots check in the equation.

Example 4

Solve: $x^4 - 2x^3 - 64x^2 + 32x + 768 = 0$

Solution

Dividing $f(x)$ by $(x - 1)$, we get the remainder: $f(1) = 735$.
Dividing $f(x)$ by $(x + 1)$, we get the remainder: $f(-1) = 675$.
Now we write the factors of $f(1)$ and of $f(-1)$:

$$f(1){:}\ 735{:}\quad 1\ \ 3\ \ 5\ \ 7\ \ 7$$
$$f(-1){:}\ 675{:}\quad 1\ \ 3\ \ 3\ \ 3\ \ 5\ \ 5$$

The equation calls for four roots. If we are not careful, we may not get the correct roots. In choosing possible roots, *we must use up all the factors of* $f(1)$ *and* $f(-1)$, except possibly some 1's.

In order to get four pairs of enveloping factors, each pair consisting of one factor from $f(1)$ and the other from $f(-1)$ that differ by 2, we rearrange the factors:

$f(1)$:	1	3	5	7	7
$f(-1)$:	1	5	3	5	9
Now we choose the possible roots:		4	−4	−6	8

We shall find that all these roots check in the equation.

Example 5

Solve: $x^5 - x^4 - 48x^3 + 88x^2 - 1152 = 0$.

Solution

Dividing $f(x)$ by $(x - 1)$, we get the remainder: $f(1) = -630$.
Dividing $f(x)$ by $(x + 1)$, we get the remainder: $f(-1) = -1500$.
Now we write the factors of $f(1)$ and of $f(-1)$, disregarding negative signs:

$f(1)$: 630:	1	2	3	3	5	7	
$f(-1)$: 1500:	1	2	2	3	5	5	5

The equation calls for five roots. In order to get five pairs of enveloping factors, we rearrange the factors:

$f(1)$:	1	2	3	3	5	7
$f(-1)$:	1	4	5	5	3	5
Now we choose the possible roots:	3	4	4	−4	−6	

We shall find that all these roots check in the equation.

Exercise 20-6

Solve the following equations:

1. $x^3 - 28x - 48 = 0$
2. $x^3 + 3x^2 - 36x - 108 = 0$
3. $x^3 - 10x^2 + 28x - 24 = 0$
4. $x^4 - 2x^3 - 28x^2 + 8x + 96 = 0$
5. $x^4 - 9x^3 + 2x^2 + 144x = 288$
6. $x^4 + x^3 - 48x^2 + 432 = 36x$
7. $x^4 - 2x^3 + 32x + 384 = 40x^2$
8. $x^4 - 6x^3 + 96x + 256 = 32x^2$

9. $x^4 - 6x^3 + 96x + 1152 = 88x^2$

10. $x^4 - 11x^3 + 15x^2 + 175x = 500$

11. $x^5 + 3x^4 - 52x^3 - 156x^2 + 576x + 1728 = 0$

12. $x^5 - 5x^4 - 37x^3 + 161x^2 + 300x - 900 = 0$

QUIZ ON CHAPTER 20, ROOTS OF EQUATIONS. FORM A.

In each of the following quadratic equations (Nos. 1–8), compute the value of the discriminant and tell what it shows about the roots of the equation: real or imaginary; equal or unequal; rational or irrational.

1. $x^2 - 7x + 10 = 0$ **2.** $2x^2 + x = 3$ **3.** $4x^2 - 20x + 25 = 0$

4. $3x^2 - 4x - 5 = 0$ **5.** $x^2 + 2x = 7$ **6.** $2x^2 = 2x - 5$

7. $5x^2 = 4$ **8.** $3x^2 - 4x = 0$

In the next two equations (Nos. 9–10), find the value of k that will give the equation equal roots.

9. $kx^2 - 8x + 2 = 0$ **10.** $3x^2 - 2x + k = 0$

In Nos. 11–13, write the equation having the two roots in each set:

11. roots: -3; 5 **12.** roots: $4i$; $-4i$ **13.** roots: $3/5$; $-1/2$

Solve the following higher degree equations:

14. $x^3 + 2x^2 - 5x - 6 = 0$ **15.** $x^3 + 3x^2 - 9x - 27 = 0$

CHAPTER TWENTY-ONE Systems Involving Quadratics

21-1 THE PROBLEM OF GRAPHING

We have seen how to solve a system of linear equations such as $3x + 2y = 6$, and $2x - 3y = 12$. The system can be solved by various algebraic methods (Chapter 14), or by graphing (Chapter 15). However, in a graphical solution, we must estimate the answers. If we graph these two equations, we shall probably estimate the answers as:

$$x = 3.2, \qquad y = -1.8 \qquad \text{(Fig. 21-1)}$$

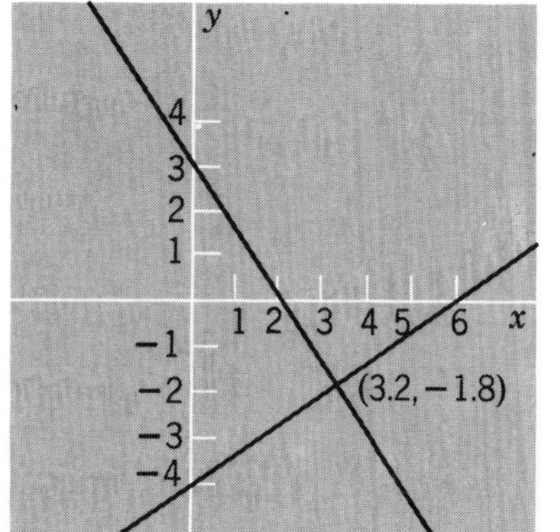

Figure 21-1

Algebraic solutions give the exact answers:

$$x = \tfrac{42}{13}, \qquad \text{and} \qquad y = -\tfrac{24}{13}.$$

If two equations are graphed carefully, the approximate coordinates of the intersection point can be read directly from the graph. The graphic method has certain advantages, but the greatest difficulty is in reading the coordinates of a point, especially when fractions are involved. In all cases, the answers read from a graph should be considered as approximate.

A system involving quadratics can also be solved graphically or by various algebraic methods. In this chapter we shall consider solving systems such as $x^2 + y^2 = 25$ and $2x + y = 0$, or a system of two quadratics, such as $2x^2 + 5y^2 = 50$, and $x^2 - y^2 = 4$.

Note especially the meaning of the following two statements:

1. *When a system is solved graphically, the coordinates of the point or points of intersection will satisfy both equations.*
2. *When a system is solved algebraically, the values obtained for x and y, as ordered pairs, will be the coordinates of the point or points of intersection.*

In graphing quadratics and higher degree equations, we run into difficulties not encountered in connection with linear equations. In graphing a linear equation two points are sufficient since two points determine a straight line. The graph of a second or higher degree equation is generally a curve of some kind, not a straight line. Then it is necessary to take some points very close together to determine the shape of the curve, especially where it curves sharply.

However, the graphs of various equations have certain characteristic forms. A knowledge of various curves will help in sketching the graph of an equation. We shall therefore first consider some second degree curves, and their characteristic shapes.

21-2 THE CIRCLE

An equation containing the second degree terms, x^2 and y^2, in which their coefficients are equal, represents a *circle*, provided there is no xy term.

Example 1

Graph the equation: $x^2 + y^2 = 25$.

Solution

We find several pairs of values for x and y that satisfy the equation. Some of these values are shown here:

$x =$	0	± 3	± 4	± 5
$y =$	± 5	± 4	± 3	0

The pair $(\pm 3, \pm 4)$ represents four points:

$$(+3, +4); (+3, -4); (-3, +4); (-3, -4)$$

When pairs of corresponding values are plotted as points on the graph and connected by a smooth curve, the result is a circle with its center at the origin $(0, 0)$, and a radius of 5 units (Fig. 21-2).

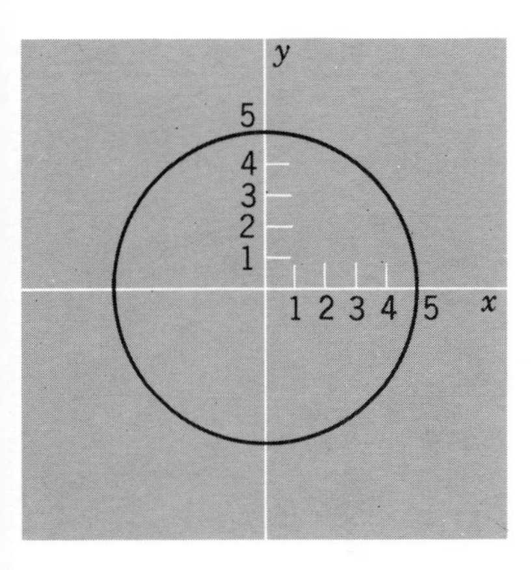

Figure 21-2

Let us see now under what conditions the graph of an equation is a circle. If we graph each of the following equations, we get a circle:

$$x^2 + y^2 = 16; \qquad x^2 + y^2 = 9; \qquad x^2 + y^2 = 20$$

Note that both x^2 and y^2 have a coefficient of 1. Then the right side of the equation represents the square of the radius. If we let r represent the radius, then the standard form of the equation of a circle with the center at the origin $(0, 0)$ is

$$x^2 + y^2 = r^2$$

Knowing the center and the radius of a circle we can draw the graph with a compass.

If the coefficients of x^2 and y^2 are equal but not 1, the equation can be changed to the standard form by dividing by this coefficient. For example, the equation, $4x^2 + 4y^2 = 49$, can be changed to the standard form by dividing both sides by 4. Then the equation becomes, $x^2 + y^2 = \frac{49}{4}$. Now we recognize a circle with center at $(0, 0)$ and radius equal to $\frac{7}{2}$.

Example 2

Write the equation of the circle with center at (0, 0) and radius equal to $2\sqrt{7}$.

Solution

Using the standard form, we get $x^2 + y^2 = 28$.

If a circle is in some other position so that its center is not at the origin, we say the circle is *translated.* If the center is at some point (h, k), the equation can be written by substituting $(x - h)$ for x, and $(y - k)$ for y in the standard form. Then we get the *standard form* for a *translated* circle: $(x - h)^2 + (y - k)^2 = r^2$.

Example 3

Write the equation of the circle whose center is at $(7, -3)$, and whose radius is equal to 5.

Solution

In this example, $h = 7$, and $k = -3$. Substituting these values in the standard form of a translated circle, we get

$$(x - 7)^2 + (y + 3)^2 = 25$$

Expanding and simplifying, we get

$$x^2 + y^2 - 14x + 6y + 33 = 0$$

Example 4

Find the center and radius of the circle: $x^2 + y^2 - 12x - 8y + 3 = 0$.

Solution

We put the equation into standard form by completing the squares in x and y. Transposing the constant term, 3, we get

$$x^2 - 12x \cdots + y^2 - 8y \cdots = -3$$

Now we add 36 and 16 to both sides to complete the squares in x and y:

$$x^2 - 12x + 36 + y^2 - 8y + 16 = -3 + 36 + 16$$

Rewriting to show squares,

$$(x - 6)^2 + (y - 4)^2 = 49$$

Now we identify the center as the point (6, 4), and the radius as 7 units. Then the circle is easily sketched (with a compass if desired).

Exercise 21-1

Sketch each of the following circles:

1. $x^2 + y^2 = 49$ **2.** $x^2 + y^2 = 16$ **3.** $x^2 + y^2 = 1$

4. $x^2 + y^2 = 40$ **5.** $4x^2 + 4y^2 = 81$ **6.** $9x^2 + 9y^2 = 64$

Write the equation of each of the following circles. Expand and simplify.

7. Center (0, 0), radius = 8 **8.** Center at (0, 0), radius = $3\sqrt{2}$

9. Center at (3, −1), radius = 7 **10.** Center at (5, 0), radius = 6

Find the center and the radius of each of these circles. Sketch the circle.

11. $x^2 + y^2 - 6x - 10y - 15 = 0$

12. $x^2 + y^2 - 8x + 12y + 43 = 0$

21-3 THE PARABOLA

Another second degree curve is the *parabola*. If an equation contains the second power of one variable and only the first power of the other, the graph, in general, is a *parabola*. Here are some examples:

$$y = x^2 - 2x - 3; \qquad x^2 = 8y + 5; \qquad y^2 = 4x$$

We show the shape of a parabola by an example.

Example 1.

Graph the equation: $y = x^2 - 2x - 3$.

Solution

Taking values of x from −2 to +4, we may use the following pairs of values:

$x =$	−2	−1	0	1	2	3	4
$y =$	5	0	−3	−4	−3	0	5

When the points are connected by a smooth curve, the result is a graph called a *parabola* (Fig. 21-3). If we take values of x less than −2 or greater than +4, the y-values will continue to increase, and the parabola will continue to extend upward.

A few characteristics of the parabola should be noted. First, unlike the circle, the *curvature* of a parabola is continuously changing. The curvature refers to the rate of change of direction. At one point the graph curves most sharply. This point is called the *vertex* of the parabola. The vertex of the parabola in Fig. 21-3 is at the point

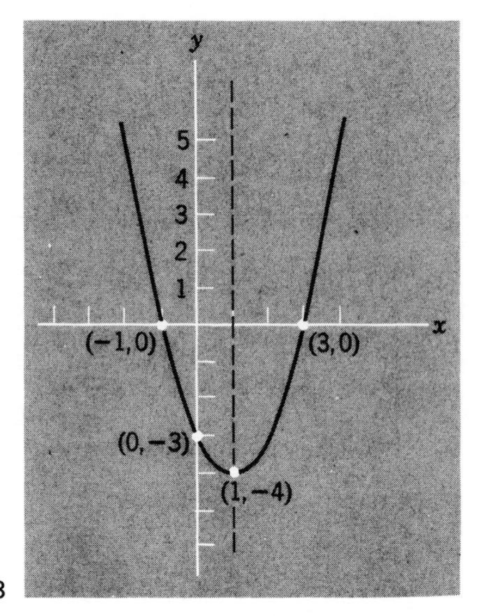

Figure 21-3

(1, −4). Another fact to note is that a straight line can be drawn through the vertex dividing the parabola into two symmetrical halves. This line is called the *principal axis* of the parabola.

A parabola is said to be in *standard position* when the vertex is at the origin (0, 0) and the principal axis is along the x-axis or the y-axis.

Example 2

Graph the equation: $y^2 = 8x$.

Solution

Since y is squared, we first assign values to y. We may use the following pairs of values:

$x =$	0	2	8
$y =$	0	± 4	± 8

These five points will give a good idea of the shape of the parabola (Fig. 21-4). This parabola is in standard position. If greater x-values are used, the parabola will continue to extend to the right. Negative x-values cannot be used for they would make y imaginary.

Note that the parabola opens toward the *right.* If the sign of x is negative, as in $y^2 = -8x$, then the parabola opens toward the *left.* If the variables are reversed, the parabola opens *upward* or *downward.* A parabola in standard position has its vertex at (0, 0) and has one of the following forms.

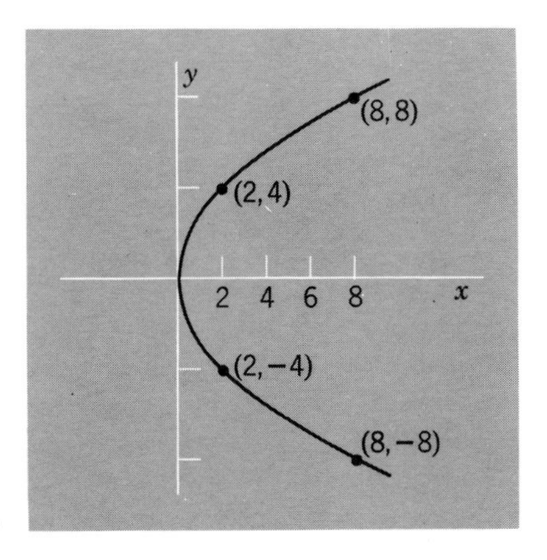

Figure 21-4

If K represents any positive constant, then

1. if $y^2 = Kx$, the parabola opens toward the right
2. if $y^2 = -Kx$, the parabola opens toward the left
3. if $x^2 = Ky$, the parabola opens upward
4. if $x^2 = -Ky$, the parabola opens downward

If the equation also contains a constant term or a first-degree term of the squared variable, the parabola is said to be *translated*, as in Example 1.

Exercise 21-2

Sketch these parabolas:

1. $y^2 = 12x$ **2.** $x^2 = 4y$ **3.** $y = x^2 - x - 2$

4. $x^2 = 4y + 3$ **5.** $y^2 = -8x$ **6.** $x^2 = -6y$

7. $y = 3 + 2x - x^2$ **8.** $x^2 = 9 - 6y$ **9.** $y^2 = 16 - 8x$

21-4 THE ELLIPSE

We have seen that this equation represents a circle: $x^2 + y^2 = 36$. Now, if the coefficients of x^2 and y^2 are different but positive, the graph of the equation is an *ellipse.* An example is the equation: $3x^2 + 5y^2 = 60$. The general form of the equation of the ellipse is: $Ax^2 + By^2 = C$, in which A, B, and C are all positive, and $A \neq B$.

The ellipse may be thought of as a flattened circle. The longest diameter is called the *major axis*, and the shortest diameter is called the *minor axis.* The *vertices* of the ellipse are at the ends of the major axis, and they represent the points of greatest curvature.

Example

Graph the equation: $2x^2 + 5y^2 = 50$.

Solution

If we set each variable in turn equal to zero and solve for the other, we get the four intercepts:

$x = 0$	± 5
$y = \pm\sqrt{10}$	0

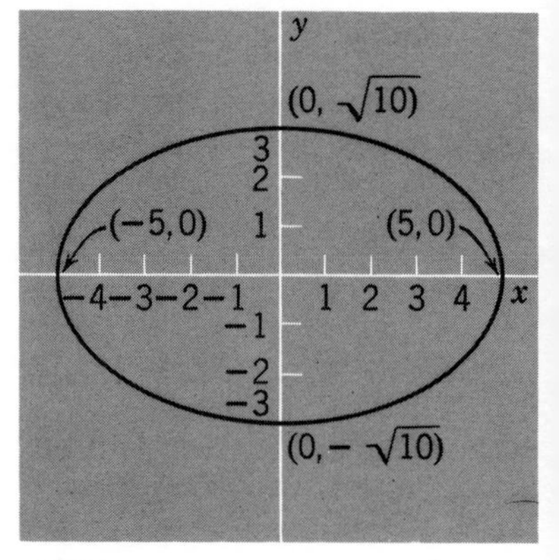

Figure 21-5

These four points are a good guide in sketching the ellipse (Fig. 21-5). Of course, a more accurate graph can be drawn by locating more points. However, with some practice, a good approximation to the true shape can be obtained by using these four points. Note that the center of this ellipse is at the origin $(0, 0)$, and the major axis is along the x-axis. The standard position of the ellipse is one in which the center is at $(0, 0)$ and the major axis is along the x-axis or the y-axis.

Exercise 21-3

Sketch each of the following equations and state the length of the major axis.

1. $4x^2 + 9y^2 = 36$ **2.** $4x^2 + y^2 = 16$ **3.** $25x^2 + 4y^2 = 100$

4. $x^2 + 2y^2 = 32$ **5.** $16x^2 + 9y^2 = 144$ **6.** $9x^2 + 25y^2 = 225$

21-5 THE HYPERBOLA

If a second degree equation contains x^2 and y^2 with coefficients of *opposite sign* (and no xy term), the graph generally is a hyperbola, another second degree curve. Here are some examples:

$$x^2 - y^2 = 16; \qquad 4x^2 - 9y^2 = 36; \qquad y^2 - 2x^2 = 18$$

The hyperbola has two branches that open in opposite directions and do not touch each other. Another characteristic of the hyperbola is that the curve approaches but never touches two fixed straight lines called *asymptotes.* An asymptote is a straight line approached by a curve. Many types of higher degree curves have asymptotes.

Example 1

Graph the equation: $4x^2 - 9y^2 = 36$.

Solution

We cannot take values of x that make y imaginary. Therefore, any x-values between -3 and $+3$ must be excluded. We may use the following pairs of approximate values:

$x =$	± 3	± 4	± 6	± 8	± 10
$y =$	0	± 1.76	± 3.46	± 4.94	± 6.36

When the points are plotted and connected by a smooth curve, the result is a curve called a *hyperbola* (Fig. 21-6). Note that the curve has two branches, one opening toward the right, the other toward the left. Note also that the curve approaches but does not touch the two diagonal lines shown in the figure. These lines are the *asymptotes* of the hyperbola.

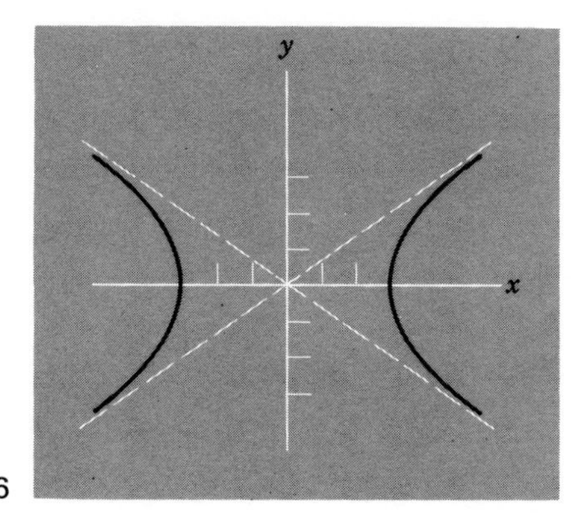

Figure 21-6

If the signs of x^2 and y^2 are reversed, and the right side is still positive, the hyperbola opens upward and downward, approaching the same asymptotes.

Another form of equation representing a hyperbola is one in which $xy = K$, where K is some constant.

Example 2

Graph the equation: $xy = 12$.

Solution

Note that neither x nor y can be zero. We may use the following values:

$x =$	1	2	3	4	6	12	24
$y =$	12	6	4	3	2	1	$\frac{1}{2}$

If we take negative values for x, then the corresponding values of y are also negative. When the points for these values are plotted, the result is a hyperbola with one branch in the first quadrant and the other in the third quadrant. The coordinate axes are asymptotes of this hyperbola. In any case, when the asymptotes of a hyperbola are perpendicular to each other, the hyperbola is said to be *equilateral* (Fig. 21-7).

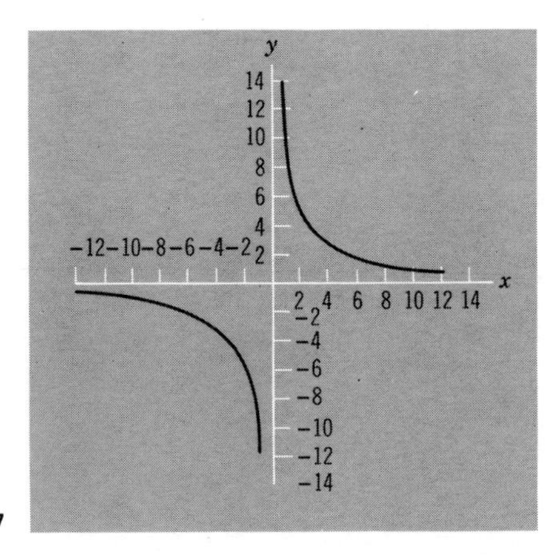

Figure 21-7

Exercise 21-4

Sketch the graph of each of the following equations:

1. $9x^2 - 16y^2 = 144$ **2.** $25x^2 - 9y^2 = 225$ **3.** $x^2 - 2y^2 = 18$

4. $9y^2 - 4x^2 = 36$ **5.** $4y^2 - 5x^2 = 80$ **6.** $3y^2 - x^2 = 48$

7. $xy = 8$ **8.** $xy = -12$ **9.** $xy = 1$

21-6 GRAPHICAL SOLUTION

To solve a system graphically, as usual we sketch the graph of each equation and then estimate the coordinates of the points of intersection. A system consisting of one quadratic and one linear equation has at most two real solutions. If the line is tangent to the curve, there is one real solution, a double point. If the line does not intersect the curve, the solutions are imaginary and cannot be shown graphically.

A system of two quadratics has at most four solutions. The solutions may all be real and different, some may be real and equal, or some or all may be imaginary.

Example 1

Solve graphically the system:

$$4x^2 + 9y^2 = 36$$

$$x - y = 2$$

Solution

The graph of the first equation is an ellipse, that of the second is a straight line (Fig. 21-8). From the figure the solutions appear to be approximately $(0, -2)$ and $(2.8, 0.8)$.

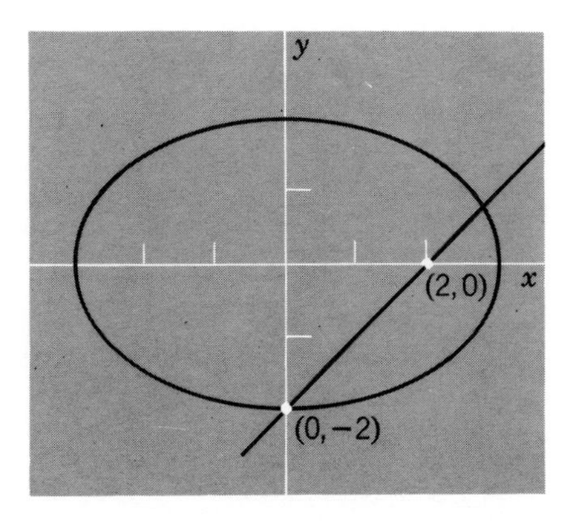

Figure 21-8

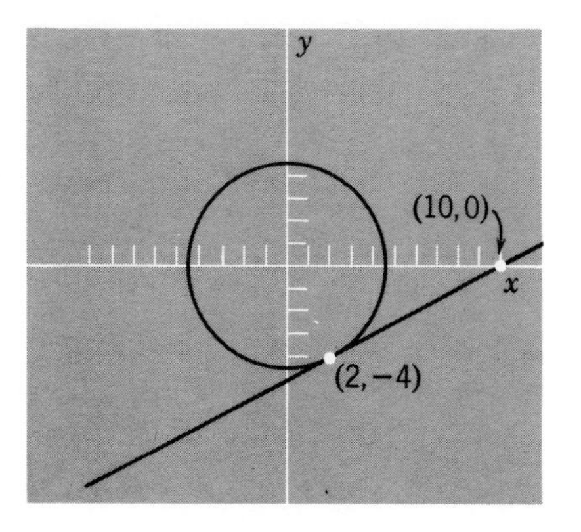

Figure 21-9

Example 2

Solve graphically:

$$x^2 + y^2 = 20$$

$$x - 2y = 10$$

Solution

Here we have the graphs of a circle and a straight line (Fig. 21-9). The line appears to be tangent to the circle at the point (2, −4), approximately.

Example 3

Solve graphically:

$$x^2 = 8y$$

$$x - 2y = 2$$

Solution

The graph of the first equation is a parabola, that of the second equation is a straight line (Fig. 21-10). Since the line does not touch the parabola, there is no real solution. Imaginary solutions cannot be shown on the graph.

Example 4

Solve graphically:

$$2x^2 + 5y^2 = 50$$

$$x^2 - y^2 = 4$$

Solution

The graph of the first equation is an ellipse, that of the second equation is a hyperbola, opening toward the right and left. The four points of intersection

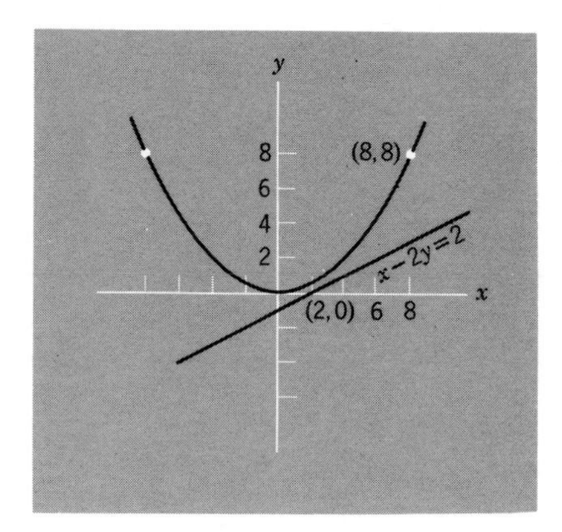

Figure 21-10

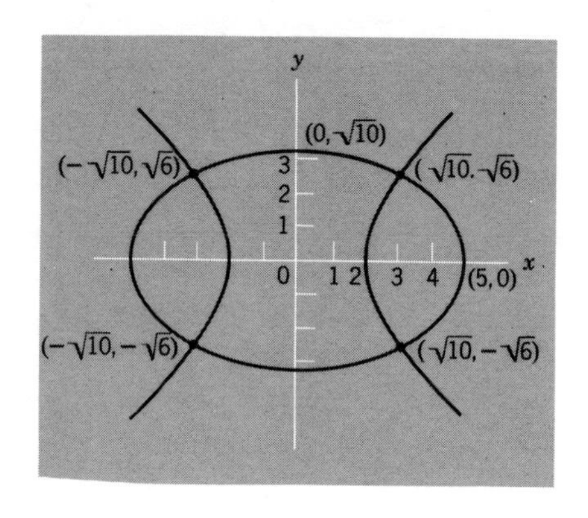

Figure 21-11

are symmetrical with respect to both axes. They appear to be approximately (± 3.2, ± 2.5) (Fig. 21-11).

Exercise 21-5

Solve the following systems graphically. Estimate to the nearest tenth.

1. $x^2 + y^2 = 9$
$2x - y = 3$

2. $x^2 + y^2 = 25$
$2x + y = 0$

3. $x^2 + y^2 = 4$
$x - 2y = 5$

4. $y^2 = 4x$
$2x - y = 4$

5. $x^2 + 2y^2 = 24$
$x - y = 2$

6. $2x^2 - y^2 = 16$
$x + 2y = 6$

7. $x^2 + y^2 = 36$
$y^2 = 4x$

8. $x^2 + y^2 = 16$
$x^2 + 9y^2 = 36$

9. $x^2 + y^2 = 36$
$2x^2 - y^2 = 4$

10. $x^2 + y^2 = 25$
$xy = 8$

11. $x^2 + y^2 = 25$
$y^2 - x = 5$

12. $4x^2 - 9y^2 = 36$
$2x - y = 4$

13. $y^2 - 2y - 4x = 7$
$2x - y = 1$

14. $x^2 + y^2 + 6x = 7$
$x^2 + y^2 = 25$

21-7 ALGEBRAIC SOLUTION

Systems involving quadratics can be solved by the same methods used in solving systems of linear equations. A system consisting of one quadratic and the other linear can best be solved by the method of substitution:

1. *Solve the linear equation for one letter.*
2. *Substitute the new expression in the quadratic.*

3. *Solve the resulting quadratic.*
4. *Find the corresponding values of the other variable by use of the linear equation.* Do not use the quadratic to find the value of the second variable.

Example 1

Solve algebraically:

$$4x^2 + 9y^2 = 36$$
$$x - y = 2$$

Solution

Solving the second equation for y, we get $\quad y = x - 2$

Substituting $(x - 2)$ for y in the quadratic, $\quad 4x^2 + 9(x - 2)^2 = 36$

Expanding and simplifying, $\quad 13x^2 - 36x = 0$

then $x = 0$, and $x = \frac{36}{13}$

From the linear equation we get the corresponding values of y.

when $x = 0$, $y = -2$; $\quad$ when $x = \frac{36}{13}$, $y = \frac{10}{13}$

These values represent the two points of intersection. (See Fig. 21-8.)

Example 2

Solve algebraically:

$$x^2 + y^2 = 20$$
$$x - 2y = 10$$

Solution

Solving the linear equation for x, $\quad x = 2y + 10$

Substituting in the quadratic, $\quad (2y + 10)^2 + y^2 = 20$

Expanding and simplifying, $\quad 5y^2 + 40y + 80 = 0$

Solving for y, $\quad y = -4$, and $y = -4$, a double root.

To find the value of x, we use the linear equation. When $y = -4$, $x = 2$. Then the point $(2, -4)$ may be called a double point, which means that the line is tangent to the circle at that point. (See Fig. 21-9.)

Example 3

Solve algebraically:

$$x^2 = 8y$$
$$x - 2y = 2$$

Solution

Solving the linear equation for y, $$y = \frac{x - 2}{2}$$

Substituting in the quadratic, we get $$x^2 = 8\left(\frac{x - 2}{2}\right)$$

Simplifying the quadratic, we get $x^2 - 4x + 8 = 0$
Solving by the formula, we get imaginary roots, $x = 2 \pm 2i$
From the linear equation we get the corresponding y values: $y = \pm i$.

To avoid fractions, the linear equation may first be solved for x. The imaginary roots indicate that the line and the parabola have no point in common. (See Fig. 21-10.)

Example 4

Solve algebraically:

$$2x^2 + 5y^2 = 50$$
$$x^2 - y^2 = 4$$

Solution

Here we can use the method of addition, or subtraction.
Multiplying the second equation by 5, $5x^2 - 5y^2 = 20$
Repeating the first equation, $2x^2 + 5y^2 = 50$
Adding the two equations, $7x^2 = 70$
Solving for x, $x = \pm\sqrt{10}$; then $y^2 = 6$ and $y = \pm\sqrt{6}$.

These values represent four solutions. Graphically they represent four real points of intersection of the graphs. (See Fig. 21-11.)

Exercise 21-6

Solve the following systems algebraically:

1. $x^2 + y^2 = 25$
$2x + y = 0$

2. $x^2 + y^2 = 20$
$2x + y = 10$

3. $x^2 + y^2 = 16$
$x - y = 6$

4. $4x^2 + 9y^2 = 36$
$x - 2y = 3$

5. $3x^2 + 2y^2 = 21$
$x - y = 2$

6. $x^2 = 4y$
$x + 2y = 4$

7. $x^2 - y^2 = 9$
$x - 2y = 3$

8. $xy = 4$
$2x + 3y = 10$

9. $xy = 6$
$3x + 2y = 12$

10. $x^2 + y^2 = 36$
$y^2 = 4x$

11. $x^2 + y^2 = 25$
$y^2 = 10 - 2x$

12. $x^2 + 4y^2 = 40$
$x^2 - 6y = 0$

13. $x^2 + y^2 = 25$
$y^2 = 3x - 3$

14. $x^2 + y^2 = 9$
$x^2 = 2y - 6$

15. $x^2 + y^2 = 21$
$4x^2 + y^2 = 36$

16. $2x^2 - 3y^2 = 12$
$x^2 + y^2 = 36$

17. $x^2 - y^2 = 8$
$xy = 3$

18. $x^2 = 2y + 16$
$x^2 + y^2 = 25$

19. $y^2 + 8x - 24 = 0$
$y^2 - 4x - 12 = 0$

20. $x^2 - 2y + 3 = 0$
$x^2 + 3y - 7 = 0$

21. $y^2 - 4x - 6y + 5 = 0$
$y - x = 1$

22. $x^2 + y^2 - 6x - 4y - 12 = 0$
$x - y = 4$

23. $x^2 + y^2 + 6x - 7 = 0$
$x^2 + y^2 = 25$

24. $y = 2x^2 - 3x - 2$
$y = 5 + x - x^2$

25. $x^2 + y^2 + 8x + 2y + 9 = 0$
$x^2 + y^2 - 4x + 2y - 15 = 0$

26. $3x^2 + y^2 - 6x - 4y - 29 = 0$
$x^2 - y^2 - 2x + 4y - 7 = 0$

27. $x^2 + y^2 = 25$
$xy = 12$

28. $x^2 - y^2 = 12$
$xy = 8$

29. $x^2 + y^2 - 2x + 6y = 15$
$x^2 + y^2 - 9 = 0$

30. $x^2 + y^2 - 4x - 8y - 5 = 0$
$x^2 - y^2 + 12 = 0$

QUIZ ON CHAPTER 21, SYSTEMS INVOLVING QUADRATICS. FORM A.

1. Where is the center and what is the radius of the circle: $x^2 + y^2 = 48$? Sketch the circle.

2. Write the equation of the circle whose center is at $(0, 0)$ and radius $= 2\sqrt{5}$.

3. Write the equation of the circle with center at $(2, -3)$ and radius $= 5$.

4. Find the center and radius of the circle:
$x^2 + y^2 + 4x - 6y - 32 = 0$.
For the following parabolas (Nos. 5–6), sketch the parabolas and state the following: coordinates of the vertex; coordinates of the focus.

5. $y^2 = 20x$;

6. $x^2 + 12y = 0$.
For the following ellipses (Nos. 7–8), sketch the figures and state the following: length of major axis; length of minor axis, coordinates of the vertices; coordinates of the foci;

7. $4x^2 + 25y^2 = 100.$
8. $9x^2 + y^2 = 36.$
9. Sketch the hyperbola: $4x^2 - 9y^2 = 36.$

Solve the following systems algebraically:

10. $x^2 + y^2 = 16$
    $2x + y = 8$
11. $x^2 + y^2 = 32$
    $y^2 = 4x$
12. $x^2 + y^2 = 25$
    $x^2 - y^2 = 11$
13. $x^2 + y^2 = 13$
    $xy = 6$
14. $x^2 + y^2 + 4x - 2y - 4 = 0$
    $x^2 + y^2 + 6x - 2y + 6 = 0$

CHAPTER TWENTY-TWO Inequalities

22-1 DEFINITIONS

We have said that an equation is a statement of equality. To express the equality we use the equal sign ($=$). If two quantities are not equal, we draw a line through the equal sign ($\neq$). This does not show, however, which is the greater quantity.

If two quantities are not equal, one of them is less than or greater than the other. A statement that one quantity is less than or greater than the other is called an *inequality*. (It is sometimes called an *inequation*). To indicate that one quantity is less than another, we use the symbol $<$. For example, $3 < 5$, is read "3 is less than 5." To indicate that one quantity is greater than another, we use the symbol $>$. For example, $5 > 3$, is read "5 is greater than 3." Note that the symbol points toward the smaller number.

We have seen how the real numbers can be shown on the number line (Fig. 22-1). On the line, a number that is to the right of another is the greater of the two. Any number is less than any number at its right. Note that 5 is at the right of 3. Then we can say: $5 > 3$; or $3 < 5$. Also, $1 > -4$; or $-4 < 1$.

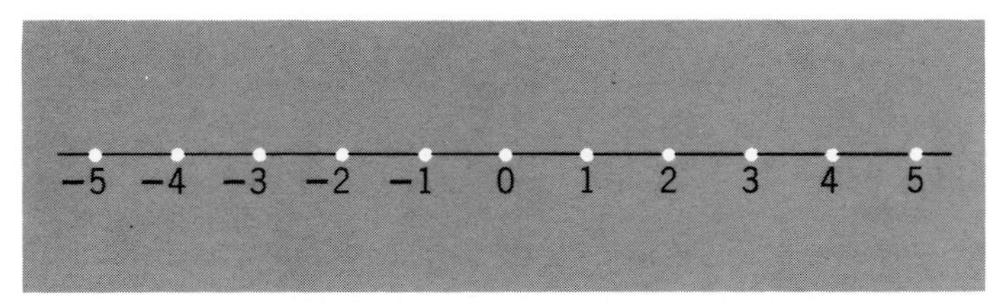

Figure 22-1

The foregoing examples are called *absolute inequalities*. They are similar to identity equations, such as $2 + 5 = 7$. On the other hand, we have *conditional inequalities* just as we have conditional equations. For example, the equation $x = 4$, is true only on the condition that the value of x is exactly 4. In a similar way, a conditional inequality depends for its truth on the value of the unknown. For example, the inequality $x < 4$ is true only if x is some value less than 4, such as 3, 2, 0, -5, and so on. In fact, x may have any value up to but not including

4. It may be 3.9999, or as close to 4 as we wish. If we wish to indicate that x may also include the value 4, we write: $x \leq 4$. This is read "x is less than or equal to 4."

22-2 SOLVING INEQUALITIES

The solution of an equation is a value of the unknown that makes the statement true. In the same way, the solution of an inequality consists of the values of the unknown that make the inequality true. The values must satisfy the inequality, the same as with an equation.

There is one important difference between the solutions of an equation and an inequality. In the solution of an equation, the value of x is one or more specific values, as in the equation, $x + 2 = 5$, which has only one solution. When we solve the equation, we pinpoint the value at exactly 3. On the other hand, in an inequality, the solution usually covers a whole range of values.

In solving equations, we use certain axioms. Most of these axioms are also true with regard to inequalities. The following are used in solving inequalities:

1. The same quantity may be added to both sides of an inequality.
2. The same quantity may be subtracted from both sides of an inequality.
3. Both sides of an inequality may be multiplied by the same positive number.
4. Both sides of an inequality may be divided by the same positive number.
5. Exception: If both sides of an inequality are multiplied or divided by a negative number, the sense of the inequality is *reversed;* that is, the *inequality sign is reversed.*

The first and second axioms are applied by transposing just as with equations. Two points should be mentioned here. To indicate that x is a positive number, we write: $x > 0$. To indicate that x is a negative number, we write: $x < 0$.

Example 1

Solve the inequality, $3x + 2 < 14$, and indicate the solution on the number line.

Solution

Transposing, $3x < 14 - 2$

Combining, $3x < 12$

Dividing both sides by 3, $x < 4$

The solution can be shown on the number line:

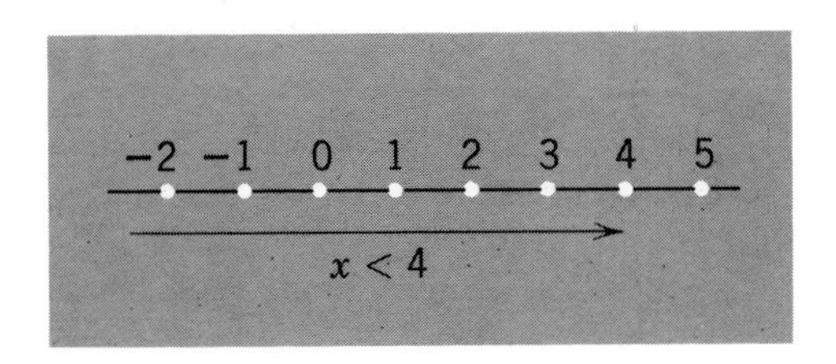

We may check the inequality with any value less than 4.

Example 2

Solve the inequality, $\frac{2x-3}{5} < \frac{3x+2}{4}$

Solution

Multiplying both sides by 20, $8x - 12 < 15x + 10$
Transposing $8x - 15x < 12 + 10$
Combining terms, $-7x < 22$
Dividing both sides by −7, $x > -\frac{22}{7}$
Note that dividing both sides by a negative number reverses the *sense* of the inequality. To check, use any value greater than $-\frac{22}{7}$, such as 0.

Exercise 22-1

Solve the following inequalities. Show the solution on the number line.

1. $\frac{2x}{3} < 2$ **2.** $\frac{3x-4}{2} > 1$ **3.** $\frac{4-x}{5} > x$

4. $\frac{x-2}{3} < x+2$ **5.** $\frac{3x}{5} < \frac{2x-1}{3}$ **6.** $\frac{2x+1}{2} > \frac{3x-4}{5}$

7. $\frac{x-2}{4} < \frac{2x+1}{3}$ **8.** $\frac{2-x}{3} < \frac{x-2}{5}$ **9.** $\frac{4x-3}{3} < \frac{4x-1}{2}$

10. $\frac{3x+5}{3} > \frac{5x-1}{4}$ **11.** $\frac{2x+4}{3} > \frac{3x-4}{2}$ **12.** $\frac{3x+5}{3} > \frac{5x+1}{6}$

22-3 GRAPHING LINEAR INEQUALITIES

We have seen that the graph of a linear equation containing two variables, x and y, is a straight line. The graph of a linear inequality containing x and y is an entire *area* on one side of a straight line. To find the area, we first write the inequality as an equation. Then we graph the equation and get a straight line. The graph of the inequality is then the area on one side of the line.

To discover the correct area representing the inequality, we select a point on either side of the line and see whether the coordinates of the point satisfy the inequality. If the inequality is satisfied, then we have the correct side of the line. If not, we take the area on the other side of the line.

Example

Show the area determined by the inequality, $x + 2y > 4$.

Solution

First, we write the equation:

$$x + 2y = 4$$

Graphing the equation, we get a straight line (Fig. 22-2). Now we select a

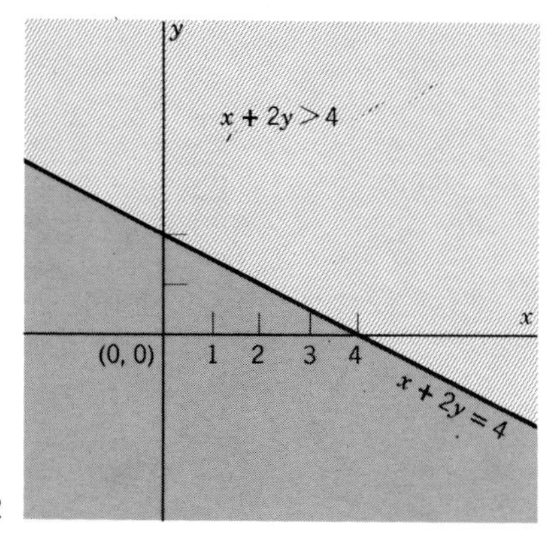

Figure 22-2

point on either side of the line. In most examples, the simplest point is the origin (0, 0). If we put the coordinates (0, 0) in the given inequality, we get.

$$0 + 0 > 4$$

Since the inequality is not satisfied, the point (0, 0) is on the wrong side of the line. Then we take the opposite side and shade the area showing the solution.

22-4 SYSTEMS OF INEQUALITIES

When we solve a system of linear equations, we get a specific point of intersection. On the other hand, when we solve a system of linear inequalities, we get a particular restricted area. Sometimes the area may be completely enclosed. Then, again, it may be limited in one or two directions and open in others.

Inequalities are coming to be recognized as increasingly important in fields of practical mathematics, for example, in economics. As an example, in locating the site for an ordnance depot, a factory, or a research center, it may not be essential to locate it at a particular point, such as, for instance, Fort Wayne, Indiana. However, it may be that because of convenience, transportation, or other conditions, it must be located in a general area, for example, west of the Appalachian Mountains, north of the Ohio River, and east of the Mississippi. The area is first restricted. This is exactly what is done in solving systems of inequalities graphically.

Example 1

Find the area determined by the solution of the two inequalities,

$$2x + y > 8 \qquad \text{and} \qquad x - y < 2$$

Solution

First, we write the inequalities as equations:

$$2x + y = 8 \qquad \text{and} \qquad x - y = 2$$

The graphs are two straight lines (Fig. 22-3). Now we use the point (0, 0) to test both inequalities for the areas. For the first, we get $0 + 0 > 8$, which is not true. For the second, we get $0 + 0 < 2$, which is true. The area determined by each inequality is shaded. Then the double shading indicates the common solution of the two inequalities.

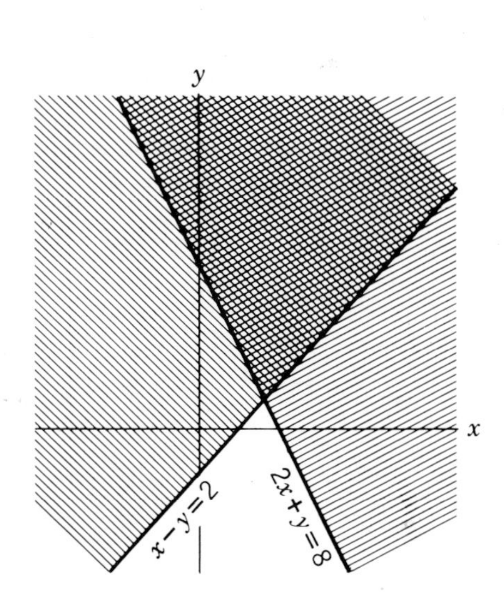

Figure 22-3

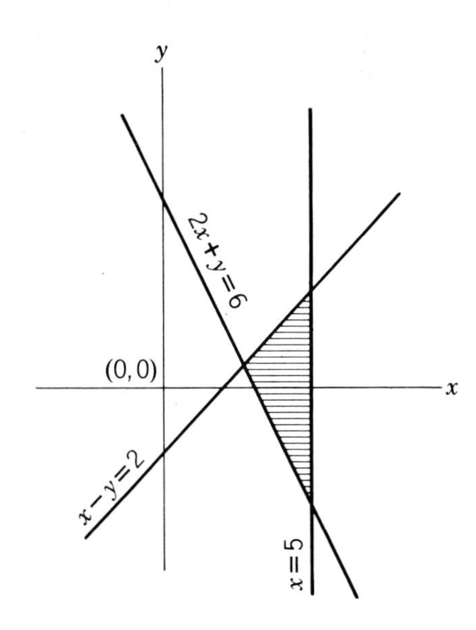

Figure 22-4

Example 2

Solve: $x - y > 2$; $2x + y > 6$; $x < 5$

Solution

We write each inequality as an equation and graph each one (Fig. 22-4). After testing each inequality with the point (0, 0), we shade the proper area for each. Then the overlapping of areas represents the common solution

Exercise 22-2

Solve these inequalities and show the area representing the solution:

1. $x + 2y > 2$ **2.** $2x + 6 > 3y$ **3.** $3x - 5y < 15$

4. $2x - y + 4 < 0$ **5.** $x < 2$ **6.** $y > -3$

Show the area determined by each of these systems of inequalities:

7. $\begin{cases} x - 3y > 6 \\ 2x + y > 4 \end{cases}$ **8.** $\begin{cases} 3x + 2y < 12 \\ 2x - 3y > 6 \end{cases}$ **9.** $\begin{cases} x + y < 4 \\ x - y < 4 \end{cases}$

10. $\begin{cases} x + 2y > 8 \\ x - 2y < 4 \\ y < 4 \end{cases}$ **11.** $\begin{cases} 2x + y > 6 \\ 2x - y < 6 \\ y < 3 \end{cases}$ **12.** $\begin{cases} x - 2y < -4 \\ x + y < 1 \\ x + 3 > 0 \end{cases}$

13. $\begin{cases} 3x - 2y < 6 \\ x + y < 3 \\ x > 0 \end{cases}$ **14.** $\begin{cases} x - y < 3 \\ y < 0 \\ x > 0 \end{cases}$ **15.** $\begin{cases} x + 3y > 3 \\ x < 0 \\ y < 4 \end{cases}$

22-5 QUADRATIC INEQUALITIES

Solving a quadratic inequality is somewhat similar to solving a quadratic equation by factoring. We transpose all terms to one side of the inequality and have a zero on the other side. If possible, we then factor the expression. The method is shown by an example.

Example 1

Solve the inequality $x^2 < 2x + 3$.

Solution

Transposing, $x^2 - 2x - 3 < 0$
Factoring, $(x + 1)(x - 3) < 0$

This statement says in effect that the product of two factors is *negative*. Now, we know that for a negative product, one factor must be negative and the other positive. There are two possibilities.

One possibility: $x + 1 < 0$ *and* $x - 3 > 0$
Solving, $x < -1$ *and* $x > 3$

If these values are to represent a solution, then both statements must be true. However, this is not possible for the same value of x. Therefore, we are left with the second possibility.

Second possibility: $x + 1 > 0$ *and* $x - 3 < 0$

Solving, $x > -1$ *and* $x < 3$

This condition is possible for any value of x between -1 and $+3$. Then this is the solution, and we can write

$$-1 < x < 3$$

This statement means that x lies between -1 and $+3$. We can check the original inequality with any value of x in this range, for example, $x = 1$.

The solution of a quadratic inequality can be determined by a study of the graph of the corresponding equation. If we set the inequality equal to y, we get an equation. For example, consider the inequality, $x^2 - 2x - 3 < 0$. If we set the function, $x^2 - 2x - 3$, equal to y, we get the equation

$$y = x^2 - 2x - 3$$

The graph of this equation is a parabola (Fig. 22-5). Now, when we write the inequality, $x^2 - 2x - 3 < 0$, this means that $y < 0$. When $y < 0$, we have the portion of the parabola below the x-axis. The graph shows this portion corresponds to the x-values between -1 and $+3$.

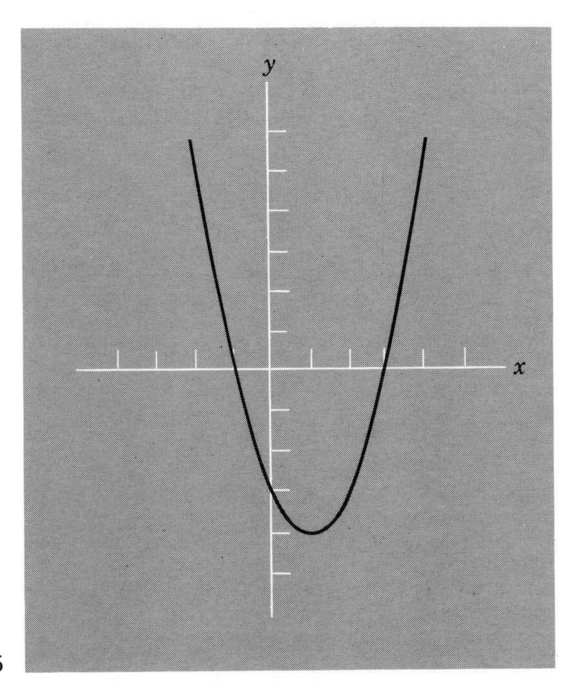

Figure 22-5

If the quadratic expression in an inequality cannot be factored, we write the corresponding equation and sketch the graph. For example, in the inequality $x^2 - 4x + 2 < 0$, the expression cannot be factored. Then we graph the equation, $y = x^2 - 4x + 2$. The portion of the curve below the x-axis represents the values where $y < 0$. Then the x-values can be determined from a study of the graph. In this example, the values of x lie between two irrational numbers: $2 + \sqrt{2}$ and $2 - \sqrt{2}$, the roots of the equation, $x^2 - 4x + 2 = 0$.

As another case, suppose we have the inequality $x^2 - 2x + 5 < 0$. If we graph the equation, $y = x^2 - 2x + 5$, we shall find that no portion of the graph lies below the x-axis. Therefore the inequality has no solution. If we solve the corresponding quadratic equation, $x^2 - 2x + 5 = 0$, we shall find the roots are imaginary.

Exercise 22-3

Solve each of the following graphically. Solve algebraically if possible.

1. $x^2 + x < 2$ **2.** $4 + 3x - x^2 > 0$ **3.** $x^2 - 4 < 0$

4. $x^2 - 5 < 0$ **5.** $x^2 - 5x < 0$ **6.** $x^2 + 2x > 3$

7. $4 - x^2 > 3x$ **8.** $x^2 - 4x + 4 < 0$ **9.** $x^2 - 2x + 4 < 0$

10. $x^2 + 2x < 7$ **11.** $x^2 + x > 2$ **12.** $2x^2 + 3 < 7x$

13. $6 + x - x^2 > 0$ **14.** $3x^2 < 5x + 2$ **15.** $4x^2 - 15 < 0$

QUIZ ON CHAPTER 22, INEQUALITIES. FORM A.

Solve the following inequalities and indicate the solution on the number line:

1. $3x - 2 < 10$ **2.** $9 - x > 3x - 1$ **3.** $4x - 3 < x - 2$

4. $7x - 5 < 4x - 5$ **5.** $6x + 11 > 5x + 9$ **6.** $4x + 1 < 6x + 4$

7. $\frac{4x + 5}{3} > 7$ **8.** $\frac{3x + 2}{4} < -1$ **9.** $\frac{4x - 1}{5} - 1 > 0$

Solve these inequalities and show the area representing the solution:

10. $2x - y > 2$ **11.** $x - 2y < 1$ **12.** $x - 1 < 0$

Show the area determined by each of these systems of inequalities:

13. $2x + y < 2$, $x - 3y > 6$ **14.** $5x - 2y > 10$, $x < 3$ **15.** $2x - y < 4$, $y > 0$

Solve these inequalities graphically and show the area determined by each:

16. $x^2 - 2x - 3 < 0$ **17.** $6 + 5x - x^2 > 0$ **18.** $x^2 < 1$

CHAPTER TWENTY-THREE Ratio, Proportion, Variation

23-1 RATIO

Much work in mathematics is concerned with comparing quantities. We can compare two quantities in various ways. For instance, if we wish to compare 8 feet with 6 feet, we can use subtraction:

$$8 \text{ feet} - 6 \text{ feet} = 2 \text{ feet}$$

That is, we can say 8 feet is 2 feet more than 6 feet.

Whenever we say one quantity is a certain amount *greater* than, or *less* than, another, we are comparing the two by *subtraction.* When we say "John is 2 inches taller than James," we are comparing by *subtraction.*

We can also compare quantities by *division.* A *comparison by division* is called the *ratio* between the two quantities. When we compare two quantities, such as 12 pounds and 4 pounds, by division we find the *ratio* between the two. For instance, the ratio of 12 pounds to 4 pounds is 12 pounds ÷ 4 pounds, which is equal to 3. Here the "3" is a pure number; it has no denomination, such as inches, feet, pounds, hours, or any other name.

When we compare quantities by subtraction, the difference between the quantities has the same denomination as the quantities themselves.

$$12 \text{ ft} - 9 \text{ ft} = 3 \text{ ft}$$

$$12 \text{ hr} - 9 \text{ hr} = 3 \text{ hr}$$

When we compare two quantities by division (that is, when we find the ratio between the two), the answer has no denomination:

$$12 \text{ ft} \div 9 \text{ ft} = \frac{4}{3} \qquad 12 \text{ hr} \div 9 \text{ hr} = \frac{4}{3}$$

A ratio is often indicated by a colon. As an example, the ratio of 6 pounds to 10 pounds is written

$$6 \text{ lb} : 10 \text{ lb}$$

This is read "6 pounds *is to* 10 pounds." The colon represents the words "*is to.*"

A ratio can also be written as a fraction:

$$6 \text{ ft} : 10\text{ft} = \frac{6 \text{ ft}}{10 \text{ ft}} = \frac{3}{5}$$

When the fraction is reduced, it becomes $\frac{3}{5}$, with no denomination.

Examples

$$\$15 : \$20 = \frac{\$15}{\$20} = \frac{3}{4}; \qquad 14 \text{ in.} : 21 \text{ in.} = \frac{2}{3}$$

As you progress further and further in mathematics, you will discover that the concept of *ratio* becomes more and more important. For example, it is the basis for a study of calculus. The idea of *ratio* should therefore be thoroughly understood. For a ratio, we have the following definition:

A ratio is a comparison of two like quantities by division.

When we wish to compare two quantities, these quantities must be of the same kind. We cannot compare, for instance, 12 pounds and 4 hours. Pounds and hours do not measure the same kind of quantity. However, two different units may be compared if they measure the same kind of quantity. We may compare 4 feet with 2 yards by first changing one so that both are expressed in the same units. The ratio of 4 feet to 6 feet (2 yards) is $\frac{2}{3}$; that is, 4 feet is two-thirds as much as 2 yards.

Exercise 23-1

Express the ratio indicated in each of the following examples, and reduce the ratio to its simplest form:

1. The ratio of \$8 to \$12

2. The ratio of 16 in. to 20 in.

3. The ratio of 24 lb to 30 lb.

4. The ratio of 21 ft to 18 ft.

5. The ratio of 5 ft to 4 yd.

6. The ratio of 2 ft to 8 in.

7. The ratio of 32 lb to 2 lb.

8. The ratio of 1 in. to 1 yd.

9. The ratio of 1 ft to 1 mile.

10. The ratio of 1320 ft to $\frac{3}{4}$ mile.

11. One circle has a diameter of 12 in., and another circle has a diameter of 6 in. What is the ratio of the radius of the first circle to the radius of the second? What is the ratio of the circumferences? What is the ratio of their areas?

12. Similar triangles are triangles that have exactly the same shape but not necessarily the same size. In similar figures all corresponding sides have the same ratio. The same is true regarding any set of corresponding lines. A certain right triangle has its sides equal to 3, 4, and 5 in., respectively. A similar triangle has a hypotenuse of 10 in. What is the ratio between the short side of the small triangle to the short side of the large triangle?

13. One square has a side of 4 in. and a second square has a side of 12 in. What is the ratio of their perimeters, diagonals, and areas?

14. On a certain house plan $\frac{1}{4}$ in. represents 1 ft of actual length of the house. What is the ratio of the distance on the plan to the actual distance on the ground?

15. On a certain map, 1 in. represents 10 miles. What is the ratio of the distances on the map to the actual distance on the earth?

16. The efficiency of an engine is defined as the ratio of the power output to the power input. Find the efficiency, stated as a percent, of an engine that requires an input of 4800 watts for an output of 3600 watts.

17. The ratio of the weight of a substance to the weight of an equal volume of water is called the *specific gravity* of the substance. Water weighs about 62.4 lb per cubic foot. Find the specific gravity of each of the following substances whose weight per cubic foot is shown: **(a)** iron: 488 lb; **(b)** lead: 710 lb; **(c)** silver: 655 lb; **(d)** gold: 1206 lb; **(e)** glass: 156 lb; **(f)** aluminium: 168 lb; **(g)** pine wood: 31.2 lb.

23-2 PROPORTION

Suppose we have the following ratios:

$$6 \text{ ft} : 9 \text{ ft} \qquad 10 \text{ min} : 15 \text{ min}$$

We find that the ratio of each is $\frac{2}{3}$. We may then write an equation stating that the two ratios are equal:

$$6 \text{ ft} : 9 \text{ ft} = 10 \text{ min} : 15 \text{ min}$$

or, as ratios

$$\frac{2}{3} = \frac{2}{3}$$

A statement of equality between two equal ratios is called a proportion.

Both ratios involved in a proportion may refer to the same denomination, as in the following example, in which both ratios refer to feet:

$$5 \text{ ft} : 15 \text{ ft} = 7 \text{ ft} : 21 \text{ ft}$$

However, the ratios may refer to different denominations as in the following:

$$6 \text{ ft} : 8 \text{ ft} = 15 \text{ lb} : 20 \text{ lb}$$

Since each ratio is equal to $\frac{3}{4}$, the two can be stated as a proportion.

You will notice that a proportion contains four terms, since the proportion consists of two ratios and each ratio has two terms. The foregoing proportions can be written as fractions:

$$\frac{5 \text{ ft}}{15 \text{ ft}} = \frac{7 \text{ ft}}{21 \text{ ft}} \qquad \frac{6 \text{ ft}}{8 \text{ ft}} = \frac{15 \text{ lb}}{20 \text{ lb}}$$

The four terms of a proportion are usually designated as *first term, second term, third term,* and *fourth term.* The first and fourth terms are called the *extremes* of the proportion; the second and third are called the *means.*

The fourth term is sometimes called the *fourth proportional* to the first three terms.

We have said that a proportion is an equation expressing an equality between two ratios. Therefore, we may treat it in the same way as any other equation. Consider the proportion

$$a : b = c : d$$

Writing it in fractional form, we have

$$\frac{a}{b} = \frac{c}{d}$$

If we multiply both sides of the equation by bd, we get

$$ad = bc$$

Notice that the term ad is the *product of the extremes* of the proportion and the term bc is the *product of the means.* From this general equation we may formulate the important *principle* of a proportion.

Principle. *In any proportion the product of the means equals the product of the extremes.*

This principle enables us to find any missing term of a proportion. For instance, suppose we have

$$5 : x = 8 : 17$$

The product of the means is $8x$. The product of the extremes is 85. Then we can write the equation

$$8x = 85$$

Solving, $\quad x = 10\frac{5}{8}$

It is possible that the second term of a proportion will be the same as the third term. This is true in the following proportion:

$$2 : 6 = 6 : 18$$

If the second and third terms are the same, that quantity is called the *mean proportional between the other two terms.* In that case we have only three *different* quantities. The fourth term is then called the *third proportional* to the first and second terms. In the foregoing example, 18 is the third proportional to 2 and 6, and 6 is the mean proportional between 2 and 18.

In the proportion

$$a : b = b : c$$

b is the mean proportional between a and c. Then $b^2 = ac$, and we have the formula

$$b = \pm\sqrt{ac}$$

Example

Find the mean proportional between 2 and 32.

Solution

We set up the proportion $2 : x = x : 32$

Then $x^2 = 64$

Solving for x, $x = \pm 8$

It will be noted that the mean proportional may be either +8 or −8.

To check the answer, we write $2 : (+8) = (+8) : 32$

Stated as fractions, $\frac{1}{4} = \frac{1}{4}$

Using −8, we have $2 : (-8) = (-8) : 32$

Stated as fractions, $-\frac{1}{4} = -\frac{1}{4}$

Exercise 23-2

Find the value of the unknown in each of these proportions:

1. $6 : x = 4 : 14$

2. $2 : 7 = 3 : x$

3. $x : 9 = 6 : 5$

4. $8 : 4.6 = 12 : 3y$

5. $36 : x = x : 4$

6. $20 : x = x : 4$

7. $2.4 : 12 = 10.8 : x$

8. $9.2 : x = 15 : 3.5$

9. $4 : (k - 2) = 13 : (2k + 1)$

10. $(2k + 3) : 3 = (k - 2) : 5$

11. $(3k - 1) : 7 = (2k + 1) : 3$

12. $(4k - 3) : 6 = (3k + 1) : 5$

13. $n : (n - 4) = (n + 4) : (n - 6)$

14. $(y - 3) : y = (y + 6) : (y + 3)$

15. Find the fourth proportional to the three numbers, 4, 6, and 5.

16. Find the mean proportional between 3 and 20.

17. Find the mean proportional between 2 and 162.

18. What is the mean proportional between $\frac{1}{9}$ and 4?

19. A recipe calls for $2\frac{1}{2}$ cups of flour for 9 servings. How many cups of flour should be used for 14 servings?

20. A certain antifreeze mixture calls for 4 quarts of antifreeze to be mixed with 3 gal of water. How many quarts of antifreeze are needed for a total solution of 18 qt?

21. According to Hooke's law, in any elastic body, the distortion is proportional to the distorting force, if kept within the "elastic limit." If a force of 5 lb will stretch a spring $3\frac{3}{4}$ in., what force will be required to stretch it 2 in.?

22. If 1 kg (2.2 lb) will stretch a spring $1\frac{3}{8}$ in., what force will be required to stretch it 1 ft at the same rate?

23. In an electric circuit, current I is proportional to voltage E. If $I = 2.4$ amp when $E = 36$ V, what voltage is required for a current of 5.6 amp?

24. A picture is 3 in. wide and 4.5 in. long. It is to be enlarged so that the length is 20 in. Find the width of the enlargement.

25. A room is 24 ft long and 16 ft wide. A floor plan of the room is 7.5 in. wide. Find the length of the floor plan.

26. If 8.25 gal of fuel are required for a drive of 180 miles, how many gallons are required for a drive of 270 miles at the same rate?

27. A city has a park 320 ft long and 240 ft wide. A map of the park is 12 in. long. Find the width of the map.

28. If 90 volts produces a current of 7.5 amperes of current in a certain circuit, what current will be produced by 120 volts in the same circuit?

23-3 PROPORTIONAL DIVISION

Sometimes it is desired to divide a given number into two or more parts such that the parts have certain ratios to each other. Consider the following example.

Example 1

A board 20 ft long is to be cut into two pieces so that the lengths of the two pieces have the ratio of 3 to 5. What should be the length of each piece?

Solution

The problem can be solved in two ways.

First method. Let x = length (in feet) of the shorter piece.
Then $20 - x$ = length (in feet) of the longer piece.

Now we set up the ratio between the two lengths, and state the value of the ratio:

$$\frac{x}{20 - x} = \frac{3}{5}$$

Solving the equation:

$$5x = 3(20 - x); \quad \text{then} \quad x = 7.5.$$

The lengths of the pieces: shorter: 7.5 ft; longer: 12.5 ft.

Second method. Let $3x$ = length (in feet) of the shorter piece.
Let $5x$ = length (in feet) of the longer piece.

We write the equation:

$$3x + 5x = 20$$

Solving,

$$x = 2.5$$

For the lengths, we have:

Shorter piece: $3(2.5) = 7.5$ (ft)
longer piece: $5(2.5) = 12.5$ (ft)

Example 2

A wire 15 ft long is to be cut into three pieces so that the lengths of the pieces have the ratio of the numbers, 2, 3, and 5. Find the length of each piece. (Use 15 ft = 180 in.)

Solution

Let $2x$ = length (in inches) of the shortest piece.

Let $3x$ = length (in inches) of the next piece.

Let $5x$ = length (in inches) of the third piece.

We write the equation: $2x + 3x + 5x = 180$; solving: $x = 18$ (in.). For the lengths of the pieces, we have: first: $2(18) = 36$ in.; second: $3(18) = 54$ in.; third: $5(18) = 90$ in.

Exercise 23-3

1. A board 12 ft long is to be cut into two pieces whose lengths have the ratio of 4 to 5. Find the length of each piece.

2. Brass is an alloy of zinc and copper, containing 2 parts of zinc to 3 parts copper. Find the amount of each metal in 24 lb of brass.

3. A mixture of antifreeze and water contains 2 parts of antifreeze and 7 parts of water. Find the amount of each in 60 gal of the mixture.

4. A certain concrete mixture calls for cement, sand, and gravel, with amounts in the ratio of 2, 5, and 9. Find the number of pounds of each in 1 ton of the concrete (1 ton = 2000 lb).

5. Chrome steel contains chromium, nickel, and steel in the proportion of 1 part of chromium, 1 part of nickel, and 23 parts of steel. Find the amount of each in 1 ton of chrome steel.

6. A feed mixture for cattle consists of wheat, corn, and oats, in the ratio of 2, 5, and 8. How much of each should be used for 1000 lb of the mixture?

7. A farmer has 360 acres of land, part planted in corn, part in oats, part in hayland, and the rest in pasture. The numbers of acres of each have the ratio of 7, 6, 2, and 3, for the uses mentioned. Find the number of acres in use for each purpose.

8. A woman has $40,000 invested, part in stocks, part in bonds, and the remainder in savings. The three investments are in the ratio of the numbers, 3, 5, and 8. Find the amount of each investment.

9. A builder agrees to build 60 homes, some low-priced, some medium-priced, and the remainder high-priced. The numbers of the three kinds are to be in the ratio of 8 to 5 to 2. Find the number of each kind to be built.

10. A university library contains 11,700 books, consisting of books on mathematics, on science, on history, and on literature. In the order mentioned, the numbers of the different kinds have the ratio to 2, 3, 5, and 8. Find the number of books of each kind in the library.

11. A store has shelf space for 600 cans of canned vegetables, all of the same size. The owner wishes to order some cans of beans, some of corn, and some of peas. He wishes to have the numbers of each kind in the ratio of 3 to 4 to 5 according to the type mentioned. How many cans of each kind should he order?

12. A company owned by 5 shareholders makes a profit of $84,000. The numbers of shares owned by the 5 shareholders have the ratios of 3, 4, 5, 7, and 9. How much of the profit should each of the owners receive?

23-4 VARIATION

In an equation, such as $y = 3x$, the quantities x and y may change and take on different values. A quantity that may vary and assume different values in a problem is called a *variable*. In the equation, $y = 3x$, the variables are x and y. A quantity that does not change, such as 3, is called a *constant*.

Variables are common in everyday life. Many quantities change, such as time of day, temperature, population, one's age, height, and weight, costs of articles, and current, voltage, and power in electric circuits.

It often happens that two variables are interdependent so that a change in one variable will cause a change in the other. Then we say that *one variable varies as the other*. In the equation, $y = 3x$, if x changes, then y also changes. Then we say that

$$y \text{ varies as } x$$

23-5 DIRECT VARIATION

A simple type of variation is one in which two variables increase or decrease at a uniform rate. As an example, suppose we drive a car at an average rate of 40 miles per hour. As the time changes, the distance also changes. The two related variables are *distance* and *time*. If we represent the distance in miles by d, and the time in hours by t, we can say that d varies as t. The relation in this case can be expressed by the equation

$$d = 40t$$

From the equation we note that d increases if t increases, and d decreases if t decreases. This is an example of *direct variation*.

As another example, consider the equation

$$y = 6x$$

In this equation, if x increases, y also increases; and if x decreases, y also decreases. The relation is seen if we list some corresponding values:

if $x =$	2	5	7	10
then $y =$	12	30	42	60

Note that any two pairs of these values form a proportion: $2:12 = 7:42$.

Variation is sometimes indicated by the symbol $\propto$: thus

$$y \propto x$$

This statement is read, "y varies directly as x." This does *not* mean that y is equal to x. In each of the following equations we can say $y \propto x$:

$$y = 7x \qquad y = 10x \qquad y = 15x$$

To avoid using the difficult variation symbol ($\propto$), we write the equation with an arbitrary positive constant K:

$$y = Kx$$

This equation does not say that y is equal to x. Instead, if K represents any positive constant, the equation means that y *varies directly as* x. The K is called the *constant of variation.* For *direct variation* we have the definition:

One quantity varies directly as another quantity when the first is equal to a positive constant times the second.

In direct variation, if one variable increases, the other also increases; if one decreases, the other decreases.

Examples of direct variation.

1. The total cost of a number of articles varies directly as the price.
2. The distance traveled in 5 hours varies directly as the speed.
3. The amount of light in a room varies directly as the number of lamps.
4. The amount of current varies directly as the electromotive force.
5. The circumference of a circle varies directly as the radius.
6. The area of a circle varies directly as the *square* of the radius.

23-6 JOINT VARIATION

It often happens that one variable depends on two or more other variables. For instance, the area of a triangle depends on the *base* and the *altitude.* The formula for the area of a triangle is

$$A = \frac{1}{2} b \cdot h$$

In this equation, if b is doubled and h remains unchanged, the area is doubled. If h is tripled and b remains constant, then the area is tripled.

Now, if the base is doubled and the altitude is tripled, the area is multiplied by 6; that is, the area varies as the *product* of the base and the altitude.

If one variable, A, varies as the *product* of two variables, b and h, then we say A varies *jointly* as b and h. *Joint* variation always implies a *product* of two variables.

If Z varies jointly as x and y, then Z varies as the *product* of x and y. For the general variation statement in this case, we write

$$Z = Kxy$$

where K is a positive constant. Note that the constant of variation, K, is included as a factor.

Exercise 23-4

By use of appropriate letters to represent quantities, express the indicated relation in each of the following statements as an equation of variation. Be sure to use a constant of variation in each equation.

1. The total cost, C, of a certain number of articles varies directly as the price, p, of each.
2. The number of miles traveled in 5 hr varies directly as the speed.
3. The amount of light in a room varies directly as the number of watts of power represented by the lamps.
4. The amount of electric current varies directly as the electromotive force, provided other factors remain constant.
5. The circumference of a circle varies directly as the radius.
6. The area of a circle varies directly as the square of the radius.
7. The interest paid on money borrowed varies jointly as the amount of money borrowed, the rate of interest, and the time for which it is borrowed.
8. The speed attained by a falling object varies directly as the time of falling. (The speed, however, reaches a limit after a certain time of falling.)
9. The distance a falling object has traversed varies as the square of the time of falling.
10. The weight of a cylinder of a particular material varies jointly as the height and square of the radius.
11. The cost of carpeting a floor varies jointly as the length and the width of the room and the cost per square yard for the carpet.
12. The weight of a sphere of a particular material varies as the cube of the radius.

23-5 INVERSE VARIATION

Suppose we must drive a distance of 120 miles. The time required to cover this distance will depend upon our speed. The two variables, *time* and *speed*, are related. However, they are related in such a way that *one decreases* as the *other increases*. The time required depends upon speed, but if the speed is *decreased*, the time required is *increased*. This is an example of *inverse variation*.

If we let r represent the rate of speed in miles per hour and t represent the time in hours, then we can relate the two variables by the equation

$$t = \frac{120}{r}$$

We can set up a table of corresponding values of r and t:

if $r =$	60	40	30	20	15
then $t =$	2	3	4	6	8

For instance, if our rate of speed is 60 miles per hour, it will require 2 hours to drive the 120-mile distance. As the *rate decreases*, the *time increases.* For *inverse variation* we have the definition:

One quantity is said to vary inversely as another when the one variable is equal to a constant divided by the second variable.

Note that a constant is also involved in inverse variation. The general equation for inverse variation is

$$y = \frac{K}{x}$$

where K represents the *constant* of variation and x and y are the variables. In inverse variation, as *one variable increases, another decreases.*

In many situations we have a combination of joint variation and inverse variation. However, in writing the general variation statement, only one constant, K, is necessary, since the K represents any positive constant.

Example

Suppose we say that a variable F varies jointly as x and y and inversely as the square of Z. The general variation equation can be written

$$F = K\frac{x \cdot y}{Z^2}$$

where K is a positive constant.
The constant K can also be written with the numerator of the fraction.

Exercise 23-5

By use of appropriate letters to represent quantities, express the indicated relationship in each of the following statements as an equation of variation. Be sure to use a constant of variation in each equation.

1. The time, t, required for a certain trip varies inversely as the speed r.

2. The number of days, d, required to erect a particular building varies inversely as the number of men, m, and the number of hours, h, per day.

3. The force of attraction between two opposite magnetic poles varies inversely as the square of the distance between them.

4. The amount of light that falls on this page varies inversely as the square of the distance of the page from the source of light, provided that the page is held at right angles to the direction of light.

5. The resistance to the flow of electricity in a conductor varies inversely as the square of the diameter of the wire.

6. The amount of water discharged from a pipe varies directly as the water pressure and the square of the radius of the pipe.

7. The volume of a gas under constant temperature varies inversely as the pressure.

8. To balance a teeter-board, if you weigh less, you sit farther from the fulcrum (the balancing point). Is this an example of direct or inverse variation?

9. The lens setting in a camera varies inversely as the square root of the shutter speed.

10. The number of vibrations of a pendulum varies inversely as the square root of the length.

23-8 FINDING AND USING THE CONSTANT OF VARIATION

Suppose we know that a variable Z varies jointly as x and y. Suppose also we know that $Z = 12$ when $x = 5$ and $y = 7$. Then it is possible to find the constant of variation.

Our *first step* is to write the general variation equation:

$$Z = Kxy$$

This equation must include K, the constant of variation.

As a *second step*, we substitute the known values for x, y, and Z and solve for K:

$$12 = K \cdot 5 \cdot 7$$

This equation can now be solved for the value of K. We get

$$K = \frac{12}{35}$$

As a *third step*, the value of the constant K is inserted in the variation equation:

$$Z = \frac{12}{35} xy$$

The result is a formula that can be used to find an unknown value of Z.

As a *fourth step*, now that we have the formula for Z, suppose we wish to find Z when $x = 15$ and $y = 8$. The formula says

$$Z = \frac{12}{35} x \cdot y$$

Substituting the new values of x and y, we get

$$Z = \frac{12}{35}(15)(8) = 41\tfrac{1}{7}$$

There are *four* principal steps in solving a problem in variation.

1. *Set up the general variation equation including the constant of variation.* (The constant may be represented by K or some other letter. However, it should not be confused with a variable.)
2. *From known values of all the variables, find the value of K.*
3. *Insert the numerical value of K in the general variation equation. The result is a formula.*
4. *Use the formula to solve the new problem involving new values of the variables.*

Example 1

Suppose L varies directly as N and inversely as the square of d. If $L = 200$, when $N = 3$ and $d = 5$, find L when $N = 8$ and $d = 10$.

Solution

Step 1. The variation equation is

$$L = K\frac{N}{d^2}$$

Step 2. Supplying the given values and solving for K,

$$200 = K\frac{3}{25}; \qquad K = \frac{5000}{3}$$

Step 3. Inserting the value of K in the general equation,

$$L = \frac{5000}{3} \cdot \frac{N}{d^2}$$

Step 4. Using the formula to find L for the new values of N and d,

$$L = \frac{5000}{3} \cdot \frac{8}{100} = \frac{400}{3} = 133\tfrac{1}{3}$$

Example 2

Suppose T varies jointly as x and the square of y and inversely as the square root of Z. If $T = 36$ when $x = 3$, $y = 8$, and $Z = 25$, find T when $x = 5$, $y = 6$, and $Z = 81$.

Solution

Step 1. The variation equation is

$$T = \frac{Kxy^2}{\sqrt{Z}}$$

Step 2. Supplying the given values and solving for K,

$$36 = \frac{K \cdot 3 \cdot 64}{\sqrt{25}}; \qquad K = \frac{15}{16}$$

Step 3. Inserting the value of K in the general equation,

$$T = \frac{15}{16} \cdot \frac{xy^2}{\sqrt{Z}}$$

Step 4. Using the formula to find T for the new values of the other variables,

$$T = \frac{15}{16} \cdot \frac{5 \cdot 36}{\sqrt{81}}; \qquad T = 18\tfrac{3}{4}$$

Exercise 23-6

Use appropriate letters for the quantities in each problem:

1. Suppose a variable, Z, varies directly as x and the square of y. If $Z = 12$, when $x = 25$ and $y = 15$, find Z when $x = 15$ and $y = 15$.

2. A variable, F, varies jointly as a and b and inversely as the square of r. If $F = 200$ when $a = 8$, $b = 6$, and $r = 40$, find F when $a = 15$, $b = 4$, and $r = 60$.

3. If Z varies directly as w and the square of x and inversely as the square root of t, find Z when $w = 20$, $x = 24$, and $t = 45$, if $Z = 5$ when $w = 15$, $x = 32$, and $t = 20$.

4. The amount of wages earned varies directly as the number of hours worked. If the amount earned in 24 hr is \$62.40, find by means of a variation equation the amount earned in 78 hr.

5. The interest paid for money borrowed varies directly as the principal, P, and the time, T. If the interest is \$168 when the principal is \$800 and the time is $3\frac{1}{2}$ yr, find the interest on \$500 for $4\frac{1}{2}$ yr at the same rate of interest.

6. The surface area, A, of a sphere varies as the square of the radius. If the area is 100π when the radius is 5, find the area when the radius is 10.

7. The weight of a sphere varies as the cube of the radius. If a sphere with a 3-in. radius weighs 100 lb, what is the weight of a sphere of the same material having a radius of 6 in.?

8. The time required to make a certain trip varies inversely as the speed. If the trip takes 7 hr at 45 mph, how long will it take at 56 mph?

9. The distance traversed by a falling object varies directly as the square of the time. If the distance is 256 ft at the end of 4 sec, find the distance fallen at the end of 5 sec. How far does the object fall in the fifth second?

10. If the distance traversed by a ball rolling down an inclined plane varies as the square of the time and if the distance is 36 in. when the time is 3 sec, find the distance traversed at the end of 6 sec. How far does it roll in the fifth second?

11. The weight of a cylinder of a certain material varies as the height and the square of the radius. If a cylinder with a radius of 4 in. and a height of 18 in. weighs 30 lb, what is the weight of a cylinder of the same material if it has a radius of 3 in. and a height of 36 in.?

12. The cost of carpeting a floor varies jointly as the length and the width of the floor. If the cost is \$180 for a room 15 ft wide and 16 ft long, what will it cost to carpet a floor 18 ft wide and 23 ft long?

13. The volume of water under constant pressure delivered by a circular pipe varies approximately as the square of the diameter. If a $\frac{1}{2}$-in. pipe delivers 30 gal of water per minute, what amount will be delivered per minute by a 3-in. pipe?

14. The resistance to the flow of electricity in a wire varies directly as the length and inversely as the square of the diameter of the wire. If the resistance is 0.26 ohm when the diameter is 0.02 in. for a wire 10 ft long, what is the resistance of a wire of the same material 5 ft long and 0.01 in. in diameter?

15. The power in an electric circuit varies directly as the resistance and the square of the current. If the power is 15 watts when the current is 0.5 amp and the resistance is 60 ohms, find the constant of variation and then find the power when the current is 3 amp and the resistance is 45 ohms.

16. The number of vibrations of a musical string is inversely proportional to the length. If a violin string approximately $12\frac{3}{4}$ in. long with a particular tension has 430 vibrations per second (approximately the tone A), what must be the length of the same string for 510 vibrations per second?

17. The number of days required to erect a certain building varies inversely as the number of men working and the number of hours per day. If the building requires 45 days to complete when 80 men work for 8 hr per day, how long will be required if 60 men work for 9 hr a day?

18. The length of time required to fill a swimming pool varies inversely as the square of the diameter of the pipe through which the water enters the pool. If it takes 5 hr for a 6-in. pipe to fill the pool, how long will it take a 3-in. pipe?

19. The force of attraction between two magnetic poles of opposite polarity varies inversely as the square of the distance between them. If the force is 80 dynes when the distance is 4 cm, what is the force when the distance is 8 cm?

20. The intensity of illumination, L, varies inversely as the square of the distance from the source of light. If the amount of light falling on a page of this book is 100 units when the book is 6 ft from the light source, how many units will fall on the page if it is held 3 ft from the source of light?

21. A student reading a book said, "I shall move nearer the lamp so that I'll be just half as far from the light. Then I'll have twice as much light on the page." What was wrong with his statement?

22. The wind pressure on a wall varies directly as the area, A, of the wall and as the square of the speed, or velocity, of the wind. If the force on a wall 12 ft wide and 18 ft long is 120 lb when the velocity of the wind is 15 mph, what is the force on a wall 10 ft wide and 20 ft long when the velocity of the wind is 25 mph?

23. The lens setting of a camera varies inversely as the square root of the shutter speed. If the speed is $\frac{1}{50}$ sec for an opening of 11.2, what should the lens opening be for a shutter speed of $\frac{1}{100}$ sec?

24. For a gas under pressure, the volume V varies inversely as the $\frac{2}{3}$

power of the pressure P. If the volume is 200 cu in. when the pressure is 64 lb per sq in., find the volume when the pressure is 125 lb per sq in.

25. The heat loss through a certain insulation material is inversely proportional to the thickness of the insulation. If the loss is 3000 BTU per hour through a thickness of 2.6 in. of insulation, find the heat loss through a thickness of 6.5 in. of insulation.

QUIZ ON CHAPTER 23; RATIO, PROPORTION; VARIATION, FORM A.

Express the following ratios as simplified fractions or whole numbers (No. 1–3):

1. \$20 to \$36 **2.** 2 ft to 3 yd **3.** 1 mile to 1 ft

4. Two circles have diameters of 4 in. and 8 in., respectively. Find **(a)** the ratio of the radius of the first circle to the radius of the second; **(b)** the ratio of their circumferences; **(c)** the ratio of their areas.

5. The side of one square is 6 in.; the side of a second square is 8 in. Find the following ratios: **(a)** the side of one square to the side of the other square; **(b)** the perimeter of one square to the perimeter of the other; **(c)** the ratio of their areas.

Find the value of the unknown in each of the following proportions:

6. $4 : x = 7 : 33$ **7.** $5 : 14 = 8 : x$

8. $2 : (h + 2) = 5 : (4h - 1)$ **9.** $4 : (k + 3) = (k - 2) : (k + 1)$

10. Find the fourth proportional to 4, 6, and 5.

11. Find the third proportional to 6 and 8.

12. Find the mean proportional between 4 and 36.

13. According to Hooke's law, a spring will be stretched such that the distance it stretches is in proportion to the force applied. If 5 lb of force will stretch a certain spring 1.5 in., how much force is required to stretch the spring 1 ft?

Write the variation statement for each of the following:

14. The power P in an electric circuit varies jointly as the resistance R and the square of the current I.

15. In an electric circuit, the current I varies inversely as the resistance R.

16. A quantity Z varies jointly as x and y. If $Z = 48$, when $x = 6$ and $y = 3$, find K the constant of variation, and then find Z when $x = 5$ and $y = 6$.

PART THREE GEOMETRY

CHAPTER TWENTY FOUR Plane Figures

24-1 THE ELEMENTS OF GEOMETRY

Geometry deals with *form* and *size*. The form and size of geometric figures continually influence our living and thinking from day to day.

Every day in countless ways we are conscious of *form*. Whenever we use expressions such as the following, we are aware of geometric form: a round plate; a square box; a long pencil; an oval table; a rectangular rug; a straight path; a curved road; a crooked trail.

We are also conscious of *size*. Whenever we use the following expressions, we have a feeling of size and measurement: a big apple; a little child; a large man; a small box; a long, narrow room; a thin board; a thick rug.

Geometry is not concerned with qualities that are not measurable. Many such qualities make life beautiful and pleasant, but they do not enter in the study of geometry, although some geometric forms are considered more beautiful than others. Geometry is not concerned with the qualities indicated in the following expressions: a red apple; a blue bird; a fierce black bear; a soft rug; a tender steak; a sweet apple; a sour lemon; a good man; a happy woman. The qualities so expressed are an essential part of our lives, but geometry is concerned instead with only *form* and *size*.

The form and size of geometric figures are determined by *points*, *lines*, and *surfaces*. A point is an *undefined element* in geometry. A true geometric point has no size—no length, width, or thickness. It indicates position or location only. We usually make a dot of some kind to indicate the location of a point, but such a dot is not a true point because it has length, width, and thickness. You can get some idea of a true point if you think of the tip of a needle just at the place where you leave the needle.

A *line* has length only. It has neither width nor thickness. A line is another undefined element in geometry, although it is sometimes defined as *the path of a moving point*. Since the point has no size, the line has no width. We often indicate a line by drawing a mark with a pen or pencil. However, such a mark is not a true line because even the finest pen or pencil line has some width and some thickness.

In geometry the word *line* is understood to refer to unlimited extent; that is, it has no end. A limited portion of a line is called a *line segment*.

A line is often named by the letter l on the line. If more than one line is under discussion in a problem, the separate lines are indicated by subscripts, such as l_1, l_2, and so on. A line segment is often indicated by a single letter or sometimes by a letter at each end.

A *straight line* (Fig. 24-1*a*) is the path formed by a moving point that always moves in the same direction. A *curved line* (Fig. 24-1*b*) is a line no part of which is straight. A *broken line* (Fig. 24-1*c*) is a line consisting of straight line segments. A *closed* curve is one that encloses a definite amount of area (Fig. 24-1*d*).

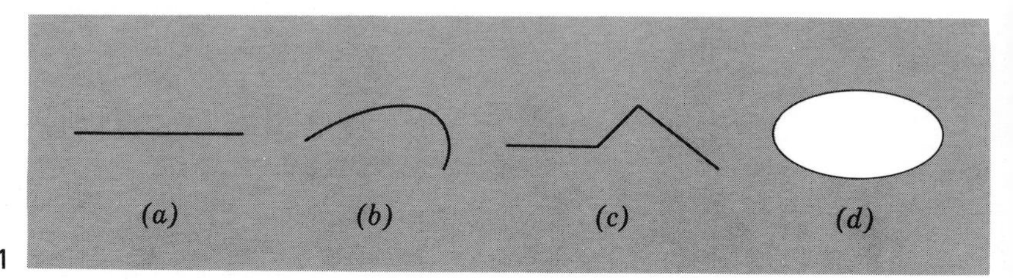

Figure 24-1

A *surface* has two dimensions, *length* and *width.* It has no thickness. When we speak of any kind of surface, such as the surface of this paper, we are not concerned with the thickness of the paper. A surface is sometimes defined as the geometric figure generated by a moving line that does not move in its own direction. A surface may be *flat* or *curved.* Fig. 24-2 shows a curved surface.

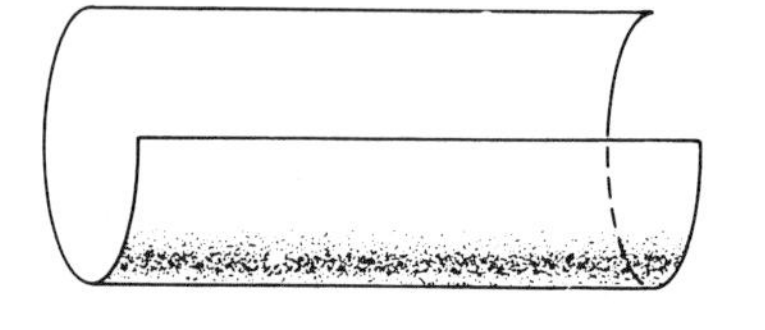
Figure 24-2

A *plane* surface is a surface usually described as "flat." More technically, a plane surface is a surface such that if a straight line has any two of its separate points anywhere in the surface, then the entire lines lies in the surface.

A *plane* figure in geometry means a figure that can be drawn on a flat surface. Such figures have no thickness, only length and width. For instance, if we think of a flat piece of paper as a plane, we can show on this flat surface several kinds of plane figures, such as a line, an angle, a triangle, a circle, a square, and other plane figures (Fig. 24-3).

Plane geometry deals with figures on a plane surface; that is, with plane figures. *Solid geometry* is the study of figures that take up space;

Figure 24-3. Plane figures.

that is, geometric solids. Solid geometry is therefore sometimes called *space geometry*.

24-2 LINES: INTERSECTING AND PARALLEL

Two straight lines in the same plane are either parallel or intersecting lines. If the lines cross each other, they are called *intersecting* lines and they intersect at one point only (Fig. 24-4*a*). If two straight lines in the same plane do not intersect no matter how far they are extended, they are called *parallel* lines (Fig. 24-4*b*). According to this definition, two parallel lines never intersect.

In space, however, two lines can be nonparallel, yet not intersect. Such lines are called *skew* lines. Two telephone wires extending in different directions, one above the other, form skew lines.

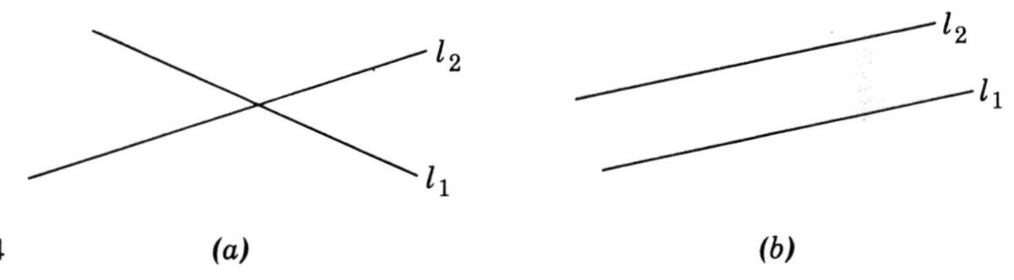

Figure 24-4 (*a*) (*b*)

24-3 ANGLE

The word *angle* is used so frequently in geometry that it is often represented by a small angle, $\angle$, as a symbol. (Plural: $\measuredangle$.)

An angle can be defined as the figure formed when two straight lines intersect. Actually, when two straight lines intersect, four angles are formed, such as the angles, 1, 2, 3, and 4 in Fig. 24-5. Two angles that have a common vertex and a common side between them are called *adjacent* angles, such as $\angle 1$ and $\angle 2$. The following sets are also pairs of adjacent angles: $\angle 2$ and $\angle 3$; $\angle 3$ and $\angle 4$; $\angle 4$ and $\angle 1$. Opposite angles, such as angles 1 and 3, are called *vertical.*

An angle is sometimes defined as *the amount of opening between two straight lines that meet at a point.* Another definition that is especially useful in most engineering is the following: *an angle is the amount of turning or rotation of a line about a point on the line* from one position to a new position. In Fig. 24-5 we can say that the angle is the amount

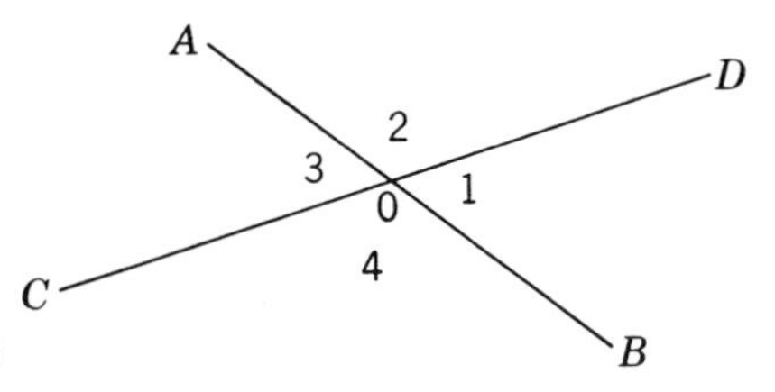

Figure 24-5

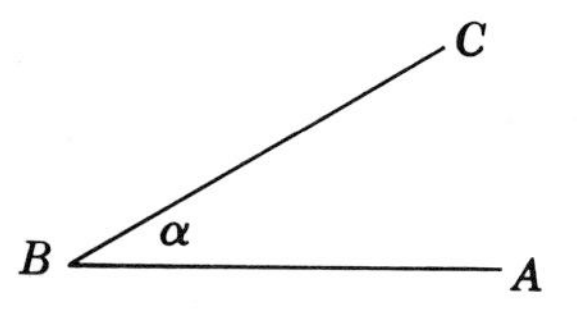

Figure 24-6

of rotation of the line AB about the point O, until the line reaches the new position, CD.

The two lines that form an angle are called the *sides* of the angle. The point where the two sides meet is called the *vertex* of the angle (Plural: *vertices*). A small curved line is often drawn between the sides of the angle near the vertex, thus $\measuredangle$.

An angle is often named by placing a capital letter at the vertex, such as $\angle B$ (Fig. 24-6). It is sometimes named with three letters, usually capital letters, such as $\angle ABC$. If three letters are used to name an angle, the vertex letter is mentioned second. The angle should be read in the same order as you would draw it without raising the pencil from the paper. An angle may also be named by a small letter inside the angle, such as $\measuredangle$. For naming angles with a single letter, we often use Greek letters, such as *alpha* (α), beta (β), *gamma* (γ), and so on. The Greek letters, theta (θ) and phi (ϕ), are also common names for angles.

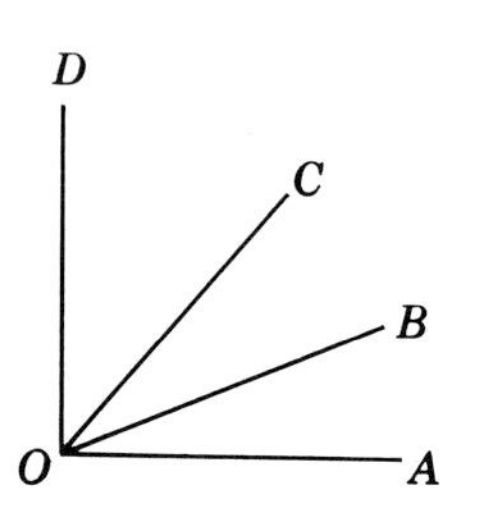

Figure 24-7

If there is any danger of confusion as to which angle is meant, the angle should be named by using three letters. In Fig. 24-7 the meaning is not clear if we simply say angle O.

24-4 PERPENDICULAR LINES

If two intersecting lines form two equal adjacent angles, the lines are said to be *perpendicular* to each other, or mutually perpendicular. The symbol, $\perp$, is used for the word *perpendicular.* In Fig. 24-8, $AB \perp CD$. (This is read *AB is perpendicular to CD.*) Also, $CD \perp AB$.

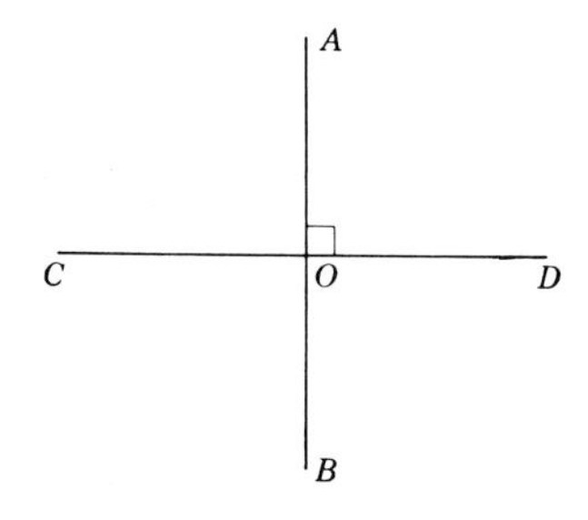

Figure 24-8

If two straight lines are perpendicular to each other, the two adjacent angles are called *right* angles. A right angle is what we often call a square corner. In Fig. 24-8, since the lines are perpendicular to each other, angle DOA and angle AOC are right angles. The other two angles are also right angles. A right angle can be indicated by the symbol ⦜.

24-5 KINDS OF ANGLES

As we have seen, if the two sides of an angle are perpendicular to each other, the angle formed is a right angle (Fig. 24-9*a*). An angle smaller than a right angle is called an *acute* angle (Fig. 24-9*b*). An angle greater than a right angle but less than the sum of two right angles is called an *obtuse* angle (Fig. 24-9*c*).

If two right angles are placed adjacent to each other, a straight line is formed by two of the sides. An angle formed as the sum of two right angles is called a *straight* angle (Fig. 24-9*d*). A straight angle forms a straight line.

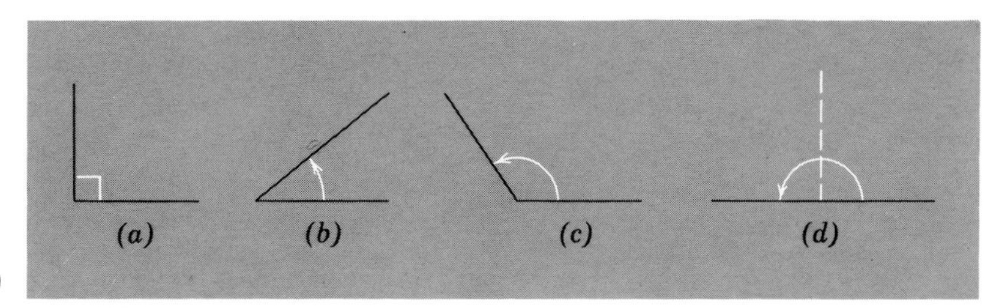

Figure 24-9

24-6 MEASUREMENT OF ANGLES

In order to measure the size of angles, the early Babylonians decided to divide a complete circle into 360 parts. (They might just as well have divided it into 100 parts or 60 parts or some other number.) Each part they called one *degree.* This system of measuring angles is still in common use. The unit of measurement, one degree (written 1°), is therefore $\frac{1}{360}$ of a complete rotation (Fig. 24-10).

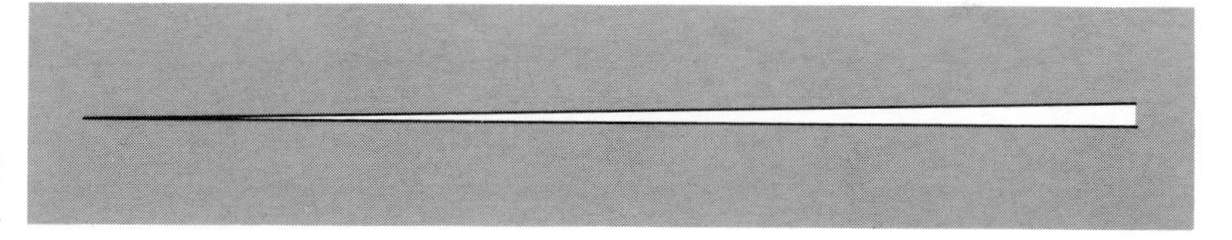

Figure 24-10. An angle of approximately one degree.

Since one rotation is called 360 degrees (360°), then one right angle is called 90 degrees (90°), and one straight angle is called 180°. An angle greater than 180° and less than 360° is called a *reflex* angle.

The size of an angle can be measured by an instrument called a *protractor.*

For very fine measurements (that is, for more accuracy), the degree is divided into 60 smaller parts or angles, called *minutes.* For extreme accuracy, such as that required in astronomy, the minute is divided into 60 parts called *seconds.* One *second* of rotation is a very small angle. In fact, an angle of one second is so small that the distance between its sides is only about $\frac{3}{10}$ of an inch at a distance of a mile from the vertex.

Two angles whose sum is equal to a right angle, or 90°, are called *complementary.* Two complementary angles need not be adjacent. An angle of 37° and an angle of 53° are complementary, since their sum is 90°. Two angles whose sum is equal to one straight angle, or 180°, are called *supplementary.* Two supplementary angles need not be adjacent. An angle of 49° and an angle of 131° are supplementary, since their sum is 180°.

Exercise 24-1

1. Draw two intersecting lines that are not perpendicular to each other. Measure the four angles formed. What can you conclude about a pair of vertical angles? What conclusion can you make concerning adjacent angles?

2. Draw two parallel lines approximately 2 in. apart. Now draw another line, called a transversal, intersecting the two parallel lines but not perpendicular to them. Measure the eight angles thus formed. What can you conclude about the eight angles formed? Mark each of four *equal* angles with the letter x and the other four angles with the letter y.

3. Show by a drawing that if a transversal is perpendicular to one of two parallel lines it is also perpendicular to the other.

4. Name five different examples of the use of parallel lines in everyday life. Which ones are necessary?

5. Name five examples of intersecting lines in everyday life. Why are they necessary?

6. Can you see ten different examples of right angles in your immediate surroundings as you look up from reading this page?

7. Name five examples of skew lines in everyday life. Why are they necessary?

8. Draw two complementary angles that are not adjacent.

9. Draw two supplementary angles that are not adjacent.

10. From a given point draw seven lines outward in different directions. Measure each of the seven angles formed and add them together. What do you conclude about the total numbers of degrees in all the angles around a point?

24-7 POLYGONS

A *polygon* may be defined as a *plane closed figure bounded by straight line segments.* If the figure is not closed, it is not a polygon. A polygon therefore contains a definite amount of area. The line segments forming the boundary are called the *sides* of the polygon. The *perimeter* of the polygon is the sum of the sides or the distance around the polygon. The points at which the sides meet are called the *vertices* of the polygon.

Polygons are named according to the number of sides. A polygon of three sides is called a *triangle* (Fig. 24-11). The prefix "tri" means three. Notice that the word triangle really means "three angles." Yet we define the triangle in terms of the sides instead of angles. Remember that in any polygon the sides must be *straight.* Perhaps no figure drawn is a perfect polygon.

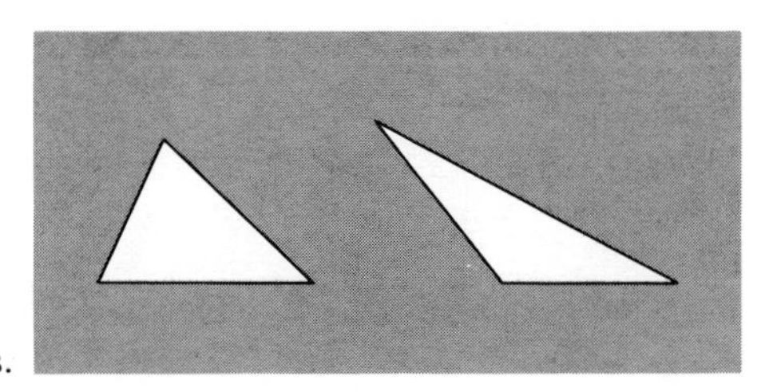
Figure 24-11. Triangles.

A polygon of four sides is called a *quadrilateral* (Fig. 24-12). The prefix "quad" means four, and "lateral" refers to side. Therefore, the word really means four sides. If a triangle were to be named after its three sides, it might be called a "trilateral." Moreover, a quadrilateral can also be named after the angles and called a "quadrangle."

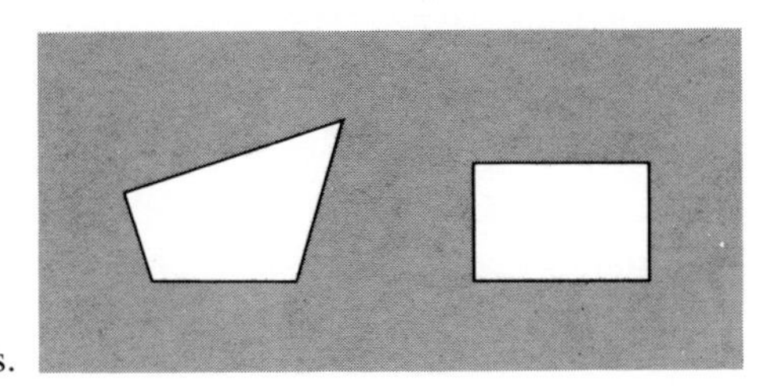
Figure 24-12. Quadrilaterals.

A polygon of five sides is called a *pentagon.* The prefix "penta" means five and "gon" is another name for angle. Fig. 24-13 shows pentagons.

Figure 24-13. Pentagons.

A polygon of six sides is called a *hexagon*; a polygon of seven sides is called a *heptagon*; a polygon of eight sides, an *octagon*; a polygon of ten sides, a *decagon*; and one of twelve sides is called a *dodecagon* (Fig. 24-14).

A *diagonal* of a polygon is a straight line segment joining any two nonadjacent vertices. A *regular polygon* is one that has all angles equal and all sides equal. Fig. 24-15 shows some regular polygons.

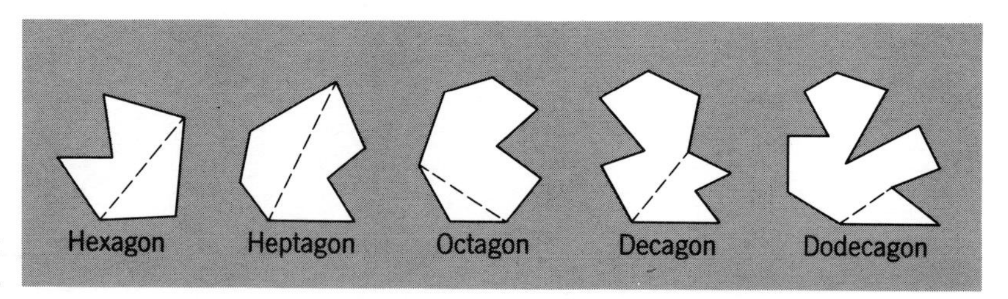

Figure 24-14. Polygons showing one diagonal in each.

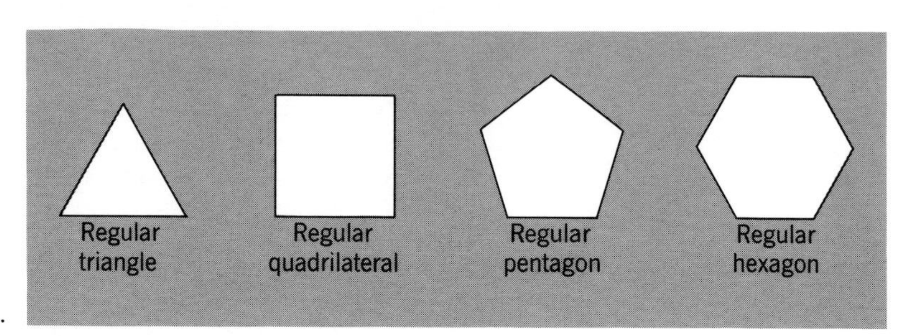

Figure 24-15. Regular polygons.

24-8 TRIANGLES (SYMBOLS)

Triangles are named according to their sides (Fig. 24-16). A triangle having all three sides equal is called an *equilateral* triangle. It can be proved that if the three sides of a triangle are equal, then all three

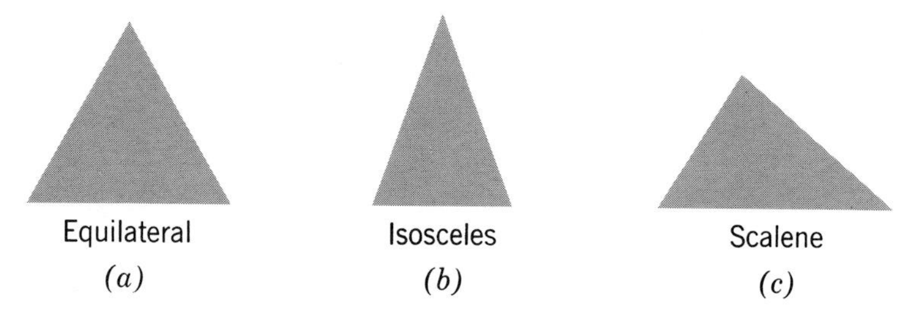

Figure 24-16. Triangles.

angles are equal; that is, an equilateral triangle is also *equiangular.* This condition is not necessarily true for other polygons. A quadrilateral may have all its four sides equal, yet not have all its angles equal.

A triangle having two sides equal is called an *isosceles* triangle. *In an isosceles triangle it can be proved that the angles opposite the equal sides are also equal. Moreover, if a triangle has two equal angles, then the sides opposite are also equal to each other.* A triangle having no two sides equal is called a *scalene* triangle.

A triangle may have angles of various sizes. However, the following statement is true with regard to the angles of all triangles:

In any triangle the sum of the three angles is equal to 180° (*two right angles, or one straight angle*).

As an example, if one angle of a triangle is 47° and a second angle is 74°, then the third angle must be 59° because the sum of the three angles must be 180°.

Triangles are also named according to their angles (Fig. 24-17). A triangle having one right angle is called a *right triangle.* Since a right

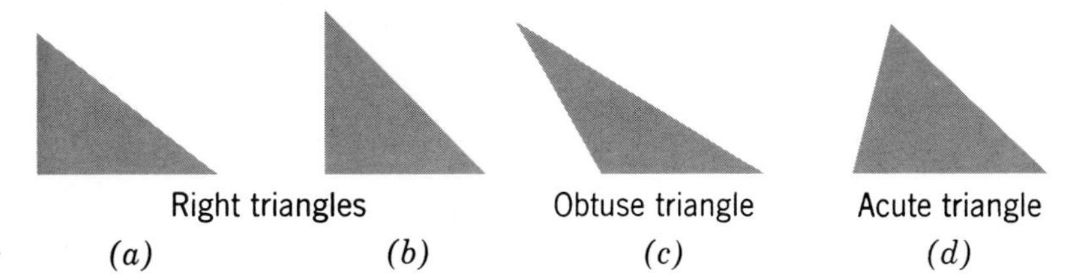

Figure 24-17. Triangles.

triangle has one right angle, the other two angles must be acute angles. Moreover, the two acute angles are complementary; that is, their sum is 90°. The two sides forming the right angle are often called the *legs* of the right triangle. The side opposite the right angle is called the *hypotenuse.* If the two legs of a right triangle are equal in length, then the triangle is also isosceles.

A triangle having one obtuse angle is called an *obtuse* triangle. A triangle all of whose angles are acute is called an *acute* triangle. Acute triangles and obtuse triangles are often called *oblique* to distinguish them from right triangles.

The *base* of a triangle is the side on which it is understood to rest. Any side of a triangle may be considered as the base. The *vertex of a triangle is* the vertex opposite the base. The *altitude* of a triangle is a straight line segment drawn from the vertex perpendicular to the base (Fig. 24-18). Since any side of a triangle may be considered as the

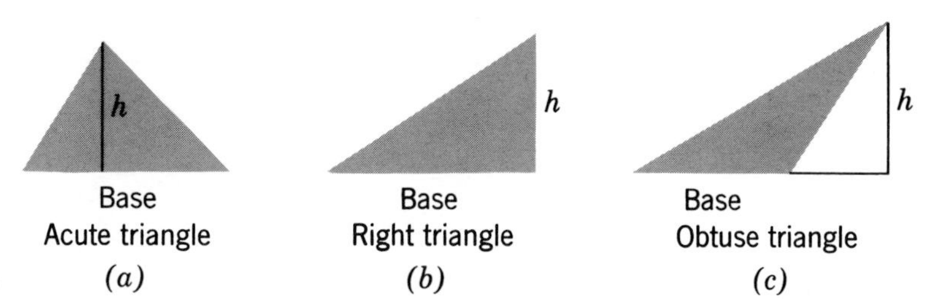

Figure 24-18. Triangles showing altitudes.

base, a triangle has three altitudes. All three altitudes meet at the same point. An altitude is often indicated by the letter h.

If one leg of a right triangle is taken as the base, then the other leg is the altitude (Fig. 24-18*b*). In an obtuse triangle two of the altitudes fall outside the triangle and must be taken to the base extended (Fig. 24-18*c*).

A *median* of a triangle is a straight line segment drawn from a vertex to the midpoint of the opposite side (Fig. 24.19). A triangle has

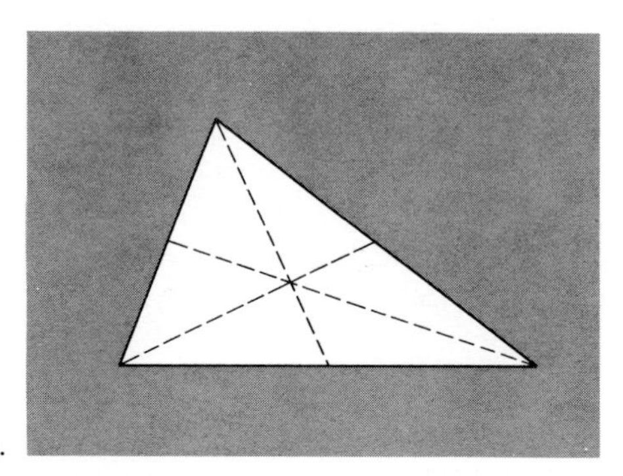

Figure 24-19. Medians of a triangle.

three medians, all of which meet at a point called the *centroid.* This point is an important one, because it is the *center of gravity*, or balancing point, of the triangle.

24-9 QUADRILATERALS

A quadrilateral may have many different shapes. Some of these are shown in Fig. 24-20.

A quadrilateral having no pairs of opposite sides parallel is called a *trapezium* (Fig. 24-20*a*).

A quadrilateral having one pair of opposite sides parallel is called a *trapezoid* (Fig. 24-20*b*). If the two nonparallel sides of a trapezoid are

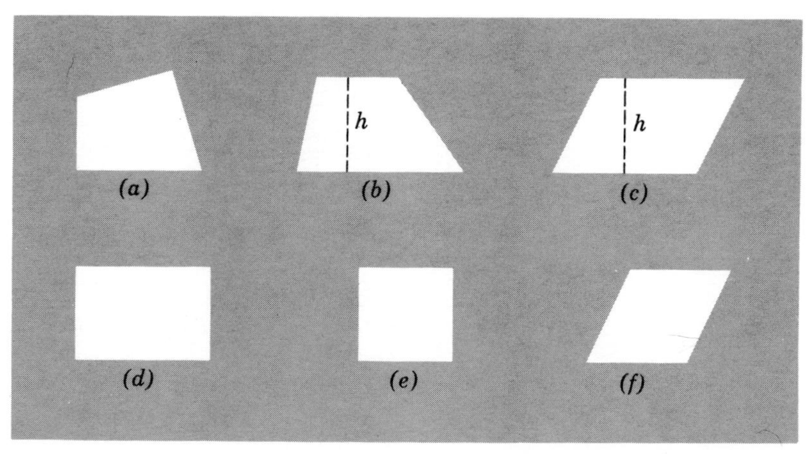

Figure 24-20. Quadrilaterals.

equal, the figure is called an *isosceles trapezoid.* The two parallel sides are called the *bases* of the trapezoid. The *altitude* of a trapezoid is the distance between the bases.

A quadrilateral having both pairs of opposite sides parallel is called a *parallelogram.* (Symbols: ▱, ▱s.) By this definition the figures (*c*), (*d*), (*e*), and (*f*) are parallelograms, because in each of these figures both pairs of opposite sides are parallel.

Now, if we focus our attention on the four parallelograms, we see that some of them have right angles. A parallelogram whose angles are right angles is called a *rectangle.* (Symbols: ▭, ▭s.) By this definition, both of the figures (*d*) and (*e*) are rectangles, since each one is a parallelogram with right angles.

If we go one step further and distinguish between the two rectangles, we notice one rectangle has all its sides equal. This figure (*e*) is a *square.* (Symbols: □, □s.) A square is defined as a rectangle having equal sides.

Notice that all rectangles are parallelograms and that they are also quadrilaterals. A square is a rectangle, a parallelogram, and a quadrilateral.

A *rhombus* is defined as a parallelogram having equal sides. By this definition, a square is also a rhombus. However, the word *rhombus* is usually restricted to mean an equilateral parallelogram that is *not* a square (Fig. 24-20*f*).

The following facts concerning quadrilaterals can be shown to be true. The student should make a drawing or sketch of a figure and try to understand clearly the meaning of each statement.

1. *The sum of all the interior angles of a quadrilateral is* 360°.
2. *Any two opposite sides of a parallelogram are equal.*
3. *If the opposite sides of a quadrilateral are equal, the figure is a parallelogram.*
4. *The opposite angles of any parallelogram are equal.*
5. *If the opposite angles of a quadrilateral are equal, the figure is a parallelogram.*
6. *The diagonals of any rectangle (including a square) are equal.*
7. *The diagonals of any rhombus (including a square) are perpendicular to each other.*

Exercise 24-2

1. Carefully draw a quadrilateral that has no two sides parallel. Measure the interior angle at each corner and find the total number of degrees in the four angles. How near is the result to 360°.
2. If you start at any point on one side of the quadrilateral and walk

around the quadrilateral until you arrive at your starting point, through how many degrees will you turn?

3. Draw a hexagon, not necessarily a *regular* hexagon. Measure the interior angle at each corner and find the total number of degrees in the six angles of the hexagon. How near is the total to 720°? It can be proved that the six interior angles of a hexagon have a total of 720°.

4. If you start at any point on a side of a hexagon and *walk* around the figure until you arrive at your starting point, through how many degrees will you turn? Through how many degrees would you turn if the figure were a decagon?

5. How many diagonals can be drawn in (a) a square? (b) A pentagon? (c) A hexagon? (d) A triangle?

6. Draw a right triangle, an acute triangle, and an obtuse triangle, and measure the three angles in each triangle. Do your measurements confirm the statement that the sum of the three angles of any triangle is 180°?

7. Draw an obtuse triangle that is also scalene. Carefully draw the three altitudes. If the altitudes are extended, do they all intersect at the same point?

8. Draw an obtuse triangle that is also isosceles. Measure the angles opposite the two equal sides. Are these angles equal?

9. Why can a triangle not have more than one obtuse angle?

10. Draw a right triangle and show the three altitudes.

11. In a right triangle why must the two acute angles be complementary?

12. Draw two right triangles, one larger than the other, but with one acute angle in each triangle equal to 35°. Notice that the angles of one triangle are equal, respectively, to the angles of the other. Such figures that have the *same shape* are called *similar* figures.

13. Draw two acute triangles with the angles of each one equal, respectively, to 42°, 65°, and 73°. Are these triangles similar?

14. Draw a quadrilateral having sides equal to 2 in., 1 in., $1\frac{1}{2}$ in., and $2\frac{1}{4}$ in., respectively. Can you draw another quadrilateral having sides of the same length but of an entirely different shape?

15. Draw a triangle having its sides equal to 2 in., $2\frac{1}{4}$ in., and $1\frac{1}{2}$ in., respectively. Can you draw another triangle having sides of the same lengths as those mentioned in this problem but of an entirely different shape?

16. Draw a quadrilateral having two opposite sides each equal to 1 in. and the two other opposite sides each equal to $1\frac{1}{2}$ in. After this is done, try to determine whether the opposite sides are also parallel.

17. Draw a quadrilateral, making the opposite sides parallel without trying to make them equal. Now measure them and determine whether they are also equal.

18. Draw a rectangle and then draw the two diagonals. Are the diagonals equal to each other? Are they perpendicular to each other?

19. Draw a rectangle much longer than it is wide. Now draw one diagonal and try to determine whether the diagonal divides the right angle into two equal angles.

20. Draw a parallelogram that is not a rectangle. Now draw the diagonals. Are the diagonals equal to each other?

21. Draw the diagonals of a square and state several facts concerning them.

CHAPTER TWENTY-FIVE Measurement of Plane Figures

25-1 THE NEED FOR MEASURING

Geometry is concerned with measurement as well as with form. It has been said that human beings first became civilized when they started to measure things. Probably the first kinds of measurement were those of length. Now we live in a world created by measurement. All scientific ideas must be translated into measurement before they can become useful. Thousands of things must be measured. Measurement is necessary whether in the making of a cake or in the building of a satellite. It is hard to realize what our lives would be like if it were not for measurement.

In the measurement of plane figures we are interested in two measurable quantities: lengths of line segments and areas of limited portions of surface. For example, we may need to find the length of a rectangle, the circumference of a circle, or the altitude of a triangle. We also may wish to know how much area is included in a rectangle, a circle, a triangle, or some other plane figure.

25-2 THE UNIT OF MEASURE

To measure any kind of quantity we must use some *unit of measure.* The unit is simply a small amount of the same kind of thing as the quantity measured.

To measure length we use a unit of length such as an inch or a foot. To measure area we use a unit of area such as a square inch or a square foot. To measure volume we use a unit of volume such as a cubic inch or a cubic foot.

To measure weight we use a unit of weight such as an ounce, a gram, or a pound. To measure time we use a unit of time, such as a minute, a second, or a day. To measure angles we use a small unit angle, the degree, or a larger unit, the radian.

In the study of electricity we also have units of measure. For current we use a unit of current called the *ampere.* For measuring electromotive force we use a unit of electromotive force called the *volt.* For electrical resistance we use a unit of resistance called the *ohm.*

In measurement we see how many times the unit is contained in the thing measured. Therefore, the unit used for measuring must be of the same kind of thing as the quantity measured. For example, we cannot measure length in pounds. We cannot measure angles in inches. We cannot measure weight in minutes. We cannot measure electric current in inches, pounds, or degrees.

25-3 PERIMETER

The *perimeter* of a polygon is the distance around a polygon. The perimeter is found by adding the lengths of the sides. For instance, if the sides of a triangle are equal to 5 inches, 7 inches, and 9 inches, respectively, the perimeter is $5 + 7 + 9 = 21$ inches. If one side of a square is 7 inches, the perimeter of the square is 28 inches.

If any polygon is equilateral (that is, having equal sides), then the perimeter can easily be found by multiplication. If we know that one side of a regular hexagon is 8 inches, then the perimeter is 48 inches.

Notice that the perimeter of any polygon is expressed in *linear units,* since the perimeter is a *length.*

25-4 AREA

Measuring a given area is done by taking a small unit of area and laying off the unit area as many times as possible on the given area. For example, if we wish to measure the area of this page, we can take a coin, such as a nickel, and see how many times the area of the page contains the small unit of area covered by the coin. The trouble with a circular unit of area is that the units do not fit together. That is the reason we use a unit of area that is rectangular. Rectangular units fit together and fill the space around a point. The same is true with a triangular or hexagonal unit. However, a square unit, such as a square inch or a square foot, has another advantage, as we shall see presently.

25-5 THE RECTANGLE

Suppose we have a rectangle 6 inches long and 4 inches wide (Fig. 25-1). To find the area of a rectangle, we usually say multiply the length by the width, or

$$(6 \text{ inches})(4 \text{ inches}) = 24 \text{ square inches}$$

The answer is correct, but does it not seem strange that we can multiply *two lines* together and get an area? If you will take a moment to analyze this problem carefully, you will probably understand much better all later problems in areas and volumes. Let us see why we get the correct answer to the problem.

Figure 25-1

We cannot measure area with a unit of linear measure such as an inch. To measure area we must use a *unit of area.* We shall use a unit that is 1 inch long and 1 inch wide. This unit is called a *square inch.*

Now, starting in one corner of the rectangle, we lay off this unit of area as many times as possible along the side of the rectangle (Fig. 25-2). We know that we can lay off six units in one row because the unit is 1 inch long and there are 6 inches in the length of the rectangle.

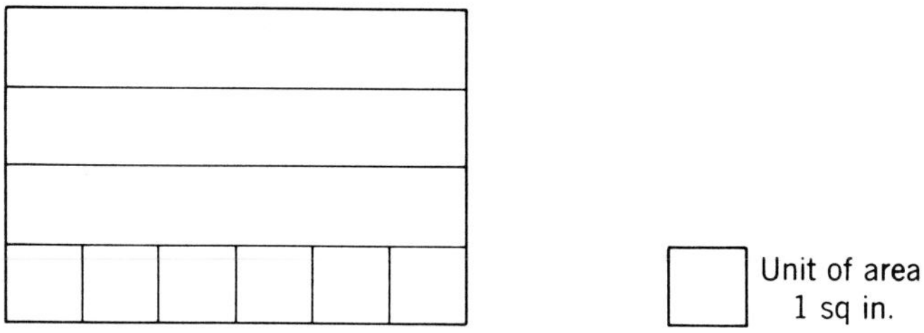

Figure 25-2

Unit of area
1 sq in.

After laying off one row of 6 units, we consider the number of rows. We see there will be 4 rows because each row is 1 inch wide and the width of the rectangle is 4 inches.

At this point we see that the total number of square inches in the rectangle can be found by multiplication:

$$(4)(6 \text{ sq in.}) = 24 \text{ sq in.}$$

The advantage of using a square unit of measure for area is that we can measure the length in *linear units* (inches), the width in *linear units* (inches), and then easily find the area by the following rule:

Rule. *Multiply the number of linear units in the length by the number of the same kind of linear units in the width. The result is the number of corresponding square units in the area.* Stated in short form the rule is

$$\text{area} = (\text{length})(\text{width})$$

If we let l represent the number of linear units in the length, let w represent the number of linear units in the width, and let A represent the number of corresponding square units in the area, we can write the rule as a formula:

$$A = lw$$

This formula can be used for all rectangles. For example, if a rectangle is 12.5 inches long and 7 inches wide, its area is 87.5 sq in. Note that the area is expressed in square units since the unit of area is a *square inch.*

The formula can also be used in reverse. For example, if a rectangle is 7 inches wide and has an area of 60 sq in., we can find the length by division: $60 \div 7 = 8\frac{4}{7}$. The answer will represent the number of linear units in the length. If we divide both sides of the formula by l, we get the formula for the width w:

$$w = \frac{A}{l}$$

If we divide both sides by w, we get the formula for the length l:

$$l = \frac{A}{w}$$

In all cases the length and width are measured in linear units, the area in square units.

The perimeter of a rectangle is equal to twice the length plus twice the width, or as a formula

$$P = 2l + 2w$$

The perimeter is of course expressed in *linear* units. Whenever we *add linear* units, we get an answer representing *linear* units. Whenever we *multiply linear* units by *linear* units, we get a product that represents an area in terms of the *square of the linear unit.*

25-6 THE SQUARE

A square is a special rectangle in which all sides are equal. The length is equal to the width (Fig. 25-3). If we represent one side of the square by the letter s, we have for the area

$$A = (s)(s) \qquad \text{or} \qquad A = s^2$$

Since all four sides are equal in length, we find the perimeter by taking 4 times the length of one side. Stated as a formula,

$$P = 4s$$

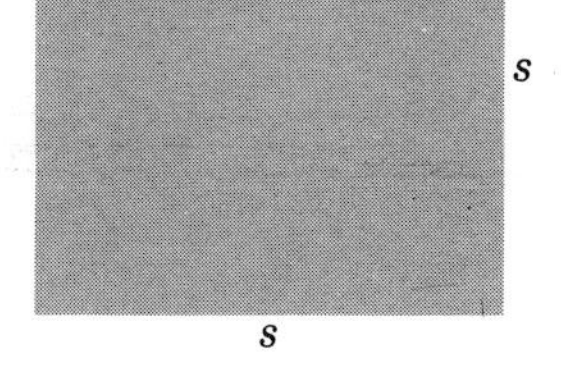

Figure 25-3

25-7 THE PARALLELOGRAM

If we know the lengths of the four sides of a parallelogram, we cannot find the area from this information alone. For instance, if we have a parallelogram with sides of 12, 8, 12, and 8 inches, respectively (Fig. 25-4), we can find its perimeter but we cannot tell the area. In order to determine the area, we must know the altitude. The altitude is the perpendicular distance between two sides. Parallelograms may have the given sides as shown and yet may have different altitudes. One side of a parallelogram, such as the 12-inch side in the example, is called its *base* rather than its *length.*

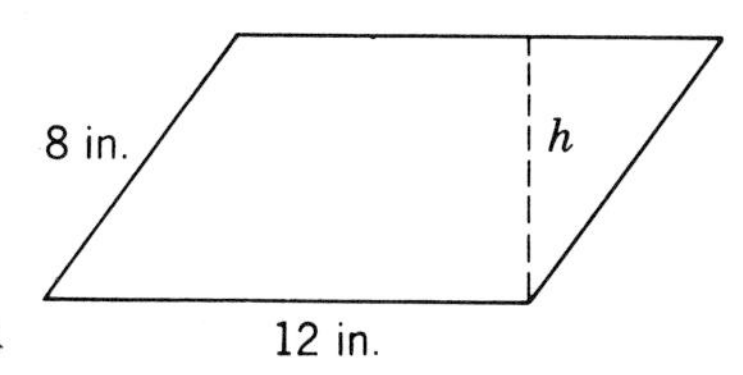

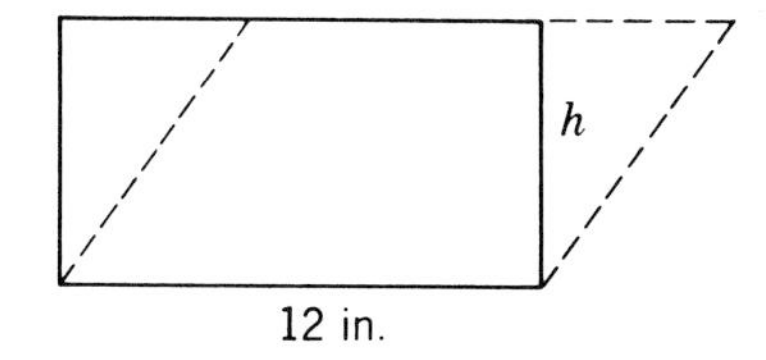

Figure 25-4

If we assume that one triangle is cut off one end of the parallelogram and attached to the other end, the result is a rectangle having the same base and the same altitude as the parallelogram. It also has the same area. Therefore, the area of a parallelogram can be found by the formula

$$A = bh$$

in which A represents the area, b represents the base, and h represents the altitude or height.

If the area and the base of a parallelogram are known, the altitude can be found by dividing the area by the base. If the area and the altitude are known, the base can be found by dividing the area by the altitude; that is,

$$h = A/b \qquad \text{and} \qquad b = A/h$$

If we know the altitude and the base of a parallelogram, we cannot find the perimeter from this information alone. Parallelograms may have the same bases and the same altitudes, yet have different perimeters.

Exercise 25-1

In each of the following, find the values indicated:

1. Rectangle: 28 in. long; 16.5 in. wide. Find area and perimeter.
2. Rectangle: 15.2 in. wide; area = 342 sq in. Find length and perimeter.
3. Rectangle: 32 in. long; area = 784 sq in. Find width and perimeter.
4. Parallelogram: base = 17.5 ft; altitude = 14.8 ft. Find area.
5. Parallelogram: area = 248.4 sq in., altitude = 13.5 in. Find base.
6. Parallelogram: sides: 15.4 in. and 10.2 in. Find perimeter.
7. Square: side = 14.3 in. Find perimeter and area.
8. Square: perimeter = 68.8 in. Find side and area.
9. Square: area = 1024 sq in. Find side and perimeter.
10. Find the cost of a piece of plywood 4 ft wide and 8 ft long at 47.5 cents per square foot.
11. Find the cost of building a driveway 60 ft long and 12 ft wide at a cost of $5.20 per square yard (1 yd = 3 ft).
12. Find the cost of paving 10 miles of road 32 ft wide at a cost of $6.50 a square yard (1 mile = 5280 ft).

13. A large rectangular hall has a floor space of 1000 sq yd. If one side measures 120 ft, find the width of the hall.

14. A field is 68 rd long and 45 rd wide. How many acres does it contain? (1 acre = 160 sq rd.)

15. A baseball diamond is a square, 90 ft between bases. What is the area of the diamond? (What part of an acre does it contain?)

16. A rectangular field contains 54 acres. If it is 120 rd long, how wide is it? Find the length of the fence needed to enclose the field.

17. A garden is 125 ft long and 84 ft wide. What part of an acre does it contain?

18. A blackboard is 26 ft 3 in. long and 3 ft 4 in. high. How many square feet does it contain?

19. A rug 13.5 ft wide and 16.8 ft long is laid in a room 15 ft wide and 18 ft long. How many square feet of the floor are not covered?

20. A rectangle is 38 in. long and 15 in. wide. Find the size of a square having the same area as the rectangle. Find the difference in the perimeters of the rectangle and the square.

21. A wire 120 in. long is to be bent so that the wire forms the four sides of a rectangle. Find the area of each rectangle formed if the following values are taken for the width:

(a) 30 in. **(b)** 24 in. **(c)** 20 in. **(d)** 15 in.
(e) 12 in. **(f)** 10 in. **(g)** 8 in. **(h)** 6 in.
(i) 4 in. **(j)** 2 in. **(k)** 1 in. **(l)** 0.5 in.

22. A rectangular field bounded on one side by a river is to be fenced on the three remaining sides with a total length of 180 rd of fence. If the length is taken as the distance along the river and the width as the distance outward from the river, find the area enclosed for each of the following widths:

(a) 10 rd **(b)** 20 rd **(c)** 30 rd
(d) 40 rd **(e)** 45 rd **(f)** 50 rd
(g) 60 rd **(h)** 70 rd **(i)** 80 rd

25-8 THE TRIANGLE

Suppose we have the triangle ABC, with base b equal to 11 inches, and altitude h equal to 8 inches (Fig. 25-5*a*).

In order to formulate a rule for the area of the triangle, we first construct another congruent triangle adjacent to the given triangle but in an inverted position (Fig. 25-5*b*). We see at once that a parallelogram is formed. The parallelogram has the same base and the same

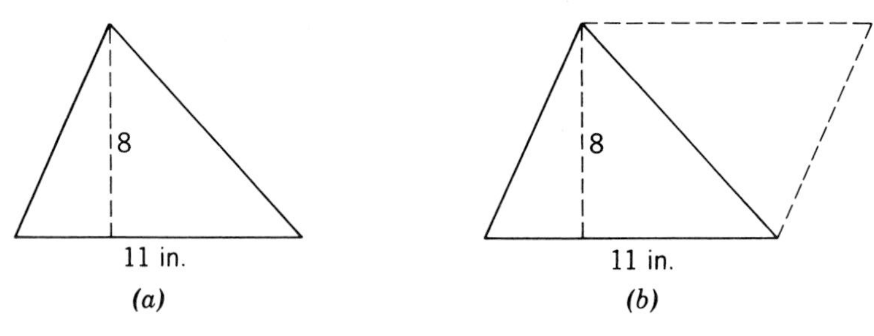

Figure 25-5

altitude as the given triangle, $b = 11$, $h = 8$. We know that the area of a parallelogram is (8)(11) or 88 square inches. For the parallelogram, we have the formula

$$A = bh$$

We also see that the area of the original triangle is exactly one-half the area of the parallelogram. Therefore, the area of the triangle is $(\frac{1}{2})(11)(8) = 44$ square inches. We can say that the area of any triangle is equal to one-half the product of the base and the altitude. This rule can be stated as a formula:

$$A = \tfrac{1}{2}bh \qquad \text{or} \qquad A = \frac{bh}{2}$$

In a right triangle either leg can be considered as the base and the other leg as the altitude. The area of a right triangle is therefore one-half the product of the two legs.

If we have several triangles, all with the same altitude, the total area may be found by computing the area of each separately, or the bases may first be added (Fig. 25-6). If we compute each of the areas

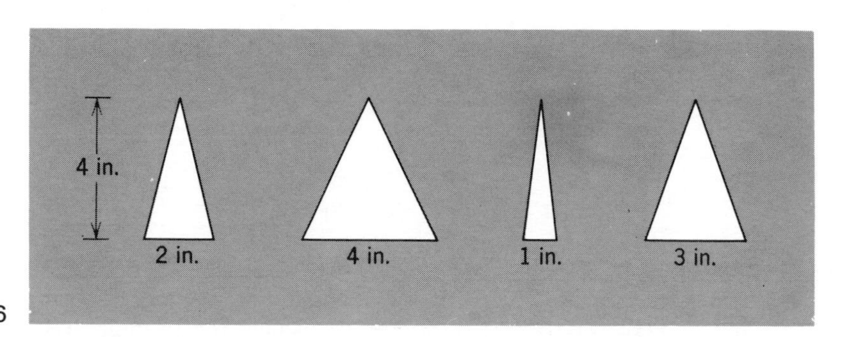

Figure 25-6

separately, we have

$$4 + 8 + 2 + 6 = 20 \text{ sq in.}$$

If we first add all the bases and then consider the figure as one large triangle with an altitude of 4 inches and a base of 10 inches, we have

$$(\tfrac{1}{2})(10)(4) = 20 \text{ sq in.}$$

In general terms let us see why the foregoing process is valid. Suppose we have any number of triangles, all having the same altitude h, and various bases, b_1, b_2, b_3, ... Adding the separate areas we have

$$\text{total area} = \tfrac{1}{2}b_1h + \tfrac{1}{2}b_2h\ \tfrac{1}{2}b_3h + \cdots$$

Factoring,

$$\text{total area} = \tfrac{1}{2}h(b_1 + b_2 + b_3 + \cdots)$$

The result shows that the bases may first be added provided the altitude is the same for all the triangles.

The formula for the area of a triangle can also be used in reverse. For example, if we know the area and either the altitude or base, we can find the unknown dimension. The formula for the area can be solved for either h or b. Since

$$A = \frac{bh}{2}$$

then

$$h = \frac{2A}{b} \quad \text{and} \quad b = \frac{2A}{h}$$

To find the perimeter of a triangle, we must know the lengths of the three sides. Then the perimeter is found simply by adding the lengths of the sides. If we let p represent the perimeter, and a, b, and c represent the sides, respectively, then we have the formula: $p = a+b+c$. For example, if a triangle has sides equal to 7 inches, 9 inches, and 10 inches, respectively (Fig. 25-7), its perimeter is given by

$$p = 7 + 9 + 10 = 26 \text{ inches}$$

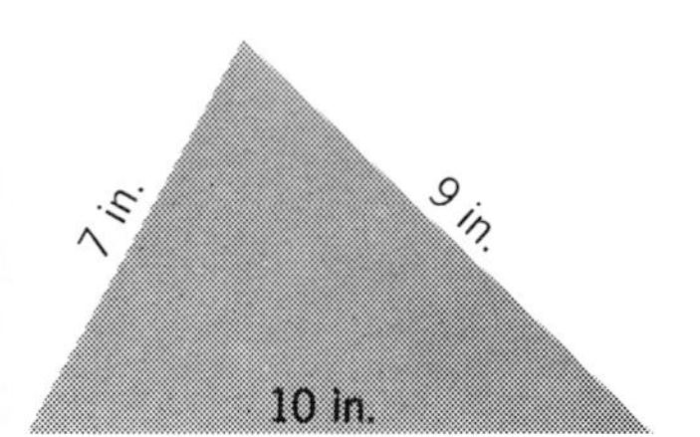

Figure 25-7

In the triangle in Fig. 25-7 since the three sides are 7, 9, and 10 inches, respectively, the shape and the size of the triangle are definitely determined. In other words, if the three sides of a triangle are given, then there is only *one shape and size* of the triangle. This is not true with respect to a quadrilateral or any other polygon. In fact, this principle forms the basis for the use of triangles in bracing. Triangular bracing produces rigidity in buildings, bridges, furniture, and so on.

Now, since the triangle in Fig. 25-7 has a definite shape and area, it should be possible to compute the area from the lengths of the three sides. The method was explained almost two thousand years ago by a mathematician named Hero (Heron). Hero showed that the area of a triangle can be computed from the three sides by the following formula. If we let a, b, and c, respectively, represent the three sides, and let s equal one-half the sum of the three sides, then the formula for the area is as follows:

$$A = \sqrt{s(s - a)(s - b)(s - c)}$$

For the foregoing triangle (Fig. 25-9) we have $a = 7$, $b = 9$, $c = 10$, $s = 13$. Therefore, in the triangle

$$A = \sqrt{13(13 - 7)(13 - 9)(13 - 10)}$$

$$= \sqrt{(13)(6)(4)(3)} = \sqrt{936} = 30.59 \text{ sq in.}$$

If we estimate the altitude of the triangle, we see that it appears to be approximately six units. Using this altitude, 6, the base, 10, and the formula, $A = \frac{1}{2}bh$, we get 30 square units as the area, which is approximately the value by the use of Hero's formula.

25-9 THE TRAPEZOID

A trapezoid is a quadrilateral with two sides parallel and the other two sides not parallel. Suppose we wish to find the area of a trapezoid having a bottom base B equal to 8 inches, a top base b equal to 5 inches, and the altitude h equal to 4 inches (Fig. 25-8*a*).

In order to formulate a rule for the area of a trapezoid, we first construct a congruent trapezoid adjacent to the given trapezoid but in an inverted position (Fig. 25-8*b*). A parallelogram is formed by the

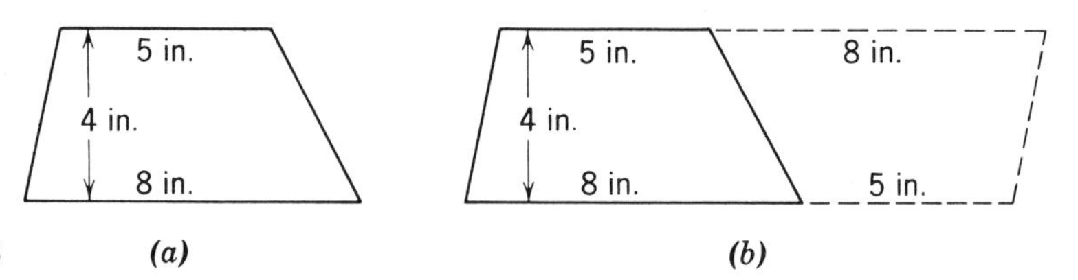

Figure 25-8 (*a*) (*b*)

two trapezoids. The parallelogram has an altitude, h, the same as the given trapezoid. The base of the parallelogram is equal to the *sum* of the bases of the trapezoid, that is, $B + b$. Therefore, the area of the parallelogram is

$$h(B + b)$$

We see that the area of the original trapezoid is exactly one-half the area of the parallelogram and therefore is given by the formula

$$A = \tfrac{1}{2}h(B+b)$$

The area of the trapezoid is $(\frac{1}{2})(4)(8 + 5) = 26$.

The formula for the area of a trapezoid can be solved for h, B, or b, to get the corresponding formula for each of these dimensions. Since

$$A = \tfrac{1}{2}h(B + b)$$

then

$$h = \frac{2A}{B + b}; \qquad B = \frac{2A}{h} - b; \qquad b = \frac{2A}{h} - B$$

The perimeter of a trapezoid can be found only if the length of each side is known. The lengths of the nonparallel sides can be computed by means of trigonometry if certain angles are known.

25-10 OTHER POLYGONS

The areas of other polygons are usually computed by dividing the polygons into triangles, rectangles, or parallelograms. The area of each separate part is computed, and then these areas are added together for the total area of the polygon.

25-11 CONVERSION OF UNITS OF AREA

It is often desirable to change the size of units of area. Let us see how this is done. We use an example from our English system of measurements.

Figure 25-9

Consider a square 1 foot long and 1 foot wide (Fig. 25-9). Such a square contains 1 square foot of area. The length of the square is 12 inches and the width is 12 inches. The formula for the area of a square is

$$A = s^2$$

Substituting numerical values, we have

$$\text{area (1 sq ft)} = 12^2 \text{ sq in.}$$

or

$$1 \text{ sq ft} = 144 \text{ sq in.}$$

The square of a linear unit is often indicated by the exponent 2 placed on the linear unit itself. For example,

$$1 \text{ sq ft} = 144 \text{ sq in.}$$

is often written

$$1 \text{ ft}^2 = 12^2 \text{ in.}^2 \qquad \text{or} \qquad 1 \text{ ft}^2 = 144 \text{ in.}^2$$

In the same way we can say,

since $1 \text{ yd} = 3$ feet, $\quad 1 \text{ sq yd} = 3^2 \text{ sq ft}$; $\quad$ or $1 \text{ yd}^2 = 9 \text{ ft}^2$;

since $1 \text{ mile} = 320$ rods, $\quad 1 \text{ sq mi} = 320^2 \text{ sq rd}$; $\quad$ or $1 \text{ mi}^2 = 102{,}420 \text{ rd}^2$;

since $1 \text{ rd} = 16.5$ ft, $\quad 1 \text{ sq rd} = (16.5)^2 \text{ sq ft}$; $\quad$ or $1 \text{ rd}^2 = 272.25 \text{ ft}^2$.

Notice that when we multiply a *linear* measure, such as inches, by the same *linear* measure, we get *square units* or units of area.

The rule is easily remembered if we think of squaring both sides of an equation. For example, suppose we have the formula,

$$1 \text{ rd} = 5.5 \text{ yd}$$

Squaring both sides, $$1 \text{ rd}^2 = (5.5)^2 \text{ yd}^2$$

Note. It is well to remember that *one square foot of area* need not be square. A rectangle 6 inches wide and 2 feet long contains one square foot of area. In the same way, one square inch of area need not be a square. A triangle having a base of $\frac{1}{2}$ inch and an altitude of 4 inches contains 1 square inch of area. *A square inch of area may have any shape, even circular.*

Exercise 25-2

In each of the following, find the values indicated:

1. Triangle: base = 35 in., altitude 26.3 in. Find area.

2. Triangle: area = 100.1 sq in.; altitude = 13 in. Find base.

3. Triangle: sides: 9 in.; 10 in.; 11 in. Find area by Hero's formula.

4. Triangle: sides: 8.1 in.; 7.1 in.; 5.8 in. Find area by Hero's formula.

5. Triangle: area = 176.3 sq in.; base = 21.5 in. Find altitude.

6. Trapezoid: B = 18.6 in.; b = 14.2 in.; altitude = 13.5 in. Find area.

7. Trapezoid: B = 35 in.; b = 24 in.; altitude = 18.2 in. Find area.

8. Trapezoid: B = 26.4 in.; b = 19.2 in.; area = 376.2 sq in. Find altitude.

9. Trapezoid: B = 50.4 in.; area = 909.5 sq in.; altitude = 21.4 in. Find b.

10. The end of a box has the shape of a trapezoid having bases of 40 in. and 28 in., respectively, and an altitude of 28 in. Find the area of the end of the box.

11. A lawn has the shape of a trapezoid. The distance between the parallel sides is 45 ft. The two parallel sides have lengths of 80 ft and 52 ft. Find the number of square yards in the lawn.

12. A railroad track cuts diagonally across a rectangular field, forming a trapezoid. The parallel sides of the field measure 120 rd and 80 rd, respectively. The distance between the two parallel sides is 64 rd. Find the number of acres in the field.

13. The gable end of a house has the shape of a triangle, 24 ft across, and the altitude of the triangle is 9.5 ft. Find the area of the gable.

14. A rectangle contains 576 sq in. Find the perimeter of each of the rectangles formed if the following values are taken for the width:

(a) 24 in.	**(b)** 18 in.	**(c)** 16 in.	**(d)** 12 in.
(e) 9 in.	**(f)** 8 in.	**(g)** 6 in.	**(h)** 4 in.
(i) 3 in.	**(j)** 2 in.	**(k)** 1 in.	**(l)** 0.5 in.

QUIZ ON CHAPTER 25, MEASUREMENT OF POLYGONS. FORM A.

1. Find the perimeter and area of a rectangle whose length is 27.3 in. and whose width is 15.7 in.

2. A certain rectangle has an area of 849 sq in. The length of the rectangle is 32.4 in. Find the width and perimeter of the rectangle.

3. A certain parallelogram has a base of 23.4 in. and altitude of 16.5 in. Find the area of the parallelogram.

4. A parallelogram has an area of 494 sq in. and an altitude of 15.2 in. Find the base of the parallelogram.

5. Find the area of a triangle whose base is 16.5 in. and altitude 12.8 in.

6. A certain triangle has an altitude of 9.5 in. and an area of 76 sq in. Find the base of the triangle.

7. A certain trapezoid has bases of 45 in. and 33 in., and an altitude of 17 in. Find its area in square inches and in square feet.

8. A trapezoid has bases of 37 in. and 28 in. The area of the trapezoid is 780 sq in. Find the altitude.

9. Find the area of a square having one side equal to 13.5 in.

10. Find the length of one side and the perimeter of a square whose area is 256 sq in.

CHAPTER TWENTY-SIX The Right Triangle

26-1 DEFINITION

A triangle having one right angle is called a *right* triangle. Since one angle is a right angle, the other two angles are acute angles.

We know that the sum of the three angles of any triangle is 180°. Since the right angle itself is equal to 90°, the sum of the two acute angles must be 180° − 90°, or 90°. Therefore, *the two acute angles of a right triangle are complementary*. This, of course, applies to a triangle in a plane.

In a right triangle the side opposite the right angle is called the *hypotenuse.* The hypotenuse happens to be the longest side. The other two sides are sometimes called the *legs* of the right triangle. Since the legs are perpendicular to each other, if one is taken as the base of the triangle, then the other is the altitude.

26-2 THE PYTHAGOREAN RULE

A right triangle has the following important characteristic:

In every right triangle the square of the hypotenuse is equal to the sum of the squares of the legs.

This principle is known as the *Pythagorean rule.* It is one of the most useful rules in all mathematics. Let us see what it means.

Suppose we have the right triangle shown in Fig. 26-1. The two legs of the triangle measure 3 and 4 inches, respectively, and the hypotenuse is 5 inches long. The sum of the two legs, 3 + 4, does not equal the

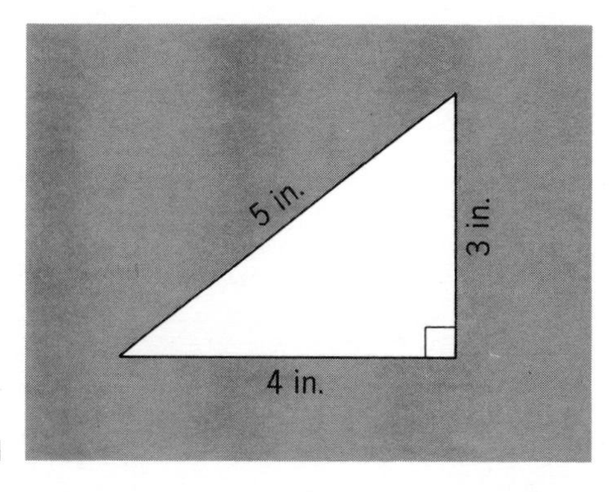

Figure 26-1

hypotenuse. However, if we square each leg and the hypotenuse, we have

$$3^2 = 9$$
$$4^2 = 16$$

Adding the squares, $5^2 = 25$

Notice that if the legs are squared and the squares are added, the sum 25 is the square of the hypotenuse.

The Pythagorean rule is named after Pythagoras, a Greek philosopher and mathematician who proved the truth of the principle about 500 B.C. The idea was known and used for specific examples before that, but no one before Pythagoras had proved the rule to be true for all right triangles. Many people often use the rule in their work without realizing why it is true. The truth of the rule has been proved in many different ways.

As an example of a proof in mathematics, we shall show why the Pythagorean rule is true. It is often called the *Pythagorean theorem*. The word *theorem* means a statement that is to be proved. In proving a theorem, we should first ask ourselves, "Is this true?" Then we proceed step by step and by careful reasoning to show that the statement must be true. We must give a reason for each statement we make. However, in the proof as given here, we shall have to make statements that are assumed to have been previously shown to be true.

One form of the proof of the Pythagorean theorem makes use of a trapezoid. Suppose we have the right triangle ABC (Fig. 26-2), with sides a, b, and c, respectively, in which c is the hypotenuse. Now we

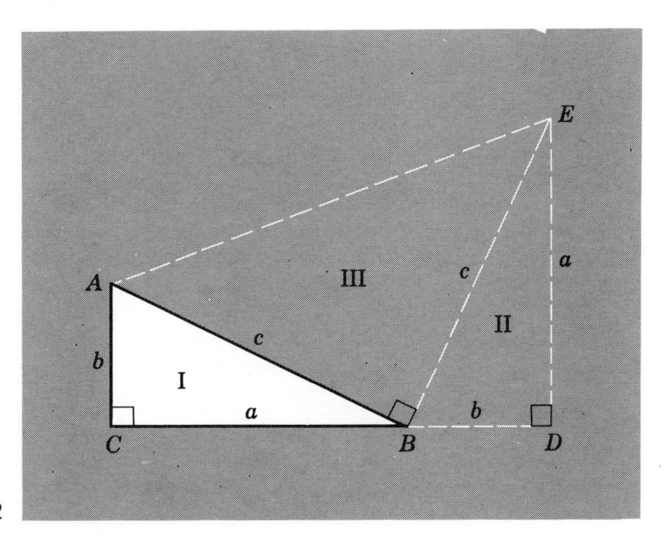

Figure 26-2

wish to show that

$$a^2 + b^2 = c^2$$

To show that this is true, we first extend side *CB* to *D*, making *BD* equal to *b*. (Construction lines are shown as broken lines.) Next, we draw *DE* perpendicular to *CD*, making *DE* equal to *a*. Finally, we draw *AE*.

Now we can say *CA* is parallel to *DE* because they are both perpendicular to the line *CD*. Then *ACDE* is a trapezoid, having bases of *a* and *b*, and an altitude of $(a + b)$. Then by the formula for the area of a trapezoid, we have

$$\text{Area of } ACDE = \tfrac{1}{2}(\text{altitude})(\text{sum of bases})$$

or

$$\text{Area} = \tfrac{1}{2}(a + b)(a + b) = \tfrac{1}{2}(a^2 + 2ab + b^2)$$

Now we see that the trapezoid is made up of three triangles. Their areas are

$$A(\text{I}) = \tfrac{1}{2}ab; \qquad A(\text{II}) = \tfrac{1}{2}ab; \qquad A(\text{III}) = \tfrac{1}{2}c^2$$

Adding the three triangles, we get a total area: $A = \frac{1}{2}(ab + ab + c^2)$. Since this total area is equal to the area of the trapezoid, we have

$$\tfrac{1}{2}(a^2 + 2ab + b^2) = \tfrac{1}{2}(ab + ab + c^2) \quad \text{or} \quad a^2 + 2ab + b^2 = 2ab + c^2$$

Subtracting $2ab$ from both sides, we get

$$a^2 + b^2 = c^2$$

The result is exactly what we set out to prove to be true.

The value of the Pythagorean rule lies in the fact that it enables us to find the hypotenuse of a right triangle if the two legs are known. For example, if the legs *a* and *b* are 5 inches and 12 inches, respectively, then the hypotenuse can be computed by the formula

$$c^2 = a^2 + b^2$$

Substituting numerical values,

$$c^2 = 5^2 + 12^2; \qquad c^2 = 169; \qquad \text{then} \qquad c = 13$$

If the hypotenuse and a leg are known, we make use of one of the following forms of the Pythagorean rule, involving subtraction:

$$a^2 = c^2 - b^2 \qquad \text{or} \qquad b^2 = c^2 - a^2$$

In using the Pythagorean rule, we shall often find that the length of an unknown side is not rational. For example, if the hypotenuse is

9 inches and one leg is 6 inches, then for the other leg we get

$$9^2 - 6^2 = 81 - 36 = 45; \qquad \text{the other leg is} \qquad \sqrt{45} = 3\sqrt{5}$$

Example 1

A telephone pole is to be braced with a wire, one end of which is fastened to the pole 32 feet above the ground and the other end to a stake in the ground 43 feet from the foot of the pole. Assuming that the ground is level and the pole is perpendicular to the ground, how long a wire will be required if 4 feet extra are needed for fastening?

Solution

In this problem we see that a right triangle is formed by the pole, the ground, and the wire. The pole and the ground form the legs of the right triangle and the wire represents the hypotenuse. Therefore, we square the given numbers and then add the squares. We get

$$\begin{aligned} 32^2 &= 1024 \\ 43^2 &= 1849 \\ \hline \text{Adding the squares,} \quad & \ 2873 \end{aligned}$$

The number 2873 represents the square of the hypotenuse. To find the hypotenuse, we take the square root of 2873:

$$\text{hypotenuse} = \sqrt{2873} = 53.6 \text{ ft (approx.)}$$

The answer is rounded off to three significant digits. Since 4 feet are needed for fastening, the length of wire required is 57.6 feet.

Example 2

One leg of a right triangle measures 23.5 in. and the hypotenuse is 31.4 in. Find the length of the other leg.

Solution

Since the hypotenuse, the longest side, is given, we square the two numbers and then find the difference between the squares. We get

$$\begin{aligned} 31.4^2 &= 985.96 \\ 23.5^2 &= 552.25 \\ \hline \text{Subtracting squares,} \quad & \ 433.71 \end{aligned}$$

The number 433.71 represents the square of the unknown leg. To find the length of the leg, we take the square root of 433.71:

$$\sqrt{433.71} = 20.8 \text{ in. (rounded off)}$$

Therefore, the other leg is approximately 20.8 in. long.

26-3 THE ISOSCELES RIGHT TRIANGLE

We have defined an isosceles triangle as one having two sides equal. If a right triangle has its two legs equal, then the triangle is also isosceles. Fig. 26-3 shows an isosceles right triangle with each of the legs equal to 6 inches, denoted by a and b, respectively.

In any isosceles triangle it can be proved that the *angles opposite the equal sides are also equal.* That is, if side a = side b, then we know that angle A is also equal to angle B.

In an isosceles right triangle the two acute angles are equal, and therefore each angle must be 45°. A right triangle of this particular shape is sometimes called a "45° right triangle."

Our problem here is to find the length of the hypotenuse. To find this length, we square the two legs, add the squares, and then find the square root of the sum. Adding the squares,

$$6^2 + 6^2 = 36 + 36 = 72$$

The number 72 represents the square of the hypotenuse. To find the hypotenuse, we take the square root of 72.

$$\sqrt{72} = \sqrt{(36)(2)} = 6\sqrt{2} \qquad \text{or} \qquad 6(1.414) = 8.484 \text{ in.}$$

An isosceles right triangle appears when we draw the diagonal of a square. Suppose we have the square $ABCD$ with diagonal d and each side s (Fig. 26-4). Then triangle BCD is an isosceles right triangle.

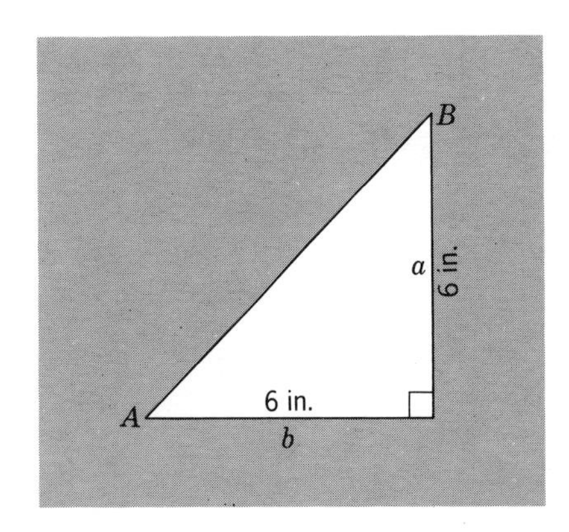

Figure 26-3

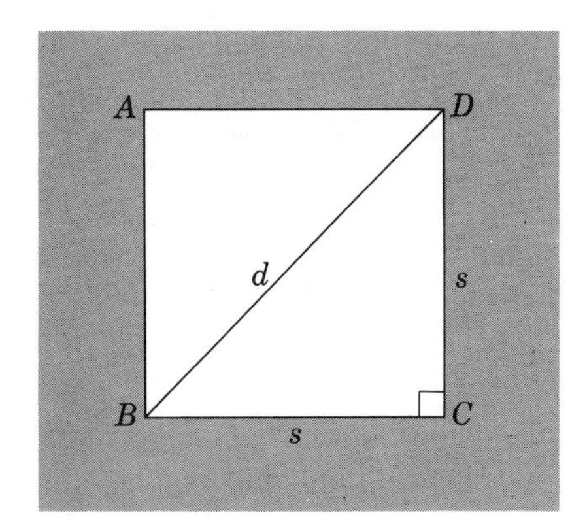

Figure 26-4

Each leg of the triangle is a side of the square, and the hypotenuse is the diagonal. Then by the Pythagorean rule, we have

$$d^2 = s^2 + s^2 \qquad \text{or} \qquad d^2 = 2s^2$$

Then

$$d = \sqrt{2s^2} \qquad \text{or} \qquad d = s\sqrt{2}$$

Stated in words: *The diagonal of any square is equal to the length of a side multiplied by the square root of* 2.

26-4 THE 30°-60° RIGHT TRIANGLE

A special triangle that occurs often in engineering mathematics is a right triangle whose acute angles are 30° and 60°, respectively. A triangle of this particular shape is often called a "30°-60° right triangle" (Fig. 26-5), Such a triangle is the kind that Plato, a famous Greek philosopher, called the "most beautiful scalene right-angled triangle." A very important fact concerning such a triangle is the following:

If any 30°-60° *right triangle, the hypotenuse is always twice as long as the shortest side.*

By use of this fact, we can find all the sides of such a triangle, provided that we know one side.

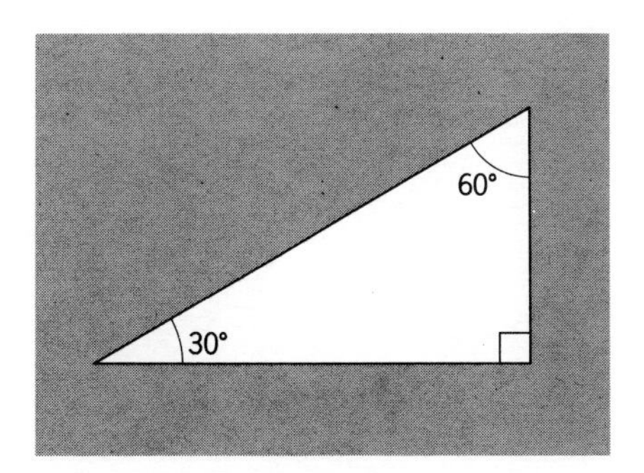

Figure 26-5

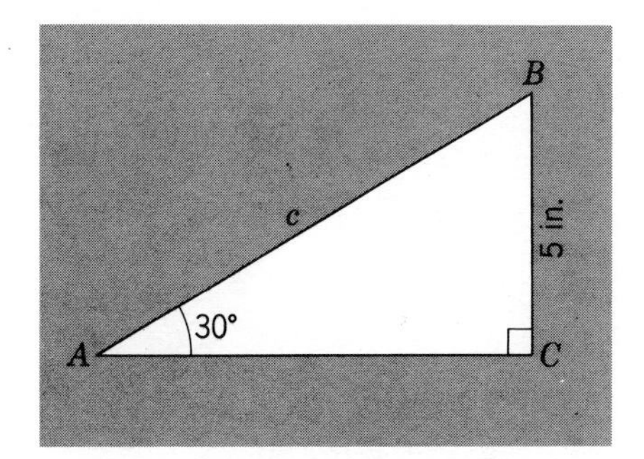

Figure 26-6

Example

Suppose we have a 30°-60° right triangle, ABC (Fig. 26-6), with the shortest leg equal to 5 in. and angle $A = 30°$. By the foregoing rule, we know that the hypotenuse is twice the side BC, or 10 in. Since the hypotenuse is 10 in. and one leg is 5 in., we can find the other leg by the Pythagorean rule. Squaring both numbers, we have

$$\begin{aligned} 10^2 &= 100 \\ 5^2 &= \underline{\ 25} \end{aligned}$$

Subtracting squares,

$$75$$

The difference, 75, represents the square of the unknown leg. Therefore, the leg is equal to the square root of 75:

$$\sqrt{75} = \sqrt{(25)(3)} = 5\sqrt{3} = 8.66 \text{ (approx.)}$$

Then area $= (\frac{1}{2})ab = 21.65$; perimeter $= 23.66$ (approx.)

26-5 THE EQUILATERAL TRIANGLE

An *equilateral* triangle is a triangle in which all three sides are equal. We have said that if two sides of a triangle are equal, then the angles opposite these sides are also equal. From this fact it follows that an equilateral triangle is also *equiangular.* Therefore, each angle in an equilateral triangle is 60° (Fig. 26-7).

With regard to the equilateral triangle, there are two problems with which we are often concerned. One is the altitude and the other is the area. We shall show how to obtain the formula for each.

Suppose we have an equilateral triangle ABC with a side equal to 8 inches (Fig. 26-7). We draw a straight line bisecting the angle C into two equal angles. The line meets the opposite side at point D. Since angle C is divided into two equal angles, each one of these is 30°.

In the figure triangle DBC is a 30°-60° right triangle. The hypotenuse of this right triangle is one side of the original equilateral triangle, which is 8 inches. Therefore, the short leg of the right triangle DBC is 4 inches.

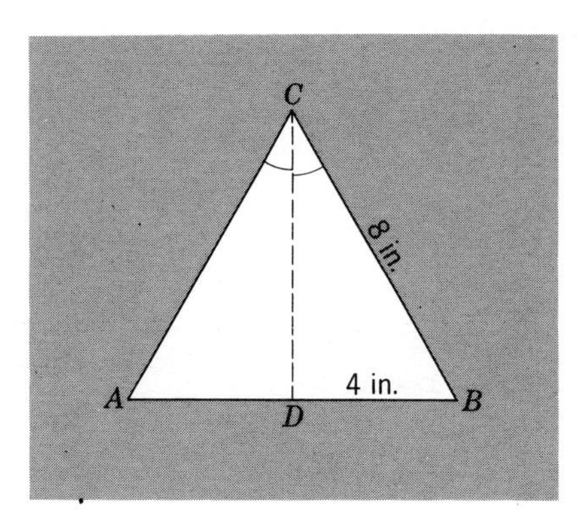

Figure 26-7

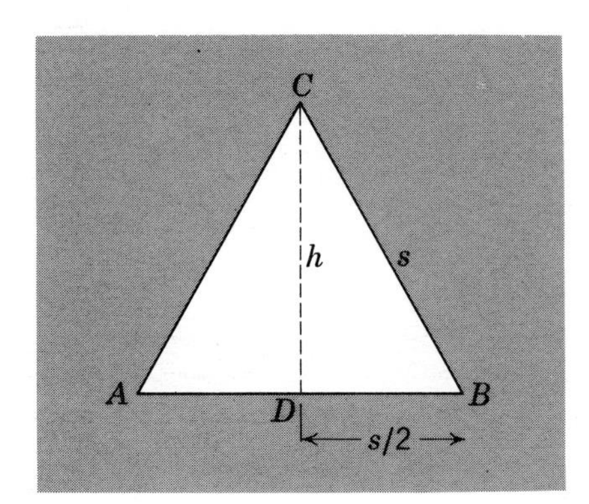

Figure 26-8

Now, in order to find the altitude of the equilateral triangle, we use the Pythagorean rule. Squaring, we have

$$8^2 = 64$$
$$4^2 = 16$$

Subtracting squares, $\quad 48$

The number 48 represents the square of the other leg, which, in this example, is the altitude of the equilateral triangle. Therefore, the altitude is equal to

$$\sqrt{48} = \sqrt{(16)(3)} = 4\sqrt{3}$$

By using a general equilateral triangle, ABC (Fig. 26-8), with one side represented by s, we can show that the altitude, h, of any equilateral triangle is equal to one-half one side multiplied by the square root of 3. In the right triangle *BCD*, the hypotenuse is *s*, and one leg is *s*/2. Squaring these two sides and subtracting, we get

$$h^2 = s^2 - \left(\frac{s}{2}\right)^2 = s^2 - \frac{s^2}{4} = \frac{3s^2}{4}; \qquad h = \sqrt{\frac{3s^2}{4}}; \qquad h = \frac{s}{2}\sqrt{3}$$

To find the area of an equilateral triangle, we simply use the formula for the area of any triangle,

$$A = \tfrac{1}{2}hb$$

Notice that the base in the general equilateral triangle is equal to *s*, or one side. Since we have the altitude, $h = \frac{1}{2}s\sqrt{3}$, we use these values to get the formula for the area:

$$A = \tfrac{1}{2}hb$$

Substituting values,

$$A = \tfrac{1}{2}(\tfrac{1}{2}s\sqrt{3})(s)$$

or

$$A = \tfrac{1}{4}s^2\sqrt{3}$$

That is, the area of an equilateral triangle is one-fourth the square of a side multiplied by the square root of 3.

Exercise 26-1

In each of the following right triangles, find the required quantity:

1. Hypotenuse = 17.2 in., one leg = 13.8 in. Find the area and the perimeter.

2. One leg = 24.3 cm, the other leg = 34.1 cm. Find the area in square inches and the perimeter in inches.

3. The legs are 3.42 and 5.13 cm, respectively. Find the area and the perimeter.

4. Hypotenuse = 26.9 in., one leg = 22.5 in. Find the area in square centimeters and the perimeter in centimeters.

5. Hypotenuse = 160 ft, one leg = 80 ft. Find the area in square yards, and the perimeter in yards.

6. How long a wire will be needed to brace a telephone pole if the wire is fastened to the pole 32 ft from the ground and to a stake in the ground 54 ft from the foot of the pole?

7. A TV broadcasting tower is braced with a cable 360 ft long. The cable is fastened in the ground at a point 270 ft from the tower. How far up on the tower is the cable fastened?

8. How long a diagonal brace is required for a gate 5 ft high and 16 ft long?

9. Two cars start at the same point, one traveling north at 45 mph and the other east at 34 mph. How far apart are they after 30 min?

10. A man rows his boat directly across a river at right angles to the current. The river is 800 ft wide. If he rows at 200 ft per min and the river flows at a rate of 150 ft per min, how fast does he travel and in what direction? How long will it take him to cross the river? How far does he travel?

11. A field is 60 rd long and 30 rd wide. A road extends diagonally from one corner to the opposite corner. How long is the road?

12. A ladder 32 ft long leans up against a wall. How high up on the wall does the ladder reach if the foot is placed 7.2 ft from the wall?

13. A ladder is 28 ft long. How far from the wall must the foot be placed in order that the ladder will just reach a window 26.4 ft from the ground.

14. The length of a rectangle is 42.3 cm. The diagonal is 53.4 cm. Find the area and perimeter of the rectangle.

15. Find the length of the diagonal of each of the following squares:

(a) A side = 13.2 in.
(b) A side = 28.7 cm
(c) A side = 3.52 m
(d) A side = 1000 m
(e) A side = 4570 mm
(f) A side = 5280 ft
(g) A side = 360 rd
(h) A side = 16.5 ft
(i) A side = 5.5 yd
(j) The perimeter = 39 in.
(k) The perimeter = 54.3 cm
(l) The perimeter = 13.2 mm
(m) The perimeter = 1 ft

16. Find the area and perimeter of each of the following 30°-60° right triangles:

(a) The shortest side = 7.4 in.
(b) The hypotenuse = 25.2 cm
(c) The hypotenuse = 485 ft
(d) The shortest side = 3.57 mm
(e) The hypotenuse = 6.73 in.
(f) The hypotenuse = 21.47 cm

17. Find the area of each of the following equilateral triangles:

(a) A side = 18 in.
(b) A side = 13.4 cm
(c) The perimeter = 42 in.
(d) The perimeter = 23.4 cm
(e) A side = 72.4 ft

QUIZ ON CHAPTER 26, THE RIGHT TRIANGLE. FORM A.

1. A right triangle has sides of 15 in. and 20 in., respectively. Find the hypotenuse, the perimeter, and the area of the triangle.
2. A right triangle has sides of 8 in. and 12 in., respectively. Find the hypotenuse, the perimeter, and the area of the triangle.
3. A right triangle has one side equal to 10 in. and the hypotenuse equal to 26 in. Find the unknown side, the perimeter, and the area.
4. The hypotenuse of a right triangle measures 23.6 in. and one side measures 15.8 in. Find the perimeter and the area of the triangle.
5. An isosceles triangle has a base of 10 in. and altitude of 12 in. Find the perimeter and the area of the triangle.
6. Find the length of the diagonal of a rectangle whose length is 10 in. and whose width is 5 in.
7. Find the length of the diagonal of a square whose side is 18 in. long.
8. Find the length of the diagonal of a square whose perimeter is 50 in.
9. An equilateral triangle has a side equal to 16 in. Find the altitude, perimeter, and area of the triangle.
10. To brace a light pole, a wire cable is fastened to the pole 45 ft above the ground. The other end of the cable is fastened to a stake in the ground 24 ft from the foot of the pole. If an extra 6 ft of cable is required for fastening at both ends, how long a cable is needed? Assume the pole is perpendicular to the ground, and the ground is level.
11. In a certain 30°-60° right triangle, the shortest side is 9 in. Find the area and the perimeter of the triangle.

CHAPTER TWENTY-SEVEN Circles

27-1 DEFINITIONS

The circle is one of the most common as well as the most useful of all geometric figures. It is used for its beauty in architecture and design. It has a practical and necessary use in the wheels of machinery in industry. In fact, the wheel, which is a circle, was one of the most important inventions in the history of civilization.

Although the circle is a very common geometric figure, it is not easy to define. We might say that the *circle is a plane closed curve such that every point on the curve is the same distance from a point within, called the center.* According to this definition, the circle is the curve itself (Fig. 27-1). For the word *circle* we use the symbols ⊙, Ⓢ.

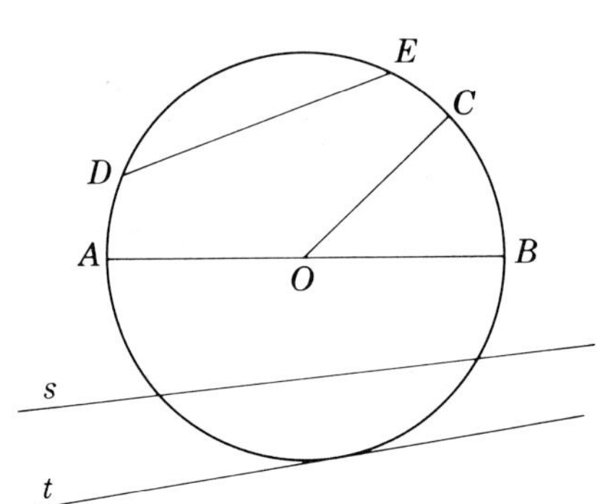

Figure 27-1

The *circumference* of a *circle* is the length of the curve. If the curve itself is called the circumference, then the circle is taken to mean the space enclosed within the circumference.

A straight line segment, such as AB, with its ends on the circumference and passing through the center, is called a *diameter* of the circle. A straight line segment, such as OC, joining the center with any point on the circumference, is called a *radius* (plural, *radii*). From these definitions, we see that a diameter is equal in length to twice the radius. If D represents the number of units of length in the diameter and r represents the number of units of length in the radius, then we have the formulas

$$D = 2r \qquad \text{and} \qquad r = \frac{D}{2}$$

Since the definition of a circle states that all the points on the circumference are the same distance from the center, it follows that *all radii of a circle are equal.* Moreover, since all radii of a circle are equal, it follows that, if the center and the radius are known, a circle can be drawn by means of an instrument called a *compass.*

A *chord* is a straight line segment, such as DE, whose ends lie on the circumference. A chord has a limited length. The longest chord that can be drawn in a circle is a diameter. A *secant line* is a straight line that cuts through the circle in two points, such as line s. A chord is a portion of a secant line. A *tangent line* is a line that touches a circle in only one point, such as line t. It has an unlimited extent.

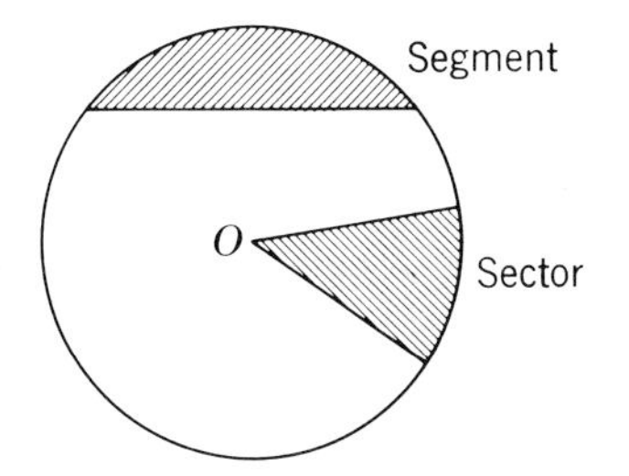

Figure 27-2

A *semicircle* is a half circle. A semicircumference is half a circumference. An *arc* is part of a circumference, such as arc BC. Any two points on a circumference divide the circumference into two arcs. If one arc is longer than the other, the longer arc is called the *major arc* and the other is called the *minor arc.* An arc can be named by the letters at the end of the arc. The symbol for the arc BC is $\widehat{BC}$. If a major arc is meant, it must be indicated by using three letters, such as $\widehat{BAC}$. The midpoint of an arc is a point that divides the arc into two equal arcs. However, an arc is also said to have a *center.* By the *center of an arc* we mean the center of the circle of which the arc is a part. For instance, the center of the arc DE is the point O (Fig. 27-1).

A *sector* of a circle is a portion of the area of the circle bounded by an arc and two radii (Fig. 27-2). A *segment* of a circle is a portion of the area bounded by an arc and a chord. *Concentric* circles are two or more circles that have the same center but different radii (Fig. 27-3).

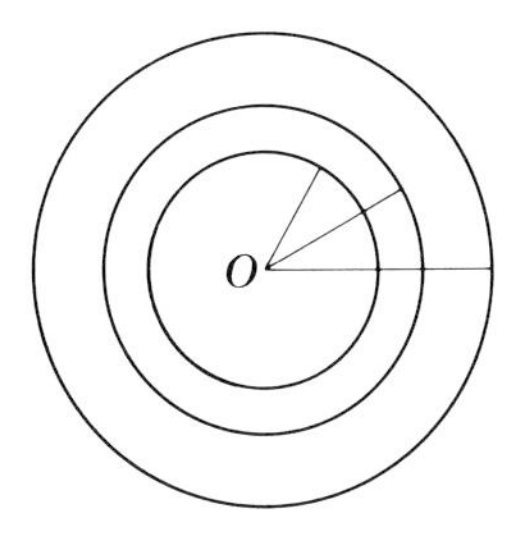

Figure 27-3

An angle formed by two radii at the center of a circle cuts off, or *intercepts,* an arc on the circumference.

We have said that an angle is sometimes defined as the amount of opening between two lines drawn from the same point. If the amount of opening is only $\frac{1}{360}$ of one complete rotation, we call the size of the angle one degree (1°). One degree is really a small angle. We can call this small angle *one angle degree* (Fig. 27-4).

An angle formed by two radii at the center of a circle cuts off, or

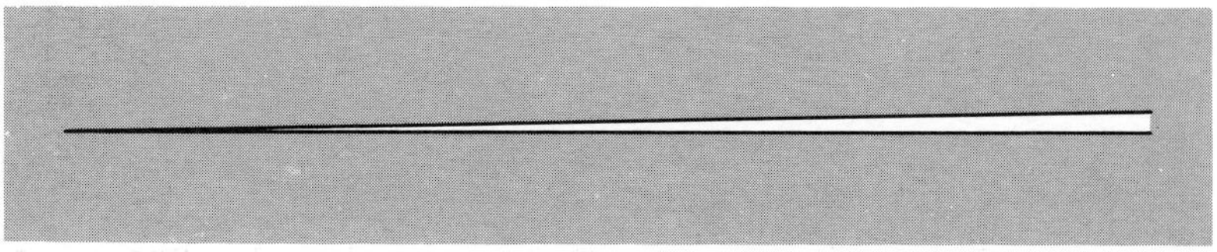

Figure 27-4. An angle of approximately one angle degree.

intercepts, an arc on the circumference. An *arc degree* is the small arc cut off the circumference of a circle *by one angle degree at the center.* Therefore, one arc degree is $\frac{1}{360}$ of the entire circumference. In order to distinguish between one *angle degree* and one *arc degree,* try to think of *one angle degree* as a *small angle* and *one arc degree* as a very *small portion of the circumference* (Fig. 27-5).

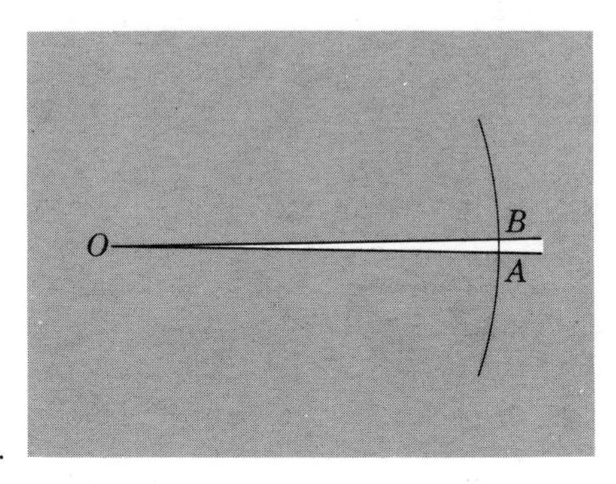

Figure 27-5. $\widehat{AB}$ is one arc degree.

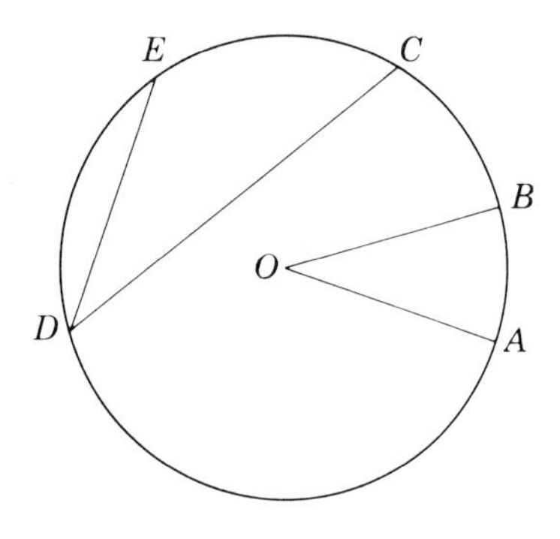

Figure 27-6

An angle formed by two radii at the center of a circle is called a *central angle.* In Fig. 27-6 angle AOB is a central angle. If a central angle contains 10 *angle degrees,* then it cuts off an arc of 10 *arc degrees* on the circumference. Whatever the size of a central angle in a circle, it will intercept an arc that contains the same number of arc degrees as the angle contains angle degrees. For this reason, we can say that in any circle a *central angle is equal, in the number of degrees, to the intercepted arc.*

We must be careful about how we make this statement. When we say a central angle of 25 angle degrees intercepts an arc of 25 arc degrees, we might be tempted to say that the angle equals the arc. Yet an angle cannot equal an arc. The two things are not comparable. It is no more correct to say that an angle equals an arc than to say, for instance, that your weight is equal to one week. Let us suppose you weigh 168 pounds. There are 168 hours in one week. How can we state any equality between these two quantities? The only thing equal about them is the number 168. We can say that the number of hours in one week is the same as the number of pounds in your weight. In any other manner the two quantities are not equal at all. In the same way, an angle cannot equal an arc. We can say, however, that the number of *angle degrees* in a central angle is equal to the number of *arc degrees* in the intercepted arc. In this manner *a central angle is said to be measured by the intercepted arc.*

An *inscribed angle* is an angle formed by two chords drawn from the same point on the circumference of a circle. In Fig. 27-6 angle CDE is an inscribed angle. *An inscribed angle is measured by one-half the intercepted arc.* This means that an inscribed angle contains one-half as many angle degrees as the number of arc degrees in the intercepted arc. If an inscribed angle intercepts an arc of 80 arc degrees, then the angle contains 40 angle degrees. If an inscribed angle contains 30 angle degrees, then it intercepts an arc of 60 arc degrees. Any angle inscribed in a semicircle intercepts an arc of 180°. The angle is therefore 90°, or a right angle.

27-2 MEASUREMENTS IN A CIRCLE: CIRCUMFERENCE

In any circle we are usually concerned with measurements that involve the diameter, the radius, the circumference, and the area. In the formulas concerning the circle we use the following notation:

C represents the number of *linear units* in the circumference.
D represents the number of *linear units* in the diameter.
r represents the number of *linear units* in the radius.
A represents the number of *square units* in the area.

We have already pointed out the fact that since the diameter is equal to twice the radius we have the two formulas

$$D = 2r \qquad \text{and} \qquad r = \frac{D}{2}$$

The circumference of any circle always has a fixed ratio to the diameter. This ratio is called π (pronounced *pie*). The value of π is approximately equal to $3\frac{1}{7}$. It is more nearly equal to 3.1416. However, you should remember that this number, 3.1416, is *not* the exact value of π. For a long time people believed that π could be written as an exact decimal or common fraction. Now it is known that the exact value of π cannot be so written. It is an unending decimal fraction. The value to 30 decimal places is

3.141592653589793238462643383280 (rounded off)

There is seldom any need for this extreme accuracy. In most problems in science π is rounded off to 3.14159 or 3.1416 (on a slide rule we use the value 3.14). As a student you should understand and use both 3.1416 and 3.14 from time to time. Then, in an actual problem in your work, you should use the value that in your judgment is the best value for that particular problem.

If we know the diameter of a circle, we can find the circumference by multiplying the diameter by the number 3.1416, or 3.14, depending on the degree of accuracy we want. Representing this number by π, we have the formula

$$C = \pi D$$

Since the diameter is twice the radius or $D = 2r$, we can substitute $2r$ for D in the formula and get, upon rearranging the values,

$$C = 2\pi r$$

The second form of the formula for the circumference, $C = 2\pi r$, is the most useful form in most mathematics and engineering.

These formulas, of course, can be used in reverse; that is, if the circumference of a circle is known and we wish to find the diameter or radius, we reverse the procedure. If we know the circumference of a circle, we divide the circumference by π to find the diameter. If the circumference of a circle is measured as 74.56 inches., then

$$D = 74.56 \div \pi = 74.56 \div 3.1416 = 23.733 \text{ in.}$$

The radius can be found by taking $\frac{1}{2}$ of the diameter.

The reverse formulas can be written as follows:

$$D = \frac{C}{\pi} \qquad r = \frac{C}{2\pi}$$

27-3 AREA OF A CIRCLE

Perhaps many people still remember from their grade school days the formula $A = \pi r^2$ for the area of a circle. This formula can be derived exactly only by more advanced mathematics. However, the meaning of the rule is not difficult to understand. Let us see why the rule is reasonable.

If we imagine a circle divided into any number of sectors, resembling triangles (Fig. 27-7), the altitude of these triangles is equal to the radius of the circle, and the sum of the bases is equal to the circumference. The total area of the triangles can then be found by the rule for the area of a triangle. The area is equal to $\frac{1}{2}$ the altitude times the sum of the many bases; or,

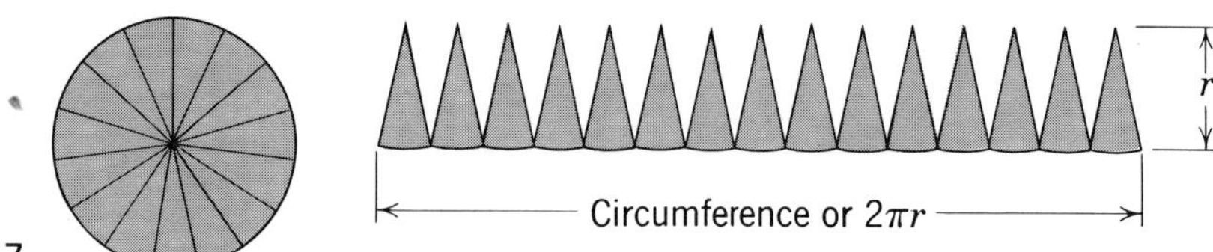

Figure 27-7

$$A = \tfrac{1}{2}(\text{radius})(\text{circumference})$$

$$A = \tfrac{1}{2}(r)(2\pi r)$$

$$A = \pi r^2$$

Of course, in this reasoning each sector is not a true triangle.

However, if we imagine that the number of sectors increases, the base of each sector becomes a smaller and smaller arc that approaches the condition of a straight line. The sum total of the bases of the sectors is still equal to the circumference of the circle.

If the area of a circle is known, the formula can be reversed, and we get

$$r^2 = \frac{A}{\pi} \qquad \text{or} \qquad r = \sqrt{\frac{A}{\pi}}$$

Another formula for the area of a circle is obtained by substituting the value $D/2$ for the radius r. The formula then becomes

$$A = \pi\left(\frac{D}{2}\right)^2 \qquad \text{or} \qquad A = \frac{\pi}{4}D^2$$

This formula can be used when the diameter is known. When the formula involving the diameter is used, then $\pi/4$ is usually taken as 0.7854. The area of a circle is therefore equal to 0.7854 times the area of the circumscribed square, or $A = 0.7854D^2$.

27-4 AREA OF A RING

By the area of a ring we mean the area between two concentric circles. It is often necessary to find such an area, as, for instance, in finding the area of a 4-foot sidewalk laid around a circular fountain, 20 ft in diameter (Fig. 27-8). If the area of the inner circle is subtracted from the area of the outer or larger circle, the remainder is the area of the ring.

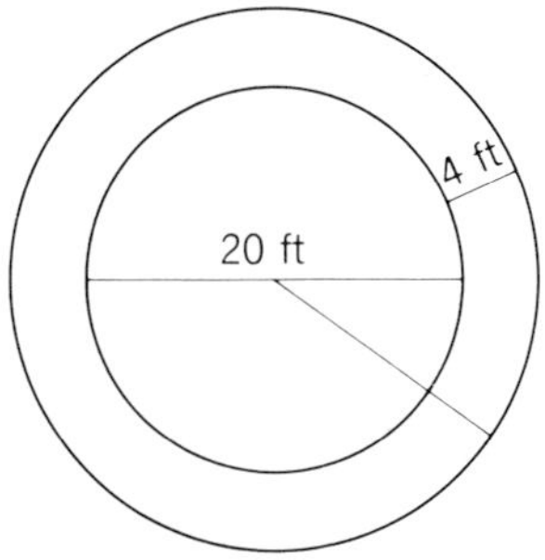

Figure 27-8

The method may be stated as a formula. We let R represent the radius of the outside or large circle, and r represent the radius of the small inner circle. Now, if we let A_L represent the area of the large circle and A_S represent the area of the small circle, we can say

$$A_L = \pi R^2 \qquad \text{and} \qquad A_S = \pi r^2$$

(area of large circle) (area of small circle)

Then the area of the ring can be stated as the difference between the areas of the two circles:

$$A \text{ (of ring)} = \pi R^2 - \pi r^2$$

The formula can be written: A (of ring) $= \pi(R^2 - r^2)$, or in factored form:

$$A = \pi(R + r)(R - r).$$

In the problem in Fig. 27-8 the radius R of the large circle is 14 ft; the radius r of the smaller circle is 10 ft. Using the formula

$$A = \pi(R^2 - r^2)$$

we have

$$\begin{aligned} A &= \pi(196 - 100) \\ &= \pi(96) \\ &= (96)(3.1416) \\ &= 301.6 \text{ (rounded off)} \end{aligned}$$

If the ring is very narrow (for instance, a ring $\frac{1}{16}$ of an inch wide around a circle 6 in. in diameter), then the approximate area of the ring can be found by a short method (Fig. 27-9).

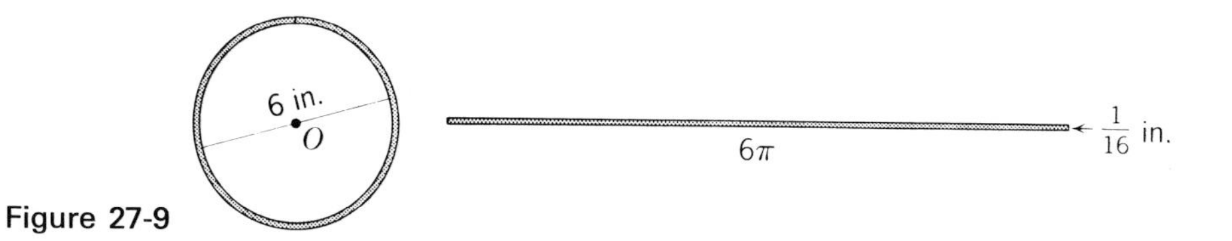

Figure 27-9

Let us assume that we cut the ring at some point and lay it out along a straight line. We can say that the ring is now a long narrow rectangle. The width of the rectangle is $\frac{1}{16}$ of an inch and the length is equal to 6π inches, the circumference of the circle. We can find the area of this rectangle by multiplying the length by the width; that is,

$$A = (\tfrac{1}{16})(6\pi) = 1.178 \text{ sq in.}$$

The error, of course, is in assuming that the ring forms a perfect rectangle. However, the difference is slight. In fact, if the area of the ring were originally computed as the difference between two circles, it would be 1.190 sq in. The short method is in error by 0.012 sq in. If the ring were still narrower, or the circle larger, then the error would be much less. If the ring were wide, as compared with the size of the circle, then the short method would result in an error too great to be useful.

27-5 AREA OF A SECTOR

The area of a sector can be computed if the number of degrees in the angle of the sector and the radius of the circle are known. We know that a central angle of 90° intercepts an arc equal in length to one-fourth of the circumference. In fact, an intercepted arc bears the same relation to the entire circumference as the central angle does to 360°. In the same way, the area of a sector has the same ratio to the entire circle as the central angle has to 360°.

As an example, suppose a sector has a central angle equal to 20° in a circle whose radius is 12 in. We know the circumference is equal to (2)(3.1416)(12) = 75.3984 in. The area of the entire circle is equal to 452.3904 sq in. The sector contains an arc that is $\frac{20}{360}$ of the circumference, or 4.1888 in. The area of the sector is $\frac{20}{360}$ of the area of the circle, or 25.1328 sq in.

27-6 AREA OF A SEGMENT

A segment is a part of a circle bounded by a chord and an arc. There is no simple accurate formula for the area of a segment of a circle. One of the simplest formulas depends on the *width* and the *height* of the segment. The width w is the length of the chord forming part of the boundary of the segment. The height h is the perpendicular distance from the midpoint of the chord to the arc. The height represents the greatest distance of the arc from the chord. A formula for a good approximation for the area of a segment is the following:

$$A = \frac{2}{3}hw + \frac{h^3}{2w}$$

Exercise 27-1

1. Find the area and the circumference of each of the following:
   (a) A round table top 42 in. in diameter.
   (b) A circle made with a compass set at 3.25 in.
   (c) A circular window $4\frac{1}{2}$ ft in diameter.
   (d) A circular mirror $27\frac{3}{4}$ in. in diameter.
   (e) A circular signboard 8 ft, 10 in. in diameter.
   (f) A circular clock dial whose radius is 4.75 in.
   (g) A skating rink 120 ft in diameter.
   (h) A coin whose diameter is 1.24 in.

2. The diameter of a drumhead is 26.3 in. Find the number of square inches in both heads.

3. A circular race track has a radius of 136. ft. How many turns does it take to run a mile?

4. On a merry-go-round you sit 16 ft from the center post. How far do you ride in making 20 complete revolutions?

5. The distance around a large tree is 11 ft, 4 in. What is the diameter of the tree?

6. A cream separator bowl makes 165 revolutions per second (rps). The diameter of the bowl is 4.5 in. What is the speed of a point on the rim of the bowl in miles per hour?

7. A bicycle wheel has a radius of 13 in. How many turns will it make in going a mile if no allowance is made for slipping?

8. An automobile tire is 28.4 in. in diameter and the wheel makes 3 revolutions per second. How far does the car move in 1 hr if no allowance is made for slipping?

9. A circular race track is 770 ft long. Find the area of the land enclosed by the track.

10. A locomotive wheel is 68 in. in diameter. If no allowance is made for slipping, how many revolutions does the wheel make per second when the train travels 60 mph?

11. A circular skating rink has a radius of 64 ft. The rink is surrounded by a 10-ft sidewalk. Find the number of square feet in the sidewalk.

12. A circular fountain having a diameter of 23 ft, 4 in. is surrounded by a cement walk 9 ft wide. Find the area of the walk in square yards.

13. The area of a circular skating rink is approximately 10,000 sq ft. What is its diameter?

14. A circular running track is $\frac{1}{8}$ mile long. How many square rods are enclosed by the track?

15. A belt pulley makes 30 rps. The pulley is 8.5 in. in diameter. If slipping is disregarded, how fast is the belt traveling in miles per hour?

16. A circle is inscribed in a square that is 16 in. on a side. Find the area of each corner of the square that is outside the circle.

17. Four concentric circles are drawn with radii of 4, 6, 8, and 10 in., respectively. Find the area of the innermost circle and the area between each of the other circles; that is, the area of each ring. (Concentric circles are circles having the same center.)

18. A certain 12-in. record plays for 15 min at a rate of $33\frac{1}{3}$ revolutions per minute. The center hole measures $\frac{9}{32}$ in. across. The distance from the edge of the hole to the outside of the recording is $5\frac{23}{32}$ in., and the distance from the edge of the hole to the inside of the recording is $2\frac{19}{32}$ in. What is the width of each groove in the record?

19. Two ventilating pipes, each 6 in. in diameter, are joined to form one single large pipe having the same capacity as the combined capacity of the two single pipes. What is the diameter of the single large pipe?

20. A square tube 8 in. on a side is to be replaced with a circular tube having the same capacity as the square tube. What is the diameter of the circular tube?

21. A circular water main 24 in. in diameter branches off into two equal circular mains having the same combined capacity as the larger main. What is the diameter of each of the smaller mains?

22. If a 24-in. circular water main branches off into 4 smaller circular mains with the same total combined capacity as the 24-in. main, what is the diameter of each of the smaller mains?

23. If a wire $\frac{1}{16}$ in. in diameter carries 6 amperes under a certain electromotive force (emf), what current will a wire $\frac{1}{32}$ in. in diameter carry under the same emf? (Current carrying capacity is proportional to the cross-sectional area.)

24. The hour hand of a clock is 10.3 cm long. How far does a point on the tip travel between 2 o'clock and 3 o'clock? How much area does the hand sweep over in 1 hr?

25. A circle 20 in. in diameter has a chord 16 in. long. How far is the center of the circle from the chord? Find the area of the segment cut off by the chord.

26. How much area can be enclosed by a wire 60 in. long if it is bent to form each of the following:

(**a**) A rectangle 4 in. wide
(**b**) A rectangle 5 in. wide
(**c**) A rectangle 8 in. wide
(**d**) A rectangle 10 in. wide
(**e**) An equilateral triangle
(**f**) A square
(**g**) A regular hexagon
(**h**) A circle

27. A company wishes to put up a fence around a circular storage tank 32 ft in diameter. If the fence is to be placed 6 ft from the tank, what will the length of the fence be? What will the area between the tank and the fence be?

28. A golfer has decided to put in a circular putting green in his back yard. What will the area of the green be if the diameter is 24.5 ft?

29. A ship's rudder has jammed and the ship is traveling in a circle with a radius of $\frac{1}{2}$ mile at the rate of 5 mph. How many minutes will it take for the ship to travel once around its circular path?

30. Three circular table tops each 4 ft in diameter are to be cut from a single board measuring 4 by 14 ft. What will the area of the scrap material be?

31. A circle is inscribed in a square that is 18 in. on a side. Find the area of each corner of the square that is outside the circle.

32. Find the area of a sector of 40° in a circle whose diameter is 12 in.

33. Find the area of a circular ring whose outside diameter is 24 in. and whose inside diameter is 23 in.

34. Four circles each having a diameter of 6 in. are drawn inside a square whose side is 24 in., so that the circles just touch one another. Find the area inside the square but outside the circles.

35. A circle is circumscribed about a square whose side is 8 in. Find the area of each portion of the circle that is outside the square.

36. One circle has a radius of 5 in. and another circle has a radius of 15 in. Find the ratio of their diameters, their circumferences, and their areas.

QUIZ ON CHAPTER 27, CIRCLES. FORM A

For each of the following circles, find the values indicated.

1. Circle: radius = 6 in. Find the diameter, circumference, and area.

2. Circle: diameter = 22 in. Find radius, circumference, and area.

3. Circle: circumference = 75π. Find diameter, radius, and area.

4. Circle: area = 900π. Find radius, diameter, and circumference.

5. A circle is inscribed inside a 10-in. square. Find the area of the four corners of the square that are outside the circle.

6. Find the area of a sidewalk 4 ft wide, built around a fountain that has a diameter of 30 ft.

7. Find the area of a 45° sector of a circle whose diameter is 32 in.

8. One circle has a diameter of 3 in., and another circle has a diameter of 6 in. Find the ratio of the radius of the first to the radius of the second, the ratio of their circumference, and the ratio of the areas.

CHAPTER TWENTY-EIGHT Geometric Solids: Prisms

28-1 POLYHEDRON

The figures we have studied up to this point do not occupy space. All can be drawn on a plane and are understood to have no thickness. *Solid geometry*, or *space geometry*, as it is sometimes called, is the study of figures that occupy space. Such figures are called *geometric solids.*

A *polyhedron* is a geometric solid bounded by planes. (The prefix "poly" means many, and "hedron" refers to faces.) The plane surfaces of a polyhedron are called *faces.* Any two faces intersect in a straight line called an *edge.* The intersection of three or more edges is called a *vertex.* Figure 28-1 shows several polyhedrons.

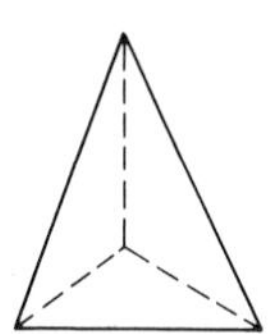
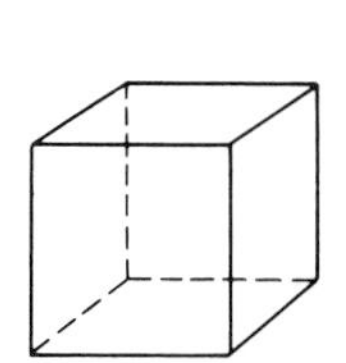
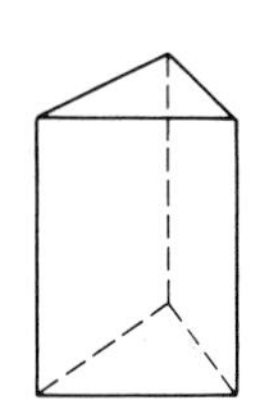
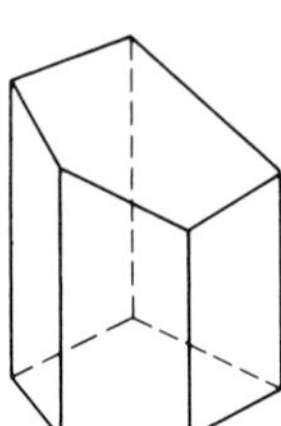

Figure 28-1. Polyhedrons.

28-2 PRISM

A *prism* is a special kind of polyhedron. In a prism two of the faces, called *bases*, are congruent and parallel polygons. The other faces, called *lateral faces*, are parallelograms formed by joining the corresponding vertices of the bases with straight lines. Figure 28-2 shows several kinds of prisms.

The bases of a prism may be triangles, quadrilaterals, pentagons, or any other kinds of polygon. Notice that the two bases, or two ends, of each prism are polygons of exactly the same size and shape. The lateral faces are parallelograms.

A *right* prism is a prism having its edges perpendicular to the bases. (Figs. 28-2*a*, *c*, *d*, *f*). If the edges are not perpendicular to the bases, the prism is called an *oblique* prism (Figs. 28-2*b*, *e*, *g*).

The most common prism is the *rectangular solid*, such as an ordinary box. Such a prism (Figs. 28-2*c*, *d*) has six faces, all of which are rectangles. A *cube* is a special rectangular solid whose faces are squares. A cube is an example of a solid called a *regular polyhedron.* In a regular polyhedron all the angles are equal and all the faces are

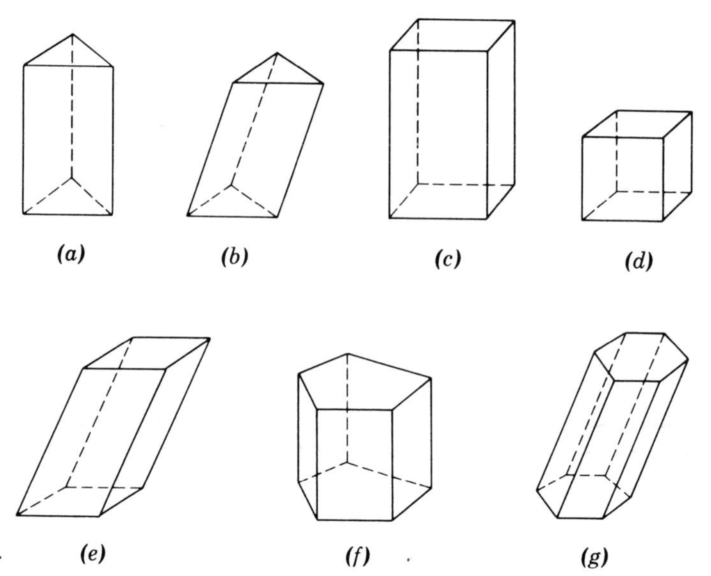

Figure 28-2. Prisms.

congruent regular polygons. There are in all only five possible regular polyhedrons.

28-3 VOLUME OF A PRISM

Volume refers to the amount of space occupied by a geometric solid. In order to understand the method of finding the volume of any prism, as well as other solids, we begin with a rectangular solid.

Suppose we have a rectangular solid 5 inches long, 3 inches wide, and 4 inches high (Fig. 28-3). We wish to find its volume. Perhaps everyone who reads this sentence knows the rule for finding the volume of a rectangular solid. To find the volume we say "multiply the length, the width, and the height together. The result is the volume."

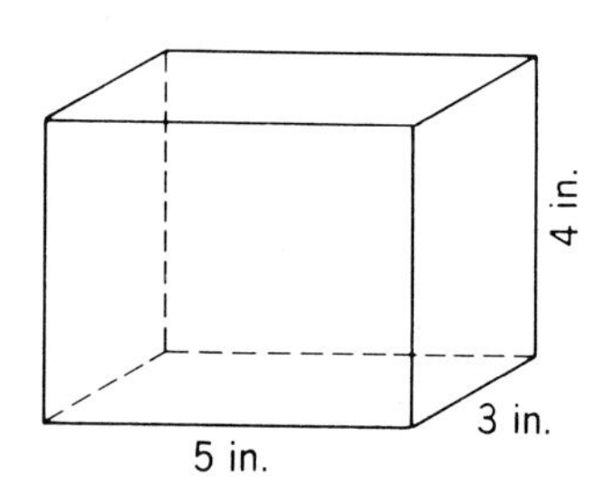

Figure 28-3

$$(5)(3)(4) = 60, \text{ volume (in cubic inches)}$$

The answer is correct, but does it not seem strange that we can multiply *lines* together and get *volume*, which occupies space? We multiply *linear* measurements, yet the result becomes *cubic units*, or volume. If you take a moment to analyze the problem, you will get a better understanding of the methods for finding the volumes of *all* kinds of prisms as well as cylinders and other solids.

To measure volume we must use a unit of volume. The most convenient unit of volume is one that has the shape of a cube. A common unit for measuring a solid of this size is a cubic inch. A *cubic inch* is a cube 1 inch long, 1 inch wide, and 1 inch high.

To measure the volume of this particular rectangular solid, we determine how many times the unit is contained in the solid. We begin by placing the units along one side at the bottom (Fig. 28-4). We find

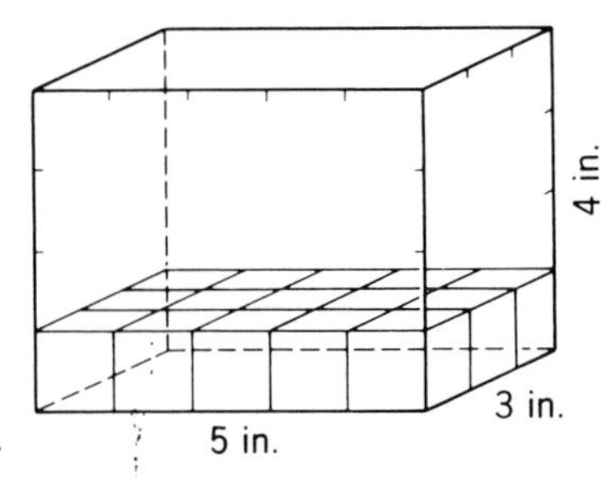

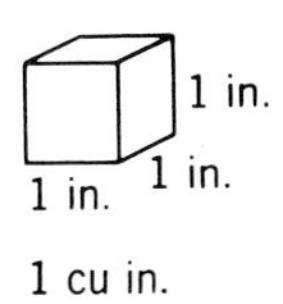

Figure 28-4

that we can place 5 cubic inch units in one row. In the bottom layer there will be 3 rows. Then the bottom layer will contain

$$(3)(5 \text{ cu in.}) = 15 \text{ cu in.}$$

Since the solid is 4 inches high, there will be 4 layers. Then the volume of the entire solid is

$$(4)(15 \text{ cu in.}) = 60 \text{ cu in.}$$

One advantage of using a cubic unit to measure volume is that we can measure the length in *linear* units, the width in *linear* units, and the height in *linear* units. When we multiply these *linear* units together, we get the number of corresponding *cubic units* of volume. The rule may be stated as a formula:

$$V = lwh$$

If the volume of a rectangular solid is known, together with two dimensions, then the unknown dimension can be found by using the formula in reverse. Solving the formula for each dimension, we have

$$l = \frac{V}{wh} \qquad w = \frac{V}{lh} \qquad h = \frac{V}{lw}$$

If we know the area of the base of a rectangular solid, together with the height, we can find the volume without knowing the length and the width. For example, if the base contains 24 square inches, we know that 24 cubic inches can be placed in 1 layer. Then, if the altitude is 15 inches, the volume is

$$(15)(24 \text{ cu in.}) = 360 \text{ cu in.}$$

This example illustrates a very useful formula for finding the volume of a prism. If we let B represent the *area of the base*, we have

$$V = Bh$$

This formula can be used to find the volume of any prism, even an oblique prism, regardless of the shape of the base. Of course, in an oblique prism, the altitude must be measured perpendicular to the bases.

28-4 LATERAL AREA OF A PRISM

The *lateral area* of a prism is found by computing the area of each side, or lateral face, and then adding these areas. If the prism is a right prism, the lateral area can be found by multiplying the height by the perimeter of the base. The *total area* is found by adding twice the area of the base to the lateral area.

In the case of a rectangular solid, which has six rectangular faces, probably the best way to get the *total* area is to find separately the area of the two sides, the two ends, and the top and bottom and then add these areas together. The total area can also be found by adding the area of one side, one end, and the bottom and then multiplying by 2.

28-5 THE CUBE

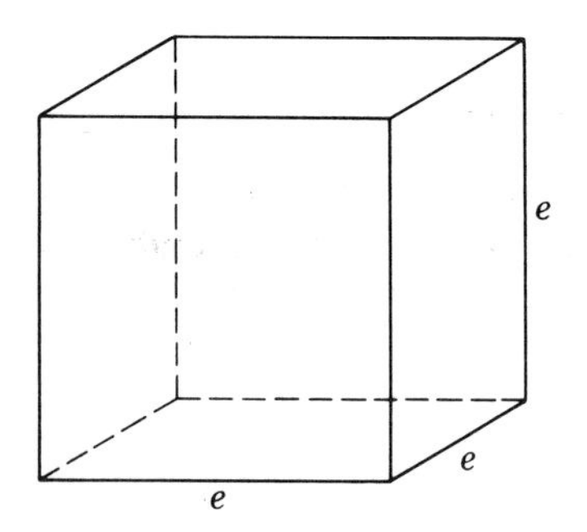

Figure 28-5. The cube.

Since a cube is a rectangular solid whose length, width, and height are equal, the formula for the volume becomes very simple. If we use the letter e to represent the length of one edge, then the formula for a rectangular solid $V = lwh$ is replaced by the formula

$$V = (e)(e)(e)$$

or

$$V = e^3$$

Each face of the cube is a square whose side is equal to the length of the edge of the cube (Fig. 28-5). The area of each face, therefore, is e^2. For the entire surface area of the six faces of the cube we have the formula

$$A = 6e^2$$

Each face has an area $\frac{1}{6}$ of the entire area of the cube.

28-6 THE VOLUME OF IRREGULAR SOLIDS

The volume of an irregular solid can be found by immersing the solid in a liquid and noting the rise of the liquid in the container. Of course, this cannot be done if the solid is soluble in the liquid or if the object is porous.

Exercise 28-1

1. A packing box is 16 in. long, 12.7 in. wide, and 9 in. deep. Find the volume and the total surface area.
2. A brick measures 8 in. by 4 in. by 2.6 in. Find the number of cubic inches in the brick.
3. A school room is 42 ft 4 in. long; 27 ft 6 in. wide, and 11 ft high. If there are 40 people in the room, find the number of cubic feet of air for each person in the room.

4. A basement is dug 36 ft long, 31.5 ft wide, and 9 ft deep. How many cubic yards of earth were removed in digging it? (1 cu yd = 27 cu ft).

5. An aquarium is 28 in. long, 16 in. wide, and 12 in. high. Find the area of the four glass walls, and the volume of the aquarium.

6. A driveway is 14.5 yd long, 10.4 ft wide, and 6 in. deep. Find the number of cubic yards of concrete in the driveway.

7. The base of a rectangular solid contains 29 sq ft. The solid is 6.5 ft high. Find the number of cubic feet in the solid.

8. A prism has a trapezoidal base whose area is 132 sq in. The prism is 4.5 ft high. How many cubic feet does it contain?

9. A room is 26 ft long, 18 ft wide, and 8.75 ft high. Find the volume of the room and the area of the four walls and ceiling.

10. A box is 29.5 in. long, 21.6 in. wide, and 16.5 in. deep. How long a string will be needed to tie the box with a cord around the box once in each of three dimensions? Find the volume of the box.

11. The edge of a cube measures 12.5 in. Find the volume of the cube and the total surface area.

12. A cube has a volume of 512 cu in. Find the edge and surface area.

13. A cube has a total surface area of 73.5 sq in. Find the length of the edge and then find the volume of the cube.

14. A rectangular solid has a volume of 1140 cu in. It is 16 in. long and 9.5 in. wide. Find the height and total surface area.

15. A rectangular solid with a square base has a volume of 338 cu in. The height of the solid is 8 in. Find the total surface area.

16. A liquid container has the shape of a cube and measures 14 in. along the edge. How many gallons will it hold? (1 gal = 231 cu in.)

17. A rectangular tank is 16 ft long, 10.5 ft wide, and 6 ft deep. How many gallons will it hold? (1 cu ft holds approximately 7.5 gal.) Find the weight of water in the full tank. (Water weighs approximately 62.4 lb per cu ft.)

18. Find the weight of a rectangular block of gold 6 in. long, 3 in. wide, and 1.75 in. thick. (The specific gravity of gold is 19.3; this means that gold weighs 19.3 times as much as water.)

19. A 300-ft ditch is 9 ft wide at the top, 5.5 ft at the bottom, and 6.25 ft deep. How many cubic yards of earth were removed in digging the ditch?

20. A swimming pool is 30 ft wide and 80 ft long. It is 3 ft deep at one end and 11 ft deep at the other end. How many gallons does it hold? How long will it take to fill the pool if water flows in at the rate of 15 gal per second?

21. The volume of Lake Mead (Hoover Dam Lake) is 29,830,000 acre-feet. (1 acre-foot is the volume needed to cover an acre 1 ft deep.) The area of the lake is 247 sq mi. What is the average depth in feet? What is the volume in gallons?

22. The area of the six faces of a rectangular solid is 384 sq in. What is the volume if the base is a square 8 in. on the edge? What is the volume if the base is 12 in. long and 4 in. wide?

23. The volume of a rectangular solid is 216 cu in. What is its surface area if it is a cube?

24. In No. 23, what is the surface area of the rectangular solid if the base is 9 in. long and 6 in. wide?

25. In No. 23, what is the surface area if the base is 12 in. long and 6 in. wide.

26. In No. 23, what is the surface area if the base is 12 in. long and 9 in. wide?

27. Our number system contains only ten digits, yet with only these ten digits we can write a number of any size, no matter how large. Archimedes, over 2000 years ago, computed the number of grains of sand in the known universe of the time to show that the number could be expressed. It may be interesting to compute the following: How many drops of water fall on a square mile when the rainfall is $\frac{1}{2}$ in. (Count 8000 drops in 1 pint and 231 cu in. in 1 gal.)

QUIZ ON CHAPTER 28, PRISMS. FORM A.

1. A rectangular box is 30 in. long, 24 in. wide, and 15 in. deep. Find the volume in cubic inches and in cubic feet. Also find the entire surface area.

2. A room is 24 ft long, 18 ft wide, and 9.5 ft high. Find the number of cubic feet and the number of cubic yards in the volume. Also find the area of the four walls and the ceiling.

3. A rectangular solid has a base of 35 sq in. The height is 18.5 in. Find the volume of the solid.

4. A prism has a square base 13.5 in. on the side. It is 24 in. high. Find the volume.

5. The edge of a cube measures 3.5 ft. Find the volume and the total surface area of the cube.
6. A cube has a volume of 64 cu in. Find the edge and the total surface area.
7. A rectangular solid has a volume of 64 cu in. It is 8 in. long and 4 in. wide. Find the height. Find the surface area and compare this area with that in No. 6.
8. A prism has a base in the shape of a trapezoid. The trapezoidal base has two parallel sides, 14 in. and 10 in., respectively. The distance between the sides of the base is 8.5 in. The prism is 24 in. high. Find the volume of the prism.
9. A rectangular box has a volume of 2664 cu in. The box is 16 in. wide and 9 in. high. Find the length and total surface area of the box.
10. A prism with a square base is 12.4 in. high and has a volume of 2790 cu in. Find the edge of the base. Find the total surface area.

CHAPTER TWENTY-NINE Cylinders

29-1 DEFINITIONS

A *cylindrical surface* is a curved surface formed by a moving straight line that moves in such a way that it is always parallel to another fixed straight line. A *cylinder* is a geometric solid bounded by a closed cylindrical surface and by portions of two parallel planes (Fig. 29-1).

The plane surface boundaries of a cylinder are called the *bases*, or the two ends, of the cylinder. If the bases are circles, then the cylinder is called a *circular cylinder*. Figures 29-1*a* and *b* show circular cylinders. Both ends of a circular cylinder are therefore equal and parallel circles.

A cylinder need not be circular. The bases may be in the shape of an ellipse or some other odd-shaped curve (Figs. 29-1*c*, *d*). However, according to the definition, both bases of the cylinder are exactly the same size and shape.

When we say *cylinder*, we usually mean a *right circular cylinder*, since this is the most common form of cylinder (Fig. 29-1*a*). The word *right* refers to perpendicularity. In a *right* circular cylinder a straight line segment connecting the centers of the circular bases is perpendicular to the bases. If this line is not perpendicular to the bases, then the cylinder is called an *oblique* cylinder (Fig. 29-1*b*).

The radius, *r*, of a circular cylinder is the radius of one of the circular ends. The altitude, or height, *h*, is the perpendicular distance between the ends of the cylinder.

If we form a right prism whose bases are *regular* polygons of three or more sides and then increase the number of sides, without limit, always keeping the base in the form of a regular polygon, the prism will become more and more nearly a cylinder. A right circular cylinder

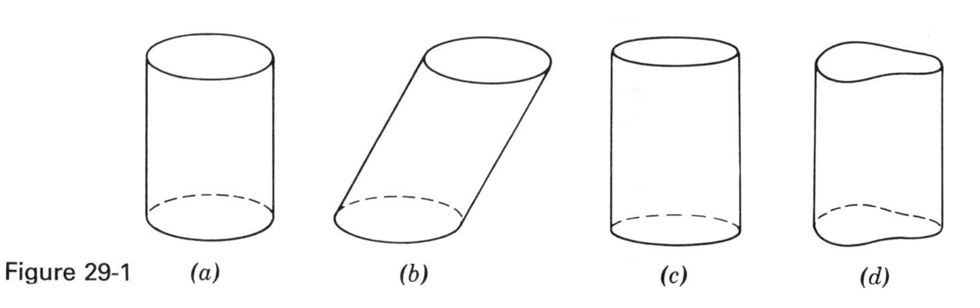

Figure 29-1 (*a*) (*b*) (*c*) (*d*)

may then be thought of as a right prism whose bases are regular polygons of an infinite number of sides (Fig. 29-2).

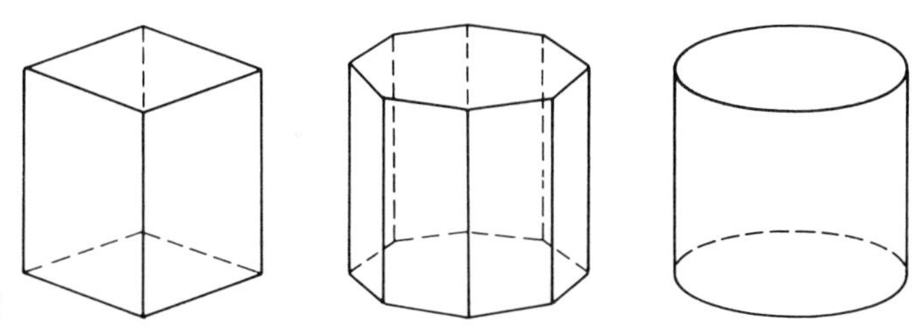
Figure 29-2

Right circular cylinders are numerous in our everyday life. They have many uses. Much of our canned food is put up in cylindrical cans. Storage tanks often have the shape of a cylinder. A round pencil is a cylinder except for the tip. We see cylinders of all sizes, from the very small, such as a shaft for a small wheel in a wrist watch, to a city's large gas storage tank, perhaps 100 ft in diameter. Some cylinders, such as a coin, have a large diameter compared with the length (height) of the cylinder. In contrast, a round wire is a cylinder whose diameter is very small compared with its length (height) (Fig. 29-3).

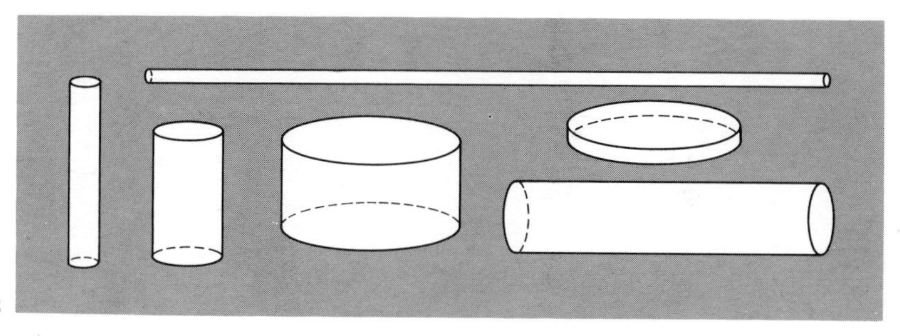
Figure 29-3

29-2 VOLUME OF A CYLINDER

Since a cylinder may be considered a prism with an infinite number of sides, then the volume of the cylinder is found by the same formula used for the volume of a prism:

$$V = Bh$$

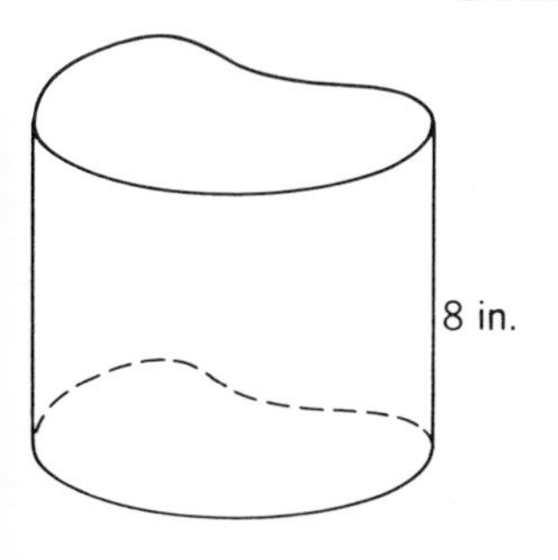

Figure 29-4

For instance, if the base of a cylinder contains 19 sq in. (Fig. 29-4), then 19 cubic inches can be placed in one layer on the bottom. If the cylinder is 8 inches high, the volume can be found by multiplying the area of the base, 19, by the height, 8. The volume is $(19)(8) =$ 152 cubic inches.

The base of a circular cylinder is a circle (Fig. 29-5). Therefore, the area of the base can be found by the formula for the area of any circle: $A = \pi r^2$. If we substitute the value πr^2 for the area of the base B in

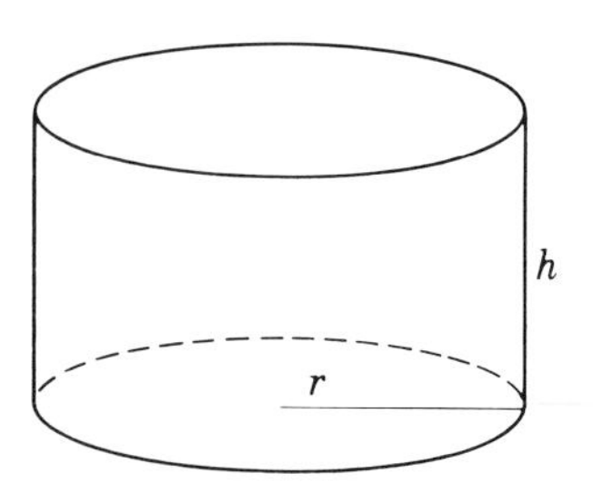

Figure 29-5

the formula $V = Bh$, we get the formula for the volume of any circular cylinder:

$$V = \pi r^2 h$$

This is an important formula. It applies to all circular cylinders, regardless of the size or shape. It also applies to oblique circular cylinders if we remember to take h as the perpendicular distance between the bases.

Notice that in the formula we are again multiplying together three linear measurements:

$$(\pi)(\text{radius})(\text{radius})(\text{height}) = \text{volume}$$

The result of the multiplication is units of volume, which has three dimensions and occupies space.

Example 1

Find the volume of a right circular cylinder whose diameter is 8.2 ft and whose altitude is 12.5 ft.

Solution

In order to use the formula, we must first find the radius r, which is one-half the diameter.

$r = \frac{1}{2}$ of $8.2 = 4.1$	radius
$r^2 = (4.1)^2 = 16.81$	square of the radius
$\pi r^2 = (3.1416)(16.81) = 52.81$	area of the base
$\pi r^2 h = (52.81)(12.5) = 660.125$ cu ft	volume

If the answer is rounded off to five significant digits, we have 660.12.

In a problem such as Example 1 we would probably assume that the given numbers are correct only to three significant digits. If the value 3.14 is used for π and the numbers are rounded off to three significant digits, as is usually done in computation by slide rule, then the computation becomes

$$r = 4.10$$
$$r^2 = 16.8$$
$$\pi r^2 = (3.14)(16.8) = 52.8$$
$$\pi r^2 h = (52.8)(12.5) = 660 \text{ cu ft}$$

If a problem is first written out in the form of the formula with the given quantities in place of the letters, the computation can often be done much more easily. This is especially true in problems containing common fractions. We shall work the next example by this method.

Example 2

Find the number of cubic inches of copper in a copper wire $\frac{3}{8}$ in. in diameter and $\frac{1}{2}$ mile long.

Solution

The wire is a cylinder whose diameter is $\frac{3}{8}$ in. and whose altitude is $\frac{1}{2}$ mile. First, the length, or altitude, $\frac{1}{2}$ mile, must be expressed in the same units as the diameter. To state $\frac{1}{2}$ mile in inches we write

$$\tfrac{1}{2}\text{ mile} = \tfrac{1}{2}(5280)(12)\text{ in.}$$

This product need not be expanded. Computation is simpler if the product is left in the form shown here.

The formula for the volume of a cylinder is

$$V = \pi r^2 h$$

In the example $r = \frac{3}{16}$ in., $h = \frac{1}{2}(5280)(12)$ in. Substituting numerical values in the formula, we get

$$V = (3.1416)\left(\frac{3}{16}\right)\left(\frac{3}{16}\right)\left(\frac{1}{2}\right)(5280)(12)$$

If the problem is written out in this form, some factors can be "canceled" in numerator and denominator.

$$V = (\overset{.3927}{\cancel{3.1416}})\left(\frac{3}{\underset{8}{\cancel{16}}}\right)\left(\frac{3}{\cancel{16}}\right)\left(\frac{1}{\cancel{2}}\right)(\overset{330}{\cancel{5280}})(\overset{\overset{3}{\cancel{6}}}{\cancel{12}})$$

$$V = 3498.957 \text{ cu in.}$$

or

$$V = 3499 \text{ cu in., rounded off to four digits.}$$

29-3 LATERAL AREA OF A CYLINDER

It is often necessary to compute the lateral area of a cylinder. By *lateral area* we mean the area of the curved surface. We may wish to paint the outside surface of a cylindrical gasoline storage tank, to find the amount of heating surface of a steam pipe, or to find the number of square inches of metal required in the manufacture of cylindrical tin cans.

In order to arrive at a formula for finding the lateral area, consider the cylinder in Fig. 29-6. Let us assume that the lateral area is cut along one side and spread out flat like a sheet of paper. The lateral area, then, has the form of a rectangle. The length of the rectangle is the circumference C of the cylinder. The width of the rectangle is the altitude, or height h of the cylinder.

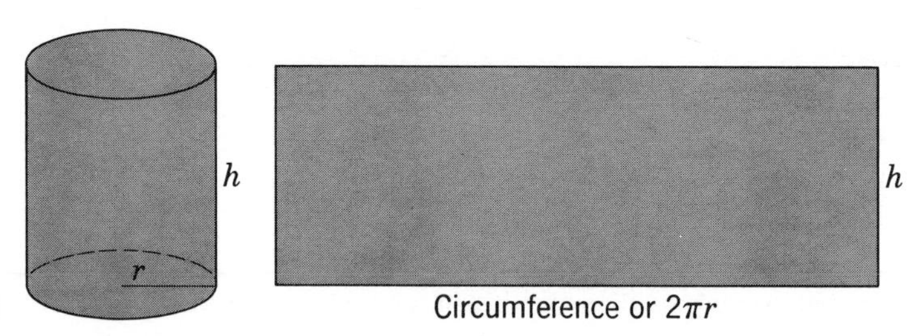

Figure 29-6

The area of a rectangle is given by the formula

$$A = (\text{length})(\text{width})$$

The lateral area (LA) of the cylinder now becomes the area of a rectangle. If we substitute the measurements of the cylinder, we get

$$LA = (\text{circumference})(\text{altitude})$$

or

$$LA = (2\pi r)(h)$$

or

$$LA = 2\pi rh$$

This formula can be used to find the lateral area of any right circular cylinder, large or small. Notice that we have in this formula the *product* of *two linear measurements.* The answer will therefore be stated in terms of *square units.*

The total surface of a cylinder must take into account not only the lateral area but also the area of both ends. If the cylinder, such as a wire, has a very small diameter compared with its length, then the area of the two ends is small enough to be negligible. However, in other cylinders the total area must include the area of both ends. Each end is a circle, and its area can be found by the formula $A = \pi r^2$. The total area is then found by the rule and formula

$$\text{total area} = \text{lateral area} + \text{area of both ends}$$

or

$$TA = 2\pi rh + 2\pi r^2$$

or, factoring,

$$TA = 2\pi r(h + r)$$

29-4 HOLLOW CYLINDER By the volume of a hollow cylinder we mean the volume of the solid, or shell, between two concentric cylinders; that is, cylinders with the

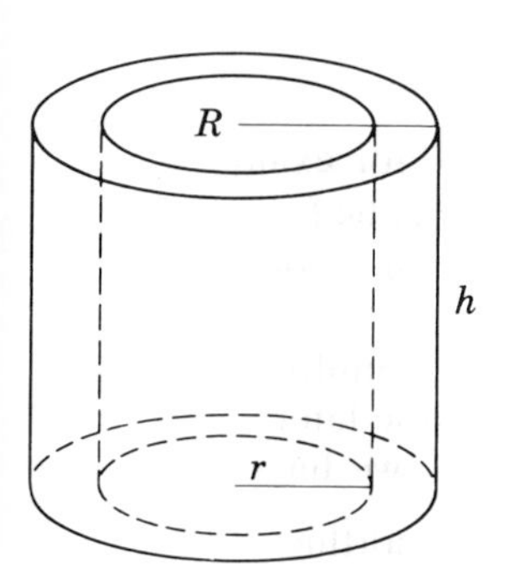

Figure 29-7

same axis but different radii (Fig. 29-7). As an example, we may wish to find the amount of metal in a hollow iron pipe.

In order to find the volume of a hollow cylinder, or shell, we can compute the volume of the large cylinder, then the volume of the small inner cylinder, or hollow, and then find the difference between the two volumes. The difference will be the volume of material in the shell.

Let us call the radius of the larger, or outer cylinder, R and the radius of the smaller, or inner cylinder, r. The altitude h is the same for both cylinders. Now, if we let V_L represent the volume of the large cylinder and V_s represent the volume of the small cylinder, we have, for the volume of each cylinder,

$$V_L = \pi R^2 h \quad \text{and} \quad V_s = \pi r^2 h$$

The volume of the shell is the difference between the two:

$$V \text{ (of shell)} = V_L - V_s$$

or

$$V \text{ (of shell)} = \pi R^2 h - \pi r^2 h$$

The formula may be written

$$V \text{ (of shell)} = \pi(R^2 - r^2)h \quad \text{or} \quad V = \pi h(R + r)(R - r)$$

Example 1

Find the volume of an iron shell 15 in. long if the inside diameter is 10 in. and the metal is 2 in. thick.

Solution

In this example the radius of the inner, or small, cylinder is 5 in. Since the metal is 2 in. thick, the radius of the large outer cylinder is 7 in. (The outer diameter is 14 in.) The height, or altitude, h is the same for both: 15 in. We use the formula

$$V \text{ (of shell)} = \pi h(R^2 - r^2)$$

Substituting the given values, we have

$$V = \pi(15)(7^2 - 5^2)$$
$$V = \pi(15)(49 - 25)$$
$$V = \pi(15)(24) = 360\pi$$
$$V = (360)(3.1416) = 1130.976 \text{ or } 1131 \text{ cu in.}$$

Warning. The student is warned that he must not subtract the small radius from the large radius before squaring the two.

In a hollow cylinder, if the metal is very thin, for example, $\frac{1}{32}$ inch thick in an 8 inch shell, the approximate volume can be found by a short method. Let us assume that the shell is cut along one side and laid out flat in the form of a rectangular solid.

The volume of the rectangular solid is found by multiplying its area by the thickness. This is the same as multiplying the lateral area of the cylinder by the thickness of the shell. Therefore, we have

$$\text{approximate volume of shell} = (\text{lateral area})(\text{thickness})$$

or

$$V\ (\text{approx.}) = (2\pi rh)(\text{thickness})$$

Example 2

Find the volume of the iron in a hollow iron cylinder having an outside diameter of 8 in. and a length of 24 in. if the metal is $\frac{1}{32}$ in. thick.

Solution

$$V\ (\text{approx.}) = (2\pi rh)(\text{thickness})$$

$$V = 2(3.1416)(4)(24)(\tfrac{1}{32}) = 18.8496 \text{ cu in.}$$

Exercise 29-1

The following containers are cylindrical unless otherwise stated:

1. A hot-water tank has a diameter of 13 in. and is 5 ft high. How many gallons will it hold? (1 gal holds 231 cu in.)

2. An oil tank is 22 ft long and has a diameter of 9.5 ft. How many gallons will it hold? (1 cu ft holds $7\frac{1}{2}$ gal.)

3. An oil can has a radius of 5 in. and is 15 in. high. Does it hold 5 gal?

4. An oil tank car is 38 ft long and 7 ft, 2 in. in diameter. How many gallons will it hold?

5. A 5-gal can has a radius of 4.8 in. Find its height.

6. Find the number of cubic inches of copper in a copper wire $\frac{3}{16}$ in. in diameter and 1 mile long. How much does it weigh if the specific gravity of the copper is 8.8?

7. How many square feet of asbestos are needed to wrap a hot-water furnace 5 ft high and 38 in. in diameter (lateral area and top)?

8. A right circular cylinder has a diameter of 24 in. and an altitude of 54 in. Find the number of square inches (and square feet) in the total surface area, including lateral area and both ends. Find the volume of the cylinder.

9. A rectangular gasoline tank on a car is 3 ft 6 in. long, 13 in. wide, and 7 in. deep. On another car the cylindrical tank is 3 ft, 2 in. long and has a diameter of 9 in. Which holds more and how much?

10. A road blacktop roller is $4\frac{1}{2}$ ft in diameter and 12 ft long. How many turns will it have to make to roll a blacktop strip *once* if the road is 28 ft wide and $\frac{1}{4}$ mile long?

11. A steam boiler has 24 flues, each 2 in. in diameter and 10 ft long. Find the total number of square inches of heating surface of the 24 flues.

12. One mile of water main is 15 in. in diameter. If water flows at the rate of 3 ft per sec, how much water flows in 1 hr?

13. A rectangular block of copper is 4 ft long and has a square cross section 6 in. on a side. What is its volume in cubic feet? What is the surface area of the block of copper? Surface area is important if we consider the exposure to the air. If the copper in this block were formed into a solid cube 1 ft long, 1 ft wide, and 1 ft high, what would the surface area be? Compare with the present surface area. Can you explain the difference? What is the surface area if this block of copper is drawn into the form of a wire $\frac{1}{4}$ in. in diameter?

14. If 10 gal of hot water are placed in a rectangular metal container 11 by 14 by 15 in., how much is the radiating surface? How much surface is exposed for radiating heat if the water is placed in a cylindrical pipe 3 in. in diameter?

15. A hollow circular concrete tube is 20 ft long. The inside diameter of the tube is 26 in. and the concrete shell is 4 in. thick. Find the number of cubic feet of concrete in the tube.

16. An iron water pipe has an inside diameter of $\frac{7}{8}$ in. If the metal is $\frac{1}{8}$ in. thick and the specific gravity of the iron is 6.8, find the weight of 100 ft of pipe.

17. Find the approximate number of cubic inches of metal in an iron shell 12 in. long with an inside diameter of 6 in., if the metal is $\frac{1}{16}$ in. thick.

18. A container labeled "1 quart" has a diameter of 4.2 in. and a height of 4.2 in. Another container marked "1 quart" has a diameter of 3.6 in. and a height of 5.7 in. Do they hold the same amount? Which one requires more metal (total area)?

19. Find the total piston displacement in an 8-cylinder engine if each cylinder has a diameter of 3 in. and a stroke of 3.8 in.

20. A manufacturer makes some quart tin cans with a diameter of 4 in. and other quart cans with a diameter of 3 in. Which size of can requires more metal for the total area? How many more square inches of metal are required for 10,000 cans of one size than for the other?

QUIZ ON CHAPTER 29, CYLINDERS. FORM A.

Use: $\pi = 3.14$, unless otherwise indicated.

1. Find the volume and total surface area of a cylindrical tin can whose diameter is 6 in. and whose altitude is 14 in.
2. Find the volume and total surface area of a cylindrical tank whose diameter is 8 ft and whose altitude is 12 ft.
3. A cylindrical oil tank has a diameter of 30 in. and an altitude of 5.5 ft. How many gallons will it hold? (1 gal holds 231 cu in.)
4. A circular cylinder has a diameter of 6.4 in. Its volume is 133.2π cu in. Find the altitude and the lateral area of the cylinder.
5. A circular cylinder whose altitude is 12 in. has a volume of 1357.2 cu in. Find the radius of the cylinder.
6. Find the number of cubic feet of copper in a copper wire $\frac{1}{3}$ mile long if the diameter of the wire is $\frac{1}{4}$ in. (1 mile = 5280 ft; use $\pi = 3.1416$).
7. Find the amount of metal in a hollow iron cylinder whose outside diameter is 18 in. and whose altitude is 12 in., if the thickness of the metal is 2 in.

CHAPTER THIRTY Pyramids and Cones

30-1 THE PYRAMID

The *pyramid* is a geometric solid that has important applications in certain topics in mathematics and engineering. It is related to the measurements in a sphere which, in turn, has certain relations to magnetism and to such problems as the storage of radioactive materials.

A *pyramid* is a polyhedron whose base is a triangle, quadrilateral, or some other polygon and whose sides, called *lateral faces*, are triangles having one vertex in common at a point opposite the base. This point is called the *apex* of the pyramid (Fig. 30-1). Notice that the base of a

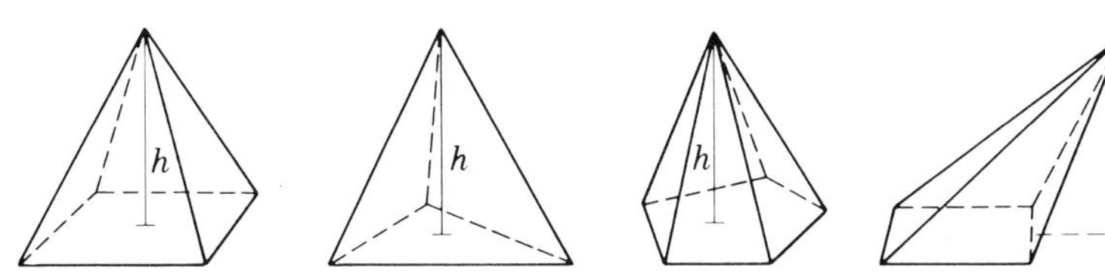

Figure 30-1

pyramid may have the shape of any polygon. The *altitude h* of a pyramid is the straight line segment from the apex perpendicular to the base.

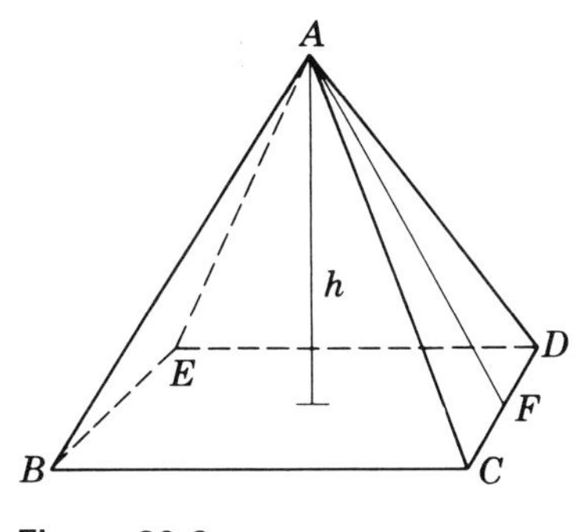

Figure 30-2

If the base of a pyramid is a regular polygon and if the apex is directly opposite the center of the base, then the pyramid is called a *regular pyramid* (Fig. 30-2). In a regular pyramid the line segment from the apex to the center of the base is the altitude.

In a regular pyramid all the lateral faces are isosceles triangles such as triangle *ABC*. The altitude of one of these triangles, *FA*, is called the *slant height* of the pyramid. The intersection, *DA*, of two lateral faces is called a *lateral edge* of the pyramid. The intersection, *BC*, of a lateral face with the base of the pyramid is called the *base edge* of the pyramid.

30-2 VOLUME OF A PYRAMID

In order to get an understanding of the volume of a pyramid, let us begin with a right prism. Suppose we begin with a rectangular solid

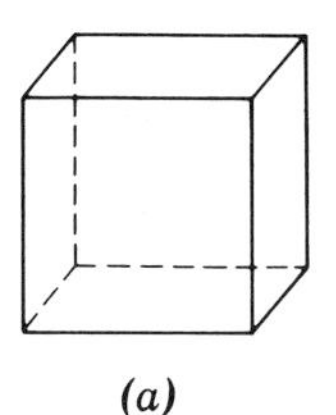
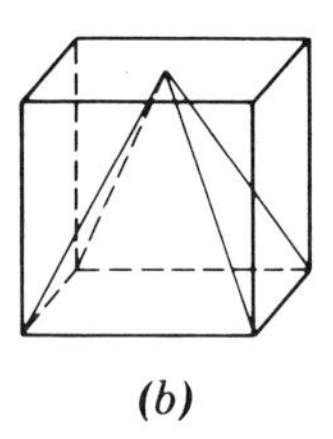
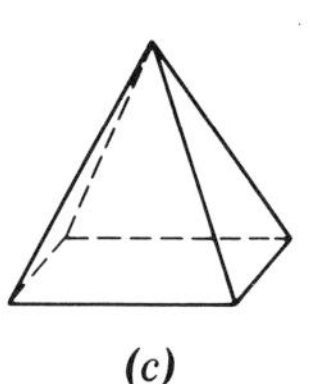

Figure 30-3 *(a)* *(b)* *(c)*

(Fig. 30-3*a*). For the volume of this prism we have the formula

$$V = Bh$$

Now, if we cut away part of the prism on each side at a slant from a point in the top down to each of the lower edges, as shown in Fig. 30-3*b*, we have a pyramid left. The question is: how much of the prism has been cut away and how much remains? The remaining pyramid is shown (Fig. 30-3*c*).

The pyramid (*c*) has the same base and the same altitude as the corresponding prism (*a*). It has been found that the volume of the pyramid is exactly one-third as much as the volume of the corresponding prism. That is, the volume of a pyramid is equal to one-third of the product of the altitude h and the area of the base B. The rule may be stated as a formula:

$$V = \tfrac{1}{3}Bh \quad \text{or} \quad V = \frac{Bh}{3}$$

The total volume of several pyramids having the same altitude h can be found by computing the volume of each pyramid separately and then adding the volumes, or the bases can first be added and the sum of these bases multiplied by the altitude. Suppose we have the pyramids shown in Fig. 30-4 with bases equal to 10, 17, 5, 13, and 18 sq in., respectively, and each pyramid having an altitude of 4 in.

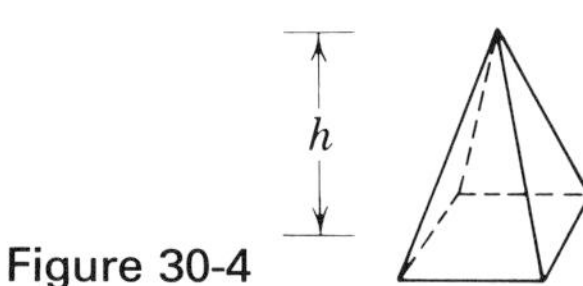

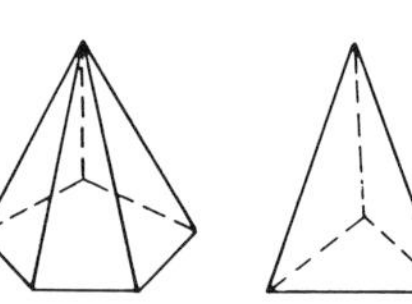
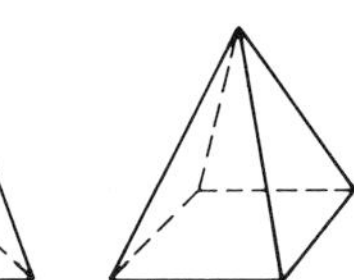
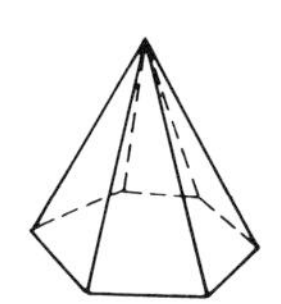

Figure 30-4

The simplest way to find the total volume of all the pyramids is to find the sum of the bases first. This is 63 sq in. Then we take the sum of the bases as one single base in the formula:

$$\text{total volume} = \tfrac{1}{3}(63)(4) = 84 \text{ cu in.}$$

30-3 LATERAL AREA OF A REGULAR PYRAMID

In a regular pyramid all the lateral faces are congruent isosceles triangles. Therefore, we can find the area of one and then multiply by the number of faces. Each isosceles triangle has a base equal to the base edge of the pyramid (Fig. 30-5). The altitude of each triangle, FA, is the slant height of the pyramid. We can find the area of a lateral face by taking one-half the product of the slant height and the base edge. Then we can find the total lateral area by multiplying by the number of faces.

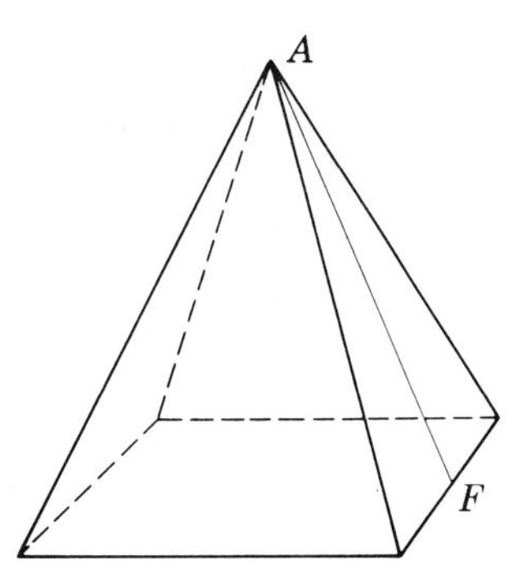

Figure 30-5

However, since all the lateral-face triangles have the same altitude, we can first find the sum of the bases, which is equal to the perimeter of the pyramid base. Then we can find the total lateral area by the following:

$$\text{total lateral area} = \tfrac{1}{2}(\text{perimeter of base})(\text{slant height})$$

or, as a formula,

$$LA = \tfrac{1}{2}ps$$

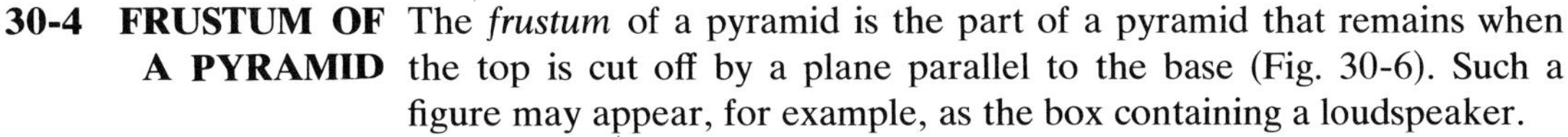

30-4 FRUSTUM OF A PYRAMID

The *frustum* of a pyramid is the part of a pyramid that remains when the top is cut off by a plane parallel to the base (Fig. 30-6). Such a figure may appear, for example, as the box containing a loudspeaker.

The top of the frustum is a polygon of exactly the same shape as the base but smaller in size. The top is sometimes called the *top base.* The two bases are often denoted by B and b, respectively.

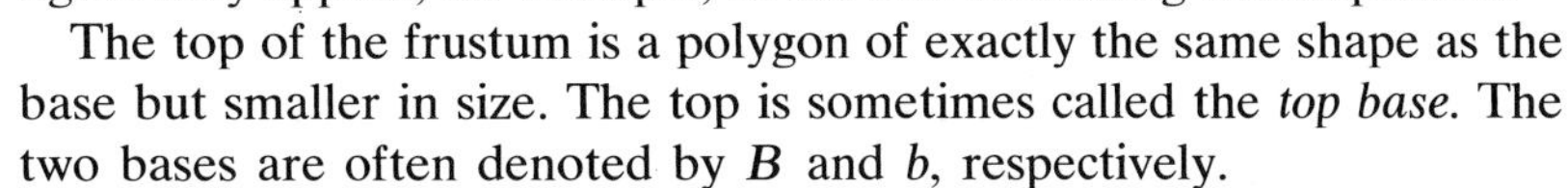

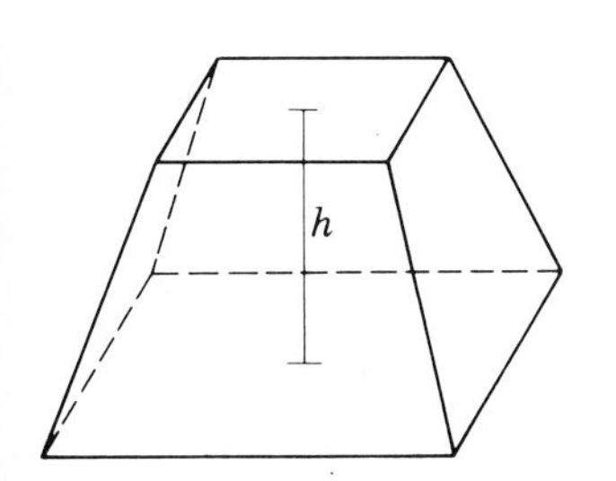

Figure 30-6

The volume of a frustum of a pyramid can be found by the following formula:

$$V = \tfrac{1}{3}h(B + b + \sqrt{Bb})$$

In the formula V represents the volume in cubic units, h represents the altitude in linear units, B the area of one base in square units, and b the area of the other base in square units.

In the frustum of a pyramid all the lateral faces are trapezoids. Therefore, we can find the lateral area by using the formula for the area of a trapezoid. Instead of finding the area of each trapezoid separately for the frustum of a regular pyramid, we find the sum of the top bases and the sum of the bottom bases of the trapezoids. These bases form the perimeters of the top base and the bottom base of the frustum. If P represents the perimeter of one base and p represents the perimeter of the other base, we use P and p in the formula for the area of a trapezoid. The altitude of each trapezoidal face is the slant height, s, of the frustum of a regular pyramid. Then, for the lateral area of a frustum, we have

$$LA = \tfrac{1}{2}s(P + p)$$

30-5 THE CONE

A cone is a geometric solid that has many applications in science and in everyday life. We have conical tents, conical speakers in radios, conical containers, ice cream cones, etc. A *cone* is a solid bounded by a plane forming the base and by a lateral curved surface that comes to a point called the *apex*. A cone might be said to be a pyramid with an infinite number of lateral faces. Fig. 30-7 shows several cones.

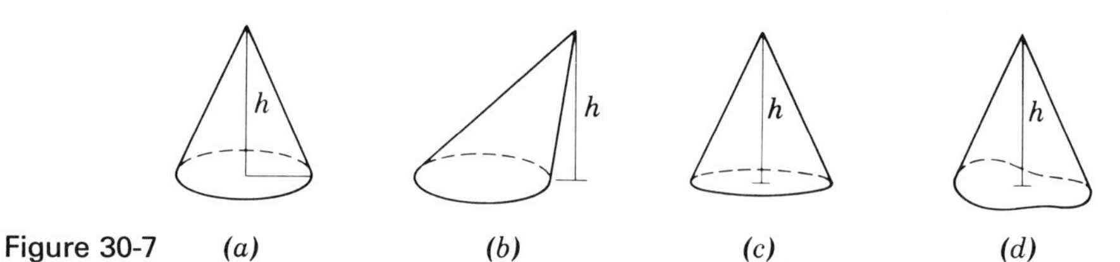

Figure 30-7 (*a*) (*b*) (*c*) (*d*)

If the base of a cone is a circle, the cone is called a *circular cone* (Figs. 30-7*a*, *b*). A cone need not be circular. The base may have other shapes, such as an ellipse or some other odd-shaped curve (Figs. 30-7*c*, *d*).

The *altitude* of a cone is a straight line segment, h, from the apex perpendicular to the base. In a *right circular cone* the base is a circle and the altitude connects the center of the circular base with the apex. Figure 30-7*a* shows a right circular cone. If a right triangle is rotated around one of its legs, a right circular cone is generated.

30-6 VOLUME OF A CONE

A cone bears the same relation to a cylinder as a pyramid bears to a prism. In order to get an understanding of the volume of a cone, let us start with a right circular cylinder (Fig. 30-8*a*). For the volume of a cylinder we have the formula

$$V = \pi r^2 h$$

If we begin at a point in the top of the cylinder and cut away part of the cylinder sloping downward to the circumference of the base (Fig. 30-8*b*), the part that remains is a cone (Fig. 30-8*c*). Now, our question is, how much of the cylinder has been cut away and how much remains?

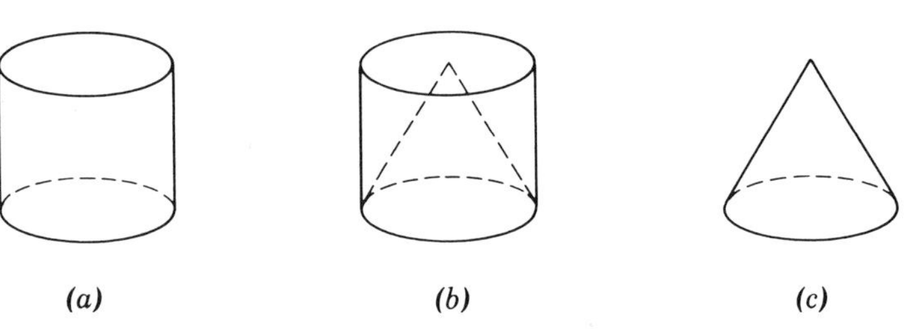

Figure 30-8 (*a*) (*b*) (*c*)

The cone has the same base and the same altitude as the corresponding cylinder (a). It has been found that the volume of a cone is exactly one-third as much as the volume of a corresponding cylinder. Therefore, the volume of the cone can be found by the formula

$$V \text{ (of cone)} = \tfrac{1}{3}\pi r^2 h \quad \text{or} \quad V = \frac{\pi r^2 h}{3}$$

30-7 LATERAL AREA OF A CONE

The lateral area of a right circular cone is found by the same formula as that used for the lateral area of a regular pyramid:

$$LA = \tfrac{1}{2}sp$$

However, for a circular cone, the perimeter is called the circumference, which is equal to $2\pi r$. The formula for the lateral area then becomes

$$LA = \tfrac{1}{2}(\text{slant height})(\text{circumference})$$

The formula reduces to

$$LA = \pi rs$$

We have said that a right circular cylinder may be thought of as a right prism in which the number of sides has increased without any limit until the lateral surface is curved. In the same way, a right circular cone may be thought of as a regular pyramid in which the number of sides of the polygon forming the base has increased without any limit until the base is a circle.

30-8 FRUSTUM OF A CONE

The *frustum* of a cone is the part that remains when the top is cut off by a plane parallel to the base (Fig. 30-9). Such a geometric form appears in many familiar objects, such as paper drinking cups, buckets, pails, containers of many kinds, dishes, lamp shades, and megaphones.

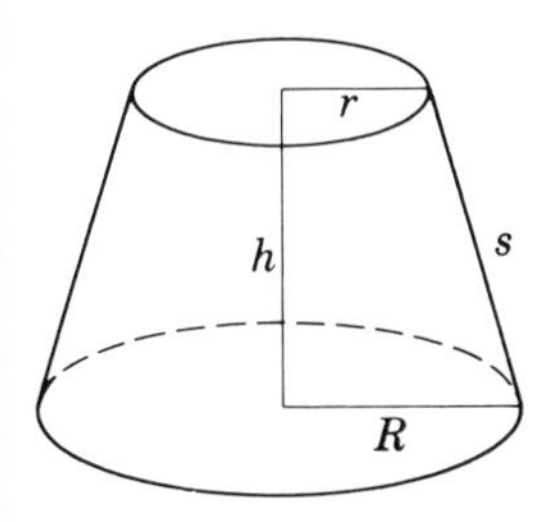

Figure 30-9

The two plane boundaries of a frustum of a cone are called the bases. In the frustum of a right circular cone one base is a smaller circle than the other. The radius of one base may be denoted by R and the radius of the other base by r. The altitude of the frustum is the perpendicular distance between the bases. The slant height s is the distance from one circumference to the other in the lateral surface.

If we denote the area of one base of the frustum of the cone by B and the area of the other base by b, then for the volume we have the same formula as that used for the volume of the frustum of a pyramid:

$$V = \tfrac{1}{3}h(B + b + \sqrt{Bb})$$

However, in the case of the frustum of a right circular cone the bases are circles and the formula can be changed to

$$V = \tfrac{1}{3}\pi h(R^2 + r^2 + Rr) \quad \text{(for the frustum of a cone)}$$

The lateral area of a frustum of a cone may be found in a way that is similar to the method used for finding the lateral area of the frustum of a pyramid. However, the perimeters are now called circumferences. Otherwise, the formula is similar. For the lateral area of the frustum of a cone, we have the formula

$$LA = \tfrac{1}{2}s(2\pi R + 2\pi r)$$

The formula reduces to

$$LA = \pi s(R + r).$$

in which s is the slant height.

Exercise 30-1

1. A rectangular prism and a corresponding pyramid both have square bases 8 in. on a side and each is 20 in. high. Find the volume of each in cubic centimeters.

2. A pyramidal monument made of granite has a base of 14 sq ft and is 18 ft high. What is the weight of the monument if granite weighs 170 lb per cu ft?

3. A pyramid contains 87 cu in. Its height is 18 in. If the base is square, find one side of the base.

4. A pyramid has a rectangular base 28×24 in. If the pyramid is 12 ft high, how many cubic feet does it contain?

5. A marble pyramid has a base 16 in. square and is 8 ft high. Find its weight if marble has a specific gravity of 2.7. (Water weighs 62.4 lb per cu ft.)

6. Imagine that you are standing 10 ft from a wall on which a picture is hung flat against the wall. The picture is 28 in. wide and 24 in. high. Your eye follows the perimeter of the picture once around. What is the volume of the geometric solid that your line of sight encloses?

7. A tent in the shape of a pyramid has a square base 9 ft on a side. The apex of the tent is 11 ft above the base. Find the number of cubic feet of air in the tent. Also find the number of square feet of canvas needed for the four sides and for the floor of the tent.

8. A pile of sand in the shape of a right circular cone has a base whose diameter is 9.2 ft? The altitude of the cone is 2.5 ft. How many cubic yards of sand are there in the pile?

9. A paper cup in the shape of a right circular cone has a diameter of 3.2 in. and an altitude of 4 in. How many cubic inches does it hold? Is this more or less than $\frac{1}{2}$ pint and how much more or less?

10. Some grain piled in one corner of a bin has the shape of a quarter cone with a radius of 5.4 ft and an altitude of 3.5 ft. How many cubic feet of grain are there in the pile?

11. Imagine that you are standing 8 ft from a wall on which hangs a circular mirror 26 in. in diameter flat against the wall. As your eye sweeps around the circumference of the mirror, what is the volume of the conical geometric solid that your line of sight generates?

12. A pyramid has a base in the shape of a triangle. The base of the triangle is 5 in. and the altitude of the triangular base is 3.5 in. Its volume is 35 cu in. Find the height or altitude of the pyramid.

13. A regular tetrahedron is a solid having four faces, all of which are equilateral triangles. This solid might be called a triangular pyramid whose base is an equilateral triangle and all of whose lateral faces are also equilateral triangles of exactly the same size and shape as the base. Find the volume of a regular tetrahedron whose base is 6 in. on a side.

14. A regular pyramid has a base that is a regular hexagon 6 in. on a side. The pyramid is 24 in. high. Where should the pyramid be divided by a plane parallel to the base so that the two parts will have the same volume?

15. A box for a radio speaker is in the shape of the frustum of a pyramid. The front opening (one base) is 18 in. square. The back (the other base) is 6 in. square. If the slant height is 11 in., find the lateral area.

16. A waste basket has the shape of the frustum of a square pyramid. The bottom is $9\frac{1}{2}$ in. square and the top 12 in. square. If the slant height is 14 in., find the volume. Another circular basket has a top 13.5 in. in diameter and a bottom 10 in. in diameter. Its slant height is 15 in. Which has the greater volume?

17. The main part of a monument is the frustum of a pyramid whose lower base is a square 36 in. on a side and whose top base is a square 12 in. on a side. The altitude of this portion of the monument is 14 ft. This frustum is surmounted by a pyramid with an altitude of 18 in. Find the weight of the monument if the stone of which it is formed has a specific gravity of 2.75.

18. A paper drinking cup has a top 3 in. in diameter and a bottom of 1.75 in. in diameter. The altitude of the cup is 3.5 in. How many cups can be filled from 6 qt of coffee?

19. One end of a megaphone has a diameter of 3 in. and the other end has a diameter of 12 in. The altitude is 24 in. How many square inches of surface are there on the outside of the megaphone?

20. Try to derive the formula for the volume of the frustum of a pyramid. To do so, imagine the completed pyramid. Let x equal the altitude of the small pyramid cut off the top. The ratio of the altitudes as well as all corresponding lines in the two pyramids is $x:(x + h)$. Now the volumes of the two pyramids (small and large) can be computed and then the top part subtracted from the large complete pyramid to get the formula for the volume of the frustum.

21. Derive the formula for the volume of the frustum of a cone.

22. The frustum of a cone has a bottom base 30 in. in diameter, a top base 16 in. in diameter, and an altitude of 24 in. Find the volume and the total surface area of the frustum.

23. A right circular cone has a diameter of 20 in. and altitude of 24 in. By using the Pythagorean rule, find the slant height of the cone. Then find the lateral area of the cone.

24. The formula for the volume of a cone is: $V = \frac{1}{3}\pi r^2 h$. Solve this formula for the altitude h, and then find h if $V = 680$ cu in., and the diameter is 12 in.

25. From No. 24, solve the formula for r, and then find the radius r of a cone whose altitude h is 25 in. and whose volume V is 2121 cu in.

26. The lateral area of a cone is 188.5 sq in. If the diameter is 12 in., find the slant height of the cone.

QUIZ ON CHAPTER 30, PYRAMIDS AND CONES. FORM A.

1. A pyramid has a rectangular base 8 in. long and 6 in. wide. The altitude of the pyramid is 12 in. Find its volume.
2. A pyramid has a base containing 48 sq in. The altitude of the pyramid is 14.5 in. Find its volume.
3. A regular pyramid has a square base 8.5 in. on the edge. The altitude of the pyramid is 16 in. Find its volume.
4. The volume of a pyramid is 870 cu in. The base is a rectangle that is 12 in. by 15 in. Find the altitude of the pyramid.
5. The frustum of a regular square pyramid has a bottom base 6 in. by 6 in., and a top base 4.5 in. by 4.5 in. The altitude of the frustum is 8 in. Find the volume.

6. A circular cone has a diameter of 15 in. and the altitude is 24 in. Find the volume of the cone.

7. The frustum of a right circular cone has a top base 5 in. in diameter and a bottom base 8 in. in diameter. The altitude of the frustum is 15 in. Find the volume of the frustum.

8. A circular cone has a diameter of 12 in. and an altitude of 8 in. By use of the Pythagorean theorem, find the slant height and then find the lateral area and total area of the cone.

9. The volume of a right circular cone is 377 cu in. The altitude is 10 in. Find the radius and the diameter of the cone. (Solve the formula for r.)

10. The volume of a right circular cone is 315 cu in. (Solve the formula for h.) Find h if the diameter of the cone is 10 in.

CHAPTER THIRTY-ONE The Sphere

31-1 DEFINITIONS

A *sphere* is a solid bounded by a curved surface such that every point on the surface is the same distance from a point within called the *center.* If a circle is rotated about one of its diameters as an axis, a sphere is generated (Fig. 31-1).

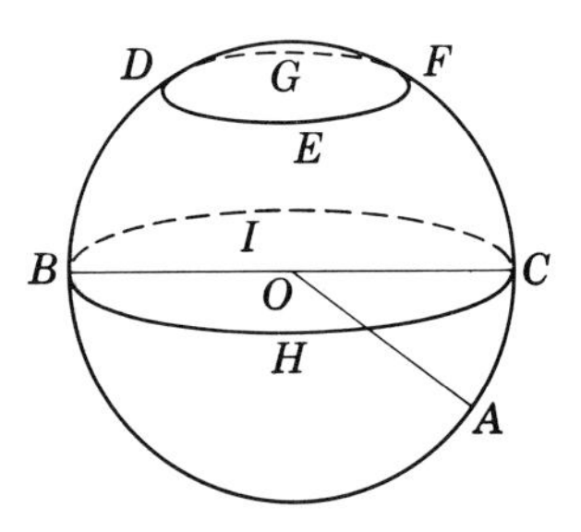

Figure 31-1

A *radius* of the sphere is a straight line segment, such as *OA*, joining the center with any point on the surface, *OA*. A *diameter* is a straight line segment through the center with its ends on the surface, *BOC*. A diameter is therefore equal to twice the radius.

If a plane cuts through a sphere, the cut surface is circular; that is, the section is a circle, *DEFG*. If a thin slice is cut off one side of a sphere, the section is a *small circle.* As the intersecting plane moves close to the center of the sphere, the sections become larger circles. The largest circle that can be cut by a plane intersecting a sphere is a section through the center of the sphere, *BHCI*. Such a circle is called a *great circle* of the sphere. The length of a minor arc of a great circle is the shortest distance between two points on the surface of a sphere.

The portion of a sphere cut off by a plane is called a *segment* of the sphere (Fig. 31-2). If two parallel planes cut through a sphere, the portion of the sphere between the planes is called a segment of two bases (Fig. 31-3).

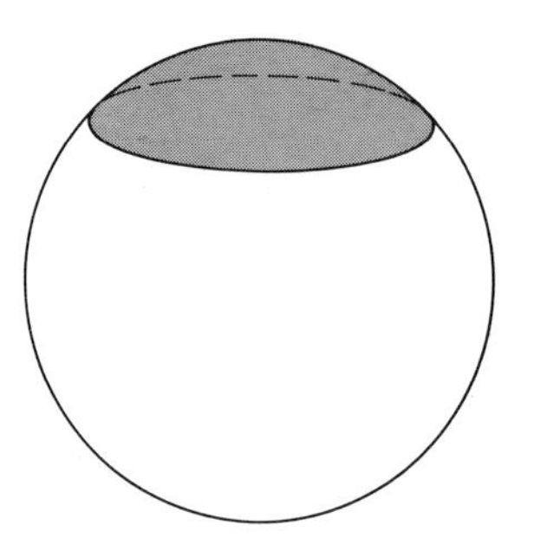

Figure 31-2

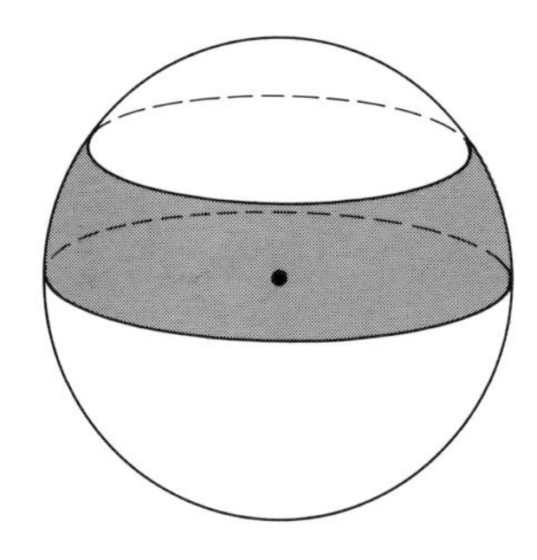

Figure 31-3

31-2 MEASUREMENTS IN A SPHERE

In connection with a sphere, there are two measurable quantities in which we are interested. They are the volume and the surface area. In stating the size of a sphere, we need state only one measurement. That measurement may be the radius, the diameter, or the circumference. If the radius of a sphere is 5 in., then the sphere is completely determined in size, and all measurable facts can be computed from this one measurement alone.

To measure the size of a sphere, we might place a plane (flat) surface on each side of the sphere so that the planes are parallel and touching the sphere. Then we measure the perpendicular distance between the planes. This distance is the diameter of the sphere. The diameter may also be measured in some instances by a set of plane calipers.

Practically, it is impossible to measure the radius of a solid sphere directly simply because we cannot get to the center. However, the formula for the surface area and the formula for the volume are usually stated in terms of the radius rather than the diameter. In almost all mathematics, the formulas for a circle and a sphere are more convenient when stated in terms of the radius rather than the diameter.

31-3 SURFACE AREA OF A SPHERE

The formula for the area of a sphere can be derived only by more advanced mathematics. However, the formula itself is not difficult to understand. It is easy to memorize and use, but you should also try to see why it is reasonable.

Suppose we have a sphere with center at point O and radius equal to 6 in. (Fig. 31-4a). If we cut the sphere in two by a plane passing through the center, we have a hemisphere (Fig. 31-4b). The hemisphere resembles a kettledrum. The plane surface of the hemisphere, $CDEF$ (the head of the kettledrum), is a great circle of the sphere. Its radius is 6 inches, the radius of the sphere.

Now we can find the area of this great circle, the head of the kettledrum. It is πr^2, or 36π.

Let us look now at the curved surface of the hemisphere, that is, the bottom of the kettledrum. This curved surface, clearly, has a greater

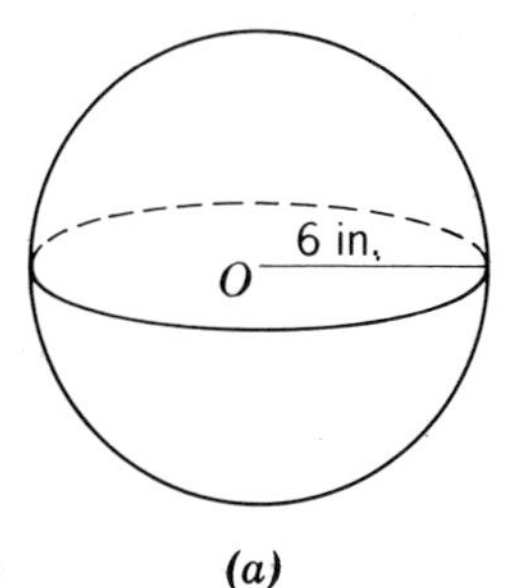

(a)

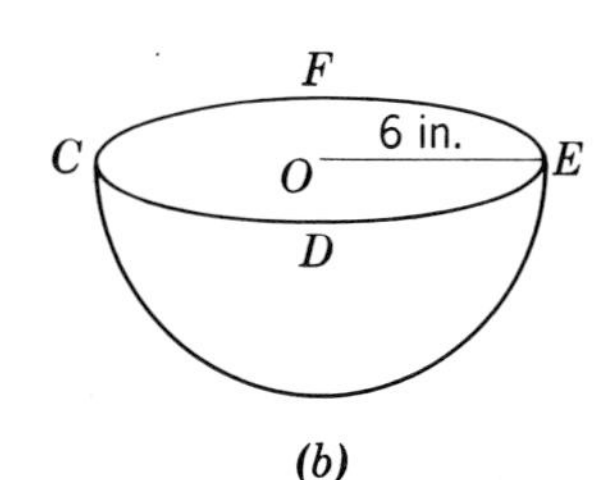

(b)

Figure 31-4

area than the flat top, the head of the drum. In fact, it can be proved that the curved surface of the hemisphere, (b), is exactly twice the area of the great circle of the sphere, or $2\pi r^2$. This fact is sometimes illustrated in the following manner.

We place a peg in the center of the circular flat top of the hemisphere. Then, starting at the peg, we wind a string of firm texture carefully around the peg, covering the flat part of the hemisphere with a single layer of string. Let us assume that 20 feet of string are required to cover the great circle, the plane surface of the hemisphere. Then we remove the string and wind it around on the curved surface in such a way that the surface is covered with a single layer of string. We discover that the 20 feet of string will cover only half the curved surface. In fact, to cover the curved surface of the hemisphere, 40 feet of string are required.

The area of the hemisphere is twice the area of the great circle. Therefore, the entire area of the whole sphere is four times the area of a great circle. The area of the great circle is given by the formula

$$A = \pi r^2$$

Therefore, the formula for the area of the entire sphere is

$$A \text{ (of sphere)} = 4\pi r^2$$

In the sphere shown in Fig. 31-4a the radius is 6 inches. For the area of the sphere, we have

$$A = 4\pi r^2 \qquad A = 4\pi 6^2 = 144\pi \qquad \text{or} \qquad 452.39 \text{ (approx.)}$$

31-4 THE UNIT SPHERE

A unit sphere is a sphere whose radius is 1 unit in length. The unit may be any measurement, 1 centimeter, 1 inch, 1 foot, and so on. The unit sphere is an important concept in many of the applications of the sphere in scientific study.

Let us suppose we have a sphere whose radius is 1 centimeter. This is a sphere about the size of a fairly large marble. The total area of the sphere is shown by the formula

$$A = 4\pi r^2$$

Since $r = 1$, we have

$$A = 4\pi 1^2$$

or

$$A = 4\pi \text{ sq cm}$$

This sphere is a unit sphere. Its entire surface area is equal to 4π sq cm, or approximately 12.57 sq cm.

Now, suppose we have straight lines or rays emanating outward in all directions from the center of the sphere and of such number that each square centimeter on the surface is pierced by only one line. Then the number of lines extending outward from the center is 4π. The idea of one and only one line from the center piercing each square centimeter on the surface of a sphere of this particular size is an important concept in the study of magnetism.

31-5 VOLUME OF A SPHERE

In trying to understand the formula for the area of a circle, we imagined the circle divided into many sectors, resembling triangles. We can think of the sphere in a similar way. A sphere may be thought of as being composed of many pyramids (Fig. 31-5). The pyramids, of course, have slightly curved bases. However, we can assume that the number of pyramids can be increased until the bases are practically flat.

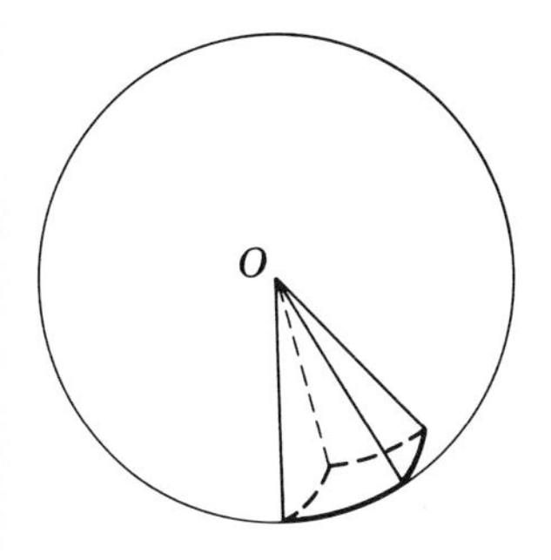

Figure 31-5

The formula for the volume of a pyramid is, as we have seen,

$$V = \tfrac{1}{3}Bh$$

In all the pyramids that make up the sphere the altitude is the same, the radius of the sphere. For many pyramids, all having the same altitude, we can say the total volume is given by the rule

$$\text{total volume of pyramids} = \tfrac{1}{3}\,(\text{sum of bases})(\text{altitude})$$

The altitude of the pyramids is the same as the radius of the sphere, r. The sum of the bases is the surface area of the sphere, $4\pi r^2$. Therefore, the volume of all the pyramids, or the total volume of the sphere, is given by the formula

$$V(\text{of sphere}) = \tfrac{1}{3}(4\pi r^2)(r) \quad \text{or} \quad V = \tfrac{4}{3}\pi r^3$$

The formula may be stated in terms of the diameter. If we replace r with its equivalent $\frac{D}{2}$, we get

$$V = \frac{\pi D^3}{6} \qquad \text{or} \qquad V = 0.5236\,D^3$$

The various formulas associated with the measurements in a sphere can also be used in reverse. Any formula may first be solved for any desired unknown value. The following examples show how this is done.

Example 1

The diameter of a sphere is 24 in. Find the radius, the circumference, the surface area, and the volume of the sphere.

Solution

We have used the formula $D = 2r$. Solving for r, $r = \frac{D}{2}$. Then $r = 12$ in. Now we find the remaining unknowns. $C = \pi D = 24\pi$ in. For the area, we have: $A = 4\pi r^2 = 576\pi = 1810$ sq in. (approx.). For the volume, we have: $V = \frac{4}{3}\pi r^3 = (\frac{4}{3})(\pi)(1728) = 2304\pi$ cu in. $= 7238$ cu in. (approx.).

Example 2

The circumference of a certain sphere is 18π in. Find the diameter, the radius, the surface area, and the volume.

Solution

Taking the formula, $C = \pi D$, we solve for D: $D = C \div \pi$. Then, diameter $= 18$ in.; radius $= 9$ in.; area $= 1017.9$ sq in.; volume $= 121.5\pi$ cu in. $= 3053.7$ cu in. (approx.).

Example 3

The circumference of a certain sphere is 39 in. Find the diameter, the radius, the surface area, and the volume of the sphere.

Solution

For the diameter, we have: $D = C \div \pi$. Here we shall use the value: $\pi = 3.1416$. Then, $D = 39 \div 3.1416 = 12.4$ in. (approx.). From the diameter, we get the radius: $r = 6.2$ in. With the known radius, we can find the surface area and the volume:

$$A = 4\pi r^2 = 4(3.1416)(6.2)^2 = 4(3.1416)(38.44) = 483 \text{ sq in. (approx.)}.$$

Also,

$$V = (\tfrac{4}{3})(3.1416)(6.2)^3 = 998.3 \text{ cu in. (approx.)}.$$

Example 4

The surface area of a certain sphere is 340 sq in. Find the radius, the diameter, the circumference, and the volume of the sphere.

Solution

Using the formula, $A = 4\pi r^2$, we solve for $r = \sqrt{A \div 4\pi}$ Then, $r = \sqrt{340 \div 4\pi} = \sqrt{340 \div 12.5664} = 5.202$ (in.)(approx.). Then, $D = 10.4$; $C = 32.7$ in.; $V = 589.8$ cu in.

Example 5

The volume of a certain sphere is approximately 310 cu in. Find the radius, the diameter, the circumference, and the surface area.

Solution

Taking the formula for volume, $V = \frac{4}{3}\pi r^3$, we solve the formula for r: first: $r^3 = 3V \div 4\pi$; then $r^3 = 930 \div 12.5664 = 74.01$ (approx.). Now we find that the cube root of 74.01 is approximately 4.2. Then $D = 8.4$ in.; $C = 26.4$ in.; $A = 22.17$ sq in.

Example 6

Find the volume of metal in a hollow iron sphere having an outside diameter of 20 in., and inside diameter of 17 in. The metal is 1.5-in. thick.

Solution

Using the formula for the volume of a sphere, we could first find the volume of the entire sphere, then compute the volume of the inner sphere (the hollow), and then subtract. The difference would be the volume of the metal in the hollow sphere.

However, let us use R to represent the radius of the larger or entire sphere, and V_L to represent its volume. We use r to represent the radius of the radius of the smaller sphere, the hollow, and V_S for its volume. Then, for the volume of the metal in the sphere, we have

$$V_L - V_S = \tfrac{4}{3}\pi R^3 - \tfrac{4}{3}\pi r^3$$

We could compute each volume separately, but instead, we factor and write

$$V_L - V_S = \tfrac{4}{3}\pi(R^3 - r^3)$$

Now, we insert the values, $R = 10$; $r = 8.5$, and get

$$V_L - V_S = \tfrac{4}{3}\pi(1000 - 614.125) = 1616 \text{ cu in. (approx.)}$$

Exercise 31-1

Find the radius, circumference, surface area, and volume of these spheres (Nos. 1–4).

1. A globe, diameter = 16 in. **2.** Iron ball, diameter = 3 in.

3. Beach ball, diameter = 22.4 in. **4.** Polo ball, diameter = 3.25 in.

Find the diameter, radius, surface area, and volume of these spheres (Nos. 5–8).

5. Tennis ball, circumference = 8.24 in.

6. Baseball, circumference = 9 in.

7. Basketball, circumference = 29.8 in.

8. Ballbearing, circumference = $\frac{5}{8}$ in.

Find the radius, diameter, circumference, and volume of these spheres (Nos. 9–12).

9. Surface area = 452.4 sq in. **10.** Surface area = 2462 sq in.

11. Surface area = 254.5 sq in. **12.** Surface area = 98.6 sq ft.

Find the radius, diameter, circumference, and surface of these spheres (Nos. 13–16).

13. Volume = 523.6 cu in. **14.** Volume = 4189 cu in.

15. Volume = 288π cu in. **16.** Volume = 2304π cu ft.

17. A glass globe is filled with water. If the globe is 12 in. in diameter, how many gallons will it hold?

18. A certain sphere holds 1 gal. Find the diameter of the sphere.

19. Find the radius and diameter of a sphere that holds 1 barrel (31.5 gal).

20. What size sphere has a volume of 1 cu ft?

21. Could you lift a sphere of gold 6 in. in diameter? (Specific gravity: 19.3.)

22. You are shown a sphere of gold 4 in. in diameter and you are told that you have a choice of taking only an outside shell $\frac{1}{2}$-in. thick or the remainder of the sphere. Which would you choose?

23. What is the diameter of a 16-lb shot if the specific gravity of the iron in the shot is 6.8?

24. Find the volume of the metal in a hollow iron sphere having an outside diameter of 24 in. and an inside diameter of 23 in.

25. Find the weight of a hollow iron sphere if the outside diameter is 16 in. and the metal is 1.5 in. thick. Specific gravity of the iron is 6.8.

26. Find the number of cubic inches of metal in a hollow metal sphere having an outside diameter of 22 in. and an inside diameter of 21 in.

27. The total surface area of a geometric solid is 384 sq in. What is its volume if the figure is a cube? What is its volume if it is a rectangular solid 12 in. long and 9 in. wide? What is its volume if the solid is a sphere? (Remember, the area is still 384 sq in.)

28. The volume of a rectangular solid is 216 cu in. What is the surface area if the solid is 12 in. long and 9 in. wide? What is the surface area if the figure is a cube? What is the surface area of a sphere containing the same volume? Can you tell from your answer why all soap bubbles are round?

QUIZ ON CHAPTER 31, THE SPHERE. FORM A.

For the given measurements of spheres, find the indicated values:

1. Diameter = 14 in.; find circumference, surface area, and volume.

2. Radius = 6 in.; find circumference, surface area, and volume.

3. Circumference = 22π in.; find diameter, surface area, and volume.

4. Circumference = 33.3 in.; find diameter, surface area, and volume.

5. Surface area = 100π sq in.; find radius, circumference, and volume.

6. Surface area = 113.2 sq in.; find radius, circumference, and volume.

7. Volume = 36π cu in.; find radius, circumference, and surface area.

8. Volume = 268 cu in.; find radius, circumference, and surface area.

CHAPTER THIRTY-TWO The Metric System of Measurement

32-1 DEFINITION

The metric system of measurement is a decimal system; that is, the system is based upon the number 10, just as our number system itself is based on 10. To change the size of the units in the metric system, all we need to do is to move the decimal point to the right or left.

In order to understand the importance of such an arrangement, let us imagine what we might do if there were 10 in. in 1 ft, 10 ft in 1 yd, 10 yd in 1 rd, 10 rd in 1 furlong, and 10 furlongs in 1 mile.

35,672 in. could be changed to
3567.2 ft
356.72 yd
35.672 rd
3.5672 furlongs
0.35672 miles

Instead, in our present method, called the English system, we must perform the following divisions:

$35{,}672 \div 12$	$= 2972.667$	number of feet
$2972.667 \div 3$	$= 990.889$	number of yards
$990.889 \div 5.5$	$= 180.16$	number of rods
$180.16 \div 40$	$= 4.504$	number of furlongs
$4.504 \div 8$	$= 0.563$	number of miles

32-2 ORIGIN OF METRIC SYSTEM

Our common English system of measurements grew up piece by piece without any definite plan. As a result there was no attempt to make it easy to convert one kind of division into another. The metric system, instead, was planned as a complete system. It was devised and set up in France in 1789 at the time of the French Revolution. At that time, the French decided to make a completely new start in setting up a logical system of measurements.

The foot measurement had originally come from a length in some way related to that of a man's foot. Instead, the French decided to start out with some measurement that was fixed in nature and not so variable as men's feet. They decided to take as a unit of length one ten-millionth of the distance from the equator to the north pole. This distance was to be the primary unit and was called one *meter* (m), meaning *measure.* The distance chosen for this length is equal to 39.37 inches in the English system, although later calculations showed that this length is not exactly one ten-millionth of the distance from the equator to the north pole. However, the relation between the English system and the metric system has been established by law, so that now

$$1 \text{ meter} = 39.37 \text{ inches}$$

The meter is therefore a little longer than the English yard of 36 inches.

32-3 UNITS OF THE METRIC SYSTEM

In the metric system all measurements of length, area, volume, and even weight are based upon the primary unit, the *meter* (m). The meter is divided into smaller divisions but not by dividing it into twelve parts as we do in the English system to change 1 foot to 12 inches. Instead, the divisions in the metric system are in *tenths.* The primary unit of length, the meter, is divided into ten parts. Each part is $\frac{1}{10}$ of a meter. The word for tenth is "deci." In fact, the word *decimal* itself means *one tenth.* Therefore, a length $\frac{1}{10}$ of a meter is called a *decimeter* (dm). A decimeter (Fig. 32-1) is equal to 3.937 inches, or approximately 4 inches.

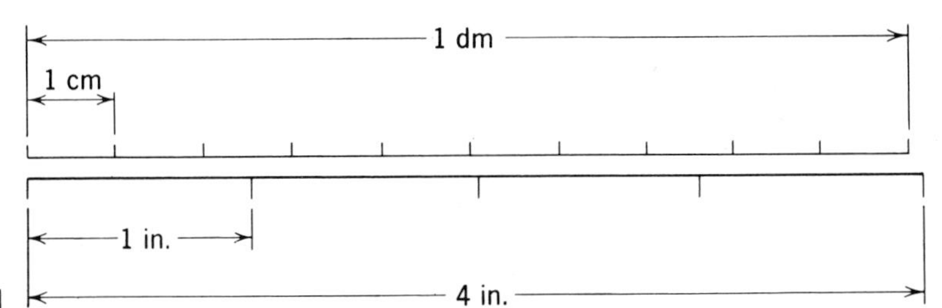

Figure 32-1

The decimeter is further divided into ten parts. Each part, about $\frac{2}{5}$ of an inch, is called a *centimeter* (cm). The prefix "centi" means one-hundredth. In fact, the word for our coin, the *cent,* means $\frac{1}{100}$ of a dollar. One centimeter is $\frac{1}{100}$ of a meter, or 0.3937 inch, which is approximately $\frac{2}{5}$ of an inch. In other words, 1 inch equals approximately $2\frac{1}{2}$ centimeters. More accurately,

$$1 \text{ in.} = 2.54 \text{ cm (approximately)}$$

The centimeter is therefore a rather small measurement. However, it is much used in scientific work.

The centimeter is further divided into ten divisions, each division called one *millimeter* (mm). The prefix "milli" means one one-thousandth. The millimeter is therefore $\frac{1}{1000}$ of a meter. It is a very small measurement, approximately $\frac{1}{25}$ of an inch.

In actual practice it always happens that some units of measurement are used more than others. For instance, the decimeter is not often used. Instead, the centimeter and the millimeter have quite common usage in our everyday speech. A 35-mm camera film is a film that is 35 millimeters wide. The sizes, 8 mm and 16 mm, are also used to refer to the width of films. The bores of cannon and of smaller firearms are often stated in terms of the metric system. An 8-mm rifle has approximately the same bore, or calibre, as a "30-30." A 150-mm cannon is a 6-inch gun.

In the explanation of the metric system up to this point we started with the primary unit, the meter, and moved toward smaller and smaller units, the decimeter, the centimeter, and down to the millimeter. Now we shall move toward larger measurements.

A distance of 10 meters is called 1 *decameter* (dkm), sometimes spelled *dekameter.* (The prefix "deca" means *ten.*) A decameter is a distance of about 32.8 feet. A distance of 10 decameters is called 1 *hectometer* (hkm). (The prefix "hecto" means 100.) A hectometer is equal to 100 meters. The decameter and the hectometer are not much used in everyday speech. For instance, a sprint of 1 hectometer is called the "hundred-meter dash."

A distance of 10 hectometers is called 1 *kilometer* (km). (The prefix "kilo" means 1000.) A kilometer is equal to 1000 meters, or approximately $\frac{5}{8}$ of a mile.

The following table is a summary of the relations between the units of length in the metric system:

$$
\begin{aligned}
10 \text{ millimeters (mm)} &= 1 \text{ centimeter (cm)} \\
10 \text{ cm} &= 1 \text{ decimeter (dm)} \\
10 \text{ dm} &= 1 \text{ meter (m)} \\
10 \text{ m} &= 1 \text{ decameter (dkm)} \\
10 \text{ dkm} &= 1 \text{ hectometer (hkm)} \\
10 \text{ hkm} &= 1 \text{ kilometer (km)} \\
100 \text{ cm} &= 1 \text{ m} \\
1000 \text{ mm} &= 1 \text{ m} \\
1 \text{ cm} &= \tfrac{1}{100} \text{ m} \\
1 \text{ mm} &= \tfrac{1}{1000} \text{ m}
\end{aligned}
$$

Notice especially the meaning of these prefixes:

deci means $\frac{1}{10}$	*deca* means 10
centi means $\frac{1}{100}$	*hecto* means 100
milli means $\frac{1}{1000}$	*kilo* means 1000
micro means $\frac{1}{1000000}$	*mega* means 1,000,000

that is,

1 decimeter	$= \frac{1}{10}$ m	1 decameter	= 10 m
1 centimeter	$= \frac{1}{100}$ m	1 hectometer	= 100 m
1 millimeter	$= \frac{1}{10000}$ m	1 kilometer	= 1000 m
1 micron	$= \frac{1}{1000000}$ m	1 megameter	= 1000000 m

In order to see the convenience of the metric system, consider the following example. Suppose we start with 1647832 millimeters.

$$\begin{aligned} 1647832 \text{ mm} &= 164783.2 \text{ cm} \\ &= 16478.32 \text{ dm} \\ &= 1647.832 \text{ m} \\ &= 164.7832 \text{ dkm} \\ &= 16.47832 \text{ hkm} \\ &= 1.647832 \text{ km} \end{aligned}$$

32-4 REDUCTION AND CONVERSION OF UNITS OF MEASUREMENT

Numbers that represent measurement are called *denominate* numbers, such as 15 feet, 48 inches, 35 centimeters and so on. In any measurement, we use a *unit* of measure, such as the *inch*, the *foot*, the *meter*, and so on.

It is often necessary to change a measurement from one unit to another, such as 8 ft to 96 inches, or 180 inches to 5 yards. Such a change within a system is called a *reduction*.

It is also often desirable to make a change from one system to another, for example, to change 15 inches to centimeters. A change from the units of one system to the units of another is called a *conversion*. In making a reduction or a conversion, we must know the relations between the units in the same system or in different systems.

Most people are familiar with the relations between the units in the English system: 1 ft = 12 in; 3 ft = 1 yd; 5280 ft = 1 mi. To change measurements from the English to the metric system, or from metric to English, we use chiefly the conversion relations:

1 meter (m) = 39.37 inches (in.)
1 inch (in.) = 2.54 centimeters (cm) (approximately)
1 kilometer (km) = 0.621 mile (or approximately $\frac{5}{8}$ mile)

The following examples show some conversions from English to metric systems.

Example 1

Change 15 in. to centimeters.

Solution

We use the relation: 1 in. = 2.54 cm.
Multiplying by 15, 15 in. = 15(2.54 cm) = 38.1 cm

Example 2

Change 80 cm to inches.

Solution

Again we use the relation: 1 in. = 2.54 cm.
In this case, we divide: $80 \div 2.54 = 31.5$ (inches)(approx.).

Example 3

Change 50 yd to meters.

Solution

First we change 50 yd to feet:

1 yard = 3 ft
Multiplying, 50 yards = 50(3 ft) = 150 ft.
Now we change to inches: 1 ft = 12 in.
Multiplying 150 ft = 150(12 in.) = 1800 in.
Now we use the relation: 1 m = 39.37 in.
Dividing by 39.37, we have $1800 \div 39.37 = 45.72$ (m) (approximately).

Example 4

Change 1350 cm to feet.

Solution

If we wish, we can first find the number of centimeters in 1 ft. Since 1 in. = 2.54 cm., then 12 in. = 12(2.54 cm) = 30.48 cm. Now we divide the number of centimeters, 1350, by 30.48, and get

$$1350 \div 30.48 = 44.3 \text{ (ft.) (approx.)}$$

Example 5

Change 20 miles to kilometers.

Solution

Using the relation: 1 km = 0.621 mile, we divide:

$$20 \div 0.621 = 32.2 \text{ (km) (approx.)}$$

To check the answer, we can change 20 miles to inches, and then divide by 39.37 to get meters, and divide by 1000 to get kilometers:

$$20 \text{ miles} = (20)(5280)(12) = 1267200 \text{ (in.)}$$

Now we divide by 39,370; $1267200 \div 39{,}370 = 32.2$ (km) (approx.).

Exercise 32-1

Change each of the following linear measurements to the form indicated. Carry decimals out to a reasonable degree.

1. 33 in. to cm
2. 95 cm to in.
3. 52.1 ft to meters
4. 221 m to ft
5. 2.25 ft to cm
6. 18.5 in. to mm
7. 42 yd to meters
8. 3000 m to yd
9. 3.42 miles to km
10. 630 mm to in.
11. 48 km to miles
***12.** 2.45 mils to in.*

13. State the length of a foot rule in centimeters.

14. A man's height is 6 ft $\frac{1}{2}$ in. State his height in centimeters.

15. The height of a table is 30 in. State the height in centimeters.

16. A desk is 44 in. long and 32 in. wide. State measurements in centimeters.

17. State the length of a 19-in. diagonal TV screen in centimeters.

18. A fish measured 2 ft 8.5 in. State the length in centimeters.

19. A woman bought 8.5 yd of a fabric. State the length in meters.

20. A boat measured 34 ft long. State the length in meters.

21. A man jogged 6.5 miles. State the distance in kilometers.

22. The height of the Sears Tower in Chicago is 1454 ft. Change to meters.

23. The length of the Brooklyn Bridge is 1595 ft. Change to meters.

24. The Mall Tunnel in The District of Columbia is 3400 ft long. Change to meters.

25. The Kentucky Derby track is 1.25 miles long. Change to kilometers.

26. A golf drive was 175 yd. Change to meters.

27. In 1976, the women's high jump record was 6 ft 3.75 in. Change to centimeters.

* 1 mil = 0.001 in.

28. In a football game, the average punt was 36.4 yd. Change to meters.

29. A certain camera has an 80-mm lens. How many inches is this?

30. A snapshot picture measures 5.72 cm by 8.26 cm. Change to inches.

31. A paper clip is made of wire 0.036 in. in diameter. How many mils is this? How many mm?

32. The thickness of some sheet metal is 0.0125 in. State the thickness in mils and millimeters.

32-5 DIMENSIONAL ANALYSIS

In the reduction or conversion of units, we often find it convenient to use a form called *dimensional analysis.* In using this form, we show the names of the units as well as the quantities themselves. Then, we perform the operations with the units in the same way as we do with the numbers or values. In this method, we "cancel" the names of the units in numerators and denominators.

We first show the method by a simple example.

Example 1

Change 15 ft to inches.

Solution

We probably first think of multiplying by 12, because there are 12 in. in 1 ft. However, let us first set up the form showing multiplication by: $\frac{12 \text{ in.}}{1 \text{ ft}}$, a fraction that is equal to 1. We write the names of the units:

$$15\,\cancel{\text{ft}} \times \frac{12 \text{ in.}}{1\,\cancel{\text{ft}}} = 180 \text{ in.}$$

In this example, the symbol "*ft*" is cancelled in the numerator and the denominator. Then the answer shows the proper denomination. Note that the multiplier is equal to 1.

Example 2

Change 6 miles to yards.

Solution

Since the result must be stated in yards, we must have "yards" as the final numerator. Then, showing the units, we write

$$\frac{\overset{2}{\cancel{6}}\,\cancel{\text{mi}}}{1} \times \frac{5280\,\cancel{\text{ft}}}{1\,\cancel{\text{mi}}} \times \frac{1\text{ yd}}{\cancel{3}\,\cancel{\text{ft}}} = 10{,}560\text{ yd}$$

The names of the units, "mi" and "ft" are "cancelled" in numerators and denominators. Then the answer has the correct name. Note that we multiply by two fractions, each equal to 1.

Example 3

Change 7700 yd to miles.

Solution

$$7700\,\cancel{\text{yd}} \times \frac{3\,\cancel{\text{ft}}}{1\,\cancel{\text{yd}}} \times \frac{1\text{ mi}}{5280\,\cancel{\text{ft}}} = 4.375\text{ mi}$$

Note that we "cancel" names of units as well as quantities in numerators and denominators. Then the answer has the correct denomination.

We can use a similar form for conversion of units from one system to another, showing the names of the units and their relations.

Example 4

Change 35 in. to centimeters.

Solution

We use the relation: 1 in. = 2.54 cm.
Of course, we might first think of simply multiplying:

$$35 \times 2.54\text{ cm} = 88.9\text{ cm}$$

However, let us show the names of the units in the following form:

$$35\,\cancel{\text{in}}. \times \frac{2.54\text{ cm}}{1\,\cancel{\text{in}}.} = 88.9\text{ cm}$$

Then we "cancel" the notation "in." for inches, and the answer shows the proper denomination, *cm*.

Example 5

Change 48 yd to meters.

Solution

We use the relations: 1 yd = 3 ft, 1 ft = 12 in., 1 m = 39.37 in. We set up the form

$$48\,\cancel{\text{yd}} \times \frac{3\,\cancel{\text{ft}}}{1\,\cancel{\text{yd}}} \times \frac{12\,\cancel{\text{in}}.}{1\,\cancel{\text{ft}}} \times \frac{1\text{ m}}{39.37\,\cancel{\text{in}}.} = 43.89\text{ meter}$$

Example 6

Change 24 miles to kilometers.

Solution

We use the relations: 1 mile = 5280 ft., 12 in. = 1 ft; 1 m = 39.37 in., 1 km = 1000 m. Then we write

$$24\ \cancel{mi} \times \frac{5280\ \cancel{ft}}{1\ \cancel{mi}} \times \frac{12\ \cancel{in.}}{1\ \cancel{ft}} \times \frac{1\ \cancel{m}}{39.37\ \cancel{in.}} \times \frac{1\ \text{km}}{1000\ \cancel{meter}} = 38.6\ \text{km}$$

Note that each multiplying fraction is equal to 1.

We can use a similar form to change units in connection with velocity.

Example 7

A speed of 45 mph is equivalent to how many feet per second?

Solution

The word "per" refers to a fraction. Then we write:

$$\frac{45\ \cancel{mi}}{\cancel{hr}} \times \frac{1\ \cancel{hr}}{60\ \cancel{min}} \times \frac{1\ \cancel{min}}{60\ \text{sec}} \times \frac{5280\ \text{ft}}{1\ \cancel{mi}} = \frac{66\ \text{ft}}{\text{sec}}$$

Here we "cancel" units as well as the quantities, and the answer shows ft/sec; that is, feet per second.

Exercise 32-2

Make the changes indicated in each of the following problems, using the method of analysis as shown in the foregoing examples (1–5) in Section 32-5.

1. 32 yd to meters
2. 64 m to yards
3. 52 ft to meters
4. 30 m to feet
5. 45 in. to centimeters
6. 1.6 ft to centimeters
7. 1.8 m to in.
8. 1.8 in. to millimeters
9. 52 miles to kilometers
10. 1000 km to miles
11. 450 mm to inches
12. 63 cm to inches
13. 15 mph to feet per second
14. 75 mph to feet per second
15. 20 mph to meters per second
16. 3.5 mph to feet per minute

32-6 SQUARE MEASURE

By a square foot we mean the amount of area contained in a square 1 ft long and 1 foot wide (Fig. 32-2). A square inch is the area equal to a square that is 1 inch long and 1 inch wide.

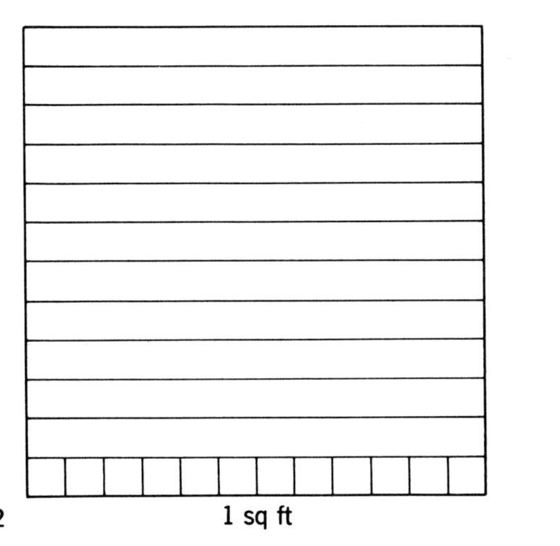

Figure 32-2

In order to find the number of square inches in 1 square foot, we lay off the square inch as many times as possible on the square foot of area. Since there are 12 inches in 1 foot, we shall have 12 square inches in 1 row and exactly 12 rows. Therefore, we have

$$1 \text{ sq ft} = 12 \times 12 \text{ sq in.} = 144 \text{ sq in.}$$

The expression is often written: $1 \text{ ft}^2 = 12^2 \text{ in.}^2$ The exponent 2 on 12 means that 12 is squared. To indicate square inches, we can write "in.^2" The 2 written as an exponent on the name of a unit indicates square units. In the same way

$$1 \text{ yd} = 3 \text{ ft}; \qquad \text{then} \qquad 1 \text{ yd}^2 = 3^2 \text{ ft}^2; \qquad \text{that is,} \qquad 1 \text{ sq yd} = 9 \text{ sq ft}$$

In the metric system we have

$1 \text{ cm} = 10 \text{ mm}$	$1 \text{ m} = 100 \text{ cm}$	$1 \text{ m} = 1000 \text{ mm}$
$1 \text{ cm}^2 = 100 \text{ mm}^2$	$1 \text{ m}^2 = 100^2 \text{ cm}^2$	$1 \text{ m}^2 = 1000^2 \text{ m}^2$

In converting square measure in the metric system to the English system, we use the same reasoning:

$1 \text{ in.} = 2.54 \text{ cm};$	$1 \text{ m} = 39.37 \text{ in.}$
$1 \text{ in.}^2 = (2.54)^2 \text{ cm}^2 = 6.45 \text{ cm}^2$	$1 \text{ m}^2 = (39.37)^2 \text{ in.}^2 = 1550 \text{ in.}^2$

Example 1

Find the perimeter in centimeters and the area in square centimeters of a sheet of writing paper, 11 in. long and 8.5 in. wide.

Solution

The length and width can first be changed to centimeters:

$$\text{length} = (11)(2.54 \text{ cm}) = 27.94 \text{ cm}; \quad \text{width} = (8.5)(2.54 \text{ cm}) = 21.59 \text{ cm}$$

The perimeter is found by adding the two sides and two ends together:

$$\text{perimeter} = (2)(27.94\ \text{cm}) + (2)(21.59\ \text{cm}) = 99.06\ \text{cm}$$

Then,

$$\text{area} = (27.94\ \text{cm})(21.59\ \text{cm}) = 603.2\ \text{cm}^2\ \text{(rounded off)}$$

To check the work, we can first find perimeter and area in English units:

$$\text{perimeter} = (2)(11\ \text{in.}) + (2)(8.5\ \text{in.}) = 39\ \text{in.}$$

Changing to centimeters:

$$(39)(2.54\ \text{cm}) = 99.06\ \text{cm (perimeter)}$$

The perimeter checks with the first method.

To check the area, we first find the area in English units:

$$\text{area} = (11\ \text{in.})(8.5\ \text{in.}) = 93.5\ \text{in.}^2$$

Now we use the relation: $1\ \text{in.}^2 = (2.54)^2\ \text{cm}^2$

Multiplying, $$(93.5)(6.45\ \text{cm}^2) = 603.075\ \text{cm}^2$$

The answer checks with the first method to three significant digits. This is the greatest accuracy that can be expected when we use the number 6.45 instead of the more accurate number: $(2.54)^2 = 6.4516$, for the number of square centimeters in 1 sq in.

If, in the foregoing example, we first change the number of inches in length and width to centimeters, we multiply *each* of the two given measurements by 2.54. Then, in the multiplication for the area, we are actually multiplying by 2.54 *twice*, or by $(2.54)^2$. The length is (11)(2.54 cm), and the width is (8.5)(2.54 cm). The area becomes

$$(11)(2.54\ \text{cm})(8.5)(2.54\ \text{cm}) = 603\ \text{cm}^2\ \text{(rounded off).}$$

If we first find the area in square inches and then convert to square centimeters, we have

$$(11)(8.5)(2.54^2\ \text{cm}^2) = 603\ \text{cm}^2$$

Note that the multiplication in both cases involves the same factors.

To convert centimeters to inches, we use the conversion factor 2.54, but we must divide instead of multiply. For example, to change 47 cm to inches, we divide by 2.54; or multiply by 1/2.54;

$$47 \div 2.54 = 18.5\ \text{(in.)}$$

We can use the form:

$$47\ \cancel{\text{cm}} \times \frac{1\ \text{in.}}{2.54\ \cancel{\text{cm}}} = 18.5\ \text{in.}$$

Example 2

A sheet of writing paper measures 32.3 cm long and 24.6 cm wide. How many square inches does it contain? Find the perimeter, in inches.

Solution

If we first find the area in cm^2, we have

$$(24.6\text{ cm})(32.3\text{ cm}) = 794.58\text{ cm}^2$$

Now we divide by 6.45 cm^2, or multiply by the reciprocal, 1/6.45 cm^2:

$$794.58\,\not{c}\!\not{m}^2 \times \frac{1\text{ in.}^2}{6.45\,\not{c}\!\not{m}^2} = 123.19\text{ in.}^2, \qquad \text{or} \qquad 123\text{ in.}^2\text{ (rounded off)}$$

If we first express the measurements in inches, we have

$$\text{length} = 32.3 \div 2.54 = 12.716\text{ (in.)}; \quad \text{width} = 24.6 \div 2.54 = 9.685\text{ (in.)}$$

For the area, we get, rounding numbers to three significant digits,

$$\text{area} = (12.7\text{ in.})(9.68\text{ in.}) = 123\text{ in.}^2\text{ (rounded off)}$$

For the perimeter, we have:

$$\text{perimeter} = (2)(32.3\text{ cm}) + (2)(24.6\text{ cm}) = 113.8\text{ cm}$$

For inches, we divide $113.8 \div 2.54 = 44.8$ (inches in perimeter).
Or, perimeter = (2)(12.7 in.) + (2)(9.68 in.) = 44.8 in.

Exercise 32-3

In Nos 1–4, find the perimeter in centimeters and the area in cm^2.

1. A rectangle, 8.6 in. by 7.2 in. **2.** A picture, 22.5 in by 18.4 in.

3. A snapshot, 4.5 in. by 3 in. **4.** A snapshot, 3.75 in by 2.75 in.

In Nos. 5–8, find the circumference in centimeters and the area in cm^2.

5. A circle: diameter = 12 in. **6.** A coin, diameter = 3.2 in.

7. Circular mirror, diameter = 5.4 in.

8. Clock face, diameter = 10.4 in.

In Nos 9–12, change measurements to centimeters and/or $cm.^2$

9. Triangle, base = 12.4 in., altitude = 9.3 in.

10. Trapezoid, bases, 14 in. and 11.4 in., altitude = 9.5 in.

11. Square, side = 16.5 in.; find perimeter and area.

12. Right triangle, legs: 8 in. and 6 in. Find hypotenuse, perimeter, and area.

13. Find the number of square meters in a floor, 18 ft by 15 ft.

14. A building lot is 84 ft long and 68 ft wide. Find the number of meters in the perimeter and the number of square meters in the area.

15. An acre lot is 16 rods long and 10 rods wide. Change these measurements to meters and then find the perimeter and the area of the lot (1 rod is equal to 16.5 ft).

16. One *hectare* is the name given to an area equal to 10,000 sq m. How much more or less than 1 acre is 1 hectare?

17. A rectangle is 26 cm long and 18 cm wide. Find the number of inches in the perimeter and the number of square inches in the area.

18. A square measures 34.5 cm on the side. Find the number of inches in the perimeter and the number of square inches in the area.

32-7 CUBIC MEASURE

The relations between measures of volume are similar to the relations between measures of area. For volumes we take the *cubes* of linear units.

$$1 \text{ ft} = 12 \text{ in}; \quad 1 \text{ ft}^3 = 12^3 \text{ in.}^3, \quad \text{or} \quad 1 \text{ cu ft} = 1728 \text{ cu in.}$$

$$1 \text{ in.} = 2.54 \text{ cm}; \quad 1 \text{ in.}^3 = (2.54)^3 \text{ cm}^3, \text{ or} \quad 1 \text{ cu in.} = 16.387 \text{ cu cm}$$

Note. In chemistry, medicine, or liquid measurement, the abbreviation for cubic centimeter is often “cc”. One thousand cubic centimeters (1000 cc) is called 1 *liter* and is slightly more than 1 quart.

Example 1

What is the volume in cubic centimeters of a rectangular container, 6 in. long, 5.5 in. wide, and 4 in. high?

Solution

We could first express each measurement in centimeters and then multiply the three measurements together. However, the work is probably simpler if we first find the volume in cubic inches.

$$(6 \text{ in.})(5.5 \text{ in.})(4 \text{ in.}) = 132 \text{ in.}^3$$

Then,

$$\begin{aligned}(132)(16.387 \text{ cm}^3) &= 2163.084 \text{ cm}^3 \\ &= 2163 \text{ cm}^3 \quad \text{(rounded off to four significant digits)}\end{aligned}$$

Example 2

Change 5 gallons to liters.

Solution

We use the relations: 1 gal = 231 in.^{3}; 1 in.^{3} = 16.387 cm^3; and 1 liter = 1000 cm^3. Then we write

$$5 \text{ gal} \times \frac{231 \text{ in.}^3}{1 \text{ gal}} \times \frac{16.387 \text{ cm}^3}{1 \text{ in.}^3} \times \frac{1 \text{ liter}}{1000 \text{ cm}^3} = 18.93 \text{ (approx) liters}$$

Exercise 32-4

1. A rectangular solid measures 8 in. by 6 in. by 5 in. Find its volume in cm^3, and the surface area in cm^2.
2. A right prism has a base of 108 sq in. and a volume of 810 cu in. Find its height (in cm), its base (in cm^2), and volume in cm^3.
3. A cube measures 6.5 in. along the edge. Find the surface area in cm^2 and the volume of the cube in cm^3.
4. A right circular cylinder has a diameter of 18 in. and an altitude of 32 in. Find the volume (in cm^3) and the surface area (in cm^2).
5. A pyramid has a square base, 13.5 in. on a side, and an altitude of 24 in. Find the number of cm^3 in the volume of the pyramid.
6. Find the number of cm^3 in the volume of a cone whose diameter is 10 in. and altitude is 24 in.
7. A sphere has a diameter of 6 in. Find the number of cm^2 in the surface area and the number of cm^3 in the volume.
8. A right prism has a square base, 15 ft on a side. The height is 18 ft. Find the number of m^2 in the volume of the prism.
9. A rectangular room measures 28 ft long, 18 ft wide, and 9 ft high. Find the volume of the room in steres (1 stere equals 1 cubic meter).
10. A large cylindrical water tank has a diameter of 32 ft and a height of 40 ft. Find the number of steres in the volume of the tank.
11. A pyramid has a rectangular base, 16 ft by 10 ft, and an altitude of 20 ft. Find the number of cubic meters in the volume of the pyramid.
12. Find the volume of a sphere whose diameter is 20 ft.

13. A jewel box measures 20 cm by 12 cm by 5 cm. Convert these measurements to inches, and then find the volume in cubic inches.

14. A suitcase has inside measurements of 72.4 cm, 57.5 cm, and 21.5 cm. Change these measurements to inches, and then find the volume in cubic inches.

32-8 METRIC SYSTEM OF WEIGHTS

In the English system the primary unit of weight is the *pound.* From this primary unit we get the *ounce*, which is 1/16 of a pound, and the *ton*, 2000 pounds.

In the metric system the primary unit of weight is the *gram.* The gram is a very small weight compared with the pound. It is equal approximately to 1/30 of an ounce. For much ordinary measurement of weight, the gram is too small. Instead, a more convenient unit is the *kilogram* (kg), which is equal to 1000 grams. A kilogram is equal to approximately 2.2 pounds.

The gram can be changed into other units in the metric system in the same way as the meter is changed, that is, by multiplying or dividing by 10 or some power of 10. One important small unit used in chemistry, medicine, and in science, is the *milligram*, which is equal to 1/1000 of a gram.

The most common conversion factor between the English system and the metric system of weight is the relation:

$$1 \text{ kg} = 2.2 \text{ lb} \quad \text{(very nearly)}$$

To change kilograms to pounds, multiply the number of kilograms by 2.2. *To change pounds to kilograms, divide the number of pounds by* 2.2.

For greater accuracy, we may use

$$1 \text{ gram} = 0.03527 \text{ ounce (oz)}; \quad 1 \text{ kilogram} = 2.2046 \text{ lb.}$$

In our common English system of measurements,* there is no relation between units of weight and units of length. There is no relation between one pound of weight and one foot of length. However, in the metric system, there is such a relation. One gram of weight

* Some people have wondered why the units of measurement in the metric system are so extremely different in size. For example, the primary unit of length, the meter, is much larger compared with the English *foot.* The size makes the unit inconvenient in much measurement in everyday life. One could not conveniently carry a meter stick in his pocket as one might a foot rule. On the other hand, the primary unit of weight, the gram, is so small compared with the pound that it is too small for use in weighing many common everyday articles. One could not conveniently buy sugar by the gram. The inconvenience of the size of the primary units may be one reason for the reluctance of many people to accept the metric system.

is defined as the weight of one cubic centimeter of pure water at a temperature of 4 degrees Celsius. At this particular temperature, water has its greatest density. At temperatures above or below this point, water expands.

Exercise 32-5

In the first 10 problems, change the measurements as indicated.

1. 1.2 lb to grams

2. 15 lb to kilograms

3. 400 kg to pounds

4. 2.4 oz to grams

5. 200 grams to ounces

6. 3000 grams to pounds

7. 0.5 oz to milligrams

8. 500 mg to ounces

9. 10 gal to liters

10. 15 liters to pints

Some containers of kitchen spices are marked with the number of ounces as well as the number of grams. The following sets of weights are marked on various cans. Do the weights agree?

11. 1.5 oz or 42.52 grams

12. 1.25 oz or 35.4 grams

13. $\frac{3}{8}$ oz or 10.63 grams

14. $\frac{1}{4}$ oz or 7.09 grams

15. 1 oz or 28 grams

16. $1\frac{1}{8}$ oz or 31.89 grams

Find the difference between:

17. 2 oz and 100 grams

18. 20 lb and 8 kg

19. 100 yd and 100 m

20. 40 cm and 15 in.

21. 70 m and 240 ft

22. 0.32 in. and 12 mm

23. 3 in.^{2} and 8 cm^2

24. 5 in.^{2} and 35 cm^2

25. 3 sq miles and 8 km^2

26. 2880 cu ft and 100 steres

27. 5 km and 3 miles

28. 1 qt and 1 liter

Change the measurement in the following common expressions to metric:

29. "A 50-yard dash"

30. "A ten-foot pole"

31. "An ounce of prevention"

32. "A pound of cure"

33. "A yard wide."

34. "I'd walk a mile"

35. "A grain of salt"
(1 lb = 7000 grains)

36. "A ten-gallon hat."

37. Sound travels 1080 ft per sec. How many meters is this?

38. The wavelength for a particular radio station is 239.6 m. How many feet and yards is this?

39. Light and radio waves travel 300,000,000 m per sec. Find the difference between this and 186,000 miles.

40. If a station broadcasts on a frequency of 1,500,000 cycles (or waves) per second (cps), how many kilocycles (kc) is this? What is the length of each wave?

41. Another station has a wavelength of 480 m. What is its frequency in kilocycles?

42. On the Apollo moon flights, the following amounts of samples were returned: Apollo 11: 21,694 grams; Apollo 12; 34,369 grams; Apollo 14: 44,500 grams; Apollo 15: 68,000 grams. Change these metric weights to pounds.

43. For Apollo 15, the earth parking orbit insertion was at the rate of 25,620 ft per sec. Change to miles per hour.

44. The second walk from the lunar module was a distance of 16.1 km. This was really the riding time in the vehicle Rover, which took 2 hours and 15 minutes. What was the speed in mph?

QUIZ ON CHAPTER 32, THE METRIC SYSTEM OF MEASUREMENT. FORM A.

In the first six problems, change the measurements as indicated:

1. 33 in. to centimeters **2.** 130 cm to inches

3. 300 yd to meters **4.** 32 ft to centimeters

5. 24 miles to kilometers **6.** 4 gal to liters

7. A rectangle measures 8 in. long and 6 in. wide. Find the length and the width expressed in centimeters and find the area in cm^2.

8. A rectangle solid measures 9 in. long, 5 in. wide, and 4 in. deep. Find the number of cubic centimeters in the volume.

9. Find the number of liters in the volume of the solid in No. 8.

10. Change a speed of 55 mph to the number of kilometers per hour.

11. If you walk at a rate of 4 mph, what is your speed in feet per minute?

PART FOUR LOGARITHMS

CHAPTER THIRTY-THREE Introduction

33-1 THE MEANING OF LOGARITHMS

Logarithms can be used to simplify computation. Long problems involving multiplication, division, powers, and roots can be made very simple by the use of logarithms. For example, consider the following problem:

$$\sqrt{\frac{(39.8)^3(0.00874)^4(986000)}{(6.49)^3(765)}} \qquad \textit{Answer: } 0.04165$$

The ordinary method of making this computation would take a very long time. We shall see that this problem can be worked by logarithms in about four minutes.

First, we must understand what is meant by a logarithm. We have seen the use of exponents in writing numbers. Here are some examples:

$$4^3 = 64 \qquad 7^2 = 49 \qquad 2^5 = 32 \qquad 10^6 = 1{,}000{,}000$$

In these examples, the exponents are 3, 2, 5, and 6. These numbers are also called *logarithms. That is, a logarithm is an exponent.*

Although logarithms are exponents, there is a slight difference in the way each is stated. Let us take a specific example:

$$5^3 = 125$$

In this example, 3 is the *exponent,* and 5 is called the *base.* Then we say 3 is the exponent on the base 5, but we say 3 is the *logarithm of* 125 *to the base* 5. Here are other examples:

since $6^2 = 36$, we say 2 is the logarithm of 36 to the base 6

since $3^4 = 81$, we say 4 is the logarithm of 81 to the base 3

In general, we have the *definition:*

$$\text{if} \qquad b^x = N$$

then x is called the *logarithm of N to the base b.*

33-2 NOTATION USED IN LOGARITHMS

The word *logarithm* is usually abbreviated "log." Now, consider a little more carefully two examples:

In the statement, $9^2 = 81$, the log of 81 is 2.
In the statement, $3^4 = 81$, the log of 81 is 4.

Note that the logarithm depends on what base is used. To make each statement clear, we must indicate the base. That is,

the logarithm of 81 is 2 when the base is 9;
the logarithm of 81 is 4 when the base is 3

In a logarithmic statement, the base is usually indicated by a subscript to the word *log*. The word *of* is omitted. Then we can write

$$\log_9 81 = 2 \qquad \text{and} \qquad \log_3 81 = 4$$

The same statements may be written in exponential form:

$$\log_9 81 = 2 \quad \text{means } 9^2 = 81; \qquad \log_3 81 = 4 \quad \text{means } 3^4 = 81$$

Any expression in logarithmic form may be written in exponential form with the same meaning as shown by these examples:

exponential form	logarithmic form
$6^2 = 36$	$\log_6 36 = 2$
$5^4 = 625$	$\log_5 625 = 4$
$10^3 = 1000$	$\log_{10} 1000 = 3$
$b^x = N$	$\log_b N = x$

Exercise 33-1

In each of the following examples find the indicated power as shown in the first two examples. Then write each in logarithmic form.

1. $5^2 = 25$ **2.** $6^3 = 216$ **3.** $12^2 =$ **4.** $11^3 =$

5. $8^4 =$ **6.** $3^5 =$ **7.** $2^{11} =$ **8.** $7^3 =$

9. $4^4 =$ **10.** $(\frac{1}{2})^3 =$ **11.** $9^3 =$ **12.** $10^5 =$

13. $(\frac{1}{3})^4 =$ **14.** $10^{-3} =$ **15.** $2^{-5} =$ **16.** $16^{\frac{1}{2}} =$

17. $8^{\frac{1}{3}} =$ **18.** $5^0 =$ **19.** $10^0 =$ **20.** $10^1 =$

Express each of the following in logarithmic form:

21. $b^x = K$ **22.** $5^x = z$ **23.** $3^x = 28$ **24.** $7^a = b$

25. $4^y = 328$ **26.** $6^t = 30$ **27.** $x^3 = y$ **28.** $5^z = 41.2$

29. $10^{1.5} = 31.63$ (approx.) **30.** $10^{3.6021} = 4000$ (approx.)

Express each of the following in exponential form:

31. $\log_8 512 = 3$ **32.** $\log_3 27 = 3$ **33.** $\log_2 512 = 9$

34. $\log_5 3125 = 5$ **35.** $\log_4 \frac{1}{2} = -\frac{1}{2}$ **36.** $\log_9 1 = 0$

37. $\log_{10} 10 = 1$ **38.** $\log_{10} 1000 = 3$ **39.** $\log_a b = c$

40. $\log_c xy = a$ **41.** $\log_{10} 0.001 = -3$ **42.** $\log_x (a + b) = y$

Find the value of x in each of the following examples:

43. $\log_4 64 = x$ **44.** $\log_5 625 = x$ **45.** $\log_3 x = 4$

46. $\log_2 x = 6$ **47.** $\log_7 x = 4$ **48.** $\log_9 x = \frac{1}{2}$

49. $\log_x 36 = 2$ **50.** $\log_x 0.25 = 2$ **51.** $\log_x 1000 = 3$

33-3 THE BASE AS A BUILDING BLOCK

Let us compare three specific examples:

$$8^2 = 64 \quad \text{then} \quad \log_8 64 = 2$$
$$4^3 = 64, \quad \text{then} \quad \log_4 64 = 3$$
$$2^6 = 64, \quad \text{then} \quad \log_2 64 = 6$$

Here we have three different logarithms of 64. The logarithm is 2, 3, or 6, depending on the base used. When the base is large, the logarithm is small; when the base is small, the logarithm is large. If we use a small base, we must use a large exponent to bring the base up to any desired value.

We might look upon the base as a sort of "building block." As an illustration, for any particular building, the number of building blocks needed will depend on the size of the blocks used. If we use a large block, or base, such as 8, we need to build it up only twice (by multiplication) to reach 64. If we use a small base, such as 2, we need to build it up six times (by multiplication) to reach 64.

33-4 ANY NUMBER EXPRESSED AS POWER OF ANOTHER NUMBER

It can be shown that any number can be expressed as a power of another number. We have expressed the number 64 as a power of 8, 4, and 2. Yet we can express 64 as a power of some other number, say, 7, if we wish. In this case, the power will not be a whole number. The second power of 7 is 49; the third power is 343. Therefore, if 64 is expressed as a power of 7, this power must be between 2 and 3. Actually, it can be shown that

$$64 = 7^{2.137} \text{ (approx.)}$$

or in log form,

$$\log_7 64 = 2.137 \text{ (approx.)}$$

Any number can be expressed as a power of 10. Consider, again, the number 64. Now, we know that $10^1 = 10$, and $10^2 = 100$. Therefore, to build the base 10 up to 64, we must use an exponent between 1 and 2. Actually,

$$64 = 10^{1.8062} \text{ (approx.)}$$

or in log form,

$$\log_{10} 64 = 1.8062 \text{ (approx.)}$$

The first question that comes to mind might be: Why bother with all this? What is the value of it? We shall soon find that it will often save time in problems involving much computation. We know that when we multiply numbers expressed as powers on the same base, we *add* the exponents. In this way *multiplication* is actually reduced to *addition.* When we wish to multiply numbers, we add their logarithms, since the logarithms are exponents. The advantage will become clearer as we get further along in our study of the subject.

33-5 COMMON LOGARITHMS: BASE 10

In computation the base 10 is used for logarithms. This base is so convenient and so common that it is not usually written as a subscript but is understood if no base is shown. Here are some numbers expressed as powers of 10, written in exponential and logarithmic form. Base 10 is understood.

Exponential form	Logarithmic form
$10 = 10^1$	$\log 10 = 1$
$100 = 10^2$	$\log 100 = 2$
$1000 = 10^3$	$\log 1000 = 3$
$64 = 10^{1.8062}$	$\log 64 = 1.8062$
$42 = 10^{1.6232}$	$\log 42 = 1.6232$
$527.4 = 10^{2.7221}$	$\log 527.4 = 2.7221$
$30000 = 10^{4.4771}$	$\log 30000 = 4.4771$

Exercise 33-2

Express each of the following in log form. Values are approximate.

1. $10^{1.9294} = 85$ **2.** $10^{2.9694} = 932$ **3.** $10^{3.3979} = 2500$

4. $10^{0.6812} = 4.8$ **5.** $10^{2.9096} = 812$ **6.** $10^{4.8921} = 78000$

7. $10^{0.7782} = 6$ **8.** $10^{1.7782} = 60$ **9.** $10^{2.7782} = 600$

Express each of the following in exponential form:

10. $\log 470 = 2.6721$ **11.** $\log 6800 = 3.8325$ **12.** $\log 59.2 = 1.7723$

13. $\log 9.15 = 0.9614$ **14.** $\log 336 = 2.5263$ **15.** $\log 75 = 1.8751$

16. $\log 248 = 2.3945$ **17.** $\log 24.8 = 1.3945$ **18.** $\log 2.48 = 0.3945$

33-6 TWO PARTS OF A LOGARITHM

It will be noted that the logarithm is often a mixed number, consisting of a whole number part and a decimal fraction part. The decimal part is, of course, rounded off, since most logarithms are not rational numbers. For example, the logarithm of 75 is 1.8751, rounded off to four decimal places.

The whole number part of a logarithm is called the *characteristic* of the logarithm. The decimal part is called the *mantissa.* For example, in the logarithm of 75, the characteristic is 1, and the mantissa is .8751. We distinguish between the two parts because they are found in different ways. To find the complete logarithm of a number we must find both characteristic and mantissa.

33-7 SCIENTIFIC NOTATION

Before finding the logarithm of a number, let us review the meaning of *scientific notation.* In science we often work with numbers that are very large or very small. For example, the distance from the earth to the sun is approximately 93,000,000 miles. Such a number is written in a special way called *scientific notation.* The number is written by placing a decimal point just to the right of the first significant digit, as 9.3. This is called the standard position for the decimal point. The number 9.3 is then multiplied by the proper power of 10. (See Sec. 16.10.)

This power of 10 is determined by counting the number of places from standard position to the decimal point in the original number. In the number 93,000,000, the decimal point is 7 places to the right of standard position. If a decimal point is not shown in the original number, it is understood to be at the right of the number. Then 93,000,000 is written in scientific notation thus:

$$93{,}000{,}000 = (9.3)(10^7)$$

A decimal fraction can be written in the same way, but the power of 10 will be negative. Example:

$$0.0000435 = (4.35)(10^{-5})$$

Exercise 33-3

Express the following numbers in scientific notation:

1. 186,000	**2.** 75,460	**3.** 4960
4. 0.00253	**5.** 0.00058	**6.** 0.000000634
7. 39.37	**8.** 3.1416	**9.** 950,000,000,000
10. 0.066325	**11.** 0.50	**12.** 0.00003665
13. 10,000,000	**14.** 300,000,000	**15.** 0.0000000073

Express the following facts using scientific notation:

16. The radius of the earth is approximately 3959 miles.

17. One inch is equal to 25,400,000 nanometers.

18. The density of dry air is 0.001293.

19. The velocity of sound is approximately 33,136 cm/sec.

20. The coefficient of expansion of gases is 0.003665.

21. The length of a wave of sodium light is approximately 0.0000005893 meters.

Express the following numbers in expanded form:

22. The constant of gravitation is approximately $(6.67)(10^{-8})$.

23. The number of electrons in 1 coulomb is approximately $(6.28)(10^{18})$.

24. The weight of an electron is approximately 10^{-28} gram.

33-8 FINDING THE CHARACTERISTIC OF A LOGARITHM

When a number is written in scientific notation, the power of 10 is the characteristic of the logarithm. Therefore, to find the characteristic of a logarithm, we count the number of places from standard position to the decimal point. For example, to find the logarithm of 5280, we see that the standard position of the decimal point is between 5 and 2. This is three places from the decimal point in the number. Therefore, the characteristic is 3. For the number 758.6, the characteristic is 2. For the number 0.000836, the characteristic is −4.

33-9 FINDING THE MANTISSA OF A LOGARITHM

To find the mantissa of the logarithm of a number we use a table of mantissas. The mantissas are decimal fractions though the decimal point is often not printed.

Mantissas have been computed by advanced mathematical formulas and in most cases are irrational numbers. They are usually rounded off to 3, 4, 5 or more places. In this book we shall use a four-place table. (See Table 1, Appendix. In this table, the base 10 is understood.)

To see how to find the mantissa of a number, let us take a specific example, say, the number 413. In the table we look in the left-hand column headed N for the first two digits, 41. From this number 41, we follow across the page until we come to the column headed 3, the third digit of the number. There we find the number 6160. This is the mantissa of the logarithm. The complete logarithm of 413 is 2.6160, including characteristic and mantissa. If the third digit of a number is 0, as in the number 410, the mantissa is found in the first column headed 0.

The mantissa of a logarithm is not affected by the position of the decimal point in the number. (This is true only when the base is 10.) The mantissa is determined only by the succession of digits in the number. For example, if the first three significant digits are 413, then the mantissa is 6160, no matter where the decimal point happens to be. The following numbers have logarithms with the same mantissa:

413 4130 41300 41.3 4.13 0.00413

Of course, the characteristics for these numbers will be different. The complete logarithms of these numbers are shown here:

$$\log 413 = 2.6160 \qquad \log 41.3 = 1.6160$$
$$\log 4130 = 3.6160 \qquad \log 4.13 = 0.6160$$
$$\log 41300 = 4.6160 \qquad \log 0.00413 = 0.6160 - 3$$

33-10 HOW TO WRITE A NEGATIVE CHARACTERISTIC

All mantissas in the table are positive. To use the table we must have a logarithm with a positive mantissa. We have seen that the logarithm of 0.00413 has the characteristic, −3, and the mantissa 0.6160. The negative characteristic cannot be written before the mantissa, as −3.6160, for this would make the mantissa also negative. The negative characteristic must be separated from the positive mantissa.

The simplest way to write a negative characteristic is to place it at the right of the mantissa and to indicate that it is negative. For example,

$$\log 0.00413 = 0.6160 - 3$$

Note that the mantissa is still positive. Another way that is often used is to write a negative characteristic in such a way that −10 is always written at the right of the mantissa. A proper adjustment is then made in the positive part of the logarithm. For example, we can write

$$\log 0.00413 = 7.6160 - 10$$

The characteristic then is partly positive and partly negative. Note that the net characteristic is still −3. In most instances we shall use the simplest form with the single characteristic, as 0.6160 − 3.

However, sometimes it is necessary to make an adjustment in the form of the characteristic. A negative characteristic can be written in any form necessary provided we compensate for any change by a positive portion. For example, the log of 0.00413 can be written in any of the following forms, as well as others:

$$0.6160 - 3 \qquad 1.6160 - 4 \qquad 5.6160 - 8$$
$$7.6160 - 10 \qquad 11.6160 - 14 \qquad 27.6160 - 30$$

In each form the net characteristic is still −3. It must be remembered that no matter what change is made, the net value of the characteristic must not be changed.

Similar changes are sometimes necessary in connection with a positive characteristic. For example,

$$\log 645 = 2.8096$$

This can be written

$$\begin{aligned} \log 645 &= 3.8096 - 1 \\ \text{or} \quad &\ 5.8096 - 3 \\ \text{or} \quad &\ 12.8096 - 10 \end{aligned}$$

In each form the net characteristic is still +2.

33-11 ANTILOGARITHMS

An *antilogarithm* (called *antilog*) is the opposite of a logarithm. To find an antilogarithm we must find the number of which a logarithm is given. As an example, the logarithm of 63400 is 4.8021. Then the antilogarithm of 4.8021 is 63400.

Now suppose we know that the logarithm of some number N is 5.8831. We are required to find the number that corresponds to this logarithm. To find the antilogarithm we use the same table, but in *reverse*.

Remember, the table contains only mantissas. Therefore we first look up the mantissa, 8831, *in the body of the table.* When we have located the mantissa, the row and column in which it is found will show the proper three consecutive digits for the required antilogarithm. The mantissa 8831 is found opposite 76 and in column 4. Then, the antilogarithm, the number N, has the digits 764, in that order.

However, now we must establish the decimal point of the number. Note that the characteristic of the logarithm is 5, which means that the decimal point is 5 places to the right of standard position. The required number N, that is, the antilogarithm, is 764000. We can check the answer by finding the logarithm of 764000.

As another example, suppose we have the following logarithm given:

$$\log A = 0.8727 - 3$$

To find the number A, the antilogarithm, we look up the mantissa and find that it corresponds to the succession of digits: 746. The characteristic, −3, indicates that the decimal point is 3 places to the left of standard position. Then we have the value: $A = 0.00746$, the antilogarithm.

Exercise 33-4

Find the logarithm of each of these numbers. Then write each as a power of 10.

1. 17200 **2.** 264 **3.** 31.5 **4.** 4.29

5. 0.697	**6.** 0.000538	**7.** 0.851	**8.** 0.000074
9. 20600	**10.** 50.4	**11.** 3100	**12.** 9830000
13. 0.0670	**14.** 0.525	**15.** 0.00491	**16.** 0.00803
17. 6	**18.** 0.0705	**19.** 0.888	**20.** 40000000
21. 1.06	**22.** 614	**23.** 0.0057	**24.** 0.000035
25. 96.9	**26.** 7.83	**27.** 0.5	**28.** 100000

Find the antilogarithm of each of the following logarithms:

29. 4.1847	**30.** 2.3729	**31.** 1.5145	**32.** 0.6821
33. 0.7513 − 2	**34.** 0.7910 − 4	**35.** 0.8987 − 1	**36.** 0.9289 − 5
37. 6.9800	**38.** 5.2788	**39.** 1.5011	**40.** 8.4800
41. 0.6201 − 6	**42.** 0.8235 − 3	**43.** 0.7490	**44.** 0.8451 − 1

33-12 INTERPOLATION

It often happens that we need to find the logarithm of a number containing four significant digits, such as 1965. However, this four-digit number is not found in the table of mantissas. The table shows the mantissa for 1960 but the next entry is for 1970. Then

$$\log 1960 = 3.2923$$
$$\log 1970 = 3.2945$$

Since the logarithm of a number increases as the number itself increases, it is reasonable that the logarithm of 1965 should lie between the logarithms of 1960 and 1970. In this example, we can safely guess that the logarithm of 1965 is about midway between the logarithms of 1960 and 1970. A good guess would be that the logarithm of 1965 is approximately 3.2934.

Interpolation may be defined as simply a good guess or estimate for a number lying between two known numbers. In making this estimate all we need to consider is often only the last two or three digits of the mantissas.

It should be mentioned that in making this estimate we are assuming that the logarithm of a number increases in direct proportion to the increase in the number itself. It is not true that logarithms increase in exactly the same ratio as the numbers, but if the difference between the two known numbers is small, the error is negligible. If the difference between two numbers is large, then the error in interpolation would be too great. For example, it is not correct to assume that the logarithm of 700 lies midway between the logarithms of 600 and 800. However, we can safely assume that the logarithm of 1965 lies approximately midway between the logarithms of 1960 and 1970.

Suppose we wish to find the logarithm of 1964. This number lies 0.4 of the way from 1960 to 1970. Then it is reasonable to assume that the logarithm lies approximately 0.4 of the way from the first logarithm to the second.

Now, we see that the difference between the two known logarithms is 22 points. We take 0.4 of this difference and get

$$(0.4)(22) = 8.8$$

The number of points is rounded off to 9, and this number is then added to the smaller logarithm. We get: $3.2923 + 0.0009 = 3.2932$.

The actual work of interpolation may be simplified by considering the mantissas only. We write down the numbers and their corresponding mantissas, as shown here. (Do not use the equal sign between the number and its mantissa.)

Number		Mantissa		
1960	→	2923	Difference is 22 points	
1964	→	?	Multiply by	0.4
				8.8
1970	→	2945	Round off to	9

The first mantissa is therefore increased by 9 points. The complete logarithm of 1964 is therefore 3.2932.

In actual computation most interpolation should be done mentally.

Interpolation is also often necessary in finding an antilogarithm. We illustrate the method by an example. Suppose we know the logarithm of some number N is 5.2655. That is, $\log N = 5.2655$. Now we wish to find the number N. Since the table contains only mantissas, we look for the mantissa 2655 in the body of the table. This particular mantissa is not in the table. We find one mantissa, 2648, slightly smaller than the given mantissa, and another mantissa, 2672, slightly larger. These two mantissas in the table correspond to the numbers 1840 and 1850, respectively.

The given mantissa, 2655, lies between the two in the table. Therefore, the antilogarithm N should lie between 1840 and 1850. Now, we try to determine what part of the distance from 1840 to 1850 we must go for the given mantissa. The difference between the mantissas in the table is 24 points. The given mantissa is 7 points above the lower one in the table. Then we can say the given mantissa is $\frac{7}{24}$ of the way from the smaller mantissa to the larger. We reduce the fraction $\frac{7}{24}$ to tenths and round the answer to 0.3. This digit is the fourth digit of the antilogarithm. The required number, the antilogarithm, then has the succession of digits, 1843. The decimal point, determined by the characteristic, is 5 places to the right of standard position. Then $N = 184300$.

To see the method more clearly, let us set up the two mantissas from the table with their corresponding antilogarithms. We also show the given mantissa for which we wish to find the antilogarithm. Read from right to left.

$$\begin{array}{ccll} \text{Number} & & \text{Mantissa} & \\ 1840 & \leftarrow & 2648 & \Big\}7 \\ ? & \leftarrow & 2655 & \Bigg\}24 \\ 1850 & \leftarrow & 2672 & \end{array}$$

The given mantissa is $\frac{7}{24}$ of the distance from the smaller to the greater. Then the unknown number N is $\frac{7}{24}$, or 0.3, of the distance from 1840 to 1850.

Exercise 33-5

Find the logarithm of each of the following numbers:

1. 126.5 **2.** 14320 **3.** 0.01683 **4.** 0.001713

5. 0.1924 **6.** 2.075 **7.** 22.95 **8.** 240100

9. 255.9 **10.** 0.02608 **11.** 0.2985 **12.** 0.0003142

13. 3.437 **14.** 371800 **15.** 0.4196 **16.** 0.04604

17. 49.61 **18.** 5087 **19.** 5.052 **20.** 5615000

21. 0.06949 **22.** 0.0007391 **23.** 0.7833 **24.** 8.006

25. 89.19 **26.** 9.538 **27.** 0.09792 **28.** 0.00009821

The following numbers are logarithms. Find the antilogarithm of each.

29. 4.1679 **30.** 6.2465 **31.** 0.2840 – 2 **32.** 0.3938 – 4

33. 5.4043 **34.** 2.4527 **35.** 0.4647 – 3 **36.** 0.4803 – 1

37. 3.5164 **38.** 1.5900 **39.** 0.6600 – 5 **40.** 0.6908 – 7

41. 0.7201 **42.** 1.7820 **43.** 0.8926 – 8 **44.** 8.9282

QUIZ ON CHAPTER 33, INTRODUCTION TO LOGARITHMS. FORM A.

In the following examples (1–8) find the indicated power and then write each in logarithmic form:

1. $7^2 =$ **2.** $8^3 =$ **3.** $2^5 =$ **4.** $5^2 =$

5. $10^4 =$ **6.** $9^0 =$ **7.** $(\frac{1}{2})^4 =$ **8.** $10^{-3} =$

Express the following in logarithmic form:

9. $a^x = y$ **10.** $4^m = n$ **11.** $5^x = 20$ **12.** $10^x = y$

13. $10^{1.6405} = 43.7$ **14.** $10^{2.4314} = 270$

Express the following in exponential form:

15. $\log_6 216 = 3$ **16.** $\log_2 128 = 7$ **17.** $\log_8 2 = \frac{1}{3}$

18. $\log_7 1 = 0$ **19.** $\log_9 9 = 1$ **20.** $\log_{10} 0.0001 = -4$

21. $\log 328 = 2.5159$ **22.** $\log 4.56 = 0.6590$

Find the following logs (use table):

23. $\log 25.36 =$ **24.** $\log 0.001984 =$ **25.** $\log 6.898 =$

Find N, the antilog, of each of the following (use table):

26. $\log N = 4.8149$ **27.** $\log N = 0.3909 - 2$

28. Find the value of x in each of the following:

(a) $\log_2 8 = x$ **(b)** $\log_3 x = 4$ **(c)** $\log_x 10000 = 4$

(d) $\log_6 6 = x$ **(e)** $\log_8 x = 0$ **(f)** $\log_x 27 = 3$

CHAPTER THIRTY-FOUR Computation by Logarithms

34-1 MULTIPLICATION

Logarithms are used to simplify computation. By use of logarithms, the processes of multiplication, division, and finding powers and roots can be made very simple. To show the use of logarithms in computation, we begin with multiplication.

In algebra we learn that $(x^3)(x^4) = x^7$; that is, when two quantities expressed as powers of the same base are multiplied together, the exponents are added. In general terms we have the rule,

$$(x^a)(x^b) = x^{a+b}$$

This rule holds true for any and all kinds of exponents, positive, negative, and fractional. When no exponent is expressed, it is understood to be 1.

Now, suppose we have the following multiplication problem:

$$(769)(874)$$

Instead of multiplying the numbers in the usual way, we may express both numbers as powers of 10:

$$769 = 10^{2.8859} \qquad \text{and} \qquad 874 = 10^{2.9415}$$

The product can now be expressed as a power of 10 by adding the exponents, or logarithms, of the numbers. The product is

$$(10^{2.8859})(10^{2.9415}) = 10^{5.8274}$$

The result means that the logarithm of the product is 5.8274. To find the product itself, (769)(874), we look up the antilogarithm of 5.8274. The antilogarithm is 672,000. This is the product of the two numbers: (769)(874) = 672,000, rounded off to three significant digits.

The work is best arranged as shown here:

$$\begin{aligned} \log 769 &= 2.8859 \\ \log 874 &= 2.9415 \\ \hline \end{aligned}$$

Adding, we get

$$\begin{aligned} \text{log of product} &= 5.8274 \\ \text{product} &= 672{,}000 \end{aligned}$$

Note especially that when we add the two logarithms, we do *not* get the product of the two numbers. Instead, we get the *logarithm of the product;* that is, we get the logarithm of (769)(874). This fact can be stated in the following equation form:

$$(\text{log of } 769) + (\text{log of } 874) = \log(769)(874)$$

Stated in words: *When we add the logarithms of two numbers, we get the logarithm of the product of the two numbers.* Note the following examples:

(a) $\log 48 + \log 85 = \log(48)(85)$, (b) $\log(25)(37) = \log 25 + \log 37$.

From these examples we may formulate the first principle of logarithms:

Principle 1. *The logarithm of the product of two or more factors is equal to the sum of the logarithms of the factors.* In general,

$$\log xy = \log x + \log y$$

Proof. To prove this principle, let

$$\log x = m \quad \text{and} \quad \log y = n$$

Then, in exponential form, $x = 10^m$ and $y = 10^n$
Multiplying, $xy = (10^m)(10^n)$ or $xy = 10^{m+n}$
Then, by definition, $\log xy = m + n$
Substituting equivalents, $\log xy = \log x + \log y$

This principle may be extended to several factors:

$$\log xyz = \log x + \log y + \log z$$

Examples

(a) $\log 4 + \log 5 + \log 7 = \log 140$; (b) $\log 2 + \log 3 = \log 6$

In *multiplication by use of logarithms*, we follow these steps:

1. *Find the logarithms of the numbers.*
2. *Add the logarithms.*
3. *Find the antilogarithm.* This is the product of the numbers.

In adding several logarithms that contain positive and negative characteristics, we add the negative characteristics separately. The positive characteristics are added with the mantissas.

Example 1

Multiply by logarithms: (57.6)(7830).

$$\begin{aligned} \log 57.6 &= 1.7604 \\ \log 7830 &= \underline{3.8938} \\ \text{log of product} &= 5.6542 \\ \text{product (or antilog)} &= 451{,}000 \quad \text{or} \quad (4.51)(10^5) \end{aligned}$$

The answer is correct to four significant digits, as can be seen if the problem is worked out by the usual method of multiplication.

Example 2

Multiply by logs: $(3.41)(0.0078)(10^2)(51.4)(0.421)$

Work:

$$
\begin{aligned}
\log 3.41 &= 0.5328 \\
\log 0.0078 &= 0.8921 - 3 \\
\log 10^2 &= 2.0000 \\
\log 51.4 &= 1.7110 \\
\log 0.421 &= 0.6243 - 1 \\
\hline
\log \text{product} &= 5.7602 - 4 \text{ (net characteristic is 1)} \\
{}^{*}\text{product} &= 57.57 \quad \text{or} \quad (5.757)(10^1)
\end{aligned}
$$

In Example 2, note that the logarithm of the product is $5.7602 - 4$. In an example of this kind the logarithm of the product may be changed to show a single characteristic: 1.7602. However, this change is not necessary. All we need to do is to observe that the net characteristic is $+1$.

Sometimes computation involves negative numbers, as in the multiplication problem, $(-3.4)(82.5)$; or in the division, $(-43.8) \div (-5.8)$. In any problem involving negative numbers, we proceed just as though the numbers were positive. After the answer has been found (by logarithms or otherwise), it is given the proper sign according to the rules for signs in algebra. That is exactly what we do if we multiply without the use of logs. In multiplying $(-3.4)(82.5)$, we get the numerical value, 280.5. Then we give it the proper sign, in this case, negative, and we have -280.5.

A problem sometimes involves addition or subtraction. Logarithms cannot be used to perform addition or subtraction of numbers. If a problem involves multiplication as well as addition or subtraction, then only the multiplication can be performed by use of logs.

Example 3

Find $(63.2)(94.3) + 534$.

* In Example 2, we find the product of the five numbers to be 57.57. If the numbers are multiplied by the usual long method, the actual product is 57.55647612. Rounding off this product to four significant digits, we get 57.56. Let us see why there is a slight difference. Mantissas, in general, are irrational numbers and have been rounded off. In some tables they are shown to five, six, or even ten places or more. Therefore, there is a slight error in the values shown in a table. Now if several mantissas are combined, these errors may be accumulated in such a way that there is a greater error in the mantissa of the answer. However, this fourth-place difference is only one point. Usually there is no difference at all.

Solution

$$\begin{aligned}
\log 63.2 &= 1.8007\\
\log 94.3 &= 1.9745\\
\text{log of product} &= \overline{3.7752}\\
\text{product, } (63.2)(94.3) &= 5960\\
\text{Now we add} &\quad\ \ \underline{534}\\
\text{and get the answer} &\quad\ 6494
\end{aligned}$$

Note that we need not find the logarithm of 534.

Exercise 34-1

Multiply the following, using logarithms.

1. (0.842)(768000)(0.0000535)
2. (3.71)(883)(0.000409)(10^3)
3. (532000)(6.07)(0.096)
4. (0.509)(86.4)(0.078)(10^{-2})
5. (6.925)(91.8)(0.0087)(10^4)
6. (372.5)(0.2673)(7.654)(10^{-4})
7. (0.001604)(2.362)(0.1867)
8. (0.02858)(327.5)(0.7894)
9. (12.63)(0.003514)(5.387)(0.8219)
10. (0.0001872)(2.606)(39910)(0.4038)
11. (106.8)(64.06)(7.128)(9.735)(10^{-6})
12. (320200)(0.4396)(7.943)(0.0008077)
13. (2307)(0.05091)(86.43)(0.0009281)
14. (32.95)(0.005841)(6.374)(7009)
15. (0.0166)(27300)(0.00389)(4.52)(69.4)(0.0808)
16. (0.128)(3.39)(0.000048)(5430)(751)(0.00009)
17. (−5.84)(63.7)(−0.493)(−90600)
18. (−72.7)(−5.32)(68.5)(144)

In the following problems, perform the multiplication by use of logs.

19. (65.2)(47.9) + 864
20. (3.96)(1280) + (25.7)(895)
21. (71.3)(6.52) − (0.845)(62.4)
22. (0.238)(8.14) − (0.0312)(0.0089)
23. A rectangular field is 87.2 rods long and 46.8 rods wide. Find the area.
24. Light travels at approximately 186,000 miles per second. How far does it travel in 8.314 min?

25. The diameter of a circle is 23.9 in. Find the circumference. (Take $\pi = 3.142$.)

26. Change 93.8 in. to centimeters. (1 in. = 2.54 cm.)

34-2 DIVISION BY LOGARITHMS

In algebra we learn that $x^8 \div x^2 = x^6$. That is, when two quantities expressed as powers of the same base are divided, the exponent of the divisor is subtracted from the exponent of the dividend. In general terms,

$$x^a \div x^b = x^{a-b}$$

The rule is true for all kinds of exponents, as shown in these examples:

$$y^8 \div y^2 = y^6 \qquad x^{4.37} \div x^{1.23} = x^{3.14} \qquad 5^{4.76} \div 5^{3.14} = 5^{1.62}$$

$$x^5 \div x^{-2} = x^7 \qquad (10^{4.786}) \div (10^{1.963}) = 10^{2.823}$$

Now suppose we have the following problem in division:

$$839{,}000 \div 23.5$$

Instead of dividing in the usual way, we first express both numbers as powers of 10:

$$839{,}000 = 10^{5.9238} \qquad \text{and} \qquad 23.5 = 10^{1.3711}$$

The quotient can now be expressed as a power of 10:

$$(10^{5.9238}) \div (10^{1.3711}) = 10^{4.5527}$$

The result means that the logarithm of the quotient is 4.5527. To find the quotient itself, $(839{,}000) \div (23.5)$, we look up the antilogarithm of 4.5527. That is, we find the number that has this logarithm. The antilogarithm is 35,700. This is the required quotient.

The work is best arranged as shown here:

$$\begin{aligned} \log 839{,}000 &= 5.9238 \\ \log 23.5 &= 1.3711 \\ \text{Subtracting, we get} \quad \text{log of quotient} &= \overline{4.5527} \\ \text{quotient} &= 35{,}700 \end{aligned}$$

Note especially that when we subtract the logarithms, we do *not* immediately get the quotient of the numbers. Instead, we get the *logarithm of the quotient;* that is, we get the logarithm of $(839{,}000) \div (23.5)$. This fact can be stated in the following equation form:

$$\log 839{,}000 - \log 23.5 = \log \frac{839{,}000}{23.5}$$

Stated in words: *When we subtract one logarithm from another logarithm, we get the logarithm of the quotient.* Note the following examples:

(a) $\log 90 - \log 18 = \log 5$, (b) $\log (592 \div 31) = \log 592 - \log 31$.

From these examples we can formulate the second principle of logarithms:

Principle 2. *The logarithm of the quotient of two numbers is equal to the logarithm of the dividend minus the logarithm of the divisor.* In general,

$$\log \frac{x}{y} = \log x - \log y$$

The proof is left to the student. It is similar to the proof of Principle 1.

In division by use of logarithms, we often run into difficulties that we do not meet in multiplication. It is often necessary to make some adjustment in the characteristic of a logarithm. It must be remembered that no matter what changes are made, the net value of the characteristic must not be changed. To show how such changes are made, we shall work out several examples in division and comment on each. Remember, *mantissas* in the table are *positive*. In division by use of logarithms, we follow these steps:

1. *Find the logarithm of the dividend and the logarithm of the divisor.*
2. *Subtract the logarithm of the divisor from the logarithm of the dividend.*
3. *Find the antilogarithm.*

Example 1

Divide 2740 by 88.1.

Work:

$$\begin{aligned} \log 2740 &= 3.4378 \\ \log 88.1 &= 1.9450 \\ \hline \text{log quotient} &= 1.4928 \\ \text{quotient} &= 31.1 \end{aligned}$$

In this example, the entire logarithms of dividend and divisor are positive. No adjustment is necessary since we can subtract the two logarithms and still get a positive mantissa.

Example 2

Divide 0.00871 by 3.19.

Work:

$$\begin{aligned} \log 0.00871 &= 0.9400 - 3 \\ \log 3.19 &= 0.5038 \\ \hline \text{log quotient} &= 0.4362 - 3 \\ \text{quotient} &= 0.00273 \end{aligned}$$

In this example, we simply subtract the mantissas as they stand and bring down the negative characteristic, -3. No adjustment is necessary since we still have a positive mantissa.

Example 3

Divide 8.98 by 21300.

Work:

$$\begin{aligned} & \quad\;\; 4 \qquad\quad -4 \\ \log 8.98 &= \not{0}.9533 \\ \log 21300 &= 4.3284 \\ \hline \log \text{quotient} &= 0.6249 - 4 \\ \text{quotient} &= 0.0004216 \end{aligned}$$

In this example, if we were to subtract the logarithms as they stand, we would get a negative logarithm; that is, the mantissa as well as the characteristic would be negative. Therefore, it is necessary to make a change in the form of the logarithm of the dividend. We do so by adding and subtracting a 4, as shown. When we then subtract, we get a positive mantissa.

There is one basic rule to follow in division: when logarithms are subtracted, *the mantissa must still remain positive.* The logarithm of the dividend must be such that the logarithm of the divisor may be subtracted without resulting in a negative mantissa.

Example 4

Divide 3.09 by 69,300.

Work:

$$\begin{aligned} & \quad\;\; 5 \qquad\quad -5 \\ \log 3.09 &= \not{0}.4900 \\ \log 69{,}300 &= 4.8407 \\ \hline \log \text{quotient} &= 0.6493 - 5 \\ \text{quotient} &= 0.0000446 \end{aligned}$$

In Example 4, it is necessary to adjust the logarithm of the dividend by adding and subtracting 5, so that the logarithm of the divisor may be subtracted.

Example 5

Divide 0.000256 by 582.

Work:

$$\begin{aligned} & \quad\;\; 3 \qquad\quad\;\; -7 \\ \log 0.000256 &= \not{0}.4082 - \not{4} \\ \log 582 &= 2.7649 \\ \hline \log \text{quotient} &= 0.6433 - 7 \\ \text{quotient} &= 0.0000004392 \quad \text{or} \quad (4.392)(10^{-7}) \end{aligned}$$

In Example 5, it is necessary to adjust the logarithm of the dividend by adding and subtracting 3. The net characteristic is still -4.

Example 6

Divide 7680 by 0.0207.

Work:

$$\begin{aligned} \log 7680 &= 3.8854 \\ \log 0.0207 &= 0.3160 - 2 \\ \hline \log \text{quotient} &= 3.5694 + 2 \\ \text{quotient} &= 371{,}000 \end{aligned}$$

In Example 6, we need not make any adjustment in the logarithm of the dividend. The negative characteristic, −2, is simply subtracted from zero, which makes +2. The net characteristic of the quotient is +5.

Example 7

Divide 0.0897 by 0.0000198.

Work:

$$\begin{aligned} \log 0.0897 &= 0.9528 - 2 \\ \log 0.0000198 &= 0.2967 - 5 \\ \hline \log \text{quotient} &= 0.6561 + 3 \\ \text{quotient} &= 4530 \end{aligned}$$

In Example 7, we subtract the negative characteristics *algebraically*.

Example 8

Divide 0.00035 by 0.0857.

$$\begin{aligned} & \quad\ 1 \qquad\ \ -5 \\ \log 0.00035 &= \not{0}.5441 - \not{4} \\ \log 0.0857 &= 0.9330 - 2 \\ \hline \log \text{quotient} &= 0.6111 - 3 \\ \text{quotient} &= 0.004084 \end{aligned}$$

In Example 8, we cannot subtract the mantissas as they stand. Therefore, we adjust the logarithm of the dividend by adding and subtracting 1.

Example 9

Divide 0.0413 by 0.0000784.

Work:

$$\begin{aligned} & \quad\ 1 \qquad\ \ -3 \\ \log 0.0413 &= \not{0}.6160 - \not{2} \\ \log 0.0000784 &= 0.8943 - 5 \\ \hline \log \text{quotient} &= 0.7217 + 2 \\ \text{quotient} &= 526.9 \quad \text{or} \quad (5.269)(10^2) \end{aligned}$$

In Example 9, we cannot subtract the mantissas as they stand. Therefore, we adjust the logarithm of the dividend by adding and subtracting 1. Subtracting the negative characteristics algebraically, we get +2.

Example 10

Compute by use of logs: $\dfrac{(362)(0.568)}{(73.5)(0.0000694)}$

Work:

We first add the logarithms of 362 and 0.568. The result is the logarithm of the numerator. Then we add the logarithms of 73.5 and 0.0000694. The result is the logarithm of the denominator. Then we subtract the logarithm of the denominator from the logarithm of the numerator.

$$\begin{aligned} \log 362 &= 2.5587 \\ \log 0.568 &= 0.7543 - 1 \\ \hline \log \text{numerator} &= 3.3130 - 1 \end{aligned} \qquad \begin{aligned} \log 73.5 &= 1.8663 \\ \log 0.0000694 &= 0.8414 - 5 \\ \hline \log \text{denominator} &= 2.7077 - 5 \end{aligned}$$

$$\begin{aligned} \log \text{numerator} &= 3.3130 - 1 \\ \log \text{denominator} &= 2.7077 - 5 \\ \hline \log \text{fraction} &= 0.6053 + 4 \\ \textit{Answer} &= 40300 \end{aligned}$$

Note that is is not necessary to find an antilogarithm until the final step. Note also that *the actual numerator or denominator does not appear in the work* and that it is nowhere necessary to combine negative and positive characteristics of the same logarithm.

Exercise 34-2

Perform the indicated operations by use of logarithms.

1. $84500 \div 96.8$ **2.** $873 \div 1.39$ **3.** $4230 \div (-78.6)$

4. $4600 \div 576$ **5.** $0.00892 \div 3.142$ **6.** $39700 \div 0.0519$

7. $60930 \div 0.7182$ **8.** $73.47 \div 0.8263$ **9.** $7.39 \div 0.0000119$

10. $0.00065 \div 9.38$ **11.** $0.000077 \div 9360$ **12.** $5.624 \div 0.000938$

13. $8.59 \div 1970$ **14.** $0.000638 \div 0.835$

15. $0.0791 \div 0.0000226$

16. $3.85 \div 965000$ **17.** $0.562 \div 0.0000214$

18. $0.47 \div 0.00000539$

19. $0.00243 \div 0.469$ **20.** $0.00037 \div 0.4508$

21. $0.0362 \div 0.0000487$

22. $\dfrac{(382)(46300)}{(7.95)(0.0857)}$ **23.** $\dfrac{(0.0059)(60.4)}{(8260)(0.0928)}$ **24.** $\dfrac{(11.54)(0.047)}{(3.65)(0.000783)}$

25. Prove Principle 2, using the general base b.

26. On a trip of 1458 miles a car averaged 16.7 miles per gallon. How many gallons of gasoline were used?

27. On a trip of 1683 miles, 95.6 gal of gasoline were used. What was the average mileage per gallon?

28. One acre contains 43,560 sq ft. A plot of land is advertised as 292,400 sq ft. How many acres are in the plot?

29. A rectange is 83.6 in. long and contains 5224 sq in. Find the width.

30. Change 54.1 meters to feet. (First change to centimeters. Then divide by 2.54, the number of centimeters in 1 in. Finally, divide by 12).

34-3 THE EXPANDED LOG FORM

We have seen by principle 1 that the *logarithm of the product* of two or more factors is equal to the *sum of the logarithms of the factors.* That is,

$$\log ab = \log a + \log b$$

The right side of the equation is called the *expanded log form.* In a similar way we can expand the logarithm of a quotient by Principle 2. That is,

$$\log \frac{m}{n} = \log m - \log n$$

Now, suppose we have a fraction containing two or more factors in numerator or denominator or both. Then, in writing the logarithm of the fraction, we can expand the log form by adding the logarithms of the numerator factors and subtracting the logarithms of the denominator factors. For example, suppose we have

$$\log \frac{ab}{xy}$$

By Principle 2, we can write this as

$$\log ab - \log xy$$

Then, by Principle 1, we can write the sum of the logs of the factors:

$$\log \frac{ab}{xy} = \log ab - \log xy = (\log a + \log b) - (\log x + \log y)$$

$$= \log a + \log b - \log x - \log y$$

We can also use the same principles in reverse to combine several logarithms into a single logarithm, as in this example:

$$\log 50 + \log 120 - \log 30 - \log 25 = \log \frac{(50)(120)}{(30)(25)} = \log 8$$

Exercise 34-3

Write each of the following in expanded log form:

1. $\log (3)(11)$ **2.** $\log abc$ **3.** $\log (15)(48)(37)$

4. $\log wxyz$ **5.** $\log (3.2)(65)(23)$ **6.** $\log (0.24)(14)(35)$

7. $\log \frac{h}{k}$ **8.** $\log \frac{3250}{23}$ **9.** $\log \frac{23}{46}$

10. $\log \frac{mn}{rs}$ **11.** $\log \frac{(6)(500)}{(24)(16)}$ **12.** $\log \frac{(2)(5)(90)}{(150)(6)(8)}$

Express each of the following as a single logarithm. Simplify if possible.

13. $\log 45 + \log 32$ **14.** $\log 2.3 + \log 500$

15. $\log 300 + \log 0.0004$ **16.** $\log h + \log k + \log t$

17. $\log 600 - \log 50$ **18.** $\log 840 - \log 12$

19. $\log 560 + \log 0.25 + \log 2$ **20.** $\log 60 + \log 8 + \log 5$

21. $\log 8 + \log 900 - \log 24$ **22.** $\log 280 - \log 56 - \log 25$

34-4 FINDING POWERS BY USE OF LOGARITHMS

In algebra, we have seen that $(x^5)^3 = x^{15}$; that is, to find the power of a power, we multiply the powers. As another example,

$$(10^{2.1342})^4 = 10^{8.5368}$$

In general terms this rule may be stated thus:

$$(x^a)^b = x^{ab}$$

To find the power of any number, we can first write the number itself as a power of 10. Then we multiply the exponent on 10 by the required power.

Suppose we have the problem:

$$(872)^3$$

Instead of multiplying the three factors together, we first express the number 872 as a power of 10:

$$872 = 10^{2.9405}$$

The answer can now be expressed as a power of 10:

$$(872)^3 = (10^{2.9405})^3 = 10^{8.8215}$$

The logarithm of the power is 8.8215. To find the answer itself, $(872)^3$, we look up the antilogarithm of 8.8215. It is 663,000,000. That is,

$$(872)^3 = 663{,}000{,}000 \qquad \text{or} \qquad (6.63)(10^8)$$

For this work we use the following form:

$$\begin{aligned} \log 872 &= 2.9405 \\ &\quad \times 3 \\ \hline \log (872)^3 &= 8.8215 \\ (872)^3 &= 663{,}000{,}000^* \end{aligned}$$

Note especially that when we multiply the logarithm by 3, we do *not* immediately get the answer to the problem. Instead, we get the *logarithm of the answer;* that is, we get the log of $(872)^3$. This fact can be stated in equation form:

$$(3)(\log 872) = \log (872)^3$$

Stated in words: *When we multiply the logarithm of a number by 3, we get the logarithm of the cube of the number.* The rule applies to any power.

(a) $(4)(\log 15) = \log 15^4$ (b) $(5)(\log 2) = \log 32$

From these examples we can formulate the third principle of logarithms:

Principle 3. *The logarithm of the power of a number is equal to the exponent times the logarithm of the number.* In general,

$$\log x^n = n \log x$$

Steps in finding powers by use of logarithms.

1. *Find the logarithm of the number to be raised to a power.*
2. *Multiply the logarithm by the exponent of the indicated power.*
3. *Find the antilogarithm.* This is the required power.

Example 1

Find $(96.7)^3$.

Work:

$$\begin{aligned} \log 96.7 &= 1.9854 \\ &\quad \times 3 \\ \hline \log (96.7)^3 &= 5.9562 \\ (96.7)^3 &= 904{,}000 \qquad \text{or} \qquad (9.04)(10^5) \end{aligned}$$

* The answer is correct to three significant digits. If the problem is done by actual multiplication, the answer is 663,054,848. When this number is rounded off to four significant digits, it becomes 663,100,000. The answer by use of logs has a difference of 1 in the fourth digit. This is because the small error in the mantissa has been multiplied by 3. A difference of 1 point in the fourth digit often happens, especially in high powers.

Example 2

Find $(93.4)^8$.

Work: $\log 93.4 = 1.9703$

$$\times 8$$

$$\log (93.4)^8 = 15.7624$$

$$(93.4)^8 = 5{,}786{,}000{,}000{,}000{,}000 \quad \text{or} \quad (5.786)(10^{15})$$

In a problem of this kind, we may realize for the first time the tremendous saving of time by use of logs. Consider the amount of work that would be involved in this problem by the long method of multiplication.

Example 3

Find $(0.00783)^4$.

Work: $\log (0.00783) = 0.8938 - 3$

$$\times 4$$

$$\log (0.00783)^4 = 3.5752 - 12$$

$$(0.00783)^4 = 0.00000000376 \quad \text{or} \quad (3.76)(10^{-9})$$

Exercise 34-4

1. $(7720)^2$ **2.** $(0.00865)^2$ **3.** $(0.399)^3$ **4.** $(43700)^3$

5. $(0.000915)^3$ **6.** $(83.4)^4$ **7.** $(0.0548)^4$ **8.** $(0.00935)^4$

9. $(539)^5$ **10.** $(0.587)^5$ **11.** $(67.3)^6$ **12.** $(0.0196)^6$

13. $(38.3)^7$ **14.** $(0.652)^8$ **15.** $(34.8)^{3.2}$ **16.** $(638)^{2.13}$

17. $(726)^3(0.48)^5$ **18.** $(6320)^4(0.0584)^3$ **19.** $(26.3)^3(0.0984)^2$

20. $(39800)^2(0.87)^5$ **21.** $(36.7)^2 + 483$ **22.** $(0.0893)^4 + 0.032$

23. $\dfrac{(286)^3(0.804)^4}{(0.00479)^2(96)^5}$ **24.** $\dfrac{(0.0286)^3(96)^5}{(8.04)^4(0.0479)^2}$ **25.** $\dfrac{(0.00078)^3(51.4)^5}{(3.19)^4(0.0607)^2}$

26. Write the proof of Principle 3, using the general base b.

27. Find the volume of a cube 3.49 feet on the edge.

34-5 FINDING ROOTS OF NUMBERS BY LOGARITHMS

Finding the square root of a number, such as the square root of 1324.69, is not an easy process in arithmetic. To find the cube root of a number is still more difficult. If we wish to find other roots of numbers, such as the seventh root of 386.52, the problem is practically impossible by the use of arithmetic alone. Such problems are easily worked by use of logarithms.

In algebra we have seen that

$$\sqrt{x^8} = x^4; \qquad \sqrt[3]{x^6} = x^2; \qquad \sqrt[3]{10^{4.761}} = 10^{1.587}$$

That is, to find the root of a number expressed with an exponent, we divide the exponent by the index of the root.

Now suppose we have the problem: $\sqrt[3]{6130}$. We can first express the radicand, 6130, as a power of 10:

$$6130 = 10^{3.7875}$$

Then the cube root can be expressed as a power of 10:

$$\sqrt[3]{6130} = \sqrt[3]{10^{3.7875}} = 10^{1.2625}$$

The logarithm of the root is 1.2625. To find the cube root itself, we look up the antilog of 1.2625. It is 18.3. This is the answer to the problem. That is,

$$\sqrt[3]{6130} = 18.3$$

For the work we use the following form:

$$\log 6130 = 3.7875$$

Dividing the log by 3, $\log \sqrt[3]{6130} = 1.2625$

$$\sqrt[3]{6130} = 18.3$$

Note especially that when we divide the log by 3, we do *not* immediately get the cube root of the number. Instead, we get the *logarithm of the cube root.* This fact may be stated in the following equation form:

$$\log \sqrt[3]{6130} = (\log 6130) \div 3; \quad \text{or} \quad \frac{\log 6130}{3}; \quad \text{or} \quad \frac{1}{3}\log 6130$$

From the example shown, we may formulate a fourth principle of logarithms:

Principle 4. *The logarithm of a root of a number is equal to the logarithm of the number divided by the index of the root.* In general,

$$\log \sqrt[r]{N} = (\log N) \div r = \frac{\log N}{r}$$

In algebra we have seen that a root may be written as a fractional power. That is,

$$\sqrt[3]{6130} = (6130)^{\frac{1}{3}}$$

Then a problem involving a root may be worked out as a power by the application of Principle 3.

Steps in finding roots by use of logarithms:

1. *Find the logarithm of the number, the radicand.*

2. *Divide the logarithm by the index of the root. This is the same as multiplying by a fractional power.*
3. *Find the antilogarithm.* This is the required root of the number.

Example 1

Find $\sqrt[5]{36900}$.

Work: $\log 36900 = 4.5670$

Dividing the log by 5, $\log \sqrt[5]{36900} = 0.9134$

$\sqrt[5]{36900} = 8.192$

If this problem is changed to the form, $(36900)^{\frac{1}{5}}$, it can be worked by the application of Principle 3. Then the logarithm is multiplied by $\frac{1}{5}$, which is equivalent to dividing by 5.

In finding the root of a decimal fraction, the entire logarithm, including mantissa and negative characteristic must be divided by the index of the root.

Example 2

Find $\sqrt[3]{0.00719}$.

Work: $\log 0.00719 = 0.8567 - 3$

dividing the log by 3, $\log \sqrt[3]{0.00719} = 0.2856 - 1$

$\sqrt[3]{0.00719} = 0.193$

In finding a root of a decimal fraction, it is often necessary to make some adjustment in the negative characteristic. Since the characteristic indicates the number of places the decimal point is to be moved from standard position, the following rule must always be observed:

Rule. *The negative characteristic must be exactly divisible by the index of the root.* After division by the index, the characteristic must be a whole number.

Example 3

Find $\sqrt[3]{0.000719}$.

Work: $\log 0.000719 = 0.8567 - 4$

At this point we notice that the characteristic is -4, which is not divisible by the index 3. Therefore, the characteristic must be adjusted so that the negative part is divisible by 3. It may be changed to -6, -9, or any other negative multiple of 3. Then we make a corresponding change in the positive portion of the logarithm. The characteristic is easily changed by adding and subtracting 2, and we get

$$\log 0.000719 = 2.8567 - 6$$
$$\log \sqrt[3]{0.000719} = 0.9522 - 2$$
$$\sqrt[3]{0.000719} = 0.08958$$

A problem involving any fractional exponent may be worked out by the application of Principle 3; that is, multiply the log by the power.

Example 4

Find $(827)^{\frac{2}{3}}$.

Work: $\log 827 = 2.9175$

$\times 2$

Multiplying the log by 2, 5.8350

and dividing by 3, $\log (827)^{\frac{2}{3}} = 1.9450$

$(827)^{\frac{2}{3}} = 88.1$

Example 5.

Find $(0.00165)^{\frac{3}{4}}$

Work: $\log 0.00165 = 0.2175 - 3$

$\times 3$

Multiplying the log by 3, $0.6525 - 9$

Before dividing by 4, we change the log to $3.6525 - 12$

Dividing the log by 4, we get $\log (0.00165)^{\frac{3}{4}}$ $0.9131 - 3$

$(0.00165)^{\frac{3}{4}} = 0.008186$

It is sometimes necessary to combine a positive mantissa with a negative characteristic into a single negative logarithm. Consider this example:

$$\log 0.00683 = 0.8344 - 3$$

or in exponential form,

$$0.00683 = 10^{0.8344-3}$$

If necessary, the negative characteristic can be combined with the positive mantissa. This is done simply by algebraic addition:

-3.0000

$+0.8344$

Adding, -2.1656

Then we can say,

$$0.00683 = 10^{-2.1656}, \quad \text{or} \quad \log 0.00683 = -2.1656.$$

The single negative number, -2.1656, is the negative logarithm of 0.00683. It is sometimes desirable to make this change, as we shall see.

The reverse change is also necessary at times. If a given logarithm is entirely negative, mantissa as well as characteristic, such as -2.3124, for example, then the antilogarithm cannot be found in the table of mantissas, since all mantissas are positive. Before we can find the antilog in the table, we must change the form in such a way that the mantissa, or decimal part, is positive. To do so, we change the form by adding and subtracting some whole number. For the negative logarithm, -2.3124, we can add and subtract the number 3, or some larger number. Then we get

$$3 - 2.3124 - 3$$

Now the positive 3 is combined with the negative logarithm, and we get

$$0.6876 - 3$$

To find the antilogarithm, we can now look up the mantissa in the table as usual. The antilogarithm is 0.004871.

Example 6

Find by logs: $(28.5)^{-3}$.

Work: $\log 28.5 = 1.4548$

Multiplying by -3, $\quad -3$

$\log (28.5)^{-3} = -4.3644$

The log is change in form to $\quad 0.6356 - 5$

$(28.5)^{-3} = 0.00004321$

Example 7

Find $(0.0548)^{2.51}$.

Work: $\log 0.0548 = 0.7388 - 2$

Before multiplying, the log is changed to -1.2612. Now this log is multiplied by 2.51, and we get a negative logarithm, -3.1656. This negative logarithm is now changed back to the form, $0.8344 - 4$. We can now look up the mantissa and find that the antilogarithm is 0.000683.

Example 8

Find

$$\sqrt{\frac{(4.68)^2(0.32)^4}{(24.3)^5(0.074)^3}}$$

Work:

$$\log 4.68 = 0.6702$$
$$\log (4.68)^2 = 1.3404$$

$$\log 0.32 = 0.5051 - 1$$
$$\log (0.32)^4 = 2.0204 - 4$$

$$\log 24.3 = 1.3856$$
$$\log (24.3)^5 = 6.9280$$

$$\log 0.074 = 0.8692 - 2$$
$$\log (0.074)^3 = 2.6076 - 6$$

$$\begin{aligned} \log (4.68)^2 &= 1.3404 \\ \log (0.32)^4 &= 2.0204 - 4 \\ \hline \log \text{numerator} &= 3.3608 - 4 \end{aligned}$$

$$\begin{aligned} \log (24.3)^5 &= 6.9280 \\ \log (0.074)^3 &= 2.6076 - 6 \\ \hline \log \text{denominator} &= 9.5356 - 6 \end{aligned}$$

$$\begin{aligned} \log \text{numerator} &= \overset{4}{\cancel{3}}.3608 - \overset{-5}{\cancel{4}} \\ \log \text{denominator} &= 3.5356 \\ \hline \log \text{fraction} &= 0.8252 - 5 \end{aligned}$$

Now we change the form of the logarithm so that the negative characteristic is divisible by 3, and get

$$\log \text{fraction} = 1.8252 - 6$$

Dividing the log by 3, $\log \text{answer} = 0.6084 - 2$

$$\text{answer} = 0.04059$$

Exercise 34-5

Compute the following by use of logarithms.

1. $\sqrt{4580}$ **2.** $\sqrt{0.00673}$ **3.** $\sqrt{703000}$ **4.** $\sqrt{0.0000486}$

5. $\sqrt[3]{74580}$ **6.** $\sqrt[3]{0.0566}$ **7.** $\sqrt[3]{0.00566}$ **8.** $\sqrt[3]{0.000566}$

9. $\sqrt[4]{1950}$ **10.** $\sqrt[4]{0.0834}$ **11.** $\sqrt[4]{0.00673}$ **12.** $\sqrt[4]{8380000}$

13. $\sqrt[5]{81560}$ **14.** $\sqrt[5]{0.0364}$ **15.** $\sqrt[5]{8730}$ **16.** $\sqrt[6]{0.6197}$

17. $\sqrt[6]{0.06197}$ **18.** $\sqrt[7]{4726}$ **19.** $\sqrt[7]{0.00923}$ **20.** $\sqrt[8]{0.00082}$

21. $(2850)^{\frac{3}{7}}$ **22.** $(277)^{\frac{3}{4}}$ **23.** $(64.6)^{\frac{2}{3}}$ **24.** $(0.00743)^{\frac{4}{7}}$

25. $(43.7)^{-2}$ **26.** $(49.8)^{-3}$ **27.** $(0.0315)^{2.71}$ **28.** $(0.0239)^{3.12}$

29. $\sqrt[3]{(63.9)^3(0.48)^2}$ **30.** $\sqrt[3]{(49.6)^2(0.037)^4}$ **31.** $\sqrt[4]{(52.3)^2(0.0029)^3}$

32. $\sqrt[4]{\dfrac{(89.4)^2(0.0076)^3}{(6.87)^5(0.58)^4}}$ **33.** $\sqrt[5]{\dfrac{(67.3)^2(0.049)^4}{(36.1)^5(0.0085)^2}}$ **34.** $\sqrt[7]{\dfrac{(9.34)^3(0.0026)^4}{(49)^5(0.00083)^2}}$

35. Find the edge of a cube whose volume is 389 cm^3.

34-6 SUMMARY OF PRINCIPLES OF LOGARITHMS

Now that we have used logarithms to find products, quotients, powers, and roots of numbers, let us summarize the principles involved. First, we state the definition of a logarithm. If any number N is stated as a power with the exponent x on a base b, then x is called the logarithm of the number N to the base b. That is,

$$\text{if} \quad b^x = N, \qquad \text{then} \qquad \log_b N = x$$

Any number except 1 or zero can be used as the base for a system of logarithms. The base of *common logarithms* is 10. This system is most convenient for computation. Common logarithms consist of two parts: *characteristic* and *mantissa*. The characteristic is an integer and is determined by the position of the decimal point. It is equal to the power of 10 for a number written in scientific notation. All mantissas shown in the table are positive. The mantissa is not affected by the decimal point in a number but is dependent only on the succession of digits. Negative numbers have no real logarithms, but logarithms themselves may be negative.

In computation by use of logarithms we use the following principles:

Principle 1. *The logarithm of the product of two or more factors is equal to the sum of the logarithms of the factors.* That is,

$$\log xy = \log x + \log y$$

This is the principle we have used in multiplication. For example, to multiply (23)(14), we (a) find the logarithms of the numbers, (b) add the logarithms, and then (c) find the antilogarithm. When we add the logarithms of the numbers, we do not get immediately the product, but the *logarithm* of the product. That is,

$$\log 23 + \log 14 = \log (23)(14)$$

As another application of this principle, $\log 5 + \log 7 = \log 35$.

Principle 2. *The logarithm of the quotient of two numbers is equal to the logarithm of the dividend minus the logarithm of the divisor.* That is,

$$\log \frac{x}{y} = \log x - \log y$$

This is the principle we have used in division. For example, to divide $(480) \div (15)$, we (a) find the logarithms of the numbers, (b) subtract the logarithm of the divisor from the logarithm of the dividend, and (c) find the antilogarithm. When we subtract the logarithms, we do not get immediately the quotient, but the *logarithm* of the quotient. That is,

$$\log 480 - \log 15 = \log \frac{480}{15}$$

As another application of this principle, $\log 54 - \log 6 = \log 9$.

Principle 3. *The logarithm of a power of a number is equal to the logarithm of the number multiplied by the exponent of the power.* That is,

$$\log x^n = n \log x$$

This is the principle we have used in finding powers. For example, to find the power, 24^3, we (a) find the logarithm of 24, (b) multiply the logarithm by 3, and (c) find the antilogarithm. When we multiply the logarithm by the exponent, we do not get immediately the power, 24^3, but the *logarithm* of the power. That is,

$$(3)(\log 24) = \log 24^3$$

As another application of this principle, $5 \log 2 = \log 32$.

Principle 4. *The logarithm of a root of a number is equal to the logarithm of the number divided by the index of the root.* That is,

$$\log \sqrt[r]{N} = (\log N) \div r, \qquad \text{or} \qquad \frac{1}{r} \log N$$

This is the principle we have used in finding roots of numbers. For example, to find $\sqrt[3]{512}$, we (a) find the logarithm of 512, (b) divide the logarithm by 3, and (c) find the antilogarithm. When we divide the logarithm by 3, we do not get immediately the $\sqrt[3]{512}$, but the *logarithm* of the root. That is,

$$\tfrac{1}{3} \log 512 = \log \sqrt[3]{512}$$

As another application of this principle, $\frac{1}{4} \log 81 = \log 3$.

The four principles of logarithms we have stated enable us to compute some logarithms from other known logarithms. In fact, in the actual computation of logarithms for a table, a few of them are computed by formulas from calculus. Then most others are computed from these few by using the four principles we have stated.

For example, suppose we know that the logarithm of 5 is 0.6990 and the logarithm of 7 is 0.8451. Knowing these, we can compute others. We can find the logarithm of 35 by use of Principle 1:

$$\log 35 = \log (5)(7) = \log 5 + \log 7 = 1.5441$$

Also, we can find the logarithm of 25 by use of Principle 3:

$$\log 25 = \log 5^2 = 2 \log 5 = 2(0.6990) = 1.3980$$

Now we can find the logarithm of 175, which is (7)(25):

$$\log 175 = \log (7)(25) = \log 7 + \log 25 = 2.2431$$

By use of Principle 2, we can find the logarithm of 1.4, which is $\frac{7}{5}$. Of course, there may be an error in the last digit because values are rounded off.

Exercise 34-6

Given: $\log 2 = 0.3010$, and $\log 3 = 0.4771$. From these two given logarithms, find the approximate value of the logarithm of each of the following numbers.

1. 6	**2.** 9	**3.** 12	**4.** 36	**5.** 81
6. 8	**7.** 32	**8.** 27	**9.** 54	**10.** 108
11. 5	**12.** 20	**13.** 40	**14.** 96	**15.** 144
16. 324	**17.** 432	**18.** 729	**19.** 512	**20.** 1728
21. 1.5	**22.** 1.8	**23.** 0.4	**24.** 1.25	**25.** 0.667

QUIZ ON CHAPTER 34, COMPUTATION BY LOGARITHMS. FORM A.

1. Multiply by use of logarithms: (483)(0.00867)(93000)(0.000206)

2. Divide by use of logarithms: $(48300) \div (0.0867)$

3. Divide by use of logarithms: $(0.00206) \div (9300)$

4. Find this power by use of logarithms: $(483)^4$

5. Find this power by use of logarithms: $(0.0867)^6$

6. Find the indicated root by use of logarithms: $\sqrt[5]{483000}$

7. Find the indicated root by use of logarithms: $\sqrt[7]{0.0867}$

Compute the following by use of logarithms:

8. $\dfrac{(3.63)(0.00527)}{(73)(0.000174)}$ **9.** $\sqrt{\dfrac{(4.52)^3(0.005)^2}{(32.4)^4(0.0703)}}$

10. Write the following in expanded log form:

(a) $\log (5)(12)(4)$ **(b)** $\log \dfrac{240}{13}$ **(c)** $\log \dfrac{(6)(7)(30)}{(8)(11)(16)}$

11. Express the following as a single logarithm. Simplify if possible.

(a) $\log 35 + \log 20$ **(b)** $\log 16 + \log 6 + \log 30$

(c) $\log 40 - \log 15$ **(d)** $\log 50 - \log 8 - \log 2$

CHAPTER THIRTY-FIVE Applications of Logarithms

35-1 SOLVING STATED PROBLEMS BY USE OF LOGARITHMS

We have seen how logarithms can facilitate computation. We shall now see how logarithms can be applied to the solving of stated word problems.

There is one point we must keep in mind in solving problems. Logarithms are in most instances irrational numbers. For use in computation, they have been rounded off to three, four, five, or more decimal places. The numbers we use, that is, those that appear in a table, are only approximate values. For this reason the answers obtained through the use of logarithm are only approximate.

Logarithms should not be used if exact answers are required. For example, the exact amount of sales of 143 automobiles at an average price of \$3754 is \$536,822. We cannot use logarithms for such a problem unless we are prepared to round off the answer to \$536,800, or \$537,000. Logarithms are usually used in connection with numbers obtained through measuring because all measurement is only approximate.

In solving problems by logarithms, we first analyze the problem and note all the steps that are involved. Then it is best to set up all the steps as a formula for the answer.

Example 1

Find the number of square centimeters in the area of a rectangle 28.7 in. long and 23.2 in. wide.

Solution

We use the formula for the area of a rectangle: $A = lw$. The product (28.7)(23.2) will show the number of square inches in the area. However, the problem calls for the number of square centimeters. Now we recall that

$$1 \text{ in.} = 2.54 \text{ cm} \qquad \text{and} \qquad 1 \text{ in.}^2 = (2.54)^2 \text{ cm}^2$$

To get the number of square centimeters, we must further multiply the product by $(2.54)^2$. Then

$$\text{area (in sq cm)} = (28.7)(23.2)(2.54)^2$$

Solving by logs,

$$\log 28.7 = 1.4579 \qquad \log 2.54 = 0.4048$$
$$\log 23.2 = 1.3655 \qquad \log (2.54)^2 = 0.8096$$
$$\log (2.54)^2 = 0.8096$$
$$\text{log of area} = 3.6330$$
$$\text{area} = 4295 \text{ cm}^2\text{, or rounded off, } 4300 \text{ cm}^2$$

Example 2

Find the sum of cubic feet in a rectangular box, 71.5 in. long, 57.6 in. wide, and 38.1 in. high.

Solution

For volume, we use the formula, $V = lwh$. If we find the product of the three given measurements, we shall have the number of cubic inches in the volume. However, the problem calls for the number of cubic feet. Then we must use the fact that 1 cu ft = 12^3 cu in. Therefore, the number of cubic inches must be divided by 12^3, or 1728. Then we write

$$\text{volume (in cu ft)} = \frac{(71.5)(57.6)(38.1)}{12^3}$$

Solving by logs:

$$\begin{aligned} \log 71.5 &= 1.8543 \\ \log 57.6 &= 1.7604 \\ \log 38.1 &= \underline{1.5809} \\ \text{log numerator} &= 5.1956 \\ \text{log denominator} &= \underline{3.2376} \\ \text{log of volume} &= 1.9580; \\ \text{volume} &= 90.78 \text{ ft}^3; \quad \text{or} \quad 90.8 \text{ ft}^3 \end{aligned} \qquad \begin{aligned} \log 12 &= 1.0792 \\ &\underline{\times 3} \\ \log (12)^3 &= 3.2376 \end{aligned}$$

In this problem the answer can be rounded off to three significant digits.

Example 3

Find the radius of a right circular cylinder if the altitude is 14.9 in. and the lateral area contains 766 sq in.

Solution

We make use of the formula for the lateral area of a right circular cylinder: $A = 2\pi rh$. First we solve the formula for r:

$$r = \frac{A}{2\pi h} \qquad r = \frac{766}{2\pi(14.9)}$$

Solving by logs:

$$\begin{aligned} \log 14.9 &= 1.1732 \\ \log \pi &= 0.4971 \\ \log 2 &= 0.3010 \\ \hline \log \text{denominator} &= 1.9713 \end{aligned} \qquad \begin{aligned} \log 766 &= 2.8842 \\ \log \text{denominator} &= 1.9713 \\ \hline \log r &= 0.9129 \\ r &= 8.182 \end{aligned}$$

The answer can be rounded off to three significant digits: 8.18.

Example 4

In a certain right triangle, ABC, the legs, a and b, are equal to 53.2 in. and 72.4 in., respectively. Find the hypotenuse c.

Solution

By the Pythagorean rule, $c^2 = a^2 + b^2$; or $c = \sqrt{a^2 + b^2}$. Substituting numerical values,

$$c = \sqrt{(53.2)^2 + (72.4)^2}$$

Solving by logs:

$$\log 53.2 = 1.7259 \qquad \log 72.4 = 1.8597$$

$$\log (53.2)^2 = 3.4518 \qquad \log (72.4)^2 = 3.7194$$

At this point, a common mistake is to add the logs of the squares. If we add these logs, we get the log of the *product* of the squares. Instead, the problem calls for the sum of the squares; that is,

$$(53.2)^2 + (72.4)^2$$

We must therefore find the antilogs:

$$\log (53.2)^2 = 3.4518 \qquad \log (72.4)^2 = 3.7194$$

$$(53.2)^2 = 2830 = a^2 \qquad (72.4)^2 = 5241 = b^2$$

Adding squares,

$$a^2 + b^2 = 2830 + 5241 = 8071$$

Now the problem reduces to finding the square root of 8071. Using logs, we get

$$\log 8071 = 3.9070; \ \log \sqrt{8071} = 1.9535; \qquad \sqrt{8071} = 89.8$$

Exercise 35-1

Solve the following problems using logarithms whenever possible.

1. Find the number of square yards in a rectangular floor that is 83.4 ft long and 52.1 ft wide.

2. Find the number of square feet in a plot of ground that is a rectangle, 48.7 yards long and 34.6 yards wide.

3. A rectangular field measures 518 ft long and 385 ft wide. How many square rods does it contain? (1 rod is equal to 16.5 ft.)

4. A rectangular solid is 18.3 in. long, 14.2 in. wide, and 7.3 in. high. What is its volume in cubic centimeters? (1 in. equals 2.54 cm.)

5. A rectangular solid 37.2 cm long, 24.6 cm wide, and 16.9 cm high. How many cubic inches does it contain?

6. A rectangular solid is 26.1 in. long, 18.7 in. wide, and 10.5 in. high. How many cubic centimeters does it contain?

7. Find the area of a circle whose radius is 23.4 in. ($A = \pi r^2$.)

8. The area of a circle is 647 square centimeters. Find the number of inches in the diameter.

9. The two legs of a right triangle measure 25.2 in. and 46.5 in., respectively. Find the hypotenuse, the perimeter, and the area of the triangle.

10. A rectangle is 28.8 in. long and 24.6 in. wide. Find the diagonal.

11. A rectangle has a length of 73.1 cm and an area of 4450 sq cm. Find the width of the rectangle and the length of the diagonal.

12. One leg of a right triangle is 27.9 in. and the hypotenuse is 54.7 in. Find the length of the other leg. Find the area of the triangle.

13. How long a cable will be required to brace a TV tower if one end of the cable is fastened to the tower 285 ft above the ground and the other end of the cable is fastened at a point on the ground level 210 ft from the tower?

14. The diameter of a right circular cylinder is 52.4 in. and the altitude is 72.3 in. How many gallons will it hold? (1 gal holds 231 cu in.)

15. A rectangular solid has a volume of 7850 cm^3, a length of 31.5 cm, and a width of 18.6 cm. Find the height.

16. A right circular cone has a volume of 5500 cm^3 and an altitude equal to 34.7 cm. Find the radius of the cone, using logs.

17. The diameter of a sphere measures 9.6 in. What is the volume of the sphere in cubic centimeters? ($V = \frac{4}{3}\pi r^3$.)

18. Find the number of centimeters in 8.2 miles.

19. If light travels 300,000,000 meters per second, find the number of miles in one light year. (Use 1 meter = 39.37 in.)

20. The formula for the distance of a freely falling body is $S = \frac{1}{2}gt^2$, in which S represents distance in feet and t represents time in seconds. If $g = 32.2$ ft per sec, per sec, how far, according to the formula, will an object fall from rest in 28.5 sec?

21. The time for one cycle of a pendulum is given by the formula

$$T = 2\pi\sqrt{L/G}$$

If $T = 1.2$ sec, $G = 32.2$, find L. (Take $\pi = 3.142$.)

22. If $R = 45$, $X_L = 73.2$, and $X_C = 48.6$, find Z in the formula

$$Z = 2\pi\sqrt{R^2 + (X_L - X_C)^2}$$

35-2 LOGARITHMS AS PARTS OF PROBLEMS

In many problems in science, a logarithm itself forms part of the problem. Then we must be sure to understand the problem as it is given. We must observe carefully the exact operations indicated.

Example 1

Find the value of x in the equation: $x = 5 + 3 \log 24$.

Solution

We first find

$$\log 24 = 1.3802; \qquad \text{then } 3 \log 24 = 4.1406$$

Now we simply add 5 to (3 log 24), and get, $5 + 4.1406 = 9.1406$. Now we have the answer to the problem: $x = 9.1406$.

Example 2

Find: $\log 68 + \log 95$.

Solution

This problem does not tell us to add 68 and 95. It does not tell us to multiply these numbers. All it says is to add the two logarithms:

$$\log 68 = 1.8325 \qquad \text{and} \qquad \log 95 = 1.9777$$

Now the problem means simply the addition of logs: $1.8325 + 1.9777 = 3.8102$. In this problem, the answer, 3.8102, is actually the logarithm of (68)(95), but we are not asked to find this product.

Example 3

Divide as indicated: $\dfrac{\log 72}{\log 4}$

Solution

This problem does not ask us to divide 72 by 4. It does not ask us to find the logarithm of the quotient, (72) ÷ (4). In the fraction shown, the numerator is not 72. It is the *logarithm* of 72, which is 1.8573. The denominator of the fraction is not 4. It is the logarithm of 4, which is 0.6021. Then the problem boils down to one in division:

$$\frac{1.8573}{0.6021}$$

At this point, some students make the mistake of subtracting. The process here indicated is division, not subtraction. It is a problem in dividing one logarithm by another logarithm; that is, (1.8573) ÷ (0.6021). By long division, we find the answer is approximately 3.084, or rounded off to 3.08. The problem in division might itself be worked by logarithms.

Exercise 35-2

By the use of the tables, find the value of each of the following:

1. $\log 37.2 + \log 49.6$

2. Work by logs: (37.2)(49.6)

3. $\log 867 - \log 21.3$

4. Work by logs: $867 \div 21.3$

5. $\log (92 + 53)$

6. $\log (123 - 38)$

7. $(\log 17.1)(\log 52.6)$

8. $\log(17.1)(52.6)$

9. $3 \log 17$

10. $6 + 5 \log 41.3$

11. $\dfrac{\log 891}{\log 13.9}$

12. $\log \dfrac{984}{23.1}$

13. $\log(63.5)^2$

14. $(\log 32.4)^2$

15. $\log 23 + \log 0.041$

16. $\log 0.045 - \log 52$

17. $8 + 3 \log 0.067$

18. $4 \log 31 - \log 0.047$

19. $\dfrac{\log 139}{3}$

20. $\dfrac{\log 0.0018}{4}$

21. $3 \log 26 + \frac{1}{2} \log 420$

22. $2 \log 4.2 - \frac{1}{3} \log 0.034$

35-3 LOGARITHMIC AND EXPONENTIAL EQUATIONS

A *logarithmic equation* is an equation containing the indicated logarithm of an unknown, such as the equation

$$5 \log x = 7.962 + \log x^2$$

An exponential equation is an equation in which the unknown appears in the exponent, such as the equation

$$6^{3x} = 594$$

A third type of equation that can be solved by use of logarithms is an equation involving an exponent in which the base is unknown, such as

$$x^{2.3} = 61.4$$

In each of the above three types of equations, our problem is to find the value of the unknown x. We shall now solve the three equations to show the procedure that used in each type.

Example 1

Solve for x: $5 \log x = 7.962 + \log x^2$.

Solution

First, we change the expression, $\log x^2$, to its equivalent, $2 \log x$, and rewrite the equation:

$$5 \log x = 7.962 + 2 \log x$$

Transposing,	$5 \log x - 2 \log x = 7.962$
Combining terms,	$3 \log x = 7.962$
Dividing both sides by 3,	$\log x = 2.654$
Taking the antilog of both sides,	$x = 450.8$

Example 2

Solve for x: $6^{3x} = 594$.

Solution

As a general rule, in solving an equation of this type, we take the logarithm of both sides, and get

$$\log 6^{3x} = \log 594$$

The equation can be written

$$3x \log 6 = \log 594$$

Dividing both sides by $(3 \log 6)$, the coefficient of x,

$$x = \frac{\log 594}{3 \log 6}$$

Substituting numerical values, $x = \dfrac{2.7738}{3(0.7782)} = 1.188$, (approx.)

The answer can be rounded off to three significant digits, 1.19. The result means that

$$6^{3(1.19)} = 594 \text{ (approx.)}$$

Example 3

Solve for x: $x^{2.1} = 83.2$.

Solution

Here again we take the logarithm of both sides of the equation:

$$\log x^{2.1} = \log 83.2$$

The equation can be written $\quad (2.1)(\log x) = \log 83.2$

Dividing both sides by 2.1, $\quad \log x = \dfrac{\log 83.2}{2.1}$

Substituting the value of log 83.2, $\quad \log x = \dfrac{1.9201}{2.1} = 0.9143$

Taking the antilog of both sides, $\quad x = 8.21$

35-4 CHANGE OF BASE OF LOGARITHMS

Sometimes it is necessary to find the logarithm of a number to a different base. As an example, suppose we wish to find the logarithm of 97 to the base 5. That is, we wish to find

$$\log_5 97$$

As in solving a problem in algebra, we let x represent the number we wish to find. Then, for this problem we can write

$$\log_5 97 = x$$

To solve for x, we first write the equation in exponential form:

$$5^x = 97$$

This is an exponential equation, a type that we have already solved. At this point we may note that x must have a value between 2 and 3. First, we take the log (base 10) of both sides of the equation:

$$\log 5^x = \log 97$$

$$\text{or, } x \log 5 = \log 97$$

Dividing both sides by (log 5), $\quad x = \dfrac{\log 97}{\log 5} = \dfrac{1.9868}{0.6990}$

Performing the division, we get $\quad x = 2.8423$

The value can be rounded off to $\quad x = 2.84$

The result means that $\log_5 97 = 2.84$ (approx.), or in exponential form, $97 = 5^{2.84}$

Note. The division itself might be done by use of logarithms.

Exercise 35-3

Solve the following equations for the unknowns:

1. $\log x + 3 \log x = 8$

2. $\log 6x + 2 \log x = 6$

3. $\log \frac{x}{14} = 1.4$

4. $\log \frac{240}{x} = 1.25$

5. $\log t^2 = \log 2t + 0.351$

6. $\log x^3 + 1.2352 = \log x$

7. $\log x^2 = 1.653 + 5 \log x$

8. $\log y + \log 4y = 5.542$

9. $\log (7x + 2) = 1 + \log x$

10. $\log (2x + 3) = \log x + \log 5x$

11. $3^x = 42$

12. $4^{2x} = 583$

13. $5^{3x-2} = 46.3$

14. $(1.5)^{2x-3} = 57.3$

15. $x^{1.7} = 2.56$

16. $x^{-1.2} = 78.9$

17. $x^{3.4} = 56.8$

18. $(5x)^{2.3} = 672$

19. Find $\log_5 57$

20. Find $\log_3 306$

21. Find $\log_8 5.47$

22. Find $\log_{2.6} 53.8$

23. Find $\log_9 5$

24. Find $\log_5 9$

25. Find $\log_6 1.73$

26. Find $\log_7 0.254$

27. $\log_x 723 = 3.25$

28. $\log_y 2.35 = -4.6$

29. Show that $\log_{10} 7 = \frac{1}{\log_7 10}$

30. Show that $\log_3 5 = \frac{1}{\log_5 3}$

35-5 NATURAL LOGARITHMS

We have seen that any number (except 1 and 0) may be used as a base for a system of logarithms. The base 10 is used almost exclusively in computation because it is more convenient than any other base. Logarithms based on 10 are called *common logarithms.*

There is another number that is often used as a base for logarithms. It is an irrational number that is equal approximately to 2.718281828459045, usually rounded off to 2.7183, or 2.718. This number is called the natural base, and is usually represented by e (or by the Greek letter epsilon ε). Logarithms based on e are called *natural logarithms.* They are sometimes called *Napierian* logarithms in honor of John Napier, the inventor of logarithms.

You may wonder why such an irrational number is used as a base for a system of logarithms. Natural logarithms are not often used for computation because they are too inconvenient. However, they force themselves into many problems in science and mathematics, particularly in physics, electronics, and calculus. This number e is also the basis of the laws of growth and decay. It is sometimes called the compound interest law.

In science it is often necessary to find the natural logarithm of a number. Of course, tables of natural logarithms have been constructed for convenience in mathematics and science. However, such tables are not essential for most work involving natural logarithms. When it is necessary to determine the natural logarithm of a number, this can be done by a simple formula. Let us see how this is done.

We have seen that the base if a sort of building block. If we wish to build any base up to a certain number, the exponent will depend on the size of the base. As an example that we have already used, the number 64 can be expressed as a power of 8, 4, 2, or any other base. The exponent of the power is equivalent to the logarithm to that particular base, as shown in the following examples. Note that when the base is large, the exponent is small; when the base is small, the exponent is large.

$$64 = 8^2 \quad \text{or} \quad \log_8 64 = 2$$

$$64 = 4^3 \quad \text{or} \quad \log_4 64 = 3$$

$$64 = 2^6 \quad \text{or} \quad \log_2 64 = 6$$

We have seen also that the base 10 may be built up to 64, although the exponent will be a mixed decimal. The power then is the common logarithm of 64. That is,

$$64 = 10^{1.8062} \quad \text{or} \quad \log_{10} 64 = 1.8062$$

Now, suppose we use the number e as a base. Recall that e is a rather small number, between 2 and 3. To build 10 up to 64 we need an exponent of 1.8062. Then, with e as a base, we need a larger exponent. Actually the required exponent is approximately 4.1589. To see the difference, compare these two statements:

$$64 = 10^{1.8062} \qquad 64 = e^{4.1589}$$

or in logarithmic form,

$$\log_{10} 64 = 1.8062 \qquad \log_e 64 = 4.1589$$

For the natural logarithm of a number, the notation "ln" is often used. Then we need not show the base. We write: $\ln 64 = 4.1589$.

To find the natural logarithm of any number, we first find the common logarithm. Then we convert this to a natural logarithm by a simple formula. We shall derive this formula, but first we need to know the natural logarithm of 10. Let us represent the natural logarithm of

10 by x. That is, we

let $\ln 10 = x$

In exponential form, this becomes $e^x = 10$

Taking the $\log_{10}$ of both sides, $\log e^x = 1$

This can be written $x \log e = 1$

Dividing both sides by ($\log e$), $x = \dfrac{1}{\log e}$

From the table of common logarithms, we find *log e* is approximately 0.4343.

Then we have $x = \dfrac{1}{0.4343} = 2.303$ (approx.)

Then we can say $\ln 10 = 2.303$ (approx.)

Now let us see how to convert a common logarithm to a natural logarithm. Let us assume that a represents the common logarithm of some number N. Then

$\log N = a$

In exponential form, $N = 10^a$

Now we take the ln of both sides, $\ln N = \ln 10^a$

This can be written $\ln N = (a)(\ln 10)$

The left side of the equation is precisely what we wish to find. The right side shows that the known common logarithm, a, must be multiplied by the natural logarithm of 10, which we have found to be 2.303. Then we have the formula:

$$\ln N = (2.303)(\log N)$$

Example 1

Find ln 41.5.

Solution

$\log 41.5 = 1.6180$

Multiplying by 2.303, $\ln 41.5 = (2.303)(1.6180) = 3.726$ (approx.)

The result means $41.5 = e^{3.726}$

Note. We must keep a few points in mind in working with natural logarithms. The whole-number part of a natural logarithm cannot be determined by inspection, as is done with common logarithms. In fact, we do not speak of characteristic and mantissa in connection with natural logarithms. Therefore, in any table of natural logarithms, the entire logarithm (whole number and decimal part) is printed in the table as a single number. If the natural logarithm is negative, then the entire logarithm is stated as negative and it is so printed in a table. For example, the natural logarithm of 0.046 is approximately -3.079.

Example 2

Find ln 0.077

Solution

$$\log 0.077 = 0.8865 - 2$$

Before multiplying, we change the logarithm to a single negative logarithm.

$$\log 0.077 = -1.1135$$

Multiplying by 2.303, $\ln 0.077 = (2.303)(-1.1135) = -2.564$ (approx.)

The reverse conversion is sometimes necessary. We may know the natural logarithm of a number and wish to find the number itself. To do so, we must find the common logarithm. Then we use the reciprocal of 2.303, which is approximately 0.4343. This number is the common logarithm of e.

Example 3

Given $\ln N = 4.119$, find N.

Solution

Multiplying by 0.4343, $\log N = (0.4343)(4.119) = 1.7889$ (approx.)
Finding the antilogarithm, $N = 61.5$

Example 4

Given $\ln X = -2.645$, find X.

Solution

Multiplying by 0.4343, $\log X = (0.4343)(-2.645) = -1.1487$ (approx.)
or, $\log X = 0.8513 - 2$
Finding antilogarithm, $X = 0.071$

The factors, 2.303 and 0.4343 should be memorized. They are actually reciprocal logarithms. That is,

$$\ln 10 = 2.303; \qquad \log e = 0.4343; \qquad \text{then } \log e = \frac{1}{\ln 10}$$

The factors are irrational numbers, but are usually rounded off as shown. They are the multipliers used in converting one form of logarithm to the other. The procedure is summarized in the following rules:

Rule 1. *To change a common logarithm to a natural logarithm, multiply the common logarithm by* 2.303 *because the logarithm must be made larger.*

Rule 2. *To change a natural logarithm to a common logarithm, multiply the natural logarithm by* 0.4343 *because the logarithm must be made smaller.*

Exercise 35-4

Find the common logarithm of each of the numbers (No. 1–12). Then change the common logarithm to the natural logarithm by use of the proper multiplier.

1. 31.7	**2.** 6.42	**3.** 2.14	**4.** 252
5. 1480	**6.** 4170	**7.** 87300	**8.** 182000
9. 0.726	**10.** 0.04	**11.** 2.718	**12.** 0.000526

Change the following natural logarithms to common logarithms by using the proper multiplier. Then find the antilogarithm for each.

13. $\ln A = 4.162$	**14.** $\ln B = 2.109$	**15.** $\ln C = 6.42$
16. $\ln D = 3.421$	**17.** $\ln E = 6.608$	**18.** $\ln F = 5.273$
19. $\ln G = 0.3001$	**20.** $\ln H = -0.3341$	**21.** $\ln J = -1.461$
22. $\ln K = -1.814$	**23.** $\ln M = -2.501$	**24.** $\ln N = -3.921$

QUIZ ON CHAPTER 35, APPLICATIONS OF LOGARITHMS. FORM A.

Solve the following problems by use of logarithms. (Use table.)

1. Find the number of cubic feet in a rectangular box, 91.2 in. long, 83.7 wide, and 71.3 in. high (1 cu. ft $= 12^3$ cu in.)

2. Find the number of cubic centimeters in a rectangular solid 20.3 in. long, 16.9 in. wide, and 9.4 in. high (1 in.$^3 = (2.54)^3$ cm^3)

3. Find the number of square inches in a circle whose diameter is 71.2 cm (1 in.$^2 = (2.54)^2$ cm^2)

4. A right circular cylinder has a radius of 6.83 in. and an altitude of 33.1 in. Find the number of cubic inches in its volume. How many gallons will it hold? (1 gal = 231 cu in.)

5. Find the number of cubic inches in the volume of a sphere whose diameter is 10.4 in. Also find the number of gallons it will hold.

6. Find the number of cubic feet in the volume of a sphere whose diameter is 6 ft. How many gallons will it hold? (1 cu ft holds 7.5 gal)

7. A right triangle has one leg = 26.4 in. The hypotenuse of the triangle is 43.7 in. long. Find the other leg and the area of the triangle.

8. Find the natural logarithms of the following numbers. First find the common logarithms.

(a) 33.5 **(b)** 2830 **(c)** 0.0597.

9. Change the following natural logarithms to common logarithms and then find the antilogarithms:

(a) $\ln A = 3.561$ **(b)** $\ln B = -2.577$

10. Solve the following equations for x:

(a) $x^{2.3} = 77$ **(b)** $\log_x 42 = 1.3$

11. Find the numerical value of the following:

(a) $2 \log 46 - (\frac{1}{2}) \log 84$ **(b)** $\dfrac{4 \log 21}{\log 3}$

(c) $(\log 40)(\log 3)$.

12. A right circular cylinder has a volume of 4720 cm and an altitude of 21.8 cm. Find the radius of the cylinder (using logarithms).

PART FIVE TRIGONOMETRY

CHAPTER THIRTY-SIX The Trigonometric Ratios

36-1 IMPORTANCE OF TRIGONOMETRY

Trigonometry is one of the most useful forms of mathematics. It has many applications in science and engineering as well as in theoretical mathematics. It is used to measure *indirectly* distances that are difficult or impossible to measure *directly*, such as the height of a flagpole or the distance across a river. It is essential in practically all engineering.

36-2 DEFINITIONS

For our purpose here, we define trigonometry as the study of *angles*. Historically, it is true that trigonometry arose out of the need to find unknown sides of triangles. It is still sometimes used for this purpose. Yet, for the many applications of trigonometry in engineering, it seems best to define trigonometry as the study of angles rather than of triangles.

What then is an angle? An angle may be defined in various ways. Let us first see what is meant by a *half-line* or a *ray*. In geometry we learn that a line is understood to have unlimited extent with no ends. However, a line drawn in one direction from a point is understood to have one end point. A line with one end point is called a *half-line* or a *ray*.

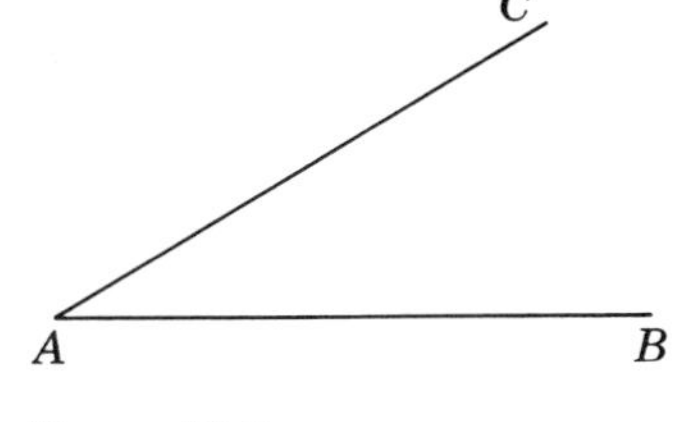

Figure 36-1

Now if two rays, AB and AC, are drawn from the same point A, they form an angle BAC (Fig. 36-1). Then we define an angle as the *figure formed* by the two rays. From this viewpoint, the angle is the *configuration* with no motion indicated. The size of the angle is determined by the *amount of opening between the rays.* The two rays are called the *sides* of the angle.

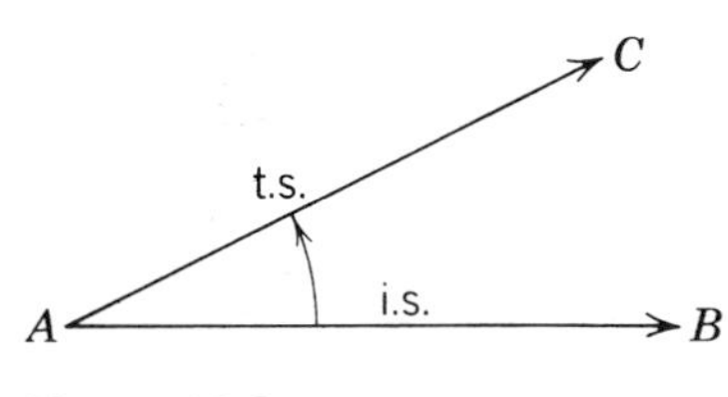

Figure 36-2

In most engineering it is necessary to define an angle as the result of motion, since much engineering involves motion. In Fig. 36-2 we begin with one ray AB, with end point at A. Now if the ray AB rotates about its end point A to the new position AC, an angle is formed. The angle is then defined as the *rotation of the ray.* AB. The size of the angle is determined by the *amount of rotation* of the ray. Note that in this definition we are dealing with only *one ray* but with *two different positions of the same ray*. The idea is like considering the angle formed by the long hand of a clock as it moves from one position to another.

From this second viewpoint, the sides of the angle are the two positions of the rotating ray. The original position, AB, is called the *initial side* (*i.s.*) of the angle. The final position, AC, is called the *terminal side* (*t.s.*) of the angle. The *vertex* of the angle is the end point around which the ray rotates. A curved arrow is used to indicate the direction and the amount of rotation. If the ray rotates in a counterclockwise direction, the rotation is called *positive.* Clockwise rotation is called *negative.*

If we consider an angle as the amount of rotation of a ray, then we may let the ray rotate to any position and as far as we wish. There is then no limit to the size of an angle.

36-3 KINDS OF ANGLES

When a ray, rotating in a positive direction, first reaches a position perpendicular to its original position, the angle generated is called a *right angle* (Fig. 36-3*a*). A right angle forms what is usually called a square corner. When the ray has reached a position so that the two sides point in opposite directions, the angle is called a *straight angle* (Fig. 36-3*b*), which is equal to two right angles. When the rotating ray reaches its original position, the angle is called *one revolution* (Fig. 36-3*c*).

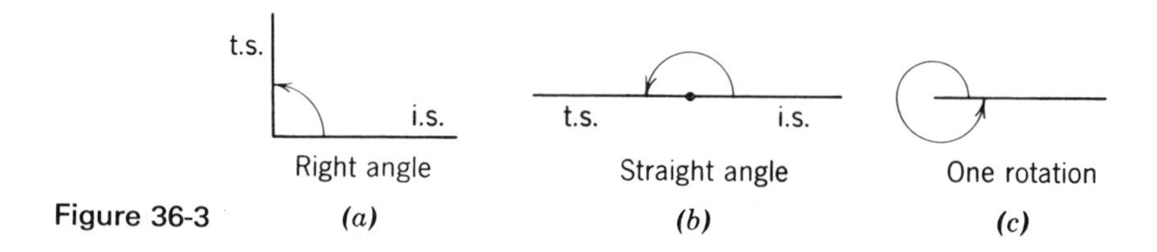

Figure 36-3 (*a*) (*b*) (*c*)

A rotation less than one right angle is called an *acute angle* (Fig. 36-4*a*). An *obtuse angle* is greater than a right angle but less than a straight angle (Fig. 36-4*b*). When three right angles are generated, the sides of the angle are again perpendicular to each other (Fig. 36-4*c*). In this case, we must be careful to call the angle *three right angles,* not simply one right angle.

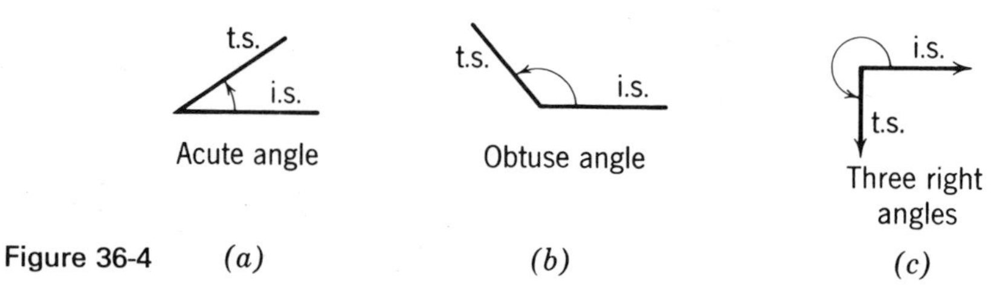

Figure 36-4 (*a*) (*b*) (*c*)

36-4 MEASUREMENT OF ANGLES

An angle is often measured in degrees. One revolution is called 360 degrees (360°). One degree (1°) is therefore $\frac{1}{360}$ of one revolution. A right angle then contains 90°; a straight angle contains 180°, or two right angles. An acute angle contains less than 90°. An obtuse angle contains more than 90° but less than 180°. For finer measurements one degree is divided into 60 minutes, and 1 minute of rotation is divided into 60 seconds.

Since an angle is the amount of rotation of a ray, there is no limit to the size of an angle. An angle of 390° is equal to one revolution and 30° more. An angle of 720° is equal to two revolutions. With this understanding of the meaning of an angle, it will make sense when an engineer speaks of an angle of 540°, 3600°, or 21,600°. An angle of 540° is equal to 1.5 revolutions or six right angles (Fig. 36-5*a*). An angle of 1800° is equal to five revolutions or 20 right angles (Fig. 36-5*b*).

Figure 36-5

36-5 STANDARD POSITION OF AN ANGLE

Angles are found in many different positions. However, angles are best studied in what is called *standard position* of the angle. Standard position refers to a certain position on the rectangular coordinate system.

To place an angle in standard position, first, place the *vertex* of the angle at the *origin* (0, 0). Then swing the angle around so that the *initial* side falls along the *positive direction* of the x-axis. An angle of 30° is in standard position in Fig. 36-6.

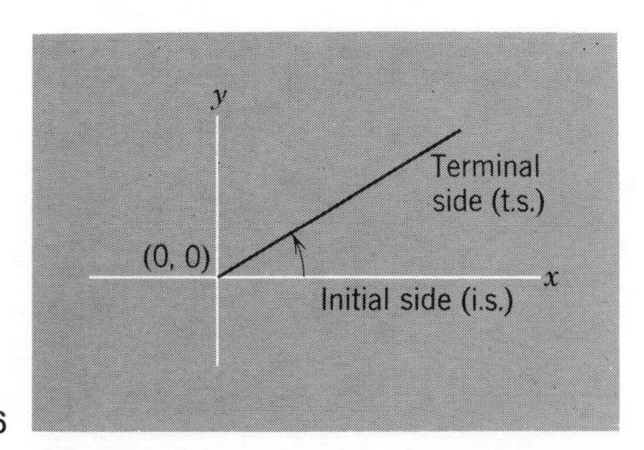

Figure 36-6

Any acute angle in standard position will have its terminal side in the first quadrant. For any obtuse angle, the terminal side will fall in the second quadrant. An angle of 210° will have its terminal side in the third quadrant. A negative angle indicates clockwise rotation. Therefore, an angle of −30° in standard position will have its terminal side in the fourth quadrant. For the standard position of any angle, the initial side must lie along the positive direction of the x-axis.

Exercise 36-1

1. Sketch the following angles (approximately to size) in standard position on the x- and y-axes. Show the initial side and the terminal side of each and show the direction and the amount of rotation by a curved arrow: 45°, 225°, 135°, 315°.
2. Do the same for the following angles (sketch these four angles on one graph): 30°, 150°, 240°, 300°.
3. Sketch the following angles on one graph. Notice that some angles have their terminal sides in the same place. Such angles are called *coterminal* angles: 60°, 420°, −300°, 120°, −240°. (A negative angle indicates clockwise rotation.)
4. Sketch the following angles in standard position, showing the initial and the terminal sides and showing the direction and amount of rotation (these need not be on the same graph). Which are coterminal? −45°, −30°, −120°, −210°, 180°, −180°, 270°, −90°.
5. On one graph show the following angles, with the use of a protractor if necessary: 10°, 20°, 30°, 40°, 50°, 60°, 70°, 80°. Now try to draw each of the following angles without using a protractor: 20°, 70°, 10°, 80°, 30°, 60°, 40°, 50°.

36-6 TRIGONOMETRIC RATIOS OR FUNCTIONS

Suppose that some angle such as A is in standard position (Fig. 36-7). Let us further suppose that the point (8, 6) lies on the terminal side of the angle. Then the abscissa of the point is 8 and the ordinate of the point is 6.

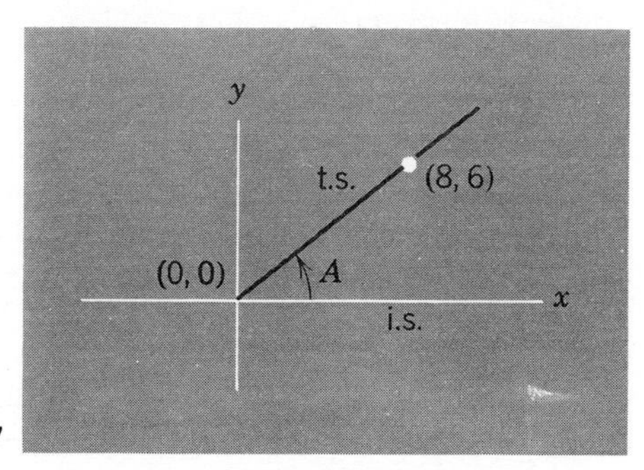

Figure 36-7

The distance from the origin out to the point is called the *radius vector* of the point. The radius vector can be computed by the Pythagorean rule. In the case of the angle A, the radius vector is 10. The three lengths, *abscissa, ordinate*, and *radius vector*, represent three distances; and the numbers 8, 6, and 10 represent the number of units in these distances, respectively.

Now, it is possible to state six and only six *ratios* between these three values: abscissa, ordinate, and radius vector. These six ratios are known as the six *trigonometric ratios* or *functions*. They have been given names as follows:

The ratio $\dfrac{\textit{ordinate}}{\textit{radius vector}}$ is called the *sine* of angle A.

The ratio $\dfrac{\textit{abscissa}}{\textit{radius vector}}$ is called the *cosine* of angle A.

The ratio $\dfrac{\textit{ordinate}}{\textit{abscissa}}$ is called the *tangent* of angle A.

The ratio $\dfrac{\textit{abscissa}}{\textit{ordinate}}$ is called the *cotangent* of angle A.

The ratio $\dfrac{\textit{radius vector}}{\textit{abscissa}}$ is called the *secant* of angle A.

The ratio $\dfrac{\textit{radius vector}}{\textit{ordinate}}$ is called the *cosecant* of angle A.

Trigonometry is concerned with these six ratios for any angle. For brevity, the ordinate is often called y, the abscissa is called x, and the radius vector is called r. The names of the ratios are abbreviated by using the first three letters, except for cosecant, which is abbreviated "csc." The word "of" is also omitted. In abbreviated form, the definitions then become

$$\frac{y}{r} = \sin A \qquad \frac{y}{x} = \tan A \qquad \frac{r}{x} = \sec A$$

$$\frac{x}{r} = \cos A \qquad \frac{x}{y} = \cot A \qquad \frac{r}{y} = \csc A$$

The names of the functions should always be pronounced as though written out in full.

The foregoing definitions hold true for any angle in standard position. If an angle is in standard position and a point on the terminal side is known, then all the ratios can be computed for that angle.

Consider again the angle A in standard position with point (8, 6) on the terminal side (Fig. 36-7). From the definitions of the trigonometric ratios, the following values can be found:

$$\sin A = \frac{y}{r} = \frac{6}{10} = 0.6000 \qquad \cos A = \frac{x}{r} = \frac{8}{10} = 0.8000$$

$$\tan A = \frac{y}{x} = \frac{6}{8} = 0.7500 \qquad \cot A = \frac{x}{y} = \frac{8}{6} = 1.3333$$

$$\sec A = \frac{r}{x} = \frac{10}{8} = 1.2500 \qquad \csc A = \frac{r}{y} = \frac{10}{6} = 1.6667$$

It must be understood that the trigonometric ratios do not indicate inches, feet, degrees, or measurements of any kind. Instead, they are pure ratios and are, therefore, numbers without any denomination.

To consider another example, suppose an angle is in standard position and the point (−7, 3) is on its terminal side (Fig. 36-8). Now we wish to find the values of the six trigonometric ratios for the angle. Let us call the angle *theta* (θ). The first thing to do is to sketch the angle in standard position (the angle should be indicated by a curved arrow).

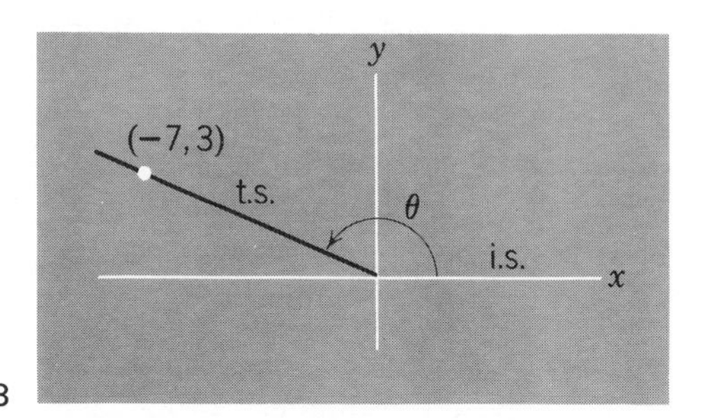

Figure 36-8

In this example the abscissa is −7, the ordinate is 3, and the radius vector, determined by the Pythagorean rule, is $\sqrt{58}$. Then the six trigonometric ratios for the angle θ are

$$\sin\theta = \frac{3}{\sqrt{58}} \qquad \tan\theta = \frac{3}{-7} \qquad \sec\theta = \frac{\sqrt{58}}{-7}$$

$$\cos\theta = \frac{-7}{\sqrt{58}} \qquad \cot\theta = \frac{-7}{3} \qquad \csc\theta = \frac{\sqrt{58}}{3}$$

The foregoing values can be changed to decimal form if it is so desired. It will be noted that some of these functions are negative. A trigonometric ratio will be negative when the two terms have opposite signs. It will be positive when it is the quotient of like signs. This is simply the application of the rule for the division of signed numbers in algebra.

Exercise 36-2

Each of the following angles, indicated by letter, is in standard position, and the given point is on the terminal side. Find the numerical and signed values of the six trigonometric functions of each angle. Consider all angles positive. Make a sketch of each angle in standard position.

Angle	Point on terminal side	Angle	Point on terminal side	Angle	Point on terminal side
A	(5, 12)	*H*	(6, 3)	*O*	(4, 5)
B	(−12, 5)	*I*	(8, 2)	*P*	(−5, −2)
C	(4, −3)	*J*	(−5, 10)	*Q*	(3, −5)
D	(−6, −8)	K	(5, 5)	*R*	(7, 1)
E	(5, −12)	*L*	(3, −3)	*S*	(5, $\sqrt{11}$)
F	(−8, 15)	*M*	(−6, 6)	*T*	($\sqrt{3}$, 1)
G	(−24, 7)	*N*	(2, −1)		

QUIZ ON CHAPTER 36, THE TRIGONOMETRIC RATIOS. FORM A.

1. Sketch these angles on the coordinate system and show the initial side, the terminal side, and the angle by a curved arrow:

(**a**) 45° (**b**) 60° (**c**) 90° (**d**) 120° (**e**) 135°

(**f**) 180° (**g**) 210° (**h**) 225° (**i**) 360° (**j**) −60°

2. Sketch these angles on the coordinate system by estimating the size of each:

(**a**) 15° (**b**) 75° (**c**) 105° (**d**) 165°

3. Sketch the following angles in standard position, showing the point on the terminal side. Then state the numerical value of the six trig functions of each angle:

Angle	Point on terminal side	Angle	Point on terminal side
A	(6, 8)	*D*	(8, 4)
B	(−15, 8)	*E*	(3, −1)
C	(12, −5)	*F*	(−4, 4)

CHAPTER THIRTY-SEVEN Tables of Trigonometric Ratios

37-1 THE TRIGONOMETRIC RATIOS AS FUNCTIONS OF AN ANGLE

The numerical values of the trigonometric ratios for any angle do not depend on the lengths of the sides of the angle but only on the *size of the angle.* For this reason, we say the ratios are *functions of the angle.*

To show that this is true, suppose we have an angle in standard position with the point (8, 6) on the terminal side (Fig. 37-1). Let us call the angle *alpha* (α). We find that the radius vector is 10 units. Then the sine of α is $\frac{6}{10}$, $\frac{3}{5}$, or 0.6000. Now we shall find that the terminal side of α passes through the point (4, 3) also. This can be proved from the fact that the right triangles formed are similar. If we take the point (4, 3), the radius vector is 5. The sine of α is $\frac{3}{5}$, the same as before.

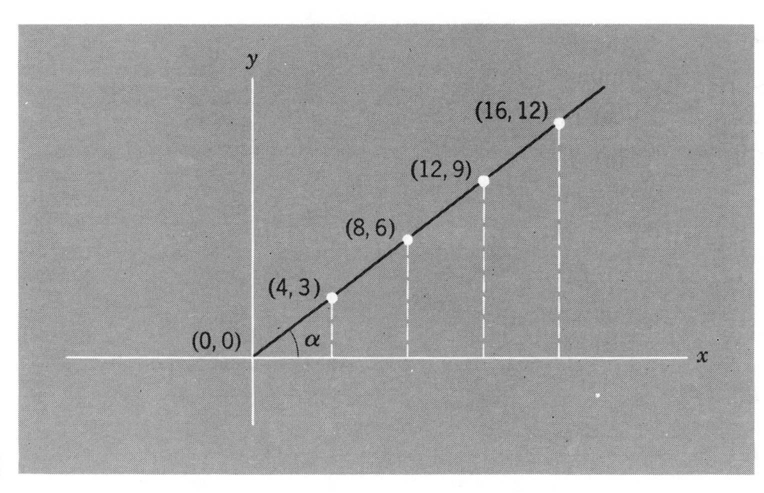

Figure 37-1

The terminal side will also pass through the points (12, 9) and (16, 12). If we compute the radius vector correctly for each point, we shall find that the sine for this particular angle (α) always has the value $\frac{3}{5}$ or 0.6000.

In fact, no matter what point is taken on the terminal side of an angle, the sine value for that angle will always remain the same. The numerical value of the sine will not change unless the angle changes in size. In the same way, all the other trigonometric ratios will also remain the same for any particular angle no matter what point is chosen on the terminal side of the angle.

Since the trigonometric ratios depend only on the size of the angle, they are called *functions of the angle.*

Now, it can be shown that the angle whose sine value is 0.6000 or $\frac{3}{5}$ is an angle of approximately 36.9°. In other words, the sine of 36.9° is approximately 0.6000 no matter what point is taken on the terminal side.

Suppose we have another angle, say, an angle of 20°, in standard position (Fig. 37-2). The trigonometric ratios of this angle will remain constant whatever point we choose on the terminal side. We take the general point (x, y) and indicate the radius vector by r. Now let us estimate the value of the sine of 20°. For the sine ratio for any angle, we have

$$\frac{\text{ordinate}}{\text{radius vector}}$$

Note in Fig. 37-2 that the ordinate is approximately $\frac{1}{3}$ of the radius vector. This ratio is true no matter where we take the point (x, y). Then we can conclude that the sine of 20° is approximately $\frac{1}{3}$. Computed by advanced formulas, the value has been found to be approximately 0.3420.

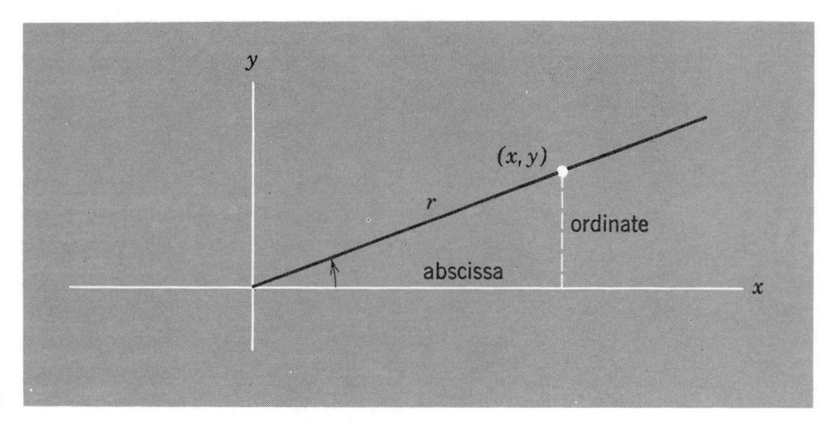

Figure 37-2

Let us in the same way estimate the value of the cosine of 20°. Note that for any position of the point (x, y), the abscissa is almost as great as the radius vector. Therefore the cosine ratio is almost 1. Actually, the cosine of 20° is approximately 0.9397.

37-2 TABLES OF TRIGONOMETRIC RATIOS

The numerical values of all trigonometric ratios have been computed for angles from zero to 90°. These values are usually listed in tabular form. The values are often given for tenths of a degree or for minutes. An angle of one minute (1′) is $\frac{1}{60}$ of a degree. An angle of one second (1″) is $\frac{1}{60}$ of a minute, or $\frac{1}{3600}$ of a degree. One of the most useful tables for technical work is one showing values of the trigonometric ratios for degrees and tenths of a degree.

In most tables the function values are arranged so that angles may be read downward at the left side and upward at the right. To find the functions of all angles from zero to 45°, find the angle in the *left-hand* column and read the function value *under* the proper heading. For angles of 45 to 90°, read *upward* at the *right of the table for the angle* and read the *footings* instead of headings.

Examples

$\sin 13.8° = 0.2385$ $\quad$ $\sin 56.4° = 0.8329$

$\cos 27.2° = 0.8894$ $\quad$ $\cos 71.8° = 0.3123$

$\tan 38.4° = 0.7926$ $\quad$ $\tan 64.9° = 2.1348$

$\cot 21.7° = 2.5129$ $\quad$ $\cot 47.7° = 0.9099$

In some problems it is necessary to use greater accuracy than tenths of a degree. In the table showing tenths of a degree we find consecutive angles such as 28.3° and 28.4°. Suppose we wish to find the functions of an angle between 28.3° and 28.4°. We may interpolate or we may use a table showing angles by minutes. Since one degree is equal to 60 minutes, then one-tenth of a degree is equal to 6 minutes. An angle of 28.3° is equivalent to an angle of 28° 18′. An angle of 28.4° is equal to 28° 24′. An angle of 31° 42′ is equal to 31.7°.

If we wish to find the sine of an angle of 28° 19′, we may use a table showing angles by minutes, in which the sine of 28° 19′ is 0.47434. If we interpolate for 28° 19′, which lies between 28.3° and 28.4°, we get a sine value of 0.47435.

Even for an angle of 24.3°, compared with the same angle stated in degrees and minutes, there is a slight difference in the values found in the five-place and the four place tables. In a table showing angles by degrees and tenths of a degree, the sine of 24.3° is shown as 0.4115. In a table showing division by minutes, the value is shown as 0.41151. The only difference is that the second value indicates greater accuracy because an angle showing minutes indicates greater accuracy than one showing only tenths of a degree.

The tables of trigonometric ratios can also be used in *reverse*; that is, when a certain function value is known, the angle itself may be found in the table. For instance, if we know that the sine of an angle is 0.3322, we can look this value up under the heading "sine." There we discover that the angle corresponding to this sine value is 19.4°.

In using the table in reverse, we must sometimes use footings instead of headings. For example, suppose we know that the tangent of angle A is 1.5340. Then we find the angle A is 56.9°.

If the exact value of a function as given is not found in the table, we may take the nearest angle shown or we may estimate by interpolation the angle lying between two known angles.

Exercise 37-1

Find the value of each of the following functions:

1. sin 25.1°	**2.** sin 40.7°	**3.** cos 17.6°	**4.** cos 31.8°
5. tan 28.2°	**6.** tan 44.9°	**7.** cot 15.4°	**8.** cot 31.3°
9. sin 48.6°	**10.** sin 60.9°	**11.** cos 53.5°	**12.** cos 76.7°
13. tan 57.3°	**14.** tan 71.8°	**15.** cot 69.6°	**16.** cot 84.9°
17. sin 1.2°	**18.** sin 8.5°	**19.** cos 1.7°	**20.** cos 7.4°
21. tan 1.3°	**22.** tan 6.8°	**23.** cot 2.6°	**24.** cot 8.9°
25. sin 0.6°	**26.** cos 0.6°	**27.** tan 0.6°	**28.** cot 0.6°
29. sin 12° 20′	**30.** cos 22° 50′	**31.** tan 73° 40′	**32.** cot 18° 10′
33. sec 6.5°	**34.** sec 66.2°	**35.** csc 15.4	**36.** csc 83.5°

Find the angle corresponding to each of the following function values:

37. $\sin A = 0.2538$	**38.** $\sin B = 0.8281$	**39.** $\cos C = 0.9178$
40. $\cos D = 0.4352$	**41.** $\tan E = 0.7212$	**42.** $\tan F = 2.267$
43. $\cot G = 3.291$	**44.** $\cot H = 0.6265$	**45.** $\sin I = 0.4730$
46. $\cos J = 0.8360$	**47.** $\sec K = 1.2200$	**48.** $\csc L = 1.6844$

37-3 ARC-FUNCTIONS OR INVERSE FUNCTIONS

At this point in our study it is necessary that we understand the terms *arcsine*, *arctangent*, and so on. Such expressions mean *angles*. The term *arc* can be used as a prefix to any of the names of trigonometric functions. Such expressions are sometimes called *inverse functions*.

To see what is meant by this notation, let us first consider the *arcsine* (usually written *arcsin*). Suppose we know that the sine of some particular angle is $\frac{2}{3}$, or 0.6667. Then we can write the equation

$$\sin \theta = \tfrac{2}{3}$$

Now, we may ask, how large is angle θ? From the table, we find that the angle θ is approximately 41.8°. This is a problem we faced in the preceding assignment.

However, there is a way of stating the size of the angle without looking it up in the table. Notice these two equations:

$$\text{if} \qquad \sin\theta = \tfrac{2}{3},$$

$$\text{then} \qquad \theta = \text{the angle whose sine is } \tfrac{2}{3}$$

You will notice that the first equation tells what "*sin θ*" equals. The second equation tells what "θ" equals.

The student will probably say, "But the second equation doesn't tell anything new. It still doesn't tell the size of the angle." That is true. Yet there are times when it is convenient to use the second form. The two equations can be shortened as follows:

$$\text{if} \qquad \sin\theta = \tfrac{2}{3}$$

$$\text{then} \qquad \theta = \arcsin\tfrac{2}{3}$$

The word *arcsine* (pronounced ark-sine) means "the angle whose sine is." A similar expression is used in connection with all the trigonometric functions. Their abbreviated forms are as follows:

arccos (pronounced ark-cosine)

arctan (pronounced ark-tangent)

arccot (pronounced ark-cotangent)

arcsec (pronounced ark-secant)

arccsc (pronounced ark-cosecant).

The *arc-functions,* often called *inverse functions,* are sometimes indicated by -1 written in the same position as an exponent. Thus arcsin x is often written $\sin^{-1} x$. If this form is used, the -1 must *not* be taken as a negative exponent. It is probably less confusing to the student to use the form *arcsin.* If it is necessary to use a negative exponent on a trigonometric function, it should be written as shown here:

$$\frac{1}{\sin x} = (\sin x)^{-1}$$

It must be understood that this expression does *not* mean *arcsin.* In our discussion here we use the forms *arcsin, arctan,* and so on.

Here are some examples showing the meaning of these expressions:

1. If $\tan A = \frac{1}{2}$, then angle $A = \arctan\frac{1}{2}$. This expression means "A is the angle whose tangent is $\frac{1}{2}$."

2. If $\sec B = \frac{5}{3}$, then $B = \text{arcsec}\,\frac{5}{3}$. This means "$B$ is the angle whose secant is $\frac{5}{3}$."

3. If we know that $\cos E = 0.7230$, then $E = \arccos 0.7230$; that is, "E is the angle whose cosine is 0.7230."

Of course, if we wish to know the size of an angle in degrees, we must look it up in the table. If $\cos E = 0.7230$, then $E = \arccos 0.7230$. From the table, we find that the angle is 43.7°; that is, $\arccos 0.7230 = 43.7°$. In the same way we find that

$$\arcsin 0.5635 = 34.3°; \qquad \arctan 1.5051 = 56.4°$$

Exercise 37-2

Find the following angles:

1. arcsin 0.3355 **2.** arcsin 0.8231 **3.** arccos 0.7157

4. arccos 0.2890 **5.** arctan 0.5704 **6.** arctan 2.023

7. arccot 0.2642 **8.** arccot 3.513 **9.** arcsin 0.9464

10. arccos 0.9982 **11.** arctan 4.370 **12.** arccot 0.3880

13. $\arcsin \frac{2}{5}$ **14.** arccos 0 **15.** $\arctan \frac{3}{4}$

16. $\operatorname{arccot} \frac{12}{5}$ **17.** arcsin 0 **18.** arctan 1

19. arccos 0.5 − arcsin 0.5 **20.** arccos 0 − arctan 1

QUIZ ON CHAPTER 37, TABLES OF TRIGONOMETRIC RATIOS. FORM A.

Find the value of each of the following functions. Use the table.

1. sin 18.6° **2.** sin 55.4° **3.** cos 22.7° **4.** cos 62.1°

5. tan 39.2° **6.** tan 71.3° **7.** cot 40.9° **8.** cot 57.5°

9. sin 4.8° **10.** cos 85.9° **11.** tan 84.7° **12.** cot 67.4°

13. sec 17.1° **14.** sec 69.3° **15.** csc 13.4° **16.** csc 83.8°

Find the following angles:

17. arcsin 0.2840 **18.** arcsin 0.7290 **19.** arccos 0.9342

20. arccos 0.5678 **21.** arctan 04515 **22.** arctan 1.739

23. arcsec 2.4300 **24.** arccsc 1.1390 **25.** arctan 3

26. Find arccos 0.8660 − .rcsin 0.3420

CHAPTER THIRTY-EIGHT The Right Triangle

38-1 DEFINITIONS OF THE TRIGONOMETRIC FUNCTIONS IN TERMS OF A RIGHT TRIANGLE

The trigonometric functions are often defined in terms of the sides of a right triangle without regard to the x and y axes. Let us see how this is done.

Consider, again, an angle, A, in standard position, with point (8, 6) on the terminal side (Fig. 38-1). The abscissa of the point is 8 units, the ordinate is 6 units, and the radius vector is 10 units. From our definitions of the trigonometric functions, we have

$$\sin A = \frac{3}{5} \qquad \cos A = \frac{4}{5} \qquad \tan A = \frac{3}{4}$$

We have seen that in this case angle A is approximately 36.9°.

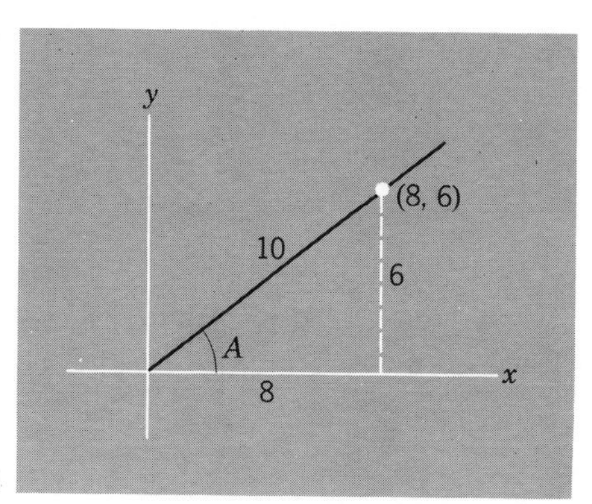

Figure 38-1

Now, suppose we erase the x-axis and the y-axis. Then we have only a right triangle with two sides and hypotenuse (Fig. 38.2). We still have the same angle A, which is still the same size. The sides of the triangle are 6 units, 8 units, and 10 units, respectively.

However, we now have no x-axis, no y-axis, and no origin. Therefore, we have no ordinate, no abscissa, and no radius vector. We can no longer define the sine of the angle as y/r. The same is true for the other functions. The definitions of the trigonometric functions must now be stated in terms of the three sides of the triangle.

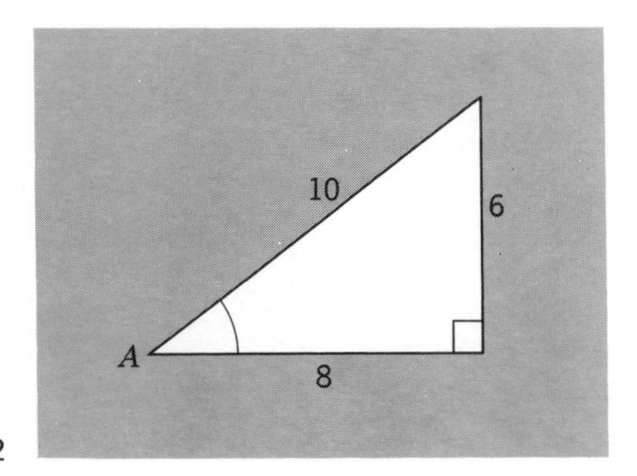

Figure 38-2

From the position of angle A, we can tell which side of the triangle was formerly the ordinate, which was the abscissa, and which was the radius vector. The hypotenuse was formerly the radius vector. To distinguish between the other two sides, we call one of them the side *opposite* angle A and the other the side *adjacent* to angle A. The side adjacent to angle A, is the side which, with the hypotenuse, forms angle A.

Our former definitions of sine, cosine, tangent, and so on, now become new definitions in terms of *hypotenuse, side opposite,* and *side adjacent.* Corresponding to our former definitions, we have

$$\sin A = \frac{\text{side opposite}}{\text{hypotenuse}} \qquad \cot A = \frac{\text{side adjacent}}{\text{side opposite}}$$

$$\cos A = \frac{\text{side adjacent}}{\text{hypotenuse}} \qquad \sec A = \frac{\text{hypotenuse}}{\text{side adjacent}}$$

$$\tan A = \frac{\text{side opposite}}{\text{side adjacent}} \qquad \csc A = \frac{\text{hypotenuse}}{\text{side opposite}}$$

The foregoing definitions hold true for any *acute* angle in a right triangle. The definitions are often abbreviated. Suppose *theta* (θ) is any acute angle in a right triangle; then the definitions of the trigonometric functions are as follows:

$$\sin \theta = \frac{\text{opp}}{\text{hyp}} \qquad \cos \theta = \frac{\text{adj}}{\text{hyp}} \qquad \tan \theta = \frac{\text{opp}}{\text{adj}}$$

$$\cot \theta = \frac{\text{adj}}{\text{opp}} \qquad \sec \theta = \frac{\text{hyp}}{\text{adj}} \qquad \csc \theta = \frac{\text{hyp}}{\text{opp}}$$

To illustrate these ratios, suppose we have a right triangle (Fig. 38-3) with sides lettered s, t, and u, in which u represents the hypotenuse and the angle θ is opposite the side s. Then, from the foregoing

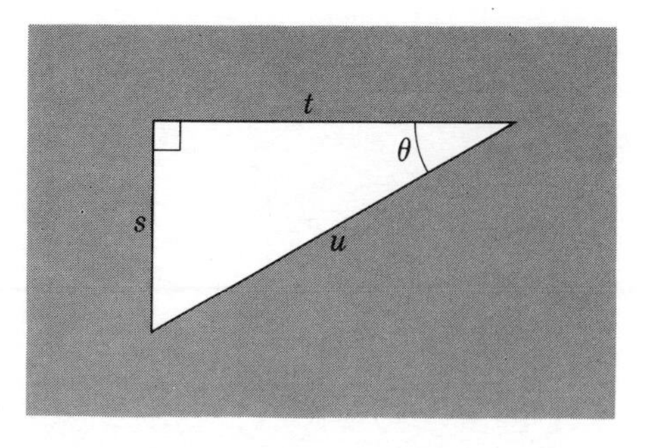

Figure 38-3

definitions, we have

$$\sin\theta = \frac{s}{u} \qquad \tan\theta = \frac{s}{t} \qquad \sec\theta = \frac{u}{t}$$

$$\cos\theta = \frac{t}{u} \qquad \cot\theta = \frac{t}{s} \qquad \csc\theta = \frac{u}{s}$$

It is important that the definitions of the trigonometric ratios be thoroughly understood and remembered. However, they should *not* be memorized in *a particular order.* Each function, by itself, should be recognized instantly.

38-2 THE FUNCTION VALUES IN ANY PARTICULAR RIGHT TRIANGLE

If we know the hypotenuse and the two sides of a right triangle, we can state the six trigonometric ratios between the sides. For instance, in the triangle shown in Fig. 38-4, suppose the sides are 5 and 12 inches, respectively, and the hypotenuse is 13 in. If we let angle R be the

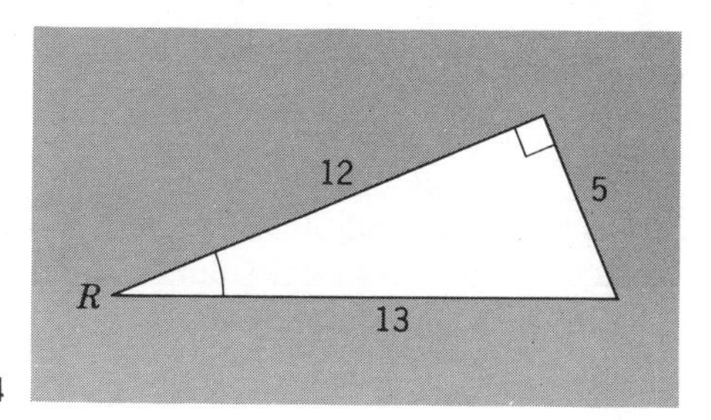

Figure 38-4

angle opposite the 5-inch side, we can state the six trigonometric ratios for angle R:

$$\sin R = \frac{\text{opp}}{\text{hyp}} = \frac{5}{13} \qquad \tan R = \frac{\text{opp}}{\text{adj}} = \frac{5}{12} \qquad \sec R = \frac{\text{hyp}}{\text{adj}} = \frac{13}{12}$$

$$\cos R = \frac{\text{adj}}{\text{hyp}} = \frac{12}{13} \qquad \cot R = \frac{\text{adj}}{\text{opp}} = \frac{12}{5} \qquad \csc R = \frac{\text{hyp}}{\text{opp}} = \frac{13}{5}$$

In a given right triangle one side may be unknown. Then it is first necessary to find the length of the unknown side. As an example, suppose we have the right triangle shown in Fig. 38-5, with sides of 2 inches and 6 inches, respectively. The hypotenuse, found by the Pythagorean rule, is $\sqrt{40}$.

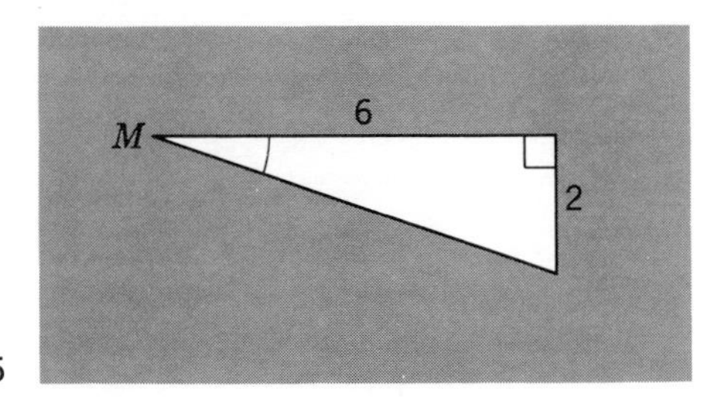

Figure 38-5

Now, using the definitions previously given, we can state the numerical values of all the trigonometric ratios for either of the two acute angles of the triangle. For instance, for angle M we have

$$\sin M = \frac{2}{\sqrt{40}} \qquad \tan M = \frac{2}{6} \qquad \sec M = \frac{\sqrt{40}}{6}$$

$$\cos M = \frac{6}{\sqrt{40}} \qquad \cot M = \frac{6}{2} \qquad \csc M = \frac{\sqrt{40}}{2}$$

These numerical values can be reduced to lower terms and also to decimal fractions if desired.

If we wish, we can also find the size of angle M. We see that the tangent of angle M is $\frac{2}{6}$, or $\frac{1}{3}$, or, in decimal form, 0.3333. From the table, we find that arctan 0.3333 is an angle of 18.4° approximately. If we wish a more accurate answer, we find that angle M equals 18°26′.

Again, it should be clearly understood that the values of the trigonometric ratios are *not measurements*, such as feet, inches, or degrees, but pure numbers. In the last example given, one side of the triangle is 2 inches and the other side is 6 inches. The tangent is the ratio of 2 inches to 6 inches, but this ratio, $\frac{1}{3}$, is not inches, but simply the pure number $\frac{1}{3}$.

Exercise 38-1

In each of the right triangles in Fig. 38-6:

1. Find the length of the unknown side.
2. Then state the value of the sine, cosine, and tangent of each of the lettered angles.

3. Finally, tell the approximate size of the lettered angle by use of the table. In finding the size of the angle, use the function involving the two *given* sides.

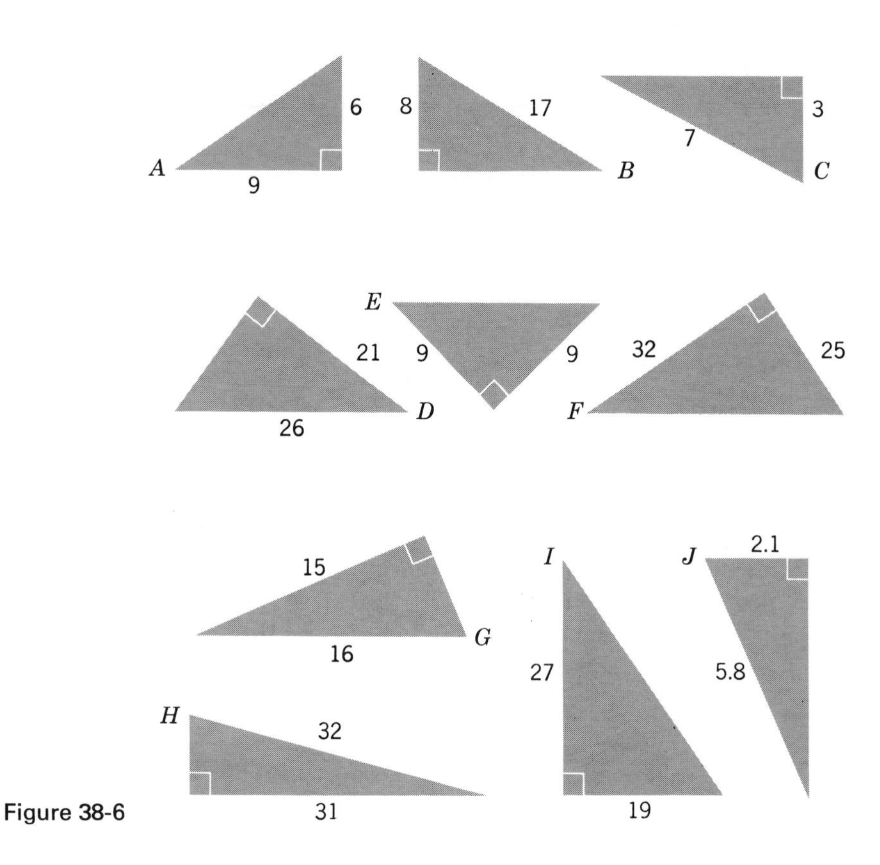

Figure 38-6

38-3 SOLVING RIGHT TRIANGLES

Every triangle has three sides and three angles. The three sides and three angles are called the six *elements* of a triangle. The three angles are often named by the capital letters, A, B, and C, respectively. Each side is named by the small letter that corresponds to the opposite angle. In a right triangle the right angle is usually called C. The hypotenuse is labeled c.

Although this is the usual method of naming the elements of a triangle, the student should not get the idea that they must be so named. An angle may be given any name we wish.

To "solve" a triangle means to find all the unknown elements. At least three of the six elements must be known, one of which must be a side. In a right triangle one angle is a right angle, or 90°. Then two more elements must be known, one of which must be a side.

In finding the unknown elements of a right triangle, we make use of the trigonometric ratios. In all cases we *set up a ratio between two sides*, one of which is known. Then we equate this ratio to the correct trigonometric function. To show how this is done, we use an example.

Example 1

Solve the right triangle, ABC, in which angle C is the right angle, angle $B = 57.1°$, and side $a = 15.2$ in. To solve the triangle, we must find the three unknown parts: angle A and sides b and c.

Solution

The first step in the solution is to make a sketch of the triangle, labeling all parts, including those given and those to be found (Fig. 38-7). Next, we set

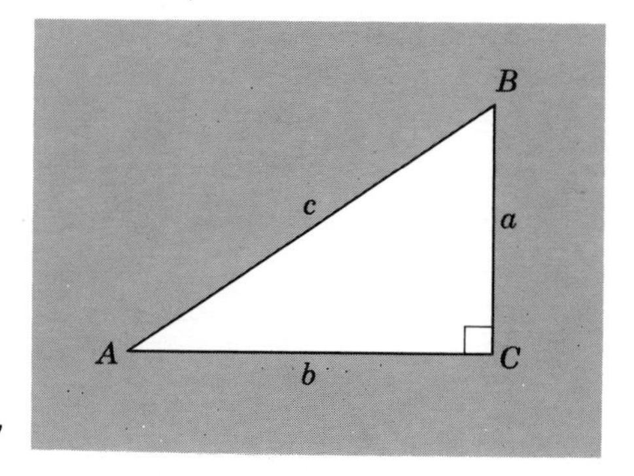

Figure 38-7

up a ratio between an *unknown side* and the *known side*:

$$\frac{b}{a}$$

Now we notice that this ratio is the tangent of angle B. This fact is then stated as an equation:

$$\frac{b}{a} = \tan B$$

Substituting known values, we get $\quad \dfrac{b}{15.2} = \tan 57.1°$

Solving the equation for b,

$$b = (15.2)(\tan 57.1°)$$
$$b = (15.2)(1.548)$$
$$= 23.50 \text{ (approx.)}$$

To find c, we could write $\quad \dfrac{c}{a} = \sec B$

Let us assume that we have no table of secants available. Then we write

$$\frac{a}{c} = \cos B$$

Solving for c in two steps, we get $$c = \frac{a}{\cos B}$$

Substituting known values, $$c = \frac{15.2}{0.5432} = 27.98 \text{ (approx.)}$$

For angle A, we have $A = 90° - 57.1° = 32.9°$

The work may be checked by the Pythagorean rule: $a^2 + b^2 = c^2$. Since the values we have used are only approximate, the two sides of the equation may not be exactly equal. However, let us ask:

does $(15.2)^2 + (23.5)^2 = (27.98)^2$?

Squaring, does $231.04 + 552.25 = 782.88$?

does $783.29 = 782.88$?

Rounded off to three significant digits, the two sides are equal, and the work may be considered correct.

Example 2

Solve the triangle ABC in which C is the right angle, $a = 34.2$ in., and the hypotenuse $c = 51.4$ in.

Solution

First we make a sketch of the triangle and label all parts. (The sketch is left for the student.) Since no acute angle is given, we must first find one of the acute angles. To do so, we set up a ratio between the given sides:

$$\frac{34.2}{51.4}$$

Now we equate this ratio to some function of one angle. We note that

$$\frac{34.2}{51.4} = \sin A; \qquad \text{which reduces to} \qquad \sin A = 0.6654$$

Then

$$A = \arcsin 0.6654 = 41.7°$$

Now we use angle A to find side b:

$$\frac{b}{51.4} = \cos A; \qquad \text{then } b = (51.4)(\cos 41.7°) = 38.4 \text{ (approx.)}$$

The answer is rounded to three places because the other two sides have only three-place accuracy. To find b, we could have used: $\frac{b}{34.2} = \cot A$. To complete the solution, we find angle B: $B = 90° - A = 48.3°$.

Note. It is suggested that all unknown sides be found by trigonometric functions and none by the Pythagorean rule. Then the Pythagorean rule can be used as a check.

Exercise 38-2

Solve the following triangles. Each triangle is understood to be lettered ABC, with angle C as the right angle, and the sides lettered with small letters that correspond, respectively, to the opposite angles. Round off answers to three significant digits (or to four if the first digit is 1).

1. $A = 24.3°$; $c = 42$ in.

2. $A = 31.6°$; $a = 24$ cm

3. $B = 25.5°$; $c = 35$ in.

4. $B = 42.3°$; $b = 64$ cm

5. $A = 63.7°$; $b = 42$ ft

6. $A = 13.8°$; $a = 54$ ft

7. $a = 21$ cm; $c = 48$ cm

8. $a = 42$ in.; $b = 48$ in.

9. $a = 48$ cm; $b = 30$ cm

10. $a = 36$ ft; $c = 60$ ft

11. $b = 45$ in.; $c = 54$ in

12. $b = 24$ cm; $c = 80$ cm

13. $A = 57.3°$; $c = 32$ cm

14. $A = 70.8°$; $a = 42$ cm

15. $A = 69.4°$; $b = 16$ ft

16. $B = 48.6°$; $c = 35$ in.

17. $B = 77.2°$; $a = 28$ cm

18. $B = 25.1°$; $b = 48$ cm

19. $b = 122$ cm; $c = 220$ cm

20. $b = 84$ cm; $a = 280$ cm

21. $a = 154$ ft; $c = 420$ ft

22. $a = 36$ cm; $b = 72$ cm

23. $A = 13.1°$; $b = 530$ ft

24. $B = 12.3°$; $c = 508$ ft

*38-4 LOGARITHMS OF FUNCTIONS

Logarithms can often be used to simplify the work in trigonometry. We have seen that solving a problem in trigonometry usually involves multiplication and division. These processes can be performed by use of logarithms. For example, suppose we have the multiplication problem:

$$a = (28.1)(\tan 72.4°)$$

The problem becomes $\quad a = (28.1)(3.152)$

By use of logarithms, we have

$$\log 28.1 = 1.4487$$
$$\log 3.152 = 0.4986$$

Adding logs, $\quad \log a = 1.9473$

Then $\quad a = 88.6$

* This section may be omitted if these tables are not available.

Now, note that in order to find the logarithm of the second factor, we must use two steps. This involves the use of two tables. First, we must look up the tangent of 72.4°, which is 3.152. Then we must find the logarithm of 3.152, which is 0.4986. Actually, what we want to find is the logarithm of the tangent of 72.4°.

Special tables have been constructed to show the logarithms of the trigonometric functions *directly*. That is, we look up the angle 72.4° and find directly, not the tangent, but the logarithm of the tangent. Using such a table, we have

$$\alpha = (28.1)(\tan 72.4°)$$

By use of logs, we have

$$\begin{aligned} \log 28.1 &= 1.4487 \\ \log \tan 72.4° &= \underline{0.4986} \\ \log a &= 1.9743; \quad a = 88.6 \end{aligned}$$

The advantage of these tables is that one table takes the place of two. It must be carefully noted that some logarithms of trigonometric functions have negative characteristics. For example, if we look up the sine of 20°, we find that it is 0.3420. The logarithm of this number is $0.5340 - 1$, or in another form, $9.5340 - 10$. In such tables, the -10 is omitted.

The following two examples are solved by use of logarithms.

Example 1

In a right triangle ABC, angle C is the right angle, $A = 68.3°$, and $b =$ 25.7 in. Find a, c, and B, using logs of the functions.

Solution

When we have sketched the triangle, we see that

$$\frac{a}{b} = \tan A; \qquad \text{or} \qquad a = (b)(\tan A)$$

Substituting values, $a = (25.7)(\tan 68.3°)$

Using logs, we have $\log 25.7 = 1.4099$

From the special table, $\log \tan 68.3° = \underline{0.4002}$

Adding logs, $\log a = 1.8101; \quad a = 64.6$ in.

To find c, we write

$$\frac{b}{c} = \cos A; \qquad \text{or,} \qquad c = \frac{b}{\cos A}$$

Substituting values,

$$c = \frac{25.7}{\cos 68.3°}$$

Using logs, we have $\log 25.7 = 1.4099$

From special table, $\log \cos 68.3° = \underline{0.5679 - 1}$

Subtracting logs for division, $\log c = 0.8420 + 1; \quad c = 69.5$ in.

We find angle B by subtraction: $B = 90° - A = 21.7°$

For a check, we may use the formula, $a^2 + b^2 = c^2$. However, if logarithms are used in the check, a better form of the Pythagorean rule is the following:

$$a^2 = c^2 - b^2; \qquad \text{or the form,} \qquad a^2 = (c + b)(c - b)$$

Taking the log of both sides, we have $\log a^2 = \log (c + b) + \log (c - b)$. Then we ask,

$$\begin{aligned}
&\text{does} \quad & 2 \log a &= \log (c + b) + \log (c - b)? \\
&\text{does} \quad & 2(\log 64.6) &= \log 95.2 + \log 43.8? \\
&\text{does} \quad & 2(1.8101) &= 1.9786 + 1.6415? \\
&\text{does} \quad & 3.6202 &= 3.6201?
\end{aligned}$$

The two sides of the equation are sufficiently close to serve as a check.

Example 2

In a right triangle ABC, angle C is the right angle, $a = 36.2$ cm, and $b = 86.1$ cm. Find A, B, and c. Do not use the Pythagorean rule to find c.

Solution

After sketching the triangle, we see that

$$\frac{a}{b} = \tan A; \qquad \text{or} \qquad \tan A = \frac{36.2}{86.1}$$

To find the value of tan A, we use logs:

$$\begin{aligned}
\log 36.2 &= \overset{2}{\not{1}}.5587 - 1 \\
\log 86.1 &= 1.9350
\end{aligned}$$

Subtracting logs, we get $\quad \log \tan A = 0.6237 - 1$

We find A directly from special table $\quad A = 22.8°; \qquad B = 67.2°$

To find c, we write

$$\frac{a}{c} = \sin A, \qquad \text{from which} \qquad c = \frac{a}{\sin A} = \frac{36.2}{\sin 22.8°}$$

Using logs, we have

$$\begin{aligned}
\log 36.2 &= 1.5587 \\
\log \sin 22.8° &= 0.5883 - 1
\end{aligned}$$

Subtracting logs,

$$\begin{aligned}
\log c &= 0.9704 + 1 \\
c &= 93.4 \text{ cm}
\end{aligned}$$

For the check, we write $2 \log a = \log (c + b) + \log (c - b)$ and ask,

$$\begin{aligned}
&\text{does} \quad & 2 \log 36.2 &= \log 179.5 + \log 7.3? \\
&\text{does} \quad & 2(1.5587) &= 2.2541 + 0.8633? \\
&\text{does} \quad & 3.1174 &= 3.1174?
\end{aligned}$$

For the check, we could also use $2 \log b = \log (c + a) + \log (c - a)$

Exercise 38-3

Solve the following right triangles by using logarithms of the trigonometric functions. Each triangle is understood to be lettered *ABC*, with angle *C* as the right angle and the sides lettered *a*, *b*, and *c*, each one opposite the corresponding angle.

1. $A = 21.9°$; $b = 24.8$ in.
2. $A = 39.2°$; $c = 340$ ft
3. $B = 38.8°$; $a = 120$ in.
4. $B = 42.3°$; $c = 204$ ft
5. $A = 58.2°$; $a = 32.2$ cm
6. $B = 64.5°$; $b = 52.4$ cm
7. $A = 73.7°$; $b = 420$ cm
8. $A = 66.2°$; $c = 34.2$ cm
9. $B = 81.8°$; $b = 23.6$ in.
10. $A = 28.5°$; $a = 24.5$ in.
11. $A = 70.2°$; $b = 21.5$ cm
12. $B = 61.5°$; $c = 86$ cm
13. $a = 265$ cm; $b = 725$ cm
14. $a = 129$ cm; $c = 153$ cm
15. $b = 58$ in.; $c = 180$ in.
16. $b = 81$ ft; $a = 75$ ft

38-5 SOLVING STATED WORD PROBLEMS

Some problems involve an angle of *elevation* or an angle of *depression*. To understand the meaning of these terms, consider a horizontal line from *A* to *B* directly. Now, suppose point *C* is above point *B*. Then, standing at point *A*, we must elevate our line of sight to look at *C*. The angle between the horizontal line *AB* and the line *AC* is called the angle of *elevation*. Now, if point *D* is a point below point *B*, then, standing at *A*, we must lower our line of sight to look at *D*. Then the angle between the horizontal line *AB* and the line *AD* is called the angle of *depression*. In this case, we can say that we must *depress* our line of sight to look at *D*.

In stated word problems we usually wish to find only one or two particular unknown quantities. All the problems we are considering here involve a right triangle. In solving any stated word problem, the following steps will serve as a guide:

1. *First, sketch a figure showing the right triangle involved.*
2. *Label all parts, showing given values and the part to be found.*
3. *Set up a ratio between two sides (at least one known).*
4. *Equate this ratio to a function of some angle.*
5. *Solve the resulting equation for the unknown value.*

Example 1

Suppose we wish to measure the height of a flagpole simply by making certain measurements from the ground. We first measure a distance, say, 120 ft, along the ground from the foot of the pole. From this point it is possible, by means of an instrument, to measure the angle of elevation of the top of the pole. Suppose the angle of elevation is 24.3°. Call the height h.

Solution

Step 1. We make a sketch of the right triangle involved (Fig. 38-8).
Step 2. We label all parts.
Step 3. Now we set up a ratio between two sides: $\frac{h}{120}$
Step 4. Express this ratio as a function of the given angle and solve the resulting equation for the unknown:

$$\frac{h}{120} = \tan 24.3^\circ \qquad h = 120 \tan 24.3^\circ$$

$$h = (120)(0.4515) \qquad h = 54.18 \quad \text{or} \quad 54.2\,\text{ft}$$

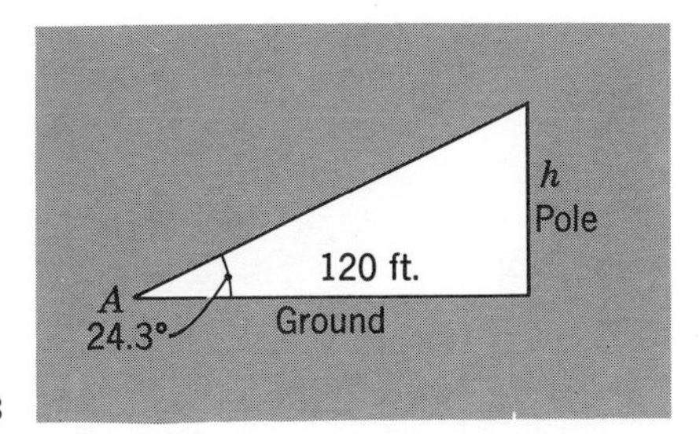

Figure 38-8

If the measurement of angle A is made from a point some distance above the level of the foot of the pole, this distance must be added to the answer in order to obtain the height of the pole. In some problems this extra distance is negligible.

Example 2

Find the angle of elevation of the sun when a flagpole 56 ft high casts a shadow 67 ft long (Fig. 38-9).

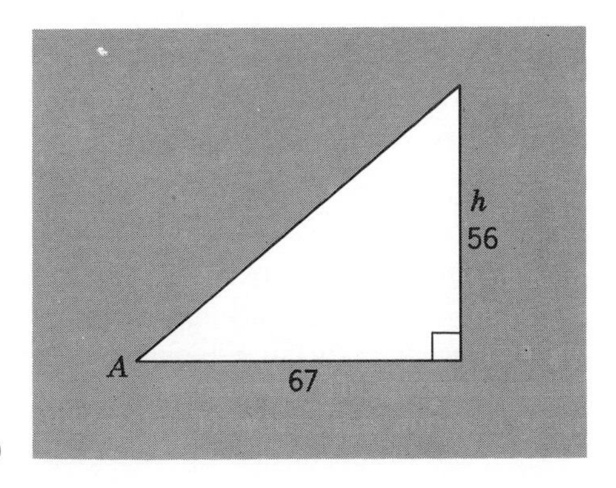

Figure 38-9

Solution

$$\frac{56}{67} = \tan A$$
$$\tan A = 0.8358$$
$$\text{angle } A = \arctan 0.8358 = 39.9°$$

Example 3

From the top of a cliff, 400 ft above the edge of a lake, a small boat is observed out on the lake at an angle of depression of 15.6°. How far is the boat from the foot of the cliff?

Solution

First, make a sketch of the right triangle in the problem and label all parts (Fig. 38-10). Next, set up the ratio $\frac{d}{400} = \cot \theta$. The angle θ is the same as

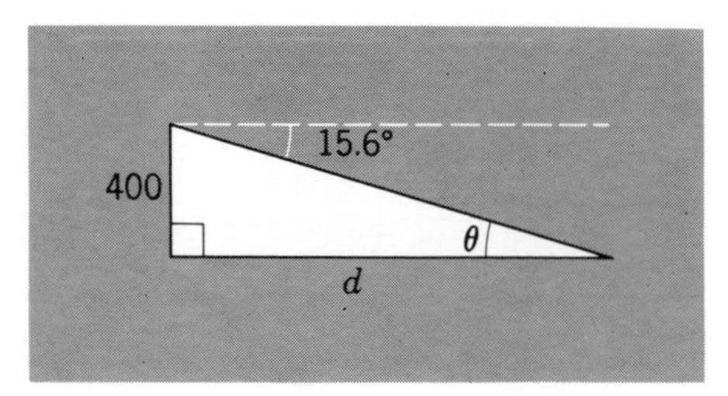

Figure 38-10

the angle of depression. Therefore,

$$\frac{d}{400} = \cot 15.6°$$

$$d = (400)(\cot 15.6°) = (400)(3.582) = 1432.8$$

The answer can reasonably be rounded off to 1430 ft.

Exercise 38-4

1. A ladder 32 ft long leans up against a vertical wall and makes an angle of 17.4° with the wall. How high up on the wall does the ladder reach? How far is the foot of the ladder from the wall?
2. An escalator from the first floor to the second floor of a building is 43.2 ft long and makes an angle of 35.7° with the floors. Find the vertical distance between the floors.
3. A road has a uniform elevation of 6.2°. Find the increase in elevation in driving one quarter of a mile along the road.

4. The shadow of a tree is 72 ft long when the sun is 43.7° elevation. Find the height of the tree.

5. An inclined railway is built to the top of a hill whose elevation is 364 ft above the level of the bottom of the railway. If the angle of elevation of the railway is 35.8°, how long is the railway?

6. A chandelier hangs on a 52-ft chain from a high ceiling. If the chandelier is pulled aside so that the chain forms an angle of 25.8° with its vertical position, how much is the chandelier raised vertically?

7. A plank 18 ft long is used to roll a barrel onto a truck. If the platform of the truck is 5.4 ft above the ground, what angle does the plank form with the ground?

8. A bridge is 24 ft above the surrounding ground level. An approach to the bridge is to be built so that the angle of elevation of the approach is not over 7.1°. How far from the bridge must the approach be started?

9. How long a shadow will be cast by the Washington Monument (height 555 ft) when the elevation of the sun is 41.7°?

10. To measure the angle of elevation of the sun, a 6-ft pole is set up vertically. Then the shadow is found to be 11 ft 3 in. long. Find the angle of elevation of the sun.

11. A 24-ft ladder is placed against a vertical wall so the foot of the ladder is 5.9 ft from the wall. What angle does the ladder make with the ground? How high on the wall does the ladder reach?

12. A telephone pole 32 ft high is braced by a wire one end of which is fastened to the top and the other end to a stake in the ground 45 ft from the foot of the pole. What angle does the brace make with the ground? How long is the brace?

13. A TV tower is braced by a wire cable fastened 20 ft below the top and to a stake in the ground 74 ft from the foot of the tower. If the brace makes an angle of 73.8° with the ground, how high is the tower? How long is the brace?

14. From the top of a 70-ft building, the angle of depression of the far side of the street is 34.7°, and to the near side, the angle is 55.2°. How wide is the street?

15. From the top of a cliff, 850 ft above the level of the sea, the angle of depression of a small boat on the water is 16.3°. Some time later the angle of depression is found to be 36.7°. How far has the boat traveled toward the cliff? How could the speed of the boat be determined?

16. Two observation posts, A and B, are 4500 ft apart. A helicopter is directly above a straight line from A to B. From A, the helicopter appears at an elevation of 54.3°, and from B the angle is 43.7°. Find the height of the helicopter above the ground.

17. Observing a tower in the distance, a man finds the angle of elevation of the top is 17.6°. He then walks directly toward the tower for a distance of 300 ft. He then finds the angle of elevation of the top to be 39.3°. How high is the tower?

18. A man sees a flagpole on top of a building. After walking 240 ft from the building, he finds that the angle of elevation of the bottom of the pole is 28.4°, and the angle of the top is 36.8°. Find the length of the flagpole.

19. Standing on top of one building, a man observes a flagpole on another building. From his position, he finds that the angle of elevation of the top of the pole is 19.5°, and the angle of depression of the bottom is 8.2°. If the pole is known to be 60 ft long, how far is it from the man?

20. Show by diagram how you might compute the distance across a river.

QUIZ ON CHAPTER 38, THE RIGHT TRIANGLE. FORM A.

1. In the right triangle at the right, find the length of the unknown side. Then write the six trigonometric functions of the angle T. Leave answers in radical and fractional form.

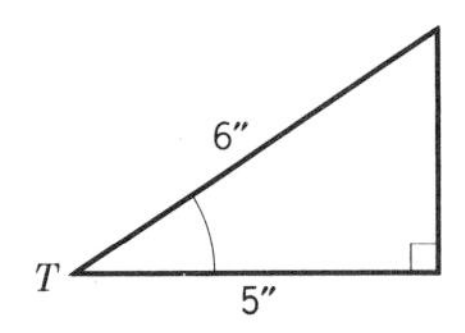

2. In a certain right triangle ABC, angle C is the right angle; angle $A = 32.7°$, and the hypotenuse $c = 5$ cm. Find B, a, and b.
3. In a certain right triangle ABC, angle C is the right angle, $a = 60$ cm, and $A = 24.9°$. Solve the triangle. (Find all unknown parts.)
4. In a certain right triangle ABC, angle C is the right angle; $a = 80$ cm, and $B = 70.6°$. Solve the triangle.
5. In a certain right triangle ABC, $a = 32$ cm and $b = 48$ cm. Solve the triangle.
6. At a distance of 35 ft from the foot of a vertical flagpole, the angle of elevation of the top of the pole is 60.4°. Find the height of the pole.
7. At a certain time of day, a flagpole 64 ft high casts a shadow 51.4 ft long. Find the angle of elevation of the sun at that time.

8. A man stands on the top of a 70-ft building. From this position, the angle of depression of the near side of the street is 62.3°, and the angle of depression of the far side is 33.1°. How wide is the street?

9. A 36-ft ladder leans up against a vertical building and forms an angle of 16.2° with the wall. How high on the wall does the ladder reach?

CHAPTER THIRTY-NINE Functions of Angles of Any Size

39-1 INTRODUCTION

Up to this time we have used a table to find the function values of any angle from zero to 90°. However, we know that angles are not limited to 90°. An angle may be any size, and it may have its terminal side in any quadrant. For example, an angle of 257° has its terminal side in the third quadrant. Our problem now is to see how we can use the same table for angles greater than 90° and for negative as well as positive angles.

To begin, we go back to our original definitions of the functions of any angle. If θ is any angle in standard position, and point (x, y) is on the terminal side, then, by definition,

$$\sin\theta = \frac{y}{r} \qquad \tan\theta = \frac{y}{x} \qquad \sec\theta = \frac{r}{x}$$

$$\cos\theta = \frac{x}{r} \qquad \cot\theta = \frac{x}{y} \qquad \csc\theta = \frac{r}{y}$$

39-2 TERMINAL SIDE IN FIRST QUADRANT

Let us first consider a positive angle less than 90°. Such an angle is called a *first-quadrant* angle. Suppose a first-quadrant angle in standard position has the point (12, 5) on its terminal side (Fig. 39-1). Call the angle *alpha* (α). Then for angle α we have

$$y = 5, \qquad x = 12, \qquad r = 13$$

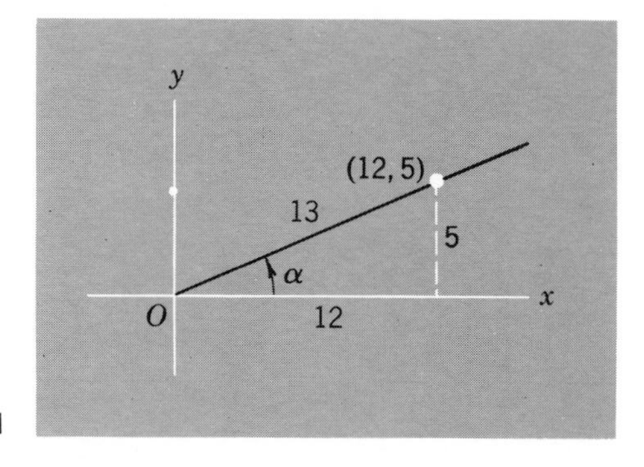

Figure 39-1

By definition, $\sin \alpha = \dfrac{y}{r} = \dfrac{5}{13}$; or 0.3846 (approx.)

From the table we find that α is approximately 22.6°.

In Fig. 39-1 note that a right triangle is formed by the line segments indicating the x, y, and r distances, respectively. This triangle may be called a "5-12-13" right triangle. The angle α is opposite the shortest side. Now keep in mind that in a right triangle of this particular shape, the smallest angle contains 22.6°. This is true no matter where this particular triangle appears.

39-3 ANGLE WITH ITS TERMINAL SIDE IN THE SECOND QUADRANT

Next consider an angle in standard position with its terminal side in the second quadrant. Such an angle is called a *second-quadrant* angle. Suppose we have a second-quadrant angle with the point (−12, 5) on its terminal side (Fig. 39-2). Let us call the angle beta (β). Note that

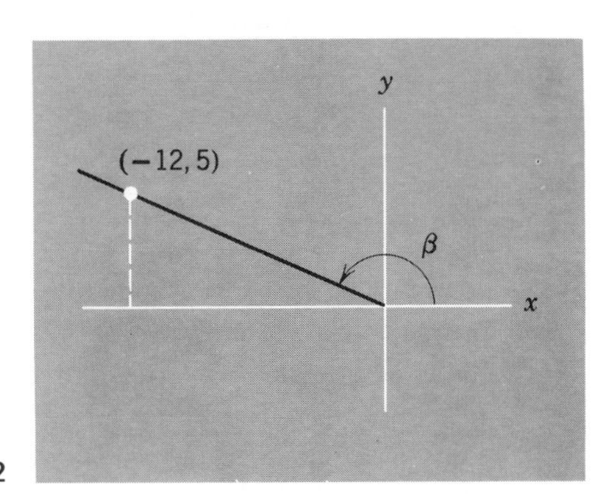

Figure 39-2

angle β is greater than 90° and less than 180°. For angle β we have

By definition, $\sin \beta = \dfrac{y}{r} = \dfrac{5}{13}$

If we show the distances represented by y, x, and r, we see the same right triangle that we saw in the first quadrant. In this "5-12-13" right triangle, we found that the smallest angle is 22.6°. Since the same triangle appears in the second quadrant, the smallest angle is 22.6°.

In Fig. 39-2, this small angle is called the *reference* angle because we can refer to it to find angle β. However, it must be remembered that this reference angle is *not* β. Yet it has the same sine value; that is, $\frac{5}{13}$. Here we see that the sine of β is the same as the sine of the reference angle. Knowing the reference angle, we find that β is 157.4°.

39-4 TERMINAL SIDE IN THIRD QUADRANT An angle in standard position with its terminal side in the third quadrant is called a *third-quadrant* angle. Suppose a third-quadrant angle has the point $(-12, -5)$ on its terminal side (Fig. 39-3). Let us

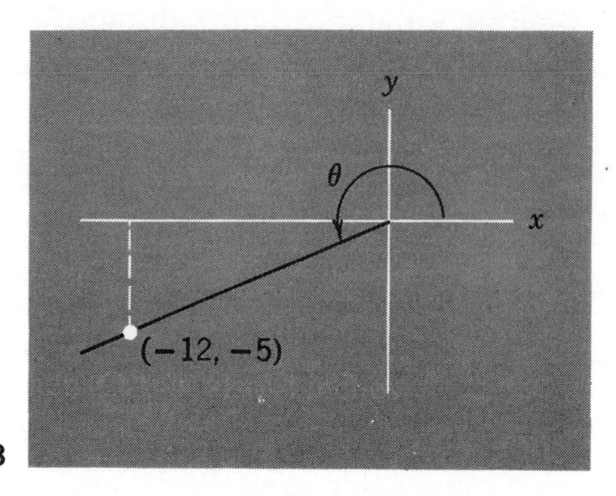

Figure 39-3

call the angle θ (theta). Note that θ is greater than 180° and less than 270°. For angle θ we have

$$y = -5, \qquad x = -12 \qquad r = 13 \qquad (r \text{ is always positive})$$

Then $$\sin\theta = \frac{y}{r} = -\frac{5}{13} \qquad \text{or} \qquad -0.3846 \text{ (approx.)}$$

Here again is the same triangle seen in the first and second quadrants. The smallest angle is therefore 22.6°. This angle, the reference angle, enables us to find θ. We see that θ is 202.6°.

It is important to keep in mind that the reference angle, 22.6°, is *not* the angle θ. The angle θ is in standard position, whereas the reference angle is *not.* Note that the sine of θ has the same *numerical value* as the sine of the reference angle. However, for any third quadrant angle, the ordinate will be negative. Since r is always positive, the sine value of a third-quadrant angle is negative. It is important that we hold to the definition that the sine of an angle is always equal to the ratio, y/r, in order that the function has the proper sign, positive or negative.

39-5 TERMINAL SIDE IN FOURTH QUADRANT An angle in standard position with its terminal side in the fourth quadrant is called a *fourth-quadrant* angle. Suppose a fourth-quadrant angle has the point $(12, -5)$ on its terminal side (Fig. 39-4). Let us call the angle ϕ (phi). If we assume that ϕ is a positive angle, it is greater than 270° and less than 360°. Then for angle ϕ we have

$$y = -5 \qquad x = 12 \qquad r = 13$$

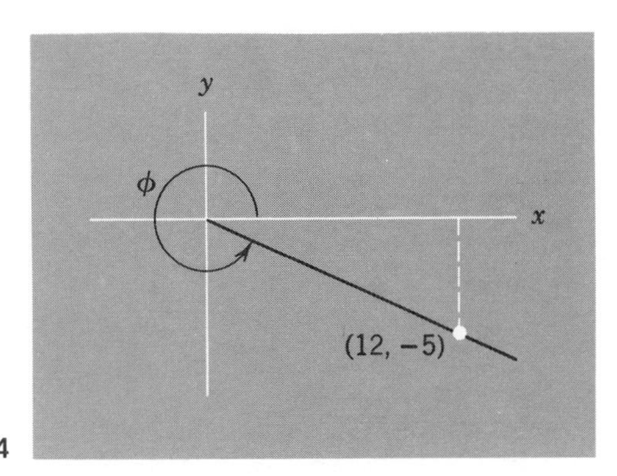

Figure 39-4

Then $$\sin \phi = \frac{y}{r} = -\frac{5}{13}$$

Here again the abscissa, the ordinate, and the radius vector form the same "5-12-13" right triangle that appeared in the other quadrants. Therefore, the smallest angle is 22.6°. We refer to this small angle to find ϕ but we must remember that the reference angle is *not* ϕ. We see that ϕ is 337.4°. Note that the sine of 337.4° has the same *numerical value* as the sine of the reference angle, 22.6°. However, the sine of ϕ is negative.

Note. We have used various letters, including θ, to refer to various angles. However, when we wish to refer to *any general angle*, we often call it θ.

39-6 REFERENCE ANGLE AND ITS USE

We have mentioned the use of the reference angle in determining the value of a trigonometric function. It is necessary to define carefully what is meant by the reference angle. For any given angle in standard position, the *reference angle* is the *acute* angle between the *terminal side and the x-axis.* It is a common mistake to consider the reference angle adjacent to the y-axis.

It will be observed that the reference angle is sometimes within the given angle and at other times outside it. This has nothing to do with its meaning or use. In Fig. 39-5 we see the angle theta (θ) ending in different quadrants. Also shown is the reference angle in each case.

We have seen that the sine of any angle ending in any quadrant has the same numerical value as the sine of the reference angle. This is true also with regard to all the other functions.

Therefore, to find a function of any angle in standard position ending in any quadrant, it is necessary only to find the *reference angle* and then to use the table to find the numerical value of the function of the reference angle. The function of the given angle will have the *same*

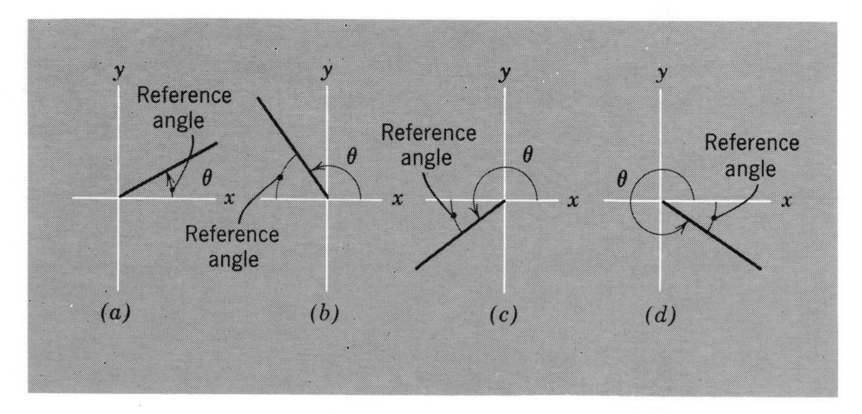

Figure 39-5

numerical value as the function of the reference angle. The final step is to determine whether the function value will be positive or negative for the given angle. This can be determined by observing the signs for x and y for the particular quadrant.

It should be remembered that the reference angle is always an *acute* angle; that is, it is less than 90°. Its functions are always positive. It might be mentioned that the reference angle is sometimes called the *related* angle.

In determining the sine, cosine, or other function of an angle, it makes no difference whether the angle is positive or negative. It makes no difference whether the angle is more or less than 360°. The only thing that determines the value of a function of any angle is the position of the terminal side.

For example, if the terminal side of any angle in standard position passes through the point $(-4, -3)$, the sine of that angle is $-3/5$, whether the angle is positive or negative, or whether it is more or less than 360°. It may be an angle of 216.9°, or −143.1°, or 1656.9°.

We need only two rules at most:

1. *The reference angle will determine the numerical value of any function.* For this we use the table of values for angles from zero to 90°.
2. *The quadrant in which the terminal side falls will determine whether the function will be positive or negative according to the signs of x and y,* since r is always positive.

To find the functions of an angle ending in any quadrant, we have the following steps:

1. *Draw the given angle in standard position, showing the terminal side and showing the direction and amount of rotation by a curved arrow.*

2. *Show the reference angle and state its size.*
3. *From the table, find the functions of the reference angle.*
4. *Give the functions of the given angle the same numerical values as the functions of the reference angle. Then give each function its proper sign, positive or negative, depending on the signs for x and y for the position of the terminal side.*

Example 1

Find sine, cosine, tangent, and cotangent of 145.7° (Fig. 39-6).

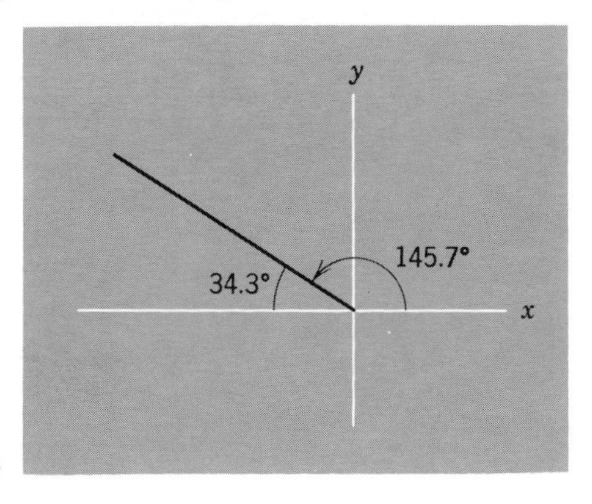

Figure 39-6

Reference angle = 34.3°.

$$
\begin{aligned}
\sin 34.3^\circ &= 0.5635 & \sin 145.7^\circ &= +0.5635 \\
\cos 34.3^\circ &= 0.8261 & \cos 145.7^\circ &= -0.8621 \\
\tan 34.3^\circ &= 0.6822 & \tan 145.7^\circ &= -0.6822 \\
\cot 34.3^\circ &= 1.4659 & \cot 145.7^\circ &= -1.4659
\end{aligned}
$$

Example 2

Find sine, cosine, and tangent of 607.3° (Fig. 39-7).

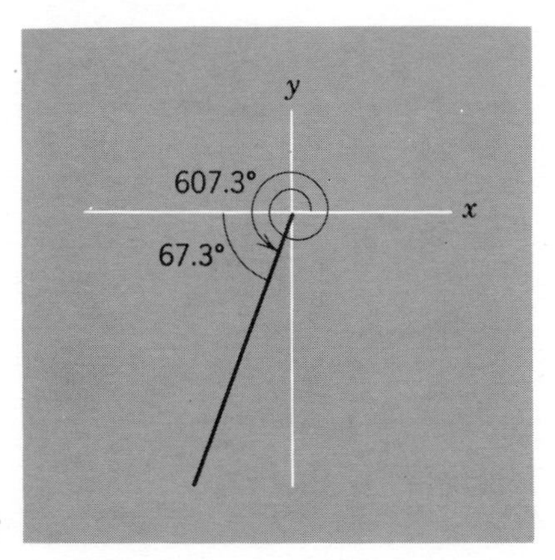

Figure 39-7

Reference angle = 67.3°.

$\sin 67.3° = 0.9225$	$\sin 607.3 = -0.9225$
$\cos 67.3° = 0.3859$	$\cos 607.3 = -0.3859$
$\tan 67.3° = 2.391$	$\tan 607.3 = +2.391$

Example 3*

Suppose we have the formula $e = 80 \sin \omega t$. This is a common formula in alternating currents in which e represents the instantaneous voltage at any particular instant, and ωt is the angle in which ω (omega) represents *angular velocity* and t represents *time*. For instance, if the angular velocity (ω) of a generator is 21,600° per sec and if the time elapsed is 0.031 sec, the angle can be found by multiplying the *angular velocity* by the *time*. Suppose we wish to find the instantaneous voltage at the end of exactly 0.031 sec.

Solution

Let us call the angle theta (θ) (Fig. 39-8). We know that

$$\theta = \omega t$$

$$\theta = (21{,}600°)(0.031) = 669.6°$$

Then we have

$$e = 80 \sin 669.6°$$

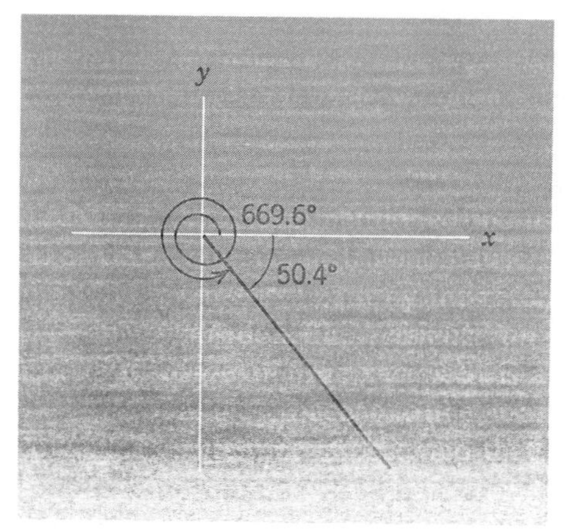

Figure 39-8

The reference angle is 50.4°.

$$\sin 50.4° = 0.7705$$

$$\sin 669.6° = -0.7705$$

Now we see that

$$e = (80)(-0.7705) = -61.64 \text{ volts}$$

* Optional. May be omitted by students unfamiliar with alternating-current theory.

Exercise 39-1

Find the sine, cosine, tangent, and cotangent of each of the following angles:

1. 156.3°	**2.** 138.1°	**3.** 164.9°	**4.** 96.7°
5. 214.8°	**6.** 192.7°	**7.** 201.5°	**8.** 232.1°
9. 248.6°	**10.** 261.4°	**11.** 344.6°	**12.** 321.3°
13. 336.2°	**14.** 285.5°	**15.** 309.7°	**16.** 298.9°
17. 377.4°	**18.** 754.2°	**19.** 473.9°	**20.** 951.6°
21. −38.5°	**22.** −332.7°	**23.** 256° 50′	**24.** 341° 20′

Find e or i in each of the following problems:

25. $e = 60 \sin 168.2°$ **26.** $i = 15 \cos 257.3°$ **27.** $i = 12 \sin 304.4°$

In the following problems* find e or i if omega (ω) is 21,600° per sec. (The angle $\theta = \omega t$.)

28. If $e = 60 \sin \omega t$, find e when $t = 0.008$ sec.

29. If $i = 12 \sin \theta$, find i when $t = 0.012$ sec.

30. If $e = 150 \cos \omega t$, find e when $t = 0.07$ sec.

39-7 FINDING AN ANGLE FROM A GIVEN FUNCTION VALUE

It is often necessary to find an angle when its function is given. In Chapter 37 we saw that the problem presents no difficulty if the angle is an acute angle (that is, less than 90°). We simply use the table of function values in reverse. We look up the function value in the proper column of the table and then find the angle opposite this value.

However, when we also wish to consider angles greater than 90°, the problem is a little more involved. We have seen that angles of 22.6° and 157.4° have the same sine value, 0.3846. Now, if we are asked what angle has a sine of 0.3846, we can say that the angle is either 22.6° or 157.4°. In fact, both answers are correct. In other words, there are two angles whose sines are equal to 0.3846.

Actually, there are other angles, both positive and negative, with the same sine value, such as −202.6° or +742.6°. However, if we confine the angles to positive angles less than 360° whose sines are equal to 0.3846, then we can state that there are two answers to the question.

In the same way (see Sections 39.4 and 39.5), we find that there are two positive angles less than 360° whose sines are equal to −0.3846. They are 202.6° and 337.4°.

We must remember that there are many angles with the same function values. However, there are always just two positive angles less than 360° with the same function values. In stating the angles for a given function value, we should always list these two angles. If only one answer is required, then additional information must be known.

Example

Find two positive angles of less than 360° whose tangent is -1.913.

Solution

The tangent of the reference angles has the same numerical value; that is, 1.913. We look for this value in the table under the heading tangent and find

$$\arctan 1.913 = 62.4°$$

We know then that the reference angle is 62.4°. Since the angles we are seeking have a negative tangent value, the angles must end in the second or fourth quadrant. Therefore, we place the reference angle in the proper position in the second and the fourth quadrants. The position of the reference angle will show the terminal side of the required angle. This is shown in Fig. 39-9.

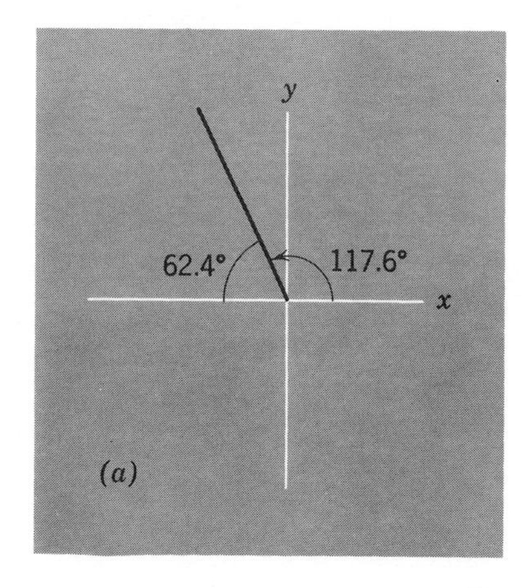

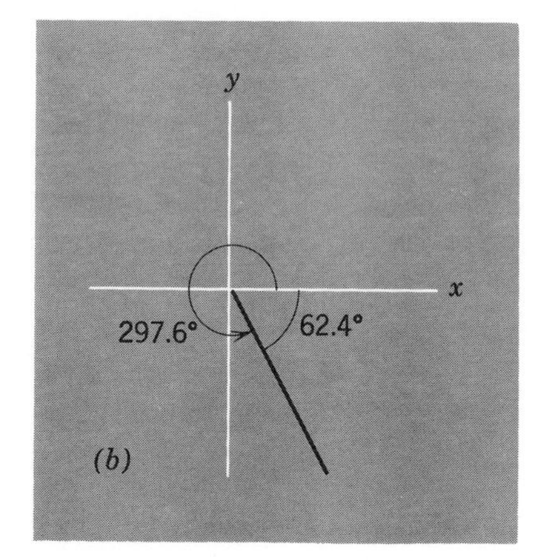

Figure 39-9

From the reference angle we can compute the size of the two required angles. They are 117.6° and 297.6°.

Now, if we have the additional information that the required angle has its terminal side in the second quadrant, then the answer is 117.6°.

Exercise 39-2

Find two positive values each less than 360° for each of the following angles:

1. arcsin (-0.4210) **2.** arcsin 0.8415 **3.** arcsin (-0.9432)

4. arccos (-0.8771) **5.** arccos 0.4321 **6.** arccos (-0.6652)

7. arctan (-0.6420) **8.** arctan 2.808 **9.** arctan (-1.889)

10. arccot (-1.1423) **11.** arccot 0.4020 **12.** arccot (-0.6950)

39-8 FINDING THE VALUES OF ALL THE FUNCTIONS OF AN ANGLE FROM ONE GIVEN FUNCTION VALUE

If the numerical value of one trigonometric function is known, the value of each of the other functions can be found from the known value. As an example, suppose we know that the sine of an angle θ is equal to $\frac{3}{5}$, or 0.6000. Let us assume that the angle is an acute angle of a right triangle, as shown in Fig. 39-10. We place the numerator, 3,

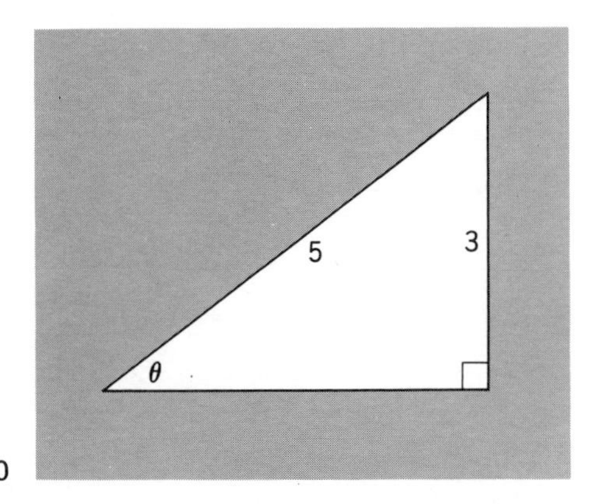

Figure 39-10

and the denominator, 5, of the fraction in the correct positions on the sides of the triangle to represent the sine ratio for θ. From these two values we can find the unknown side by the Pythagorean rule.

$$\sqrt{5^2 - 3^2} = \sqrt{25 - 9} = \sqrt{16} = 4$$

Now, since we know the length, 4, of the side adjacent to angle θ, we can write the value of each of the six functions. For example, $\cot \theta = \frac{4}{3}$.

If the value of a function is given as a decimal fraction or a mixed decimal, we simply write this value over the denominator 1 and then proceed as in the foregoing example. For instance, if we know that the secant of an angle ϕ is 2.31, we can write this number as 2.31/1. We sketch a right triangle and call one of the angles ϕ. We place the two numbers, 2.31 and 1, on the hypotenuse and side adjacent, respectively, so that the ratio, 2.31/1 will represent the secant of ϕ. The length of the unknown side, the side opposite ϕ, can now be found by use of the Pythagorean rule.

$$\sqrt{2.31^2 - 1^2} = \sqrt{5.3361 - 1} = \sqrt{4.3361} = 2.08$$

Knowing the two sides and the hypotenuse of the right triangle, we can write the value of each of the six functions of ϕ.

Exercise 39-3

From each of the following given function values, find the five other function values for the same angle. Consider only the acute angle in each case. The angle itself need not be found, except in No. 25.

1. $\sin A = \frac{5}{13}$ **2.** $\cos B = \frac{8}{17}$ **3.** $\tan C = \frac{3}{4}$

4. $\cot D = 1$ **5.** $\sin E = 0.5$ **6.** $\cos F = 0.4$

7. $\tan G = \sqrt{2}$ **8.** $\cot H = \sqrt{3}$ **9.** $\sec I = \sqrt{10}$

10. $\sin J = 0.25$ **11.** $\cos K = 0.28$ **12.** $\csc L = 2.1$

13. $\tan M = \frac{5}{3}$ **14.** $\cot N = \frac{15}{8}$ **15.** $\sec P = \frac{3}{2}$

16. $\csc Q = \dfrac{x}{3}$ **17.** $\sin R = \dfrac{4}{x}$ **18.** $\cos S = \dfrac{x}{5}$

19. $\tan T = \dfrac{5}{x}$ **20.** $\sec U = \dfrac{3x}{2}$ **21.** $\sin V = \dfrac{2t}{3}$

22. Find sin arctan $\frac{4}{3}$. (First show the angle whose tangent is $\frac{4}{3}$. Remember that arctan $\frac{4}{3}$ is an angle. Then find the sine of this angle.)

23. Find tan arcsin $\frac{1}{2}$. **24.** Find cot arctan $\frac{5}{8}$.

25. Find arctan 1 − arcsin 0.5. (First find the angles. Then subtract angles.)

QUIZ ON CHAPTER 39, FUNCTIONS OF ANGLES OF ANY SIZE. FORM A.

Find the sine, cosine, tangent, and cotangent of each of these angles:

1. 155.8° **2.** 211.6° **3.** 317.5° **4.** 379.4°

5. 248.7° **6.** 102.2° **7.** 278.1° **8.** 407.1°

9. Find e: $e = 120 \sin 316.3°$. **10.** Find i: $i = 15 \cos 218.4°$

Find two angles less than 360° for each of the following values:

11. $\sin A = 0.4003$ **12.** $\cos B = 0.8203$ **13.** $\tan C = 0.6322$

14. $\cot D = 1.3079$ **15.** $\sin E = -0.6730$ **16.** $\cos F = -0.3600$

17. $\tan G = -2.125$ **18.** $\cot H = -0.3000$

Find all the six functions of each angle for the given value:

19. $\sin A = \frac{3}{5}$ **20.** $\cos B = \frac{1}{2}$ **21.** $\tan C = \frac{15}{8}$ **22.** $\cot D = \sqrt{3}$

CHAPTER FORTY Trigonometric Identities

40-1 MEANING OF IDENTITY

There are certain relations between the trigonometric functions that are always true. Such relations are called *trigonometric identities.*

In algebra we learned that an identity is an equation that is true for all values of a variable. For example, take the equation

$$3x + 2x \equiv 5x$$

This equation is true no matter what value we give to x. It is an identity. An identity is often indicated by three parallel lines instead of two, as in the usual *equal* sign.

Compare the foregoing equation with the following:

$$3x + 2 = 17$$

This equation is true only on the condition that $x = 5$. It is a *conditional equation.*

In trigonometry certain relations are true for all values of an angle. Other relations are conditional equations and call for a solution. As an example, let us ask the question:

$$\text{does} \qquad \sin\theta = \cos\theta?$$

The equation is true if θ is 45° or 225°, as well as for some angles greater than 360°. Yet it is *not* true for all values of θ. For example, if θ is equal to, say, 20°, then the statement is not true. Therefore the equation is *not* an identity.

However, there are certain relations in trigonometry that are always true regardless of the size of the angle. For example, there is a definite relation between the sine and the cosecant of any particular angle, no matter what the size of the angle. There are certain definite relations between all the trigonometric functions whatever the size of the specific angle. These relations are called *identities.* A few identity relations should be understood and memorized. You will find the trigonometric identities very helpful in simplifying much of the work in mathematics.

40-2 RECIPROCAL RELATIONS

Certain relations in trigonometry are called *reciprocal relations.* Let us recall the meaning of *reciprocal.*

In arithemetic and algebra we learned that the reciprocal of a number is *1 divided by the number.* The reciprocal of 5 is $\frac{1}{5}$; the reciprocal of -7 is $\frac{1}{-7}$. The reciprocal of $\frac{2}{5}$ is $1 \div \frac{2}{5}$, which is $\frac{5}{2}$.

When we say that "the sine of an angle, θ, is the reciprocal of the cosecant of the same angle, θ," we mean that the following relation exists:

$$\sin\theta = \frac{1}{\csc\theta}$$

To make the relation still clearer, let us take a specific example. Suppose θ is an angle of 18.2°. Then the statement means

$$\sin 18.2^\circ = \frac{1}{\csc 18.2^\circ}$$

Let us see if this is true when we substitute numerical values. The cosecant of 18.2° is approximately 3.2017. Then the expression

$$\frac{1}{\csc 18.2^\circ} \quad \text{means} \quad \frac{1}{3.2017}$$

If we work out the division, $1 \div 3.2017$, we get 0.3123, rounded off to four digits. From the table we find that the sine of 18.2° is 0.3123. Therefore, we see that

$$\frac{1}{\csc 18.2^\circ} = \sin 18.2^\circ$$

Let us consider again the relation

$$\sin\theta = \frac{1}{\csc\theta}$$

We have seen that this relation is true for a particular angle, that is, if θ is 18.2°. However, this does *not* make the statement an identity. To be an identity, it must be true for *all values* of θ. Our problem now is: can this statement be shown to be true for all angles?

We might go on to show that it is true for some other values of the angle, such as 61.2°, 47.3°, or any other particular angle. However, any so-called "proof" of a trigonometric identity by the use of a specific example is not a proof at all but only an illustration. To prove an identity, we must show it to be true for *all* angles.

This is not so difficult as it might seem. An identity can be proved by using a *general* angle rather than a specific angle. If we use a general angle, this general angle represents all specific angles. Then, if the statement can be proved to be true for the general angle, it will be true for all angles.

Let us see why the relation is true for all angles. We might phrase the statement as a question and ask, "For any general angle, θ, is it true always that

$$\sin\theta = \frac{1}{\csc\theta}\text{?"}$$

Suppose we take a general angle in standard position, as shown in Fig. 40-1. We call the angle θ. We already know from the definitions that

$$\sin\theta = \frac{y}{r} \qquad \csc\theta = \frac{r}{y}$$

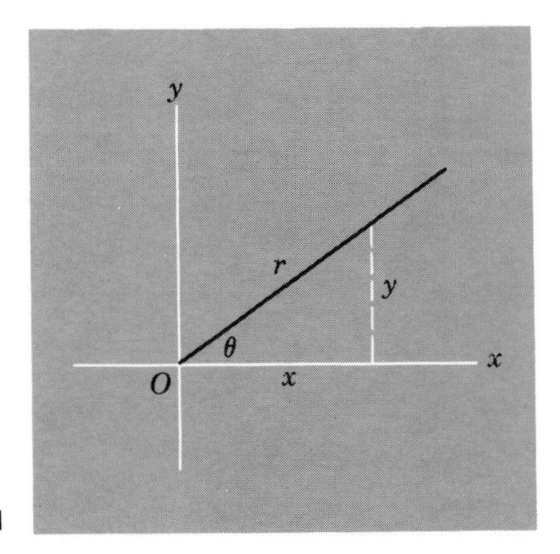

Figure 40-1

If we work out the meaning of $\frac{1}{\csc\theta}$, we have

$$\frac{1}{\csc\theta} = \frac{1}{r/y} = 1 \div \frac{r}{y} = 1 \cdot \frac{y}{r} = \frac{y}{r}$$

But we already know that the ratio y/r is the sine of θ. Therefore, we see that the relation is true no matter what the size of θ. Then we can say that the relation is an identity; that is,

$$\frac{1}{\csc\theta} \equiv \sin\theta \qquad \text{(for all values of } \theta\text{)}$$

If any particular relationship has been proved to be true for all angles, it can be checked by showing that it is true for some specific angle. For instance, since we have proved that the foregoing relation is identically true for all angles, we can show that it is true for some particular angle, such as 15°, 21.4°, or 27°19′. If the statement fails for a single angle, it is not an identity. Remember, however, that the values in the table are only approximate.

Here is a list of the six reciprocal relations. We have proved the first one. Now, using a general angle, try to show that the rest are also identities.

1. $\sin\theta \equiv \dfrac{1}{\csc\theta}$ 2. $\cos\theta \equiv \dfrac{1}{\sec\theta}$ 3. $\tan\theta \equiv \dfrac{1}{\cot\theta}$

4. $\cot\theta \equiv \dfrac{1}{\tan\theta}$ 5. $\sec\theta \equiv \dfrac{1}{\cos\theta}$ 6. $\csc\theta \equiv \dfrac{1}{\sin\theta}$

Note. It must be understood that, in the identities, we must omit cases in which a denominator is zero. For example, in the following identity, we must exclude the case where the angle θ is 90°:

$$\tan\theta = \frac{1}{\cot\theta}$$

In this case, if $\theta = 90°$, we get

$$\tan 90° = \frac{1}{0},$$

which is undefined, because we get division by zero. In the same way, in the identity,

$$\csc\theta = \frac{1}{\sin\theta}$$

we must exclude the case where $\theta = 0$, because the result would be division by zero.

Exercise 40-1

Check each of the foregoing identities by showing it to be true for some particular angle. Remember that the values in the table are only approximate. You may use a different angle to check each identity. Then prove that each is true for all angles by using a general angle.

40-3 COFUNCTION RELATIONS

Other identities are the so-called "co-function" relations. There is a definite relation between a function of any angle and the cofunction of that angle. For instance, there is a definite relation between the sine and the cosine of any angle, regardless of the size of the angle.

We see from the table that the sine of 20° is the same as the cosine of 70°. Also the sine of 31.6° is the same as the cosine of its complement, 58.4°; that is, *the sine of any angle is always equal to the cosine of the complementary angle.* (Complementary angles are two angles whose sum is 90°.)

The same is true with regard to the other functions. The tangent of 15° is the same as the cotangent of 75°. The secant of 34° is the same as the cosecant of 56°. To summarize, we can say that *any function of an angle is equal to the cofunction of the complementary angle.* It is the relation between cofunctions that enables us to arrange a table of values reading angles from top to bottom on one side and from bottom to top on the other side.

Probably the best way to understand the reason for this cofunction relation is to make use of a right triangle and the corresponding definitions. In the right triangle shown in Fig. 40-2, we have, by definition.

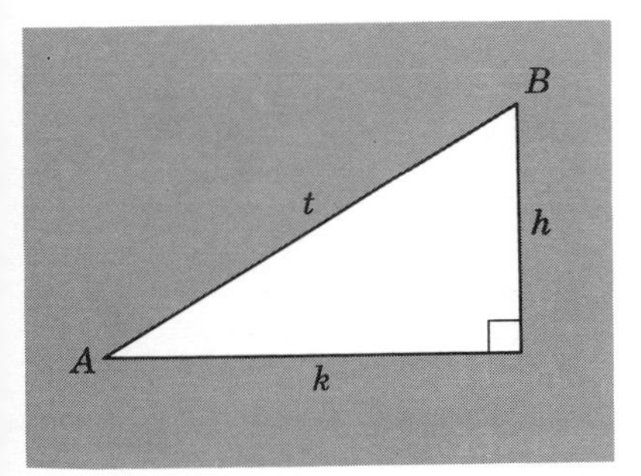

Figure 40-2

$$\sin A = \frac{\text{opposite}}{\text{hypotenuse}} = \frac{h}{t} \qquad \cos A = \frac{\text{adjacent}}{\text{hypotenuse}} = \frac{k}{t}$$

For angle B we have

$$\sin B = \frac{\text{opposite}}{\text{hypotenuse}} = \frac{k}{t} \qquad \cos B = \frac{\text{adjacent}}{\text{hypotenuse}} = \frac{h}{t}$$

Here we see that the sine of A has the same value as the cosine of B, regardless of the size of either angle. Also, $\cos A = \sin B$.

We know that angles A and B are complementary; that is, $A + B = 90°$. Therefore, $A = 90° - B$ and $B = 90° - A$.

Since	$\sin A = \cos B$
then	$\sin A = \cos(90° - A)$
In the same way	$\cos A = \sin(90° - A)$

Here are some examples showing this confunction relation:

$$\sin 42° = \cos 48° \qquad \tan 23.6° = \cot 66.4°$$

$$\cos 25.2° = \sin 64.8° \qquad \sec 31.8° = \csc 58.2°$$

The relation between the cofunctions might be better understood if we realize that the present names were not always used for the cofunction relations. For instance, the word *cosine* has not always been used for this particular ratio. For some time the cosine ratio was called the *complementary sine.* Instead of $\sin 20° = \cos 70°$, this relation was expressed as $\sin 20° =$ complementary $\sin 70°$. In time, the words "complementary sine" led to the shortened form, *cosine.*

Exercise 40-2

Supply the proper term to complete each statement:

1. $\sin 23° = \cos$____ **2.** $\cos 31° = \sin$____

3. $\tan 42° = \cot$____ **4.** $\cot 16° = \tan$____

5. $\sec 62° = \csc$____ **6.** $\csc 53° = \sec$____

7. $\cos 63.7° = \sin\_\_\_\_$

8. $\sin 72.4° = \cos\_\_\_\_$

9. $\cot 81.4° = \tan\_\_\_\_$

10. $\tan 69.6° = \cot\_\_\_\_$

11. $\cos 45° = \sin\_\_\_\_$

12. $\sin 90° = \cos\_\_\_\_$

13. $\sin 14° = \_\_\_\_ 76°$

14. $\cot 34° = \_\_\_\_ 56°$

15. $\sin \theta = \cos\_\_\_\_$

16. $\tan \phi = \cot\_\_\_\_$

17. $\sec \alpha = \csc\_\_\_\_$

18. $\cos \beta = \sin\_\_\_\_$

19. $\sin (90° - M) = \cos\_\_\_\_$

20. $\cot (90° - A) = \tan\_\_\_\_$

21. $\sin 120° = \cos\_\_\_\_$

22. $\cos 240° = \sin\_\_\_\_$

23. $\csc 10° = \dfrac{1}{\sin\_\_\_\_}$

24. $\cos 40° = \dfrac{1}{\sec\_\_\_\_}$

25. $\tan 35° = \dfrac{1}{\_\_\_\_ 35°}$

26. $\sin 15° = \dfrac{1}{\_\_\_\_ 15°}$

27. $\sin 30° = \dfrac{1}{\sec\_\_\_\_}$

28. $\csc 25° = \dfrac{1}{\cos\_\_\_\_}$

29. $\tan 20° = \dfrac{1}{\_\_\_\_ 70°}$

30. $\sec 35° = \dfrac{1}{\_\_\_\_ 55°}$

40-4 IDENTITIES INVOLVING SQUARES

There are three important identities involving squares of functions. These should be thoroughly understood and then memorized. Here, again, we should not only know these identities but we should see that they are true for any particular angle as well as for a general angle.

To indicate the square of the sine of an angle θ, we can write

$$(\sin \theta)^2$$

However, in most cases the exponent 2 is written next to the word *sine*. To indicate the square of the sine of θ, we write

$$\sin^2 \theta$$

No parentheses are needed when this form is used. The same notation is used to indicate any power of any trigonometric function. The only exception is that the reciprocal of a trigonometric function cannot be indicated by -1 in this position.

Now let us consider a particular angle, say, 28.3°. The sine of 28.3° is approximately 0.4741, and the cosine is approximately 0.8805. Suppose we wish to find the squares of these values.

The square of the sine value is $(0.4741)^2 = 0.22477081$

The square of the cosine value is $(0.8805)^2 = \underline{0.77528025}$

Adding the squares of sine and cosine, we get 1.00005106

Now we might guess that the sum of these squares would be exactly 1 if we had the exact values for the sine and cosine. This guess is correct.

In fact, the following relation can be shown to be true for any value of θ:

$$\sin^2 \theta + \cos^2 \theta \equiv 1$$

This relation can be checked for any particular angle. However, if we use the values for a particular angle shown in the table, we find that the sum is not exactly 1. That is because the values in the table are only approximate.

The actual proof that the foregoing relation is an identity is derived by taking a *general* angle, θ, in standard position, as shown in Fig. 40-3. Let us state the relation as a question: is it always true that $\sin^2 \theta + \cos^2 \theta = 1$?

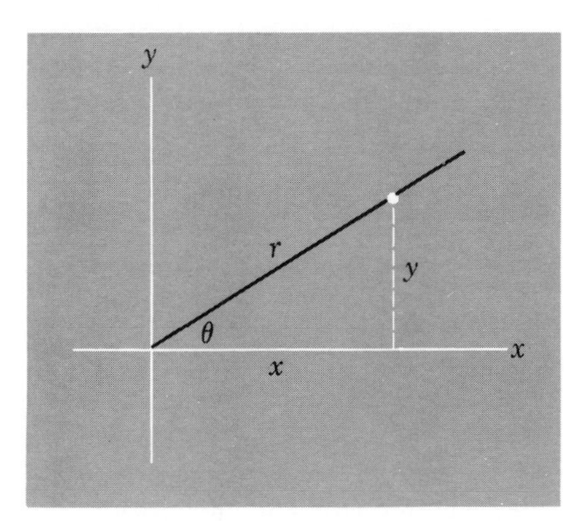

Figure 40-3

We begin with a statement that we know to be true always; that is, by the Pythagorean rule, we know that

$$y^2 + x^2 \equiv r^2$$

We may divide both sides of the identity by anything we wish except zero. Suppose we divide both sides of the equation by r^2. We get

$$\frac{y^2}{r^2} + \frac{x^2}{r^2} \equiv \frac{r^2}{r^2}$$

In the resulting equation note that the first term is actually the square of the sine, the second term is the square of the cosine, and the right side of the equation is equivalent to 1. This proves that

$$\sin^2 \theta + \cos^2 \theta \equiv 1$$

Other relations involving squares of functions are

$$\tan^2 \theta + 1 \equiv \sec^2 \theta; \qquad \cot^2 \theta + 1 \equiv \csc^2 \theta$$

These identities can be proved in the same way as the preceding identity.

The proofs can be shown by starting with the Pythagorean relation that is true for any angle:

$$y^2 + x^2 \equiv r^2$$

If we divide both sides of the equation by r^2, we get the first of the identities. If we divide both sides by x^2, we are led to the second. If we divide by y^2, we get the third identity.

The method shown is the general proof for these three identities. However, we should also show that the identities hold true for some particular angle. For instance, we might show that the identity $\tan^2 \theta + 1 = \sec^2 \theta$ is true if θ is an angle of, say, 25.2°.

It should be noted that the foregoing identities are often seen in a slightly different form. For instance, instead of

$$\sin^2 \theta + \cos^2 \theta = 1$$

we may have

$$\sin^2 \theta = 1 - \cos^2 \theta; \qquad \text{or} \qquad \cos^2 \theta = 1 - \sin^2 \theta$$

The difference between two squares may also be written in factored form.

40-5 TWO SPECIAL RELATIONS

Two other important relations should be understood and memorized. They are (for any angle, θ)

$$\frac{\sin \theta}{\cos \theta} \equiv \tan \theta \qquad \frac{\cos \theta}{\sin \theta} \equiv \cot \theta$$

The proofs for these identities can be derived by taking a general angle, θ, and using the following definitions:

$$\sin \theta \equiv \frac{y}{r} \qquad \cos \theta \equiv \frac{x}{r} \qquad \tan \theta \equiv \frac{y}{x}$$

These should also be shown to be true for some particular angle.

Exercise 40-3

1. Write out a list of the 17 general identities mentioned in this chapter.
2. Check each one of the 17 identities by using a particular angle, say, $\theta = 32.8°$, 64.7°, or some other angle.

40-6 PROVING IDENTITIES

An identity, as we have said, is an equation that is true for all values of the variable. A *trigonometric identity* is an equation that is true for all values of an angle.

To prove that an equation is an identity, we need only show that the equation can be transformed into one of the known identities we have already shown to be true. In some cases this is simple. In others the identity may be more involved and therefore more difficult to prove.

Given an equation in trigonometry, we might prove that it is an identity in one of three ways:

1. Transform the left side of the equation and try to make it exactly like the right side.
2. Transform the right side of the equation and try to make it exactly like the left side.
3. Transform both sides into new forms so that the new equation shows one of the known identities.

There is no rule that will always lead to success in proving an identity easily. One rule that may help when other attempts fail is to change all terms into sines and cosines. If an equation can be transformed in such a way that it will show one of the known identities, its identity is considered proved.

As a general rule, in proving an identity, we transform only one side of the equation and leave the other side unchanged. We usually transform the side containing the more complicated expression.

Example 1

Show that the following equation is an identity:

$$\frac{\sin x}{\tan x} = \cos x$$

Solution

The left side can be changed as follows:

$$\frac{\sin x}{\tan x} = (\sin x)\left(\frac{1}{\tan x}\right)$$

$$= (\sin x)(\cot x) = (\sin x)\left(\frac{\cos x}{\sin x}\right) = \cos x$$

Example 2

Show that this equation is an identity:

$$\frac{(\sin x)(\sec x)}{\tan^2 x} = \cot x$$

Solution

We change the left side as follows:

$$\frac{(\sin x)(\sec x)}{\tan^2 x} = (\sin x)(\sec x)\left(\frac{1}{\tan^2 x}\right)$$

$$= (\sin x)\left(\frac{1}{\cos x}\right)(\cot^2 x) = \left(\frac{\sin x}{\cos x}\right)\left(\frac{\cos^2 x}{\sin^2 x}\right) = \frac{\cos x}{\sin x} = \cot x$$

Example 3

Show that the following equation is an identity:

$$\frac{(\csc^2 x)(\cos^2 x)}{\cot^2 x} = 1$$

Solution

We can change the left side as follows:

$$(\csc^2 x)(\cos^2 x)\left(\frac{1}{\cot^2 x}\right)$$

$$= \left(\frac{1}{\sin^2 x}\right)(\cos^2 x)(\tan^2 x) = \left(\frac{\cos^2 x}{\sin^2 x}\right)\left(\frac{\sin^2 x}{\cos^2 x}\right) = 1$$

Example 4

Show that the following equation is a identity:

$$\sin x = \frac{\sec x}{\cot x + \tan x}$$

Solution

We change the right side of the equation as follows:

$$\frac{\sec x}{\cot x + \tan x} = \frac{\sec x}{\dfrac{1}{\tan x} + \tan x} = \frac{\sec x}{\dfrac{1 + \tan^2 x}{\tan x}}$$

$$= \frac{(\sec x)(\tan x)}{1 + \tan^2 x} = \frac{(\sec x)(\tan x)}{\sec^2 x} = \frac{\tan x}{\sec x} = (\tan x)\left(\frac{1}{\sec x}\right)$$

$$= \left(\frac{\sin x}{\cos x}\right)(\cos x) = \sin x, \qquad \text{(which is the left side of the equation)}$$

Example 5

Show that the following equation is an identity:

$$\tan x + \cot x = (\sec x)(\csc x)$$

Solution

We change the left side as follows:

$$\tan x + \cot x = \frac{\sin x}{\cos x} + \frac{\cos x}{\sin x} = \frac{\sin^2 x + \cos^2 x}{(\cos x)(\sin x)}$$

$$= \frac{1}{(\cos x)(\sin x)} = \left(\frac{1}{\cos x}\right)\left(\frac{1}{\sin x}\right) = (\sec x)(\csc x)$$

Note. This identity might have been shown by changing all functions to sines and cosines:

$$\frac{\sin x}{\cos x} + \frac{\cos x}{\sin x} = \left(\frac{1}{\cos x}\right)\left(\frac{1}{\sin x}\right)$$

Multiplying both sides by $(\sin x)(\cos x)$, we get

$$\sin^2 x + \cos^2 x = 1$$

However, in most instances, one side of the equation should be left unchanged.

Exercise 40-4

Prove each of the following identities or show that it is not an identity:

1. $\sin A \csc A = 1$

2. $\tan B \cos B = \dfrac{1}{\csc B}$

3. $\dfrac{1}{\sin^2 \theta} - 1 = \dfrac{\cos^2 \theta}{\sin^2 \theta}$

4. $\cos^2 M = 1 - \tan^2 M$

5. $\sin^2 x(\cot^2 x + 1) = 1$

6. $\dfrac{\sin A}{\csc A} + \dfrac{\cos A}{\sec A} = 1$

7. $\tan^2 \phi - \sin^2 \phi = \sin^2 \phi \tan^2 \phi$

8. $(\sin x)(\cot x)(\sec x) = 1$

9. $(\sin A)(1 + \cot^2 A) = \csc A$

10. $\sin t = \csc t - (\cos t)(\cot t)$

11. $\csc^2 Y \sec^2 Y = \csc^2 Y + \sec^2 Y$

12. $\csc^4 A - \cot^4 A = \csc^2 A + \cot^2 A$

13. $(\cot x)(\csc x) = \dfrac{\cos x}{1 - \cos^2 x}$

14. $\dfrac{1 + \sin \theta}{1 - \sin \theta} = \dfrac{\csc \theta + 1}{\csc \theta - 1}$

15. $\sin x \cos x (\tan x + \cot x) = 1$

16. $2 - \sin^2 \phi - \cos^2 \phi = \sec^2 \phi - \sin^2 \phi \sec^2 \phi$

Simplify the following expressions. State each one in the simplest form possible without fractions.

17. $\sec x \cot x$ **18.** $\tan B \csc^2 B$ **19.** $\sec^3 \phi \cot \phi$

20. $\dfrac{\cot^3 \theta}{\csc^5 \theta}$ **21.** $\dfrac{\tan^2 M}{\sec^3 M}$ **22.** $\dfrac{1 + \tan^2 x}{\tan^2 x}$

23. $\dfrac{\cos^2 A}{\sin^3 A}$ **24.** $\dfrac{\sin^3 B}{\cos^5 B}$ **25.** $\dfrac{\tan^3 A}{\sec^2 A}$

26. $\tan^2 \theta \csc^2 \theta$ **27.** $\cot^3 Z \sec^2 Z$ **28.** $\tan^4 T \csc^3 T$

29. $\dfrac{\sec^3 \phi}{\tan^3 \phi}$ **30.** $\dfrac{\csc^3 \phi}{\cot^2 \phi}$ **31.** $\dfrac{\tan^5 \theta}{\sec^3 \theta}$

QUIZ ON CHAPTER 40. TRIGONOMETRIC IDENTITIES. FORM A.

Using the definitions of the trigonometric ratios, in terms of x, y, and r, prove the following identities:

1. $\sin \theta = \dfrac{1}{\csc \theta}$ **2.** $\cos \theta = \dfrac{1}{\sec \theta}$ **3.** $\tan \theta = \dfrac{1}{\cot \theta}$

4. $\sec \theta = \dfrac{1}{\cos \theta}$ **5.** $\dfrac{\sin \theta}{\cos \theta} = \tan \theta$ **6.** $\dfrac{\cos \theta}{\sin \theta} = \cot \theta$

Fill the blanks with the correct angle or term:

7. $\sin 21° = \cos$____ **8.** $\cos 43° = \sin$____ **9.** $\tan 68° = \cot$____

10. $\cot 81° = \tan$____ **11.** $\sec 15° = \csc$____ **12.** $\csc 52° = \sec$____

13. $\sin 15° = \dfrac{1}{\csc\_\_\_\_}$ **14.** $\cos 25° = \dfrac{1}{\sec\_\_\_\_}$

15. $\sin 20° = \dfrac{1}{\_\_\_\_75°}$ **16.** $\cos 35° = \dfrac{1}{\_\_\_\_65°}$

17. Show why: $\sin^2 \theta + \cos^2 \theta = 1$

18. Prove: $\cot^2 \theta - \cos^2 \theta = (\cos^2 \theta)(\cot^2 \theta)$

CHAPTER FORTY-ONE Radian Measure

41-1 DEFINITION OF RADIAN

The size of an angle is usually measured in degrees. Angles may also be measured in right angles. For instance, we can say that the size of an angle is 2 right angles, 3.5 right angles, or 0.24 right angles. Since most tables give angles in degrees, we usually express angular measurement in degrees before looking up the values of the functions in a table; that is, the sine of 0.3 right angles is the same as the sine of 27°, which is 0.4540.

Angles may also be measured in *radians*. The radian is so important in work in trigonometry and in all engineering that we should understand clearly the meaning of the term.

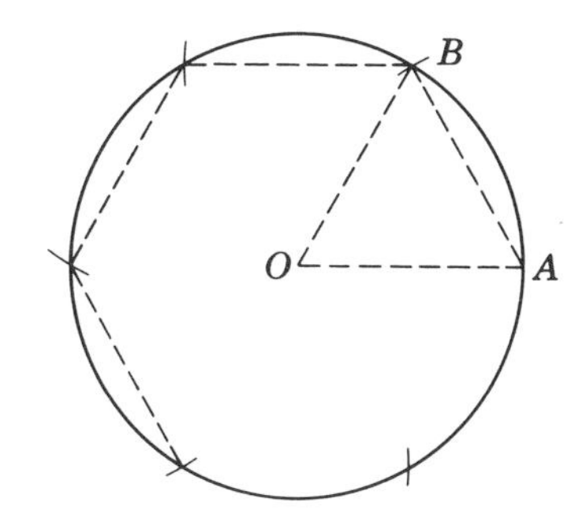

Figure 41-1

It happens that a radian is an angle of approximately 57.3°; that is, it is a little less than 60°. This will be better understood when we learn the definition of the term. When we say that one radian is an angle of approximately 57.3°, that is a fact, but it is *not* a definition.

Let us see, then, what we mean by the term *radian*. Suppose that we have a circle of any convenient size, such as the circle with center point O (Fig. 41-1). Starting at any point on the circumference, such as A, let us mark off arcs on the circumference, using the same radius as the radius of the circle; that is, the compass is set at length OA, so that $AB = OA = OB$.

If the six points are joined successively and then each point is joined to the center of the circle by straight lines, six equilateral triangles will be formed. The triangles are also equiangular.

The six angles at the center will add up to 360°, or one complete revolution. Then each angle at the center will be 60°. Therefore, there will be exactly six such angles at the center.

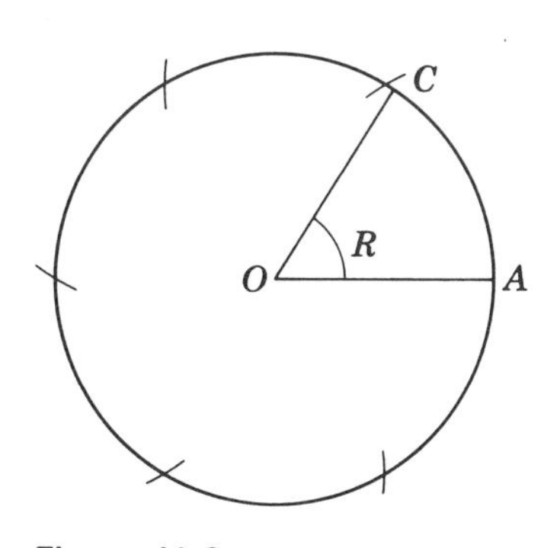

Figure 41-2

Now, if we look at the *arc* AB, we realize that it is slightly longer than the *chord* AB. Whatever the size of the circle, a 60° angle at the center will cut off on the circumference an arc slightly greater than the radius of that circle.

Suppose we wish to have an angle at the center of a circle so that the intercepted arc will be *exactly as long as the radius of the circle.* In Fig. 41-2 let us assume that the angle R at the center does cut off an arc, AC, that is equal in length to the radius r. Then *angle* R *is* defined as one *radian.*

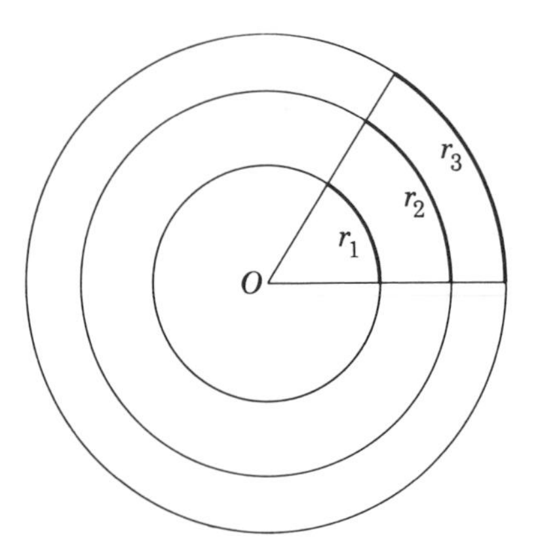

Figure 41-3

We see that angle R must be slightly less than 60°. Actually, one radian is approximately 57.3°. Remember, the radian is the *angle* which, if placed at the center of the circle, will cut off an arc equal in length to the radius.

A radian need not be at the center of any circle. It may be found anywhere, just as any other angle may be found anywhere. No matter where it is seen, a radian is approximately 57.3°.

The size of the circle does not affect the size of a radian. In Fig. 41-3 we see an angle of one radian at the center of concentric circles. In each circle the intercepted arc is exactly equal in length to the radius of that circle.

41-2 MEASUREMENT OF ANGLES IN RADIANS

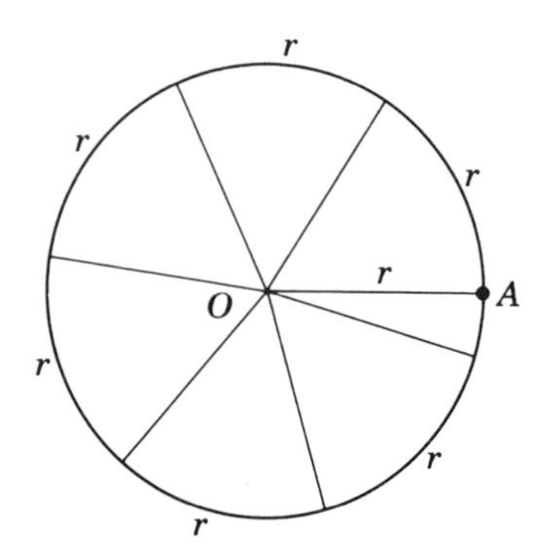

Figure 41-4

In 180° there are a little over three radians. Actually, 180° is equal to approximately 3.14159 radians, or *exactly* π radians. One revolution, or 360°, is equal to 2π radians. Let us see why this is so.

Suppose we have a circle with center at O, as shown in Fig. 41-4. Let us lay off the radius successively on the circumference starting at any point, A. For each radius length that we lay off on the circumference, we have one radian at the center. We know that the circumference is always 2π times the radius.

If	$C = 2\pi \cdot \text{radius},$
then	$360° = 2\pi$ radians
Therefore,	$180° = \pi$ radians

In radian measure we often omit the word *radian*; that is, instead of saying $180° = \pi$ radians, we say $180° = \pi$.

Exercise 41-1

Fill in the blanks in the following. Work out each answer by *inspection*. For instance, if $\pi = 180°$, then $2\pi = 360°$. State radians in terms of π.

1. $3\pi =$ ___° **2.** $4\pi =$ ___° **3.** $\frac{\pi}{2} =$ ___°

4. $\frac{\pi}{3} =$ ___° **5.** $\frac{\pi}{6} =$ ___° **6.** $8\pi =$ ___°

7. $10\pi =$ ___° **8.** $\frac{2}{3}\pi =$ ___° **9.** $\frac{3}{2}\pi =$ ___°

10. $180° =$ ___ (radians) **11.** $135° =$ ___ (radians)

12. $150° = \_\_\_$ (radians) **13.** $18° = \_\_\_$ (radians)

14. $15° = \_\_\_$ (radians) **15.** $720° = \_\_\_$ (radians)

41-3 CHANGING FROM DEGREES TO RADIANS AND FROM RADIANS TO DEGREES

In the preceding exercises we can change radians to degrees and degrees to radians rather easily. However, the conversion of one kind of measure to the other is often more difficult. Let us formulate two rules for making the changes.

We start with $180° = \pi$ radians

Dividing both sides of this equation by 180, we get

$$1° = \frac{\pi}{180} \text{ radians}$$

This means that any angle stated in degree measure can be changed to radian measure by multiplying by $\pi/180$.

Rule 1. *To change degrees to radians, multiply the number of degrees by* $\frac{\pi}{180}$. The result is the number of radians.

Example 1

Change 86.4° to radians.

Solution

$$86.4 \cdot \frac{\pi}{180} = 1.508 \text{ radians.}$$

The foregoing rule is sometimes simplified by using the number 57.3, the approximate number of degrees in one radian. We may divide the number of degrees by 57.3 to obtain the number of radians.

$$86.4 \div 57.3 = 1.508, \text{ number of radians}$$

To change radians to degrees, we begin, again, with the equation

$$\pi \text{ radians} = 180°$$

Dividing both sides of the equation by π, we get the equation

$$1 \text{ radian} = \frac{180°}{\pi}$$

Rule 2. *To change radians to degrees, multiply the number of radians by* $\frac{180}{\pi}$. The result is the number of degrees.

Example 2

Change 2.4 radians to degrees.

Solution

$$(2.4)\frac{180^\circ}{\pi} = 137.5^\circ$$

Here, again, the work may be simplified by using the number 57.3. We have then

$$2.4 \text{ radians} = 2.4 \times 57.3^\circ = 137.5^\circ$$

Note. In the examples shown, the answers have been rounded off to four significant digits.

There is one point that must be clearly understood with reference to any number used in connection with a trigonometric function. When we say, for instance, $\sin\theta$, we think of θ as an angle. The angle is usually stated in degrees. For example, suppose we have the equation

$$\sin 45^\circ = 0.7071$$

Note that the number 45 after the trigonometric function states the angle in *degrees.* However, when any number is used in any way to indicate an angle, then, *if the angle is not definitely stated as degrees,* it must be taken to mean *radians.* For instance, the expression, sin 1, means the sine of one radian; that is, sin 1 means

$$\sin 1 \text{ radian} = \sin 57.3^\circ = 0.8415$$

On the other hand, $\sin 1^\circ = 0.01745$.

We must be especially careful with regard to the number π. The number π taken by itself is the irrational number approximately equal to 3.14159265358979323846264338328O, or, less accurately, to 3.1416. But when π, or any other number, is used in such a way that an angle is indicated, then the number must be taken to mean *radians* unless it is definitely stated as degrees.

As an example, the expression $\sin\pi$ means $\sin\pi$ radians, which is the same as $\sin 180^\circ$, which is zero; that is,

$$\sin\pi = \sin 180^\circ = 0$$

However,

$$\sin\pi^\circ = \sin 3.14^\circ = 0.0548 \text{ (approx.)}$$

As another example,

$$\sin\frac{\pi}{3} = \sin\frac{3.1416}{3} = \sin 1.0472 \text{ radians} = \sin 60^\circ = 0.866$$

The expression $\sin 2\pi$ means the sine of 360°, which is zero. However, $\sin 2\pi°$ is simply the sine of 6.28°, which is approximately equal to 0.10939.*

Exercise 41-2

Change the following degree measure to radian measure:

1. 45°	**2.** 210°	**3.** 225°	**4.** 540°
5. 330°	**6.** 1080°	**7.** −90°	**8.** 25°
9. 38.2°	**10.** 152°	**11.** 21600°	**12.** 13.5°
13. 270°	**14.** −120°	**15.** 112.5°	**16.** 4.5°

Change the following *radian measure* to degree measure:

17. 5π	**18.** 2	**19.** 3.4	**20.** 1.32
21. 10	**22.** 120	**23.** 0.015	**24.** 6.2832

Find the value of each of the following:

25. $\sin \frac{\pi}{2}$	**26.** $\cos \frac{\pi}{12}$	**27.** $\sin \frac{\pi}{3}$
28. $\tan \frac{\pi}{6}$	**29.** $\cot \frac{\pi}{4}$	**30.** $\cos \frac{2\pi}{3}$
31. $\tan \frac{3\pi}{4}$	**32.** $\sec \frac{\pi}{5}$	**33.** $\sin \pi°$
34. $\sin 1.3$	**35.** $\cos 1.2$	**36.** $\tan 0.4$
37. $2 + \sin 3$	**38.** $\pi + \cos \pi$	**39.** $3 + \tan 2$
40. $\frac{\pi}{4} - \tan \frac{\pi}{4}$	**41.** $\frac{\pi}{2} + \sin \frac{\pi}{2}$	**42.** $\left(\sin \frac{\pi}{4}\right)^2$

†41-4 THE SOLID ANGLE

Radian measure is extended to the measurement of *solid angles.* The solid angle is an important concept in science. The concept is useful in connection with such problems as studio lighting, the magnetic force between masses, the storage of fissionable materials, and in other ways. To see what is meant by a solid angle, let us first repeat the definition of a plane angle.

* In some work in mathematics we consider the trigonometric functions of numbers rather than of angles. However, at present we are concerned only with angles.

† This section may be omitted without any discontinuity in the study of trigonometry.

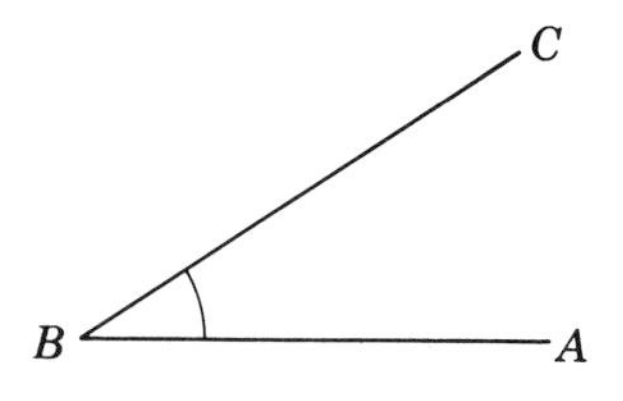

Figure 41-5

In trigonometry we find it convenient to define a *plane angle* as the amount of rotation of a ray about its end point. However, for our present purpose, let us go back to an earlier definition and say that a *plane angle* is the *amount of opening between two lines* drawn from a point (41-5).

Now let us extend the term *angle* to three dimensions. A *dihedral* angle is a three-dimensional figure formed by two intersecting planes. When we try to represent three-dimensional figures on a plane, we run into trouble. The paper is flat, yet we must try somehow to make a drawing that will represent three dimensions. The result is that we draw figures that are optical illusions. They only *look like* the figures they represent.

Suppose we fold a piece of cardboard and open up the fold as we open the pages of a book. We try to show the result in Fig. 41-6. The figure formed is a *dihedral* angle. (The word *hedral* refers to *side* or *face*.) The two planes forming the angle are called *faces*. The two faces meet in a *line* called the *vertex* of the angle.

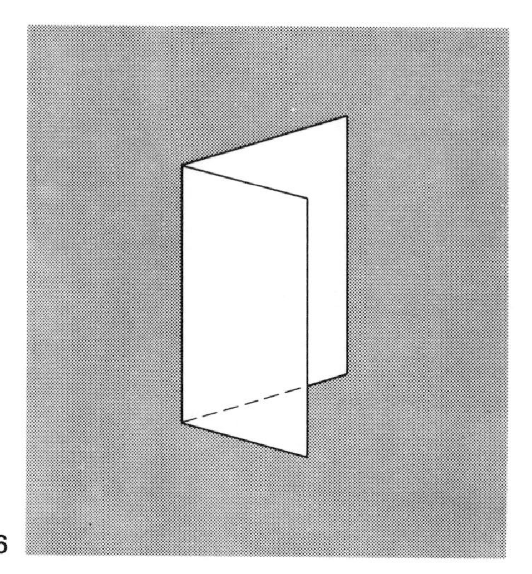

Figure 41-6

We often measure a plane angle in degrees. A dihedral angle can also be measured in degrees. To measure the angle we measure the plane angle between two straight lines in the faces of the dihedral angle from the same point in the vertex (edge) and perpendicular to the vertex. For example, if the two faces are perpendicular to each other, we say the dihedral angle is 90°.

Now, let us go one step further. Suppose we have three or more planes meeting at a point, such as the apex of a pyramid (Fig. 41-7). The figure formed by these planes outward from their point of intersection will be three-dimensional and, in general, will be pyramidal in

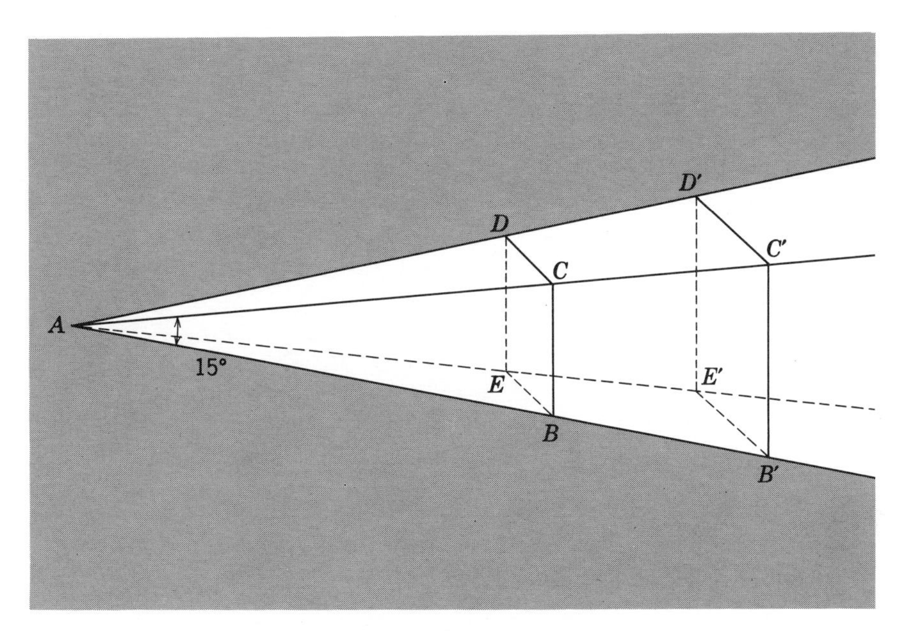

Figure 41-7

shape. Then the *amount of opening* of the planes at the apex is called a *solid angle.*

We have said that a plane angle can be called the amount of opening between two lines drawn from a point. In a similar way, a solid angle is the *amount of opening between planes* or any surrounding surfaces coming to a point. Moreover, there may be any number of planes or other surfaces forming the solid angle. (A solid angle may be denoted by the symbol $\underline{/\theta/}$.)

In Fig. 41-7, we have four planes meeting at point A, the apex of the pyramid. Our question now is: How shall we measure the solid angle? It is not the sum of the plane angles at A, nor is it the sum of the dihedral angles.

Let us take points B, C, D, and E on the edges of the pyramid, respectively, so that these points are equidistant from A. Then the area of the polygon $BCDE$ is a measure of the size of the solid angle. To be strictly correct, we should assume that the surface $BCDE$ is curved so that every point on the surface is equidistant from A.

If we take four other points, B', C', D', E', all equidistant from A, the area of this surface, $B'C'D'E'$, is also a measure of the same solid angle. This surface has a greater area, but the solid angle is still the same size.

The unit for measuring solid angles is called a *steradian.* If we think of a solid angle as being located at the center of a sphere, then we consider the amount of area that the central solid angle intercepts on the surface of the sphere. Then we have the following definition of a *steradian.*

Definition. *One steradian is a solid angle of such size that when placed at the center of a sphere, it intercepts on the surface of the sphere an area exactly equal to the square of the radius* (Fig. 41-8).

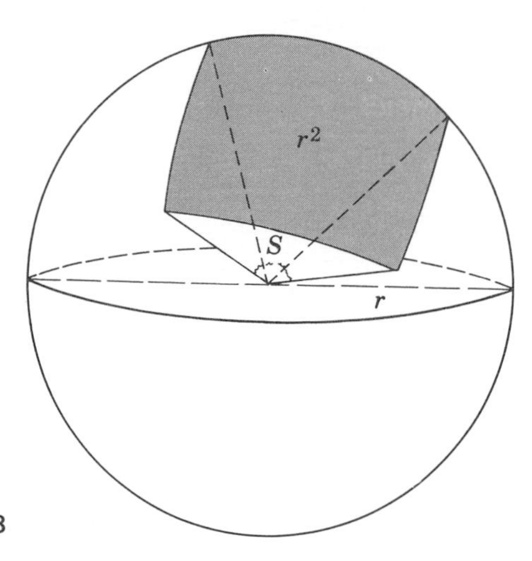

Figure 41-8

If the radius of a sphere is 10 inches, then one steradian at the center will intercept an area of 100 square inches on the surface of the sphere. This area, in fact, may be any shape—square, triangular, circular, or irregular. In general, one steradian at the center of a sphere of radius r intercepts an area equal to r^2 on the surface.

The surface area of any sphere is given by the formula: $A = 4\pi r^2$. Since one steradian at the center intercepts an area of r^2 on the surface, the total surface of the sphere must be associated with 4π steradians. Therefore, the entire solid angle from any point in space, or the total amount of opening in all directions from a point, is 4π steradians.

For measuring *plane angles,* the radian is a rather large unit. Instead, we often use a smaller unit, the *degree.* In the same way, the steradian is a rather large unit for measuring *solid angles.* It is possible to use a smaller unit. Let us see how this is done.

We have seen that an angle of one degree is called an *angle degree* (Section 27.1). Also, one angle degree at the center of a circle intercepts an arc called one *arc degree* on the circumference.

Consider now a regular square pyramid with its apex at the center of a sphere, and of such size that each plane angle at the apex is 1° (Fig. 41-9). Each plane angle of 1° will intercept one arc degree on a great

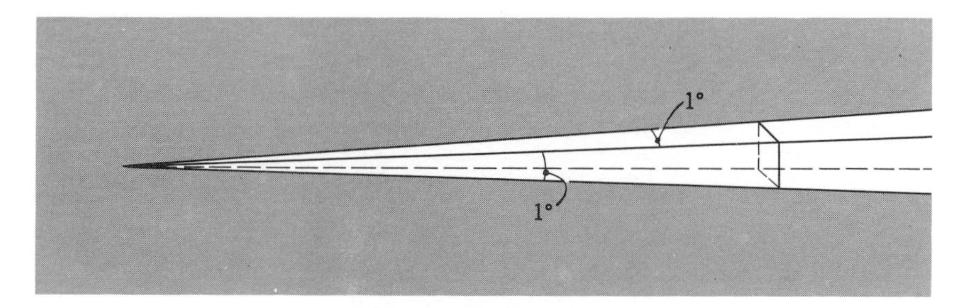

Figure 41-9

circle of the sphere. We call the solid angle or amount of opening at the apex of this pyramid one *square degree.* (This is analogous to our definition of a *square inch* as a unit of area.) Then we have the following definition of a square degree.

Definition. *One square degree is a solid angle equivalent to the opening at the apex of a regular square pyramid each of whose plane angles at the apex is one angle degree.*

We already have the following formulas:

$$\pi \text{ radians} = 180° \qquad 1 \text{ radian} = \frac{180°}{\pi}$$

Therefore, the radius is equal in length to $180/\pi$ arc degrees.
Now we derive the following formulas. Since

$$r = \frac{180}{\pi} \text{ arc degrees in length,} \qquad r^2 = \left(\frac{180}{\pi}\right)^2 \text{ square arc degrees}$$

For the area of a sphere,

$$A = 4\pi r^2 = 4\pi\left(\frac{180}{\pi}\right)^2 \text{ square arc degrees}$$

Then

$$4\pi \text{ steradians} = \frac{129600}{\pi} \text{ square degrees}$$

Therefore, the entire solid angle about any point in space is $\dfrac{129600}{\pi}\,\text{deg}^2$.

The size of a solid angle is an important consideration in the size of a TV or movie screen. There is an optimum size of the solid angle for viewing a TV or movie if viewing is to be most comfortable. Sometimes it is a simple matter to compute the size of the solid angle. Suppose you sit directly before a home movie screen that is rectangular in shape. Let us assume that the plane angle between your lines of sight to the left side and the right side is 15° (Fig. 41-10). That is, the

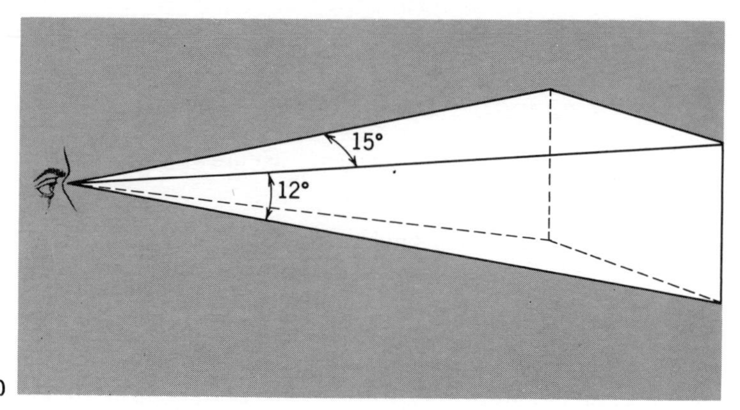

Figure 41-10

horizontal spread of your eyesight to the screen is 15°. Suppose also that the vertical spread of your eyesight to the top and bottom of the screen is 12°. Then, the screen is rectangular, the solid angle, $\underline{/A/}$, is given by

$$\underline{/A/} = (15°)(12°) = 180 \text{ deg}^2$$

If one is seated about 7 ft from a TV screen that is 15 in. high and 19 in. wide, the solid angle in the viewing is approximately 132 sq degrees. If one moves up to a distance of 6 ft from the TV, the solid angle is approximately 179 sq degrees. Let us see how the size of a solid angle can be computed.

Example 1

Suppose you sit directly in front of a home movie screen that is 4 ft wide and 3 ft high. You sit 15 ft from the screen with the center of the screen on a level with your eyes. What is the size of the solid angle from your eyes to the perimeter of the screen?

Solution

We take the top of the screen at 1.5 ft above eye level and the bottom of the screen at 1.5 ft below eye level (Fig. 41-11). Now let us say 2α is the plane angle between the lines of sight to the top and bottom of the screen. Then we have an angle of α above and α below eye level. Then

$$\tan \alpha = 1.5/15 = 0.1000 \qquad \text{or} \qquad \alpha = 5.7° \text{ (approx.)}; \qquad \text{then } 2\alpha = 11.4°$$

Note that we cannot take 3/15 as the tangent of 2α.

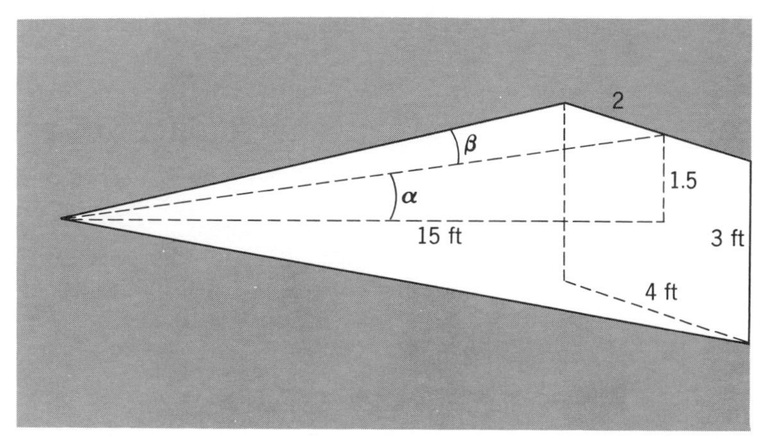

Figure 41-11

In a similar manner we can find the horizontal angle between the left and right edges of the screen. Calling this angle 2β, we have

$$\tan\beta = 2/15 = 0.13333 \qquad \text{or} \qquad \beta = 7.6^\circ \text{ (approx.)}; \qquad \text{then } 2\beta = 15.2^\circ$$

Now we have a vertical spread of 11.4° and a horizontal spread of 15.2°. Then the solid angle, $\underline{/A/}$, is found by

$$\underline{/A/} = (11.4^\circ)(15.2^\circ) = 173.3 \text{ deg}^2 \text{ (approx.)}$$

Approximate solution. We can get an approximation to the size of the solid angle without using trigonometry if we consider every point on the screen to be 15 ft from our eyes (which of course is not true). Then the number of square degrees can be computed as a 12-sq-ft portion of the entire surface of a sphere having a radius of 15 ft. That is, we take

$$\frac{12}{4\pi 225}\left(\frac{129600}{\pi}\right) = \frac{12}{900}\left(\frac{129600}{\pi^2}\right) = 175.1 \text{ (approx.)}$$

The formula is approximately: $\underline{/A/} = \left(\frac{\text{area}}{d^2}\right)(3280)$, where d represents the distance from the viewer to the object.

Example 2

A right circular cone has an altitude of 20 in. and the diameter of its base is 8 in. Find the solid angle at the apex of the cone.

Solution

Let α be the plane angle between the altitude and an element of the cone. Then

$$\tan\alpha = 4/20 = 0.20000; \qquad \alpha = 11.31^\circ; \qquad 2\alpha = 22.62^\circ$$

Now if the base were a square with 8 in. on a side the solid angle would be

$$(22.62°)^2 \qquad \text{or} \qquad 511.7\ \text{deg}^2$$

However, we know the area of a circle is approximately 0.7854 times the area of a circumscribed square. Then, for the solid angle of the cone, we have (0.7854) (512.6) = 401.9 deg^2, the solid angle at the apex of the cone. Using the approximation formula, we get 412.2 deg^2 (approx.).

Note that the result from using geometry instead of trigonometry turns out to be slightly greater in each case than the true size of the solid angle. This is because in using geometry, we neglect the fact that the surface we are considering is a plane rather than the curved surface of a sphere. The error becomes greater as the viewed area increases in size with reference to the distance from the viewer. If we take the maximum distance across any viewed area as compared with the distance from the viewer, then, if this ratio is less than 1/4, there is little error through use of the approximate formula.

Exercise 41-3

Use trigonometry to solve the following problems.

In the first six of the following exercises, find the size of the solid angle from a point directly in front of the center of the object:

1. A picture, 8 ft wide and 6 ft high, from a distance of 20 ft.
2. A rectangular picture, 6 by 4 ft, from a distance of 20 ft.
3. A TV screen, 24 in. wide and 20 in. high, from 12 ft away.
4. A movie screen, 24 ft wide and 15 ft high, from 60 ft away.
5. A circular picture, 36 in. in diameter, from 12 ft away.
6. A speck of uranium, 4 ft from a ball of material 8 in. in diameter.
7. Find the size of the solid angle in No. 1, if the line of sight is perpendicular to the picture at one corner of the picture.
8. Work No. 4 if you sit with your eyes on a level with the bottom of the screen and 4 ft to the right of center.
9. How far from a movie screen, 15 ft high and 24 ft wide, should you sit for a solid angle of 240 square degrees if you sit directly before the center?
10. What should be the size of a TV screen, with a ratio of width to height of 6 to 5, if you wish to sit 6 ft from the TV and the solid angle is to be 270 square degrees?
11. How far should the speck of uranium in No. 6 be placed from the other ball of material if the solid angle is not to exceed 36 square degrees?

QUIZ ON CHAPTER 41, RADIAN MEASURE. FORM A.

The following numbers indicate the number of radians in each angle. Change to degree measure.

1. $\frac{\pi}{4}$ **2.** 5π **3.** $\frac{3}{4}\pi$ **4.** 9π

5. $\frac{5\pi}{6}$ **6.** $\frac{\pi}{8}$ **7.** $\frac{\pi}{10}$ **8.** 20π

9. 2.5 **10.** 1.8 **11.** 5.6 **12.** 120π

Change the following angles in degree measure to radian measure:

13. 30° **14.** 225° **15.** 25° **16.** 300°

17. 10° **18.** 15° **19.** 240° **20.** 450°

21. 900° **22.** 550° **23.** 31.4° **24.** 21,600°

Find the value of each of the following:

25. $\sin\frac{3\pi}{2}$ **26.** $\cos\frac{\pi}{3}$ **27.** $\tan\frac{5\pi}{4}$ **28.** $\cot 1.2$

29. $3 + \sin 1$ **30.** $2\pi + \cos \pi$ **31.** $\frac{\pi}{6} + \sin\frac{\pi}{6}$

32. A picture 6 ft by 4 ft hangs on a wall. You sit 10 ft from the picture with your eyes on the level of the center of the picture. What is the size of the solid angle from your eyes to the perimeter of the picture?

CHAPTER FORTY-TWO Special Angles

42-1 THE 30°-60° RIGHT TRIANGLE

If we wish to know any function of an angle, we usually look up the value in a table. Yet it is often desirable to find the value of a trigonometric function without the use of a table. This is especially true with regard to certain angles.

In the case of most angles the computation of the numerical values of the functions is a complicated process. However, for a few special angles we can compute the values rather easily. For instance, one such angle is a 30° angle. Let us see, then, how we might find the sine, cosine, and other functions of a 30° angle without using a table.

Suppose we have a 30° angle in standard position (Fig. 42-1). We take a point, P, on the terminal side of the angle and show the ordinate. The triangle formed by the ordinate, the abscissa, and the radius vector has one acute angle of 30°. Therefore, the other acute angle of the triangle is 60°. Here we have what is called a "30°-60° right triangle."

In geometry it is proved that *in any* 30°-60° *right triangle the hypotenuse is always exactly twice the length of the shortest side.* Therefore, in Fig. 42-1 side AP is one-half the length of side OP.

If we assume that AP is 4 inches, then OP is 8 inches. Whatever length we assume for AP, the hypotenuse OP will always be twice as long. Therefore, the sine of 30° is $\frac{1}{2}$, or 0.5000. We may just as well

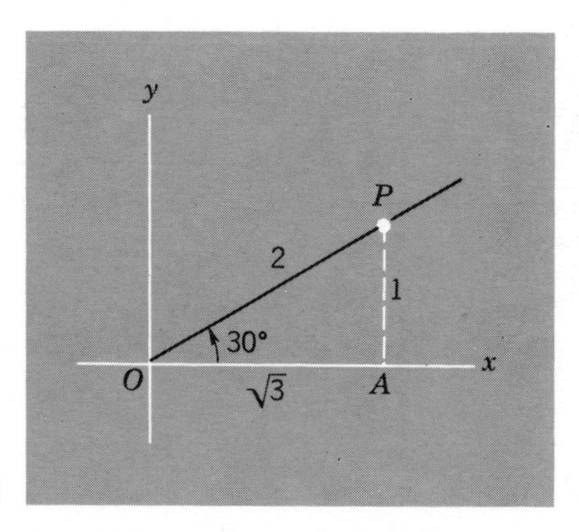

Figure 42-1

assume that the length of AP is 1 unit and that the length of OP is equal to 2 units. The sine of 30° is still $\frac{1}{2}$, or 0.5000.

To find the values of the other functions of a 30° angle, we place the numbers 1 and 2 on the triangle in the proper places to represent those lengths. The third side of the triangle is then computed by the Pythagorean rule. It is $\sqrt{3}$. These numbers, 1, 2, and $\sqrt{3}$, are shown in proper positions on the triangle. We must be sure to use the number 1 for the shortest side of the triangle and the number 2 for the hypotenuse.

We can now express the six functions of an angle of 30° or $\pi/6$ radians. From the definitions of the functions, we have

$$\sin\frac{\pi}{6} = \sin 30° = \frac{y}{r} = \frac{1}{2} \qquad \cos\frac{\pi}{6} = \cos 30° = \frac{x}{r} = \frac{\sqrt{3}}{2}$$

$$\tan\frac{\pi}{6} = \tan 30° = \frac{y}{x} = \frac{1}{\sqrt{3}} \qquad \cot\frac{\pi}{6} = \cot 30° = \frac{x}{y} = \sqrt{3}$$

$$\sec\frac{\pi}{6} = \sec 30° = \frac{r}{x} = \frac{2}{\sqrt{3}} \qquad \csc\frac{\pi}{6} = \csc 30° = \frac{r}{y} = 2$$

These values can be changed to decimal form if desired.

If we have a 60° angle in standard position (Fig. 42-2), the same 30°-60° triangle will appear in the sketch. We can use the same numbers, 1, 2, and $\sqrt{3}$, to find the values of the functions of a 60° angle. For instance, by definition, the sine of 60° = $y/r = (\sqrt{3})/2$.

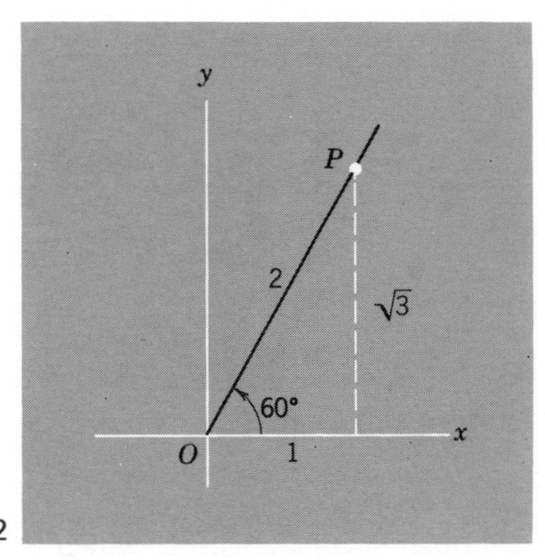

Figure 42-2

The same 30°-60° right triangle will appear for several other angles in standard position. As an example, suppose we wish to find the functions of an angle of 240°, which is equal to $4\pi/3$ radians.

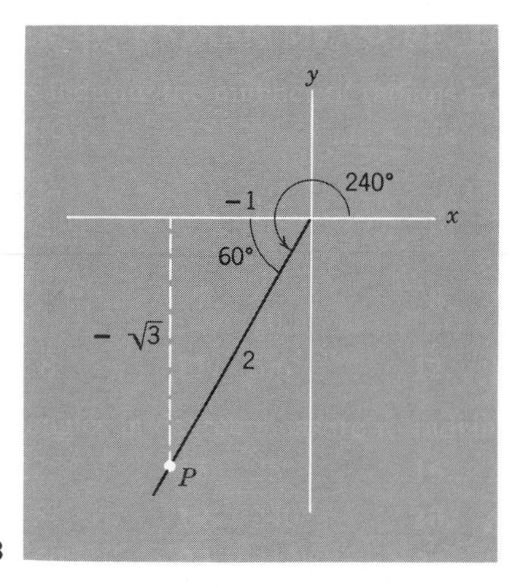

Figure 42-3

We sketch the 240° angle in standard position (Fig. 42-3). When we draw a vertical line to indicate the ordinate, we see a 30°-60° right triangle in the sketch. In this case the reference angle is 60°. We place the numbers, 1, 2, and $\sqrt{3}$ in the proper positions on the right triangle. Moreover, we must indicate whether these values are positive or negative for the 240° angle. The signs of the numbers will depend on the signs of x and y for the point on the terminal side. The hypotenuse is always positive.

For the 240° angle, $x = -1$, $y = -\sqrt{3}$, and $r = 2$. Then we have the following values for the angle of 240°, or $(4\pi/3)$ radians:

$$\sin\frac{4\pi}{3} = \sin 240° = \frac{y}{r} = \frac{-\sqrt{3}}{2} \qquad \cos\frac{4\pi}{3} = \cos 240° = \frac{x}{r} = \frac{-1}{2}$$

$$\tan\frac{4\pi}{3} = \tan 240° = \frac{y}{x} = \frac{-\sqrt{3}}{-1} \qquad \cot\frac{4\pi}{3} = \cot 240° = \frac{x}{y} = \frac{-1}{-\sqrt{3}}$$

$$\sec\frac{4\pi}{3} = \sec 240° = \frac{r}{x} = \frac{2}{-1} \qquad \csc\frac{4\pi}{3} = \csc 240° = \frac{r}{y} = \frac{2}{-\sqrt{3}}$$

These values can be changed to decimal form and each one given its proper sign.

42-2 THE 45° RIGHT TRIANGLE

Another special angle for which we can easily compute the functions is a 45° angle. Suppose we sketch a 45° angle in standard position (Fig. 42-4). Take point P on the terminal side. In the triangle AOP one acute angle is 45°. Therefore, the other acute angle is also 45°. Therefore, $OA = AP$. The triangle formed is called a "45° right

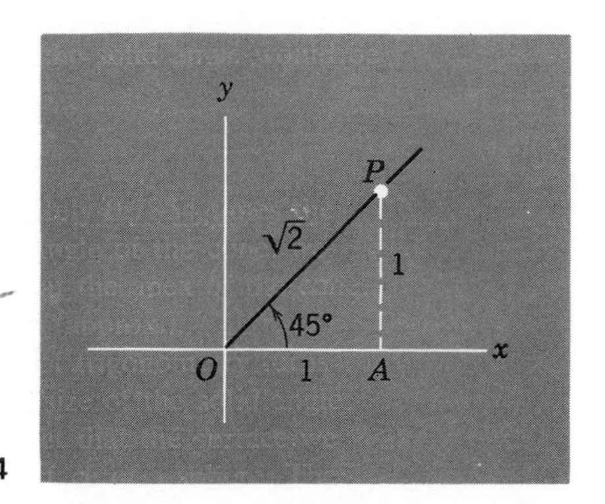

Figure 42-4

triangle." If we assume that AP is 4 inches, then OA is also 4 inches and $OP = 4\sqrt{2}$ inches.

We may assume that AP is equal to 1 unit. Then $OA = 1$, and $OP = \sqrt{2}$. The numerical values of the ratios will not change, regardless of the length we assume for OA. Using the numbers 1, 1, and $\sqrt{2}$ for the two sides and the hypotenuse, respectively, we can compute all the values of the functions of the 45° angle, which is equal to $(\pi/4)$ radians:

$$\sin\frac{\pi}{4} = \sin 45° = \frac{y}{r} = \frac{1}{\sqrt{2}} \qquad \cos\frac{\pi}{4} = \cos 45° = \frac{x}{r} = \frac{1}{\sqrt{2}}$$

$$\tan\frac{\pi}{4} = \tan 45° = \frac{y}{x} = 1 \qquad \cot\frac{\pi}{4} = \cot 45° = \frac{x}{y} = 1$$

$$\sec\frac{\pi}{4} = \sec 45° = \frac{r}{x} = \sqrt{2} \qquad \csc\frac{\pi}{4} = \csc 45° = \frac{r}{y} = \sqrt{2}$$

We can use the numbers 1, 1, and $\sqrt{2}$ to find the values of the functions of any angle whenever a 45° angle appears in the sketch of the given angle in standard position. This will be true for an angle of 135°, an angle of 225°, an angle of 315°, or for any angle, positive or negative, whose terminal side falls in one of the corresponding positions.

We must be sure to place the numbers 1, 1, and $\sqrt{2}$ in their proper positions on the right triangle in the sketch. Moreover, we must indicate whether the numbers are positive or negative for the given angle. This will depend upon the signs of x and y for the point on the terminal side.

As an example, let us take an angle of $(-45°)$ in standard position (Fig. 42-5). (Parentheses are used to enclose a negative angle.) We take any point P, on the terminal side and show the ordinate length. The triangle formed is a 45° right triangle. We place the numbers 1, 1,

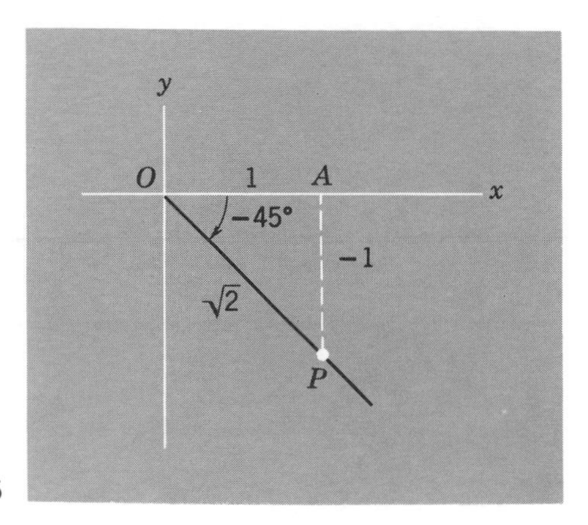

Figure 42-5

and $\sqrt{2}$ on the sides of the triangle in proper positions and with the correct signs. Since point P lies in the fourth quadrant, the abscissa is positive and the ordinate is negative. The radius vector is always positive.

From the values shown, we can write all the function values for $(-45°)$, or $(-\pi/4)$ radians. The values can be reduced to decimal form if desired.

$$\sin\left(-\frac{\pi}{4}\right) = \frac{y}{r} = \frac{-1}{\sqrt{2}} \qquad \tan\left(-\frac{\pi}{4}\right) = \frac{y}{x} = \frac{-1}{+1} \qquad \sec\left(-\frac{\pi}{4}\right) = \frac{r}{x} = \frac{\sqrt{2}}{+1}$$

$$\cos\left(-\frac{\pi}{4}\right) = \frac{x}{r} = \frac{+1}{\sqrt{2}} \qquad \cot\left(-\frac{\pi}{4}\right) = \frac{x}{y} = \frac{+1}{-1} \qquad \csc\left(-\frac{\pi}{4}\right) = \frac{r}{y} = \frac{\sqrt{2}}{-1}$$

42-3 QUADRANTAL ANGLES

We now consider a few angles that require special attention. They are the *quadrantal* angles whose terminal sides fall on one of the coordinate axes, such as 0°, 90°, 180°, 270°, 360°. The functions of such angles may be found in the following manner.

Sketch the angle in standard position, showing the terminal side along the axis. Indicate the angle by a curved arrow. Assume some point on the terminal side. Then express the abscissa, the ordinate, and the radius vector of the point. The function values can then be written from the definitions.

Example

Find the six trigonometric functions of 180°, or π radians.

First, we sketch the angle in standard position (Fig. 42-6). Now we assume some point, say, $(-6, 0)$, on the terminal side. Then, for this angle of

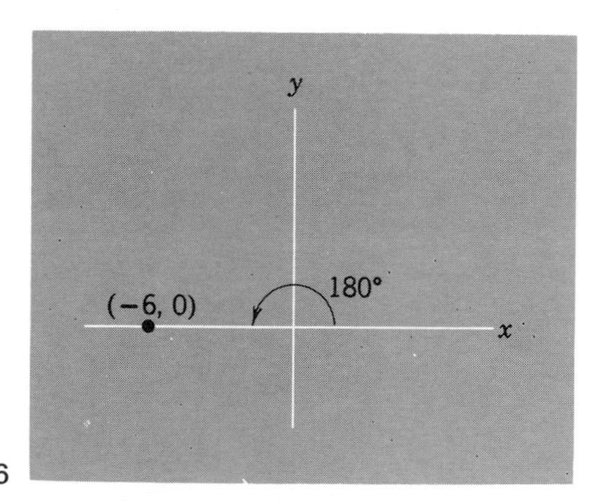

Figure 42-6

180°, we have

$$x = -6 \qquad y = 0 \qquad r = 6$$

From these values the numerical values of all the functions of 180° can be written. Remember to show the proper sign for each function. If a zero appears in the denominator of a fraction, the function is not defined for the angle. The functions of 180°, or π radians are as follows:

$$\sin \pi = \frac{y}{r} = \frac{0}{6} = 0 \qquad \cot \pi = \frac{x}{y} = \frac{-6}{0} \quad \text{(not defined)}$$

$$\cos \pi = \frac{x}{r} = \frac{-6}{6} = -1 \qquad \sec \pi = \frac{r}{x} = \frac{6}{-6} = -1$$

$$\tan \pi = \frac{y}{x} = \frac{0}{-6} = 0 \qquad \csc \pi = \frac{r}{y} = \frac{6}{0} = \text{(not defined)}$$

Exercise 42-1

Without the use of a table, find the numerical value of each of the six trigonometric functions of each of the following special angles. Leave values in radical and fractional form. Also sketch each angle in standard position and state each angle in radian measure.

1. 120° **2.** 135° **3.** 150° **4.** 210°

5. 225° **6.** 300° **7.** 315° **8.** 0°

9. 90° **10.** 270° **11.** 360° **12.** −90°

13. −135° **14.** −60° **15.** 495° **16.** 990°

Find each of the following by use of special angles:

17. $\sin \frac{\pi}{6} + \cos \frac{\pi}{3}$ **18.** $\cos \frac{2\pi}{3} + 2 \cos \pi$ **19.** $\sin \frac{5\pi}{6} + \cos \frac{4\pi}{3}$

20. $\cos^2 \frac{\pi}{4} - \cos \pi$ **21.** $\tan \frac{2\pi}{3} \sec \frac{2\pi}{3}$ **22.** $\tan \frac{5\pi}{6} + \sec \frac{5\pi}{6}$

23. $\sec^2 \frac{5\pi}{4} - \tan^2 \frac{5\pi}{4}$ **24.** $\sin^2 \frac{3\pi}{2} + \cos^2 \frac{3\pi}{2}$ **25.** $\tan \frac{3\pi}{2} \cot \frac{3\pi}{2}$

26. $\sec^2 (-45°) - \tan^2 (-45°)$ **27.** $\sin^2 300° \cos^4 300°$

28. $\sec^3 150° \csc^3 150°$ **29.** $\csc^2 225° \cot^3 150°$

30. $(\sin 150°)(\cos 240°) - (\cos 150°)(\sin 240°)$

31. $(\cot 30°)(\csc 60°) + (\tan 60°)(\sec 45°)$

32. If $\phi = \arctan(-1)$, find two positive values of ϕ less than 360°.

33. Find $\log_e \tan^2 120°$.

34. Find $\log_e \sec 45° + \log_e \csc 135°$.

35. Find $\log_e \cot 210° + \log_e \csc 150°$.

36. Find $\log_e \sin 120° - \log_e \tan 240°$.

42-4 TRIGONOMETRIC EQUATIONS

In Chapter 40 we saw the meaning of a trigonometric identity. An identity is an equation that is true for all values of the angle θ, just as an identity in algebra is true for all values of an unknown, x.

Now, a trigonometric equation may be true for some values but not for all values of the angle θ. Then it is not an identity but a *conditional equation.* Just as a conditional equation in algebra calls for a *solution*, so does a conditional equation in trigonometry.

Consider the following conditional equation in algebra: $x^2 - 2x - 8 = 0$. The equation is true only if $x = -2$ or $x = 4$. These are solutions. To solve a conditional equation, we find values of the unknown that make the equation true.

In trigonometry we have a similar situation. For example, suppose we have the equation,

$$\sin \theta = \cos \theta$$

This is not an identity because it is not true for all values of θ. Now we try to find the solutions, that is, the values of θ that make the equation true. If we divide both sides of the equation by $\cos \theta$, we get

$$\frac{\sin \theta}{\cos \theta} = 1$$

We can replace the left side of the equation by its equivalent, $\tan \theta$, and get

$$\tan \theta = 1$$

There are an infinite number of angles θ whose tangent is 1. However, we confine the solutions to values of θ from 0 to 360°, but not including 360°, because the values for 360° are the same as those for zero. Then,

$$\text{if} \quad \tan\theta = 1$$

$$\theta = 45° \quad \text{and} \quad 225°$$

Checking these values, we find that they both satisfy the original equation.

Example 1

Find all values of θ from 0° to but not including 360° such that

$$\sin 3\,\theta = \tfrac{1}{2}$$

Solution

First, we seek values of the entire angle, 3θ, having a sine value equal to $\frac{1}{2}$. The only angles less than 360° satisfying this condition are angles of 30° and 150°. Then we first say

$$\text{if} \quad \sin 3\theta = \tfrac{1}{2}$$

$$\text{then} \quad 3\theta = 30° \quad \text{and} \quad 150°$$

However, these values are not the solutions to the problem. The problem calls for values of θ, not only 3θ. As values of θ, we get

$$\theta = 10° \quad \text{and} \quad 50°$$

There are also other values that satisfy the equation. To get all values of θ, we go around the circle three times for 3θ and get

$$3\theta = 30°;\ 150°;\ 390°;\ 510°;\ 750°;\ 870°$$

$$\text{then} \quad \theta = 10°;\ 50°;\ 130°;\ 170°;\ 250°;\ 290°$$

Example 2

Solve the following equation: $\cos^2\theta = \frac{1}{4}$.

Solution

Taking the square root, $\cos\theta = \pm\frac{1}{2}$. Since we can take positive and negative values, we get

$$\theta = 60°;\ 120°;\ 240°;\ 300°$$

Exercise 42-2

Solve the following equations for all values of the unknown from zero up to but not including 360°.

1. $\sin\theta = \frac{1}{2}$ **2.** $\sin^2\theta = 1$ **3.** $\cos^2 2\phi = 1$

4. $\tan^2 \theta = 1$ **5.** $\tan^2 \phi = 3$ **6.** $\sin 3\phi = 1$

7. $\cos 3x = 0$ **8.** $\sin^2 4x = \frac{3}{4}$ **9.** $\tan 5x = -1$

10. $\sec 2\phi = 1$ **11.** $\csc^2 \theta = 4$ **12.** $\cot^2 3x = 0$

13. $\sin \theta = \sqrt{3} \cos \theta$ **14.** $3 \cos 2x = 2 - \cos 2x$

15. $4 \cos^2 3x - 3 = 0$ **16.** $\tan^2 \theta - \tan \theta = 0$

17. $\sin^2 \theta - 2 \sin \theta = 0$ **18.** $3 \sin^2 \theta - 2 \sin \theta = 1$

19. $\sin^2 \theta + \sin \theta = 2$ **20.** $\sin^2 2x - 3 \cos^2 2x = 0$

QUIZ ON CHAPTER 42, SPECIAL ANGLES. FORM A.

In Nos. 1–18, sketch each of the angles in standard position. Then, without a table or calculator, find the six functions of each special angle:

1. 60° **2.** 210° **3.** 225° **4.** 270°

5. 300° **6.** −30° **7.** −120° **8.** −180°

In Nos. 9–16, find the value of each function given:

9. $\sin \frac{\pi}{6}$ **10.** $\cos \frac{\pi}{4}$ **11.** $\tan \frac{\pi}{3}$ **12.** $\sin^2 \frac{2\pi}{3}$

13. $\cos^2 \frac{3\pi}{4}$ **14.** $\cot \frac{7\pi}{4}$ **15.** $\sec \frac{2\pi}{3}$ **16.** $\csc \frac{\pi}{2}$

Solve the following equations for all values of the angle θ from zero up but not including 360°; that is, $0\theta < 360°$.

17. $\sin \theta = -(\frac{1}{2})$ **18.** $\cos 2\theta = \frac{1}{2}$ **19.** $\tan 3\theta = 1$

20. $\sin 2\theta = 1$ **21.** $\sin^2 \theta = \frac{1}{2}$ **22.** $\tan^2 2\theta = 1$

CHAPTER FORTY-THREE Composite Angle Formulas

43-1 PREASSIGNMENT

In order to understand more clearly the subject of this chapter, the student should first work out the following assignment:

1. Copy in *table form* the sine for each of the following angles:

$$0°, 10°, 20°, 30°, 40°, 50°, 60°, 70°, 80°, 90°$$

Then answer the following questions:

(a) As the angle increases from 0 to 90°, does the sine value increase or decrease?
(b) If an angle is doubled, is its sine doubled?
(c) Is the sine of 80° equal to the sine of 50° plus the sine of 30°; that is, does $\sin 80° = \sin 50° + \sin 30°$?
(d) If one angle is one-half another angle, is its sine also one-half as much?
(e) Do angles have the same ratio as their sines? For example, if one angle is three-fifths of another angle, is its sine value also three-fifths as great?

2. Copy in table form the cosines of the angles in No. 1.
(a, b, c, d, e) Answer the same questions for the cosine.

3. Copy in table form the tangents of the same angles.
(a, b, c, d, e) Answer the same questions for the tangent.

43-2 SUM AND DIFFERENCE FORMULAS

From the preceding exercises a few facts should be noted. First, we note that the *sine of the sum* of two angles is *not* equal to the *sum of the sines* of the two angles. For instance, the sine of 50° cannot be found simply by adding the sine of 20° to the sine of 30°.

As another example, if we wish to find the sine of 75°, we cannot find it simply by adding the sine of 45° to the sine of 30°. However, it is sometimes convenient and desirable to compute the functions of some angle, such as the sine of 75°, from the functions of other angles.

Let us make one point clear. In most problems, if we wish to find the functions of any angle, we look up the values in a table. Yet it is possible to find the functions of some angles by using functions of other angles that can be computed *without a table*. Let us see how this is done.

We have seen that we can easily find the functions of 45° and 30° without the use of a table. Now, it happens that the functions of 75° can be computed from the functions of 45° and 30°. Remember, the sine of 75° *cannot* be found simply by adding the sines of the other two angles. Instead, it can be shown that

$$\sin 75° = (\sin 45°)(\cos 30°) + (\cos 45°)(\sin 30°)$$

or

$$\sin (45° + 30°) = (\sin 45°)(\cos 30°) + (\cos 45°)(\sin 30°)$$

By using the values of the sines and cosines of 30° and 45°, we can find the sine of 75°. The following values can be found withut a table:

$$\sin 45° = \frac{\sqrt{2}}{2} \qquad \sin 30° = \frac{1}{2} \qquad \cos 45° = \frac{\sqrt{2}}{2} \qquad \cos 30° = \frac{\sqrt{3}}{2}$$

Substituting numerical values in the foregoing equation, we get

$$\sin 75° = \frac{\sqrt{2}}{2} \cdot \frac{\sqrt{3}}{2} + \frac{\sqrt{2}}{2} \cdot \frac{1}{2}$$

$$= \frac{\sqrt{6}}{4} + \frac{\sqrt{2}}{4} = \frac{\sqrt{6} + \sqrt{2}}{4}$$

$$= \frac{2.449 + 1.414}{4} = \frac{3.863}{4} = 0.966 \text{ (approx.)}$$

Let us state the rule in words.

Rule. *The sine of the sum of two angles is equal to the sine of the first angle times the cosine of the second angle, plus the cosine of the first angle times the sine of the second angle.*

If A and B represent any two angles, the rule says:

$$\sin (A + B) = (\sin A)(\cos B) + (\cos A)(\sin B) \quad \text{(Formula 1)}$$

This is called the formula for the *sine of the sum of two angles.* It can be used to compute the sine of any angle that can be expressed as *the sum of any two of the special angles we have studied.* For example, we can use the formula to compute the sine of 165°, since 165° may be called (120° + 45°).

Note. The formula is an identity and can be used for any angle if the angle is expressed as the sum of two other angles. For instance, it could be used to find the sine of 68° by combining the functions of 47° and 21° in the proper manner. The formula might be checked for any combination of angles. However, it is not used generally, except with reference to a combination of the special angles we have mentioned in Chapter 42.

We shall now show how this formula is derived. Let us start with any two general angles. A and B, as shown in Fig. 43-1.*

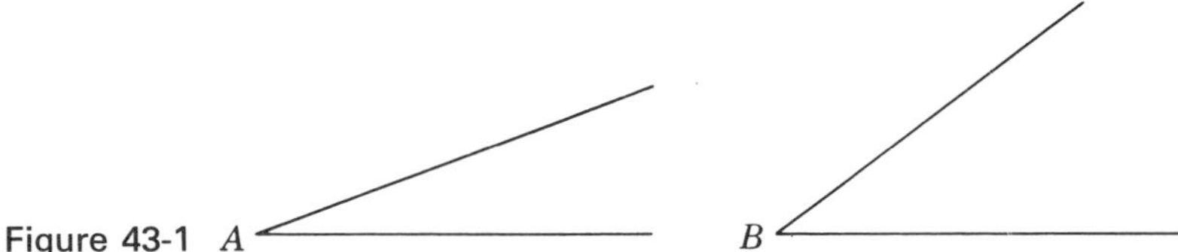

Figure 43-1

To show the sum of the two angles, $A + B$, we add them geometrically and place the sum in standard position. In Fig. 43-2.

$$A + B = \text{angle } COD$$

The sine of $(A + B)$ is the same as the sine of angle COD.

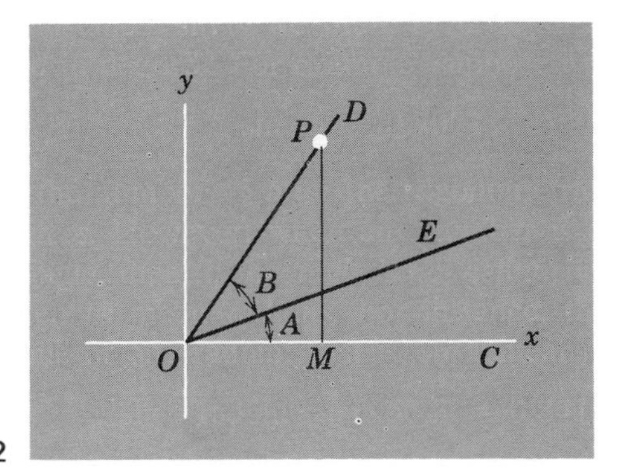

Figure 43-2

To express the sine of angle COD, take a point P on the terminal side OD. Draw a line PM perpendicular to the x-axis. We then have

$$\sin(A + B) = \frac{MP}{OP}$$

Our problem now becomes a question of expressing the two lengths, MP and OP, as functions of the angles, A and B. To do so, we first draw three other line segments. (Fig. 43-3).

1. Draw PQ perpendicular to OE.
2. Draw QN perpendicular to the x-axis.
3. Draw QR perpendicular to MP.

* The derivation shown here is geometric in nature. The formula may also be derived by the analytic method which depends on the use of the formula for the distance between two points.

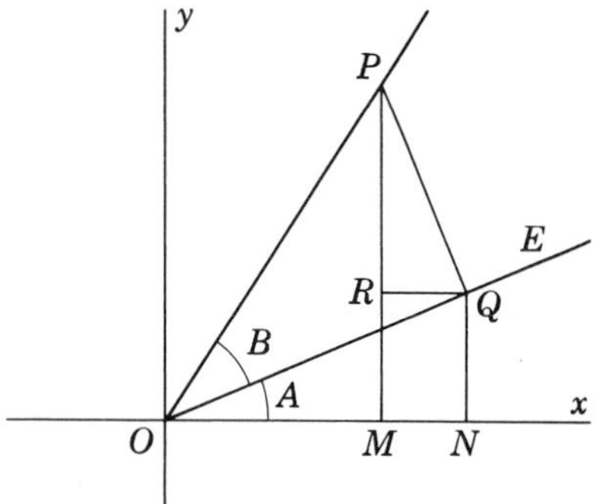

Figure 43-3

In Fig. 43-3 we see that MR is equal to NQ. Also angle RPQ is equal to the angle A, since their sides, respectively, are perpendicular, right to right, and left to left. Let us go back to the definition of the sine of $(A + B)$.

$$\sin(A+B) = \frac{MP}{OP}$$

Now, since $MP = MR + RP$, we can write

$$\sin(A+B) = \frac{MR+RP}{OP}$$

Separating the right side into two fractions, we get

$$\sin(A+B) = \frac{MR}{OP} + \frac{RP}{OP}$$

or

$$\sin(A+B) = \frac{NQ}{OP} + \frac{RP}{OP}$$

Sine and cosine functions may now be substituted for the fractions on the right side of the equation. Note that

$$\frac{NQ}{OQ} = \sin A; \qquad \text{therefore, } NQ = OQ \sin A$$

$$\frac{RP}{PQ} = \cos A; \qquad \text{therefore, } RP = PQ \cos A$$

If we replace NQ with its equal, $OQ \sin A$, and replace RP with its equal, $PQ \cos A$, in the formula, we get

$$\sin(A+B) = \frac{OQ \sin A}{OP} + \frac{PQ \cos A}{OP}$$

From Fig. 43-3 we note that

$$\frac{OQ}{OP} = \cos B \qquad \text{and} \qquad \frac{PQ}{OP} = \sin B$$

By substitution of equals, the final formula becomes

$$\sin(A+B) = \sin A \cos B + \cos A \sin B \qquad \text{(Formula 1)}$$

(The parentheses are usually omitted on the right side of the equation.)

To derive the formula for the cosine of $(A + B)$, we also use Fig. 43-3. To express this function, we first write

$$\cos(A+B) = \frac{OM}{OP}$$

From the figure we see that $OM = ON - MN$ and $MN = RQ$. The final formula becomes

$$\cos (A + B) = \cos A \cos B - \sin A \sin B \quad \text{(Formula 2)}$$

This is the formula for the *cosine* of the *sum* of two angles. The derivation is left as an exercise for the student.

In demonstrating that the foregoing formulas hold true, we have used two angles whose sum is less than 90°; that is, the terminal side of the sum $(A + B)$ still falls in the first quadrant. The angles and the figure we have used do not apply to the cases in which the terminal side of the sum falls in other quadrants. A formula derived from one particular situation cannot always be assumed to be true for a different situation. However, these formulas can be proved true for all cases by using two angles whose sum falls in each of the four quadrants. The complete proof requires four different figures. The student should try to show that the formulas are true for each of the four quadrants.

Other formulas refer to the *difference* between two angles. In the exercises in the preassignment at the beginning of this chapter, it will be noted that we cannot find the sine of the *difference* between two angles simply by subtracting the sine of the two angles. For instance, the sine of 15° cannot be found simply by subtracting the sine of 30° from the sine of 45°.

However, the sine, as well as the other functions, of 15° can be found by the formulas for the functions of the *difference* between two angles. The *difference* formulas can be derived directly from the foregoing *sum* formulas. First, it is necessary to see the relation between functions of negative and positive angles of equal magnitude.

First, we ask: What is the relation between the sine of a negative angle and the sine of the corresponding positive angle of equal magnitude? Let us take a positive angle θ ending in the first quadrant. Then the corresponding negative θ of the same magnitude ends in the fourth quadrant (Fig. 43-4).

We take a point (a, b) on the terminal side of θ. Then the corresponding point on the terminal side of $(-\theta)$ is $(a, -b)$. The radius vector r is the same for the positive and the negative angle. Then we have

$$\sin (+\theta) = \frac{b}{r} \quad \text{and} \quad \sin (-\theta) = \frac{-b}{r}$$

In general, $$\sin (-\theta) = -\sin (+\theta)$$

That is, *the sine of a negative angle is the negative of the sine of the corresponding positive angle.* The statement is true for angles ending in any quadrant.

For the cosine of the angles, $+\theta$ and $-\theta$, we note that the abscissa

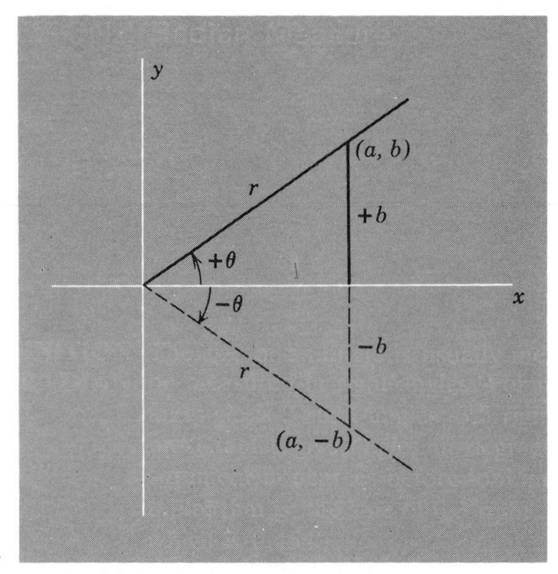

Figure 43-4

has the same sign for the negative as for the positive angle. Therefore,

$$\cos(+\theta) = \frac{a}{r} \quad \text{and} \quad \cos(-\theta) = \frac{a}{r}$$

In general,

$$\cos(-\theta) = +\cos(+\theta)$$

That is, the *cosine of a negative angle has the same sign as the cosine of the corresponding positive angle.* This statement can also be shown to be true for all values of θ.

The student should work out the rules for the remaining functions showing the relations between positive and negative angles.

The formula for the sine of the *difference* between two angles can now be derived by substituting $(-B)$ for $(+B)$ in Formula 1. We get

$$\sin(A - B) = \sin A \cos B - \cos A \sin B \qquad \text{(Formula 3)}$$

The formula for the cosine of the *difference* between two angles may be derived by substituting a negative angle $(-B)$ for the $+B$ in the formula for the cosine of the sum of two angles. We get

$$\cos(A - B) = \cos A \cos B + \sin A \sin B \qquad \text{(Formula 4)}$$

The plus sign (+) on the right side of the equation is reasonable if we think of the angle $(A - B)$ as being a small angle whose cosine is therefore comparatively large. Again, we must remember that the formula is true for angles of any size.

To derive the formula for the tangent of $(A + B)$, we begin with the identity

$$\tan\theta = \frac{\sin\theta}{\cos\theta}$$

By using Formulas 1 and 2, we get

$$\tan(A + B) = \frac{\sin(A + B)}{\cos(A + B)} = \frac{\sin A\cos B + \cos A\sin B}{\cos A\cos B - \sin A\sin B}$$

By dividing the entire numerator and denominator of the last fraction by the quantity $\cos A \cos B$, we get the formula for the tangent of the sum of two angles:

$$\tan(A + B) = \frac{\tan A + \tan B}{1 - \tan A\tan B} \qquad \text{(Formula 5)}$$

In the same way we can derive the formula for $\tan(A - B)$:

$$\tan(A - B) = \frac{\tan A - \tan B}{1 + \tan A\tan B} \qquad \text{(Formula 6)}$$

The formulas for the cotangent of the sum and the difference of two angles can be derived in a similar manner.

Note. The foregoing formulas are identities. Therefore they apply to the combinations of any angles, whatever the size. However, in practical use they have little value. If we wish to find the functions of an angle such as 75° or 15°, we look up the values in a table. Yet it should be remembered that many formulas turn out to be important later in our study of trigonometry and other mathematics. Many formulas are necessary in deriving other simpler formulas that have practical use in solving problems.

Example 1

Combine Formulas 1 and 3 in such a way as to get a formula for the product of the sine of one angle times the cosine of another angle. That is, the formula for $(\sin A)(\cos B)$. Then show that the resulting formula is true for the values: $A = 45°$ and $B = 15°$.

Solution

We write Formulas 1 and 3, using A and B for the two angles:

$$\sin(A + B) = \sin A\cos B + \cos A\sin B$$

$$\sin(A - B) = \sin A\cos B - \cos A\sin B$$

Adding: $\sin(A + B) + \sin(A - B) = 2\sin A\cos B$

Dividing by 2, $\sin A\cos B = \frac{1}{2}[\sin(A + B) + \sin(A - B)]$ (Formula 7)

Now we have a formula for the product of the sine of one angle times the cosine of another angle. We could check the result by using any two angles for A and B. Let us use: $A = 45°$ and $B = 15°$.

Using values: $(\sin 45°)(\cos 15°) = \frac{1}{2}[\sin 60° + \sin 30°)$

We get $(0.7071)(0.9659) = \frac{1}{2}[0.8660 + 0.5000)$

$$0.6830 = 0.6830$$

Note that the right side of the formula is easier to find by special angles.

If we use another set of angles, such as $A = 50°$ and $B = 20°$, the formula is true:

$$(\sin 50°)(\cos 20°) = \frac{1}{2}[\sin 70° + \sin 30°)$$

If we subtract Formula 3 from Formula 1, the result is essentially the same, with angle A and angle B interchanged.

Example 2

Use Formulas 1 and 3 to derive a formula for the difference between the sines of two angles; for example, $\sin x - \sin y$. The result is a very useful formula in calculus. Check with angles: $x = 75°$; $y = 15°$.

Solution

Writing the Formulas 1 and 3:

$$\sin(A + B) = \sin A \cos B + \cos A \sin B$$

$$\sin(A - B) = \sin A \cos B - \cos A \sin B$$

Subtracting, $\sin(A + B) - \sin(A - B) = 2\cos A \sin B$

Note that the left side represents the difference between two sines. To simplify the formula, we shall represent $(A + B)$ by the letter x, and represent $(A - b)$ by the letter y; that is,

$$\text{let} \quad x = A + B \quad \text{and} \quad \text{let} \quad y = A - B$$

Adding the two, we get: $x + y = 2A$; then $A = (\frac{1}{2})(x + y)$
Subtracting the two, $x - y = 2B$; then $B = (\frac{1}{2})(x - y)$
Now we replace $(A + B)$ with x; and $(A - B)$ with y. We get

$$\sin x - \sin y = 2\cos(\tfrac{1}{2})(x + y)\sin(\tfrac{1}{2})(x - y) \qquad \text{(Formula 8)}$$

This is now the formula for the difference between the sines of two angles. We shall check the formula by using the angles: $x = 75°$; $y = 15°$.

$$\sin 75° - \sin 15° = 2\cos 45° \sin 30°$$

Substituting values; $0.9659 - 0.2588 = 2(0.7071)(0.5000)$

Combining values: $0.7071 = 0.7071$

This Formula 8 is used in calculus to find the derivative of the sine of an angle.

Exercise 43-1

Find the following by use of special angles and Formulas 1–8:

1. sin 75°, using 45° and 30°

2. sin 165°, using 135° and 30°

3. cos 75, using 45° and 30°

4. cos 105°, using 60° and 45°

5. sin 15°, using 45° and 30°

6. sin 15°, using 135° and 120°

7. cos 15°, using 60° and 45°

8. cos 15°, using 135° and 120°

9. tan 75°, using 45° and 30°

10. tan 15°, using 60° and 45°

11. Derive the formula for the tangent of a negative angle.

12. Derive the formula for the cotangent of the sum of two angles.

13. Combine Formulas 2 and 4 to get a formula for $(\cos A)(\cos B)$.

14. Using Formulas 1 and 3, derive a formula for $(\sin x + \sin y)$.

15. Check Formula 7 by using the angles 60° and 10°.

16. Check Formula 8 by using the angles 70° and 30°.

Find the instantaneous current i or instantaneous voltage e in the following:

17. $i = 20 \sin 75°$

18. $e = 120 \cos 105°$

19. $i = (\sin 60°)(\cos 20°)$

20. $e = \sin 70° - \sin 10°$

43-3 DOUBLE-ANGLE FORMULAS

Another fact to be observed from the preassignment exercises is that if we *double an angle*, we do *not* thereby *double the sine value*; that is, the sine of 80° is *not* simply twice the sine of 40°. The tangent of, say, 36° is *not* exactly twice the tangent of 18°. In fact, none of the function values is exactly proportional to the corresponding angle.

However, it is possible, by using the functions of a particular angle, to compute the functions of an angle twice as large. For such computations we have the so-called "double-angle" formulas. These formulas can be derived from Formulas 1, 2, and 5 by assuming that angle B is equal to angle A in each case; then $A + B = 2A$. We get the following double-angle formulas:

$$\sin (A + A) = \sin A \cos A + \cos A \sin A$$

or

$$\sin 2A = 2 \sin A \cos A \qquad \text{(Formula 9)}$$

Also

$$\cos 2A = \cos^2 A - \sin^2 A \qquad \text{(Formula 10)}$$

and

$$\tan 2A = \frac{2 \tan A}{1 - \tan^2 A} \qquad \text{(Formula 11)}$$

As examples of the application of Formula 9, we have

$$\sin 60^\circ = 2(\sin 30^\circ)(\cos 30^\circ) = (2)\cdot\frac{1}{2}\cdot\frac{\sqrt{3}}{2} = \frac{\sqrt{3}}{2}$$

Since the formulas are identities, they are true for all angles. For instance,

$$\begin{aligned}\sin 36.4^\circ &= (2)(\sin 18.2^\circ)(\cos 18.2^\circ)\\ &= (2)(0.3123)(0.9500) = 0.5934\end{aligned}$$

43-4 HALF-ANGLE FORMULAS

Still another fact is to be noted from the preassignment: we cannot obtain the functions of half an angle simply by taking half the function value of a given angle. For instance, the sine of 20° is not simply one-half the sine of 40°. Yet it is sometimes desirable to compute the functions of one-half a given angle. To do so, we use the so-called "half-angle" formulas.

To derive the formula for the sine of half a given angle, we begin with Formula 10:

$$\cos 2A = \cos^2 A - \sin^2 A$$

In this formula angle A represents any angle. Then $2A$ represents twice the angle A. Therefore, we can say that angle A is one-half angle $2A$.

First, we recall the identity $\cos^2 A = 1 - \sin^2 A$. For $\cos^2 A$ in Formula 10 we substitute $1 - \sin^2 A$. The formula

$$\cos 2A = \cos^2 A - \sin^2 A$$

then becomes

$$\cos 2A = 1 - \sin^2 A - \sin^2 A$$

or

$$\cos 2A = 1 - 2\sin^2 A$$

Transposing,

$$2\sin^2 A = 1 - \cos 2A$$

Dividing by 2,

$$\sin^2 A = \frac{1 - \cos 2A}{2}$$

Solving,

$$\sin A = \sqrt{\frac{1 - \cos 2A}{2}}$$

If we replace angle $2A$ with any other angle, say, θ, then angle A is equal to $\theta/2$, or one-half θ. In terms of θ, we have the formula for the sine of half an angle:

$$\sin\frac{\theta}{2} = \pm\sqrt{\frac{1 - \cos\theta}{2}} \qquad \text{(Formula 12)}$$

By the use of this formula, the sine of 15° may be computed from the cosine of 30°. Since the formula is an identity, it applies to the half of *any* angle.

Summary of half-angle formulas:

$$\sin\frac{\theta}{2} = \pm\sqrt{\frac{1-\cos\theta}{2}} \qquad \text{(Formula 12)}$$

$$\cos\frac{\theta}{2} = \pm\sqrt{\frac{1+\cos\theta}{2}} \qquad \text{(Formula 13)}$$

$$\tan\frac{\theta}{2} = \pm\sqrt{\frac{1-\cos\theta}{1+\cos\theta}} \qquad \text{(Formula 14)}$$

Formula 14 may be written in two other forms:

$$\tan\frac{\theta}{2} = \frac{\sin\theta}{1+\cos\theta} = \frac{1-\cos\theta}{\sin\theta} \qquad \text{(Formula 15)}$$

The student should try to work out the derivation of each of these forms.

43-5 APPLICATION OF FORMULAS TO PARTICULAR SITUATIONS

Formulas 1 to 12 must be understood to be true for all values of an angle. In other words, they are identities. When we say, as in Formula 9, $\sin 2A = 2\sin A\cos A$, we must understand that the formula is true for all values of the angle A.

This means, of course, it is true for any *particular* angle. Now, we might wish to check the formula when $A = 20°$. The sine of 20° is approximately 0.3420; the cosine of 20° is approximately 0.9397. Therefore, the sine of 40° should be approximately

$$\sin 40° = (2)(0.3420)(0.9397)$$

The product of the three factors on the right is 0.6427548, or, rounded off, 0.6428. If we look up the sine of 40°, we find it is given as 0.6428 to four places. This example shows that the formula is true for this one angle.

However, when we show that a formula is true for a particular angle, the example should not be considered as a proof but only an illustration. An illustration by example is not a proof at all. If the application of a formula turns out to be correct in a particular case, the example itself does not prove the formula to be always true.

Perhaps the chief value of trying out a formula in a particular example is that it helps us to understand the full meaning of the formula. An example or illustration is valuable in that it shows the application of the formula to a special situation. The example may be considered as a check on the formula. If the formula fails in only one instance, then it is not an identity.

Example

From the triangle in Fig. 43-5 show that

$$\cos 2A = \cos^2 A - \sin^2 A$$

To understand our problem better, let us state it in the form of a question: in Fig. 43-5,

$$\text{does} \quad \cos 2A = \cos^2 A - \sin^2 A?$$

Perhaps the best way is to work out the right side of the equation first.

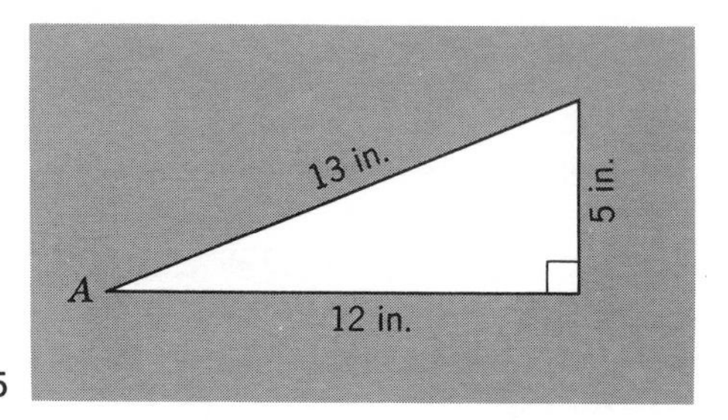

Figure 43-5

From the figure we see that

$$\cos A = \frac{12}{13} \qquad \cos^2 A = \frac{144}{169}$$

$$\sin A = \frac{5}{13} \qquad \sin^2 A = \frac{25}{169}$$

The right side of the equation becomes

$$\cos^2 A - \sin^2 A = \frac{144}{169} - \frac{25}{169} = \frac{119}{169} = 0.70414$$

Now we shall determine whether the cos $2A$ is equal to this value. First, we must find angle A. Angle A is arctan $\frac{5}{12}$, or approximately 22.6°. Then angle $2A$ is approximately 45.2°. The table shows cos 45.2° to be approximately 0.7046. The two sides of the equation are not exactly equal, because only approximate values were used, but they are sufficiently close.

If we compute angle A more accurately, we shall find it to be about 22° 37.2′. Then angle $2A$ is 45° 14.4′. The table shows the cosine of this angle as 0.70414 by interpolation. The two sides of the equation are then more nearly equal.

Exercise 43-2

Using the double-angle formulas, find the following:

1. sin 60°, using a 30° angle

2. cos 60°, using a 30° angle

3. tan 60°, using a 30° angle

4. sin 90°, using a 45° angle

Using the half-angle formulas, find the following:

5. sin 15°, using a 30° angle

6. cos 15°, using a 30° angle

7. tan 15°, using a 30° angle

8. sin 22.5°, using a 45° angle

9. cos 75°, using a 150° angle

10. tan 22.5°, using a 45° angle

11. sin 67.5°, using a 135° angle

12. cos 67.5°, using a 135° angle

13. Find sin 7.5° without a table. First find sin 15° and cos 15°; then take 7.5° as half of 15°.

14. A right triangle has sides equal to 3 in., 4 in., and 5 in., respectively. Angle A is the acute angle opposite the 3-in. side. From this triangle, show that Formula 11 is correct, using angle A.

15. A right triangle has sides equal to 5 in., 12., and 13 in., respectively. Angle R is the acute angle opposite the 12-in. side. From this triangle show that Formula 12 is correct, using angle R.

16. Using angle R in No. 15, show that Formula 13 is correct.

17. Derive the two forms of Formula 15.

QUIZ ON CHAPTER 43, COMPOSITE ANGLE FORMULAS. FORM A.

Without a table or calculator, find the values indicated by use of special angles and the sum and difference formulas:

1. sin 75°, using 45° and 30°

2. cos 165°, using 135° and 30°

3. tan 105°, using 60° and 45°

4. sin 15°, using 60° and 45°

5. cos 15°, using 45° and 30°

6. tan 15°, using 60° and 45°

7. sin 150°, using 120° and 30°

8. cos 120°, using 90°and 30°

Using the double-angle formulas, find the following:

9. sin 120°, using a 60° angle

10. cos 90°, using a 45° angle

Using the half-angle formulas, find the following:

11. sin 15°, using a 30° angle

12. cos 22.5°, using a 45° angle

13. tan 30°, using a 60° angle

14. sin 75°, using a 150° angle

15. Check Formula 7 by using the angles: $A = 75°$; $B = 45°$.

16. Check Formula 8 by using the angles: $x = 105°$; $y = 15°$.

CHAPTER FORTY-FOUR Solving Oblique Triangles

44-1 TWO KINDS OF OBLIQUE TRIANGLES

An oblique triangle is any triangle that is *not* a right triangle. The are two kinds of oblique triangles. An *acute triangle* is a triangle in which all *three* of its angles are *acute* (Fig. 44-1). An *obtuse triangle* is a triangle with *one obtuse* angle. In an obtuse triangle, two of the angles are acute (Fig. 44-2).

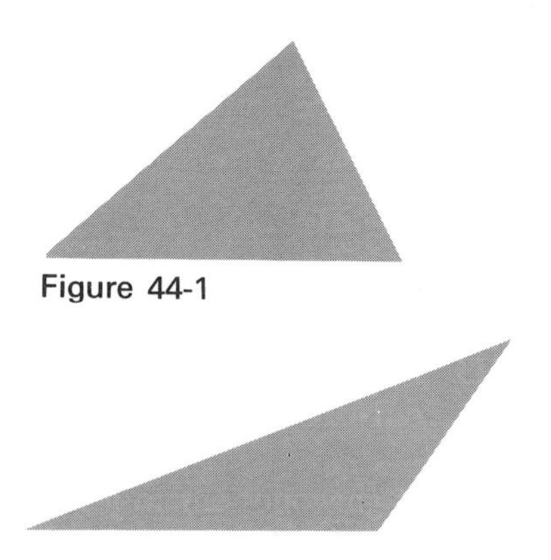

Figure 44-1

Figure 44-2

The three sides and the three angles of any triangle are called the six *elements* of a triangle. The three angles are often indicated by capital letters, A, B, and C. The sides are usually indicated by small letters, a, b, and c. Each side is denoted by a small letter that corresponds to the *opposite* angle. Thus side a is opposite angle A.

Oblique triangles can be solved indirectly by using the regular trigonometric functions and by drawing extra line segments. However, such triangles are often more conveniently solved by using the *sine law* or the *cosine law*.

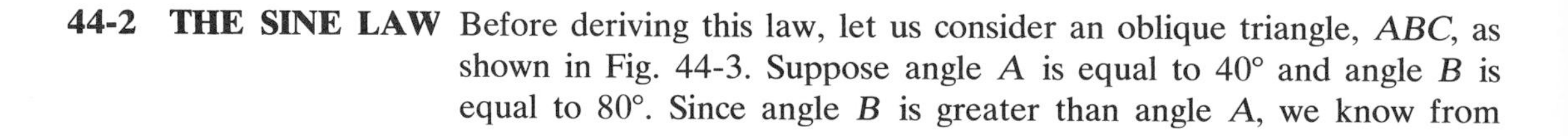

44-2 THE SINE LAW

Before deriving this law, let us consider an oblique triangle, ABC, as shown in Fig. 44-3. Suppose angle A is equal to 40° and angle B is equal to 80°. Since angle B is greater than angle A, we know from

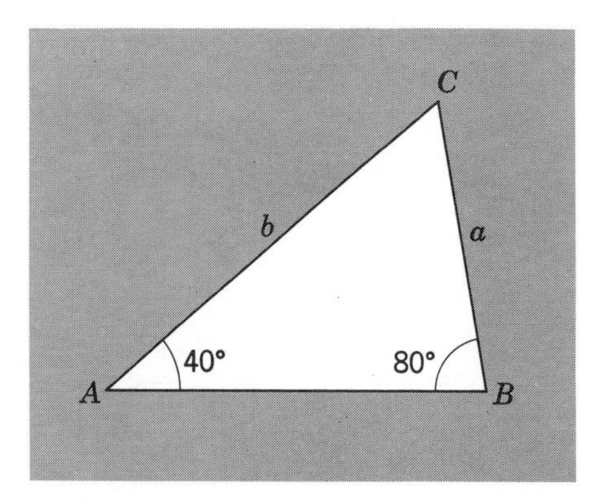

Figure 44-3

plane geometry that side b is greater than side a. In plane geometry we prove the following theorem:

If two angles of a triangle are unequal, the sides opposite these angles are also unequal and the greater side is opposite the greater angle.

Now, since angle B is twice angle A, we might be inclined to say, at first guess, that side b is also twice side a. However, this is not the case. The sides of a triangle do *not* have the same *ratio* as the opposite angles.

Instead, the ratio of side b to side a will always be the same as the ratio of the *sines* of the opposite angles. In other words,

$$\text{the ratio } \frac{b}{a} = \text{the ratio } \frac{\sin B}{\sin A}$$

This is the *sine law*. The law is usually stated in another form:

$$\text{Form 1.} \quad \frac{a}{\sin A} = \frac{b}{\sin B} = \frac{c}{\sin C}$$

$$\text{Form 2.} \quad \frac{\sin A}{a} = \frac{\sin B}{b} = \frac{\sin C}{c}$$

In words, the *sine law* may be stated as follows: *Any side of a triangle has the same ratio to the sine of the opposite angle as any other side has to the sine of its opposite angle.*

Either of the foregoing forms of the sine law may be used. In solving an oblique triangle by this law, one form is usually more convenient than the other, depending upon the information given in the problem.

The sine law is easily derived from a triangle. We draw any triangle ABC (Fig. 44-4) with sides a, b, and c. Draw a line segment from any angle, say, angle C, perpendicular to the opposite side, c. Then two right triangles are formed. We call the altitude h. From the definitions

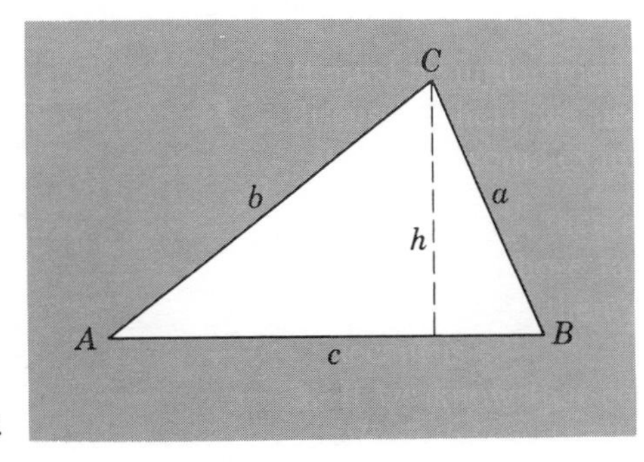

Figure 44-4

of the trigonometric functions we have

$$\frac{h}{b} = \sin A \qquad \text{and} \qquad \frac{h}{a} = \sin B$$

or

$$h = b \sin A \qquad \text{and} \qquad h = a \sin B$$

Here we have two values of h. Equating these values, we have

$$b \sin A = a \sin B$$

Dividing both sides of the equation by the quantity $\sin A \sin B$,

$$\frac{b}{\sin B} = \frac{a}{\sin A}$$

If, instead, we divide both sides of the equation by ab, we get the second form

$$\frac{\sin A}{a} = \frac{\sin B}{b}$$

Since angle A and angle B represent any two angles of the triangle, we can call one of them angle C and get the complete form of the sine law.

In using the sine law, we set up an equation between the two fractions that involve the given parts and the parts to be found. It is usually best to begin with the part to be found. It will be observed that in order to use this law at least one angle must be given.

Example 1

In a certain oblique triangle angle $A = 40.7°$, angle $C = 78.9°$, and $a = 14.2$ in. Find side c.

Solution

First we make a sketch of the triangle and label the parts (Fig. 44-5). Next, we set up a statement of the sine law involving the parts given and the part

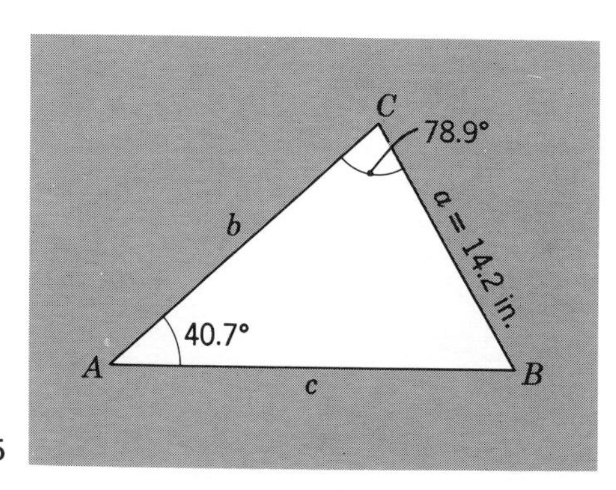

Figure 44-5

to be found. We begin with the part to be found.

$$\frac{c}{\sin C} = \frac{a}{\sin A}$$

Solving the equation for c, we get $$c = \frac{a \sin C}{\sin A}$$

Substituting the given values, $$c = \frac{(14.2)(\sin 78.9°)}{\sin 40.7°}$$

Taking sine values from the table, we get $$c = \frac{(14.2)(0.9813)}{0.6521}$$

$$c = 21.37 \quad \text{or} \quad 21.4$$

The actual computation can be performed by means of logarithms, if desired.

Example 2

In a oblique triangle, $C = 73.4°$, $b = 21.3$ in., and $c = 27.6$ in. Solve the triangle completely. This means, find A, B, and a.

Solution

To find angle B, we set up the sine law involving the angles, B and C. We begin with angle B, because it is unknown.

$$\frac{\sin B}{b} = \frac{\sin C}{c}$$

Then $$\sin B = \frac{b \sin C}{c}$$

Substituting values, we have $$\sin B = \frac{(21.3)(\sin 73.4°)}{27.6}$$

From this, we get $$\sin B = \frac{(21.3)(0.9583)}{27.6} = 0.7396$$

$$\text{angle } B = 47.7°$$

Since we now know two angles of the triangle, we find the third by using the fact that the sum of the angles of a triangle is 180°. Then $A = 58.9°$.

To find side a, we use the form $$\frac{a}{\sin A} = \frac{c}{\sin C}$$

Substituting values, we get $$a = \frac{(27.6)(0.8563)}{0.9583} = 24.66, \quad \text{or} \quad 24.7$$

Note. It might be pointed out here that there may be two possible answers for angle B. Since $\sin B = 0.7396$, we might note there are two angles whose sines are 0.7396. They are 47.7° and the obtuse angle 132.3°. However, in this case we must discard the second value since the sum of B and C would exceed 180°. In the next example, we find that there are two possible answers, each of which can be considered correct. This type of problem is called the *ambiguous* case.

Example 3

Solve the triangle in which $A = 34.5°$; $b = 60$, and $a = 36$.

Solution

To find B, we use the form
$$\frac{\sin B}{b} = \frac{\sin A}{a}$$

or
$$\sin B = \frac{(b)(\sin A)}{a}$$

Inserting numerical values,
$$\sin B = \frac{(60)(0.5664)}{36} = 0.9440$$

Then B has the two possible values: $B = 70.7°$ and $109.3°$. For these values of B, we have two values of $C : C = 74.8°$ and $36.2°$.

Using the form
$$\frac{c}{\sin C} = \frac{a}{\sin A}$$

we get
$$c = \frac{(36)(0.9650)}{(0.5664)} = 61.3$$

and
$$c = \frac{(36)(0.5906)}{(0.5664)} = 37.5$$

The two solutions are shown in Fig. 44-6.

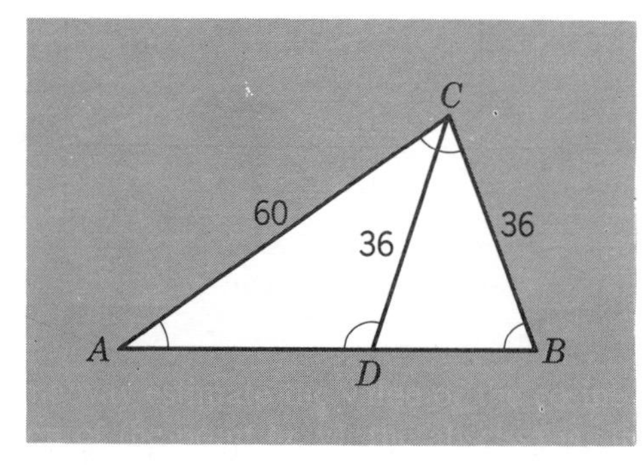

Figure 44-6

Exercise 44-1

Use the sine law to solve the triangles with the following given parts. Linear measurements are assumed to be in corresponding linear units, such as inches, centimeters, feet, and so on. Round off angles to the nearest tenth of a degree and linear measures to three significant digits (or four if the first digit is 1).

1. $A = 38.2°$; $B = 67.1°$; $b = 42$

2. $A = 26.9°$; $B = 70.7°$; $a = 36$

3. $A = 72.3°$; $C = 84.5°$; $c = 72$

4. $A = 21.8°$; $C = 81.4°$; $a = 24$

5. $A = 52.1°$; $C = 57.3°$; $b = 32.4$

6. $A = 86.2°$; $B = 49.5°$; $c = 43.2$

7. $B = 41.7°$; $C = 78.8°$; $a = 25.2$

8. $A = 76.6°$; $C = 38.5°$; $b = 50.4$

9. $A = 21.9°$; $B = 120.8°$; $c = 45$

10. $A = 106.2°$; $C = 20.5°$; $b = 22.5$

11. $B = 68.2°$; $a = 53$; $b = 65$

12. $A = 41.7°$; $a = 25.2$; $c = 32.4$

13. $C = 74.5°$; $b = 17$; $c = 24$

14. $B = 79.1°$; $a = 14$; $b = 18$

15. $B = 112.7°$; $b = 75$; $c = 27$

16. $A = 125.7°$; $a = 54$; $b = 21$

17. $A = 37°$; $a = 12$; $b = 16$

18. $B = 41°$; $b = 72$, $c = 80$

19. $A = 35°$; $a = 24$; $b = 32$

20. $C = 41°$; $b = 40$; $c = 30$

21. $B = 43°$; $a = 60$; $b = 36$

22. $B = 17.5°$; $a = 18$; $b = 8$

23. An observer at A wishes to find the distance to a house at point B. From his line of sight toward B, he turns 74° to his right and walks 200 ft. He then finds he must turn left 117° to look toward B. How far is he from B?

24. One gun is located at point A. Another gun is at B, which is 5 miles directly east from A. They both wish to hit a target at C. From A the direction to the target is 31° north of east. From B the target is 67° north of east. For what firing range should the guns be set?

25. From a point P near a river we wish to find the distance to a point R on the opposite side of the river. We turn our steps to the right 72°, and walk 180 ft in a straight line. From our new position we find we must turn left through 124° to see the point R Find the distance from P to R.

44-3 THE COSINE LAW

In some problems involving oblique triangles, we may not be able to use the sine law because of a lack of sufficient information. In such cases it may be possible to use the *cosine law*. This is another convenient method for solving some oblique triangles.

For example, suppose we have the triangle shown in Fig. 44-7 in which we have given two sides and the included angle: $b = 7$ in., side $c = 9$ in., and angle $A = 58°$. If we try to use the sine law, we find that no matter how we set up the equation we shall have two unknowns. Therefore, the sine law cannot be used to solve the triangle.

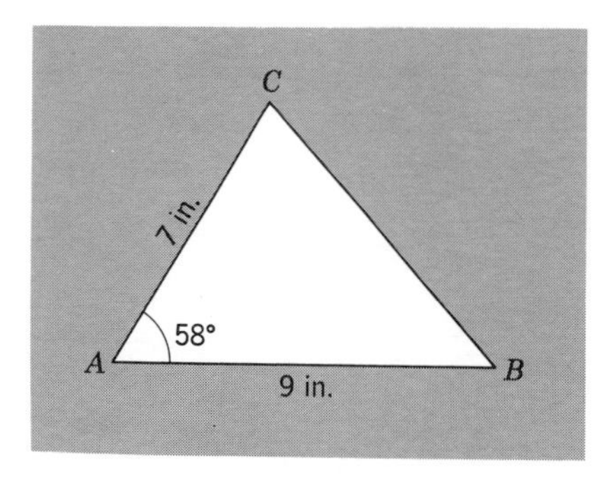

Figure 44-7

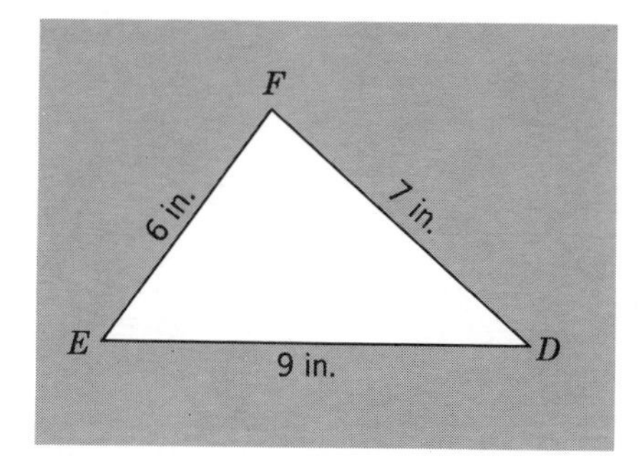

Figure 44-8

As another example, suppose we have the three sides of a triangle *DEF* shown in Fig. 44-8; $d = 6$ in., $e = 7$ in., and $f = 9$ in. Here, again, we find that we cannot use the sine law because no angle is given. In such problems we can use the cosine law.

To derive the cosine law, we begin with a general triangle such as *ABC* in Fig. 44.9. Let us suppose that sides b and c are known and that angle A is also known. We draw a perpendicular from C to the opposite side at point D. Call the line segment h. The distance AD we may call x. Then the distance DB is $c - x$.

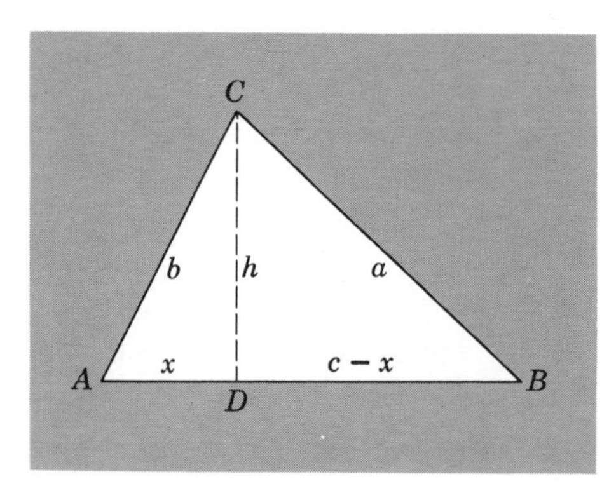

Figure 44-9

We now write two equations involving h and then eliminate the h between the two equations. By the Pythagorean rule, we get the following equations:

$$b^2 = h^2 + x^2$$
$$a^2 = h^2 + (c - x)^2$$

Expanding the second equation, we get $\quad a^2 = h^2 + c^2 - 2cx + x^2$
Rewriting the first equation, $\quad b^2 = h^2 \qquad\qquad + x^2$

Subtracting (eliminating h), $\quad a^2 - b^2 = c^2 - 2cx$
Transposing, $\quad a^2 = b^2 + c^2 - 2cx$

The last of the foregoing equations is solved for a^2 and should enable us to find a. The right side of the equation involves only known quantities b and c, except for the factor x. If we can also eliminate this factor, all of the right side of the equation will be known. From the original figure we see that

$$\frac{x}{b} = \cos A \qquad \text{or} \qquad x = b \cos A$$

In the final equation we replace x with its equivalent, $b \cos A$. The result is the *cosine law:*

$$a^2 = b^2 + c^2 - 2bc \cos A$$

In this formula every term and factor on the right side of the equation is known. Therefore, we can find a.

It should be understood that the letters a, b, and c refer to any of the sides of the triangle. Therefore, the formula may be stated in three forms:

$$a^2 = b^2 + c^2 - 2bc \cos A$$
$$b^2 = a^2 + c^2 - 2ac \cos B$$
$$c^2 = a^2 + b^2 - 2ab \cos C$$

The *cosine law* may be stated as follows: *The square of any side of a triangle is equal to the sum of the squares of the other two sides minus twice the product of those two sides times the cosine of the included angle.*

If each of the three forms of the cosine law is solved for the cosine of the angle, we get three useful alternate forms:

$$\cos A = \frac{b^2 + c^2 - a^2}{2bc}; \cos B = \frac{a^2 + c^2 - b^2}{2ac}; \cos C = \frac{a^2 + b^2 - c^2}{2ab}$$

The cosine law is used for two kinds of problems: (1) when two sides and the included angle are given, and (2) when the three sides are given.

Example 1

In a certain triangle, $a = 17$ cm, $c = 21$ cm, $B = 56.1°$. Find b.

Solution

Since we are to find side b, we write the formula for b^2:

$$b^2 = a^2 + c^2 - 2ac \cos B$$

Substituting values, we get

$$b^2 = 289 + 441 - (2)(17)(21)(0.5577) = 331.8$$

$$b = 18.22 \text{ (approx.)}$$

Now if we wish to find A and C, we can use the sine law.

Example 2

Find the angles of the triangle in which $a = 10$; $b = 12$; $c = 18$.

Solution

We find all angles by use of the alternate forms of the cosine law:

$$\cos A = \frac{b^2 + c^2 - a^2}{2bc} = \frac{144 + 324 - 100}{432} = 0.8519; \quad \text{then } A = 31.6°.$$

$$\cos B = \frac{a^2 + c^2 - b^2}{2ac} = \frac{100 + 324 - 144}{360} = 0.7778; \quad \text{then } B = 38.9°$$

$$\cos C = \frac{a^2 + b^2 - c^2}{2ab} = \frac{100 + 144 - 324}{240} = -0.3333; \quad \text{then } C = 109.5°$$

Check: $31.6° + 38.9° + 109.5° = 180.0°$. Note that if the cosine of the angle is negative, then the angle is *obtuse*, and the figure is an *obtuse triangle*.

Note. The sine law or the cosine law should *not* be used to solve a right triangle. Of course, the formulas hold true for all triangles, but the work is simpler if these laws are not used for right triangles.

Exercise 44-2

Use the consine law and the sine law to solve the triangles having the following given parts. Linear measurements are assumed to be in corresponding linear units, such as inches, feet, centimeters, and so on. If three sides are given, find all angles by the cosine law. Round off angles to the nearest tenth of a degree, and linear measures to three significant digits (or four if the first digit is 1).

1. $A = 50.9°$; $b = 45$; $c = 40$ **2.** $C = 58.2°$; $a = 56$; $b = 64$

3. $B = 42.1°$; $a = 38$; $c = 35$ **4.** $A = 63°$; $b = 24$; $c = 35$

5. $B = 104.3°$; $a = 52$; $c = 32$ **6.** $A = 118.1°$; $b = 12$; $c = 20$

7. $C = 146.3°$; $a = 21$; $b = 12$ **8.** $B = 114.9°$; $a = 40$; $c = 35$

9. $a = 20$; $b = 25$; $c = 28$ **10.** $a = 17$; $b = 19$; $c = 11$

11. $a = 21$; $b = 22$; $c = 25$ **12.** $a = 42.7$; $b = 53.4$; $c = 38.1$

13. $a = 20$; $b = 15$; $c = 28$ **14.** $a = 15$; $b = 22$; $c = 32$

15. $a = 22$; $b = 15$; $c = 10$ **16.** $a = 14$; $b = 22$; $c = 30$

17. $B = 23.5°$; $a = 17$; $c = 30$ **18.** $C = 36.2°$; $a = 28$; $b = 40$

19. $A = 47.7°$; $b = 35$; $c = 14$ **20.** $B = 23.8°$; $a = 17$; $c = 30$

21. An airline pilot flies from an airfield at A directly east 200 miles; then directly northeast for 160 miles. How far and in what direction is he then from the airfield?

22. A pilot flies from a town at B directly west 80 miles, then 50 miles in a direction 65° north of west. How far and in what direction must he fly to return to the town at B!

23. A hunter leaves his camp and walks directly west 6.5 miles, and then turns and walks 4 miles in a direction 40° south of east. He then decides to walk directly to his camp. How far and in what direction must he walk?

24. From town A a straight road leads to town B, a distance of 4 miles. From B another road leads directly to C, a distance of 3.5 miles. The angle between the two roads is 70°. A road is to be built directly from A to C. Find the distance from A to C and the direction with respect to AB.

25. A pilot flies from an airfield 200 miles in a direction 40° east of north. He then flies 300 miles in a direction 20° south of east. How far and in what direction must he then fly to return directly to the airfield?

26. A pilot flies 180 miles at a direction 15° south of east from his home airport. Then he turns to a direction 60° north of east and flies 240 miles. How far and in what direction must he then fly to return directly to his home port?

27. A man walks from his home directly west for a distance of 400 ft. Then he turns to a direction 70° south of west and walks for a distance of 350 ft. How far and in what direction must he then walk to return directly to his home?

28. A triangular park is to be enclosed with three fences. Two sides of the park are 140 ft and 125 ft, respectively, with an angle of 80°

between them. Find the total length of fence required to enclose the park.

29. A triangular plot of land is enclosed with fences. The three sides of the plot measure 200 meters, 180 meters, and 170 meters, respectively. Find the angle at each corner.

30. A motor boat that travels 8 mph in still water starts on one side of a river and travels downstream at an angle of 71° with the river bank. If the current of the river is 3 mph, how far does the boat travel in crossing the river?

31. A motor boat whose rate is 10 mph in still water must cross a river from point A on one side of the river to a point B directly opposite on the other side. If the current of the river is 4 mph, in what direction must the boat be steered? How long will it take to cross the river? (The river is 3 miles wide.)

QUIZ ON CHAPTER 44, SOLVING OBLIQUE TRIANGLES. FORM A.

In an oblique triangle, lettered ABC, with sides, a, b, and c, each side corresponding to the opposite angle, find the indicated values from the given values (or tell if there is no solution):

1. $A = 58°$; $B = 73°$; $a = 40$ cm. Find C, b, and c.

2. $A = 37°$; $B = 74°$; $c = 20$ cm. Find C and a.

3. $B = 43°$; $b = 15$ cm; $a = 24$ cm. Find A and C.

4. $A = 47°$; $a = 15$ cm; $c = 20$ cm. Find C (two answers).

5. $C = 115°$; $B = 42°$; $b = 30$ cm. Find A and c.

6. $A = 60°$; $b = 15$ cm; $c = 18$ cm. Find a.

7. $a = 10$ cm; $b = 13$ cm; $c = 15$ cm. Find A.

8. $B = 118°$; $a = 20$ cm; $c = 15$ cm. Find b.

CHAPTER FORTY-FIVE Vectors*

45-1 ADDITION: ARITHMETIC AND ALGEBRAIC

Quantities may be added in various ways. When we first begin the process of addition of numbers and quantities in the first grade, we learn that 4 plus 3 equals 7; $4 added to $3 are $7; 4 miles plus 3 miles equals 7 miles. In fact, all through grade school we never think of 4 added to 3 as being anything but 7. This is *arithmetic addition.*

When we get to the study of algebra and positive and negative numbers, we learn that a distance of 4 miles added to a distance of 3 miles may not always result in 7 miles. For instance, if we walk 4 miles *east* and then 3 miles *west,* we are only 1 mile from our starting point. Here the result of our walking has been to take us only 1 mile.

In algebra we show direction by a plus sign (+) and a minus sign (−). Walking 4 miles *east* may be indicated by +4, and 3 miles *west* may be indicated by −3. The addition of a + 4 to a − 3 results in +1. *Algebraic addition is the addition involving positive and negative numbers.*

Algebraic addition of signed numbers may be understood even more clearly by considering two forces operating at the *same time.* Let us suppose we can row a boat at a rate of 4 miles per hour (mph) in still water. In 1 hour we row a distance of 4 miles.

Now, suppose we row with this same effort in a river in which the water flows at a rate of 3 mph. If we row *downstream,* our rate of travel with reference to the shore will be 7 mph. In 1 hour we row 4 miles and the river carries us 3 miles in the *same direction.* The result of the two forces and motions is that we shall move 7 miles downstream. This may be expressed as (+4) plus (+3) equals +7.

Suppose we reverse our rowing and row upstream with the same force as before. That is, our rowing rate is still 4 mph. However, the force of the current has a downstream effect of 3 mph. The net result is that after 1 hour we shall have moved upstream only 1 mile. To add these two forces and motions, we use algebraic addition: $(+4) + (-3) = +1$.

* In electrical engineering the word *phasor* is used in place of *vector* because a difference in time between two magnitudes such as voltages represents a difference in phase rather than in direction.

So we see that adding 4 miles and 3 miles does not necessarily result in a total of 7 miles. To get the net result we must consider the direction of the motions.

45-2 VECTOR QUANTITIES

As we have seen, to describe some quantities completely it is necessary to state their *direction* as well as their *magnitude.* By *magnitude,* we mean the size or the amount of the quantity.

A *vector* is a quantity that involves both *magnitude* and *direction.* It may be a distance, velocity, force, or other quantity operating in a particular direction. For example, a distance of 4 miles can be described more completely by stating the direction as east. When we say an object moves with a velocity of 20 feet per second, we may need to add that the direction is north. To describe the effect of a force of 50 pounds, it may be necessary to state the direction in which the force operates.

To understand more clearly the meaning of a vector, let us look at some quantities that do not involve direction. Such quantities are called *scalar* quantities. Scalars contain *magnitude* only, such as 10 dollars, a dozen apples, 5 gallons, and so on. These quantities have nothing to do with direction. The abstract numbers themselves, such as 2, 6, 10, are also scalars.

45-3 REPRESENTATION OF VECTORS BY LINE SEGMENTS

A vector is conveniently represented by a line segment with an arrowhead at one end. The arrowhead indicates the *direction* of the vector. The length of the arrow represents the *magnitude.*

The arrow itself is often called the vector. To be strictly accurate, the vector is the force or velocity or distance itself and is only *represented* by the arrow. When we refer to the arrow as the vector, we should understand that the vector is the quantity itself, not the arrow.

Let us represent by two arrows the two vectors, *rowing downstream* at a rate of 4 mph, and the *river flowing* at a rate of 3 mph. We draw the arrows pointing in the same direction (Fig. 45-1). The lengths of the arrows represent the magnitudes, 4 and 3, respectively; that is, one line segment is 4 units long, the other is 3 units long.

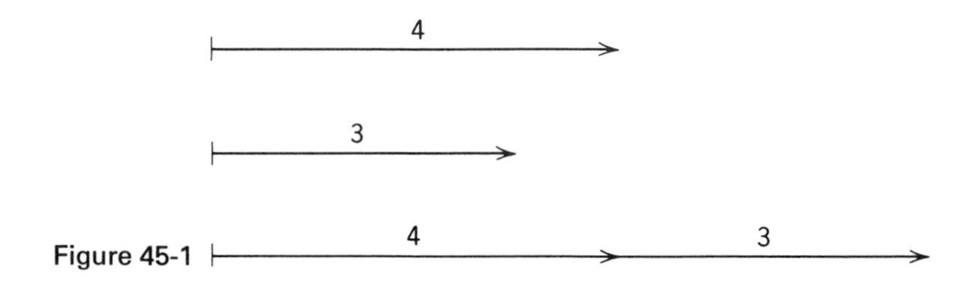

Figure 45-1

The addition of the two vectors may be represented by the sum of the two line segments. One arrow is drawn first. Then the second arrow is drawn from the head of the first, in the proper direction and of proper length. The sum of the arrows is 7 units from the starting point, which represents 7 mph in the proper direction.

Two vectors operating in *opposite directions*, such as 4 mph *upstream* and 3 mph *downstream*, may be represented by arrows pointing in opposite directions (Fig. 45-2). The sum of the two vectors by algebraic addition is $+1$, as we have seen. Algebraic addition may be shown

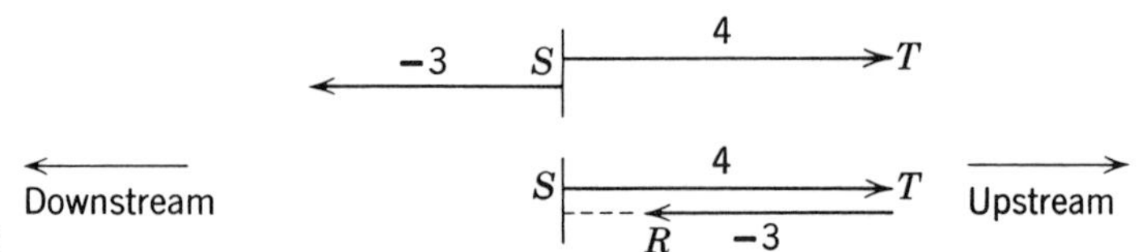

Figure 45-2

here also by combining the two arrows. We first draw the arrow, *ST*, in the proper direction from the starting point. Then, from the head of the first arrow, *T*, we draw the second arrow in the proper direction and of correct length.

The head of the second arrow, *R*, represents the algebraic sum of the two vectors. It is one unit from the starting point. The result can be shown by a third vector, 1 unit long, representing the sum. This vector is shown by a broken line.

There is one point that must be clearly understood with reference to the addition of vectors. If we are rowing upstream at a rate of 4 mph and the current has a force of 3 mph downstream, our actual motion is somewhat different from that indicated by the two arrows. The arrows appear to indicate that, while we are rowing, the river remains still and waits for us to move 4 miles upstream in 1 hour. Then, when we have arrived at that point, we wait for the river to carry us 3 miles downstream. This does not happen.

Instead, we have the two forces operating at the *same time.* As a result, our actual motion is only 1 mile upstream in 1 hour. Yet it is proper to show the two motions by arrows correctly drawn; and the *result* of the two motions will be that indicated in the sketch, that is, 1 mile upstream.

45-4 GEOMETRIC ADDITION

As a third situation, suppose we row directly *across* the river. Let us suppose the river is 4 miles wide and the water runs downstream at 3 mph. Here we must use a different kind of addition. If we row directly across the river at *right angles to the current,* in one hour we shall have crossed the river but the current will have carried us 3 miles downstream. Again we can represent graphically by arrows the two vector forces operating at right angles to each other (Fig. 45-3). These forces are indicated in the diagram by the line segments *ST* and *TR*.

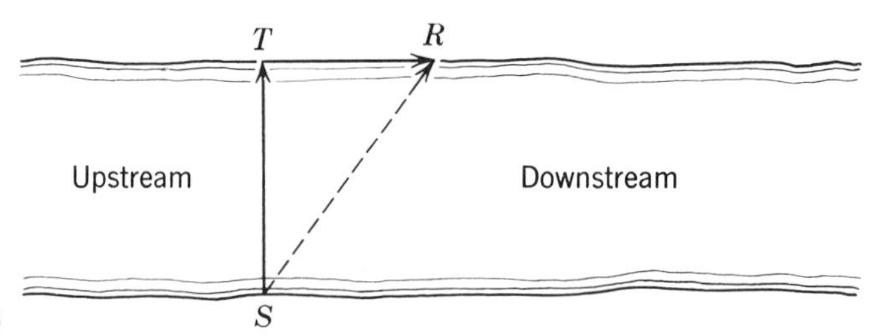

Figure 45-3

Our actual path of travel is represented by the arrow *SR*. This line segment represents our *direction* of travel and also the actual *distance* traveled.

Again, our actual travel is not exactly as indicated by the arrows, in the diagram. It is important to consider carefully just what happens in this case, since we have two forces operating at right angles to each other and at the *same time.* The arrows appear to indicate that the river waited one hour for us to go directly across and, after that, carried us downstream 3 miles. If this had happened, then the actual distance traveled would have been 7 miles. But such is not the case.

Our actual travel is represented by the diagonal line *SR*. By the Pythagorean rule, the length of this path is 5 miles. Here we cannot use arithmetic or algebraic addition. We must use *geometric* addition. The one vector, 4 miles in one direction, plus the second vector, 3 miles at right angles, equals a third vector, 5 miles in a diagonal direction. Again, we see that a vector has both magnitude and direction. Here, 4 miles + 3 miles = 5 miles.

Geometric addition is sometimes called *vectorial* addition. Geometric addition, or vectorial addition, can be done only by showing the vectors by arrows in a diagram.

45-5 RESULTANT AND COMPONENTS

A *resultant* vector is the vector that results from the addition of two or more vectors. In the diagram showing our path across the river we add the two vectors, 4 miles and 3 miles. The *resultant* is the sum of these two vectors. The *result* of the two forces is the *resultant* vector, 5 miles, in the *direction* indicated by the diagonal *SR* (Fig. 45-4).

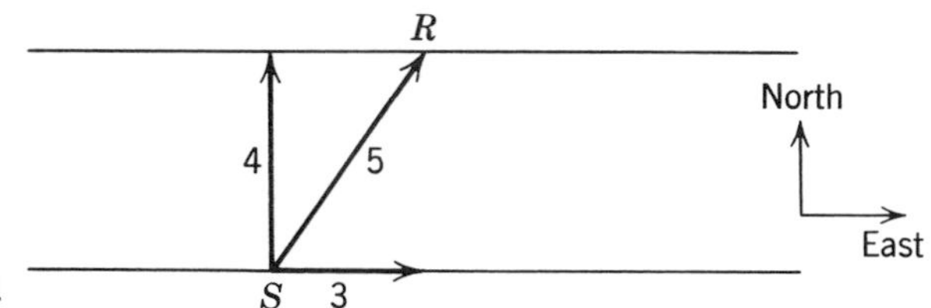

Figure 45-4

Component vectors are the vectors that are added to form the resultant. In the example given let us assume that the river runs *east* and our rowing is *north*. We can say that the two components are 4 miles *north* and 3 miles *east*. The *resultant* is 5 miles in a direction somewhat east of north. The *direction* can be indicated more accurately, as will be shown presently.

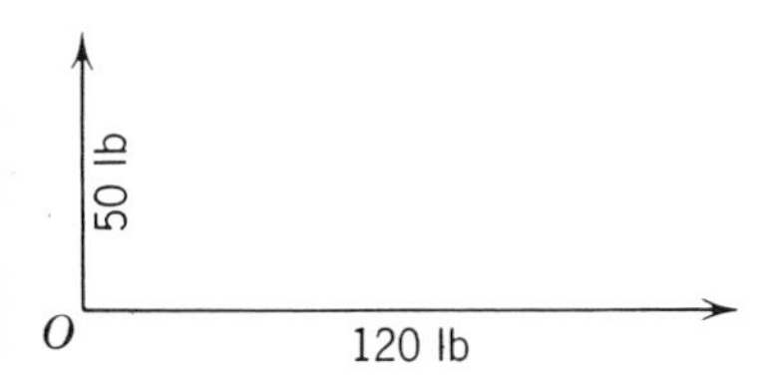

Figure 45-5

Let us consider another example. Suppose we have an object at point O (Fig. 45-5). Suppose that a force of, say, 140 pounds is required to move the object. Let us also suppose that a man pulls east with a force of 120 pounds and his son pulls north with a force of 50 pounds. The two forces, 120 pounds *east* and 50 pounds *north*, are the two component vectors. They contain magnitude and direction.

Our question now is, "Will the object move, and, if so, in what direction?" The resultant of the two forces, 120 pounds and 50 pounds, is not 170 pounds.

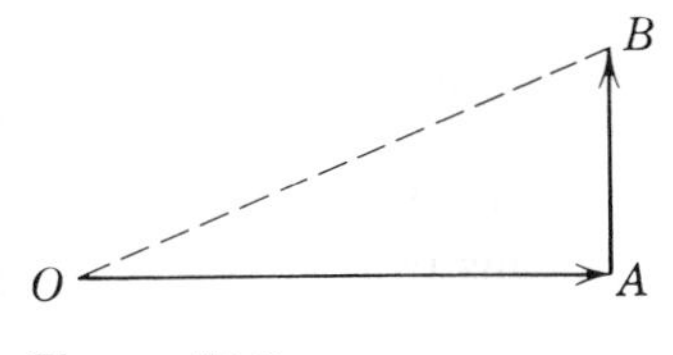

Figure 45-6

In order to determine the resultant of the two forces, we add them graphically, or vectorially. There are two methods by which the resultant may be found. We may draw one arrow first, representing the proper direction and distance from the point O. Then we draw the second arrow from the head of the first, observing proper direction and distance (Fig. 45-6).

By another method we draw each arrow in the proper direction and length from point O, such as OA and OB (Fig. 45-7). Then we complete a rectangle with these two line segments as sides. The diagonal OC of the rectangle represents the direction and magnitude of the resultant. By the Pythagorean rule, this diagonal is 130. The resultant force is therefore 130 pounds. The direction of the resultant force will also be indicated by the diagonal. The resultant shows that the object will not move.

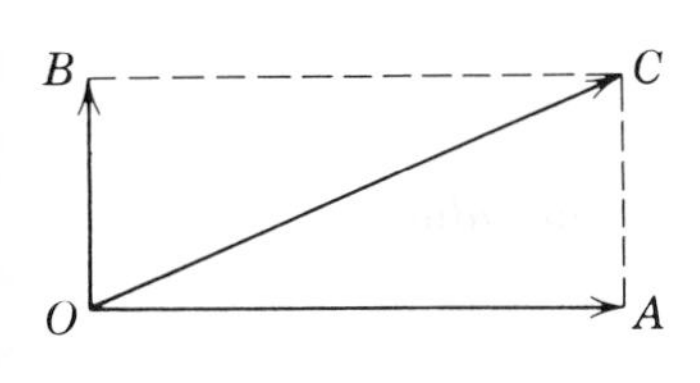

Figure 45-7

We can now state the direction as being 22.6° north of east. This angle is found by reducing its tangent $\frac{50}{120}$ to the decimal 0.4167. The angle corresponding to this tangent is 22.6°.

The complete resultant vector has a magnitude of 130 pounds in a direction 22.6° north of east. The most compact way of indicating this resultant vector is by use of this notation:

$$130\underline{/22.6^\circ}$$

45-6 VECTORS REPRESENTED ON THE COORDINATE SYSTEM

A set of component vectors and their resultant can be shown conveniently on the common system of rectangular coordinates. The two forces in the foregoing example are shown in Fig. 45-8.

Starting at the origin (0, 0) the 120-pound vector is shown by an arrow along the x-axis. The 50-pound vector is shown by an arrow along the positive y-axis. The two forces are thus represented by

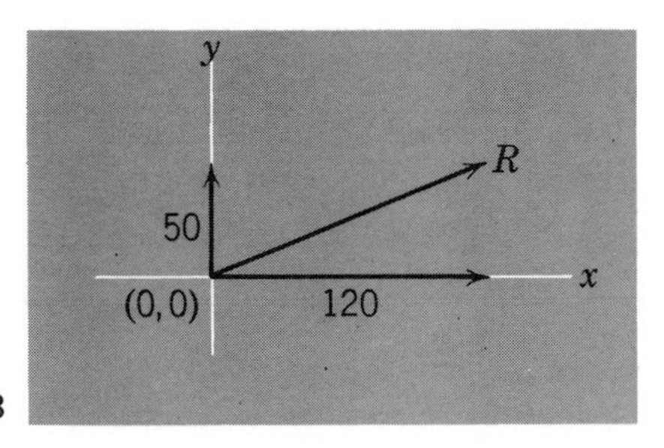

Figure 45-8

abscissa and ordinate, respectively. These two coordinates represent the component vectors. The vectorial sum of the two components is the resultant, denoted by R.

45-7 FINDING THE RESULTANT OF TWO COMPONENTS

Let us suppose we have two component vectors represented by the abscissa and ordinate, respectively. We may call one vector the x-component and the other vector the y-component. Our problem now is to find the resultant. Remember, resultant means two things: *direction* and *magnitude.*

Suppose we have an x-component equal to -8, and a y-component equal to $+6$, shown on the coordinate system (Fig. 45-9). By the Pythagorean rule, the *magnitude* of the resultant is 10. If we indicate the magnitude by the small letter r, we have the formula

$$r = \sqrt{x^2 + y^2}$$

The magnitude is sometimes called the *absolute value* of the vector.

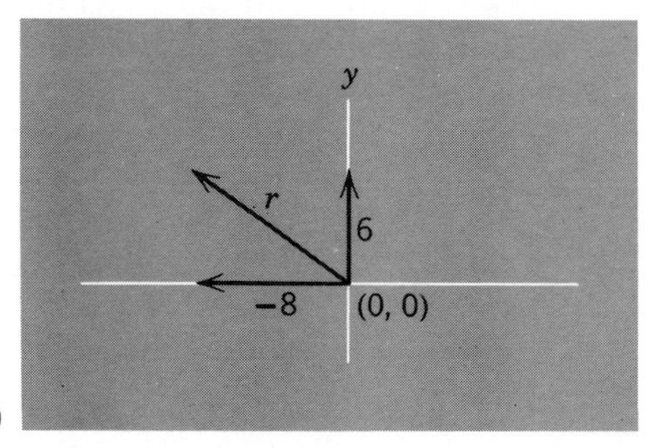

Figure 45-9

In order to state the resultant completely, we must also tell its *direction.* Using the tangent value $\frac{6}{8}$ and disregarding the negative sign, we find the *reference* angle is 36.9°. However, this is not the angle of the resultant. The angle of the resultant, which must be measured from the right-hand direction of the x-axis, is 143.1°. The entire resultant can be written in this compact form:

$$R: 10\underline{/143.1^\circ}$$

You will note that the angle of the resultant is arctan y/x.

45-8 FINDING THE COMPONENTS OF A GIVEN VECTOR Let us suppose that we have given a particular vector that has been obtained as the sum of two components. If we know the resultant vector, including its magnitude and direction, our problem is to find the two components that will produce the given resultant vector.

Example

Suppose we have the given resultant vector $60\underline{/20^\circ}$. Our problem is to find the components.

Solution

Let us represent this vector on the coordinate system (Fig. 45-10). The magnitude is 60 units, and the angle is 20°. The given vector is represented

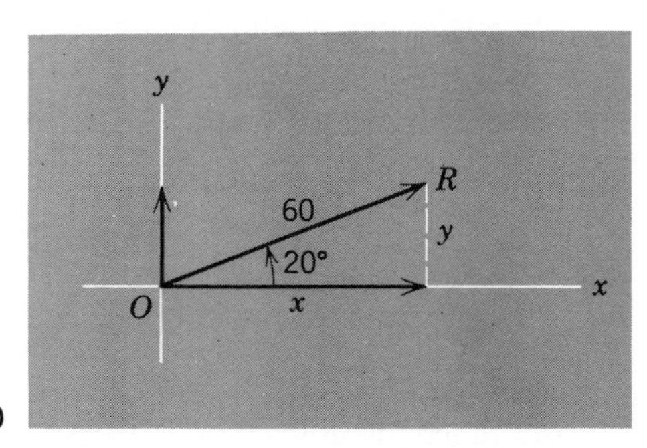

Figure 45-10

by the arrow OR. We denote the components by x and y, respectively. In the diagram we see that

$$\frac{y}{60} = \sin 20^\circ$$

Solving for y, we get

$$y = 60 \sin 20^\circ$$

$$y = 60(0.3420) = 20.52 \text{ (approx.)}$$

In the same way, we see that $\frac{x}{60} = \cos 20^\circ$

$$x = 60 \cos 20^\circ$$

$$x = 60(0.9397) = 56.38 \text{ (approx.)}$$

Now we have the components: $y = 20.52$; $x = 56.38$

To check the values for x and y, we can use the Pythagorean rule; that is, $x^2 + y^2$ should equal the square of 60. Since all the values are only approximate, the check will not show the two *exactly* equal.

The solution of this type of problem in more general terms may be shown by starting with a resultant stated in general terms (Fig. 45-11).

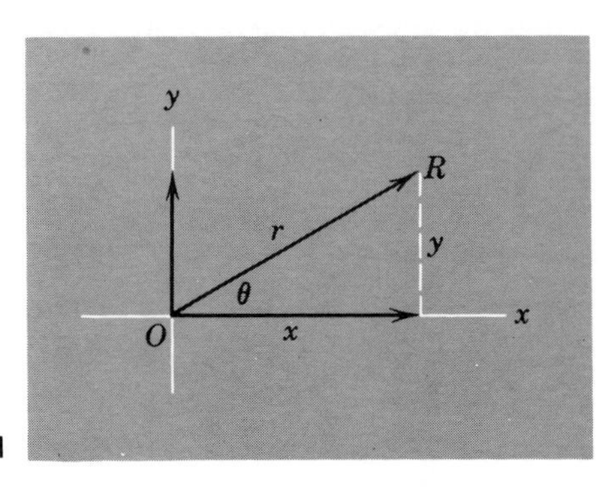

Figure 45-11

If we have the given resultant $r\angle\theta$, where r is the given distance and θ is the angle, our solution becomes

$$\frac{y}{r} = \sin\theta, \quad \text{from which we get} \quad \boxed{y = r\sin\theta}$$

$$\frac{x}{r} = \cos\theta, \quad \text{from which we get} \quad \boxed{x = r\cos\theta}$$

The two formulas, $y = r\sin\theta$ and $x = r\cos\theta$, can be used to find the components of any given vector, regardless of its magnitude or direction. We must be sure to observe the correct algebraic signs for the functions used.

Exercise 45-1

Find the resultant for each of the following sets of components. (Resultant includes both magnitude and direction.)

1. $x = 8$, $y = 5$

2. $x = -15$ $y = 11$

3. $x = -12$, $y = -7$

4. $x = 5.2$ v, $y = -2.1$ v

5. $x = -3.21$ v, $y = 5.32$ v

6. $x = -23.2$, $y = -4.13$

7. $x = 1.24$, $y = -5.32$

8. $x = 2.45$, $y = 5.13$

9. $x = 15$ ohms, $y = 5$ ohms

10. $x = 0.5$ ohms, $y = 50$ ohms

11. $x = 7.21, \quad y = 1.32$ **12.** $x = -32.4, \quad y = 124.0$

13. $x = 8.41, \quad y = -0.92$ **14.** $x = 5.31, \quad y = -5.12$

15. $x = 42.3, \quad y = -1.42$ **16.** $x = -2.46, \quad y = -9.40$

Find the components of each of the following:

17. $80\underline{/28.3°}$ **18.** $32\text{v}\underline{/169.8°}$

19. $50\text{v}\underline{/74.2°}$ **20.** $4.2 \text{ ohms}\underline{/-38.4°}$

21. $6 \text{ amp}\underline{/148.8°}$ **22.** $23.4 \text{ lb}\underline{/58.1°}$

23. $150 \text{ v}\underline{/204.9°}$ **24.** $385 \text{ miles}\underline{/208.3°}$

25. $70 \text{ ohms}\underline{/123.4°}$ **26.** $57.4 \text{ mph}\underline{/152.8°}$

27. $12\underline{/235.3°}$ **28.** $42.6\text{v}\underline{/493°}$

29. $15\underline{/287.6°}$ **30.** $34.7\underline{/1432.6°}$

31. $25\underline{/300.7°}$ **32.** $1.65\underline{/73.2°}$

33. If a plane is set to fly at 230 mph directly east and a wind is blowing from the north at 25 mph, find the direction of travel and the distance traveled in 1 hr.

34. If two forces of 360 and 155 lb, respectively, are exerted at right angles to each other, find the magnitude and direction of the resultant force. Show by diagram.

35. A motor boat is steered directly across a river at 12 mph. The speed of the river is 3.5 mph. What is the actual speed and direction of the boat through the water? Show by diagram.

36. What single force and direction would have the same effect as a pull of 25 lb east and 15 lb north?

37. A man pulls with a force of 160 lb in a direction of 41° north of east. What are the east and north components of the force?

38. A plane travels 220 miles in a direction south of east by 31°. What are the east and south components of the distance traveled?

39. A plane is headed due east; the wind is blowing from the north. At 12 noon the plane passes over town *A*, and at 1 P.M. it passes over town *B*. The towns are 295 miles apart. Town *B* is 10° south of east from *A*. Find the speed of the plane and the speed of the wind.

40. A pull of 52 lb is exerted in a direction 64° south of west. Find the west and south components of the force.

45-9 GRAPHICAL ADDITION OF ANY TWO OR MORE VECTORS

It is often necessary to add vectors of any direction and magnitude. For instance, suppose we have the two vectors

$$A:50\underline{/23.4^\circ} \qquad \text{and} \qquad B:20\underline{/71.2^\circ}$$

These vectors can be represented graphically (Fig. 45-12)

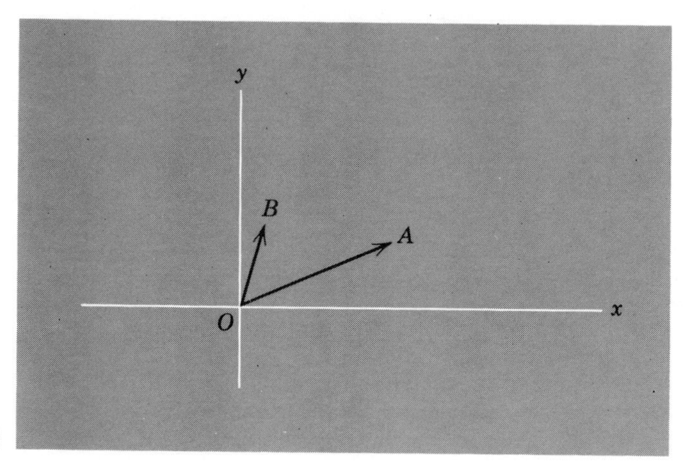

Figure 45-12

If vector A is a force of 50 pounds in a direction indicated by 23.4° and vector B is a force of 20 pounds in a direction indicated by 71.2°, our problem then might be stated thus: what are the magnitude and the direction of the resultant force?

Graphically, we can add the vectors by completing a parallelogram on the two arrows, OA and OB (Fig. 45-13). The resultant vector is represented by the diagonal arrow, OR. This arrow shows the direction

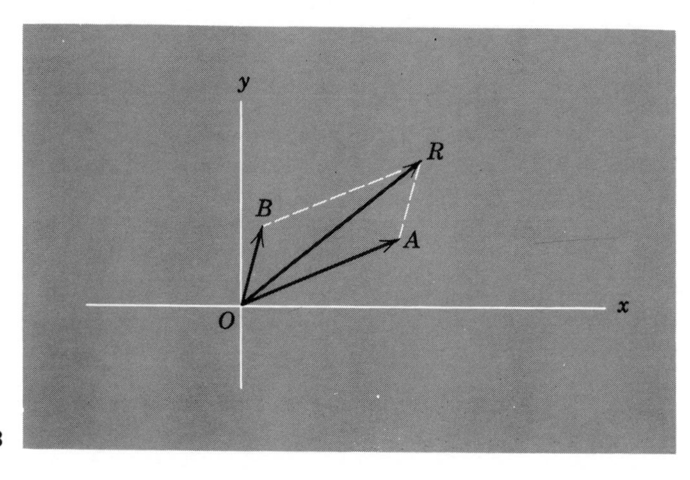

Figure 45-13

and magnitude of the resultant. If we make careful measurements of line segments and angles, we can also measure very closely the magnitude and direction of the resultant. This method is called *graphical* or *geometric* addition of vectors.

45-10 ADDITION OF VECTORS BY ALGEBRA AND TRIGONOMETRY

If we wish to obtain a more accurate measure of the sum of the two vectors, we add them by means of algebra and trigonometry. To add two or more vectors of any magnitude and direction, we follow this general procedure:

1. *Find the x-component and the y-component of each vector.*
2. *Add the x-components algebraically. The result will be the x-component of the resultant.*
3. *Add the y-components algebraically. The result will be the y-component resultant.*
4. *Combine the two components of the resultant by vectoral addition. Show the result as a magnitude and direction.*

Example

Add the vectors $A: 50\underline{/23.4^\circ}$ and $B: 20\underline{/71.2^\circ}$.

Solution

First we find the components of each vector. Let us indicate the components of A by x_1 and y_1 and the components of B by x_2 and y_2. By formula, $x = r\cos\theta$ and $y = r\sin\theta$. Then

$$\begin{aligned} x_1 &= 50\cos 23.4^\circ & y_1 &= 50\sin 23.4^\circ \\ &= 50(0.9178) & &= 50(0.3971) \\ &= 45.890 & &= 19.855 \\ x_2 &= 20\cos 71.2^\circ & y_2 &= 20\sin 71.2^\circ \\ &= 20(0.3223) & &= 20\ (0.9466) \\ &= 6.446 & &= 18.932 \end{aligned}$$

The x-components are now added algebraically to form the x-component of the resultant. The y-components are added to form the y-component of the resultant. The subscript R indicates resultant.

$$\begin{aligned} x_1 &= 45.890 & y_1 &= 19.855 \\ x_2 &= \underline{\ \ 6.446} & y_2 &= \underline{18.932} \\ x_R &= 52.336, \text{ or } 52.34 & y_R &= 38.787, \text{ or } 38.79 \end{aligned}$$

We now find the magnitude of the resultant vector by the Pythagorean rule. By squaring the components, adding these squares, and finding the square root of the sum, we find that the magnitude of the resultant is

$$r = \sqrt{(52.34)^2 + (38.79)^2} = 65.15$$

The angle of the resultant is arctan 38.79/52.34 = 36.5°. The resultant is written

$$65.15\underline{/36.5^\circ}$$

45-11 VECTORS AS COMPLEX NUMBERS: OPERATOR-*j*

The components of a vector may be written in the form of a complex number: $a + bj$. The x-component becomes the real part of the complex number, and the y-component becomes the coefficient of j. The letter j is often placed before its coefficient, as $j5$. For instance, if the x-component of a vector is 17 and the y-component is -8, they may be written $17 - j8$. This form is known as "operator-j" notation. Such notation is common in the study of electric circuits.

If the components of vectors are written in operator-j notation, their addition becomes simply the addition of complex numbers (see Section 19-8). The procedure is shown in the following example.

Example 1

Add the vectors A: $40\underline{/68.8^\circ}$ and $60\underline{/205.7^\circ}$.

Solution

We indicate the components by subscripts.

$$\begin{aligned} x_1 &= 40 \cos 68.8^\circ & y_1 &= 40 \sin 68.8^\circ \\ &= 40\ (0.3616) & &= 40\ (0.9323) \\ &= 14.464 & &= 37.292 \\ x_2 &= 60 \cos 205.7^\circ & y_2 &= 60 \sin 205.7^\circ \\ &= 60\ (-0.9011) & &= 60\ (-0.4337) \\ &= -54.066 & &= -26.022 \end{aligned}$$

We now write these vectors as complex numbers.

$$\begin{aligned} A&:\ \ \ 14.464 + j37.292 \\ B&:\ -54.066 - j26.022 \\ \text{Resultant,}\quad R&:\ -39.602 + j11.270 \end{aligned}$$

The magnitude and direction of the resultant are now computed.

$$r = \sqrt{(-39.602)^2 + (11.270)^2} = 41.17$$

The angle of the resultant is arctan $11.270/-39.602$. Disregarding the negative sign, we find the reference angle is 15.9°. Since the resultant appears in the second quadrant, the angle of the resultant is 164.1°. The resultant is written

$$41.17\underline{/164.1^\circ}$$

Example 2

Add the following four vectors:

$$A: 80\underline{/24.4^\circ} \qquad B: 50\underline{/131.4^\circ} \qquad C: 70\underline{/233.8^\circ} \qquad D: 60\underline{/317.9^\circ}$$

Solution

$$
\begin{array}{llll}
A: x_1 = 80 \cos 24.4^\circ = 72.856 & & y_1 = 80 \sin 24.4^\circ = 33.048 \\
B: x_2 = 50 \cos 131.4^\circ = -33.065 & & y_2 = 50 \sin 131.4^\circ = 37.505 \\
C: x_3 = 70 \cos 233.8^\circ = -41.342 & & y_3 = 70 \sin 233.8^\circ = -56.490 \\
D: x_4 = \quad \cos 317.9^\circ = 44.520 & & y_4 = 60 \sin 317.9^\circ = -40.224
\end{array}
$$

The vectors are now written in operator-j notation and added.

$$
\begin{array}{ll}
A: & 72.856 + j33.048 \\
B: & -33.065 + j37.505 \\
C: & -41.342 - j56.490 \\
D: & 44.520 - j40.224 \\
\hline
R: & 42.969 - j26.161
\end{array}
$$

We compute the magnitude and direction from the components of the resultant. By the Pythagorean rule, the magnitude is equal to $\sqrt{x^2 + y^2}$.

$$r = \sqrt{(42.97)^2 + (-26.16)^2} = 50.31$$

To find the angle of the resultant, we first find the reference angle, which is arctan 26.16/42.97. In finding the reference angle, we disregard the negative sign. The reference angle is 31.3°. Since the resultant falls in the fourth quadrant, the angle of the resultant is 328.7°.

In the case of a vector falling in the fourth quadrant, the angle is sometimes written as a negative angle. This should be done only in the case of a fourth-quadrant angle. The answer to the foregoing problem is

$$R: 50.31\angle 328.7^\circ \qquad \text{or} \qquad 50.31\angle -31.3$$

Exercise 45-2

Add the following sets of vectors:

1. $A: 80\angle 59.2^\circ$ $\quad B: 60\angle 194.3^\circ$
2. $A: 20\angle 14.2^\circ$ $\quad B: 75\angle 231.6^\circ$
3. $A: 70\angle 31.8^\circ$ $\quad B: 40\angle 153.7^\circ$
4. $A: 20\angle 143.3^\circ$ $\quad B: 30\angle 257.4^\circ$
5. $A: 25\angle 51.3^\circ$ $\quad B: 35\angle 26.8^\circ$ $\quad C: 50\angle 165.2^\circ$
6. $A: 34\angle 68.7^\circ$ $\quad B: 42\angle 243.8^\circ$ $\quad C: 55\angle 327.6^\circ$
7. $A: 45\angle 15.3^\circ$ $\quad B: 23\angle 147.2^\circ$ $\quad C: 67\angle 235^\circ$ $\quad D: 54\angle 312^\circ$
8. $A: 22.5\angle 57.3^\circ$ $\quad B: 31.2\angle 114^\circ$ $\quad C: 43/218^\circ$ $\quad D: 22.1\angle -13.2^\circ$
9. $A: 24\angle 26.3^\circ$ $\quad B: 36\angle 116.7^\circ$ $\quad C: 52\angle 246^\circ$ $\quad D: 23\angle -23.6^\circ$

10. $A: 30\angle 16.2^\circ$ $B: 50\angle 72.6^\circ$ $C: 40\angle 131.7^\circ$
    $D: 60\angle 212^\circ$ $E: 10\angle 293.1^\circ$

11. $A: 32\angle 0^\circ$ $B: 14.3\angle 42.3^\circ$ $C: 38\angle 261^\circ$

12. $A: 65\angle 90^\circ$ $B: 85\angle 207.3^\circ$ $C: 72\angle -68^\circ$

13. $A: 52\angle 35.2^\circ$ $B: 73\angle 180^\circ$ $C: 54\angle 297^\circ$

14. $A: 23\angle 71.4^\circ$ $B: 31\angle 158.2^\circ$ $C: 21\angle 270^\circ$

15. $A: 36\angle 0^\circ$ $B: 42\angle 90^\circ$ $C: 27\angle -90^\circ$

16. Find the resultant of two forces, one of 300 lb at an angle of 20° north of east, and another of 200 lb at 70° north of east.

17. Find the result of two forces, one of 50 lb at 80° north of east, and the other force of 70 lb at an angle of 40° north of west.

18. A plane flies from its home port for 150 miles at 15° north of east, and then 180 miles in a direction of 65° north of west. Find the distance and the direction from the home port.

19. Find the resultant of two forces, one of 80 lb acting in a direction 75° north of east, and the other of 140 lb in a direction 50° south of west.

20. A force of 300 lb is applied to an object in a direction 20° north of east. At what angle must a force of 240 lb be applied so that the resultant force will be directly east?

21. The resultant of two forces acting on an object is a force of 40 lb in a direction of 70°. One force is 50 lb at an angle of 20° north of east. Find the other component force.

22. Three forces are acting on an object. Two of the forces are 30 lb at an angle of 40° north of east, and 50 lb at an angle of 30° north of west. What force is required at an angle of 10° west of south in order that the resultant force be zero?

45-12 POLAR COORDINATES

To locate a point in the plane we have used the *rectangular system* consisting of the x and y axes drawn at right angles. Then we locate the point by two numbers showing the distances from the axes. For example, the point (8, 6) has the coordinates, $x = 8$ and $y = 6$ (Fig. 45-14).

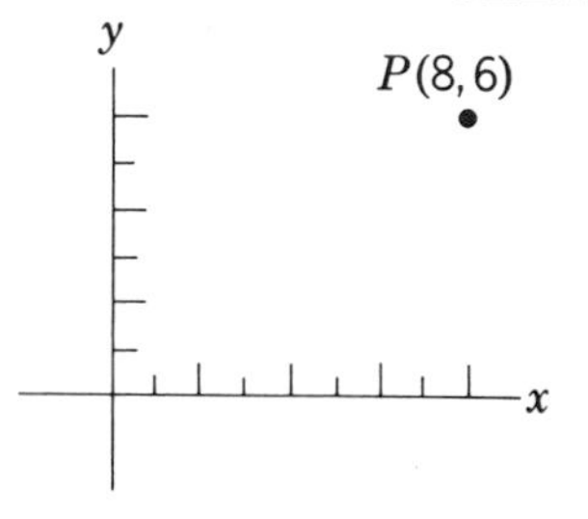

Figure 45-14

A point may be located by using what is called the *polar coordinate* system. In this system we locate the point by stating its *distance* from a fixed point called the *pole*, and its *direction* with reference to a fixed line called the *polar axis* (Fig. 45-15).

In going from the origin to the point (8, 6) by rectangular coordinates, we must travel 14 units. However, if we use polar coordinates,

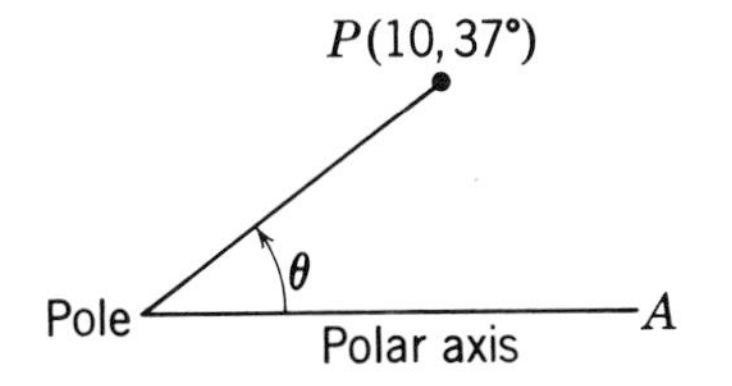

Figure 45-15

we first turn to a direction 37° from the polar axis, and then travel only 10 units to get to the point *P*.

The two numbers, distance 10 units and angle of 37°, are also written in parentheses: (10, 37°). These numbers are the *polar coordinates* of the point. *This is exactly what is implied in writing a vector as*: $10\underline{/37^\circ}$. For this reason, the expression, $10\underline{/37^\circ}$, is called the *polar form* of a vector.

In general, the location of a point in polar coordinates is denoted by (r, θ), where θ indicates the angle and r indicates the distance from the pole.

45-13 TRIGONOMETRIC FORM OF A COMPLEX NUMBER

We have seen that a vector can be written as a complex number in the rectangular form, $a + jb$. For example, the vector with components, $x = 4$ and $y = 3$, is written $4 + j3$. If we take the components, x and y, of a vector in general terms, then the complex number becomes: $x + jy$.

We have also seen that the x and y components of the vector, $r\underline{/\theta}$, can be found by the formulas:

$$x = r \cos \theta \qquad \text{and} \qquad y = r \sin \theta$$

Now using the quantities, $(r \cos \theta)$ and $(r \sin \theta)$, as components, we can write the complex number as

$$r \cos \theta + jr \sin \theta$$

or in factored form,

$$r(\cos \theta + j \sin \theta)$$

This is called the *trigonometric form of a complex number.* Note that this is still in rectangular form since the components are shown. The magnitude (or absolute value) of the vector is r, and the angle (or amplitude) is θ. The values of r and θ are the same as in the polar form, $r\underline{/\theta}$.

Example 1

Given the components of a vector, $x = 1$ and $y = \sqrt{3}$, write the complex number in rectangular form; then find the polar and trigonometric forms.

Solution

In the algebraic form, we have

$$1 + j\sqrt{3}$$

Now, using the formulas, $r = \sqrt{x^2 + y^2}$, and $\theta = \arctan y/x$, we get

$$r = \sqrt{1 + 3} = 2 \text{ and } \theta = \arctan \sqrt{3} = 60^\circ$$

Now that we know r and θ, we can write the polar form, $2\underline{/60^\circ}$. In the trigonometric form, we have $2(\cos 60^\circ + j \sin 60^\circ)$.

Example 2

Given the complex number, $10(\cos 20^\circ + j \sin 20^\circ)$, change to polar form and to algebraic form.

Solution

The given form is equivalent to the polar form, $10\underline{/20^\circ}$, since $r = 10$ and the angle $\theta = 20^\circ$. For the components, we get

$$x = 10 \cos 20^\circ = 9.397; \qquad y = 10 \sin 20^\circ = 3.420$$

With these components we can now write the algebraic form: $9.397 + j3.420$

Example 3

Given the polar form of a vector: $20\underline{/150^\circ}$; change to the trigonometric form and to the algebraic form.

Solution

Since r and θ are known, we can immediately write the trigonometric form:

$$20(\cos 150^\circ + j \sin 150^\circ)$$

For the algebraic form, we find the components by simply expanding the trigonometric form:

$$20\left(-\frac{\sqrt{3}}{2} + j0.5\right) = -10\sqrt{3} + j10$$

45-14 MULTIPLICATION OF VECTORS

We have seen that the addition of vectors is rather simple when they are expressed in rectangular form as complex numbers. For example, suppose we have the two vectors, $A = 3 + j2$ and $B = 1 + j4$. Then the sum, $A + B$, is $4 + j6$.

We have also seen that vectors may be added graphically when they are expressed as complex numbers. We plot each number on the graph, draw the arrows representing the vectors, and then complete a parallelogram on the two arrows.

However, there is no simple way to multiply vectors graphically. We can, of course, multiply the two complex numbers algebraically and then plot the point that represents their product. If we have the two vectors, $A = 3 + j2$ and $B = 1 + j4$, then by algebraic multiplication, we get the product: $AB = (3 + j2)(1 + j4) = -5 + j14$. The two vectors and the point that represents their product AB are shown on the graph (Fig. 45-16). The magnitude of the product is $\sqrt{221}$. The

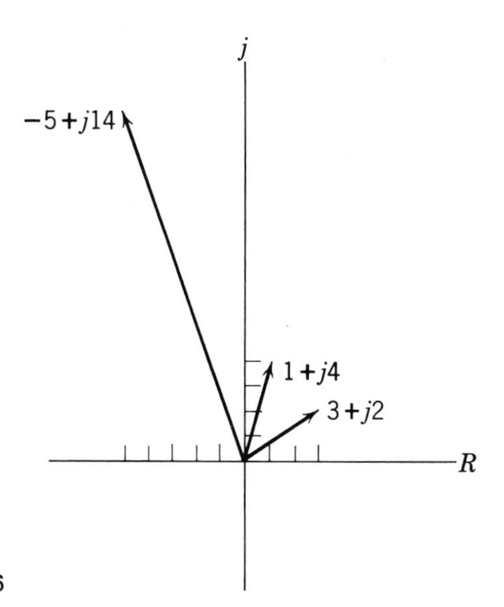

Figure 45-16

angle of the product is arctan $(-\frac{14}{5})$, which is approximately 109.7°.

The multiplication of two vectors is comparatively simple when the vectors are expressed in polar form. We now state the rule, and then later we shall show why it is true.

Rule. *For the product of two vectors in polar form, we multiply the magnitudes and add the angles.* Let us apply the rule to the foregoing vectors A and B.

For A, the magnitude is $\sqrt{13}$, and the angle is approximately 33.7°.

For B, the magnitude is $\sqrt{17}$, and the angle is approximately 76°. The product of the magnitudes is $(\sqrt{13})(\sqrt{17}) = \sqrt{221}$. This is exactly the magnitude of the product of the vectors as found by algebraic multiplication.

The sum of the angles is $33.7° + 76° = 109.7°$. Although these are approximate values of the angles, we can also show that the angle of the product is exactly correct. Calling the angles θ and ϕ for the vectors A and B, respectively, we have

$$\tan\theta = \frac{2}{3} \qquad \text{and} \qquad \tan\phi = 4$$

Now, using the formula for the tangent of the sum of two angles, we get

$$\tan(\theta + \phi) = \frac{\tan\theta + \tan\phi}{1 - (\tan\theta)(\tan\phi)} = \frac{\frac{2}{3} + 4}{1 - (\frac{2}{3})(4)} = \frac{2 + 12}{3 - 8} = -\frac{14}{5}$$

The result is exactly the tangent of the product, AB, or $-5 + j14$. That is, the angle of the product is $\theta + \phi$, the sum of the angles of the vectors.

In general terms, for vectors in polar form, $(r\underline{/\theta})(s\underline{/\phi}) = rs\underline{/\theta + \phi}$. The rule applies equally to the trigonometric form:

$$[r(\cos\ \theta + j\ \sin\ \theta)][s(\cos\ \phi + j\ \sin\ \phi)] = (r)(s)[\cos\ (\theta + \phi) + j\ \sin\ (\theta + \phi)]$$

Rule. *The product of two vectors in trigonometric form has a magnitude equal to the product of the magnitudes and an angle equal to the sum of the angles of the vectors.*

Now let us see why this rule is true in general terms, using two general complex numbers: $r(\cos A + j \sin A)$ and $s(\cos B + j \sin B)$. We write the product and expand and simplify the result:

$$\begin{aligned}
&[r(\cos A + j \sin A)][s(\cos B + j \sin B)] \\
&= (r)(s)(\cos A\ j \sin A)(\cos B + j \sin B) \\
&= rs(\cos A \cos B + j \sin A \cos B\ j \cos A \sin B + j^2 \sin A \sin B) \\
&= rs[\cos A \cos B - \sin A \sin B + j(\sin A \cos B + \cos A \sin B)] \\
&= rs[\qquad \cos (A + B) \qquad + j \qquad \sin (A + B) \qquad] \\
&= rs[\cos (A + B) + j\ \sin\ (A + B)]
\end{aligned}$$

Note that the magnitude, rs, of the product is the product of the magnitudes of the vectors. The angle of the product is equal to $A + B$, the sum of the angles of the vectors.

The proof of the rule may be shown also by taking the two general complex numbers: $A = a + jb$ and $B = c + jd$. For the vector A, the magnitude is $\sqrt{a^2 + b^2}$ and the angle is arctan b/a. For the vector B, the magnitude is $\sqrt{c^2 + d^2}$ and the angle is arctan d/c. If we multiply the two complex numbers, we get a product of $(ac - bd) + j(bc + ad)$. Using these values, we can show that the magnitude of the product is equal to the product of the magnitudes, and the angle of the product is equal to the sum of the angles of the vectors. The proof is left to the student.

Example 1

Multiply: $(8\underline{/50°})(5\underline{/70°}) = 40\underline{/120°}$

Example 2

$[4(\cos 60° + j \sin 60°)][7(\cos 80° + j \sin 80°)] = 28(\cos 140° + j \sin 140°)$.

45-15 DIVISION OF VECTORS

Since division is the reverse of multiplication, we have the rule: *To divide two vectors in polar form, divide the magnitudes and subtract the angles.*

$$\text{In general, } (r\angle\theta) \div (s\angle\phi) = \frac{r}{s}\angle\theta - \phi$$

Examples

(a) $(24\angle 210^\circ) \div (6\angle 75^\circ) = 4\angle 135^\circ$, (b) $(60\angle 40^\circ) \div (12\angle 70^\circ) = 5\angle -30^\circ$,
(c) $[10(\cos 80^\circ + j\sin 80^\circ)] \div [2(\cos 15^\circ + j\sin 15^\circ)] = 5(\cos 65^\circ + j\sin 65^\circ)$.

45-16 A POWER OF A VECTOR IN POLAR FORM

The rule for finding a power of a vector in polar or trigonometric form comes directly from the rule for a product. For the product $(6/50^\circ)(6\angle 50^\circ) = 36\angle 100^\circ$, we can say $(6\angle 50^\circ)^2 = 36\angle 100^\circ$.

Rule. *To find a power of a vector in polar or trigonometric form, raise the magnitude to the indicated power and multiply the angle by the exponent.* In general,

$$(r\angle\theta)^n = r^n\angle n\theta.$$

Examples

(a) $(4\angle 150^\circ)^2 = 16\angle 300^\circ$,
(b) $(2\angle 30^\circ)^6 = 64\angle 180^\circ$.
In trigonometric form, the rule becomes

$$[r(\cos\theta + j\sin\theta)]^n = r^n(\cos n\theta + j\sin n\theta)$$

This rule is known as DeMoivre's theorem. As an example

$$[5(\cos 30^\circ + j\sin 30^\circ)]^3 = 125(\cos 90^\circ + j\sin 90^\circ)$$

45-17 FINDING THE ROOT OF A VECTOR IN POLAR FORM

To find a root of a vector in polar form, we can call the root a fractional power. Then we use the rule for a power. For example,

$$\sqrt{36\angle 100^\circ} \text{ can be written } (36\angle 100^\circ)^{\frac{1}{2}}$$

By the power rule, we get

$$(36\angle 100^\circ)^{\frac{1}{2}} = 36^{\frac{1}{2}}\angle(\tfrac{1}{2})(100^\circ) = 6\angle 50^\circ$$

Stated in another way: *To find a root of a vector in polar form, we first find the root of the magnitude. The result is the magnitude of the root. Then, for the angle of the root, we divide the original angle by the index of the root.* For example, to find the cube root of $125\underline{/90^\circ}$, we have

$$\sqrt[3]{125\underline{/90^\circ}} = \sqrt[3]{125}\underline{\bigg/\frac{90^\circ}{3}} = 5\underline{/30^\circ}$$

By the power rule, we get

$$125^{\frac{1}{3}}\underline{/(\tfrac{1}{3})(90^\circ)} = 5\underline{/30^\circ}$$

Actually, there are two other cube roots of $125\underline{/90^\circ}$. To find the angle for one other root, we add 360° to the given angle and then divide by 3. We get (450°) ÷ 3 = 150°. For the third cube root, we add another 360° to the original angle and get (810°) ÷ 3 = 270°. The magnitude is 5 for each root. Then the three cube roots of $125\underline{/90^\circ}$ are

$$r_1 = 5\underline{/30^\circ}; \qquad r_2 = 5\underline{/150^\circ}; \qquad r_3 = 5\underline{/270^\circ}$$

If we plot the points representing these three cube roots, we shall find that they are evenly spaced, separated by 120°, on a circle whose radius is 5 (Fig. 45-17). Note that in finding the three cube roots, we begin with the given angle of the vector, and then add 360° twice.

In finding the cube roots of any vector, we first find the cube root of the magnitude. To find the angle of the root, we begin with the given angle and divide it by 3 for the first root. We then add 360° twice, dividing by 3 after each addition of 360°. To find the angles of the fourth roots of a vector, we begin with the given angle of the vector and divide it by 4. We then add 360° three times, dividing by 4 after each addition of 360°.

In all cases in finding all the indicated roots of a given vector, we begin with the given angle and divide it by the index of the root for the first root of the vector. We then add 360° to the original angle a

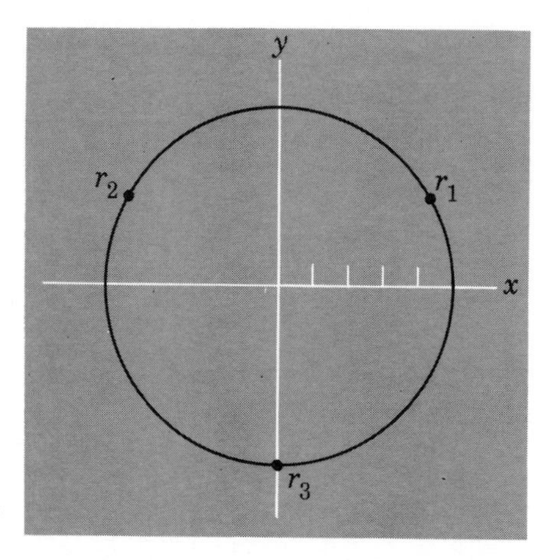

Figure 45-17

number of times that is one less than the index of the root. If we add 360° too many times, the roots will simply be repeated. For example, for a cube root of $125\underline{/90^\circ}$, if we add 360° three times, we get the root $5\underline{/390^\circ}$, which is the same as r_1. The rule is often stated as follows in general terms:

$$\sqrt[n]{r/\theta} = \sqrt[n]{r}\,\underline{\Big/\frac{\theta + 2\pi k}{n}}, \text{ where } k = 0, 1, 2, \ldots \text{ to } k = n - 1$$

If any complex number in rectangular form is changed to polar form, the operations involving multiplication, division, powers, and roots are usually much simpler.

Example 1

Find the following power: $(-1 - j\sqrt{3})^3$.

Solution

To find the polar form, we find r by the formula, $r = \sqrt{x^2 + y^2}$. In this example, $r = \sqrt{1 + 3} = 2$. Angle $\theta = \arctan \sqrt{3}$. However, θ is a third quadrant angle and is therefore 240°. Then we have the polar form:

$$2\underline{/240^\circ}$$

For the third power, we get

$$(2\underline{/240^\circ})^3 = 8\underline{/720^\circ}$$

In trigonometric form, we have

$$8(\cos 720^\circ + j \sin 720) = 8 + j0$$

Now, if we were to expand the power in algebraic form, we should get 8.

Example 2

Find the cube roots of -8.

Solution

The number -8 represents the polar form, $8\underline{/180^\circ}$, or in trig. form,

$$8(\cos\ 180^\circ + j \sin 180^\circ)$$

Then for the cube roots, we have the magnitude 2. For one root we take $\frac{1}{3}$ of 180°. Adding 360° twice and dividing by 3 each time, we get the three roots:

$$r_1 = 2\underline{/60^\circ}; \qquad r_2 = 2\underline{/180^\circ}; \qquad r_3 = 2\underline{/300^\circ}$$

In complex form, the three roots are:

$$r_1 = 1 + j\sqrt{3}; \qquad r_2 = -2 + j0; \qquad r_3 = 1 - j\sqrt{3}$$

Example 3

Find the cube roots of $j8$.

Solution

The number $j8$ can be written: $8\underline{/90^\circ}$, or in trigonometric form,

$$8(\cos 90^\circ + j \sin 90^\circ)$$

Then for the three cube roots of $j8$, we get:

$$2\underline{/30^\circ}; \qquad 2\underline{/150^\circ}; \qquad 2\underline{/270}$$

These can also be written in the form of complex numbers.

Exercise 45-3

Write the polar and trigonometric forms of vectors with the following components:

1. $x = 12, y = 5$ **2.** $x = -15, y = 8$ **3.** $x = -\sqrt{3}, y = 1$

Write the following vectors in polar and algebraic form:

4. $20(\cos 135^\circ + j \sin 135^\circ)$ **5.** $15(\cos 210^\circ + j \sin 210^\circ)$

Multiply as indicated; write the result in algebraic form:

6. $(8\underline{/60^\circ})(5\underline{/30^\circ})$

7. $[15(\cos 50^\circ + j \sin 50^\circ)][4(\cos 40^\circ + j \sin 40^\circ \;]$

Multiply algebriacally and then change each to polar form and multiply:

8. $(4 + j4\sqrt{3})(-2\sqrt{3} - j2)$ **9.** $(3 + j4)(5 + j12)$

Divide as indicated:

10. $(40\underline{/60^\circ}) \div (5\underline{/20^\circ})$ **11.** $(30\underline{/20^\circ}) \div (3\underline{/80^\circ})$

12. $(10\underline{/270^\circ}) \div (2\underline{/90^\circ})$ **13.** $\dfrac{20(\cos 120^\circ + j \sin 120^\circ)}{4(\cos 30^\circ + j \sin 30^\circ)}$

14. $\dfrac{-8j}{\sqrt{3} + j}$ (change to polar form)

Find the indicated powers:

15. $(2\underline{/60^\circ})^4$ **16.** $[4(\cos 30^\circ + j \sin 30^\circ)]^3$ **17.** $(2\underline{/45^\circ})^8$

Change to polar form and find the following powers:

18. $(1 + j\sqrt{3})^6$ **19.** $(\sqrt{2} + j\sqrt{2})^6$ **20.** $(\sqrt{3} + j)^9$

Find the following roots:

21. Cube roots of 8 **22.** Square roots of j

23. Fourth roots of $-8 - j8\sqrt{3}$

24. If the current in a circuit is $6 - j2$ amps and the impedance is $2 + j3$ ohms, find the absolute value of the voltage.

25. If the voltage in a circuit is $6 - 3j$ volts and the impedance is $2 + j$ ohms, find the absolute value of the current.

45-18 PYTHAGOREAN NUMBERS

We have seen that the power of any vector has a magnitude raised to the indicated power. Then the square of a vector has a magnitude equal to the square of the magnitude of the vector itself.

This rule enables us to find rather easily a set of three integers called a "Pythagorean triple," such as a, b, and c, that satisfy the equation:

$$a^2 + b^2 = c^2$$

One set of such numbers consists of the integers 3, 4, and 5. Such a set of integers is called a Pythagorean triple (or Pythagorean numbers), because they can represent the two legs and the hypotenuse, respectively, of a right triangle.

Now, consider the vector in rectangular form, as a complex number:

$$3 + 2j$$

The magnitude of this vector is $\sqrt{13}$. Then the square of this vector has a magnitude of 13. Squaring the complex number, we get

$$(3 + 2j)^2 = 9 + 12j + 4j^2 = 5 + 12j$$

Note that the magnitude of the result is 13.

In the graph of the number $5 + 12j$, the numbers 5, 12, and 13, form a right triangle. The real part is

$$3^2 - 2^2 = 5$$

The coefficient of the imaginary unit is

$$(2)(3)2) = 12$$

The magnitude of the result is

$$3^2 + 2^2 = 13$$

Then a right triangle is formed by the real part, the imaginary part, and the magnitude, the triangle having sides of 5, 12, and 13.

Now let us see if this is true for the general vector as a complex number, $a + bj$. The magnitude of this vector is $\sqrt{a^2 + b^2}$. Squaring the vector, we get

$$(a + bj)^2 = a^2 + 2abj + b^2j^2$$

which reduces to

$$a^2 - b^2 + 2abj$$

The real part of the result becomes

$$a^2 - b^2$$

one side of a right triangle. The coefficient of j becomes

$$2ab$$

another side of the triangle. The magnitude becomes

$$a^2 + b^2$$

the hypotenuse of the triangle. These three quantities form the sides of a right triangle.

As another example, consider the complex number $7 + 4j$. The magnitude of this vector is $\sqrt{65}$. Then the square has a magnitude of 65. In the square of this number, the real part becomes

$$7^2 - 4^2 = 33$$

the imaginary part has coefficient of

$$(2)(7)(4) = 56$$

the magnitude of the result becomes

$$7^2 + 4^2 = 65$$

Then the three numbers, 33, 56, and 65, form a Pythagorean triple.

We need not write the complex number itself. We simply take any two integers and from these we compute a set of Pythagorean numbers. For example, let us take the numbers 12 and 7. Then we get the following:

for one side of a right triangle, we have	$12^2 - 7^2 = 95$
for another side of the triangle, we have	$(2)(12)(7) = 168$
for the hypotenuse, we have	$12^2 + 7^2 = 193$
As a check, we have	$95^2 + 168^2 = 193.^2$

Exercise 45-4*

Find a Pythagorean triple, using each of the following pairs of

* Optional.

integers. Reduce the result to lowest terms.

1. 7; 2	**2.** 8; 5	**3.** 9; 4	**4.** 9; 5	**5.** 12; 5
6. 7; 5	**7.** 6; 4	**8.** 13; 8	**9.** 14; 11	**10.** 17; 11

QUIZ ON CHAPTER 45, VECTORS. FORM A.

Find the resultant of the following components and write in polar and trigonometric form:

1. $x = 12; y = 5$ **2.** $x = -8y; y = 15$

3. $-6 - 4j$ **4.** $8 - 3j$

Find the components of the following vectors and write in the form of a complex number:

5. $40\underline{/20^\circ}$ **6.** $50\underline{/201^\circ}$

7. Find the components and write in the form of a complex number:

$$40(\cos 145^\circ + j \sin 145^\circ)$$

8. Multiply as indicated: $(6\underline{/70^\circ})(5\underline{/80^\circ})$

9. Divide as indicated: $40\underline{/210^\circ} \div 8\underline{/60^\circ}$

10. Find the indicated power and write as a complex number: $(2\underline{/80^\circ})^3$

11. Find the three cube roots of $64\underline{/135^\circ}$

12. Add: $30\underline{/73^\circ} + 80\underline{/320^\circ}$

CHAPTER FORTY-SIX Graphs of the Trigonometric Functions

46-1 SELECTION OF VALUES

We recall that in graphing an algebraic equation, $y = f(x)$, we find sets of values for x and y. As an example, consider the equation

$$y = x^2 - 3x - 4$$

In this equation we take any arbitrary value for x and find the corresponding value of y. For instance, if we take $x = 2$, then $y = -6$; if $x = 3$, $y = -4$; and so on. These pairs of values are plotted as the coordinates of points on the rectangular coordinate system.

If we have an equation involving a trigonometric function, we proceed in the same way. Suppose we wish to graph the equation

$$y = \sin x$$

We take any arbitrary values for x, such as 1, 2, 3, and so on, and find the corresponding values of y. These pairs of values are then plotted as points on the coordinate system.

However, in graphing the trigonometric functions, we run into difficulties not encountered in purely algebraic functions. For instance, in a trigonometric equation, such as $y = \sin x$, the values of x must be in *radians*. If we take the value $x = 1$, then we have $y = \sin 1$. Here the 1 must be taken as 1 *radian;* that is,

$$y = \sin 1 = \sin 57.3° = 0.8415$$

In a trigonometric equation it is usually easier to take the values of x in terms of π radians. If this is done, the numbers of degrees can be rational numbers. For instance,

$$\sin \frac{\pi}{6} = \sin 30° = 0.5000$$

In plotting the points for the true trigonometric curves on the rectangular coordinate system, we must remember two things:

1. The units on the x-axis and the y-axis must be equal in length.
2. The units on the x-axis must represent radians.

It is true that we usually find the values of x by looking up the angle in degrees, since most tables give angles by degrees. However, if we are to get the true and proper shape of the curves, the scales on the axes must be laid off according to the foregoing rules. In actual practice, it will be somewhat easier to graph a trigonometric function if the scales are modified slightly; but if this is done, the result will not be the true shape of the curve.

46-2 THE SINE CURVE

To graph the equation $y = \sin x$, we first set up a table of pairs of values. The corresponding numbers of degrees are shown in parentheses.

	x	y		x	y
0	0	0	(210°)	$\frac{7\pi}{6}$	−0.5000
(30°)	$\frac{\pi}{6}$	0.5000	(225°)	$\frac{5\pi}{4}$	−0.7071
45°	$\frac{\pi}{4}$	0.7071	(240°)	$\frac{4\pi}{3}$	−0.8660
(60°)	$\frac{\pi}{3}$	0.8660	(270°)	$\frac{3\pi}{2}$	−1.0000
(90°)	$\frac{\pi}{2}$	1.0000	(300°)	$\frac{5\pi}{3}$	−0.8660
(120°)	$\frac{2\pi}{3}$	0.8660	(315°)	$\frac{7\pi}{4}$	−0.7071
(135°)	$\frac{3\pi}{4}$	0.7071	(330°)	$\frac{11\pi}{6}$	−0.5000
(150°)	$\frac{5\pi}{6}$	0.5000	(360°)	2π	0
(180°)	π	0			

All of the foregoing values of the sine can be found without a table by the method of special angles. If we wish to find the sine of angles at smaller intervals, such as 10° or 5°, we can use a table of values and plot as many values as desired.

To plot the foregoing values as points, we first show the x-axis and the y-axis on rectangular coordinate paper. We lay off equal units on the x-axis and the y-axis and number the points 1, 2, 3, and so on. For the sine curve the coordinate will never be greater than *one* (1).

Since the angles are usually found in terms of π radians, we locate the point π on the x-axis approximately $3\frac{1}{7}$ units to the right of the origin. This point represents an angle of π radians, or 180°. Using this

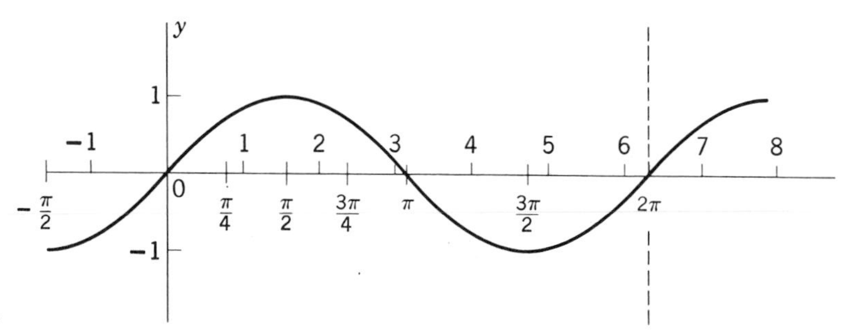

Figure 46-1. $y = \sin x$.

point as a guide, we lay off the common fractional values of π, such as $\pi/6$, $\pi/4$, and $\pi/3$ (Fig. 46-1).

The values of x (that is, the angle in radians) and the corresponding values of y (that is, the sine values) are then plotted as points on the graph. If we join the points by a smooth curve, we get the true sine curve or the graph of the equation, $y = \sin x$. For values greater than 360°, the curve repeats the same pattern.

46-3 THE GRAPH OF THE EQUATION $y = \sin 2x$

In finding pairs of values for x and y in the equation $y = \sin 2x$, we must be careful to double the value of the angle x before finding the sine value. If we take the value of x as $\pi/6$, or 30°, then we must multiply this value by 2 before finding the value of y. To find the value of y, we have, for $x = 30°$, $y = \sin (2)(30°) = \sin 60° = 0.8660$.

In general, whenever a value of x is taken in this equation, this value must be multiplied by 2 before the sine value is found. Some of the angles from 0 to 2π and the corresponding sine values are shown in the table. The graph for the equation $y = \sin 2x$ is shown in Fig. 46-2. Note that the curve repeats itself after 180°.

x	y ($\sin 2x$)
0	0
15°	0.50000
30°	0.8660
45°	1.0000
60°	0.8600
90°	0
120°	−0.8660
135°	−1.0000
150°	−0.8660
180°	0
210°	0.8660
225°	1.0000

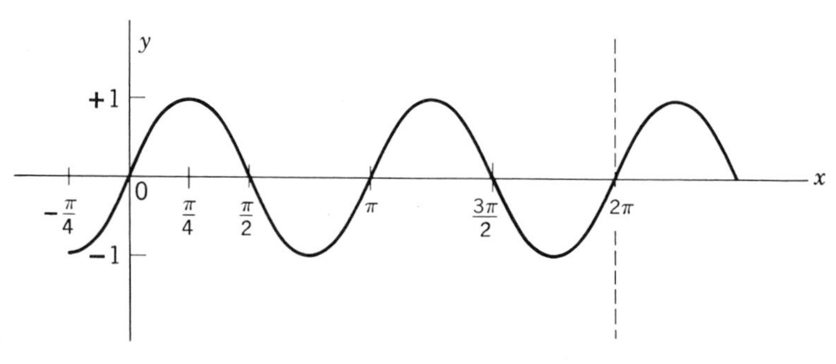

Figure 46-2. $y = \sin 2x$.

46-4 CYCLE; FREQUENCY; PERIOD

In the curves we have graphed, $y = \sin x$ and $y = \sin 2x$, we note that after a certain interval on the x-axis, the curve repeats the same pattern. One complete pattern of a curve is called a *cycle.* In the curve, $y = \sin x$, we get one cycle from 0 to 2π. In the curve, $y = \sin 2x$, we get two cycles in this interval. If we were to graph $y = \sin 3x$, we should get three cycles in this interval.

The number of cycles in the interval from 0 to 2π is called the *frequency*. In the curve, $y = \sin 2x$, the frequency is 2, which indicates two cycles. In the equation, $y = \sin 3x$, the frequency is 3. In general, in the graph of $y = \sin nx$, the coefficient n represents the frequency or number of cycles in an interval of 2π.

The *period* of a curve is the interval or distance along the x-axis required for one cycle. Note that the curve, $y = \sin x$, requires an interval of 2π for one cycle. The period of this curve is therefore 2π. The curve $y = \sin 2x$ has two cycles from 0 to 2π. One cycle requires only an interval of π. Therefore the period is π. In the graph of $y = \sin 3x$, the frequency is 3, which indicates three cycles. For one cycle we need an interval of only $\frac{1}{3}$ of 2π, or $\dfrac{2\pi}{3}$. In general, for the equation, $y = \sin nx$, the frequency is n, and the period is $\dfrac{2\pi}{n}$.

Then we have the following reciprocal relations between frequency, f, and period p:

$$f = (2\pi)\left(\frac{1}{p}\right) \qquad \text{and} \qquad p = (2\pi)\left(\frac{1}{f}\right)$$

46-5 THE GRAPH OF THE EQUATION $y = 2 \sin x$

In finding pairs of values for x and y in the equation $y = 2 \sin x$, we first take any value for x and find the sine of that angle. Then we multiply this sine value by 2. We do *not* double the angle. For instance, if we take $x = 30°$ (that is, $\pi/6$), we find that the sine of the angle is 0.5000. Then we multiply this value by 2 to find the value of y. When the angle is 90°, the sine value is 1, but this value is then doubled, so $y = 2$.

Some of the values of x and the corresponding values of y are shown in the table. The graph of the equation $y = 2 \sin x$ is shown in Fig. 46-3, with values of x from $x = 0$ to $x = 2\pi$.

x	y
0	0
30°	1.0000
45°	1.4142
60°	1.7320
90°	2.0000
120°	1.7320
135°	1.4142
150°	1.0000
180°	0
210°	−1.0000
225°	−1.4142

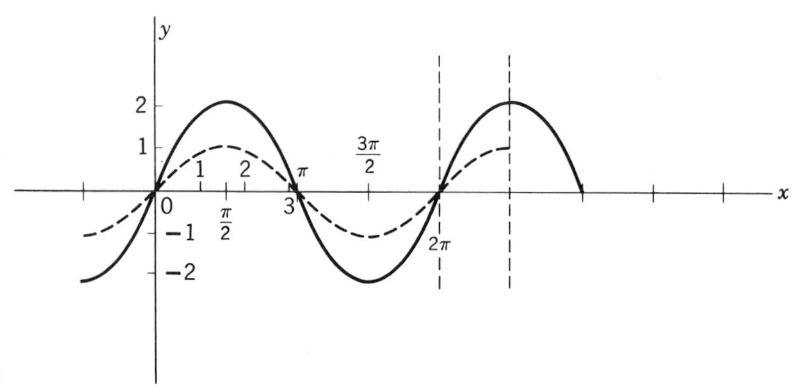

Figure 46-3. $y = 2 \sin x$. (The broken line shows $y = \sin x$.)

Note that in the interval from 0 to 2π there is only one complete cycle of the curve, since the frequency factor is 1. However, the curve has a greater distance from the x-axis. The ordinate y at one time reaches a value of 2.

46-6 AMPLITUDE FACTOR

In the equation $y = 2 \sin x$, whatever value we assign to x, we first find the sine value and then double this value. At every point on the curve the ordinate, or y-value, extends twice as far as that for the curve of $y = \sin x$. The height of the curve, or the greatest distance from the x-axis, is called the *amplitude* of the curve. The amplitude of $y = 2 \sin x$ is twice the amplitude of the sine curve, $y = \sin x$.

If we were to graph the equation $y = 3 \sin x$, we should find that the amplitude is three times that of the sine curve. For this reason, the 3 is called the *amplitude factor.* In any equation of the form $y = m \sin x$ the coefficient m is the amplitude factor because it determines the greatest height or amplitude of the curve. Moreover, for any particular value of x, the curve has a y-value m times the y-value of the sine curve itself.

If we graph the equations $y = 3 \sin x$, $y = 4 \sin x$, or $y = m \sin x$, where m is any constant, we get only one complete cycle from $x = 0$ to $x = 2\pi$. Note that the frequency factor is 1 in each case. The frequency factor is not affected by the amplitude of the curve.

If we were to graph the equation $y = 30 \sin 20x$, we should get a curve of 20 cycles over a 360° interval. The amplitude factor is 30, which means that the curve reaches a point 30 units above and below the x-axis. Such a curve is called a *sinusoidal* curve, since it resembles the sine curve. For such curves it is often convenient to compress the scales in the y-direction.

46-7 THE COSINE CURVE

In finding pairs of values for x and y in the equation $y = \cos x$, we proceed as we do for the sine curve. We take any convenient values for x and then compute the values for y. Some of these values are shown in the table. Note especially the following values:

$$\cos 0 = 1$$

$$\cos \frac{\pi}{2} = 0$$

$$\cos \pi = -1$$

x	y
0	1.0000
30°	0.8600
45°	0.7071
60°	0.5000
90°	0
120°	−0.5000
135°	−0.7071
150°	−0.8660
180°	−1.0000

If we take values of x from 0 to 2π and then plot these values carefully, the result is the cosine curve. The graph is shown in Fig. 46-4. Note that the amplitude is 1 and that the frequency is also 1. In general appearance, the cosine curve resembles the sine curve. The period is 2π.

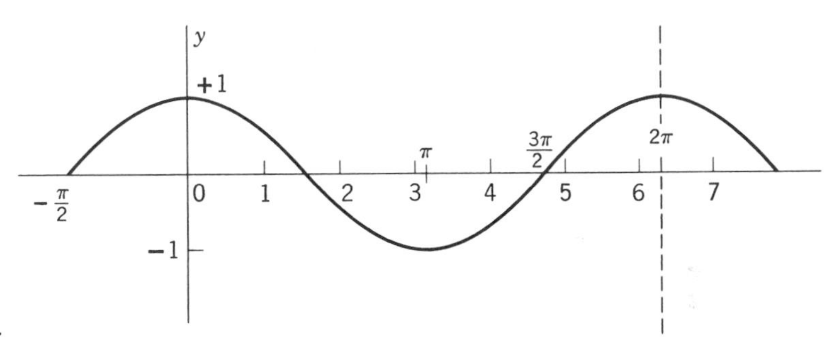

Figure 46-4. $y = \cos x$.

46-8 PHASE DIFFERENCE

In the graph of the equation $y = \cos x$ we get one cycle in the interval $x = 0$ to $x = 360°$. In the graph of the equation $y = \sin x$ we also get one cycle in the same interval. If we continue the sine and cosine curves beyond 360°, we find that the two curves have the same shape. However, their positions are separated by a 90° interval, or difference. This difference in position along the x-axis is called "phase difference." A phase difference between two curves represents a shift in a curve to the right or left.

46-9 OTHER TRIGONOMETRIC CURVES

In graphing the curves for the tangent, the cotangent, the secant, and the cosecant, we should note the following points:

1. For some values of x the function values do not exist. They increase or decrease without limit. The graphs in such cases

approach straight lines but do not cross them. A straight line approached by a curve in this way is called an *asymptote.* For instance, the tangent of 90° does not exist. It can be said to be infinitely great. The tangent value continues to increase as x approaches 90°. Note these tangent values as the angle approaches 90°. The tangent of 89° is approximately 57.3; the tangent of 89° 59′ is approximately 3438; the tangent of 89° 59′ 59″ is approximately 206,000. Just beyond 90° the tangent is a large *negative* number.

2. The tangent and cotangent curves cross the x-axis. The secant and cosecant curves do not cross the x-axis, because their smallest numerical value is 1.

3. The cosecant curve touches the sine curve at 90° and 270°, because cosecant is the reciprocal of the sine. When the sine value is 0, the cosecant value is infinite.

4. The secant curve touches the cosine curve at 0, 180°, and 360°, because the secant is the reciprocal of the cosine. When the cosine value is 0, the cosecant value is infinite.

For the graph of the tangent curve, $y = \tan x$, we may use the corresponding values shown in the list at the right. The values from 180° to 360° are the same as from 0° to 180°.

x	y
0°	0
30°	0.58
45°	1.0
60°	1.73
90°	∞
120°	−1.73
135°	−1.0
150°	−0.58
180°	0

To find the values from the table, we take angles stated in degrees, rather than in terms of radians. However, the units along the x-axis must be in terms of radians, as usual, taking $\pi = 180°$.

Figure 46-5 shows the graph of the tangent curve, $y = \tan x$, by a solid line. The cotangent curve, $y = \cot x$, is shown by a broken line. Asymptotes are shown by dotted lines.

Note that the tangent curve and the cotangent curve have two cycles from 0 to 2π. That is, the period of these curves is π, or 180°.

For the graph of the cosecant curve, $y = \csc x$, we may use the corresponding values shown in the list at the right. The values from 180° to 360° are negative but have the same numerical values as from 0° to 180°.

x	y
0	∞
30°	2.0
45°	1.4
60°	1.2
90°	1.0
120°	1.2
135°	1.4
150°	2.0
180°	∞

Figure 46-6 shows the graph of the cosecant curve, $y = \csc x$, by a solid line. The sine curve, $y = \sin x$, is shown by a broken line. Asymptotes are shown by dotted lines.

Note that at 90°, both the sine and the cosecant have the value 1. At 0°, the sine value is 0. Since

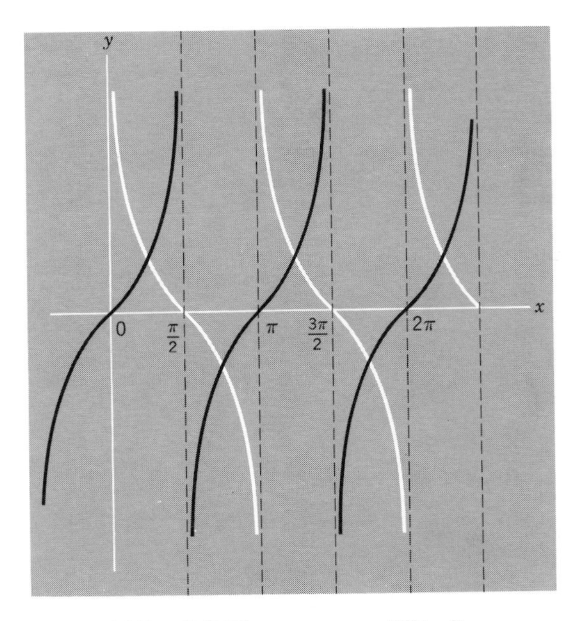

Figure 46-5. *Solid line:* $y = \tan x$. *White line:* $y = \cot x$. *Dotted lines:* asymptotes.

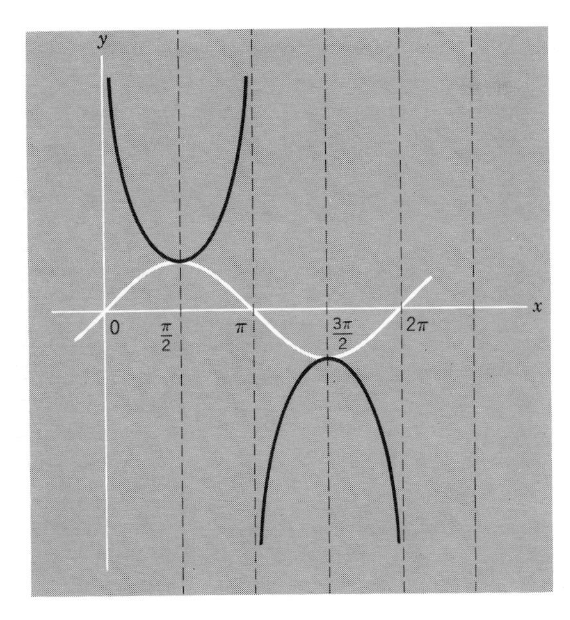

Figure 46-6. *White line:* $y = \sin x$. *Solid line:* $y = \csc x$. *Dotted lines:* asymptotes.

the sine and the cosecant of an angle are reciprocals, that is, $\csc x = 1/\sin x$, then, for the csc 0, we get the expression, $\csc 0 = 1/0$, which is undefined, and usually indicated by the infinity sign, ∞. We might think of the cosecant curve as being the inverted form of the sine curve. The same relation exists between the curves of the cosine and the secant of an angle. Note that these curves have only one cycle from 0 to 2π. That is, the period is 2π. The graphs of the cosine and secant are shown in Fig. 46-7.

46-10 COMPOUND CURVES

In some cases it is desirable to find the curve that represents the sum of two or more functions. For example, we may wish to find the curve for the equation

$$y = \sin x + \sin 2x$$

One way is to graph each function separately on the same form. Then, by means of a compass, we measure the ordinate value of one curve at any point on the graph and then add this value to the ordinate of the other curve.

Another way is to set up a table of values for y that correspond to the selected x values. In this case, the y values must represent the sum, $\sin x + \sin 2x$. For any particular value of x, we first find the *sine of x*,

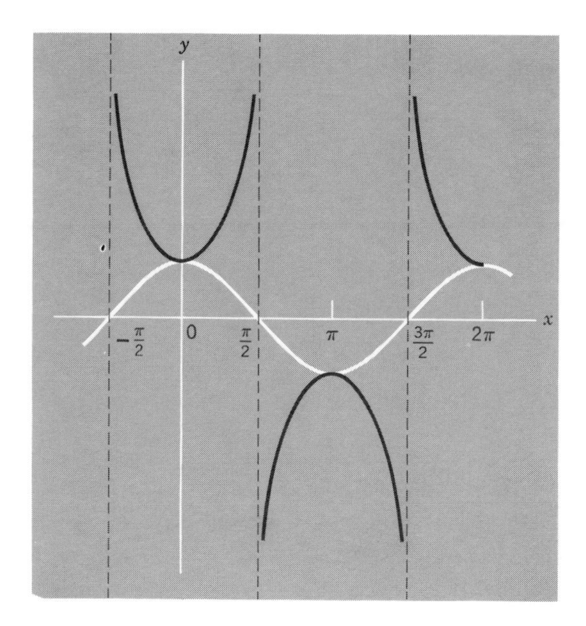

Figure 46-7. *White line:* $y = \cos x$. *Solid line:* $y = \sec x$. *Dotted lines:* asymptotes.

then the *sine of* $2x$, and then add these two to get the proper value of y. This method has been used to graph the foregoing equation.

For graphing the equation, $y = \sin x + \sin 2x$, we may use the following sets of corresponding values: (values have been rounded off):

x	0	10°	20°	30°	45°	50°	60°	70°	80°	90°
$\sin x$	0	0.17	0.34	0.50	0.71	0.77	0.87	0.94	0.98	1.0
$\sin 2x$	0	0.34	0.64	0.87	1.0	0.98	0.87	0.64	0.34	0
y	0	0.51	0.98	1.37	1.71	1.75	1.74	1.58	1.32	1.0

x	100°	110°	120°	130°	135°	140°	150°	160°	170°	180°
$\sin x$	0.98	0.94	0.87	0.77	0.71	0.64	0.50	0.34	0.17	0
$\sin 2x$	−0.34	−0.64	−0.87	−0.98	−1.0	−0.98	−0.87	−0.64	−0.34	0
y	0.64	0.30	0	−0.21	−0.29	−0.34	−0.37	−0.30	−0.17	0

The graph of this equation is shown in Fig. 46.8. In most instances, the graph of a curve may not require so many pairs of values. The general shape of the curve can usually be determined by taking certain particular values, showing highest points, lowest points, and values where the curve crosses the x-axis.

In Fig. 46-8, the graphs of $y = \sin x$, and $y = \sin 2x$ are shown by broken lines. The graph of the sum of these two, $y = \sin x + \sin 2x$, is shown by a solid line.

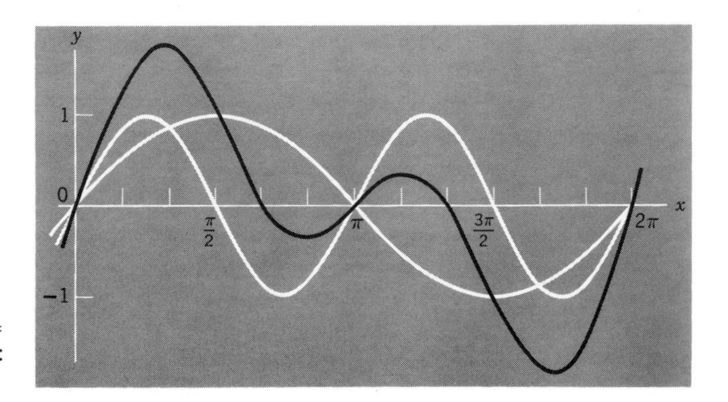

Figure 46-8. *Broken lines:* $y = \sin x$ and $y = \sin 2x$. *Solid line:* $y = \sin x + \sin 2x$.

46-11 GRAPHS OF INVERSE TRIGONOMETRIC FUNCTIONS

The graphs of the inverse trigonometric functions are similar in shape to the graphs of the trigonometric functions, with the x and y axes reversed. Consider the equation,

$$y = \arcsin x$$

This equation means that y is an angle whose sine is x. The equation may be stated in the form

$$x = \sin y$$

Now, if we graph the equation $x = \sin y$, we get the sine curve along the y-axis instead of along the x-axis. We shall find that the graphs of the inverse functions have the same relation to the y-axis as the graphs of the trigonometric functions have to the x-axis.

For the values of the angles (that is, the y-values), we restrict these values between certain limits. In the equation

$$y = \arcsin x$$

There is a relation between x and y. The x-values are limited from -1 to $+1$, because the sine values of angles do not lie outside these limits. However, for each value of x, there are many values of the angle y. For example,

$$\text{if} \qquad y = \arcsin 0.5$$

then the angle y may have many values, such as 30°, 150°, 390°, −210°, and many more.

In order that y may be a function of x, there should be only one value of y for each value of x. That is the definition of a function. For this reason we restrict the value of $\arcsin x$ to angles from −90° to +90°; that is, from $-\pi/2$ to $+\pi/2$. Then there will be only one value of *arcsin* x for each value of x.

Values of the angle within the restricted range are called *principal* values. Principal values are usually indicated by capitalizing the names.

For example,

$$\text{if} \qquad y = \text{Arcsin } 0.5$$
$$\text{then} \qquad y = 30°$$

the principal value of the angle. In the same way,

$$\text{Arcsin}(-0.5) = -30°$$

For the Arccos x, we take the principal values of the angle from 0 to π; that is, 0 to 180°. Then

$$\text{if} \qquad y = \text{Arccos}(-0.5)$$
$$\text{then} \qquad y = 120°$$

the principal value. Then there is only one principal value of the angle whose cosine is -0.5.

The following principal values are taken for the inverse functions:

Arcsin x from $-\pi/2$ to $+\pi/2$; Arccos x from 0 to π;
Arctan x from $-\pi/2$ to $+\pi/2$; Arccot x from 0 to π;
Arcsec x from 0 to π; Arccsc x from $-\pi/2$ to $+\pi/2$

The graphs of the inverse trigonometric functions are shown in Figures 46-9 through 46-14. The principal values are shown by solid lines.

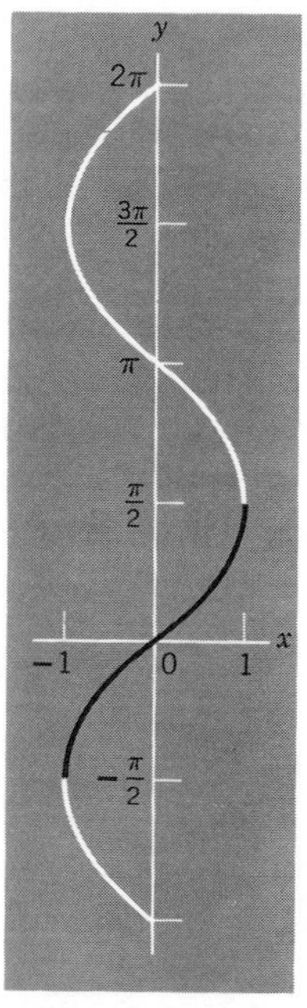

Figure 46-9. $y = \arcsin x$. *Solid line:* $y = \text{Arcsin } x$.

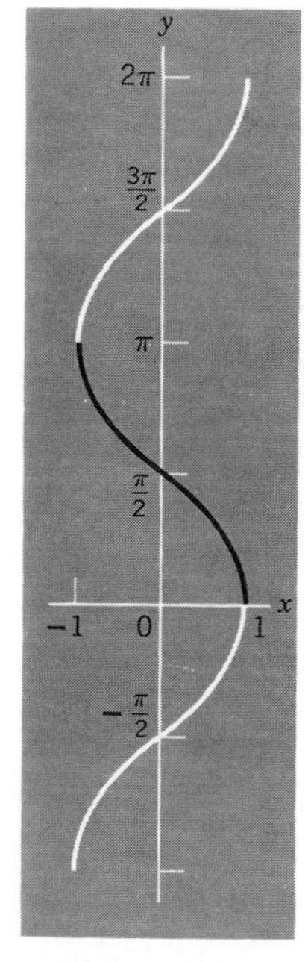

Figure 46-10. $y = \arccos x$. *Solid line:* $y = \text{Arccos } x$.

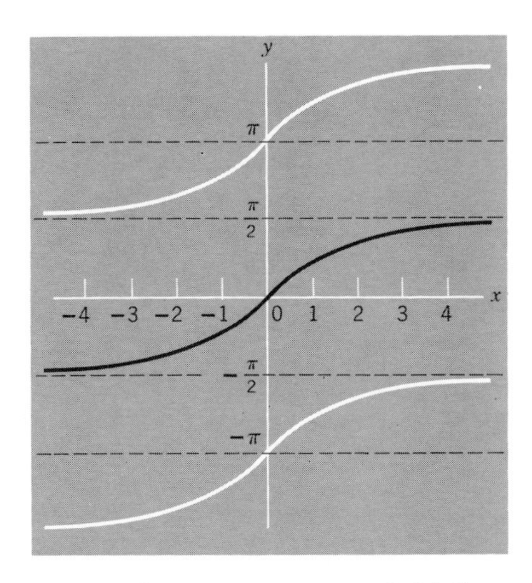

Figure 46-11. $y = \arctan x$. *Solid line*: $y = \text{Arctan}\, x$.

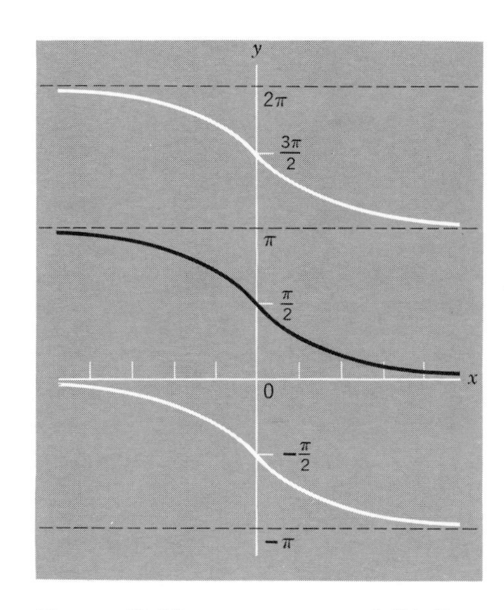

Figure 46-12. $y = \text{arccot}\, x$. *Solid line*: $y = \text{Arccot}\, x$.

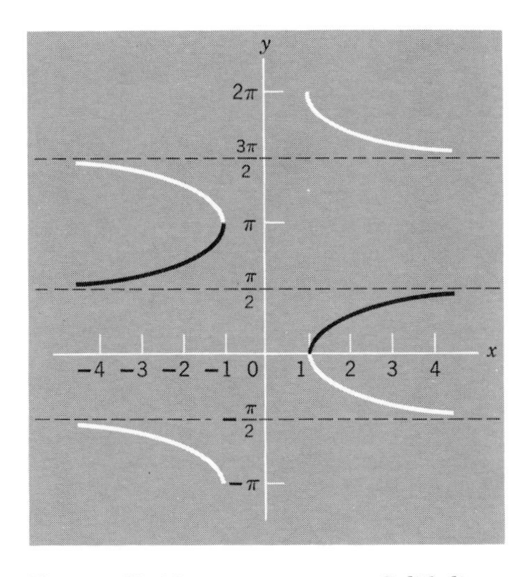

Figure 46-13. $y = \text{arcsec}\, x$. *Solid line*: $y = \text{Arcsex}\, x$.

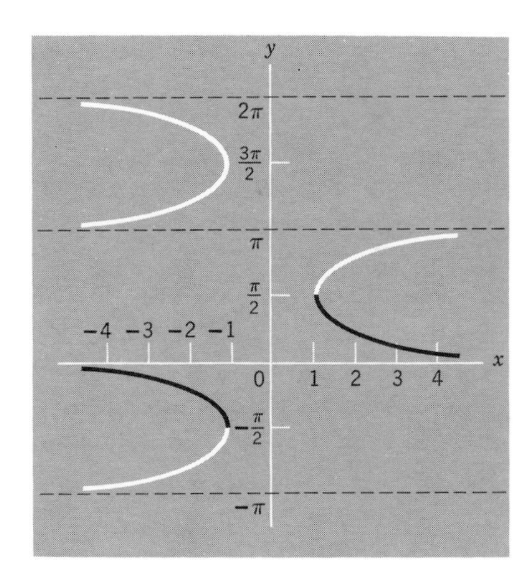

Figure 46-14. $y = \text{arccsc}\, x$. *Solid line*: $y = \text{Arccsc}\, x$.

Exercise 46-1

Graph the following equations on rectangular coordinate paper:

1. $y = \sin x$

2. $y = \cos x$

3. $y = \tan x$

4. $y = \cot x$

5. $y = \sec x$

6. $y = \csc x$

7. $y = \sin 3x$

8. $y = 3 \sin x$

9. $y = \cos 4x$

10. $y = 3 \cos 2x$

11. $y = 1 + \sin x$

12. $y = x + \sin x$

13. $y = \sin (90° + x)$

14. $y = \cos (x + 45°)$

15. $y = \sin (x - 30°)$

16. $y = \sin (x + 15°)$

17. $y = \cos \left(x - \frac{\pi}{2}\right)$

18. $y = \cos (x + 60°)$

19. $y = \sin x + \cos x$

20. $y = \sin x + \sin 3x$

21. $y = \sin x + \sin 2x + \sin 3x$

22. $y = \sin x + \sin 3x + \sin 5x$

23. $y = \arcsin x$

24. $y = \arccos x$

25. $y = \arcsin 2x$

26. $y = \arccos 2x$

27. $y = \arctan x$

28. $y = \text{arccsc}\, x$

PART SIX CALCULUS

CHAPTER FORTY-SEVEN Introduction to Calculus

47-1 CONSTANTS AND VARIABLES

In calculus we are concerned with two kinds of quantities, *variables* and *constants.* A *variable* is a quantity whose value may change in a particular problem. In the equation, $2x + 3y = 12$, the letters x and y represent variables. A *constant* is a number or quantity whose value does not change in a particular problem. An *absolute constant* is one whose value never changes, such as the numbers 3, -4, π, and so on. An *arbitrary constant* is a quantity whose value is assumed to remain constant in a particular problem. In other problems an arbitrary constant may have another value. In the equation, $y = mx + b$, x and y represent variables; m and b represent constants.

47-2 FUNCTION

If two variables are so related that for each value of one there is a corresponding value of the other, then we say one variable is a *function* of the other. The relationship is often expressed by an equation, such as, $y = 3x + 7$. In this equation, for each value of x there is a corresponding value of y. Then we say that y is a function of x. A function implies dependence between variables. One variable, usually x, is called the *independent* variable, then the other variable is called the *dependent* variable. Of course, the role of dependent and independent variable may be reversed.

A function may be expressed in other ways, such as by a table of corresponding values of related variables, as in this table showing temperature:

Hour (P.M.)	1	2	3	4	5	6	7	8
Temperature (°)	68	69	71	71	73	72	69	67

For each hour there is a corresponding temperature. Then we say the temperature is a function of the hour (time). However, in this case, the relation cannot usually be expressed by an equation.

In most problems, the function is expressed by an equation, such as

$$y = x^2 + 2x + 5; \qquad s = t^2 + 5t - 3; \qquad P = 20i^2; \qquad A = \pi r^2$$

Each of these equations shows an expressed relation between two variables.

There is a way to define a function of x without using another letter. We can make use of the following operational definition:

Definition. *A function of x is any expression that contains the variable x.*

As an example, the expression, $3x + 7$, is a function of x. For every value of x, there is a corresponding value of the function. If $x = 6$, the function has the value 25. In like manner,

$$t^2 + 5t \text{ is a function of } t; \qquad i^2 + 4i - 5 \text{ is a function of } i$$

In most instances we use another letter, such as y, to represent a function of x. For example, we say, $y = 3x + 7$. A function of x is often represented by the symbol, $f(x)$. Then we write: $f(x) = 3x + 7$. The equation is read:

"the f function of $x = 3x + 7$," or simply, "f of $x = 3x + 7$."

It should be remembered that the function of x is not the symbol used to represent it. Whether we use y or $f(x)$, or some other symbol, the function of x is the expression itself, such as $3x + 7$, and *not* the symbol we use to represent it. We can use any letter to represent a function of x, or t, or any other variable, such as: $g(x) = x^2 + 5$; $h(t) = t^2 - t$; $k(v) = v^3 + 2v$.

The expression $f(x)$ to represent a function of x is a convenient notation when we wish to indicate the value of a function. To indicate the value of a function when x is equal to, say, 5, we substitute 5 for x in the symbol $f(x)$ as well as in the function itself. Then, if

$$f(x) = x^2 + 3x + 2, \qquad f(5) = 42; \qquad f(0) = 2; \qquad f(-1) = 0$$

Exercise 47-1

Evaluate the following functions:

1. If $f(x) = x^2 - 2x - 3$, find $f(1)$; $f(3)$; $f(-1)$; $f(3x)$; $f(2x - 1)$.
2. If $f(t) = t^2 + 5t + 2$, find $f(3)$; $f(-2)$; $f(t^2)$; $f(0)$; $f(x)$.
3. If $f(r) = \pi r^2$, find $f(3)$; $f(7)$; $f(-4)$; $f(1)$; $f(2r)$.
4. If $f(i) = i^2 + 2i + 3$, find $f(2)$; $f(-2)$; $f(0)$; $f(1.5)$; $f(2.1)$.
5. If $g(x) = 3 - 2x - x^2$, find $g(0)$; $g(1)$; $g(-1)$; $g(3)$; $g(x^2)$.
6. If $k(t) = t^3 + 2t^2 - 5t$, find $k(0)$; $k(1)$; $k(-1)$; $k(4)$; $k(2t)$.
7. If $f(x) = x^2 + 5x + 2$, and $g(x) = 2x - 3$, find the following:

   (a) $f(x) + g(x)$; (b) $f(2) + g(0)$;
   (c) $f(x) - g(x)$; (d) $f(3) - g(-1)$

8. If $f(x) = \dfrac{x^2 - 9}{x - 3}$, and $g(x) = x + 3$, find (a) $f(5)$ and $g(5)$; (b) $f(2)$ and $g(2)$; (c) $f(3)$ and $g(3)$; (d) Is $f(x) = g(x)$?

9. If $f(x) = \log_{10}(x + 95)$, find $f(5)$; $f(905)$.

10. If $f(x) = \sin x$, find $f\left(\dfrac{\pi}{2}\right)$; $f(1)$.

11. If $g(t) = \cos t$, find $g(\pi)$; $g(2)$.

12. If $f(x) = 10^x$, find $f(1)$; $f(3)$; $f(-4)$.

13. Graph the function: $y = 3x - 2$, using $x = -3, -2, -1, 0, 1, 2, 3, 4$.

14. Graph the function: $f(x) = x^2 - 2x - 3$, for integral values of x, from $x = -3$ to $x = 4$. (First let y equal the function.)

47-3 INCREMENTS

Calculus is the study of changes in variables. If two variables are so related that one is a function of the other, then a change in one variable will usually result in some change in the other. Calculus is concerned with the relation between changes in related variables.

Any change in a variable is called an *increment.* As used here, an increment refers to any change, positive, negative, or zero. If the temperature is 58° at 8 A.M. and 64° at 11 A.M., then the increment in temperature is 6°. If the temperature is 72° at 3 P.M. and 72° at 5 P.M., then the increment in temperature is zero. If the current in an electric circuit changes from 4.8 amperes to 4.3 amperes, the increment in current is −0.5 amperes.

An increment is often denoted by the Greek letter *delta* (Δ), placed before the variable. An increment in x is written "Δx," and read "delta x." This expression does not indicate a product but only a change in x. As a variable changes from one value to another, the increment is computed by taking the *second value minus* the *first value.*

We often denote first and second values by subscripts, 1 and 2, respectively. If $x_1 = 6$ and $x_2 = 9$, then $\Delta x = x_2 - x_1 = 9 - 6 = 3$. If temperature T changes from −15° to −7°, the increment is positive:

$$\Delta T = T_2 - T_1 = (-7) - (-15) = +8$$

That is, a change from −15° to −7° represents a *rise* in temperature.

The increment in a function is found in the same way. If one variable y is a function of another variable x, then for any increment in x, there will be a corresponding increment, positive, negative, or zero, in the function. The increment in the function is represented by Δy, or $\Delta f(x)$. As an example, consider the function

$$y = x^2 + x$$

If x changes from 4 to 7, the function y changes from 20 to 56. Then

$$\Delta x = x_2 - x_1 = 7 - 4 = 3; \qquad \Delta y = y_2 - y_1 = 56 - 20 = 36$$

47-4 RELATION BETWEEN INCREMENTS: AVERAGE RATE OF CHANGE

We are now concerned with the relation between the increment of a variable and the increment of its function. Calculus is concerned specifically with the *ratio* of the increments in this order:

$$\frac{\Delta y}{\Delta x} \quad \text{or} \quad \frac{\Delta f(x)}{\Delta x}$$

In the function, $y = x^2 + x$, as x changes from 4 to 7, the function y changes from 20 to 56. Then $\Delta x = 3$ and $\Delta y = 36$. For the desired ratio, we have

$$\frac{\Delta y}{\Delta x} = \frac{36}{3} = 12$$

The number 12 represents the *average rate of change* in the function as x changes from 4 to 7. That is, the function changes 12 units for each change of 1 unit in x, when x changes from 4 to 7. If the increment in x is different from 3, then, generally, the ratio between the increments may be different.

The average rate of change of one variable with respect to another is an important concept in many problems in everyday life, as well as in science. For example, suppose the odometer of a car shows a mileage of 6530 miles at 1 P.M. and 6740 at 6 P.M. the same day. To find the average velocity, or average rate of change of distance with respect to time, we first find the increments. Using t for time, and s for distance, we have

$$\Delta t = t_2 - t_1 = 6 - 1 = 5 \text{ hours};$$

$$\Delta s = s_2 - s_1 = 6740 - 6530 = 210 \text{ miles}$$

For average velocity, we have

$$\frac{\Delta s}{\Delta t} = \frac{210}{5} = 42 \text{ mph}$$

That is, the average rate of change of distance is 42 miles for each change of 1 hour in time.

Note. It must be remembered that the average rate of change does not tell the instantaneous rate of change at a particular instant. We shall see later how to find the *instantaneous rate of change* by calculus.

Exercise 47-2

1. If the odometer of a car shows 6845 miles at 8 A.M. and 7130 miles at 2 P.M. the same day, find average velocity per hour during the interval.

2. The speedometer of a car shows 43 mph at 3 P.M., and 63 mph 40 sec later. Find the average acceleration (average rate of change of velocity) during the 40-sec interval.

3. If the temperature was 58° at 5 P.M., and 46° at 11 P.M. the same day, find the average rate of change in temperature per hour.

4. The barometric pressure was 30.08 in. at 4 P.M., and 29.63 at 7 P.M. Find the average rate of change in pressure per hour.

5. An object falls according to the formula, $s = 16t^2$, where s represents distance in feet, and t represents time in seconds. Find the average velocity in the interval from $t = 4$ to $t = 7$.

6. An object rolling down an incline has rolled 16 ft at the end of 2 sec, and 144 ft at the end of 6 sec. Find average velocity.

7. Find the average rate of change in the area of a circle for each change of 1 in. in the radius when the radius changes from 5 in. to 8 in.

8. If an electric current changes from 6.2 amperes to 5.6 amperes in $\frac{1}{2}$ sec, what is the average rate of change in current per second?

9. The power in an electric circuit is given by the formula, $P = 60i^2$. Find the average change in power P (watts) for each change of 1 ampere in current i, when i changes from 2 amperes to 4.5 amperes.

10. Towns A and B are 150 miles apart. In a trip from A to B a car averaged 50 mph. On the return trip the car averaged 30 mph. What was the average velocity for the entire round trip?

Find the average rate of change for each of the following functions for the given values of the independent variable.

11. $y = x^2$; $x = 6$ to $x = 9$

12. $y = x^2 + 2x$; $x = 3$ to $x = 7$

13. $i = t^2 + 3t$; $t = 2$ to 2.4

14. $v = 3t - 4$; $t = 6.5$ to $t = 8$

15. $v = t^2 - 2t$; $t = 3.6$ to 4.2

16. $s = t^2 + 5t$; $t = 4$ to 5.2

17. $y = \frac{12}{x}$; $x = 2$ to $x = 4$

18. $y = \frac{72}{x^2}$; $x = 3$ to 6

19. $P = 20i^2$; $i = 6$ to 7.5

20. $I = \frac{600}{R}$; $R = 10$ to 20

47-5 LIMITS Before we try to determine instantaneous rate of change, it is necessary to understand the meaning of a *mathematical limit.* In mathematics, the word *limit* does not have exactly the same meaning as it does in everyday use. When we say the speed limit on a highway is 60 mph, then a speed of 60 mph must not be exceeded, *but it may be reached.*

On the other hand, a mathematical limit is a constant that can be approached as close as we wish, but it *cannot be reached.* Suppose a variable x starts at zero and moves toward some constant, say, 8, by steps. For each step, the variable x moves half the remaining distance to 8 (Fig. 47-1). Then for the successive steps, x reaches the following

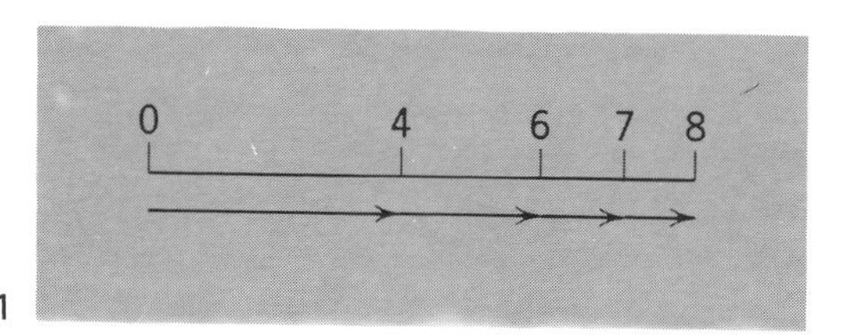

Figure 47-1

points: 4, 6, 7, $7\frac{1}{2}$, $7\frac{3}{4}$, $7\frac{7}{8}$, and so on. Notice the following facts:

1. The variable gets closer and closer to the constant 8.
2. It can get as close to 8 as we wish by taking enough steps.
3. It will never reach 8.
4. The difference, $8 - x$, becomes small and smaller and nearer to zero.

Under these conditions we say that x approaches 8 as a limit. To indicate this fact we use an arrow to denote *approaches,* and write

$$x \to 8$$

Moreover,

$$(8 - x) \to 0$$

To indicate that 8 is the limit of the variable x, we write

$$\lim_{x \to 8} x = 8$$

The expression is read: "The limit of x is 8 as x approaches 8." Also

$$\lim_{x \to 8} (8 - x) = 0$$

Definition. *If a variable x approaches a constant K in such a way that the variable x can be made as close to K as we wish but not equal to K, then the constant K is called the limit of the variable x.*

Notice especially the following facts. The limit K is a constant. It does not approach anything. The variable x can be made as close to the constant K as we wish, but it is never equal to the constant K. The difference, $K - x$, approaches zero as a limit, but is not equal to zero.

A variable may increase or decrease without approaching a limit. For example, if x represents the positive integers, then it has the values, 1, 2, 3, 4, and so on. Then there is no limit to the value of x, and it becomes infinitely large. We indicate this condition by an arrow and the infinity sign:

$$x \to \infty$$

However, in this case, the arrow should not be read "approaches" because infinity cannot be approached. Moreover, the difference, $\infty - x$, does not approach zero as a limit. The expression should be read, "x becomes infinitely large," or "x increases without limit."

We have seen what is meant by the limit of a variable. Now let us see what is meant by the *limit of a function*. Consider the function, $x^2 + 1$. Let us see what happens to this function as x approaches some limit, say, 3. We are not concerned with the value of the function when x is exactly 3. We have emphasized that when x approaches 3 as a limit, it never reaches 3, yet it can get as close to 3 as we wish. We get the following corresponding values of the function:

if $x =$	0	1	2	2.5	2.8	2.9	$2.99 \to 3$
then $x^2 + 1 =$	1	2	5	7.25	7.84	9.41	$9.9401 \to ?$

From the table it appears that as x approaches 3, the function gets nearer and nearer to 10. In fact, in this case, if x is exactly 3, then the function value is 10. However, we are interested in the limit of the function as x approaches 3, not when x equals 3. In this example, as x approaches 3, the function, $f(x) = x^2 + 1$, approaches $f(3)$, which is 10.

To find the limit of a function, $f(x)$, we may think all we need to do is to substitute the limit of x in the function. This is true in most instances, but sometimes the limit of a function cannot be found by substitution.

Example 1

Find the limit of $f(x) = x^2 + 3x + 5$, as x approaches 2.

Solution

The limit can be found by simple substitution. Then

$$\lim_{x \to 2} (x^2 + 3x + 5) = 4 + 6 + 5 = 15$$

Example 2

Find the limit of the function, $f(x) = \dfrac{x^2 - 4}{x - 2}$, as x approaches 2.

Solution

If we substitute 2 for x, we get $f(2) = \frac{0}{0}$, which is meaningless. However, if x is any quantity except 2, we can reduce the fraction by dividing numerator and denominator by the quantity, $x - 2$, and get $(x + 2)$. Then

$$\lim_{x \to 2} \frac{x^2 - 4}{x - 2} = \lim_{x \to 2} (x + 2) = 4$$

In this example, if we take values of x closer and closer to 2, we shall find that the value of the function gets closer and closer to 4. For example, when $x = 1.99$, the function has the value, 3.9601.

There are two special types of limits that cannot be found by simple substitution. They are shown by these examples:

$$\text{(a)} \lim_{x \to 0} \frac{6}{x} \qquad \text{and} \qquad \text{(b)} \lim_{x \to \infty} \frac{6}{x}$$

In these examples, substitution has no meaning. In the first example, by substitution, we get $\frac{6}{0}$, which has no meaning. To see whether a limit exists, we can take x-values closer and closer to zero. Then we shall find that the function becomes larger and larger, and therefore it has no limit. Then we write

$$\lim_{x \to 0} \frac{6}{x} = \infty \qquad \text{(becomes infinitely large)}$$

Consider the second example, (b): $\lim_{x \to \infty} \frac{6}{x}$. In this case, we take larger and larger values of x, letting x increase without limit. Then we shall find that the function becomes smaller and smaller and approaches zero as a limit. Then we write

$$\lim_{x \to \infty} \frac{6}{x} = 0 \qquad \text{(zero is a limit)}$$

Exercise 47-3

Evaluate the following limits. Tell when a limit does not exist.

1. $\lim_{x \to 5} (x + 4)$ **2.** $\lim_{x \to 4} (2x - 3)$ **3.** $\lim_{x \to 3} (x^2 + 4x - 1)$

4. $\lim_{x \to 3} (x^2 - 1)$ **5.** $\lim_{x \to 0} (x^2 + 3)$ **6.** $\lim_{x \to -2} (3x^2 - 5x + 1)$

7. $\lim_{x\to 3} \frac{x^2 - 9}{x - 3}$
8. $\lim_{x\to -2} \frac{x^2 - 4}{x + 2}$
9. $\lim_{x\to 4} \frac{x^2 - 4x}{x - 4}$
10. $\lim_{x\to\infty} (x + 5)$
11. $\lim_{x\to\infty} (x^2 + 2)$
12. $\lim_{x\to\infty} (x^2 - 3x + 1)$
13. $\lim_{x\to\infty} \frac{12}{x}$
14. $\lim_{x\to\infty} \frac{x}{2}$
15. $\lim_{x\to 2} \frac{x - 2}{x + 6}$
16. $\lim_{x\to 0} \frac{x^2 + 2}{x^2 - 5}$
17. $\lim_{x\to 1} \frac{x^2 + x - 2}{x^2 + 2x - 3}$
18. $\lim_{x\to 2} \frac{3x^2 - 2x - 8}{x - 2}$

47-6 INSTANTANEOUS RATE OF CHANGE

We have seen how to compute the average rate of change of a function. As an example, for an object in "free-fall" the distance s is given approximately by the formula, $s = 16t^2$. When $t = 3$ sec, $s = 144$ ft; when $t = 5$ sec, $s = 400$ ft. Then $\Delta t = 2$ sec, and $\Delta s = 256$ ft. Then average velocity is equal to $\Delta s/\Delta t$, or 128 ft/sec during the 2-sec interval.

Now suppose we wish to know the exact velocity at a particular instant, say, when $t = 3$ sec. For an object in "free-fall" the velocity is continuously changing. At the instant when $t = 3$ sec, the velocity is not the same as when $t = 3.1$ sec. To get the *instantaneous velocity* we cannot take the average between two values. It would seem that we must take $\Delta t = 0$ and $\Delta s = 0$. But then we should have $\Delta s/\Delta t = 0/0$, which is meaningless.

As another example, let us take the power formula. In an electric circuit suppose $P = 20i^2$. When $i = 4$ amps, $P = 320$ watts; when $i = 7$ amps, $P = 980$ watts. Then $\Delta i = 3$ amps; $\Delta P = 660$ watts. Then the average rate of change in power with respect to current i is

$$\frac{\Delta P}{\Delta i} = \frac{660 \text{ watts}}{3 \text{ amps}} = 220 \text{ watts per amp}$$

The result means that power changes at an average of 220 watts for each change of one ampere in current. However, now we may wish to know the rate of change of power at a particular instant, that is, *instantaneous rate of change.*

In many problems in science it is important to know the rate of change of a variable at some particular instant. If we wish to know the force of a 5-lb weight falling to the ground from a height of 100 ft, we must know the *instantaneous velocity* at the exact instant of impact. Even in driving a car, in the case of a collision, the damage done depends on the instantaneous velocity at the time of impact, not on the average velocity for the trip.

To see further the meaning of the instantaneous rate of change of a function and how it may be determined, let us consider the simple function, $y = x^2$. Let us begin with the values, $x = 4$ and $y = 16$. Beginning with these initial values, we let x take on an increment Δx which becomes smaller and smaller. We shall let the increment Δx approach zero and see what happens to Δy and to the ratio, $\Delta y/\Delta x$. The following table shows the results:

Δx	new x $x + \Delta x$	new y $y + \Delta y$	Δy	$\frac{\Delta y}{\Delta x}$
3	7	49	33	11
2	6	36	20	10
1	5	25	9	9
0.5	4.5	20.25	4.25	8.5
0.1	4.1	16.81	0.81	8.1
0.01	4.01	16.0801	0.0801	8.01
↓			↓	↓
0			0	?

From the table, as Δx approaches zero, and as Δy approaches zero, the ratio, $\Delta y/\Delta x$, approaches some constant. It appears from the table that the *limit of the ratio*, $\Delta y/\Delta x$, is 8. In this particular problem, this is the correct limit. This limit is called the *derivative.*

47-7 THE DERIVATIVE

The derivative of a function is so important that it must be clearly understood. The entire course of differential calculus is the study of the derivative and its uses.

In the foregoing example, the fifth column in the table shows the various values of the ratio $\Delta y/\Delta x$, but these values are not the derivative. The derivative is the *limit of this ratio* as the increment Δx approaches zero. The derivative does not approach a limit. It *is* the limit of the ratio $\Delta y/\Delta x$. Then, for the derivative, we have this definition:

Definition. *The derivative of a function, $y = f(x)$, is the limit of the ratio $\Delta y/\Delta x$ as the increment Δx approaches zero as a limit.* That is, the

$$\text{derivative} = \lim_{\Delta x \to 0} \frac{\Delta y}{\Delta x}; \qquad \lim_{\Delta x \to 0} \frac{\Delta f(x)}{\Delta x}$$

Two variables are involved in the derivative, the independent variable and the dependent variable. Stated in another way, *the derivative of a function is the limit of the ratio of the increment of the dependent variable to the increment of the independent variable as the latter increment approaches zero as a limit.*

To denote the derivative, we often use other symbols rather than the form showing the limit. For the derivative, the following notations are used:

$$\frac{dy}{dx}; \quad y'; \quad f'(x); \quad D_x; \quad D$$

47-8 FINDING THE DERIVATIVE: DELTA METHOD

There are four distinct steps in finding the derivative by the method suggested in the foregoing example. We illustrate the four steps by using the same function.

Example 1

Find the derivative, dy/dx, of the function $y = x^2$.

Solution

Step 1. We substitute the quantity, $x + \Delta x$, for x, and $y + \Delta y$ for y, in the equation:

$$y + \Delta y = (x + \Delta x)^2$$

Expanding,

$$y + \Delta y = x^2 + 2x(\Delta x) + (\Delta x)^2$$

Step 2. To get Δy alone on one side of the equation, we subtract the original function,

$$y \quad = x^2$$

and get

$$\Delta y = \quad 2x(\Delta x) + (\Delta x)^2$$

Step 3. To get the desired ratio, we divide both sides of the equation by Δx:

$$\frac{\Delta y}{\Delta x} = 2x + \Delta x$$

Step 4. As a final step, we let Δx approach zero. On the left side we state the result as a limit, which is the derivative. On the right side we get the value of the derivative:

$$\text{derivative} = \lim_{\Delta x \to 0} \frac{\Delta y}{\Delta x} = 2x$$

or in another form,

$$\frac{dy}{dx} = 2x$$

The result means that the derivative is twice the initial value of x. Recall that in the original example, we began with the initial value

$x = 4$, and we found that the derivative came to be 8, which is twice the original value of x. If we were to begin with $x = 6$, we should find that the derivative would equal 12.

The foregoing method of finding the derivative is by various names, such as the *delta method*, the *four-step rule*, the *increment method*, and others. This rule is basic because it is essentially the definition of the derivative. We shall discover many simple formulas for finding derivatives, but the delta method should be thoroughly known and memorized. All formulas for derivatives are based on the delta method. The process of finding a derivative is called *differentiation*. To *differentiate* a function means to find the derivative.

Summary of the Delta Method for Finding the Derivative

1. *Substitute $(x + \Delta x)$ for x, and $(y + \Delta y)$ for y. Expand if necessary.*
2. *Subtract the original function to get Δy alone on one side.*
3. *Divide both sides of the equation by Δx to get the ratio, $\Delta y/\Delta x$.*
4. *Let Δx approach zero as a limit. Then the left side of the equation is the definition of the derivative, and the right side shows its value.*

To evaluate a derivative for a particular value of the independent variable, we substitute the given value in the derivative.

Example 2

Find the derivative, dy/dx, of the following function and find the numerical value of the derivative at $x = 5$.

$$y = x^2 + 3x - 4$$

Solution

As a first step, we substitute $(y + \Delta y)$ for y, and $(x + \Delta x)$ for x:

$$y + \Delta y = (x + \Delta x)^2 + 3(x + \Delta x) - 4$$

Expanding, $$y + \Delta y = x^2 + 2x(\Delta x) + (\Delta x)^2 + 3x + 3(\Delta x) - 4$$

We subtract, $$y = x^2 + 3x - 4$$

Subtracting, $$\Delta y = 2x(\Delta x) + (\Delta x)^2 + 3(\Delta x)$$

Dividing by Δx, $$\frac{\Delta y}{\Delta x} = 2x + \Delta x + 3$$

Then $$\lim_{\Delta x \to 0} \frac{\Delta y}{\Delta x} = 2x + 3$$

The result is the derivative and can be written:

$$\frac{dy}{dx} = 2x + 3$$

For the value at $x = 5$, we have $$\frac{dy}{dx} = 13$$

Example 3

Find the derivative, ds/dt, of the following function and then find the value of the derivative for $t = 0.5$: $s = 3t^2 - 5t + 4$

Solution

Step 1. Substituting,

$$s + \Delta s = 3(t + \Delta t)^2 - 5(t + \Delta t) + 4$$

Expanding, $$s + \Delta s = 3(t^2 + 2t(\Delta t) + (\Delta t)^2) - 5t - 5(\Delta t) + 4$$

or, $$s + \Delta s = 3t^2 + 6t(\Delta t) + 3(\Delta t)^2 - 5t - 5(\Delta t) + 4$$

Subtract, $$s = 3t^2 - 5t + 4$$

We get $$\Delta s = 6t(\Delta t) + 3(\Delta t)^2 - 5(\Delta t)$$

Dividing by Δt, $$\frac{\Delta s}{\Delta t} = 6t + 3(\Delta t) - 5$$

As Δt approaches zero, $\dfrac{ds}{dt} = 6t - 5$; (the derivative).

For the value when $t = 0.5$, we have $ds/dt = 3 - 5 = -2$.

Example 4

Find the derivative, dy/dx, of the following function, and then evaluate the derivative for $x = 5$.

$$y = \frac{6}{x - 2}$$

Solution

Substituting, $$y + \Delta y = \frac{6}{x + \Delta x - 2}$$

Subtracting the original function:

$$\Delta y = \frac{6}{x + \Delta x - 2} - \frac{6}{x - 2}$$

Here we must combine the fractions at the right; the common denominator is the product of the two denominators. We get

$$\Delta y = \frac{6(x - 2) - 6(x + \Delta x - 2)}{(x - 2)(x + \Delta x - 2)}$$

or,

$$\Delta y = \frac{6x - 12 - 6x - 6(\Delta x) + 12}{(x - 2)(x + \Delta x - 2)}$$

This reduces to

$$\Delta y = \frac{-6\Delta x}{(x - 2)(x + \Delta x - 2)}$$

Dividing by Δx,

$$\frac{\Delta y}{\Delta x} = \frac{-6}{(x - 2)(x + \Delta x - 2)}$$

Now let Δx approach zero, we get

$$\frac{dy}{dx} = \frac{-6}{(x - 2)^2}$$

the derivative. For the value, $x = 5$, we have $dy/dx = -\frac{2}{3}$.

Exercise 47-4

By the delta method, find the expression for the derivative of each of the folowing functions:

1. $y = x^2 + 4$ **2.** $y = 3 - x^2$ **3.** $y = 2x^2 + 3x + 4$

4. $y = x^3$ **5.** $y = x^2 + 4x - 5$ **6.** $y = 3x^2 - 2x + 1$

7. $y = 7 - 6x - x^2$ **8.** $s = t^2 + 3t - 2$ **9.** $i = t^2 - 3t + 5$

10. $P = 15i^2$ **11.** $q = t^2 - 4t + 6$ **12.** $V = 8r^3$

13. $y = \frac{6}{x}$ **14.** $y = \frac{5}{x + 3}$ **15.** $y = \frac{4}{2 - 3x}$

Find the expression for the derivative of each of the following and evaluate the derivative for the given value of x or t:

16. $y = 5 + 3x - x^2$; $(x = -2)$ **17.** $y = x^2 - 5x + 5$; $(x = 3.5)$

18. $i = t^2 + 4t - 3$; $(t = 1)$ **19.** $i = 1 - 3t - 2t^2$; $(t = 3)$

20. $y = \frac{10}{x}$; $(x = 4)$ **21.** $i = \frac{12}{t + 1}$; $(t = 2)$

22. If $A = \pi r^2$, find dA/dr when $r = 8$.

23. If $P = 20i^2$, find the rate of change of power P with respect to current i when $i = 3$ amperes.

24. If $s = 16t^2$, find the instantaneous rate of change of s when $t = 4$.

47-9 DERIVATIVE BY FORMULA

Up to this point we have used the four-step rule, or delta method, to find the derivative. This rule is basic because it simply applies the definition of the derivative. Moreover, it is used to derive all formulas for the derivative. However, the delta method is often long and tedious. We shall see that there are several simpler rules for differentiation. These rules are themselves derived by the delta method.

We first derive the so-called *power rule*, one of the most useful of all formulas. To do so, we use the delta method. We have seen that

$$\text{if} \quad y = x^2, \quad \text{then} \quad \frac{dy}{dx} = 2x$$

Now let us take a higher power of x, say, $y = x^4$, and see what happens. Using the delta method, we have

Step 1. $y + \Delta y = (x + \Delta x)^4$

or, $y + \Delta y = x^4 + 4x^3(\Delta x) + 6x^2(\Delta x)^2 + 4x(\Delta x)^3 + (\Delta x)^4$

$y = x^4$

Step 2. $\Delta y = 4x^3(\Delta x) + 6x^2(\Delta x)^2 + 4x(\Delta x)^3 + (\Delta x)^4$

Step 3. $\dfrac{\Delta y}{\Delta x} = 4x^3 + 6x^2(\Delta x) + 4x(\Delta x)^2 + (\Delta x)^3$

Step 4. Now we let Δx approach zero. Each term on the right except the first contains the factor Δx, and therefore these terms approach zero as a limit. The left side of the equation becomes the expression for the derivative, and we get

$$\frac{dy}{dx} = \lim_{\Delta x \to 0} \frac{\Delta y}{\Delta x} = 4x^3$$

In the derivative, $4x^3$, note especially the following facts:

1. *The coefficient of the term is 4, which was the original exponent on x.*
2. *The original exponent on x has been decreased by 1.*

The result in the foregoing example shows exactly what happens when we find the derivative of a power on a variable. We get a similar result whatever the original power on x. If we were to begin with the function, $y = x^8$, and go through all four steps, we should find that:

$$\frac{dy}{dx} = 8x^7$$

The power rule may be stated as a formula. If n represents any power on x, and

$$\text{if } y = x^n, \quad \text{then} \quad \frac{dy}{dx} = nx^{n-1}$$

This power rule is one of the most useful rules. It enables us to find the derivative very quickly. The rule applies to all kinds of exponents, positive, negative, and fractional. It applies to any variable raised to a power.

Examples

(1) If $y = x^{15}$ then $\frac{dy}{dx} = 15x^{14}$

(2) If $y = x^{-2}$, $\frac{dy}{dx} = -2x^{-3}$

(3) If $i = t^{\frac{1}{3}}$, $\frac{di}{dt} = (\frac{1}{3})t^{-\frac{2}{3}}$

(4) If $s = t^4$, $\frac{ds}{dt} = 4t^3$

Differentiation of a function can be conveniently indicated by the symbol, "d/dx," called the *derivative operator.* To see how this notation is used, consider the function, x^5. To find the derivative of x^5, we usually first let y represent the function: $y = x^5$; then we say, $dy/dx = 5x^4$.

Now, instead of using y to represent the function, we can indicate differentiation by placing the symbol, d/dx, before the function to be differentiated:

$$\frac{d}{dx}(x^5) = 5x^4$$

The symbol, d/dx, is called the *derivative operator* and indicates that the quantity following the symbol is to be differentiated.

The derivative operator is sometimes denoted by a single symbol, D_x, in which the subscript x represents the independent variable. The subscript may be omitted if there is no doubt as to the independent variable. Examples:

$$D_x(x^9) = 9x^8 \qquad D(x^4) = 4x^3 \qquad D_t(t^7) = 7t^6$$

The following *special rules* for differentiation are simply special applications of the power rule. They enable us to find derivatives quickly. They can all be derived by the delta method.

1. *The derivative of the product of a constant and a variable is equal to the constant times the derivative of the variable.*

Example

If $y = 7x^4$, $\frac{dy}{dx} = 7(4x^3) = 28x^3$.

That is, when the variable has a constant coefficient, we
(a) *multiply the coefficient by the exponent on the variable,* then
(b) *decrease the exponent by 1.*

In general, if c is any constant and if $y = cx^n$, then $dy/dx = cnx^{n-1}$. Using the derivative operator, we have

$$\frac{d}{dx}(cx^n) = cnx^{n-1}$$

2. *The derivative of the product of a constant and a variable to the first power is equal to the constant alone.* That is,

$$\text{if } y = 5x, \qquad \text{then } \frac{dy}{dx} = 5; \qquad \text{or} \qquad \frac{d}{dx}(5x) = 5$$

In general,

$$\text{if } y = cx, \qquad \text{then } \frac{dy}{dx} = c; \qquad \text{or} \qquad \frac{d}{dx}(cx) = c$$

3. *The derivative of a constant is zero.* That is, if $y = 8$, $dy/dx = 0$.

4. *The derivative of a variable with respect to itself is 1.* Examples:

$$\text{If } y = x, \qquad \frac{dy}{dx} = 1; \qquad \frac{d}{dx}(x) = 1; \qquad \text{if } y = t, \qquad \frac{dy}{dt} = 1$$

5. *A multinomial may be differentiated term by term.* Example:

$$\text{If } y = 5x^3 + 4x^{-2} + 8x^{\frac{1}{4}} - 3x + 2,$$

$$\frac{dy}{dx} = 15x^2 - 8x^{-3} + 2x^{-\frac{3}{4}} - 3$$

47-10 HIGHER DERIVATIVES

It is sometimes desirable to find what we call *higher derivatives.* The derivative of a function of x is also generally a function of x. For example,

$$\text{if } y = 5x^3 + 4x^2 - 3x + 2, \qquad \text{then} \qquad y' = \frac{dy}{dx} = 15x^2 + 8x - 3$$

Now we may wish to find the *derivative of the derivative.* The result is called the *second derivative* of the original function. The second derivative can be indicated in various ways. If y is a function of x, we can use any of the following symbols for the second derivative:

$$\frac{d}{dx}\left(\frac{dy}{dx}\right); \qquad \frac{d^2y}{dx^2}; \qquad y''; \qquad f''(x); \qquad D^2(y)$$

The third and higher derivatives are denoted in a similar manner.

Example

If $y = x^4 - 3x^3 - x^2 + 9x - 5$, find the first three derivatives of the function; then evaluate each derivative for $x = 0$, and for $x = 1$.

Solution

$$\frac{dy}{dx} = 4x^3 - 9x^2 - 2x + 9; \qquad \frac{d^2y}{dx^2} = 12x^2 - 18x - 2;$$

$$\frac{d^3y}{dx^3} = 24x - 18.$$

If we use $f(x)$ to represent the function, we can represent the derivatives by $f'(x)$; $f''(x)$; and $f'''(x)$. This notation is convenient when we wish to indicate the values of the derivatives. For the first derivative, when $x = 0$, we write $f'(0)$. Then, for the derivatives and their values we have

$$f'(x) = 4x^3 - 9x^2 - 2x + 9; \quad \text{then } f'(0) = 9; \quad f'(1) = 2$$

$$f''(x) = 12x^2 - 18x - 2; \quad \text{then } f''(0) = -2; \quad f''(1) = -8$$

$$f'''(x) = 24x - 18; \quad \text{then } f'''(0) = -18; \quad f'''(1) = 6$$

Exercise 47-5

Find the instantaneous rate of change of y with respect to x when x has the indicated value. Use the power rule.

1. $y = x^2 - 5x + 2, (x = 1)$

2. $y = 3x^2 - 7x + 3, (x = 2)$

3. $y = 2x^3 + x^2 - 3x - 7, (x = 4)$

4. $y = 4x^5 - 7x^4 + 3x^3 - x, (x = -2)$

5. $y = 15x^3 - 8x^4 - x^5, (x = -1)$

6. $y = 3x^{-2} - 6x^{\frac{1}{2}}, (x = 4)$

7. $y = \frac{2}{x}, (x = 3)$

8. $y = x^{\frac{2}{3}}, (x = 8)$

In the next four problems (Nos. 9–12), find the derivative of the function with respect to the independent variable, and evaluate as indicated.

9. $s = 2t^3 - 6t^2 + 3t, (t = 2)$

10. $i = 3t^2 + 2t + 5, (t = 3)$

11. $A = 8\pi r + 2\pi r^2, (r = 4)$

12. $V = (\frac{4}{3})\pi r^3, (r = 6)$

Find the following:

13. $\frac{d}{dx}(3x^4 - 5x^2 + 3)$

14. $\frac{d}{dt}(t^3 - 5t^2 + 8t)$

15. $\frac{d}{dw}(80w - 2w^2)$

Find the second and third derivatives of each of the following (Nos. 16–18):

16. $y = x^4 - 5x^3 + 3x^2$

17. $y = x^{\frac{3}{2}} + 4x^{\frac{1}{3}}$

18. $y = t^{-2} - 5t^{-1}$

19. If $f(x) = 3x^2 - 4x - 2$, find $f'(x)$; $f'(1)$; $f'(0)$.

20. If $f(r) = r^2 - 5r + 1$, find $f'(r)$; $f'(1.5)$; $f'(0)$.

21. If $g(t) = t^2 + 5t - 3$, find $g'(t)$; $g'(-2)$; $g'(4)$.

22. In each of these functions, for what values of x is y' equal to zero?

(a) $y = x^2 - 6x + 5$; (b) $y = x^3 - 12x$;
(c) $y = 2x^3 - 9x^2 + 12x - 5$

23. If $f(x) = x^3 - 3x^2 - 9x + 10$, for what values of x is $f'(x)$ equal to zero?

24. If $s = t^3 - 6t^2 + 12t - 5$, for what values of t is ds/dt equal to zero?

25. Find the instantaneous rate of change of the area of a circle with respect to the radius when the radius is 5 in.

26. In a certain electric circuit, $P = 30i^2$. Find the rate of change of power P with respect to current i when i equals 2.5 amps.

27. In a certain electric circuit, $P = E^2/R$, and $E = 80$ volts. Find the rate of change of power P with respect to resistance R, when $R = 200$ ohms.

QUIZ ON CHAPTER 47, INTRODUCTION TO CALCULUS. FORM A.

Evaluate the following functions:

1. If $f(x) = x^2 + 3x - 5$, find $f(1)$; $f(0)$; $f(-1)$.

2. If $f(t) = 3 + 4t - t^2$, find $f(0)$; $f(1)$; $f(a)$.

3. If $g(x) = \log_2 (x + 11)$, find $g(-3)$; $g(5)$.

4. If $f(x) = \sin x$, find $f(\frac{\pi}{6})$; $f(\pi)$.

5. If $f(x) = 4^x$, find $f(0)$; $f(3)$; $f(-2)$.

6. If the temperature changes from 84° to 56° in the time from 3 P.M. to 11 P.M., find the average rate of change per hour.

7. If an electric current changes from 8.3 amps to 7.1 amps in 3 sec, find the average rate of change of current per second.

8. Find the average rate of change of y with respect to x, if $y = x^2 + 3x$, and x changes from 3 to 5.

9. Find the following limits:

(a) $\lim_{x\to 2} (x^2 + 3x - 1)$ **(b)** $\lim_{x\to 4} \frac{x^2 - 16}{x - 4}$ **(c)** $\lim_{x\to\infty} \frac{8}{x}$

(d) $\lim_{x\to 0} \frac{6}{x}$ **(e)** $\lim_{x\to\infty} (x + 3)$

10. Find the derivative, dy/dx, and evaluate for the given value: (use delta method)

$$y = x^2 - 3x - 2; \quad (x = 3)$$

11. Find the derivative and evaluate for the given value: (use delta method):

$$y = \frac{10}{x}; \quad (x = 4)$$

12. Find the derivatives by formula:

(a) $y = 2x^3 - 5x^2 + 3x - 4$ **(b)** $i = t^2 + 4t + 7$

(c) $y = 8t^{-3}$ **(d)** $y = 12x^{\frac{1}{2}}$

13. Given, $y = x^3 + 6x^2 - 8x + 5$, find $y'(0)$; $y''(0)$.

CHAPTER FORTY-EIGHT The Derivative: Applications

48-1 DISTANCE; VELOCITY; ACCELERATION

We have already referred to *velocity as the rate of change of distance.* To find average velocity between two points, we measure the distance between them and then measure the time that elapses in passing from one point to the other. Suppose a car passes point A and then 5 sec later it passes point B (Fig. 48-1). Suppose A and B are 200 ft apart.

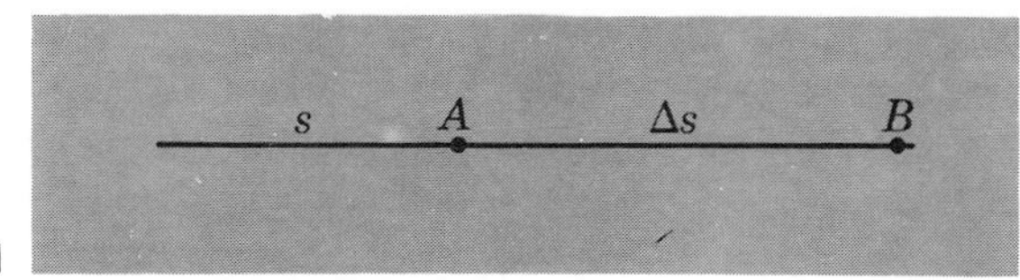

Figure 48-1

If s represents the distance covered before A, then the distance AB we call Δs, the increment in distance. The time 5 sec is Δt, the increment in time. Dividing, we get

$$\text{average velocity} = \frac{\Delta s}{\Delta t} = \frac{200 \text{ ft}}{5 \text{ sec}} = 40 \text{ ft per sec}$$

Now we are concerned with *instantaneous velocity* at a point. To get instantaneous velocity at A, we let Δt get smaller and approach zero. Then Δs also approaches zero. The instantaneous velocity at A will be the limit of the ratio, $\Delta s/\Delta t$, as Δt approaches zero. That is,

$$\text{instantaneous velocity } v = \lim_{\Delta t \to 0} \frac{\Delta s}{\Delta t}$$

Note that this limit is the derivative of s with respect to t:

$$v = \frac{ds}{dt} = s'; \qquad \text{or} \qquad v = f'(t)$$

Example 1

An object in free-fall within a reasonable distance from the earth's surface falls approximately according to the formula:

$s = 16t^2$; (distance s in ft, velocity v in ft/sec; t in seconds)

How far and what will be the velocity at the end of 5 seconds?

Solution

For distance s, we have $s = 16t^2$ (ft)
For the velocity, we have

$$v = \frac{ds}{dt} = 32 \text{ (ft/sec)}$$

At the end of 5 seconds,

$$s = 400 \text{ ft.}; \qquad v = 160 \text{ ft/sec}$$

Example 2

An object is dropped from a plane at a height of 3600 feet. Find the velocity of the object when it hits the earth. (neglect air resistance).

Solution

First, we must find the time of the fall, in seconds. Since the distance is 3600 feet, we set $s = 3600$, and have the equation

$$3600 = 16t^2$$

Solving for t,

$$225 = t^2; \qquad \text{then} \qquad t = 15 \text{ seconds}$$

For the velocity, we have

$$v = \frac{ds}{dt}; \qquad \text{then} \qquad v = 32t$$

When time $t = 15$ seconds,

$$v = (32)(15) = 480; \quad \text{(ft/sec; velocity at impact)}$$

Example 3

A ball is projected directly upward from the top of a building. Its distance s (in feet) from the ground at any time t (in seconds) is given by the formula

$$s = 64 + 96t - 16t^2$$

Find the height of the building, the initial velocity of the ball, and the maximum height reached by the ball (neglect air resistance).

Solution

We have the formula for distance s: $s = 64 + 96t - 16t^2$.
For velocity, we take the derivative of distance: $v = 96 - 32t$.
When the ball leaves the top of the building, we take time $t = 0$. Then, also we have the initial velocity of the ball. For $t = 0$, we have

$$s = 64 + 0 - 0 = 64 \text{ (ft)}; \qquad \text{height of building}$$

$$v = 96 - 0 = 96 \text{ (ft/sec)}; \qquad \text{initial velocity}$$

When the ball reaches its maximum height, its velocity is zero. Therefore, we set

$$96 - 32t = 0; \qquad \text{solving for } t, \text{ we get } t = 3 \text{ (seconds)}$$

The result means the ball reaches its maximum height at the end of 3 seconds. To find the height, we use 3 for t in the formula for s, and get

$$s = 64 + 288 - 144 = 208 \text{ (ft, maximum height)}$$

Example 4

A particle moves according to the formula, $s = t^2 - 6t + 5$ (s in feet, t in seconds). Find the position (distance from zero), and the velocity when $t = 0$; 1 sec; and 4 sec. When and where is the particle at rest?

Solution

Since

$$s = t^2 - 6t + 5 \qquad \text{(formula for distance } s\text{)}$$

$$v = \frac{ds}{dt} = 2t - 6 \qquad \text{(formula for velocity } v\text{)}$$

At $t = 0$, $s = 5$ ft; $v = -6$ft/sec; at $t = 1$, $s = 0$; $v = -4$ ft/sec. At $t = 4$, $s = -3$ ft; $v = 2$ ft/sec. When the particle is at rest, $v = 0$; that is, $2t - 6 = 0$; then $t = 3$ sec. When $t = 3$ sec, $s = -4$ ft.

Acceleration is a *change in velocity*. Then *instantaneous acceleration* is the *instantaneous rate of change of velocity*. Let us first consider *average acceleration*. Suppose two types of cars are each traveling at 10 ft per sec. Some time later each one is traveling at 70 ft per sec. Each car has increased its velocity by 60 ft per sec. However, one car required 20 sec, the other only 5 sec to make the change. If we represent the changes in velocity and time by Δv and Δt, respectively, then average acceleration is given by the ratio, $\Delta v/\Delta t$. The following table shows Δv and Δt, and the ratio of the increments:

	v_1	v_2	Δv	Δt	$\Delta v/\Delta t$
First car	10	70	60 ft/sec	20 sec	3 ft/sec^2
Second car	10	70	60 ft/sec	5 sec	12 ft/sec^2

The last column shows average acceleration during the time interval. To get instantaneous acceleration at a point, we let Δt approach zero as a limit. Then Δv also approaches zero. Then the *limit of the ratio*, $\Delta v/\Delta t$, we call *instantaneous acceleration a*.

$$\text{Instantaneous acceleration } a = \lim_{\Delta t \to 0} \frac{\Delta v}{\Delta t} = \frac{dv}{dt} = v'$$

We have seen that $v = ds/dt$, the first derivative of distance s. Then

$$a = \frac{dv}{dt} = v' = \frac{d}{dt}\left(\frac{ds}{dt}\right) = \frac{d^2s}{dt^2} = s'' = f''(t)$$

To summarize:

1. *Velocity is the first derivative of distance s, that is,* $v = ds/dt$.
2. *Acceleration is the first derivative of velocity v, and the second derivative of distance s.*

Example 5

A particle moves according to the formula, $s = t^3 - 6t^2 + 9t - 5$ (s in feet, t in seconds.) Find the position, velocity, and acceleration when $t = 0$, and when $t = 2$. When and where is the particle at rest?

Solution

Given:

$$s = t^3 - 6t^2 + 9t - 5$$

then

$$v = 3t^2 - 12t + 9$$

$$a = 6t - 12$$

When $t = 0$, $s = -5$ft; $v = 9$ ft/sec; $a = -12$ ft/sec^2
When $t = 2$, $s = -3$ ft; $v = -3$ ft/sec; $a = 0$
The particle is at rest when $v = 0$; that is, $3t^2 - 12t + 9 = 0$. Solving, we get, $t = 1$ and $t = 3$. When $t = 1$, $s = -1$ ft; when $t = 3$, $s = -5$ ft

Exercise 48-1

In each of the first 8 exercises below, find distance s, velocity v, and acceleration a, when $t = 0$ and when $t = 1$ (s in feet, t in seconds).

1. $s = 4t + 5$ **2.** $s = 3t^2 - 2t + 4$ **3.** $s = 4t^2 + 3t - 2$

4. $s = t^2 - 5t - 4$ **5.** $s = 8 + 3t - t^2$ **6.** $s = t^3 - 6t^2 + 15t$

7. $s = 2t^4 - 3t^3 + 4t^2 - 9t - 5$ **8.** $s = t^4 - 2t^3 + 5t^2 - 3t + 6$

9. A particle moves according to the formula, $s = t^3 - 9t^2 + 15t - 4$. Find the position, velocity, and acceleration when $t = 0$, and when $t = 4$ sec. When and where is the particle at rest? Show the motion on a horizontal line.

10. A particle moves according to the formula, $s = 2t^3 - 5t^2 + 4t - 6$. Find the position and acceleration when the particle is at rest.

11. A ball is projected upward from the top of a building, and its distance from the ground is given by the formula: $a = 80 + 96t - 16t^2$. Find the following: (a) height of the building;

(b) initial velocity of the ball; (c) number of seconds it takes to reach the highest point; (d) maximum height reached by the ball; (e) velocity when it returns to the ground.

12. A ball is projected directly upward from the top of a building, and its distance from the ground is given by the formula: $s = 90 + 80t - 16t^2$. Find the facts asked for in No. 11.

13. A small bore rifle bullet is fired directly upward from ground level, and its distance from the ground is given by the formula, $s = 640t - 16t^2$ (s in feet, t in seconds). Neglecting air resistance, how high will the bullet rise? What is the muzzle velocity of the bullet?

14. A high-powered rifle bullet is fired directly upward from ground level, and its distance from the ground is given by the formula, $s = 2640t - 16t^2$. Neglecting air resistance, how high will it rise (how many miles)? What is the muzzle velocity of the bullet?

15. A bomb is dropped from a plane at a height of 19,600 ft above the surface of the earth. Find the distance and the velocity at the end of 10 sec. Find the velocity of the object at the time of impact with the earth.

16. Near the surface of the moon, an object falls approximately according to the formula: $s = 2.6t^2$. Find the distance and velocity at the end of 20 sec.

17. An object is released at a height of 585 ft above the surface of the moon. Find the velocity of the object when it hits the moon's surface.

18. The motion of a particle is given by the formula: $s = 10 - 9t + 6t^2 - t^3$. Describe the position, the velocity, and the acceleration at $t = 0$. When and where is the particle at rest? What is the acceleration when $v = 0$?

48-2 DERIVATIVE AS THE SLOPE OF A CURVE

The derivative has an important geometric interpretation. If we graph a function $y = f(x)$, we shall find that the derivative happens to be equivalent to the slope of the curve at any point on the curve. Let us see why this is so.

For the slope of a straight line, we take two points on the line and then define the slope by the ratio, *rise/run*. However, in considering the slope of a curve we run into difficulties. The slope of a curve is continuously changing. We cannot take two points on the curve and use the same ratio. Our problem is to define the slope of a curve at a single point. If we draw a line tangent to the curve at the point, then we define the slope of the curve at a point as follows:

Definition. *The slope of a curve at a single point on the curve is defined as the slope of the tangent line at that point.*

To show that the slope of a curve at a point is equivalent to the derivative, we draw the graph representing the general function $y = f(x)$ (Fig. 48-2). Let $P(x, y)$ be any point on the curve. Now we let

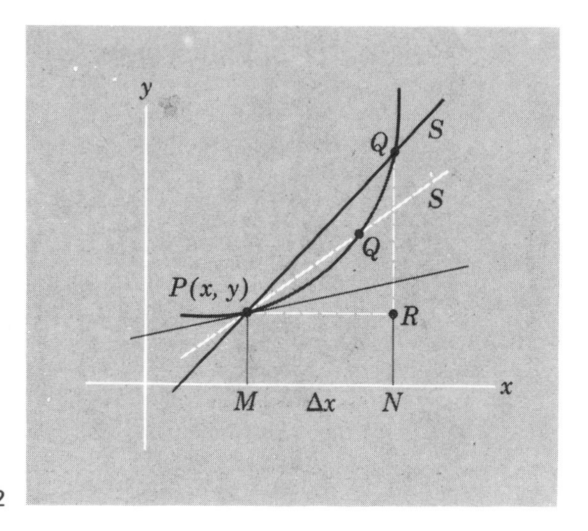

Figure 48-2

x take on an increment, $\Delta x = MN$. Then y changes by some increment, $\Delta y = RQ$. The secant line s through P and Q has the slope:

$$m = \frac{\text{rise}}{\text{run}} = \frac{\Delta y}{\Delta x} = \frac{RQ}{PR}$$

Now we let point Q move along the curve toward P, so that Δx and Δy become smaller and smaller. Until Q reaches P, the secant line will always have the slope $\Delta y/\Delta x$. The limiting position of the secant line is the tangent line at P. As Δx and Δy approach zero, the slope of the secant line approaches the slope of the tangent line. Since $\Delta y/\Delta x$ approaches dy/dx, the derivative, we get

$$\frac{dy}{dx} = \text{slope of the tangent}$$

Since the slope of the tangent is equivalent to the slope of the curve, we have

$$\frac{dy}{dx} = \textit{slope of the curve at any point on the curve}$$

Example 1

Given, the function, $y = x^2 - 4x - 5$, find (a) the slope of the curve at the point $(3, -8)$; (b) the equation of the tangent at $(3, -8)$.

Solution

The graph is shown in Fig. 48-3. Since

$$y = x^2 - 4x - 5$$

$$\frac{dy}{dx} = 2x - 4$$

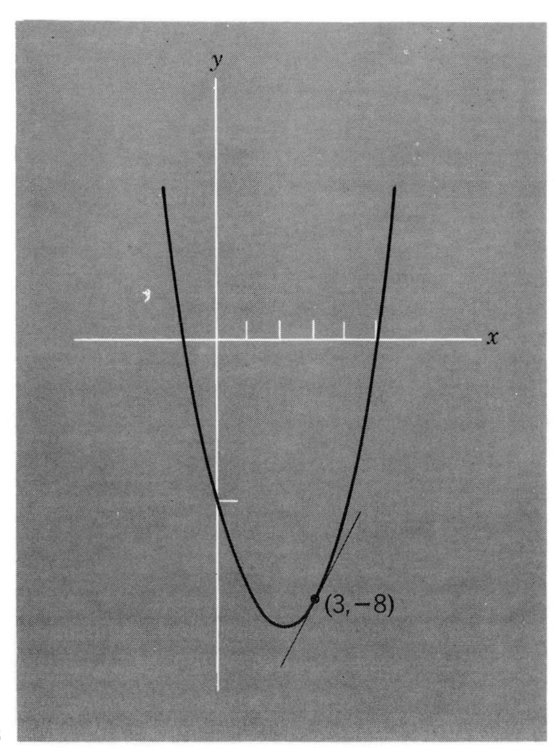

Figure 48-3

(a) The formula for the slope is: $m = 2x - 4$. Where $x = 3$, $m = 6 - 4 = 2$.
(b) The tangent has the same slope as the curve at the point $(3, -8)$. Using the slope $= 2$ and the point $(3, -8)$, we get the equation of the tangent:
$2x - y = 14$

Example 2

Where does the curve, $y = x^2 - 6x + 8$, have a horizontal tangent?

Solution

For a horizontal tangent, the slope must be zero. Then we have

$$\frac{dy}{dx} = 2x - 6 = m; \qquad \text{then } 2x - 6 = 0; \qquad x = 3$$

Therefore the curve has a slope of zero at $(3, -1)$.

Example 3

Given the function, $y = x^3 - 3x^2 - 9x + 12$, find the points on the graph where $m = 0$. Also find the equation of the tangent where $x = 2$.

Solution

Since	$y = x^3 - 3x^2 - 9x + 12$
Differentiating,	$dy/dx = 3x^2 - 6x - 9$
For $m = 0$, we set	$3x^2 - 6x - 9 = 0$
Solving for x,	$x = -1$, and $x = 3$

To find y-values, we find that when $x = -1$, $y = 17$; when $x = 3$, $y = -15$. The slope is zero then at the points $(-1, 17)$ and $(3, -15)$.

Where $x = 2$, $m = -9$, and $y = -10$. Equation of tangent: $9x + y = 8$.

Exercise 48-2

For the first four functions, find the slope at the given points.

1. $y = x^2 - 2x - 3$; at $(2, -3)$; $(-1, 0)$; where $x = 1.5$.
2. $y = x^3 - 2x^2 - 5x + 4$; at $(1, -2)$; $(0, 4)$; $(-1, 6)$.
3. $y = 2x^3 + 4x^2 - 5x - 3$; at $(1, -2)$; $(-1, 4)$; $(0, -3)$.
4. $y = 2x^3 - 3x^2 - 12x + 12$; at $(0, 12)$; $(1, -1)$; where $x = 1.5$.

In each of the next six functions (Nos. 5–10), write the equation of the tangent to the curve at the given point.

5. $y = x^2 - x - 2$; at $x = 1$
6. $y = 2x^2 - 5x - 1$; at $x = 2$
7. $y = 4 - 2x - x^2$; at $x = -1$
8. $y = x^2 - 2x - 6$; at $x = 0$
9. $y = x^2 - x + 3$; at $(1, 3)$
10. $y = 3x^2 - 2x + 6$; at $(1, 7)$

In each of the next eight functions (Nos. 11–18), find the point where $m = 0$.

11. $y = x^2 - 4x - 3$
12. $y = 5 + 2x - x^2$
13. $y = 2x^2 - 5x + 3$
14. $y = 3x^2 - 4x - 2$
15. $y = x^3 - 3x^2 - 9x + 10$
16. $y = 4 + 15x + 6x^2 - x^3$

17. $y = 2x^3 - 3x^2 - 36x + 8$ **18.** $y = 6 + 24x - 9x^2 - 2x^3$

19. If $y = x^3 - 3x^2 - 12x - 6$, find the equation of the tangent where $x = -1$.

20. If $y = 7 + 15x + 6x^2 - x^3$, find the equation of the tangent at $x = 2$.

21. If $i = t^2 - 3t + 2$, find the value of i when $di/dt = 0$.

22. If $i = t^3 - 9t^2 + 15t + 5$, find the value of i when $di/dt = 0$.

48-3 MAXIMUM AND MINIMUM POINTS AND VALUES

Consider the graph of a general function, $y = f(x)$ (Fig. 48-4). To estimate the slope of the curve at various points we draw a short line tangent to the curve at each point. Then the slope of the curve is equal to the slope of the tangent at that point.

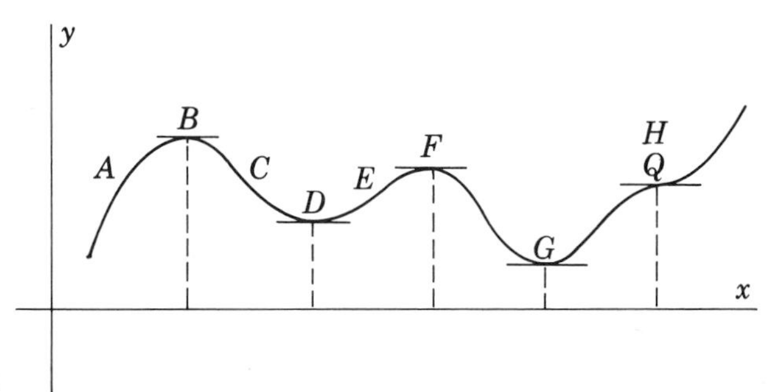

Figure 48-4

A point such as B is called a *maximum point.* A maximum point is a point that is higher than a nearby point on either side, like the top of a hill. At a maximum point, the function has a *maximum value.* A curve may have more than one maximum point. Each is then called a *relative maximum.* An *absolute maximum* is a point higher than any other point on the entire curve.

A point such as D is called a *minimum* point. A minimum point is a point that is lower than a nearby point on either side, like the lowest point in a valley between two hills. At a minimum point, the function has a *minimum value.* Note that this curve has two *relative maximum* and two *relative minimum* points, but no absolute maximum or minimum point.

At one point on the curve, at H, the slope is zero. However, this point is neither a maximum nor a minimum. Then we may form these conclusions:

(a) At a maximum or a minimum point, the slope must be zero; $dy/dx = 0$.

(b) A zero slope does not guarantee a maximum or a minimum point.

Note four facts about a maximum point (B): (1) at a maximum point, $m = 0$; (2) a little to the left, as at A, m is positive; (3) a little to the right, as at C, m is negative; (4) as the curve passes through a maximum point, the slope m changes from *positive to negative.* These facts are used to determine whether or not a point is a maximum.

Note four facts about a minimum point (D): (1) at a minimum point, $m = 0$; (2) *a little to the left, as at* C, m is negative; (3) a little to the right, as at E, m is positive; (4) as the curve passes through a minimum point, the slope m changes from *negative to positive.* These facts are used to determine whether or not a point is a minimum.

To find a maximum or a minimum point on a curve, we first set the derivative equal to zero and solve for x. This will show the x-values of points where the slope m is zero. Such values are called *critical values.* However, a point where the slope is zero may or may not be a maximum or minimum. The point must be tested.

To test a point where $m = 0$, we take one x-value a little to the *left* and another x-value a little to the *right* of the critical value. Then we determine the sign of the slope at either side of the critical value. As the curve passes through the critical point (where $m = 0$),

1. *If the slope changes from (−) to (+), the point in question is a minimum.*
2. *If the slope changes from (+) to (−), the point is a maximum.*
3. *If the sign of the slope does not change, the point is neither maximum nor minimum.*

Example 1

Find any maximum and/or minimum points on the curve:

$$y = x^2 - 6x + 4$$

Solution

Differentiating, we get $dy/dx = 2x - 6 = m$. For a maximum or minimum, m must equal zero. Then $2x - 6 = 0$; $x = 3$. When $x = 3$, $y = -5$. Then the point $(3, -5)$ may be a maximum or minimum. As a test, we take an x-value on either side of $x = 3$. We can use the values, $x = 2$ and $x = 4$, and find the sign of the slope for these values.

At $x = 2$, $m = 4 - 6 = (-)$; at $x = 4$, $m = 8 - 6 = (+)$. Since the slope changes from *negative to positive,* the point $(3, -5)$ is a *minimum.* The function then has a minimum value of -5.

Example 2

Find any maximum and/or minimum points on the curve,

$$y = x^3 - 3x^2 - 9x + 12$$

Solution

Differentiating, $dy/dx = 3x^2 - 6x - 9$. We set $3x^2 - 6x - 9 = 0$; solving, $x = -1$ and $x = 3$. These x-values represent the points, $(-1, 17)$ and $(3, -15)$. These may be maximum or minimum points. To test the value, $x = -1$, we use $x = -2$, and $x = 0$.

At $x = -2$, $m = 12 + 12 - 9 = (+)$; at $x = 0$, $m = 0 - 0 - 9 = (-)$. Since the slope changes from $(+)$ to $(-)$, the point $(-1, 17)$ is a maximum. To test the value, $x = 3$, we use $x = 0$ and $x = 4$.

At $x = 0$, $m = (-)$; at $x = 4$, $m = 48 - 24 - 9 = (+)$. Then the point $(3, -15)$ is a minimum. The graph is shown in Fig. 48-5. The vertical scale has been compressed. This is done often when the function has large values.

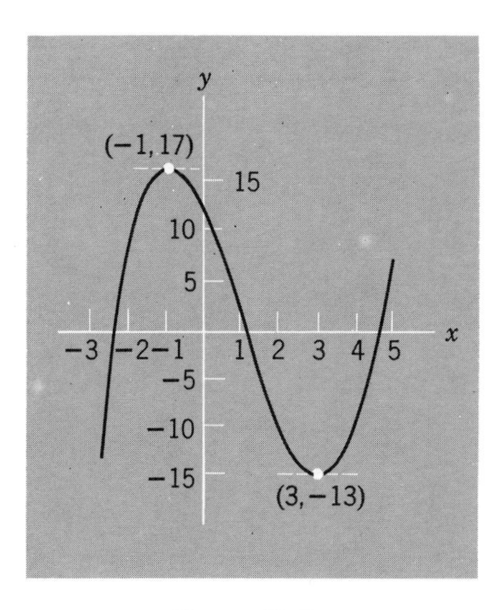

Figure 48-5

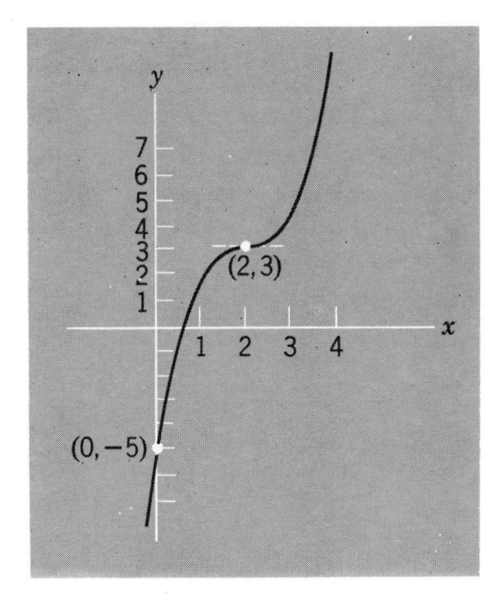

Figure 48-6

Example 3

Find any maximum and/or minimum points on the curve,

$$y = x^3 - 6x^2 + 12x - 5$$

Solution

Differentiating, $dy/dx = 3x^2 - 12x + 12$. We set $3x^2 - 12x + 12 = 0$; solving, $x = 2$, and 2, a double value. The corresponding y-value is 3. To test the point $(2, 3)$, we use $x = 1$, and $x = 3$. At $x = 1$, $m = (+)$; at $x = 3$, $m = (+)$. Since the slope does not change in sign, the point $(2, 3)$ is neither maximum nor minimum (Fig. 48-6). In this example, a point such as $(2, 3)$ is called an *inflection* point. An inflection point is a point at which the sense of curvature changes, from a left turn to a right turn, or from right to left. *At such a point, the second derivative is equal to zero.*

Maximum and minimum points may be tested sometimes by what is called the *second derivative test.* Since the derivative, dy/dx, is equal to the slope m, then the second derivative, d^2y/dx^2, is the *rate of change of slope,* or dm/dx. At a maximum point the slope is zero, but it is *decreasing* in value, and d^2y/dx^2 is *negative.* At a minimum point the slope is zero, but it is *increasing* in value, and d^2y/dx^2 is *positive.* Therefore, where $m = 0$,

if d^2y/dx^2 *is negative, the point is a maximum;*
if d^2y/dx^2 *is positive, the point is a minimum;*
if d^2y/dx^2 *is zero, this test fails.*

In Example 2 above, $d^2y/dx^2 = 6x - 6$. Then

at $x = -1$, $d^2y/dx^2 = (-)$; therefore, $(-1, 17)$ is a maximum point,

at $x = 3$, $d^2y/dx^2 = (+)$; therefore, $(3, -15)$ is a minimum point.

In Example 3, the second derivative test fails.

Exercise 48-3

Find any maximum and/or minimum points on each of the following curves. Sketch each curve. Test each critical point by using an x-value on either side. Try the second derivative test for some.

1. $y = x^2 - 4x + 1$

2. $y = x^2 + 2x - 3$

3. $y = 6 + 2x - x^2$

4. $y = 4 + 3x - x^2$

5. $y = x^2 + 5x + 4$

6. $y = 2 + x - x^2$

7. $y = 2x^2 - 6x + 5$

8. $y = 5 + 4x - 3x^2$

9. $y = x^3 - 9x^2 + 15x + 6$

10. $y = x^3 + 6x^2 + 9x + 2$

11. $y = x^3 - 12x + 9$

12. $y = 2x^3 - 9x^2 - 12$

13. $y = 2x^3 - 15x^2 - 36x - 32$

14. $y = 4 - 24x - 15x^2 - 2x^3$

15. $y = 6 - 24x - 9x^2 - x^3$

16. $y = 18 + 9x + 3x^2 - x^3$

17. If $s = 60 + 96t - 16t^2$, find when and where s is a maximum.

18. If $s = 6 + 15t - 2.5t^2$, find when and where s is a maximum.

19. If the current i in a circuit is given by the formula, $i = t^2 - 4t + 10$, find the maximum current i.

20. If the electric charge q is given by the formula, $q = t^2 - 6t + 12$, find the maximum charge, q.

48-4 PROBLEMS IN MAXIMA AND MINIMA

In many practical problems it is necessary to find a maximum or a minimum value of a function. We have seen how to find and test such values on a graph. However, in many practical problems we can find maximum or minimum values without the use of a graph. If we can write the proper equation showing the relation between a variable and its function, we can find the derivative, set it equal to zero, and solve. We can do this without a graph. At first a graph may help to understand the meaning of maximum and minimum values, but in many problems a graph is not necessary.

Moreover, in many practical problems it is not necessary to test the answers. The nature of the problem will usually determine whether an answer represents a maximum or minimum.

The following steps may serve as a guide in solving problems:

1. Make a sketch of the figure for the problem, if possible.
2. Set up a general equation that is always true for all values of the variables. This general equation is often simply a formula.
3. If the formula contains more than one independent variable, eliminate all but one by finding a relation between the independent variables.
4. Find the derivative with respect to the independent variable.
5. Set the derivative equal to zero and solve for the independent variable.
6. Find the corresponding value of the function.

Example 1

A farmer wishes to enclose part of his land along a river to form a rectangular field. No fence is needed along the river. If he has 180 rods of fencing, find the dimensions of the field for a maximum area.

Solution

We make a sketch of the field along the river, and indicate the width and length in general terms, l and w (Fig. 48-7). Since the area is to be a maximum, we write the equation for area:

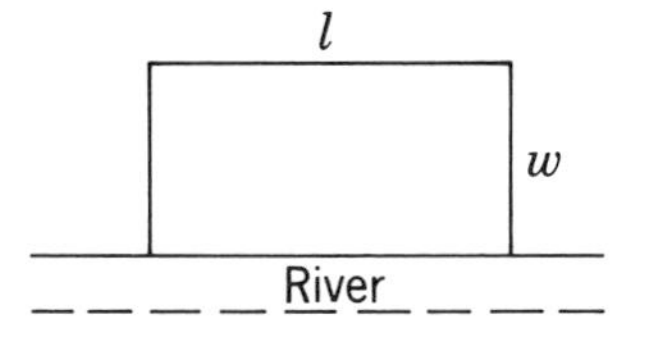

Figure 48-7

$$A = lw$$

Here we have two independent variables. We eliminate one by expressing it in terms of the other. We note that

$$l + 2w = 180; \qquad \text{or} \qquad l = 180 - 2w$$

Substituting in the formula, we get

$$A = w(180 - 2w); \qquad \text{or} \qquad A = 180w - 2w^2$$

Differentiating, we get

$$dA/dw = 180 - 4w$$

For a maximum or minimum, we set

$$180 - 4w = 0; \qquad \text{then} \qquad w = 45$$

Then the width is 45 rods; the length is 90 rods; area = 4050 sq rods. To show that the answer represents a maximum and not a minimum, all we need to do is to note that any other measurements will give a smaller area. In a practical problem, we may get two or more answers for the unknown. In some problems, either answer will be acceptable. However, in many practical problems, only one answer can be accepted. The nature of the problem will show that certain answers must be discarded. This is shown in the next example.

Example 2

A rectangular container is to be formed from a sheet of metal 32 in. long and 20 in. wide by cutting a square out of each corner and turning up the edges. Find the size of the square to be cut out so that the container will have the greatest volume. Find its volume.

Solution

We make a sketch of the sheet of metal and indicate the square to be cut out of each corner (Fig. 48-8). If we take squares of various sizes, the volume will vary. For one particular size of square, the resulting container will have a maximum volume. We let x represent the edge of the square (in inches). Then the container will have the following dimensions:

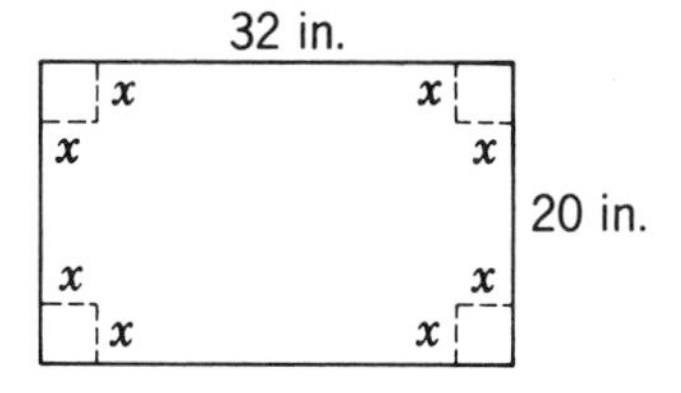

Figure 48-8

$$\text{length} = 32 - 2x; \qquad \text{width} = 20 - 2x; \qquad \text{height} = x$$

Taking the formula for the volume of any rectangular container, $V = lwh$, we have

$$V = (32 - 2x)(20 - 2x)(x); \qquad \text{or} \qquad V = 640x - 104x^2 + 4x^3$$

Differentiating, we get

$$\frac{dV}{dx} = 640 - 208x + 12x^2$$

For a maximum or minimum, we set

$$640 - 208x + 12x^2 = 0$$

Solving, we get two values of x:

$$x = \frac{40}{3} \qquad \text{and} \qquad x = 4$$

The edge of the square could not be $\frac{40}{3}$ because the sheet of metal is only 20 in. wide. Then the correct answer is 4 in. For the volume we get 1152 cu in. We can conclude that 1152 cu in. is a maximum and not a minimum because any other measurement would give a smaller volume.

Exercise 48-4

1. A farmer has 200 rods of fence to be used to enclose a rectangular field, both ends and both sides. The field is to have two extra cross fences from side to side forming three plots. What dimensions of the field will give the entire field a maximum area?

2. What should be the dimensions of the field in No. 1 if the field is to have one additional fence lengthwise through the middle of the field forming six plots?

3. A man wishes to fence in his yard so that it shall contain 180 sq yd. If the cost is 50 cents per yard along one end and two sides, and 75 cents per yard along the end facing the street, what should be the dimensions for the minimum cost?

4. A sheet of aluminum is 15 in. long and 12 in. wide. A square is to be cut out of each corner and the sides turned up to form a rectangular container. Find the dimensions of the container for a maximum volume.

5. A rectangular box without a top has a square base. If the total surface area (outside) is 432 sq in., what are the dimensions of the box for a maximum volume?

6. A rectangular box without a top is twice as long as it is wide. If the total outside surface area is 192 sq in., find the dimensions of the box for a maximum volume.

7. The strength of a beam with a rectangular cross section is proportional to the width and to the square of the depth (height). What are the cross-sectional dimensions of the strongest beam that can be sawed from a log 12 in. in diameter?

8. The current i in an electric circuit is given by the formula, $i = 6t - t^3/3$ (t is time in seconds; i is current in amperes). Find any maximum and/or minimum current. What is the voltage if the resistance R is 120 ohms?

9. In an electric circuit, the voltage E is given by: $E = 120t^3 - 90t$. Find any maximum and/or minimum voltage.

10. The electric charge q transmitted in a circuit varied according to the formula: $q = 3t^4 - t^3$. Find any maximum and/or minimum current, i.

11. Find the volume of the largest right circular cylinder that can be cut from a right circular cone whose diameter is 24 in. and whose altitude is 16 in.

12. A printer offers to print 1000 circulars at 30 cents each. If the

number of circulars is over 1000, the price per circular is 2 cents less for each hundred over 1000. What number of circulars will give the printer the maximum amount of sale?

48-5 DIFFERENTIALS

The *differential* of a variable is a useful concept in calculus. The differential of x is denoted by the symbol, dx. The differential of the function is denoted by dy, or by the symbol, $df(x)$. Similarly, the differential of t is dt; the differential of s is ds. The differential involves the derivative, but the two should not be confused.

To see what is meant by the differential, suppose we graph the general function, $y = f(x)$ (Fig. 48-9). We take any point $P(x, y)$ on the curve. For this point the value of x is the segment OA. The value of y is AP.

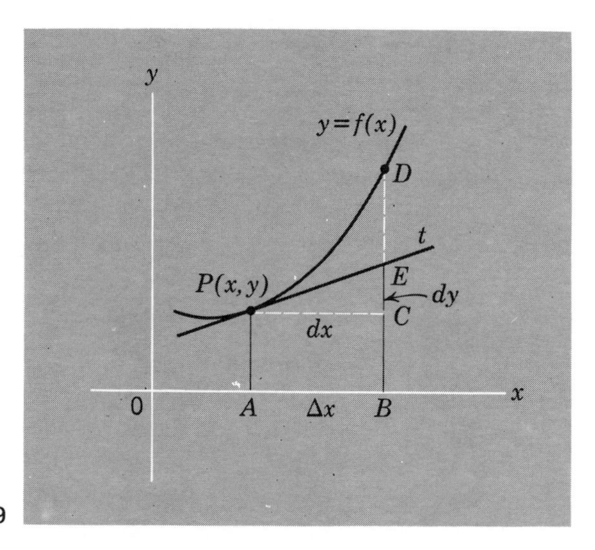

Figure 48-9

Now we let x take on an increment, $\Delta x = AB$. Then y changes by some increment, $\Delta y = CD$. We draw the tangent t to the curve at P, intersecting CD at E. The line segment PC represents Δx. This segment is also equivalent to dx, the differential of x. That is, $\Delta x = dx$.

The differential of y, that is, dy, is represented by the line segment CE. Note especially that dy does not equal Δy. However, if dx is small, then dy is very nearly equal to Δy. Now we state the slope of the tangent line in terms of differentials:

$$m = \frac{\text{rise}}{\text{run}} = \frac{CE}{PC} = \frac{\text{differential of } y}{\text{differential } x}$$

If we use the symbols, dy and dx, for the differentials, we have, for the slope of the tangent,

$$m = \frac{dy}{dx}$$

Here we think of dy and dx as separate quantities. Then the slope of the tangent is equal to the quotient: $dy \div dx$.

We have seen that the slope of the tangent at P is equal to the *derivative.* Since the slope is also equal to the *quotient of the differentials*, we can say:

$$\frac{dy}{dx} = \text{the derivative}; \qquad \text{or} \qquad \frac{dy}{dx} = \frac{dy}{dx}$$

Here we think of the left side of the equation as a fraction showing the division of differentials, dy and dx, considered as separate quantities. The right side is not a fraction but instead the expression for the derivative, the limit of a ratio, a single quantity. Since the left side is a fraction, we can multiply both sides of the equation by the denominator dx, and get

$$dy = \left(\frac{dy}{dx}\right) dx$$

That is, dy, the differential of y, is equal to the *derivative times the differential of* x. This is the algebraic definition of the differential of y.

Definition. *The differential of the function* y *is defined as the derivative of the function multiplied by the differential of* x. That is,

$$dy = (\text{derivative})\, dx; \qquad \text{or} \qquad df(x) = f'(x)\, dx; \qquad \text{or} \qquad dy = y'\, dx.$$

We have seen that the quotient, $dy \div dx$, is equivalent to the derivative, dy/dx. Therefore, hereafter whenever the derivative, dy/dx, appears, we may consider it as the quotient of two differentials, if we wish. Moreover, the quotient of any two differentials may be considered as a derivative.

Note that dy, the differential of the function y involves the derivative. To find dy, we find the derivative and multiply it by dx.

Example 1

If $y = x^3 + 5x^2 + 2x - 3$, evaluate dy when $x = 6$ and $dx = 0.2$.

Solution

By definition, $dy = (\text{derivative})\, dx$

then $dy = (3x^2 + 10x + 2)\, dx$

Evaluating, we get $dy = (108 + 60 + 2)(0.2) = 34$

The differential dy is used to find the approximate change in a function when it might be diffcult to find Δy, the actual change. The advantage of the differential lies in the fact that it is much easier to compute. If the change in x, that is, dx, is small, then dy may be satisfactory for the change in y. The differential of y represents the change that would have occurred in y if the rate of change at P had continued over the interval dx.

In any function $y = f(x)$ if x changes by some increment, Δx, then the function y changes by some actual increment Δy. From Fig. 48-9, note:

1. dy is not exactly equal to Δy;
2. if dx is small, then dy is very nearly equal to Δy.

Example 2

Find the actual change, Δy, and the approximate change, dy, in the following function as x changes from 8 to 8.02: $y = x^2 + 5x + 9$.

Solution

When $x = 8$, $y = 113$. To find Δy, we must compute the value of the function when $x = 8.02$. We get

$(8.02)^2 + 5(8.02) + 9 = 113.4204$;

$$\text{then} \qquad \Delta y = 113.4204 - 113 = 0.4204$$

To find dy, the differential of y, we have

$$dy = (\text{derivative})\, dx.$$

Then $$dy = (2x + 5)(dx) = (16 + 5)(0.02) = 0.42$$

Note that dy differs from Δy by only 0.0004. The result means that as x changes by 0.02 from 8 to 8.02, the approximate change in y is 0.42, whereas the actual change is 0.4204. Note expecially two facts:

1. the computation for Δy is much more difficult than for dy;
2. for practical purposes, the value of dy is sufficiently close to Δy (when Δx is relatively small).

Example 3

In a certain circuit, the power P is given by $P = 35i^5$. If the current i changes from 5 to 4.96 amps, what is the approximate change in power P, and what is the approximate power for the new current value?

Solution

First we note that when $i = 5$, $P = 875$. Also, $di = -0.04$ amps. For approximate change in power, we have

$$dP = \left(\frac{dP}{di}\right)(di) = 70i\, di$$

Substituting numerical values, we get

$$dP = (70)(5)(-0.04) = -14 \text{ watts}$$

For the approximate new power, we get $875 - 14 = 861$ watts. If we compute the actual power using the current, 4.96, we get 861.056 watts.

Note. Recall that we have used the symbol, "d/dx," to indicate the derivative. This symbol is called the *derivative operator.* In a similar manner, we indicate the differential of a function by the symbol "d" before a function. Note the similarity and the difference in the *derivative operator* and the *differential operator d*:

derivative operator: $\quad \dfrac{d}{dx}(x^2 + 3x) = 2x + 3$

differential operator: $\quad d(x^2 + 3x) = (2x + 3)\,dx$

Exercise 48-5

1. If $y = x^4$, find dy and Δy as x changes from 5 to 5.03.

2. If $y = 2x^2 - 3x + 4$, find dy and Δy as x changes from 6 to 5.95.

3. If $y = x^3 + 4x^2 + 3x + 6$, find dy as x changes from 3 to 2.998.

4. A particle moves according to the formula:

$$s = t^3 - 2t^2 + 10t + 3.$$

Find the approximate change in s when t changes from 4 sec to 4.06 sec.

5. A freely falling object falls according to the formula, $s = 16t^2$. How far will it fall in 5 sec? How far in the next 0.2 sec? (Use differentials.)

6. One formula for power P in a circuit is: $P = 30i^2$. Find the approximate change in power when current i changes from 3.6 to 3.64 amps.

7. One formula for power is: $P = E^2/R$. If the resistance R is 20 ohms, find the approximate change in power when E changes from 120 to 119.5 volts.

8. Find by differentials the approximate change in the area of a circle when the radius changes from 20 to 20.3 in. What is the actual change?

9. Find the approximate area of a circular ring with an inside diameter of 16 in. and an outside diameter of 16.4 in.

10. Find the approximate change in the volume of a sphere when its radius changes from 20 in. to 21 in.

11. What is the approximate number of cubic inches of iron in a hollow iron sphere if the outside diameter is 24 in. and the shell is $\frac{1}{4}$-in. thick?

12. Differentials may be used to compute approximate errors. In measuring the diameter of a circle, the diameter is stated as being 24 in. If the maximum error in this measurement is ± 0.05 in., what is the approximate maximum error in the area of the circle?

13. The diameter of a steel ball is measured as 4.2 in. with a tolerance of ± 0.005 in. What is the approximate maximum error in the volume?

14. An electric circuit containing a resistance of 40 ohms has a current measured at 3.40 amps with a possible error of 0.005 amps. What is the approximate maximum error in power?

15. In a circuit containing a resistance of 40 ohms, the voltage is measured as 110 ± 0.4 volts. What is the approximate maximum error in power?

16. What is the approximate error in the volume of a cube if the edge is measured as 15 in. with a possible error of 0.4 in.

17. A steel ball is measured as having a diameter of 1.75 in. with a possible maximum error of ± 0.02 in. What is the approximate maximum error in the volume?

18. An electric circuit containing a resistance of 60 ohms has a current measured as 4.5 amps, with a possible error of ± 0.05 amps. What is the approximate maximum error in power? This error is what percentage of the power as computed with 4.5 amps? (This is percentage error.)

QUIZ ON CHAPTER 48, THE DERIVATIVE: APPLICATIONS. FORM A.

1. A ball is projected directly upward from the top of a building. Its distance from the ground at any time t is given by the formula: (s in ft): $s = 80 + 64t - 16t^2$. Find (a) the height of the building; (b) the initial velocity; (c) the maximum height reached by the ball.

2. A particle moves according to the formula:

$$s = t^3 - 9t^2 + 24t + 10 \text{ (s ln ft, t in sec).}$$

When and where is the particle at rest?

3. Find the slope of the following curve at the point $(2, -3)$, and write the equation of the tangent at the point: $y = x^2 - 3x - 1$.

4. Find any maximum and/or minimum points of the curve: $y = x^3 + 3x^2 - 9x - 4$.

5. In an electric circuit, the current i is given by the formula: $i = t^2 - 6t + 12$. Find i when $di/dt = 0$.

6. In a circuit, $q = 2t^3 - t^2$. Since current $i = dq/dt$, find any maximum or minimum value of q.

7. A man has 120 rods of fence to enclose a rectangular field with one extra cross fence from side to side. Find the dimensions of the field if the area is a maximum.

8. If $y = x^2 + 3x$, find the actual change Δy, as x changes from 4 to 4.2. Then find the approximate change in y, dy, by using differentials.

9. In the function, $y = x^3 - x^2 + 5x - 6$, find by differentials the approximate change in y as x changes from 6 to 6.3.

10. In an electric circuit, power P is given by the formula: $P = 24i^2$. Find dP, the approximate change in power as i changes from 8 to 7.8 amps.

11. In the formula for power, $P = E^2/60$, if E is measured as 40 volts with a possible error of ± 0.5, find the maximum approximate error in power.

12. Find by differentials the approximate area of a ring whose outside diameter is 5 in. and inside diameter is 4.7 in.

CHAPTER FORTY-NINE Derivatives of Algebraic Functions

49-1 POWERS OF FUNCTIONS

We have seen the power rule applied to powers of single terms:

$$\text{if } y = x^5, \frac{dy}{dx} = 5x^4; \qquad \text{if } y = t^5, \frac{dy}{dt} = 5t^4; \qquad \text{if } y = u^5, \frac{dy}{du} = 5u^4$$

In the case of powers on a single variable, we take the derivative of the function with respect to the independent variable. In general terms,

$$\text{if } y = x^n, \frac{dy}{dx} = nx^{n-1}$$

Now, suppose we have a power of a function of x, such as $x^2+3x)^5$. We might ask: Can the power rule be applied on the function $(x^2 + 3x)$ raised to the fifth power in the same way as a on a single term x^5? That is,

$$\text{if } y = (x^2 + 3x)^5, \qquad \text{does} \qquad \frac{dy}{dx} = 5(x^2 + 3x)^4$$

We shall see that the result is correct as far as it goes, but there is an additional factor. First let us see the meaning of a *function of a function.* Let us take again the example,

$$\text{if } y = u^5, \qquad \text{then} \qquad \frac{dy}{du} = 5u^4$$

Here we have the derivative of y with respect to u as the independent variable.

Now, let us suppose in this example that u itself is some function of x, such as $(x^2 + 3x)$. Since y is a function of u, and u is a function of x, then y is actually a function of x through u. Then y is a function of a function.

Our problem now is to find the derivative of y, not only with respect to u but also with respect to x as the basic independent variable. That is, we wish to find dy/dx, not only dy/du. Let us see how this can be done.

Consider again the increments in the variables. For any increment in x, we shall have corresponding increments in u and in y. We denote the increments by Δx, Δu, and Δy, respectively. As Δx approaches zero, Δu and Δy will also approach zero, if the functions are continuous. As long as the increments are anything but zero, we can say

$$\left(\frac{\Delta y}{\Delta u}\right)\left(\frac{\Delta u}{\Delta x}\right) = \frac{\Delta y}{\Delta x}$$

As the increments approach zero, each fraction will approach a limit, which is a derivative, and we get

$$\left(\frac{dy}{du}\right)\left(\frac{du}{dx}\right) = \frac{dy}{dx}$$

This is the so-called chain rule for the derivative of a function of a function. That is, the derivative of y with respect to u times the derivative of u with respect to x is equal to the derivative of y with respect to x.

Now, let us take the problem,

$$y = (x^2 + 3x)^5$$

If we take the function $(x^2 + 3x)$ as u, then the problem is essentially

$$y = u^5 \qquad \text{and} \qquad \frac{dy}{du} = 5u^4$$

To get dy/dx, we multiply both sides of the equation by du/dx, and get

$$\left(\frac{dy}{du}\right)\left(\frac{du}{dx}\right) = 5(u)^4\left(\frac{du}{dx}\right)$$

The result means that when we have applied the power rule on the u function, we must follow this up with (du/dx), the derivative of u with respect to x. In this example, since $u = (x^2 + 3x)$, then $du/dx = 2x + 3$. This is the extra factor we need, and we get

$$\text{if } y = (x^2 + 3x)^5$$

$$\frac{dy}{dx} = 5(x^2 + 3x)^4(2x + 3)$$

Rule. *To find the derivative of a power of a function, apply the power rule on the function, then multiply the result by the derivative of the function.* The following notation may help to emphasize the rule:

$$\text{if } y = (\text{function of } x)^n$$

$$\text{then } \frac{dy}{dx} = (n)(\text{function of } x)^{n-1}(\text{derivative of the function})$$

As a formula, $\quad$ if $y = u^n$, $\quad$ then $\quad \dfrac{dy}{dx} = (n)(u)^{n-1}\left(\dfrac{du}{dx}\right)$

Note. It is wlll to think of the complete power rule as being applied to the power on x alone. For example,

$$\text{if } y = x^5, \qquad \text{then} \qquad \frac{dy}{dx} = 5(x)^4\left(\frac{dx}{dx}\right)$$

But since $dx/dx = 1$, we usually omit dx/dx, and write simply, $5x^4$.

Example 1

If $y = (x^2 + 7x + 4)^6$, $\quad dy/dx = 6(x^2 + 7x + 4)^5(2x + 7)$.

Example 2

If $i = (t^3 - 5t^2 + t)^3$, $\quad di/dt = 3(t^3 - 5t^2 + t)^2(3t^2 - 10t + 1)$.

Example 3

If $y = \sqrt{x^2 - 4x + 7}$, evaluate dy/dx at $x = 3$.

Solution

First we write the function as a power: $y = (x^2 - 4x + 7)^{\frac{1}{2}}$. Differentiating, we get

$$\frac{dy}{dx} = \left(\frac{1}{2}\right)(x^2 - 4x + 7)^{-\frac{1}{2}}(2x - 4)$$

The result can be simplified to,

$$\frac{dy}{dx} = \frac{x - 2}{(x^2 - 4x + 7)^{\frac{1}{2}}}$$

For the value, $x = 3$, the derivative is equal to $\frac{1}{2}$.

Example 4

If $y = \dfrac{8}{\sqrt{x^2 - 9}}$, find the slope of the curve at $x = 5$.

Solution

First we write the function: $y = 8(x^2 - 9)^{-\frac{1}{2}}$. Differentiating, we get

$$\frac{dy}{dx} = -4(x^2 - 9)^{-\frac{3}{2}}(2x) = \frac{-8x}{(x^2 - 9)^{\frac{3}{2}}}$$

For the slope of the curve at $x = 5$, we have

$$m = \frac{dy}{dx} = -\frac{5}{8}$$

Exercise 49-1

Find the derivatives with respect to the independent variable.

1. $y = (4x - 3)^5$ **2.** $y = (2 - 5x)^6$ **3.** $y = (x^2 - 3x + 4)^8$

4. $i = (t^2 - 6t)^4$ **5.** $s = (t^2 + 4t)^2$ **6.** $i = (t^3 - 2t^2 + 1)^3$

7. $q = (4t - t^2)^5$ **8.** $i = (3 - 2t^3)^3$ **9.** $y = (1 - 3t - t^2)^4$

10. $\frac{d}{dx}(x^3 - 2x^2)^6$ **11.** $\frac{d}{dt}(6t^2 - 3t)^{\frac{1}{3}}$ **12.** $\frac{d}{dx}(x^4 - 5x^3 - x)^7$

13. $D_x(5 - x^2 - x^3)^8$ **14.** $D_t(3t^2 + 4t)^{\frac{1}{2}}$ **15.** $D(x^3 + 3x^2 - 4)^{\frac{1}{2}}$

Evaluate the derivative for the given value of x or t:

16. $y = (4t - t^2)^{\frac{1}{2}}, (t = 2)$ **17.** $y = \sqrt{5x - 6}, (x = 3)$

18. $i = (2t - 4)^{\frac{1}{3}}, (t = -2)$ **19.** $q = (t^2 + 6t)^{\frac{1}{2}}, (t = 2)$

20. $y = \dfrac{2}{\sqrt{x^2 + 9}}, (x = 0)$ **21.** $y = \dfrac{6}{(x^3 - 7)^{\frac{2}{3}}}, (x = -1)$

49-2 PRODUCT RULE

It is often necessary to differentiate the product of two functions, as

$$y = (x^2 + 3x)(5x^2 - 2)$$

In this problem as in some others, we could expand the product into a single polynomial before differentiating. Then the function becomes

$$y = 5x^4 + 15x^3 - 2x^2 - 6x$$

Now we could differentiate term by term as in any polynomial.

However, in many problems the multiplication of two factors is very difficult or impossible. We therefore need a rule for differentiating a product. The product rule can be derived by the delta method.

In the product above, let us call the two functions u and v, respectively:

$$u = x^2 + 3x \qquad \text{and} \qquad v = 5x^2 - 2$$

Then we have essentially, $y = (u)(v)$

Since u, v, and y are all functions of x, then for any increment Δx, we shall have corresponding increments Δu, Δv, and Δy. Then by the delta method, we get $y + \Delta y = (u + \Delta u)(v + \Delta v)$

Expanding, $y + \Delta y = uv + u(\Delta v) + v(\Delta u) + (\Delta u)(\Delta v)$

Subtracting, $y = uv$

we get $\Delta y = u(\Delta v) + v(\Delta u) + (\Delta u)(\Delta v)$

Dividing by Δx, $\dfrac{\Delta y}{\Delta x} = (u)\left(\dfrac{\Delta v}{\Delta x}\right) + (v)\left(\dfrac{\Delta u}{\Delta x}\right) + (\Delta u)\left(\dfrac{\Delta v}{\Delta x}\right)$

As Δx approaches zero, we get the formula,
$$\frac{dy}{dx} = (u)\left(\frac{dv}{dx}\right) + (v)\left(\frac{du}{dx}\right) + 0$$

In words, the formula says: *The derivative of the product of the two functions, u and v, is equal to the first function u times the derivative of the second function v, plus the second function v times the derivative of the first function u.*

The derivative of a product should be written in exactly the order shown:

1. *Write the first function exactly as it appears.*
2. *Multiply this by the complete derivative of the second function.*
3. *After the plus sign, write the second function exactly as given.*
4. *Multiply this by the derivative of the first function.*

Example 1

If $y = (x^2 + 3x)(5x^2 - 2)$, evaluate $\frac{dy}{dx}$ at $x = 1$.

Solution

$$\frac{dy}{dx} = (x^2 + 3x)(10x) + (5x^2 - 2)(2x + 3)$$

At $x = 1$, we get $\frac{dy}{dx} = (1 + 3)(10) + (5 - 2)(2 + 3) = 55$

Example 2

If $y = (x^2 - 2)^4(4x + 1)^3$, evaluate $\frac{dy}{dx}$ at $x = -1$.

Solution

If we call the two functions u and v, respectively, we have

$$u = (x^2 - 2)^4 \quad \text{and} \quad v = (4x + 1)^3$$

Note that the two factors are powers. We must be careful to take the complete derivative of each factor by the power formula. Then we get

$$\frac{dy}{dx} = (x^2 - 2)^4(3)(4x + 1)^2(4) + (4x + 1)^3(4)(x^2 - 2)^3(2x)$$

Here we identify $(\quad u \quad)\left(\quad \frac{dv}{dx} \quad\right) + (\quad v \quad)\left(\quad \frac{du}{dx} \quad\right)$

If we factor the quantity, $(4)(x^2 - 2)^3(4x + 1)^2$, out of each term, we get the derivative in the form, $(4)(x^2 - 2)^3(4x + 1)^2(11x^2 + 2x - 6)$. For the value, $x = -1$, the derivative is equal to (-108).

Exercise 49-2 (Use the product rule for this exercise.)

Find the derivative of each of the following:

1. $y = (4x - 3)(x^2 - 5)$
2. $y = (x^2 + 3)(4x - x^3)$
3. $y = (2 - x^2)(x^3 - x^2 + 1)$
4. $y = (x^2 - 5x)(3x^2 - 4x + 2)$
5. $y = (2x^3 - 6x^2)(x^2 + 2x - 5)$
6. $y = (4x^2 - 3x + 5)(3x^2 - 2x)$
7. $i = t^3(3 - t^2)^4$
8. $i = t^2(t^3 + 4t - 1)^3$
9. $y = (x + 4)^5(x - x^2)^3$
10. $y = (x^2 - x)^4(6 + x^2)^{\frac{1}{2}}$
11. $i = \dfrac{4 - t^2}{(t^2 - 9)^2}$
12. $q = \dfrac{2t - 3}{(t^2 - 4t)^{\frac{1}{2}}}$

Evaluate the derivative for each for the given value of x or t.

13, $y = x^3(x^2 - 5); (x = 2)$
14. $y = x^2(x^2 + 4x + 1); (x = -1)$
15. $y = (1 - x)^2(x^3 - x)^4;$ $(x = -1)$
16. $y = (x^2 + 1)(x^2 - 4)^3; (x = 1)$
17. $y = x^3(5 + x^2)^{\frac{1}{2}}; (x = 2)$
18. $y = (x^2 - 3)^4(x^3 - 3x)^3; (x = 2)$
19. $s = t^{\frac{3}{2}}(2x - t^2)^{\frac{2}{3}}; (t = 4)$
20. $i = (t^2 + 1)^3(6t - 1)^{\frac{1}{3}}; (t = 0)$
21. $y = \dfrac{x^2}{(x^2 - 5)^{\frac{1}{2}}}; (x = 3)$
22. $s = \dfrac{t^2 - 3}{(3t^2 + 4)^{\frac{1}{2}}}; (t = 2)$

Note. Nos. 11, 12, 21, and 22 may be written as products by moving the denominator to the numerator and changing the sign of the exponent.

49-3 QUOTIENT RULE

We need a special rule for finding the derivative of the quotient of two functions, as in

$$y = \frac{x^3 + 2}{x^2 + 5}$$

In some problems of this kind, we might use the product rule by shifting the denominator to the numerator, but in most instances the quotient rule is more convenient. The quotient rule can be derived by the delta method. If we represent the numerator by u and the denominator by v, we can represent the above expression by

$$y = \frac{u}{v}$$

Since u and v are functions of x, then for any increment Δx, we shall have the corresponding increments, Δu, Δv, and Δy. By the delta method we can get the formula:

$$\frac{dy}{dx} = \frac{(v)(du/dx) - (u)(dv/dx)}{v^2}$$

In words, the formula says: *The derivative of the quotient of two functions, u/v, is equal to the denominator v times the derivative of the numerator, du/dx, minus the numerator u times the derivative of the denominator, dv/dx, all divided by the square of the denominator, v^2.* We have these steps:

1. *First, write the denominator of the fraction just as it appears, v.*
2. *Multiply this by the derivative of the numerator, du/dx.*
3. *After the minus sign, write the numerator just as it appears, u.*
4. *Multiply this by the derivative of the denominator, dv/dx.*
5. *Finally, write the square of the denominator as the denominator of the derivative.*

Example 1

If $y = \dfrac{x^3 + 2}{x^2 + 5}$, evaluate the derivative for $x = 1$.

Solution

$$\frac{dy}{dx} = \frac{(x^2 + 5)(3x^2) - (x^3 + 2)(2x)}{(x^2 + 5)^2} = \frac{x^4 + 15x^2 - 4x}{(x^2 + 5)^2}$$

At $x = 1$, the value of the derivative becomes $\frac{1}{3}$.

Example 2

If $y = \dfrac{x^2 + 2}{(x^2 - 1)^{\frac{1}{2}}}$, for what values of x is $m = 0$?

Solution

$$\frac{dy}{dx} = \frac{(x^2 - 1)^{\frac{1}{2}}(2x) - (x^2 + 2)(\frac{1}{2})(x^2 - 1)^{-\frac{1}{2}}(2x)}{x^2 - 1}$$

If we multiply numerator and denominator of the fraction by $(x^2 - 1)^{\frac{1}{2}}$, we get

$$\frac{dy}{dx} = \frac{(x^2 - 1)(2x) - x(x^2 + 2)}{(x^2 - 1)^{\frac{3}{2}}} = \frac{x^3 - 4x}{(x^2 - 1)^{\frac{3}{2}}}$$

Setting dy/dx equal to zero, we get: $x = 0$, $x = 2$, and $x = -2$. The value $x = 0$ must be discarded since this value makes y imaginary.

In some problems it may be easier to find dx/dy, rather than dy/dx. Then we find dy/dx by *inversion*, using the relation:

$$\frac{1}{dx/dy} = \frac{dy}{dx}$$

This inverse relation can be proved by the delta method.

Example 3

If $x = \dfrac{3y}{(y^2+5)^{\frac{1}{2}}}$, evaluate $\dfrac{dy}{dx}$ for $y = 2$.

Solution

$$\frac{dx}{dy} = \frac{(y^2+5)^{\frac{1}{2}}(3) - (3y)(\frac{1}{2})(y^2+5)^{-\frac{1}{2}}(2y)}{y^2+5}$$

Multiplying numerator and denominator by $(y^2+5)^{\frac{1}{2}}$, and simplifying the result, we get

$$\frac{dx}{dy} = \frac{15}{(y^2+5)^{\frac{3}{2}}}; \qquad \text{then} \qquad \frac{dy}{dx} = \frac{(y^2+5)^{\frac{3}{2}}}{15}$$

Evaluating at $y = 2$, we get

$$\frac{dx}{dy} = \frac{5}{9}; \qquad \text{then} \qquad \frac{dy}{dx} = \frac{9}{5}$$

Exercise 49-3 (Use the quotient rule for this exercise.)

Find dy/dx in each of the first twelve exercises:

1. $y = \dfrac{3x}{2x+5}$ **2.** $y = \dfrac{2x-1}{x^2-1}$ **3.** $y = \dfrac{x^2}{\sqrt{4x-1}}$

4. $y = \dfrac{3x-2}{x^2}$ **5.** $y = \dfrac{1-x}{\sqrt{x^2-5}}$ **6.** $y = \dfrac{x^2-2}{\sqrt{3-x^2}}$

7. $y = \dfrac{\sqrt{1-x^2}}{x^3}$ **8.** $y = \dfrac{(2x-1)^3}{x^3+4}$ **9.** $y = \dfrac{2+3x}{(3-x^2)^{\frac{1}{2}}}$

10. $x = (2-y^2)^3$ **11.** $x = (y^2-4y)^{\frac{1}{2}}$ **12.** $x = (y^2)(y^3-7)^{\frac{2}{3}}$

Evaluate dy/dx in exercises Nos. 13–16:

13. $x = (y^2-3)(y^2+5); (y = 2)$ **14.** $x = y^2(1-y^2)^{\frac{1}{3}}; (y = 3)$

15. $x = \dfrac{3y-1}{(y-3)^2}; (y = 1)$ **16.** $x = \dfrac{3y^2}{\sqrt{5-y^2}}; (y = -1)$

Evaluate the derivative for the given value of t:

17. $i = \dfrac{2t^2}{t^2 - 5}$; $(t = 1)$

18. $q = \dfrac{(t^2 - 3)^{\frac{1}{2}}}{t^3}$; $(t = 2)$

19. $i = \dfrac{3t + 4}{(1 - t^2)^2}$; $(t = -2)$

20. $s = \dfrac{t^2 - 3}{\sqrt{4 - t^2}}$; $(t = 0)$

49-4 IMPLICIT DIFFERENTIATION

As we have seen, a function is often expressed by an equation, such as

$$y = 5x - 4; \qquad y = x^2 - 2x - 3$$

Note that each equation is solved for y. If an equation is solved for y, we say that y is an *explicit function* of x. The function has the form: $y = f(x)$.

An equation may indicate a functional relationship between x and y in another form, as in

$$2x + 3y = 18$$

This equation is not solved for y. It does show that y is a function of x, but the relation is implied. In this form, we say y is an *implicit function* of x. Here are some examples of implicit functions:

$$3x - 4y = 13; \qquad x^2 + 3y^2 + 2y = 3; \qquad x^3 + xy^2 - y^3 = 4$$

Up to this time we have found derivatives of explicit functions only. An explicit function is more convenient for finding a derivative. An implicit function can sometimes be changed to an explicit function by solving for y. For example, the equation $2x + 3y = 18$, can be changed to

$$y = -\frac{2}{3}x + 6; \qquad \text{then} \qquad \frac{dy}{dx} = -\frac{2}{3}$$

However, changing an implicit to an explicit function is often difficult and sometimes impossible, as in the equation: $x^3 - x^2y - y^4 + xy^3 = 6$.

Implicit functions can be differentiated by a method called *implicit* differentiation. In using this method we do not solve the equation for y. Instead, we differentiate each term with respect to x. There are two steps:

1. *Differentiate each term as it stands.*
2. *Solve the resulting equation for dy/dx.*

In differentiating implicitly two points must be observed:

(a) The variable y to any power is treated in the same way as any other variable that is a function of x. For example,

$$\frac{d}{dx}(u^3) = 3u^2\left(\frac{du}{dx}\right); \qquad \frac{d}{dx}(y^3) = 3y^2\left(\frac{dy}{dx}\right)$$

(b) A product of two factors involving x and y, such as x^2y^4, must be differentiated by the product rule. For example,

$$\frac{d}{dx}(x^2y^4) = (x^2)(4y^3)\left(\frac{dy}{dx}\right) + (y^4)(2x)\left(\text{note that } \frac{dx}{dx} \text{ is omitted}\right)$$

Example 1

If $x^2 + y^2 = 13$, find the equation of the tangent at $(-2, -3)$.

Solution

Differentiating implicitly, $2x + 2y\left(\frac{dy}{dx}\right) = 0$; $\frac{dy}{dx} = -\frac{x}{y}$. At $(-2, -3)$, $\frac{dy}{dx} = m = -\frac{-2}{-3} = -\frac{2}{3}$. Equation of tangent: $2x + 3y = -13$.

Note. In implicit differentiation, the derivative will usually contain the variable y. If desired, the derivative may be expressed entirely in x-terms, by referring to the original equation. However, this change is not usually necessary since the form containing y is satisfactory for evaluation.

Example 2

If $x^2 - 2xy - y^2 - 9x - 3y = -8$, find the equation of the tangent at the point $(2, -1)$.

Solution

Differentiating, $$2x - 2\left(x\frac{dy}{dx} + y\right) - 2y\frac{dy}{dx} - 9 - 3\left(\frac{dy}{dx}\right) = 0$$

Removing parentheses, $$2x - 2x\frac{dy}{dx} - 2y - 2y\frac{dy}{dx} - 9 - 3\frac{dy}{dx} = 0$$

Transposing and solving for dy/dx, $$\frac{dy}{dx} = \frac{2x - 2y - 9}{2x + 2y + 3}$$

At $(2, -1)$, $\frac{dx}{dy} = -\frac{3}{5}$. Equation of tangent: $3x + 5y = 1$.

Example 3

If $x^2y - xy^2 - 2y^3 = 4$, evaluate dy/dx for the point $(-2, 1)$.

Solution

Differentiating, $$x^2\frac{dy}{dx} + y(2x) - \left[(x)(2y)\left(\frac{dy}{dx}\right) + y^2\right] - 6y^2\frac{dy}{dx} = 0$$

Removing parentheses, $$x^2\frac{dy}{dx} + 2xy - 2xy\frac{dy}{dx} - y^2 - 6y^2\frac{dy}{dx} = 0$$

Solving, $$\frac{dy}{dx} = \frac{y^2 - 2xy}{x^2 - 2xy - 6y^2}; \quad \text{at } (-2, 1), \quad \frac{dy}{dx} = \frac{5}{2}$$

Exercise 49-4

Differentiate implicitly and find the slope at the given point:

1. $2x + 5y = 11; (-2, 3)$
2. $x^2 + y^2 = 25; (-4, -3)$
3. $x^2 - y^2 = 16; (5, -3)$
4. $x^2 + xy - y^2 = 10; (4, -1)$
5. $2x^2 + 5y^2 = 38; (3, 2)$
6. $5x^2 - 3y^2 = 17; (2, -1)$
7. $3xy - 4x + 3y = 11; (-2, -1)$
8. $x^2 + y^3 = 1; (3, -2)$
9. $x^3 + 4y^2 = 28; (-2, 3)$
10. $x^2y - xy^2 + 4 = 0; (1, -2)$
11. $x^3 - y^3 - 7xy = 0; (4, 2)$
12. $3x^2 - 2xy - 6y^2 = 15; (3, -2)$
13. $x^2 + y^2 + 3x - 5y = 0; (-3, 5)$
14. $2x^2 - 3y^2 - 5x + 2y = 13; (-2, -1)$
15. $x^{\frac{2}{3}} + y^{\frac{2}{3}} = 13; (8, 27)$
16. $x^{\frac{2}{3}} - x^{\frac{1}{3}}y^{\frac{1}{3}} + y^{\frac{2}{3}} = 7; (27, 8)$
17. $3x^2 - 4xy - 2y^2 - 3x + 5y + 6 = 0; (1, 2)$
18. $x^2 - 3xy + y^2 - 4x - y - 8 = 0; (-1, -3)$

49-5 RELATED RATES

It is often desirable to know *how fast* a variable is changing. The derivative by itself does not indicate speed of changes. It refers to the rate of change between variables, but it does not tell how fast the changes take place. For example, if $y = (\frac{1}{2})x - 4$, then $dy/dx = \frac{1}{2}$. The derivative, $\frac{1}{2}$, means that the change in y is one-half *as much as* the change in x. It refers to the ratio between changes, not to the speed of those changes.

If we wish to know *how fast* a variable is changing, we must refer to *time*. Then we have what is called a *time-rate* of change. We have already used time-rates in connection with distance, velocity, and acceleration. In the example above, the derivative, $\frac{1}{2}$, does not tell how fast the variables are changing. However, if x is changing at 6 units per second, then we can say that y is changing at 3 units per second; that is, y is changing $\frac{1}{2}$ as fast as x. We are then concerned with *rates of change with respect to time.*

Time-rates of change of x and y are denoted as follows:

(a) *The time-rate of change of x is denoted by dx/dt.*
(b) *The time-rate of change of y is denoted by dy/dt.*

Since the variables, x and y, are understood to be related, their time-rates are also related. Then the time-rates are called *related time-rates*, or simply *related rates*.

To find a time-rate of change, we differentiate with respect to *time*. In any problem involving time-rates, we must consider not only the variables x and y themselves, but also their time-rates. If x and y are the variables, then we must consider four variables: x; y; dx/dt; dy/dt.

A problem usually involves finding an unknown time-rate, although any of the variables may be found if the other three are known. Note that the ratio between the time-rates is equal to the derivative:

$$\frac{dy/dt}{dx/dt} = \frac{dy}{dx}$$

Example 1

If $y = x^2 - 3x + 7$, find dy/dt when $x = 3$, and $dx/dt = 4$ units per sec.

Solution

Differentiating with respect to time,

$$\frac{dy}{dt} = 2x\left(\frac{dx}{dt}\right) - 3\left(\frac{dx}{dt}\right)$$

Inserting known quantities, we get

$$\frac{dy}{dt} = (2)(3)(4) - 3(4) = 12 \text{ units per sec}$$

Warning. *Do not insert particular quantities before differentiating.*

Example 2

If $2x^2 - y^2 = 14$, find dy/dt when $x = 3$, $y = 2$, $dx/dt = 5$ per sec.

Solution

Differentiating, $$4x\frac{dx}{dt} - 2y\frac{dy}{dt} = 0$$

Solving for $\frac{dy}{dt}$, $\quad \frac{dy}{dt} = \frac{2x}{y}\left(\frac{dx}{dt}\right)$; $\quad$ then $\quad \frac{dy}{dt} = \frac{6}{2}(5) = 15$ units per sec

Numerical values may be inserted immediately after differentiating. Then we get

$$(4)(3)(5) - (2)(2)\left(\frac{dy}{dt}\right) = 0; \qquad \text{then} \qquad \frac{dy}{dt} = 15$$

Example 3

In the equation, $2x^3 + 3y^3 = 19$, how fast is y changing if x is changing at a rate of 6 units per sec when $x = 2$ and $y = 1$?

Solution

Differentiating, $$6x^2\left(\frac{dx}{dt}\right) + 9y^2\left(\frac{dy}{dt}\right) = 0$$

Inserting given values at once, we get $$(6)(4)(6) + (9)(1)\left(\frac{dy}{dt}\right) = 0$$

Solving for $\frac{dy}{dt}$, we get $$\frac{dy}{dt} = -16 \text{ units per sec}$$

The negative value indicates that y is changing negatively at 16 per units per sec while x is increasing at 6 units per sec.

Example 4

In a certain electric circuit, power P (in watts) is given by the formula, $P = 30i^2$, in which i represents current in amperes. How fast is the power changing at the instant when the current is 2 amps and the current is changing at 0.05 amps per sec?

Solution

Differentiating with respect to time, $$\frac{dP}{dt} = 60i\left(\frac{di}{dt}\right)$$

Inserting given values, $$\frac{dP}{dt} = (60)(2)(0.05) = 6$$

The result means that the power is changing at a rate of 6 watts per sec.

Example 5

A ladder 39 ft long leans up against a vertical wall. The foot of the ladder is being pulled away from the wall at a rate of 0.5 ft per sec. How fast is the top descending when the foot of the ladder is 15 ft from the wall?

Solution

We show the ladder in any general position (Fig. 49-1). Now we let x equal the number of feet from the wall to the foot of the ladder at any time. Let y equal the distance, in feet, from the ground to the top of the ladder. Now we write an equation that is always true: $x^2 + y^2 = 39^2$. Since the ladder is being pulled out, x is changing, and we have,

$$\frac{dx}{dt} = 0.5$$

Differentiating with respect to time, we get

$$2x\left(\frac{dx}{dt}\right) + 2y\left(\frac{dy}{dt}\right) = 0$$

When $x = 15$, $y = 36$, by the Pythagorean rule.

Inserting numerical values, we get $\quad (2)(15)(0.5) + (2)(36)\left(\frac{dy}{dt}\right) = 0$

Solving for $\frac{dy}{dt}$, we get, $\quad \frac{dy}{dt} = -2.083$ ft per sec

The negative sign indicates that y is decreasing, since the top is descending.

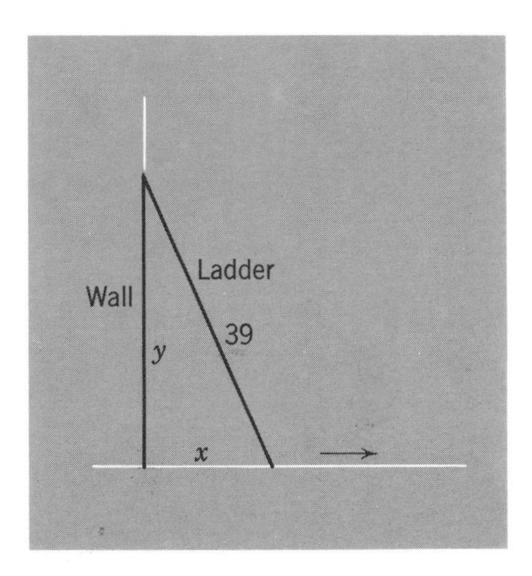

Figure 49-1

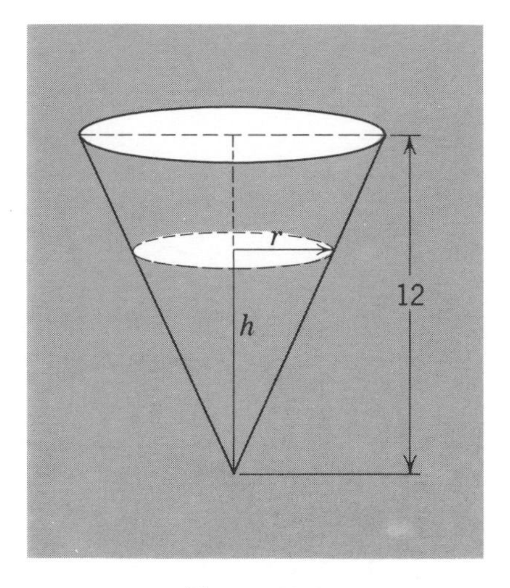

Figure 49-2

Example 6

A tank has the shape of an inverted cone whose altitude is 12 ft. The diameter of the top is 12 ft. Water is running out through a hole in the bottom at 24 cu in. per sec. How fast is the surface of the water descending when the water is 3 ft deep?

Solution

We sketch a figure showing the diameter and altitude of the tank (Fig. 49-2). For the volume of the water in the tank at any time, we let

$$h = \text{altitude}; \qquad r = \text{radius}$$

For the volume of water we have

$$V = (\tfrac{1}{3})\pi r^2 h$$

We must eliminate one of the variables, either r or h. Since we wish to find the rate of change in h, we eliminate r. We note that

$$\frac{r}{h} = \frac{6}{12}; \qquad \text{then} \qquad r = \frac{h}{2}$$

Substituting, we get

$$V = \left(\frac{1}{3}\right)(\pi)\left(\frac{h^2}{4}\right)(h); \qquad \text{or} \qquad V = \frac{\pi}{12}h^3$$

Differentiating with respect to time,

$$\left(\frac{dV}{dt}\right) = \left(\frac{\pi}{4}\right)(h^2)\left(\frac{dh}{dt}\right)$$

Now we have the following known quantities:

$$h = 36 \text{ in.}; \qquad \frac{dV}{dt} = -24 \text{ in.}^3\text{/sec}$$

Substituting values, we get $\quad -24 = \frac{\pi}{4}(1296)\left(\frac{dh}{dt}\right)$

then $\quad \frac{dh}{dt} = -\frac{2}{27\pi}$

The result is approximately -0.024 in. per sec, for dh/dt.

Exercise 49-5

1. If $x^2 - 2y^2 = 1$, how fast is y changing when $x = 3$, $y = 2$, and $dx/dt = 6$ units per sec? Show that the ratio of the time-rates is equal to the derivative.

2. If $2x^3 + y^3 = 17$, find dy/dt when $x = 2$, $y = 1$, and $dx/dt = 0.5$ in. per sec.

3. If $x^2y - xy^2 = 6$, find dy/dt when $x = 3$, $y = 1$, and $dx/dt = 6$ units per sec.

4. If $2x^2 + 5y^2 = 37$, find dx/dt when $x = 3$, $y = -1$, and $dy/dt = 12$ units per sec.

5. If $P = 80i^2$, how fast is power P changing (in watts per second) when i is equal to 4 amps, and i is changing at 0.05 amps per sec?

6. If $P = E^2/20$, how fast is power P changing when voltage E is 120 volts and E is changing at 0.6 volts per sec?

7. A ladder 35 ft long leans up against a vertical wall. The foot of the ladder is being pulled from the wall at a rate of 0.6 ft per sec. How fast is the top descending when the foot of the ladder is 21 ft from the watt?

8. A ladder 30 ft long leans up against a vertical wall. The foot of the ladder is being pushed toward the wall at a rate of 0.5 ft per sec. How fast is the top rising when the top of the ladder is 15 ft from the ground?

9. A tank in the shape of an inverted circular cone has an altitude of 12 ft. The circular top has a diameter of 16 ft. If water is running into the tank at 36 cu ft per sec, how fast is the surface of the water rising when the water is 4 ft deep?

10. Sand is falling through a hole onto a level surface forming a conical pile having an altitude always equal to one-half the radius of the base. If the sand is falling at a rate of 2.5 cu ft per min, how fast is the altitude increasing when the radius of the pile is 3 ft?

11. The volume of a sphere is increasing at a rate of 0.6 cu in. per sec. How fast is the radius of the sphere increasing when the diameter is 18 in.? How fast is the surface of the sphere increasing?

12. If the edge of a cube is changing at a rate of 0.2 in. per sec, how fast is the volume changing when the edge is 12 in.?

13. If the side of a square is increasing at $\frac{1}{4}$ cm per minute, how fast is the area changing when the side is 16 cm.?

14. One car leaves a particular point and travels directly east at an average of 33 mph. At the same time a second car leaves the same point and travels directly north at an average of 44 mph. How fast are the cars separating at the end of 2 hours?

15. In No. 14, if the second car starts out a $\frac{1}{2}$ hour after the first car has left, how fast are the two separating when the second car has traveled 2 hours?

16. A tank in the shape of a paraboloid with the vertex at the bottom has a circular top 6 ft in diameter. The tank has a total depth of 8 ft. If water is running into the tank at a constant rate of 3.6 cu ft per minute, how fast is the surface of the water rising when the water is 3 ft deep?

17. A man 6 ft tall starts walking from a point directly under a street light that hangs 20 ft above the street. He walks directly away from the light at a steady rate of 3 mph. How fast is his shadow lengthening when the man has walked 30 ft?

18. Boyle's law states that with a constant temperature the volume of gas under pressure varies inversely as the pressure p. If the volume is 120 cm^3 when the pressure is 2.5 lb per cm^2, how fast is the volume changing when the pressure is 12 lb per cm^2 and the pressure is changing at 1.5 lb per sq cm?

19. Kinetic energy K is given by the formula, $K = (\frac{1}{2})mv^2$, where mass m is in grams, velocity v is in centimeters per second, and energy K is in ergs. If $m = 20$ grams and $v = 10$ cm/sec, how fast is K changing when velocity v is changing by 3 cm/sec^2?

20. Electrical resistance in a certain conductor is given by the formula, $R = 40 + 0.05T^2$, where T represents temperature in degrees Celsius (°C). If T is increasing at a rate of 0.08 C° per second, find the rate of change in the resistance R when $T = 40°$ C. (Note the difference in meaning between degrees Celsius and Celsius degrees.)

QUIZ ON CHAPTER 49, DERIVATIVES OF ALGEBRAIC FUNCTIONS. FORM A.

Find the derivatives of the following functions with respect to the independent variables:

1. $y = (4 - x^2)^5$ **2.** $y = (x^3 - 4x^2 + 1)^4$

3. $s = (t^3 - 5t)^6$ **4.** $y = (3x^2 - 4x)^{\frac{1}{2}}$

5. $i = (4 + 3t - t^2)^3$ **6.** $q = (3 - t^2)^{-3}$

Find the derivatives by the product rule:

7. $y = (x^2 + 3x)(4 - x^3)$ **8.** $y = x^3(3 - x^2)^4$

Find the derivatives by the quotient rule:

9. $y = \dfrac{x^2 - 3}{4x - x^3}$ **10.** $y = \dfrac{3x + 2}{\sqrt{6 - x^2}}$

11. In the following function, first find dx/dy then dy/dx by inversion.

$$x = (y^3 - 4)^2$$

12. Find the slope of the curve, $3x^2 - 4xy - 2y^2 = 18$, at the point $(2, -3)$ (by implicit differentiation).

13. In the function, $y = x^2 - 3x - 7$, how fast is y changing when $x = 5$, and x is changing at 2 units per second (related time-rates).

14. In the function, $x^2 + 2y^2 = 33$, how fast is y changing when $x = 5$, $y = 2$ and x is changing by 0.5 units per second (related time-rates).

15. In the equation for power, $P = 15i^2$, how fast is power P changing when the current is 2 amps and changing at 0.25 amps per second?

16. A ladder 26-ft long leans up against a vertical wall. If the foot of the ladder is being pulled away from the wall at a steady rate of 3 in. per second, how fast is the top of the ladder descending when the foot of the ladder is 10 ft from the wall?

CHAPTER FIFTY Integration

50-1 INTEGRATION AS ANTIDIFFERENTIATION

We have seen that the process of finding a derivative or differential is called *differentiation*. We now come to the *inverse process* of finding the function itself when the *derivative* (or *differential*) is given. This inverse process is called *integration* or *antidifferentiation*.

Integration has many uses in mathematics. We have seen that the derivative of distance s is equal to velocity v. Now, if we know the velocity of an object in motion, we can integrate and get the formula for distance s. If the slope of a curve is given, we can integrate and find the equation of the curve itself. If we know di/dt, the rate of change of current, we can find the equation for the current i by integration.

To see clearly the meaning of integration, consider an example. Suppose we have the function, $y = x^5$. Then $dy/dx = 5x^4$; or $dy = 5x^4\,dx$. Now, if we know that the derivative of a function is $5x^4$, or in differential terms, $5x^4\,dx$, we wish to find the function itself. This is the problem of integration. In this example, we might immediately know that the function itself is x^5.

To indicate integration, we use the symbol, $\int$, an elongated S. The symbol is written before the expression to be integrated. The quantity to be integrated is called the *integrand*. The integrand is usually a *differential* rather than only the derivative, and it is written immediately after the integral sign. Whether the integrand is the differential or only the derivative, the answer is the same. For example, we write

$$\int 5x^4 = x^5, \qquad \text{but more often} \qquad \int 5x^4\,dx = x^5$$

The integral sign means, "Find the function whose derivative is $5x^4$, or whose differential is $5x^4\,dx$." The answer, x^5, is called the *integral*. The factor, dx, is included in the integrand because it has important uses, as we shell see later.

To find the integral, we can sometimes recognize the integrand as the differential of some known function. However, for most integration, we need some specific steps to follow.

50-2 THE POWER RULE FOR INTEGRATION

One of the most useful rules for integration is the *Power rule.* Let us see how this rule comes about. Consider a problem in differentiation:

$$\frac{d}{dx}(8x^3) = 24x^2, \qquad \text{or} \qquad d(8x^3) = 24x^2\,dx$$

Note carefully the steps in differentiation. Then we shall reverse these steps, since integration is the reverse of differentiation:

1. To differentiate, *we multiply the coefficient by the exponent on* x.
2. *Then we decrease the exponent by* 1.

Now suppose we have the derivative $24x^2$, or the differential, $24x^2\,dx$, and wish to find the integral, that is, the function itself. Our problem is

$$\int 24x^2\,dx$$

We reverse the differentiating process. Then, for integration, we have:

1. *Increase the exponent by* 1, *and get* $24x^3$.
2. *Divide the entire quantity by the new exponent, and get*
   $24x^3/3 = 8x^3$

Note that the factor dx disappears. In general terms, we have the formula, the power rule:

$$\int Kx^n\,dx = \frac{Kx^{n+1}}{n+1} \text{ (where } K \text{ is any constant) } (n \neq -1)$$

The power rule for integration applies to any variable. Examples:

$$\int 12x^2\,dx = 4x^3; \qquad \int 20t^3\,dt = 5t^4; \qquad \int 48u^5\,du = 8u^6$$

Note that the integrands contain the differentials, dx, dt, and du. In each integrand, the variable to be integrated has associated with it the corresponding differential. Each such differential indicates the *variable of integration.* To integrate the power, $20t^3\,dt$, the integrand must contain the factor dt. For example, there is no way to integrate the following:

$$\int 12y^2\,dx$$

The factor dx indicates the variable of integration. The only way to perform the integration here is to replace y with some function of x

that is known to be equal to y. If we know that $y = 2x$, then we can rewrite the integration and have

$$\int 12(4x^2)dx = \int 48x^2\,dx = 16x^3$$

50-3 THE INDEFINITE INTEGRAL

Note that the following functions have the same derivative:

$$\frac{d}{dx}(x^5) = 5x^4; \quad \frac{d}{dx}(x^5 + 3) = 5x^4; \quad \frac{d}{dx}(x^5 - 7) = 5x^4$$

For all these functions, the derivative is $5x^4$ because the derivative of a constant is zero. Now, if we have the integrand, $5x^4\,dx$, the integral will contain x^5, but *it may also contain some constant term*. The integral may be any of these, as well as others: $x^5 + 7$; $x^5 - 8$; $x^5 + \frac{1}{2}$.

To show that the integral may also contain some constant term, we say the integral is x^5 *plus some possible constant*. This indefinite constant is indicated by the letter C, or some other letter understood to be a constant. The constant may be positive, negative, fractional, or zero. Then we write

$$\int 5x^4\,dx = x^5 + C$$

Here, C is called the *constant of integration* and should always be included in a problem of this kind. The answer is called the *indefinite integral*, because C may be any constant. The indefinite integral represents an entire family of functions. To find the indefinite integral,

1. *Increase the exponent by* 1.
2. *Divide by the new exponent.*
3. *Add C, the constant of integration.*

Formula:

$$\int Kx^n\,dx = \frac{Kx^{n+1}}{n+1} + C$$

(Note: $n \neq -1$)

The following special rules are simply applications of the power rule.

1. Since $d/dx(6x) = 6$, and $d(6x) = 6\,dx$, then $\int 6\,dx = 6x + C$. *That is, the integral of a constant is equal to the constant times the variable to the first power.* For example,

$$\int 7\,dt = 7t + C; \quad \int 3\,dy = 3y + C; \quad \int 4\,di = 4i + C$$

2. *The integral of a differential is the first power of the variable:*

$$\int dx = x + C; \qquad \int dy = y + C;$$

$$\int dv = v + C; \qquad \int dA = A + C$$

Actually, in this case the coefficient of each differential is 1. Then this rule is covered in Rule (1).

3. *The power rule applies to all kinds of exponents* (except -1):

$$\int 12x^{-5}\,dx = -3x^{-4} + C; \qquad \int 6x^{\frac{1}{2}}\,dx = 4x^{\frac{3}{2}} + C$$

4. *A multinomial may be integrated term by term.* Only one C is needed.

$$\int (6x^2 + 8x^{-3} - 12x^{\frac{1}{3}} - 1)dx = 2x^3 - 4x^{-2} - 9x^{\frac{4}{3}} - x + C$$

Example 1

The slope of a curve is given by the formula, $m = 6x - 5$. Find the general equation of the curve itself.

Solution

To get the equation of the curve, we integrate the slope:

$$\int (6x - 5)dx = 3x^2 - 5x + C$$

Letting y represent the function, we get

$$y = 3x^2 - 5x + C$$

Now we recognize the equation as that of a parabola.

Example 2

The velocity of an object in motion is given by the formula, $v = 3t^2 - 2t + 4$. What is the general equation for distance s?

Solution

$$s = \int (3t^2 - 2t + 4)dt = t^3 - t^2 + 4t + C$$

Example 3

The acceleration of an object in motion is given by: $a = 12t + 3$. Since $a = dv/dt$, find the general equation for velocity v.

Solution

$$v = \int (12t + 3)dt = 6t^2 + 3t + C$$

Example 4

In a circuit, the rate of change of current is: $di/dt = 4t - 3$. What is the formula for the current i?

Solution

$$i = \int (4t - 3)dt = 2t^2 - 3t + C$$

Exercise 50-1

Find the indefinite integrals. Don't forget the constant C.

1. $\int 6\,dx$ **2.** $\int 6x\,dx$ **3.** $\int 6x^3\,dx$ **4.** $6\int x^2\,dx$

5. $\int dq$ **6.** $\int dw$ **7.** $\int 5t\,dt$ **8.** $\int 3y\,dy$

9. $\int 8t^3\,dt$ **10.** $8\int t^3\,dt$ **11.** $\int 6x^{-3}\,dx$ **12.** $\int 5t^{-2}\,dt$

13. $\int (t^3 - t)\,dt$ **14.** $\int (i^2 + 5i)\,di$ **15.** $\int (x^{-4} - x^{-2})\,dx$

16. $\int (8t^{\frac{1}{2}} - 3t^{\frac{1}{2}})\,dt$ **17.** $\int (4t^{-3} - 3t^{-\frac{1}{2}})\,dt$ **18.** $\int (8x - 3)x^2\,dx$

19. $\int (2x - 3)^2 x\,dx$ **20.** $\int (2x - 3)^3 x^2\,dx$ **21.** $\int (x - 3)(x + 5)\,dx$

22. $\int \frac{1}{x^2}\,dx$ **23.** $\int \frac{6}{x^3}\,dx$ **24.** $\int \frac{10}{x^6}\,dx$

25. $\int (x^{-3} + 6x^{-2} - 3x^2 - x + 2)dx$

26. $\int (8x^7 + 4x^5 - 2x^3 + x^2 + 1)\,dx$ **27.** $\int y\,dx$, if $y = x^2 - 3x$

28. $\int i\,dt$, if $i = 2t - t^2$

29. If the slope of a curve is given by the formula, $m = 3x - 5$, find the general equation of the curve. Name the curve.

30. If acceleration of an object is given by the formula, $a = 8t - 1$, find the general equation for velocity v.

31. If velocity v is given by the formula, $v = 4t^2 - 3t + 5$, find the general equation for distance s.

32. In finding the integral, $\int 6$, what is the difficulty?

33. If q represents electric charge, then current i is defined as dq/dt. If $i = 3t^2 + 4t + 3$, find the general equation for q.

50-4 THE PARTICULAR INTEGRAL

When we integrate a function, we get another function, the integral function. Then we add the constant C to indicate the indefinite integral. For example,

$$\int (2x - 5)\,dx = x^2 - 5x + C$$

Whatever the value of C, we can represent the indefinite integral by y, and write,

$$y = x^2 - 5x + C$$

Note that this equation represents a parabola, in fact, a family of parabolas.

Now, if we specify that the parabola passes through the point $(-2, 6)$, this point will determine the particular parabola, and also the value of C. Since the point must satisfy the equation, we substitute the coordinates of the point in the general equation, $y = x^2 - 5x + C$, and get

$$6 = 4 + 10 + C; \qquad \text{solving, we get } C = -8$$

The equation of this particular parabola becomes: $y = x^2 - 5x - 8$. The function, $x^2 - 5x - 8$, is called the *particular integral*, and may be indicated by the equation: $P(x) = x^2 - 5x - 8$; or $y = x^2 - 5x - 8$.

Example 1

In the following problem, find the particular integral if the integral itself is equal to 6 when $x = -1$:

$$\int (3x^2 - 4x - 5)\,dx$$

Solution

Integrating, $\int (3x^2 - 4x - 5)\,dx = x^3 - 2x^2 - 5x + C$

When $x = -1$, and the integral is equal to 6, we have the following values:

$$-1 - 2 + 5 + C = 6; \qquad \text{solving, } C = 4$$

Then the particular integral is the function, $P(x) = x^3 - 2x^2 - 5x + 4$.

Example 2

The slope of a curve is given by the formula, $m = 3 - 4x$. Find the equation of the curve if it passes through the point (1, 3). Draw the curve.

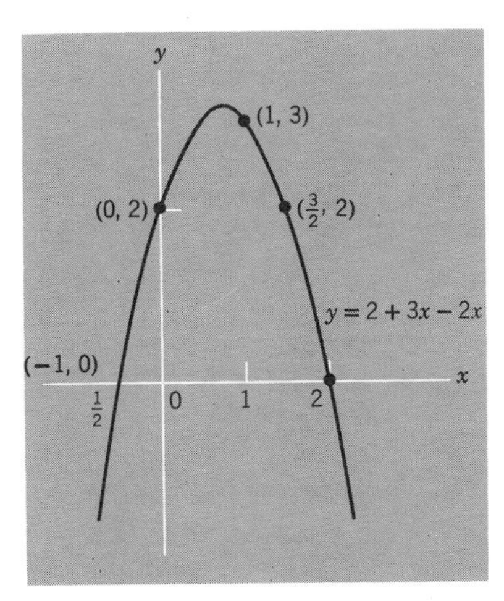

Figure 50-1

Solution

Integrating, $\qquad \int (3 - 4x)\,dx = 3x - 2x^2 + C$

The general equation can be written:

$$y = 3x - 2x^2 + C$$

However, we cannot draw the curve, until we know the value of C. Substituting the coordinates of the point (1, 3), we get

$$3 = 3 - 2 + C; \qquad \text{solving,} \quad C = 2$$

Then the particular equation of the parabola becomes:

$$y = 2 + 3x - 2x^2$$

The drawing is shown in Fig. 50.1.

Example 3

The velocity of a particle in motion is given by: $v = 3t^2 - 8t + 5$. Find the formula for distance s if $s = 30$, when $t = 2$

Solution

Integrating, $\qquad s = \int (3t^2 - 8t + 5)\,dt = t^3 - 4t^2 + 5t + C$

When $s = 30$ and $t = 2$, we get $30 = 8 - 16 \geqq 10 + C$; $\quad$ then $C = 28$

Then the particular equation for distance is $\qquad s = t^3 - 4t^2 + 5t + 28$

Example 4

A ball is thrown directly upward from the top of a 90-ft building with an initial velocity of 64 ft per sec. If the acceleration due to the earth's gravity is 32 ft/sec^2 *downward*, find the particular equations for velocity v and distance s. Then find s when $t = 5$ sec.

Solution

Taking acceleration downward, we have $\quad a = -32$

since $a = dv/dt$, we have $\qquad v = \int (-32)\,dt = -32t + C_1$

Since $v = 64$ when $t = 0$, we find that $C_1 = 64$. Then the particular equation for velocity becomes: $v = 64 - 32t$

Integrating for s, we get $s = \int (64 - 32t)\,dt = 64t - 16t^2 + C_2$

Since $s = 90$ when $t = 0$, $C_2 = 90$, and we get $s = 90 + 64t - 16t^2$

Using this formula, when $t = 5$, we get $s = 90 + 320 - 400 = 10$ ft

Exercise 50-2

Find the particular equation of the curve whose slope is given by each of the functions of x in Nos. 1–8, if the indicated point lies on the curve.

1. $m = (2x + 5)$; pt(2, 8)

2. $m = 3x - 4$; pt(−2, 5)

3. $m = 3x^2 - 2x + 1$; pt(2, 6)

4. $m = 3 - 8x - 6x^2$; pt(−1, 5)

5. $m = 8x^3 - 3x^2 - 6x$; pt(2, −1)

6. $m = x^3 - 2x^2 - 5x + 7$; pt(2, −3)

7. $m = 4x^3 - 8x + 2$; pt(2, −5)

8. $m = 5 + 2x - 6x^2 - x^3$; pt(−1, −5)

9. If $di/dt = 4t - 5$, find the equation for i, if $i = 3$ amps when $t = 1$ sec.

10. If $di/dt = 3 - 5t$, find the equation for i, if $i = 2$ amps when $t = 1.5$ sec.

11. If acceleration $a = 8$ ft per sec^2, find equation for velocity v, if $v = 30$, when $t = 2$. Then find the equation for distance s, if $s = 40$ ft when $t = 2$ sec.

12. If $a = -16$ ft per sec^2, find the equations for velocity and distance s if $v = 30$ ft per sec and $s = 50$ ft when $t = 0$.

13. A rifle bullet is fired directly upward from ground level. The muzzle velocity of the bullet is 2400 ft per sec. If the acceleration due to earth's gravity is 32 ft per sec^2 *downward*, find the equations for velocity v and distance s. Find the maximum height of the bullet.

14. The moon's gravity is such that the acceleration is approximately 5.2 ft per sec^2 downward. If an object is thrown directly upward from the moon's surface with an initial velocity of 78 ft per sec, find the particular equations for the velocity v and the distance s. How high will the object rise?

15. A man who can high jump 6.8 ft on earth has an initial velocity of approximately 20.8 ft per sec when he leaves the ground. How high could he jump on the moon with the same initial velocity?

16. The gravity on Jupiter is such that acceleration due to gravity is about 83.2 ft per sec^2 downward. How high could the man in No. 15 jump on Jupiter?

50-5 THE DEFINITE INTEGRAL

To understand the meaning of the definite integral, consider again the indefinite integral:

$$\int (2x + 5)\,dx = x^2 + 5x + C$$

We have seen how to find the value of C. For example, if this integral is equal to 9 when $x = 1$, then $C = 3$, and the particular integral becomes

$$P(x) = x^2 + 5x + 3$$

Now we come to the *definite integral*. One of the most important uses of calculus is in finding the *difference* between two integral values. For example, in the particular integral, $P(x) = x^2 + 5x + 3$, when $x = 6$, $P(6) = 69$; when $x = 4$, $P(4) = 39$. The difference between the two integral values is $69 - 39 = 30$. The difference, 30, is called the *definite integral.*

Definition. *The definite integral is the difference between two integral values.*

In the particular integral, $P(x) = x^2 + 5x + 3$, if we show the substitution of separate values, we have:

$$P(6) = 36 + 30 + 3; \qquad P(4) = 16 + 20 + 3$$

Subtracting,

$$\begin{aligned} P(6) - P(4) &= (36 + 30 + 3) - (16 + 20 + 3) \\ &= 36 + 30 + 3 - 16 - 20 - 3 = 30 \end{aligned}$$

Note especially that in combining the terms, the $+3$ *and the* -3 *cancel each other. This means that we get the same answer if we omit the constant* 3. *Therefore, in finding a definite integral, the constant* C *can be omitted.*

The two values of x, such as 6 and 4, used in the integral are called the *limits of integration.* The greater algebraic value is called the *upper limit*; the smaller is called the *lower limit.* To indicate a definite integral, we place the two limits near the integral sign, as shown here:

$$\int_4^6 (2x + 5)\,dx$$

In finding a definite integral we first integrate, omitting the constant of integration C. We rewrite the limits near the integral. We find the value of the integral at the *upper limit*, and then *subtract* the value of the integral at the *lower limit*. Example:

$$\int_4^6 (2x+5)\,dx = x^2+5x\Big]_4^6 = (36+30)-(16+20) = 30$$

If a and b represent the lower and upper limits, respectively, of integration, then, in general terms, the definite integral is

$$\int_a^b f(x)\,dx = F(x)\Big]_a^b = F(b)-F(a)$$

The definite integral may have a value that is positive, negative, or zero. Note that the value does not depend on what particular variable is used.

Example 1

$$\int_2^3 (2x-5)\,dx = x^2-5x\Big]_2^3 = 9-15-4+10 = 0$$

Example 2

$$\int_{-2}^1 (4t+3)\,dt = 2t^2+3t\Big]_{-2}^1 = 2+3-8+6 = 3$$

Example 3

$$\int_1^5 (2x-7)\,dx = x^2-7x\Big]_1^5 = 25-35-1+7 = -4$$

Example 4

$$\int_{-3}^2 (3t^2-4t-5)\,dt = t^3-2t^2-5t\Big]_{-3}^2 = 20$$

Exercise 50-3

Evaluate the following definite integrals:

1. $\int_1^4 (4x-3)\,dx$ **2.** $\int_1^4 (4y-3)\,dy$ **3.** $\int_1^4 (4t-3)\,dt$

4. $\int_{-2}^3 (2x+5)\,dx$ **5.** $\int_{-2}^3 (2t+5)\,dt$ **6.** $\int_0^3 (3x^2-4x)\,dx$

7. $\int_0^2 (3t-t^2)\,dt$ **8.** $\int_{-2}^{-1} (8t-6t^2)\,dt$ **9.** $\int_1^4 (3-x-x^2)\,dx$

10. $\int_{-3}^{1} (2v - v^2)\, dv$ 11. $\int_{5}^{10} (4q^3 - 3q^2)\, dq$ 12. $(3x^2 - 4x + 7)\, dx$

13. $\int_{1}^{64} (x^{\frac{1}{3}} - x^{\frac{1}{2}})\, dx$ 14. $\int_{1}^{4} (x^{\frac{1}{2}} - x^{-\frac{3}{2}})\, dx$ 15. $\int_{-1}^{2} (6x^2 + 8x - 9)\, dx$

16. $\int_{1}^{8} (t^{\frac{2}{3}} + t^{\frac{1}{3}})\, dt$ 17. $\int_{1}^{3} (x^{-2} - x^{-3})\, dx$ 18. $\int_{50}^{100} (6x + 2)\, dx$

19. $\int_{1}^{2} y^2\, dx;\ (y = 2x)$ 20. $\int_{-1}^{2} x^2 dy;\ (x = 2y + 3)$

21. $\int_{1}^{3} y\, dx;\ \left(y = \frac{x^2}{4}\right)$ 22. $\int_{0}^{4} x\, dy;\ (x^2 = 4y)$

23. $\int_{-2}^{4} (x_2 - x_1)\, dy;\quad x_2 = \frac{y}{2} + 2;$ and $x_1 = \frac{y^2}{4}$

24. $\int_{-2}^{4} (y_2 - y_1)\, dx;\quad y_2 = \frac{x}{2};$ and $y_1 = \left(\frac{x^2}{4}\right) - 2$

50-6 THE POWER RULE APPLIED TO FUNCTIONS

The power rule for integration can be applied to the power of a function u, provided the integrand contains du, the differential of the function on which the power rule is to be applied. We have seen that

$$\int u^n\, du = \frac{u^{n+1}}{n+1} + C$$

If we consider u as a function of x, then before the power rule can be applied on the function u, the integrand must contain not only dx but also the complete differential du. Consider the following problem:

Example 1

$$\int (x^2 + 3)^5\, dx$$

Here we identify the function, $u = x^2 + 3$. To apply the power rule on this function, we need the complete differential, $du = 2x\, dx$, in the integrand. Since the factor $2x$ is lacking, we cannot apply the power rule. On the other hand, suppose we have the problem:

Example 2

$$\int (x^2 + 3)^5 2x\, dx$$

Here we identify the following: $u = x^2 + 3$, then $du = 2x\, dx$. Since the integrand contains the complete differential, $2x\,dx$, we immediately apply the power rule on the function u and get

$$\int (x^2 + 3)^5 2x\,dx = \frac{(x^2 + 3)^6}{6} + C$$

Note that the differential, $2x\,dx$, disappears in the process of integration.

If the integrand lacks only a *constant factor* for a complete differential du, then this factor can be supplied as in the following example:

Example 3

$$\int (x^3 + 2)^6 x^2\,dx$$

We identify the following: $u = x^3 + 2$, then $du = 3x^2\,dx$. Since we need the differential, $3x^2\,dx$, in the integrand, we supply the constant factor, 3, and its reciprocal as follows:

$$\int (\tfrac{1}{3})(3)(x^3 + 2)x^2\,dx$$

Now the factor $(\frac{1}{3})$ may be moved to the left of the integral sign and we get

$$(\tfrac{1}{3})\int 3(x^3 + 2)^6 x^2\,dx$$

Now the integrand contains the complete differential, $3x^2\,dx$. In the process of integration, this differential disappears, but the factor $(\frac{1}{3})$ outside the integral sign remains, and we get

$$\frac{1}{3}\int 3(x^3 + 2)^6 x^2\,dx = (\tfrac{1}{3})\frac{(x^3 + 2)^7}{7} + C = \frac{(x^3 + 2)^7}{21} + C$$

To summarize: *If the integrand lacks only a constant factor for a perfect differential, this factor may be supplied and its reciprocal written outside the integral sign. The differential under the sign disappears, but the reciprocal outside remains as a part of the integral.*

Warning. *A variable factor cannot be supplied in this manner.*

Exercise 50-4

Supply any necessary constant and integrate by the power rule:

1. $\int (x^2 - 5)^3 x\,dx$

2. $\int (1 - x^2)^5 x\,dx$

3. $\int (x^4 + 2)^6 x^3\,dx$

4. $\int 3x(x^2 + 1)^4\,dx$

5. $\int (x^2 + 6x + 1)^3 (x + 3)\,dx$

6. $\int (3 + 4x - x^2)^2 (x - 2)\,dx$

7. $\int (4x^2)(x^3 - 4)^5\,dx$

8. $\int (x^3 - 6x^2 + 3)^4(x^2 - 4x)\,dx$

9. $\int 5(x^2 + 8x)^{\frac{1}{2}}(x + 4)\,dx$

10. $\int (2 + 6x - x^2)^{\frac{1}{3}}(x - 3)\,dx$

11. $\int 2(t^3 - 6t)^4(t^2 - 2)\,dt$

12. $\int 3(4s^2 - 6s)(4s - 3)\,ds$

13. $\int \frac{(x - 2)\,dx}{\sqrt{4x - x^2}}$

14. $\int \frac{(6 - 3x)\,dx}{\sqrt[3]{x^2 - 4x + 3}}$

15. $\int_0^2 x(x^2 + 1)^3\,dx$

16. $\int_0^2 (3x^2 + 4)^{\frac{1}{2}}x\,dx$

17. $\int_1^3 (3x^2 - 2)^{\frac{1}{2}}x\,dx$

18. $\int_0^3 (x^2 - 1)^{\frac{1}{3}}x\,dx$

19. $\int_0^2 (4 - x^2)^{\frac{1}{2}}x\,dx$

20. $\int_{-2}^0 (2 - x^3)^2x^2\,dx$

QUIZ ON CHAPTER 50, INTEGRATION. FORM A.

Find the indefinite integral of each of the following (Nos. 1–9):

1. $\int 7dt$ **2.** $\int 24x^3\,dx$ **3.** $\int 6y^{-4}\,dy$

4. $\int 8x^{\frac{1}{2}}\,dx$ **5.** $\int (x^2 - 3x)\,dx$ **6.** $\int (i^3 - 4i^2)\,di$

7. $\int \frac{6}{x^3}\,dx$ **8.** $\int (x(x^2 + 3)^4\,dx$ **9.** $\int x^2(4 - x^3)^5\,dx$

10. If the slope of a curve is given by the formula, $m = 4x - 3$, find the general equation of the curve, and then find the particular equation if the curve passes through the point $(-2, -7)$.

11. If the slope of a curve is given by the formula, $m = 3x^2 - 6x + 5$, find the particular equation of the curve if it passes through the point $(-1, 4)$.

12. If the formula for acceleration is given by the formula, $a = 6t - 5$, find the particular equation for velocity v, if $v = 20$ when $t = 0$.

13. A ball is projected directly upward from the top of a 90-ft building with an initial velocity of 60 ft/sec. If the acceleration due to gravity is -32 ft/sec^2, find the particular equations for velocity v, and distance s.

14. Find the definite integral: $\int_{-1}^{3}(3x^2 - 5x + 2)\,dx$.

CHAPTER
FIFTY-ONE Applications of Integration

51-1 AREA

We are familiar with the formula for the area of a rectangle, $A = lw$. However, finding the area bounded by a curve is not so simple. Consider the area bounded by the curve, $x^2 = 4y$, the x-axis, and the line $x = 4$ (Fig. 51-1). To estimate the area we can divide it into vertical strips. Then, considering the strips as approximate rectangles, we can estimate the area of each strip and then add these areas. Yet the result will not be the exact area under the curve. However, the exact area can often be found by the definite integral.

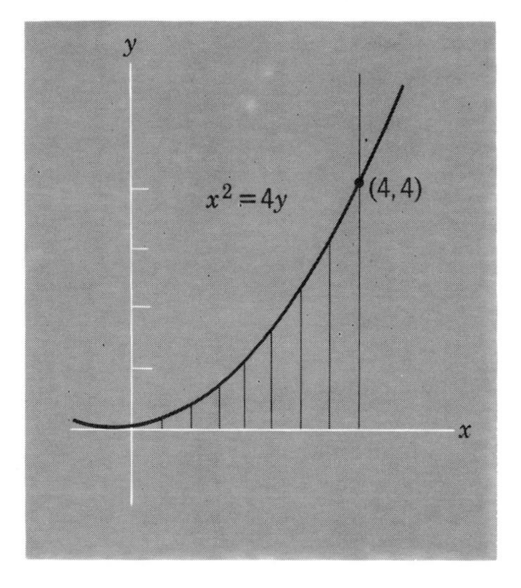

Figure 51.1

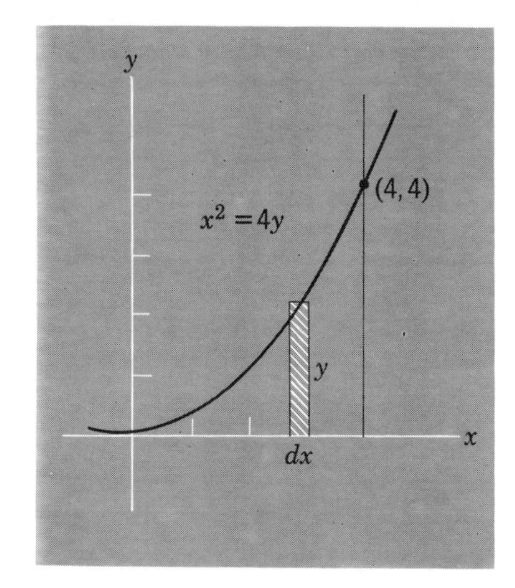

Figure 51.2

One of the strips of area is shown in Fig. 51-2. This strip is called an *element of area* and is denoted by dA, as the *differential of area.* Now, assuming the strip is a rectangle, the length is y, the ordinate of the curve. The width we call Δx or dx. Then the area of the element is

$$dA = (y)(dx)$$

Of course, the element shown is not an actual rectangle.

But now we go one step further. We imagine that dx becomes smaller and smaller and approaches zero. The number of strips increases without limit, becoming infinite. Then *the sum of an infinite number of strips is equal to the exact area under the curve.*

At this point we discover one of the most important facts of calculus. The infinite number of strips can be added up by the definite integral, evaluated between the x-limits of the curve. We need show only one strip. That is,

$$dA = \text{the element}; \qquad \text{then, exact area} = \int \text{the element}$$

In this example, $dA = y\,dx$, then, exact area $= \int_0^4 y\,dx$

Before we can integrate, we must express y as a function of x. In this example, $x^2 = 4y$, then $y = x^2/4$. Replacing y with $x^2/4$, we have

$$\text{area} = \int_0^4 \left(\frac{x^2}{4}\right) dx = \frac{x^3}{12}\bigg]_0^4 = \frac{16}{3}, \text{ the exact area}$$

For the area between two curves, we follow the same general procedure. We imagine the area divided into an infinite number of vertical or horizontal strips. We show one of these strips as the element of area. The important thing is to express the element of area correctly. The area will always be

$$dA = (\text{length})(\text{width})$$

For a vertical strip the length will be: *top* y − *bottom* y; or $y_2 - y_1$.
For a horizontal strip the length will be: *right* x − *left* x; or $x_2 - x_1$.

Example 1

Find the area between the curves, $y^2 = 4x$, and $2x - y = 4$.

Solution

We sketch the curves (Fig. 51-3) and imagine the enclosed area divided into horizontal strips, one of which is shown. This is the element of area, dA. The length is $(x_2 - x_1)$; the width is dy. Then

$$dA = (x_2 - x_1)dy$$

To find the limits of integration, we solve the two equations as a system and get the points $(1, -2)$ and $(4, 4)$. Now the infinite number of strips must be added up from $y = -2$ to $y = 4$. Then

$$A = \int_{-2}^{4} (x_2 - x_1)\,dy$$

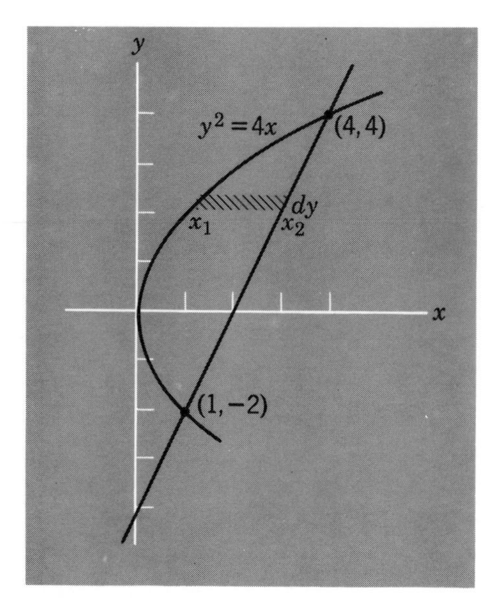

Figure 51.3

We must replace x_2 and x_1 with the equivalent functions of y. Solving each equation for x, we get

$$x_2 = \frac{y}{2} + 2; \quad x_1 = \frac{y^2}{4}$$

Then $$A = \int_{-2}^{4} \left(\frac{y}{2} + 2 - \frac{y^2}{4}\right) dy = \frac{y^2}{4} + 2y - \frac{y^3}{12}\Big]_{-2}^{4} = 9$$

Note. Some thought must be given to the matter of the proper element of area. Sometimes the element may be either vertical or horizontal. However, sometimes it must be taken only one way. It must be taken so that either end of the element will always be on the same curve, wherever it is taken. In Example 1, above, if we take the element vertically, the bottom of the element in some positions will not be on the straight line.

Example 2

Find the area between the curves: $x^2 + 8 = 4y$; $x + 8 = 2y$

Solution

We sketch the curves: (Fig. 51.4). In this example, we imagine the area divided into vertical strips, one of which is shown. This strip is dA, the element of area. For dA, the length is

$$\text{top } y - \text{bottom } y$$

or

$$y_2 - y_1$$

The width of the element is dx. Then, for the area of the element, we have

$$dA = (y_2 - y_1)\,dx$$

Now we must replace y_2 and y_1 with the equivalent functions of x. Solving the equations for y, we get

$$y_2 = \frac{x}{2} + 4; \qquad y_1 = \frac{x^2}{4} + 2$$

Then, the length of the element becomes

$$y_2 - y_1 = \left(\frac{x}{2} + 4\right) - \left(\frac{x^2}{4} + 2\right) = \frac{x}{2} + 2 - \frac{x^2}{4}$$

For the element of area, we have

$$dA = \left(\frac{x}{2} + 2 - \frac{x^2}{4}\right\}dx$$

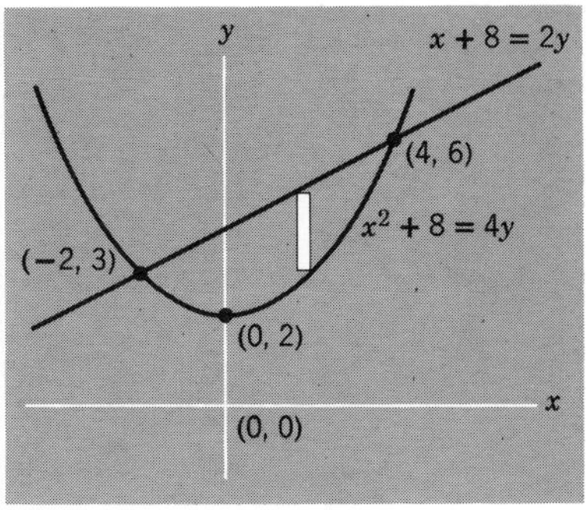

Figure 51.4

To find the limits of integration, we solve the two equations as a system and get the points of intersection: $(-2, 3)$ and $(4, 6)$.

Using the x-limits, we have, for the area,

$$A = \int_{-2}^{4} \left(\frac{x}{2} + 2 - \frac{x^2}{4}\right) dx = \frac{x^2}{4} + 2x - \frac{x^3}{12}\Big]_{-2}^{4} = 9$$

Example 3

Find the total area bounded by the x-axis and the straight line, $x + 2y = 6$, from $x = 0$ to $x = 10$ (Fig. 51-5).

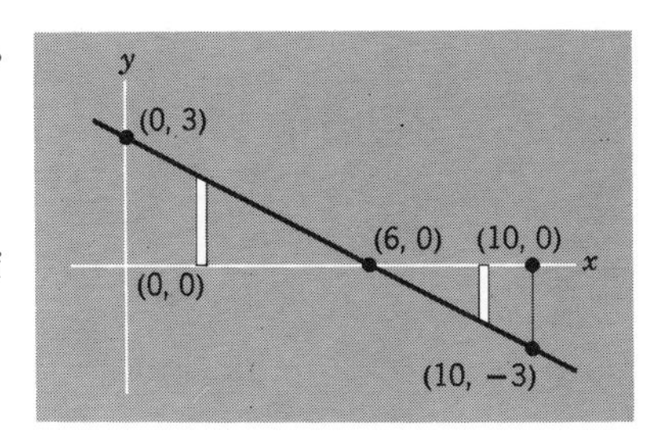

Figure 51.5

Solution

In this example, the total area must be computed in two parts. A portion of the area lies above the x-axis, and part below. We take a vertical element, whose area is

$$dA = y\,dx$$

Solving the equation for y, we get

$$y = 3 - \frac{x}{2}$$

Then we indicate the integral,

$$\int \left(3 - \frac{x}{2}\right) dx = 3x - \frac{x^2}{4}$$

If we use the limits, 0 and 10, we get the definite integral: 5 sq units. However, since a portion of the area lies below the x-axis, this area subtracts from the area above. To get the total area, we must integrate first between the limits, $x = 0$ to $x = 6$, because this portion lies above the x-axis. For this portion we get the area of 9 sq units. For the portion below the x-axis, we get the area: -4 sq units. Since no area can be negative, we must call this a positive area, and get a total area of 13 sq units.

Exercise 51-1

In Nos. 1–6, find the area under the curve between the indicated limits:

1. $y = 2x$, from $x = 2$ to $x = 4$. **2.** $x^2 = 8y$, from $x = 0$ to $x = 4$

3. $y = x^3$, from $x = 0$ to $x = 2$ **4.** $y = x^4$, from $x = 0$ to $x = 2$

5. $y = 4x - x^2$; $x = 0$ to $x = 4$ **6.** $y = 9 - x^2$; $x = -3$ to $x = +3$

7. Find the area between the y-axis and the curves, $y^2 = 4x$; $y = 4$.

8. Find the area between the y-axis and the curve, $x = 2y - y^2$.

9. Find the total area bounded by the curve, $y = x^3 - 2x^2 - 5x + 6$, and the x-axis, from $x = -2$ to $x = 3$. This area must be computed in two portions, one above the x-axis and the other below.

10. Find the total area bounded by the curve, $y = x^3$, and the x-axis from $x = -1$ to $x = +1$.

In Nos. 11–20, find the area bounded by the two curves:

11. $y^2 = 4x$; $x^2 = 4y$ **12.** $y = x^4$; $y = x^2$

13. $y^2 - 4x = 12$; $y^2 + 8x = 24$ **14.** $y^2 + x = 1$; $y^2 = 4x + 16$

15. $y^2 = 4x$; $y = x$ **16.** $y = 4x - x^2$; $y = x$

17. $x^2 = 2y$; $2x - y = 0$ **18.** $y = x^3$; $4x - y = 0$

19. $x^2 = 4y$; $x + 4 = 2y$ **20.** $y = x^3$; $y = x^2$

51-2 VOLUME

If an area is rotated about an axis, it generates a *solid of revolution.* Consider the area bounded by the curves, $x = 2y$, $y = 0$, and $x = 8$ (Fig. 51-6). If this area is rotated about the x-axis, a solid of revolution

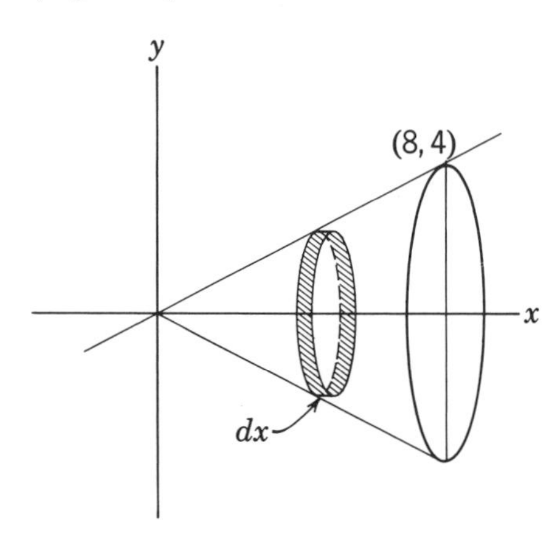

Figure 51.6

is formed, in this case, a cone. To find the volume, we could use the formula for the volume of a cone. However, we shall find the volume by calculus to illustrate the method.

Imagine the volume divided vertically into slices, one of which is shown. This slice is an *element of volume,* a disc, resembling approximately the shape of a coin or thin cylinder. The volume of the element is then given by the formula for the volume of a cylinder: $V = \pi r^2 h$. In this example, the radius r is equal to y, the ordinate of the line (or curve). The thickness is dx, which is the height of the cylinder. Then the volume of the element is

$$dV = \pi y^2\,dx$$

Now we imagine that the number of discs becomes infinite and that dx approaches zero. Then the infinite number of discs (or elements) may be added up by the definite integral, evaluated between x-limits:

$$V = \pi\int_0^8 y^2\,dx$$

Before we integrate we must replace y^2 with the equivalent function of x. From the line, $x = 2y$, we get $y^2 = x^2/4$. Then

$$V = \pi\int_0^8 \frac{x^2}{4}\,dx = \pi\frac{x^3}{12}\bigg]_0^8 = \frac{128}{3}\,\pi$$

In taking a slice of a solid revolution, the disc may resemble a washer, a disc with a hole at the center. Consider the area between the curves, $y^2 = 4x$ and $y = x$ (Fig. 51-7). If this area is rotated about the x-axis, the result is a geometric solid as shown. Now, if we take a slice at right angles to the axis of revolution, the slice resembles a washer. Then the volume is found by computing the volume of the solid disc and subtracting the volume of the hole. If R and r represent the large and the small radii, respectively, we have

$$dV = \pi R^2\,dx - \pi r^2\,dx$$

From the equations, R is the y of the parabola, and $R^2 = 4x$; r is the y of the line, and $r^2 = x^2$. Then $dV = \pi(4x - x^2)\,dx$, and we get

$$V = \pi\int_0^4 (4x - x^2)\,dx = \pi\left[2x^2 - \frac{x^3}{3}\right]_0^4 = \frac{32}{3}\,\pi$$

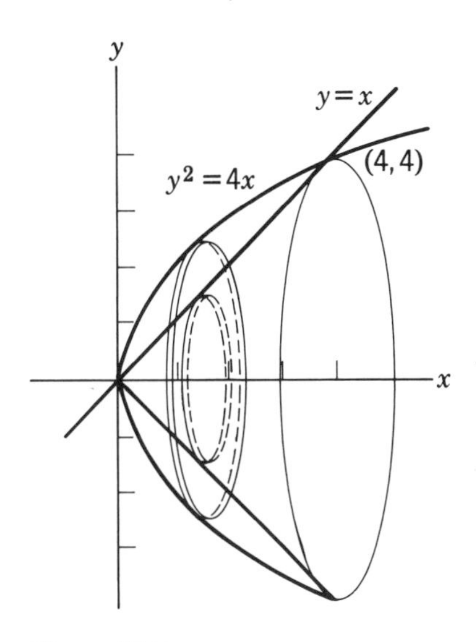

Figure 51.7

The volume of a solid of revolution may also be found by an alternate method, called the *shell method.* In this case we imagine the volume divided into shells, each having its axis as the axis of the solid of revolution. Figure 51.8 shows the same solid indicated in Fig. 51.6. A typical shell is called the element of volume, dV, as shown in the figure. This shell is a hollow cylinder. Now we use the formula for the

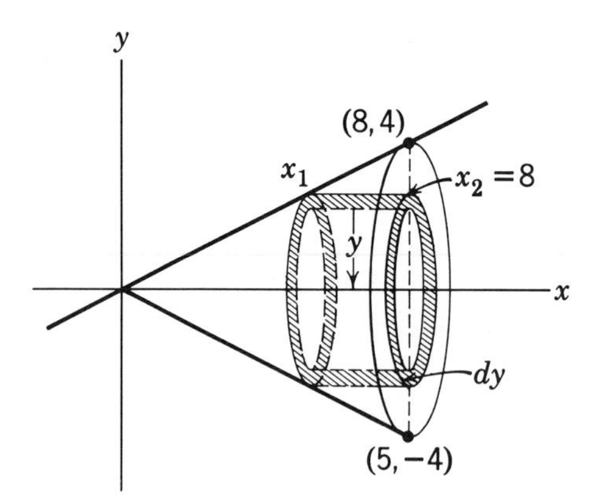

Figure 51.8

approximate volume of a hollow cylinder:

$$V = 2\pi rh(\text{thickness})$$

In this example, $r = y$, $h = x_2 - x_1$, and thickness $= dy$. Then

$$dV = 2\pi y(x_2 - x_1)\,dy; \qquad \text{or} \qquad \text{equivalently,}$$

$$dV = 2\pi(8y - 2y^2)\,dy$$

Now we add up the infinite number of shells by the definite integral and get

$$V = 2\pi\int_0^4 (8y - 2y^2)\,dy = 2\pi\left[4y^2 - \frac{2y^3}{3}\right]_0^4 = 2\pi\left(64 - \frac{128}{3}\right)$$

$$= \frac{128}{3}\pi$$

Exercise 51-2

Find the volume of the solid of revolution formed by rotating the given area about the indicated axis. Use the disc method for Nos. 1–7.

1. Area bounded by: $y^2 = 4x$; $x = 4$; $y = 0$; 1st quadrant; x-axis.

2. Area bounded by: $x^2 = 4y$; $y = 0$; $x = 4$; 1st quadrant; x-axis.

3. Area bounded by: $y^2 = 2x$; $y = 4$; $x = 0$; 1st quadrant; y-axis.

4. Area between: $y^2 = 4x$; $y = x$; y-axis.

5. Area bounded by: $y = x^3$; $x = 2$; $y = 0$; x-axis.

6. Area between: $y = 4x - x^2$; $y = x$; x-axis.

7. Area bounded by: $x = 2y$; $x = 8$; $y = 0$; y-axis.

8–12. Work the following by the shell method: Nos. 1, 2, 3, 4, 7.

51-3 WORK Another type of practical problem is one involving work. *Work* is defined as the *product of a force and a distance* through which the force acts. Force alone does not constitute work. No work is done unless the force acts through some distance. Work is measured in *foot-pounds*, or in *inch-pounds*. In the metric system it may be measured in *centimeter-grams*.

The work done by a force of 20 lb acting through a distance of 4 ft is equal to 80 foot-pounds (ft-lb). The work done by a force of 3 lb through a distance of 5 in. is 15 inch-pounds (in.-lb). If you weigh 150 lb and walk up a stair leading to a floor 16 ft above, you do 2400 ft-lb of work.

If a force is constant, the amount of work done can be found simply by multiplying the force by the distance. As a formula, work = Fs. However, if the force or the distance is a *variable*, the work can be computed by the definite integral.

Example 1

A spring is stretched 3 in. by a 6-lb force. Find the work done in stretching the spring whose normal length is 20 in. to a length of 28 in.

Solution

First we must find the spring constant, or spring modulus. The spring constant is the force required to stretch a spring 1 in. or 1 ft. According to Hooke's law, the distortion of an elastic body is proportional to the distorting force. Therefore, if a stretch of 3 in. requires a force of 6 lb, a stretch of 1 in. requires a force of 2 lb. This is the spring constant for this particular spring. Now, for any length of stretch, x, the force required is $2x$ pounds. As we stretch the spring some small distance dx, the work done in this case is (force)(distance), or $(2x)\,(dx)$ (Fig. 51-9). This is called the *element of*

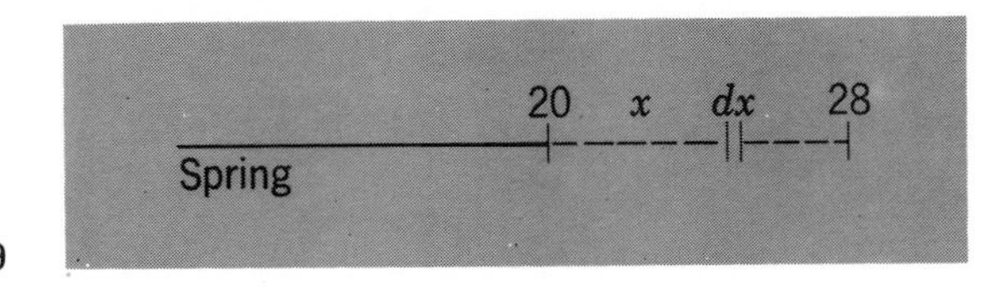

Figure 51.9

work; that is, $d(\text{work}) = 2x\,dx$. Now, when the spring is stretched from a length of 20 in. to 28 in., x changes from 0 to 8. Then we add up the elements of work over this interval, and get

$$\text{work} = \int_0^8 2x\,dx = x^2\Big]_0^8 = 64 \text{ in.-lb}$$

In the foregoing problem, the spring constant can be stated as 24 lb per ft. Then we get

$$\text{work} = \int_0^{2/3} 24x\,dx = 12x^2\Big]_0^{2/3} = \frac{16}{3} \text{ ft-lb}$$

Example 2

A chain 60 ft long hangs from a windlass on which it is to be wound. Find the work done in winding up 40 ft of the chain; (Weight: 4 lb/ft)

Solution

At the beginning the entire chain must be raised. After part of it has been raised, the remaining part requires less force. Then *force* is a variable. We let x = number of feet remaining at any time to be lifted. Then $4x$ = force required to lift it at that point. Now we let dx represent the small distance through which $4x$ pounds acts. Then the element of work, dw, is equal to the (force)(distance), or $dw = (4x)(dx)$. For the total amount of work done we add up the elements of work through the limits of x, and get

$$\text{work} = \int_{20}^{60} 4x\,dx = 2x^2\Big]_{20}^{60} = 6400 \text{ ft-lb}$$

Example 3

A cylindrical tank 20 ft long and 12 ft in diameter stands on end. It is full of water, which weights 62.4 lb per cu ft. Find the work done in pumping the water to a level 5 ft above the top of the tank.

Solution

We take a horizontal slice of water, a disc (Fig. 51–10). This slice is to be raised through a distance h, a variable. We state the volume and weight of this slice. The radius is 6 ft, and the thickness is dh. Then

$$dV = \pi(36)(dh); \qquad \text{and} \qquad d(wt) = (62.4)(36)(\pi)(dh)$$

This is the force required to lift this element of volume. The distance through which the forces acts is h. Then, for the element of work, we have

$$d(\text{work}) = (62.4)(36)(\pi)(h)(dh)$$

To find the amount of work done, we add up the elements of work between the h-limits, and get

$$\text{work} = (36)(62.4)(\pi)\int_{5}^{25} h\,dh = 673{,}920\pi \text{ ft-lb}$$

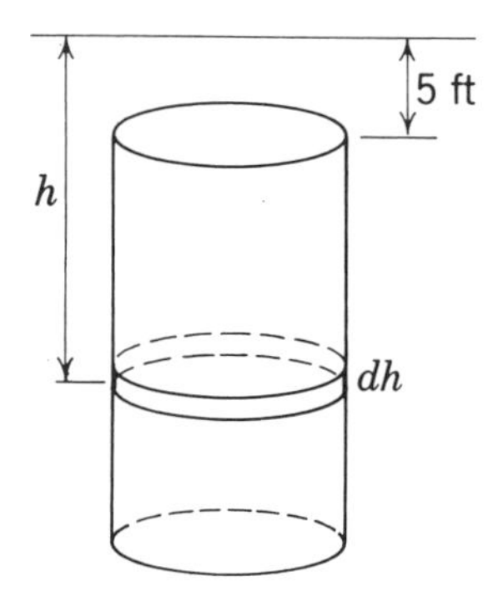

Figure 51.10

Exercise 51-3

1. A spring 30 in. long requires a force of 6 lb to stretch it 2 in. Find the work done in stretching it to a length of 38 in.
2. A spring 20 in. long is stretched 3 in. by a 2-lb force. Find the work done in stretching it to a length of 26 in.
3. A spring 40 in. long requires a force of 5 lb to stretch it 4 in. Find the work done in stretching it 10 in.

4. A spring 20 in. long is stretched 2.4 in. by a 4-lb force. Find the work done in stretching it from a length of 20 in. to a length of 25 in.

5. A spring requires a force of 3.5 lb to stretch it 3.375 in. Find the work done in stretching it from a normal length of 16 in. to 21.5 in.

6. A chain 60 ft long weighs 4 lb per ft. Find the work done in winding up the first half of the chain if it hangs on a windlass. Find the work done in winding up the second half.

7. A chain, 60 ft long, weighs 4 lb per ft. It hangs from a windlass on which it is to be wound. A 50-lb weight is attached to the bottom. Find the work done in winding up 20 ft of the chain.

8. If the chain in No. 7 is 40 ft long and weighs 3 lb per ft and has an 80-lb weight attached to the bottom, find the work done in raising the weight 30 ft.

9. A safe weighing 800 lb is fastened to the bottom end of a 20-ft chain hanging from a windlass. The chain weighs 6 lb per ft. Find the work done in raising the safe 12 ft.

10. A chain 80 ft long, weighing $\frac{1}{2}$ lb per ft, hangs from a windlass and has a 30-lb weight attached to the bottom end. Find the work done in raising the weight 60 ft.

11. A cylindrical tank has a diameter of 8 ft and is 14 ft high. It is to be filled with water from a supply that is 6 ft below the bottom of the tank. Find the work done in pumping the tank full of water.

12. Find the work done in No. 11 if the tank is to be only half filled.

13. A tank in the shape of a right circular cone has the apex at the bottom. The diameter of the top is 12 ft and the tank is 12 ft. high. Find the work done in filling the tank through a hole in the apex at the bottom if the water supply is at the level of the bottom of the tank.

14. If the tank in No. 13 is full of water, find the work done in pumping the water to a level 4 ft above the top of the tank.

15. If the tank in No. 13 is empty and is to be filled through a hole in the bottom, find the work done in filling the tank if the water supply is 4 ft below the bottom of the tank.

16. The velocity of a falling object in free-fall near the earth's surface is given by the formula: $v = 32t$. This represents the derivative of the distance s. By integrating and evaluating between limits, find the distance fallen in the interval from $t = 3$ sec to $t = 5$ sec.

17. An object is projected upward along a frictionless inclined plane with an initial velocity of 20 ft/sec. If the acceleration due to gravity is -6 ft/sec^2, find the distance covered by the object in the interval from $t = 1$ sec to $t = 3$ sec.

18. The velocity of an object projected directly upward from the earth's surface is given by the formula: $v = 96 - 32t$. Find the distance covered in the interval from 2 sec to 4 sec.

19. In an electric circuit, the current i is defined as the rate of transport of charge q (in coulombs). Then we have the formula, $i = dq/dt$. The current i is measured in amperes. If $i = 4t + 3$, where t is in seconds, find the total transport of charge q in the interval from $t = 4$ to 6 sec.

20. If $i = 6t^2$, in amperes, find q in the interval from 2 to 5 sec.

21. In the formula, $L(di/dt) = 3 - 4t$, if $L = 2$, find the current between the limits, $t = 1$ to $t = 4$.

22. Find the area under the curve, $y^2 = 4x$, from $x = 0$ to $x = 9$. Then find the *average height* of the area.

QUIZ ON CHAPTER 51, APPLICATIONS OF INTEGRATION. FORM A.

1. Find the area in the first quadrant bounded by the curves $y^2 = 8x$, the x-axis, and the line $x = 2$.

2. Find the area in the first quadrant bounded by the curves $y^2 = 2x$, the y-axis, and the line $y = 4$.

3. Find the area between the curves $y^2 = 6x$ and $x = y$.

4. Find the area between the curves $y = x^3$ and $y = x^2$.

5. Find the area bounded by the curves $y^2 = 2x$ and the line $x = 2y$.

6. If the area bounded by the curves $y^2 = 2x$, the x-axis, and the line $x = 8$ is rotated about the x-axis, find the volume of the solid generated. Use the disc method.

7. If the area bounded by the curve $x^2 = 4y$, the x-axis, and the line $x = 4$ is rotated about the y-axis, find the volume of the solid generated. (Use the shell method.)

8. A certain spring 3-ft long has a spring constant of 16 lb/ft. How much work is done in stretching the spring to a length of 4.5 ft?

9. A chain weighing 1/4 lb/ft is 50 ft long. It hangs from a windlass on which it is to be raised. If a weight of 20 lb is attached to the lower end, how much work is done in raising the weight a distance of 40 ft?

10. A cylindrical tank has a circular diameter of 6 ft and a height of 10 ft. How much work is done in filling the tank if the water supply is 4 ft below the bottom of the tank? (1 cu ft. of water weighs 62.4 lb/cu ft.)

11. The velocity of an object in motion is given by the formula $v = 50 - 6t$. By integration between limits, find the distance s covered in the interval from $t = 1$ sec to $t = 3$ sec.

12. If the current in an electric circuit is given by the formula $i = 3t^2 + 2t + 3$, find the total transport of charge q from $t =$ 2 sec to $t = 5$ sec.

CHAPTER FIFTY-TWO Transcendental Functions

52-1 TRIGONOMETRIC FUNCTIONS

Some functions, such as the trigonometric, the logarithmic, and the exponential, are nonalgebraic and are called *transcendental functions.* In Chapter 46 we were introduced to the trigonometric curves. To determine the slope of these curves involves the derivative. Moreover, we may need to find the area under the curves between limits, which involves the definite integral. It is therefore necessary to understand the differentiation and integration of the trigonometric functions. The same is true regarding the logarithmic and the exponential functions.

Although the necessary derivation is rather involved, it can be shown that

$$\text{if } y = \sin x, \qquad \text{then} \qquad dy/dx = \cos x$$

This result can be seen by a study of the sine and cosine curves. The curve, $y = \sin x$, passes through the origin at a 45° angle. Then its slope at $x = 0$ is equal to 1. This is the value of cos 0. That is, we shall find that the value of the cosine curve at any point is equal to the slope of the sine curve at that point. Note the following:

$$\text{at } \sin 0,\ m = \cos 0 = 1; \qquad \text{at } \sin 90°,\ m = \cos 90° = 0$$

$$\text{at } \sin 180°;\ m = \cos 180° = -1; \qquad \text{at } \sin 270°,\ m = \cos 270° = 0$$

If we plot the slope of the sine curve, we get the cosine curve.

If u is some function of x, and if $y = \sin u$, we have

$$dy/du = \cos u; \qquad \text{and} \qquad dy/dx = (\cos u)(du/dx)$$

That is, *the derivative of the sine of an angle u is equal to the cosine of the angle u, times the derivative of the angle.*

Examples

(a) If $y = \sin 3x, \dfrac{dy}{dx} = (\cos 3x)(3)$, *or* $3 \cos 3x$

(b) If $y = \sin (5 - x), \dfrac{dy}{dx} = [\cos (5 - x)][-1] = -\cos (5 - x)$

(c) *If* $y = \sin t^2$, find y''.

For the first derivative, we have $y' = (\cos t^2)(2t)$.

To find the formula for the derivative of the cosine, we first make use of the trigonometric identity, $\cos u = \sin(90° - u)$. Then, by the formula for the derivative of the $\sin u$,

$$\text{if} \quad y = \cos u, \quad \text{we write} \quad y = \sin(90° - u)$$

$$\text{then} \quad \frac{dy}{dx} = [\cos(90° - u)][-1]\left[\frac{du}{dx}\right] = -[\cos(90° - u)]\left[\frac{du}{dx}\right]$$

Now we use the same identity in reverse and get

$$\frac{dy}{dx} = -[\sin u]\left[\frac{du}{dx}\right]$$

That is, *the derivative of the cosine of an angle is the negative of the sine of the same angle times the derivative of the angle.*

Examples

(a) If $y = \cos 5x$, $dy/dx = -5 \sin 5x$

(b) If $y = \cos(10 - x)$, $dy/dx = -[\sin(10 - x)][-1] = +\sin(10 - x)$

(c) If $y = \sin x \cos x$, find y'

Here we must use the product rule. Then

$$y' = (\sin x)(-\sin x) + (\cos x)(\cos x) = \cos^2 x - \sin^2 x$$

For the derivative of the tangent of an angle, we first change $\tan u$ as shown.

$$\text{If} \quad y = \tan u,$$

we write

$$y = \frac{\sin u}{\cos u}$$

By the quotient rule,

$$\frac{dy}{dx} = \frac{(\cos u)(\cos u) - (\sin u)(-\sin u)}{\cos^2 u} = \frac{\cos^2 u + \sin^2 u}{\cos^2 u}$$

The numerator is equal to 1, and we get

$$dy/du = 1/\cos^2 u = \sec^2 u$$

Then

$$dy/dx = \sec^2 u \, (du/dx)$$

Example

$$\text{If} \quad y = \tan 6x, \quad \text{then} \quad \frac{dy}{dx} = 6 \sec^2 6x$$

In the same manner we can derive the formulas:

$$\frac{d}{dx}(\cot u) = -\csc^2 u \frac{du}{dx}$$

$$\frac{d}{dx}(\sec u) = (\sec u)(\tan u)\left(\frac{du}{dx}\right); \qquad \frac{d}{dx}(\csc u) = -(\csc u)(\cot u)\left(\frac{du}{dx}\right)$$

The forms of the derivatives of the trigonometric functions should be well known because they must be recognized when we come to the integration of trigonometric functions.

For the *derivative of a power of a trigonometric function*, we apply the same rule used for the derivative of a power of an algebraic function. For example, we have seen that

$$\text{if} \quad y = (x^2 + 3x)^5, \quad \text{then} \quad dy/dx = 5(x^2 + 3x)^4(2x + 3)$$

Example

If $y = \sin^5 3x$, find dy/dx. Evaluate dy/dx for $x = \pi/9$.

Solution

The derivative of a power of a trigonometric function involves three steps:

power rule:	$5(\sin^4 3x)$
derivative of sine:	$5(\sin^4 3x)(\cos 3x)$
derivative of angle:	$5(\sin^4 3x)(\cos 3x)(3)$

At $x = \pi/9$, $3x = 60°$. Then dy/dx has the value, 135/32.

The three steps in finding the derivative of a power of a trigonometric function as indicated in this example should always be followed in exactly the order shown. We summarize the steps.

1. *Apply the power rule with regard to the exponent.*
2. *Find the derivative of the trigonometric function.*
3. *Finally, find the derivative of the angle.*

Exercise 52-1

Find the derivative of each of the following:

1. $y = \sin 5x$ **2.** $y = \cos 3x^2$ **3.** $y = \tan 4t$

4. $y = \cot 7t$ **5.** $y = \sec 2x$ **6.** $y = \csc x^3$

7. $i = \sin 20t$ **8.** $i = \sin t^2$ **9.** $y = \sin (5 - 3x^2)$

10. $y = \cos (3 - x)$ **11.** $y = \cos 50t$ **12.** $i = \cos (3t - 2)$

13. $\frac{d}{dt}(50 \sin 6t)$ **14.** $\frac{d}{dt}(30 \cos 20t)$ **15.** $\frac{d}{dt}(40 \sin 377t)$

16. $\frac{d}{dt}(100 \cos 40t)$ **17.** $\frac{d}{dt}(20 \sin \omega t)$ **18.** $\frac{d}{dt}[20 \sin (2\pi f)t]$

19. $i = 30 \sin 20t + 15 \cos 20t$ **20.** $e = 20 \sin 50t - 30 \cos 50t$

21. Find the slope of the curve, $y = \sin x$, at $x = 0$; at $x = \frac{2\pi}{3}$; at $x = \frac{\pi}{2}$.

22. Find the slope of the curve, $y = \sin 2x$, at $x = 0$. Where is $m = 0$?

23. Find the slope of the curve, $y = \cos 3x$, at $x = 0$; at $x = \frac{\pi}{6}$.

24. Find the slope of the curve, $y = \tan x$, at $x = 0$; at $x = \frac{\pi}{6}$.

25. In an electric circuit the current is given by the formula; find di/dt when $t = \pi/4$:

$$i = I_{max} \sin (t + \phi); \qquad I_{max} = 20 \text{ amps}; \qquad \phi = 15°$$

Find the derivative of each of the following:

26. $y = \sin^3 6t$ **27.** $i = \cos^5 2t$ **28.** $y = \tan^4 3t$

29. $y = \cot^6 2t$ **30.** $y = \sec^3 4t$ **31.** $y = \csc^4 2t$

32. $y = t^2 \sin 3t$ **33.** $y = t^3 \cos 2t$ **34.** $y = \sin 3t \cos 3t$

Find the derivative and evaluate for the given value of θ.

35. $y = \sin^3 2\theta, \left(\theta = \frac{\pi}{3}\right)$ **36.** $y = \cos^4 \theta, \left(\theta = \frac{\pi}{4}\right)$

52-2 LOGARITHMIC FUNCTIONS

To graph a logarithmic function, such as $y = \ln x$, we must omit $x = 0$ and all negative values of x, since zero and negative numbers have no real logarithms. We may use the following values for graphing, $y = \ln x$:

$x = 0.5$	1	2	e	4	7	10
$y = -0.7$	0	0.7	1	1.4	1.9	2.3

The graph is shown in Fig. 52-1. The graphs of $y = \log_{10} x$, and $y = \log_2 x$, are shown on the same graph and have similar shapes. As

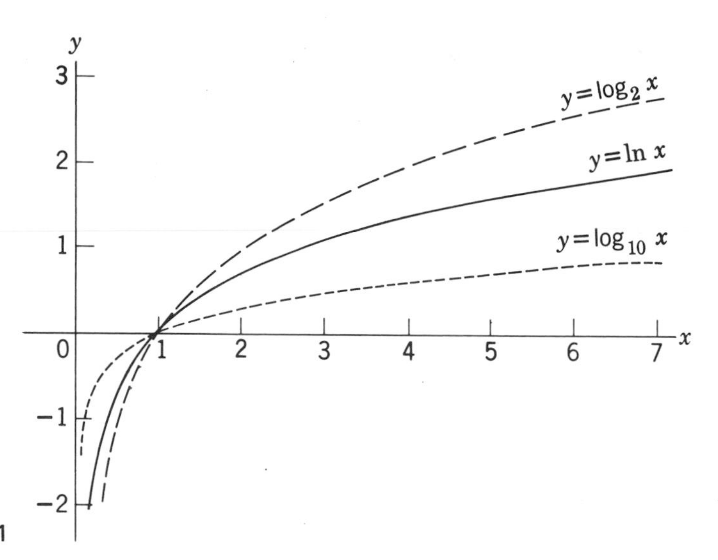

Figure 52-1

the value of x approaches zero, the curves approach the y-axis negatively as an asymptote.

The formula for the derivative of a logarithmic function can be derived by the delta method. Whatever base is used, the derivative will involve the limit, e, the natural base. With base e, and if u is some function of x, then

$$\text{if } y = \ln u, \qquad \frac{dy}{dx} = \frac{du/dx}{u}$$

Stated in words: For base e, the derivative of a logarithmic function is equal to a fraction having the function u as the denominator, and its derivative, du/dx, as the numerator.

Example 1

$$\text{If } y = \ln(x^2 + 3x + 2), \qquad \text{then} \qquad \frac{dy}{dx} = \frac{2x + 3}{x^2 + 3x + 2}$$

Example 2

$$\text{If } y = \ln \tan 3x, \qquad \text{then} \qquad \frac{dy}{dx} = \frac{3\sec^2 3x}{\tan 3x} = 3 \sec 3x \csc 3x$$

If the base of the logarithm is not e but is some other base b, we have

$$y = \log_b u; \qquad \text{then} \qquad \frac{dy}{dx} = \frac{du/dx}{u}(\log_b e)$$

Stated in words: For any other base b, the derivative of the logarithm is the same as for base e, with the additional factor, $\log_b e$.

Example 3

If $y = \log_{10}(x^2 + 3x + 2)$, $\quad \dfrac{dy}{dx} = \dfrac{2x + 3}{x^2 + 3x + 2}(\log_{10} e)$.

Note. $\log_{10} e = 0.4343$ (approximately).

Exercise 52-2

Find the derivative of each of the following: ($\log_{10}$ is shown as *log*).

1. $y = \ln(x^2 + 5)$ **2.** $y = \ln(x^3 + x^2)$ **3.** $y = \ln(2 - 5x - x^2)$

4. $y = \log(x^2 + 3)$ **5.** $y = \log(7 - 4x)$ **6.** $y = \log(x^3 + 5x^2)$

7. $y = \ln x^4$ **8.** $y = \ln(3x - 2)^5$ **9.** $y = \ln(5 - x^2)^3$

10. $y = \ln \sin 3t$ **11.** $y = \ln \cos t^2$ **12.** $y = \ln \tan 2t^3$

13. $y = \ln \sin^2 5t$ **14.** $y = \ln \cos^3 2t$ **15.** $i = \ln \tan^4 3t$

16. $y = \ln \dfrac{3x}{x^2 + 4}$ **17.** $y = \ln \dfrac{6}{3x - 5}$

18. $y = \ln(x^2 - 3)(2x - 5)$

19. Find the derivative of each of these and compare results:

$y = \ln x$; $\quad y = \ln 2x$; $\quad y = \ln 5x$; $\quad y = \ln 10x$; $\quad y = \ln cx$

20. Find the derivative of each of these and compare results:

$y = \ln x$; $\quad y = \ln x^2$; $\quad y = \ln x^5$; $\quad y = \ln x^8$; $\quad y = \ln x^n$

21. Find the slope of the curve, $y = \ln x$, where it crosses the x-axis.

22. Find the slope of the curve, $y = \log_{10} x$, where it crosses the x-axis.

23. Find the derivative of $y = x(\ln x) - x$, and simplify.

24. Find the derivative of $y = \ln(\sec x + \tan x)$, and simplify.

Evaluate the following derivatives for the indicated value of x:

25. $y = \ln(x^2 + 3)$, $(x = -1)$ **26.** $y = \ln \sin x$, $(x = \pi/4)$

27. Where does the curve, $y = \ln(x^2 - 4x + 5)$, have $m = 0$?

52-3 EXPONENTIAL FUNCTIONS

Since the exponential function is the inverse of the logarithmic function, the graph of the function, $y = e^x$, is similar in shape to the graph of $y = \ln x$, with the x and y axes reversed. The same is true regarding the inverses:

$$y = \log_{10} x, \quad \text{and} \quad y = 10^x$$

$$y = \log_2 x, \quad \text{and} \quad y = 2^x$$

The graphs of these exponential functions are shown in Fig. 52-2.

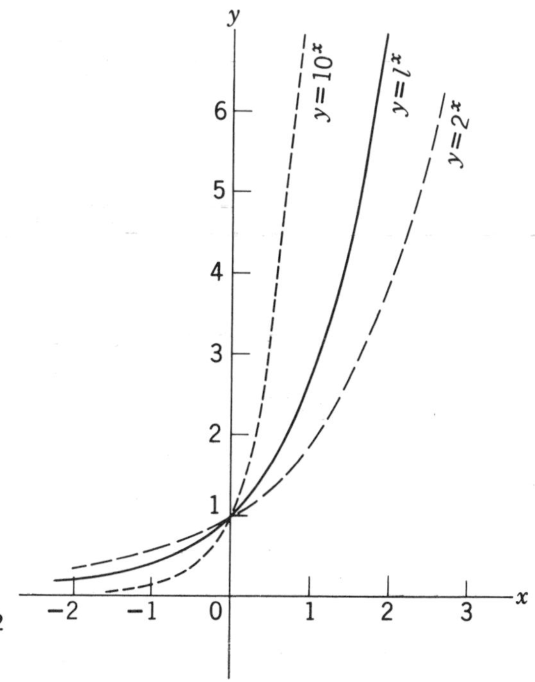

Figure 52-2

If the base of the exponential function is e, and if u is some function of x, then

$$y = e^u, \quad \text{and} \quad \frac{dy}{dx} = (e^u)\left(\frac{du}{dx}\right)$$

In words: The derivative of an exponential function, with base e, is the function itself times the derivative of the exponent.

Example 1

If $y = e^{3x}$, $\frac{dy}{dx} = (e^{3x})(3)$, or $3e^{3x}$

Example 2

If $y = e^{5x-x^2}$, $\dfrac{dy}{dx} = (e^{5x-x^2})(5 - 2x)$

Example 3

If $i = e^{\sin 4t}$, $\dfrac{di}{dt} = (e^{\sin 4t})(\cos 4t)(4)$

If the base of an exponential function is some number, b, other than e, we have

$$y = b^u, \qquad \text{and} \qquad \frac{dy}{dx} = (b^u)\left(\frac{du}{dx}\right)(\ln b)$$

In words: For a base b other than e, the derivative of an exponential function is the same as for base e, with the additional factor, ln *b, the natural logarithm of the base.*

Example 4

If $y = 10^{3x}$, $\dfrac{dy}{dx} = (10^{3x})(3)(\ln 10)$; $\ln 10 = 2.3026$ (approx.)

If the base e is used, the final factor becomes $\ln e$, which is 1. The use of base e in exponentials as well as in logarithms makes the final factor unnecessary.

Exercise 52-3

Find the derivatives of the following functions:

1. $y = e^{6x}$ **2.** $y = e^{-7x}$ **3.** $y = e^{x^2+3x}$

4. $i = e^{-10t}$ **5.** $q = e^{-x^2}$ **6** $y = e^{x^2-4x+2}$

7. $i = e^{4t^2}$ **8.** $i = e^{3-5t}$ **9.** $i = e^{t^2-2t-3}$

10. $y = 10^{5x}$ **11.** $y = 4^{3x}$ **12.** $y = 2^{3x-4}$

13. $i = e^{\sin 2t}$ **14.** $y = e^{\cos 3x}$ **15.** $y = e^{\tan 4x}$

16. $i = t^2e^{2t}$ **17.** $i = t^3e^{-3t}$ **18.** $i = t^4e^{3t+2}$

19. $x = t^{-2}e^{-t}$ **20.** $q = e^{3t} - e^{-3t}$ **21.** $y = e^{5t} + e^{-5t}$

22. $y = e^x \ln x$ **23.** $y = e^x \sin 5x$ **24.** $i = e^{-2x} \cos 2x$

25. $i = e^{-3t}(\sin 20t + \cos 20t)$ **26.** $q = e^{-4t}(\sin 15t - \cos 15t)$

27. $i = e^{-20t}(\sin 30t - \cos 30t)$ **28.** $q = e^{-10t}(\sin 40t + \cos 40t)$

Find the second derivative of each of the following:

29. $y = e^{3x^2}$ **30.** $y = e^{-4x^3}$ **31.** $y = t^2e^{-3t}$

32. Find the first five derivatives of $y = e^x$, and of $y = e^{-x}$.

33. If $y = e^{-x}$, find $\dfrac{dy}{dx} + y$ **34.** If $y = 4e^{-5x}$, find $y'' + 5y'$

35. Find the second derivative of No. 27.

36. Find q'' in No. 28.

37. Find the slope of the curve, $y = e^x$, where it crosses the y-axis.

38. Find the slope of the curve, $y = 10^x$, where it crosses the y-axis.

39. For the curve, $y = e^x$, find y and the slope of the curve where $x = 1$.

40. For the curve, $y = e^{-x}$, find y and the slope of the curve where $x = 1$.

41. For the curve, $y = \frac{1}{2}(e^x + e^{-x})$, find y and the slope where $x = 0$.

42. For the curve, $y = xe^x - e^x$, find the slope where $x = 2$.

52-4 INTEGRATION OF TRANSCENDENTAL FUNCTIONS

The following integrals involving the trigonometric functions follow directly from the derivatives:

$$\int \sin u\, du = -\cos u + C \qquad \int \cos u\, du = \sin u + C$$

$$\int \sec^2 u\, du = \tan u + C \qquad \int \csc^2 u\, du = -\cot u + C$$

$$\int \sec u \tan u\, du = \sec u + C \qquad \int \csc u \cot u\, du = -\csc u + C$$

The following integrals can be derived:

$$\int \tan u\, du = \ln \sec u + C \qquad \int \cot u\, du = \ln \sin u + C$$

$$\int \sec u\, du = \ln (\sec u + \tan u) + C \qquad \int \csc u\, du = -\ln (\csc u + \cot u) + C$$

It must be remembered that the integrand must always contain the complete factor, du, before the integration can be performed. A constant factor can be supplied as usual and its reciprocal written outside the integral sign.

Example 1

Find $\int \sin 5x\, dx$.

Solution

The derivative of $\cos 5x$ is $(-\sin 5x)(5)$. Therefore, we supply the factor, -5, and write its reciprocal outside the integral sign. We get

$$\int \sin 5x\, dx = -\frac{1}{5}\int -5 \sin 5x\, dx = -\tfrac{1}{5}\cos 5x + C$$

Example 2

Find $\int \sin^4 2x \cos 2x\, dx$.

Solution

To apply the power rule on $\sin 2x$, we must have the complete differential of $\sin 2x$, which is $2(\cos 2x)(dx)$. We supply the constant 2, and get

$$\frac{1}{2}\int 2 \sin^4 2x \cos 2x\, dx = \frac{1}{2}\left(\frac{\sin^5 2x}{5}\right) + C = \tfrac{1}{10}\sin^5 2x + C$$

We have seen that

$$\frac{d}{dx}(\ln u) = \frac{du/dx}{u}$$

Therefore, if a fraction has a *numerator that is the perfect derivative of the denominator*, then the integral is the *natural logarithm of the denominator*. Again, if a constant factor is lacking, it may be supplied as usual.

Example 3

$$\int \frac{2x\, dx}{x^2 + 3} = \ln (x^2 + 3) + C$$

Example 4

$$\int \frac{(x - 3)\, dx}{x^2 - 6x + 2} = \frac{1}{2}\int \frac{2(x - 3)\, dx}{x^2 - 6x + 2} = \tfrac{1}{2}\ln (x^2 - 6x + 2) + C$$

Example 5

$$\int \cot x\, dx = \int \frac{\cos x\, dx}{\sin x} = \ln \sin x + C$$

Example 6

$$\int \tan^2 5x\, dx = \int (\sec^2 5x - 1)\, dx = \tfrac{1}{5}\tan 5x - x + C$$

We have seen that the derivative of an exponential function contains the same exponential function. For example,

$$\frac{d}{dx}(e^u) = (e^u)\left(\frac{du}{dx}\right)$$

Therefore, the integral of an exponential function is the same exponential function. However, before the integration can be done, the integrand must contain the perfect *derivative of the exponent.* If a constant is lacking, it can be supplied as usual.

Example 6

$$\int e^{4x}\,dx$$

Solution

The integrand must contain the derivative of the exponent. We need the constant 4, which can be supplied. Then

$$\int e^{4x}\,dx = \frac{1}{4}\int 4e^{4x}\,dx = \tfrac{1}{4}e^{4x} + C$$

Let us consider the function, e^{4x}. It is interesting to note that when we differentiate, we get 4 as a *factor.* When we integrate, we get 4 as a *divisor*:

$$\frac{d}{dx}(e^{4x}) = 4e^{4x}; \qquad \int e^{4x}\,dx = \frac{e^{4x}}{4} + C$$

Example 7

$$\int_0^1 t^2e^{t^3}\,dt = \frac{1}{3}\int_0^1 3t^2e^{t^3}\,dt = \tfrac{1}{3}e^{t^3}\Big]_0^1 = \tfrac{1}{3}(e^1 - e^0) = 0.573$$

Example 8

Find $\int \sin 3xe^{\cos 3x}\,dx$.

Solution

We need the derivative of the exponent, which is $-3\sin 3x$. We supply the factor, -3; then

$$\int \sin 3xe^{\cos 3x}\,dx = -\frac{1}{3}\int -3\sin 3xe^{\cos 3x}\,dx = -\tfrac{1}{3}e^{\cos 3x} + C$$

Exercise 52-4

Integrate as indicated:

1. $\int \cos 4x\,dx$
2. $\int \sin 7x\,dx$
3. $\int \sin^4 2x \cos 2x\,dx$
4. $\int \sec^2 5x\,dx$
5. $\int \cos^3 3x \sin 3x\,dx$
6. $\int \tan^2 6x \sec^2 6x\,dx$
7. $\int \frac{x\,dx}{x^2 - 3}$
8. $\int \frac{(x - 3)\,dx}{x^2 - 6x + 2}$
9. $\int \frac{(2 - x)\,dx}{x^2 - 4x + 1}$
10. $\int \frac{4\,dx}{5x - 2}$
11. $\int \frac{\sec^2 3x\,dx}{\tan 3x}$
12. $\int \frac{(6x - 3x^2)\,dx}{x^3 - 3x^2 + 2}$
13. $\int e^{-5x}\,dx$
14. $\int xe^{-x^2}\,dx$
15. $\int \sin 2xe^{\cos 2x}\,dx$
16. $\int_0^{\pi/6} \cos 2x\,dx$
17. $\int_0^2 e^{-x}\,dx$
18. $\int_0^9 \frac{4}{x + 1}\,dx$
19. $\int \tan x\,dx$
20. $\int \cot^2 3x\,dx$
21. $\int \sec 4x \tan 4x\,dx$
22. $\int \csc^2 8x\,dx$
23. $\int \tan^4 x\,dx$
24. $\int \csc 2x \cot 2x\,dx$
25. Find the area under the curve, $y = \sin 2x$, from $x = 0$ to $x = \frac{\pi}{2}$.
26. Find the area under the curve, $y = \cos x$, from $x = 0$ to $x = \frac{\pi}{4}$.
27. Find the area under the curve, $y = \tan x$, from $x = 0$ to $x = \frac{\pi}{3}$.
28. Find the area under the curve, $y = e^x$, from $x = -1$ to $x = +1$.
29. Find the area under the curve, $y = \frac{1}{2}(e^x + e^{-x})$ from $x = -1$ to $x = +1$.
30. Find the area between the curves, $y = \sin x$ and $y = \cos x$, from $x = 0$ to $x = \frac{\pi}{4}$.

52-5 AVERAGE VALUE OF VOLTAGE OR CURRENT

The voltage and current in an AC electric circuit follow the form of the sine curve or the cosine curve. The value of voltage or current reaches a peak or maximum value represented by the maximum height of the curve.

One problem is to find the average value of the voltage or current. To find this value, we first find the total area under the curve from one zero value to the next. Then we find the average height of the curve by dividing the total area by the length of the base line.

Example 1

Find the total area under the sine curve from $x = 0$ to $x = \pi$; then find the average height of the curve.

Solution

For the total area we have the integral:

$$A = \int_0^{\pi} \sin x\, dx = -[\cos x]_0^{\pi} = -[-1 - 1] = 2$$

Then the total area under the sine curve is 2 square units from $x = 0$ to π.
For the average height of the curve, we divide by the base line, π, and get

$$2 \div \pi = 0.6366, \qquad \text{or} \qquad 0.637 \text{ (approx.)}$$

The result means that the average value of voltage or current in an AC electric circuit is 0.637 times the maximum value. That is,

$$E_{av} = 0.637E_{max}$$

where E_{av} represents average voltage, and E_{max} represents maximum voltage.

Example 2

The instantaneous voltage in a certain AC circuit is given by

$$e = 150 \sin \theta$$

Find the average voltage.

Solution

The maximum voltage occurs when the angle is equal to $\pi/2$, at the time when $\sin \theta$ is equal to 1. Then maximum voltage = 150 volts. For the average voltage, we have

$$E_{av} = (0.637)(150 \text{ volts}) = 95.4 \text{ volts}$$

Example 3

If the average voltage in an AC circuit is 70 volts, find the maximum voltage.

Solution

We have the equation: $\qquad 0.637E_{max} = 70 \text{ volts}$

Dividing both sides by 0.637, $E_{max} = 70 \div 0.637 = 110$ volts

Then the maximum voltage in the circuit is 110 volts.

Example 4

In an AC electric circuit, the instantaneous current i is given by the formula:

$$i = 30 \cos 10t$$

Find the average current.

Solution

For the average current, we have the formula

$$I_{av} = 0.637 I_{max}$$

The maximum current is 30 amps at a time when the angle $10t = 0$. For the average current we have

$$I_{av} = (0.637)(30) = 19.1$$

Then the average current is 19.1 amps.

52-6 EFFECTIVE VALUE (RMS) OF VOLTAGE OR CURRENT

In measuring voltage or current in an AC electric circuit, most meters show effective values, based on power producing the same heating effect as in DC circuits. Since power involves the *square* of voltage or current, the effective value is found by squaring the expressions for voltage or current.

To find the effective value, we must first find the value of the integral

$$\int_0^{\pi} \sin^2 x \, dx$$

Then we divide the integral by the base line, π, to get the average (or mean) of this value. Finally, we take the square root of the result. The effective value is therefore often called the "root-mean-square" (or *RMS*) value.

Example 5

Find the formula for the effective value of voltage or current in an AC circuit.

Solution

We first find the value of the integral

$$\int_0^{\pi} \sin^2 x \, dx$$

Here we must use the trigonometric identity:

$$\sin^2 x = \tfrac{1}{2}(1 - \cos 2x)$$

Then

$$\int_0^\pi \sin^2 x\,dx = \frac{1}{2}\int_0^\pi (1 - \cos 2x)\,dx$$

$$= \frac{1}{2}\left[\int_0^\pi dx - \int_0^\pi \cos 2x\,dx\right] = \tfrac{1}{2}[x - \tfrac{1}{2}\sin 2x]_0^\pi = \frac{\pi}{2}$$

Dividing by π, the base line, we get

$$\frac{\pi}{2} \div \pi = \tfrac{1}{2}$$

Taking the square root of the result, we get $\sqrt{1/2} = 0.707$ (approx.). The result means that the *RMS*, or effective, value of voltage or current is found by multiplying the maximum value by 0.707. That is,

$$E_{rms} = 0.707 E_{\max}$$

For the effective value of current we have the formula: $I_{rms} = 0.707 I_{\max}$

Example 6

In a certain AC circuit, the instantaneous voltage is given by the formula:

$$e = 160 \sin 10t$$

Find the *RMS* (effective) voltage.

Solution

We use the formula:

$$E_{rms} = 0.707 E_{\max}$$

In the circuit, we note that the maximum voltage is 160 volts. Then we have

$$E_{rms} = (0.707)(160 \text{ volts}) = 113 \text{ volts}$$

Example 7

If a meter shows a voltage of 120 volts in an AC circuit, find the maximum voltage. Then find the average voltage.

Solution

We have the equation:

$$0.707 E_{\max} = 120$$

Dividing by 0.707, we get

$$E_{\max} = (120) \div (0.707) = 170$$

Then the maximum voltage is 170 volts. To find the average voltage, we have

$$E_{av} = 0.637E_{max}$$

or

$$E_{av} = (0.637)(170 \text{ volts}) = 108 \text{ volts}$$

Example 8

The average current i in a certain AC circuit is 18 amps. Find the *RMS* (effective) value of the current.

Solution

First we find the maximum current, I_{max}, we have the equation

$$0.637I_{max} = 18$$

Dividing by 0.637,

$$I_{max} = (18) \div 0.637 = 28.3$$

Since the maximum current is 28.3 amps, we take

$$I_{rms} = (0.707)(28.3) = 20 \text{ (amperes)}$$

Exercise 52-5

1. In an AC circuit, the voltage is given by the formula, $e = 60 \sin \theta$. What is the maximum voltage? Find the effective (*RMS*) value and the average value of the voltage.
2. The average value of voltage in a certain circuit is 80 volts. Find the maximum voltage, and then find the average voltage.
3. The effective value of the current in a certain circuit is 25 amps. Find the maximum current and then find the average current. What is the ratio of the *RMS* value to the average value? What is the reverse ratio?
4. A voltmeter shows a voltage of 110 volts in a circuit. What is the maximum voltage? The average voltage?
5. The effective (*RMS*) current in a certain circuit is 260 milliamperes. Find the maximum current and the average current in the circuit.
6. The instantaneous current i in a certain circuit is given by $i = 45 \cos \theta$. What is the maximum current? Find the *RMS* value and the average current in the circuit.

QUIZ ON CHAPTER 52, TRANSCENDENTAL FUNCTIONS. FORM A.

Find the derivative of each of the following (1–20) with respect to the independent variable:

1. $y = 8 \sin 3x$ **2.** $y = 10 \cos 20t$ **3.** $y = \tan x^2$

4. $y = 12 \cot 5x$ **5.** $y = 5 \sec 8x$ **6.** $y = 10 \csc 3x$

7. $i = 20 \sin 10t$ **8.** $e = 120 \cos 6t$ **9.** $y = x^2 \sin 20x$

10. $y = x^3 \cos 10x$ **11.** $y = \sin^2 4x$ **12.** $y = \cos^3 6x$

13. $y = \tan^2 5x$ **14.** $y = \sec^3 4x$ **15.** $y = \ln (x^2 + 3x + 2)$

16. $y = \ln (2t + 7)$ **17.** $y = 6e^{3x}$ **18.** $y = 8e^{-4x}$

19. $y = x^3 e^{5x}$ **20.** $y = x^2 \sin 7x$ **21.** $y = \sin 4x$ (find y'')

Integrate as indicated:

22. $\int 5 \cos 3x\, dx$ **23.** $\int 3x \sin x^2\, dx$ **24.** $\int 10 \sec^2 3t\, dt$

25. $\int 8 \csc^2 5x\, dx$ **26.** $\int \frac{(2x + 3)\, dx}{x^2 + 3x + 2}$ **27.** $\int \frac{3x\, dx}{4 - x^2}$

28. $\int 4\, e^{3t}\, dt$ **29.** $\int \sec x \tan x\, dx$ **30.** $\int \tan 3x\, dx$

31. Find the area under the cosine curve, $y = \cos x$, from $x = 0$ to $\pi/2$.

32. Find the area under the sine curve, $y = \sin x$, from $x = 0$ to π. Also, find the average height of the curve.

33. Evaluate the following integral between limits: $\int_0^1 e^t\, dt$.

34. In an AC circuit, $E_{max} = 130$ volts. Find E_{av} and E_{rms}.

35. In an AC circuit, $I_{av} = 13.7$ amps. Find I_{max} and I_{rms}.

36. In an AC circuit, $E_{rms} = 85$ volts. Find E_{max} and E_{av}.

CHAPTER FIFTY-THREE Number Bases

53-1 OUR COMMON DECIMAL SYSTEM

In this age of computers everyone should have some idea of the meaning of number bases. We have mentioned the fact that our common system of numbers makes use of only ten digits: 0, 1, 2, 3, 4, 5, 6, 7, 8, and 9. Each of these symbols was invented by early man in counting objects.

When counting continued to one more than nine, it was decided to lump all these together and write "1" for the bunch, with a zero at the right of the 1. When we write 10 for ten, the 1 represents one bunch of ten, and the zero shows that there are no extra ones. Then we say that *ten* is the *base* of our number system. It is therefore called a *decimal system*, the word *decimal* referring to *ten*. When we write a number such as 37, we mean 3 bunches of ten, and 7 ones more.

When we count to 99 and then count one more, we get ten bunches. However, we have no symbol for ten, so we put them all together into a larger bunch and write 1, followed by two zeros: 100. Here, the 1 represents 1 bunch of the second order, or 10^2. For numbers written in base ten, the second digit from the right represents tens. Then, moving toward the left, the places represent successive powers of ten.

Many electronic computers use only two digits: 0 and 1. We shall see how any number can be expressed by using only these two digits. First, let us look at a particular base other than ten; for example, base *six*.

53-2 A PARTICULAR BASE: SIX

Suppose we decide to use only six digits: 0, 1, 2, 3, 4, and 5. When we count a number one more than 5, we put these six objects together and call them one *bunch*. We write 1 for the bunch, and then 0 for no extra ones. Then the symbol for six becomes: 10. In this case, we say the number is written in base *six*.

Note that in base six, we have no digit for *six*, the base. This is true for any base we might use. In base ten (our common base) we have no single symbol for *ten*. However, in our explanation here, for convenience, we shall use our usual digit to represent the base. For example, to refer to base *six*, we shall call it *base 6*, although in this base there is no single digit for *six*. In writing a number in any particular base, we often indicate the base by a *subscript.* For base *ten*, we shall use the subscript *10.*

As an example in the use of base 6, consider the number of fingers on two hands. We call this number *ten* and write it 10. Now, if we wish to write this number in base 6, we would say that we have *one bunch* of 6, and 4 ones more. Then in base 6, we would write the number of fingers on two hands as

$$14_6$$

The subscript indicates the base.

Now consider the number of days in the month of September. This number we call thirty and usually write it 30 (in base ten). The symbol means 3 bunches of ten each and none over. Suppose now that we wish to write this number in base 6. We first see how many bunches of 6 we can get in 30. Using our common notation (base *ten*), we divide 30 by 6 and get 5, with no remainder. This means there are 5 bunches of 6 with none over. Then the number of days in September written in base 6 becomes

$$50_6$$

As another example, suppose we wish to write the number 27 (base ten) in base 6. We divide 27 by 6 and get 4 with a remainder of 3. Then the number 27 (base ten), becomes, in base 6,

$$43_6; \qquad \text{that is,} \qquad 27_{10} = 43_6$$

Let us go one step further. There are 36 inches in 1 yard. Suppose we wish to write this number in base 6. When we count off 5 bunches of six each, we have another six left, which makes another bunch. Then we have six bunches. But we have no digit for six. So we put the six bunches together and call it a bunch of the second order. Now we write 1 for the bunch of the second order, and follow it with two zeros. Then

36 in base *ten* becomes 100 in base *six*

That is,

$$36_{10} = 100_6 \quad \text{(we use the subscript 10 for base } ten\text{)}$$

53-3 CHANGING A NUMBER FROM BASE TEN TO ANY OTHER BASE

The foregoing examples lead directly to the following rule for changing a number from base *ten* to any other base:

Rule 1. *To change a number (base ten) to any other base, divide the number (base ten) by the new base successively as many times as possible and write down all remainders, including zeros. Then the last quotient and the remainders in reverse order are the digits of the number in the desired base.*

Example 1

Change the following numbers (base *ten*) to the base indicated: (a) 679 to base 6; (b) 477 to base 5; (c) 3115 to base 8; (d) 13 to base 2.

Solution

We divide each number by the base. Remainders are shown R.

(a)	R	(b)	R	(c)	R	(d)	R
$6\overline{)679}$		$5\overline{)477}$		$8\overline{)3115}$		$2\overline{)13}$	
$6\overline{)113}$	$-\ 1$	$5\overline{)\ 95}$	$-\ 2$	$8\overline{)\ 389}$	$-\ 3$	$2\overline{)\ 6}$	$-\ 1$
$6\overline{)\ 18}$	$-\ 5$	$5\overline{)\ 19}$	$-\ 0$	$8\overline{)\ 48}$	$-\ 5$	$2\overline{)\ 3}$	$-\ 0$
3	$-\ 0$	3	$-\ 4$	6	$-\ 0$	1	$-\ 1$

Answer: 3051_6 ***Answer:*** 3402_5 ***Answer:*** 6053_8 ***Answer:*** 1101_2

53-4 CHANGING A NUMBER TO BASE TEN FROM ANY OTHER BASE

To change a number to base *ten* from any other base, let us first consider the meaning of *place value* in any base. When we write a number such as 3724 (base *ten*), each of the digits, 3, 7, 2, and 4 have the following meanings. The first digit at the right, 4, represents single units. The second digit from the right, 2, represents 10's. The next digit, 7, represents 10^2's, or 100's. The next digit, 3, represents 10^3's, or 100's. The places have the following values:

$$\overline{10^3}\quad\overline{10^2}\quad\overline{10}\quad\overline{\text{units}}$$

Briefly, each place starting at the right represents successive powers of the base.

In a similar manner, when we write a number in another base, the digits have place values that represent successive powers of the base. For example, in base 6, the places have the following values:

$$\overline{6^3}\quad\overline{6^2}\quad\overline{6}\quad\overline{\text{units}}$$

Now let us see the meaning of a number written in base 6, say, 2543_6. The digit 3 at the extreme right represents the number of units. The next digit, 4, represents the number of 6's. The next digit, 5, represents the number of 6^2's, or 36's. The next digit, 2, represents the number of 6^3's, or 216's. Now we can compute the value of the number in base ten:

The digit at the right represents units:	3
The digit 4 means 4(6), or	24 (base 10)
The digit 5 means 5(36), or	180 (base 10)
The digit 2 means 2(216), or	432 (base 10)
Adding these values, we get	639_{10} (base 10)

Rule 2. *To change a number to base ten from any other base, write the digit at the extreme right as the number of units. Then, moving toward the left, multiply each digit by successive powers of the base. Finally, add the results.*

Note that Rules 1 and 2 involve opposite operations. The first involves division; the second involves multiplication.

53-5 COMPUTATION IN ANY BASE

Computation can be performed in any base. However, in all cases, we take into account the base used. For example, in base 6, $4 + 5 = 13$. Also, $14 - 5 = 5$. When we borrow a 1, as in subtraction, we must remember that the 1 borrowed represents the base. The following examples show some computation in base 6. The principles involved are the same for all bases. Each remainder in division represents the base.

1. Add	**2.** Subtract	**3.** Multiply	**4.** Divide
324	514	55	4/5512
435	−345	44	1245
1203	125	352	
		352	
		4312	

Check subtraction by addition; check division by multiplication. We can also check answer by changing all numbers to base ten. Then we have, for the above:

1. Add (base 10)	**2.** Subtract	**3.** Multiply	**4.** Divide:
124	190	35	4/1268
167	137	24	317
291	53	*Ans.* 980	

Exercise 53.1

Change the following numbers (base ten) to the base indicated:

1. 1963 to base 6 **2.** 3699 to base 6 **3.** 7605 to base 6

4. 2711 to base 5 **5.** 5983 to base 7 **6.** 1267 to base 3

7. 6115 to base 8 **8.** 3638 to base 4 **9.** 43 to base 2

The following numbers are written in the base indicated. Change to base ten.

10. 345_6 **11.** 415_6 **12.** 2021_3 **13.** 2301_4 **14.** 3142_5 **15.** 256_7

The following numbers are written in base 6. Perform the indicated operations.

16. 34 + 45 **17.** 345 + 452 **18.** 53 − 25 **19.** 432 − 145

20. (3)(54) **21.** (45)(35) **22.** 312 ÷ 4 **23.** 2432 ÷ 5

53-6 THE BINARY SYSTEM

The *binary* system uses only two digits, 0 and 1. Then we say the base is *two*. This is the system used by many electronic computers because the electric switches in a computer are either ON or OFF. Then the digit 1 can represent the ON position, and the digit 0 can represent the OFF position.

In the binary system we have no symbol for two. Instead, two units are combined into one bunch and written: 10. In this number, the 1 represents one bunch of two. However, we shall often use the symbol 2 to represent the base of the binary system. For three units, we write: 11. The 1 at the right represents 1 unit. The 1 at the left represents two units. To express four units, we write: 100. Here the 1 represents two bunches of two units each, or a bunch of the second order. Then five is written: 101.

The following table shows the relation between the decimal system (base ten) and the binary system (base two) for some numbers:

decimal	binary	decimal	binary	decimal	binary
1	1	7	111	13	1101
2	10	8	1000	14	1110
3	11	9	1001	15	1111
4	100	10	1010	16	10000
5	101	11	1011	17	10001
6	110	12	1100	18	10010

The binary system requires many more digits to express a number. However, this is no obstacle for electronic computers in which switches can go *on* and *off* thousands of times in a second.

To change a number from *decimal* (base ten) to *binary* (base two), we follow Rule 1 and divide the number (base ten) by 2 successively as many times as possible and write down all remainders, including zeros. Then the remainders in reverse order are the digits of the number in binary notation.

Example 1

Change the number 23 (base ten) to binary notation.

Solution

	Rem.
2/23	
2/11	− 1
2/ 5	− 1
2/ 2	− 1
2/ 1	− 0
0	− 1

Then the number in binary notation becomes: 10111. Actually, the last step in division might be omitted if we take the final quotient as the first digit of the number.

To change a number *from binary notation to decimal notation*, we must consider the place value of each digit. In binary notation, the places have the following values:

2^8	2^7	2^6	2^5	2^4	2^3	2^2	2^1	units

or

256	128	64	32	16	8	4	2	units

Example 2

Change the binary number, 1011011, to decimal (base ten) notation.

Solution

The 1 at the extreme right:	1
The next 1 represents 2:	2 (base 10)
The next digit represents 4:	0 (base 10)
The next digit represents 8:	8 (base 10)
The next digit represents 16:	16 (base 10)
The next digit represents 32:	0 (base 10)
The last digit represents 64:	64 (base 10)
Adding, we get	91 (base 10)

To check the result, we can change the number 91 back to binary notation by dividing by 2, as in Example 1.

Computation can be performed with numbers in binary notation. However, we must remember a few elementary facts: $1 + 1 = 10$; $10 + 1 = 11$; $10 + 10 = 100$; $10 - 1 = 1$; $100 - 1 = 11$; $(11)(11) = 1001$.

To check computation, we can change the numbers to decimal (base ten) form and then perform the operations. The following examples show some computation with binary numbers.

Add the following numbers (binary notation):

(a)	(b)	(c)	(d)
1011	1101	1110	101101
101	1001	1011	110110
10000	10110	11001	1100011

Changed to decimal notation, the foregoing numbers become:

11	13	14	45
5	9	11	54
16	22	25	99

Subtract the bottom number from the top:

(a)	(b)	(c)	(d)
1101	11101	101011	111011
111	1110	11011	11101
110	1111	10000	11110

Changed to decimal notation, the foregoing numbers become:

13	29	43	59
7	14	27	29
6	15	16	30

Subtraction can be performed by changing the sign of the subtrahend and then adding. However, then the resulting digit at the left must be omitted and added to the number. To change the sign of a number, we simply change 1 to 0, and 0 to 1. Work the next example by this method.

Example 3

Subtract: 110111 − 101101, by changing the sign of the subtrahend and adding.

Solution

Changing sign of the subtrahend, we add:

```
  110111
  010010
 -------
 X001001
      →1
 -------
    1010
```

Adding the numbers, we get X001001 (the left-hand 1 crossed out).
Now we shift the left-hand digit 1, and add it: →1
The result shows the remainder when subtracting: 1010

Example 4

Subtract: 1101101 − 1001001 by the method shown.

Solution

We change the sign of the subtrahend,

```
  1101101
  0110110
 --------
 X0100011
       →1
 --------
   100100
```

Adding, X0100011 (the left-hand 1 crossed out).
Now we shift the left 1, and add it: →1
Result: 100100

Checking this example by changing to base ten, we have: $109 - 73 = 36$.

Example 5

Multiply: (10111)(101)

Solution

The multiplication is shown at the right.

$$\begin{array}{r} 10111 \\ 101 \\ \hline 10111 \\ 10111 \\ \hline 1110011 \end{array}$$

Changing to decimal notation, we have

$$\begin{array}{r} 23 \\ 5 \\ \hline 115 \end{array}$$

Example 6

Divide: (110101) ÷ (101).

Solution

The division is shown at the right:
Note that the quotient represents *ten* (decimal notation), and the remainder represents *three.*
Changing to base *ten* notation, we have

$$53 \div 5 = 10, \text{ remainder } 3$$

$$\begin{array}{r} 1010 \\ 101\overline{)110101} \\ 101 \\ \hline 110 \\ 101 \\ \hline 11 \end{array}$$

Exercise 53-2

Change these numbers from binary form to decimal (base ten) form:

1. 10111 **2.** 11101 **3.** 1010101 **4.** 10001110

5. 10110101 **6.** 100101101 **7.** 110101011 **8.** 1100110011

Change these numbers from decimal to binary notation:

9. 35 **10.** 27 **11.** 43 **12.** 75 **13.** 113 **14.** 147

Add these numbers in binary notation:

15. $\begin{array}{r} 1101 \\ 1011 \\ \hline \end{array}$ **16.** $\begin{array}{r} 1001 \\ 1101 \\ \hline \end{array}$ **17.** $\begin{array}{r} 1110 \\ 1001 \\ \hline \end{array}$ **18.** $\begin{array}{r} 1010 \\ 1001 \\ \hline \end{array}$ **19.** $\begin{array}{r} 1110 \\ 1111 \\ \hline \end{array}$

20. $\begin{array}{r} 1110011 \\ 1011010 \\ \hline \end{array}$ **21.** $\begin{array}{r} 11010110 \\ 1101011 \\ \hline \end{array}$ **22.** $\begin{array}{r} 11011101101 \\ 10010011011 \\ \hline \end{array}$

Subtract the bottom number from the top:

23. $\begin{array}{r} 10101 \\ 1101 \\ \hline \end{array}$ **24.** $\begin{array}{r} 11011 \\ 10101 \\ \hline \end{array}$ **25.** $\begin{array}{r} 1101011 \\ 110110 \\ \hline \end{array}$ **26.** $\begin{array}{r} 1101010 \\ 1011011 \\ \hline \end{array}$

Multiply:

27. (1101)(101) **28.** (10111)(1011) **29.** (11011)(1101)

Divide:

30. (110111) ÷ (101) **31.** (11101101) ÷ (1011)

53-7 THE OCTAL NUMBER SYSTEM

Another number system used by some computers is the *octal* system. In this system the base is *eight*, as the name suggests. In the octal system the digits used are 0, 1, 2, 3, 4, 5, 6, and 7. There is no digit for eight. Instead, when we count to one more than 7, we lump the eight objects together into one bunch with no ones over. We write *eight* as 10. Then, as an example, for the number of fingers on both hands, we write 12, which means one bunch of eight and two units more.

When we count to 63, we have 7 bunches and 7 ones, which is 77. Then one more makes another bunch, and we write: 100. In this case, the 1 represents one bunch of (8)(8), or 64 (base ten).

The following table shows the relation between decimal and octal notation for some numbers:

decimal	octal	decimal	octal	decimal	octal
1	1	7	7	13	15
2	2	8	10	14	16
3	3	9	11	15	17
4	4	10	12	16	20
5	5	11	13	17	21
6	6	12	14	18	22

To change a number from *decimal* (base ten) notation to *octal* notation, we divide the number (base ten) by 8 as many times as possible, and show all remainders including zeros. Then the last quotient and the remainders in reverse order show the digits in the octal (base eight) notation.

Example 1

Change the number 354 (base ten) to base 8 (octal) notation.

Solution

```
     Rem.
8 /3541
8  /442 - 5
8   /55 - 2
      6 - 7
```

Then the number in octal notation is 6725_8

We need not divide the quotient 6 by 8,
Instead, we begin the answer with this quotient.

We use the symbol 8 as a subscript to denote the base.

To change a number from *octal* notation to *decimal* notation, we must consider the place value of each digit. In octal notation, the places have the following values:

8^4	8^3	8^2	8	units
4096	512	64	8	units

Example 2

Change the number 245_8 (octal) to decimal notation.

Solution

For the units, we have	5
The digit 4 means (4)(8):	32 (base 10)
The digit 2 means (2)(64):	128 (base 10)
Adding, we get	165 (base 10)

Example 3

Change the octal number 5627_8 to decimal notation:

Solution

For the units, we have	7
The digit 2 means (2)(8):	16 (base 10)
The digit 6 means (6)(64):	384 (base 10)
The digit 5 means (5)(512):	2560 (base 10)
Adding, we get	2967 (base 10)

To check the result, we can change the number back to octal (base 8) by dividing by 8, as in Example 1.

In computation with octal numbers, we must remember a few elementary facts:

$$3 + 5 = 10; \qquad 14 - 7 = 5; \qquad (5)(7) = 43; \qquad 22 \div 3 = 6$$

In subtraction, any time a "1" is borrowed, it must be considered as 8. The following examples show some computation with octal numbers:

Addition:

(1)	2.	3.	4.
576	465	756	564
643	573	647	456
1441	1260	1625	1242

We can check by changing the numbers to decimal (base 10) notation.

Then the above examples become:

	(1)	(2)	(3)	(4)
	382	309	494	372
	419	379	423	302
Adding,	801	688	917	674

Subtract the bottom number from the top: (numbers are in *octal* notation):

1. 623	**2.** 725	**3.** 652	**4.** 751	**5.** 61325
356	467	274	275	25476
245	236	356	454	33627

The answers can be checked by addition, or by changing all numbers to base 10.
Multiply: (base 8): (67)(475) = 42303. Multiply: (3)(437) = 1535
Divide: (base 8): 6456 ÷ 7 = 742.
Change to base 10: 3374 ÷ 7 = 482

There is a convenient relation between the binary system and the octal system. This comes about because of the fact that $8 = 2^3$.

The digits in any octal number represent a series of three digits in the binary system. If we write the numbers from *one* through *seven* in octal and binary form, we have

octal digits:	1	2	3	4	5	6	7
binary form:	001	010	011	100	101	110	111

In any binary number, if we separate the digits in groups of three starting at the right, we can recognize the digits in the octal system. For example, consider the number: (binary):

11101110

Starting at the right, we separate the digits into groups of three.

11 101 110

Then each group of three represents a digit in the octal system.

11 101 110
Then we have the octal digits: 3 5 6

In octal notation, the number is: 356_8. The result can be checked by changing each to decimal notation. We get 238 (base10) for each.

We can also use the foregoing principle in reverse. The digits of an octal number can be changed immediately to binary digits. For example, in the number, 3745 (octal), each digit can be written in binary form: 3 = 011; 7 = 111; 4 = 100; 5 = 101. Then, in binary form, we have the number:

$$3745_8 = 011111100101 \text{ (binary)}$$

Exercise 53-3

Change each of the following numbers (base ten) from decimal to octal form:

1. 473 **2.** 5436 **3.** 3065 **4.** 21073 **5.** 98304

Change each of these octal numbers (base 8) to decimal notation (base 10):

6. 563_8 **7.** 3047_8 **8.** 4357_8 **9.** 6305_8 **10.** 350741_8

Perform the indicated operations with these octal numbers:

11. Add: $\begin{array}{r} 475 \\ \underline{367} \end{array}$ **12.** Subtract $\begin{array}{r} 734 \\ \underline{253} \end{array}$ **13.** Multiply $\begin{array}{r} 435 \\ \underline{27} \end{array}$ **14.** Divide $4637 \div 5$

15. Change these numbers from octal to binary notation: (these are in octal):

(a) 374 (b) 563 (c) 21376 (d) 20763

16. Change these binary numbers to the octal notation:

(a) 11101100 (b) 111000101 (c) 10110111 (d) 10011001 0011

53-8 THE HEXADECIMAL NUMBER SYSTEM

Computers sometimes use a system called the *hexadecimal* system of numbers. This system uses *sixteen* digits. The first ten are the same as in the common decimal system, 0 through 9. To provide for the extra six digits, we often use the first six letters of the alphabet, *A* through *F*. These letters then represent the following numbers:

A—ten	C—twelve	E—fourteen
B—eleven	D—thirteen	F—fifteen

When we count to one more than fifteen (that is, sixteen), we write the number as: 10. Here the "1" represents sixteen units. Any digit in second place from the right represents *sixteens*. For example, the number 43 in base *sixteen*, is the same as the number 67 in base *ten*. The number 3D (base sixteen) means: $(3)(16) + 13 = 51$, in base *ten*.

To change a number from decimal (base *ten*) notation to base sixteen, we use Rule 1, which we have used in connection with other bases. We divide the number (base ten) by 16 successively as many times as possible, noting the remainders and the final quotient. Then the final quotient and the remainders in reverse order represent the digits of the number in base *sixteen*. We shall sometimes represent base *sixteen* as 16.

Example 1

Change the number 742 (base ten) to base sixteen (16).

Solution

Dividing, we get the remainders and final quotient: 6, 14, 2. Now we note that 14 is represented by E. Then, taking the final quotient and the remainders in reverse order, we get the number: 2E6 (base sixteen).

Example 2

Change 68397 (base ten) to base sixteen.

Solution

Dividing successively by 16, we get the remainders: 13, 2, 11, 0, and final quotient 1. Taking these numbers in reverse order, we get the number in base sixteen: 10B2D.

To change a number from base *sixteen* to base *ten*, we have the following place values from base sixteen:

16^4	16^3	16^2	16	units

or,

65536	4096	256	16	units

Example 3

Change the following hexadecimal number (base sixteen) to decimal (base ten) form: A39D

Solution

The digit D means units:	13	(base ten)
The digit 9 means (9)(16):	144	(base ten)
The digit 3 means (3)(256):	768	(base ten)
The digit A means (10)(4096):	40960	(base ten)
Adding, we get	41885	(base ten)

The following examples show some simple computation with hexadecimal numbers;

Add: E + F = 1D; C + D = 19; 9 + 7 = 10; 98C + ED7 = 1863;
Subtract: 10 − 8 = 8; 15 − C = A; 2A − 1B = F; C34 − 58B = 6A9;
Multiply: (7)(F) = 69; (9)(8) = 48; (6)(23) = D2;
Divide: (54) ÷ (4) = 15; A8 ÷ C = E

Exercise 53-4

1. Change the following numbers (base ten) to hexadecimal notation:
   (a) 437 (b) 719 (c) 3243 (d) 8673 (e) 8947 (f) 29387
2. Change these hexadecimal (base 16) numbers to decimal (base ten) form:
   (a) 395 (b) 1487 (c) 8A5 (d) A3F (e) BE9 (f) 3A4BF
3. Multiply: (a) (D)(13) (b) (B)(2A)
4. Divide: (a) A8 ÷ 6 (b) 324 ÷ 1C

CHAPTER FIFTY-FOUR Electronic Hand Calculators

54-1 INTRODUCTION

In these days, electronic hand calculators are in quite common use, especially by students of technology, but also among grade and high school students. These calculators have in many instances replaced the slide rule because they afford a quick and efficient as well as a more accurate method for many mathematical computations.

Calculators are of many types. Some provide for more forms of computation than others. Let us briefly describe two general types. Our explanation of the use of calculators will be chiefly concerned with these two types.

Type 1. This is the simpler type. It has chiefly the following keys:

Digits: [0] through [9]; the decimal point: [.].
Basic arithmetic operations: addition: [+]; subtraction: [−]; multiplication: [×]; division: [÷].
Square of a number: [x^2]; square root of a number: [$\sqrt{x}$].
Reciprocal: [$1/x$]; clear key: [C].

It may have two or three other miscellaneous keys.

Type 2. In addition to all the basic keys of Type 1, the Type 2 calculator has these additional keys, and perhaps a few others:

Trigonometric functions: [sin]; [cos]; [tan].
Logarithmic functions: common logs: [log]; natural log: [ln].
For storage of information: [STO].
Recall of stored information: [RCL].
Power key: [y^x]; also π key: [π].
Key for conversion between degrees and radians.

As a first rule, know your calculator and what it can do. Most calculators have detailed and special instructions, which accompany the calculator. Yet many operations are the same on all.

A word of warning may be in order. Watch for errors in your computation. One example will show one type of error in the use of the calculator. One student had the problem; 7 × 9. On the calculator he got the answer: 56. He wrote down 56, never realizing he had hit the "8" key instead of the "9." A good rule to follow is to do the computation a second time. Above all, know the basic arithmetic combinations for addition and multiplication.

54-2 ADDITION AND SUBTRACTION

The operations of addition and subtraction are done in the order in which they occur in a problem. Parentheses must be observed. These operations are basically the same on all calculators. In many instances the equal sign (=) need not be used after each separate operation. It is used at the end of a sequence to show the answer. In some particular problems, the equal sign must be used after certain operations, as we shall see.

In showing the steps in a solution, we shall enclose in brackets each key to be used in the sequence. The multiplication sign is shown as [×]; the decimal point is shown as [.].

Example 1

Find: 20.8 + 5.62 − 3.4

Solution

The following sequence shows the keys used:

[2][0][.][8][+][5][.][6][2][−][3][.][4][=]

The answer is displayed: 23.02

Example 2

Find: 35 − 4.6 − 13.5

Solution

Sequence: [3][5][−][4][.][6][−][1][3][.][5][=]
Answer: 16.9

Example 3

Find: 10 − 7.2 − 8

Solution

Sequence: [1][0][−][7][.][2][−][8][=]
Answer: −5.2.

Example 4

Find: 11 − (2 + 5)

Solution

Removing the parentheses, we have: 11 − 2 − 5, showing that the 2 and the 5 must be subtracted. Sequence: [1][1][−][2][−][5][=]
Answer: 4.

Example 5

Find: 25 + (13 − 6)

Solution

Sequence: [2][5][+][1][3][−][6][=] *Answer:* 32

54-3 MULTIPLICATION AND DIVISION

If only multiplications and/or divisions occur in the same problem, these operations are performed in the order in which they occur, but sometimes may be reversed.

Example 6

Find (6)(5) ÷ 2

Solution

Sequence: [6][×][5][÷][2][=] *Answer:* 15

Example 7

Find 4 ÷ 3 × 9

Solution

Sequence: [4][÷][3][×][9][=]; we get the answer: 11.999999. If we reverse the multiplication and division, we get
Sequence: [4][×][9][÷][3][=]; we get the exact answer: 12.

If additions and/or subtractions occur together with multiplications and/or divisions in the same problem, the operations involving multiplications and/or divisions are performed first, and then additions and subtractions in order.

Example 8

Find: 6 + 5 × 8 ÷ 2

Solution

Sequence: [5][×][8][÷][2][+][6][=] *Answer:* 26.
In this example, if we take the numbers in the order shown, we get the wrong answer. Then the sequence is: [6][+][5][×][8][÷][2][=]. We get the answer: 44, which is wrong.

Example 9

Find: $34-8\times6\div3$

Solution

Sequence: [8][×][6][÷][3][−][3][4][=]
For the answer, we get −18. However, note, that 34 is greater the (8)(6) ÷ 3. Therefore, the answer should be positive. We change it to positive: +18.

Some calculators have a key to change the sign of a number. However, all we need to do, is to note that the answer should be positive.

Example 10

Find: $6 - 20 \div 4 \times 3$

Solution

Sequence: [2][0][÷][4][×][3][−][6][=] *Answer:* 9.
Note that in the problem itself, the first portion is positive, (+6), whereas the second portion is negative. As we set up the sequence, the first portion is positive, the second is negative, −6. Then we change the answer to a negative; −9.

Example 11

Find: $35 - 6.4 \times 2.5 \div 0.8$

Solution

Sequence: [6][.][4][×][2][.][5][÷][.][8][−][3][5][=].
We get the answer: −15. We know the answer must be positive: +15.

Example 12

Find $20 + (3)(5) - 12$.

Solution

Sequence: [3][×][5][+][2][0][−][1][2][=] *Answer:* 23.

54-4 SQUARES OF NUMBERS

If a problem involves the square of a number, this operation should be performed first. This is true for all calculators. However, some can use a slightly different approach.

Example 13

Find: $(8)(15^2)$

Solution

Sequence: [1][5][x^2][×][8][=] *Answer:* 1800.
This problem might be done as follows: [8][×][1][5][×][1][5][=]. Note that the multiplication still involves (15)(15).
On Type 2 calculators, the sequence could be: [8][×][1][5][x^2][=]. That is, the calculator finds 15^2 independently of the 8. However, on Type 1 calculators, we would get the wrong answer because the first operation would be (8)(15), which is 120. Then we would have 120^2, or 14400 (wrong).

In general, if a square of a number appears in a sequence, this operation should be done first. Then you cannot go wrong.

Example 14

Find: $42 - 5^2$.

Solution

Sequence: [5][x^2][−][4][2][=] *Answer:* −17.
Note, however, that we have changed the signs of the two portions of the problem in setting up the sequence. Therefore, the answer should be +17.

Example 15

Find: $\frac{1}{9 + 4^2}$ (Type 1 calculator).

Solution

Sequence for denominator: [4][x^2]+[9][=].
We want the reciprocal, however, so we add the key $1/x$, and have

[4][x^2][+][9][$1/x$][=] *Answer:* 0.04.

Example 16

Find $\frac{6}{5 + 3^2}$ (Type 1 calculator).

Solution

After computing the denominator, we can divide it by 6, and then find the reciprocal. Then we have the sequence:

[3][x^2][+][5][÷][6][1/x][=] *Answer:* 0.42857.

If a Type 2 calculator is used for Examples 15 and 16, an extra equal sign [=] is necessary immediately before the reciprocal sign [1/x].

Example 17

Find: $12^2 + 5^2$.

Solution

To find the sum of two squares or their difference, the sequence depends on the particular type of calculator. In this example, we can always use the following sequence. First: [1][2][x^2][=].
Now we write down the answer, 144, and clear [C]. Then we use the sequence:

[5][x^2][+][1][4][4][=] *Answer:* 169.

However, for Type 2 calculators, we can use the sequence:

[1][2][x^2][+][5][x^2][=].

On such calculators, the "5" will be squared independently of what has gone before. If we use the latter sequence for Type 1 calculators, we would get the *wrong* answer: 22201. This is because we would first get $12^2 = 144$. Then we would get $144 + 5 = 149$. Finally, we would get $(149)^2 = 22201$.

Example 18

Find: $25^2 - 7^2$.

Solution

For Type 2 calculators, we use the sequence:

[2][5][x^2][−][7][x^2][=] *Answer:* 576.

Such calculators will find 7^2 independently of any preceding operation.
However, for all calculators, we can use the following sequence:
First, [7][x^2][=]; we write down this answer, 49, and clear [C].
Next, we have [2][5][x^2][−][4][9][=] *Answer:* 576.
Using this method, note that we first find 7^2 because this quantity is to be subtracted from 25^2.

Example 19

Find $\sqrt{225} - \sqrt{36}$

Solution

For Type 2 calculators, we use the sequence:

[2][2][5][$\sqrt{x}$][−][3][6][$\sqrt{x}$][=] *Answer:* 9.

For Type 1 calculators we would have to use: [3][6][$\sqrt{x}$][=] *Answer:* 6.
Then: [2][2][5][$\sqrt{x}$][−][6][=] *Answer;* 9.
As usual for this method, we must clear [C] after the first answer, 6.

54-5 THE PYTHAGOREAN RULE

Example 20

Find: $\sqrt{15^2 + 8^2}$

Solution

For a Type 2 calculator, we use the sequence:

[1][5][x^2][+][8][x^2][=][$\sqrt{x}$][=] *Answer:* 17.

However, in this example, the equal sign key [=] must be used immediately after the last [x^2]; that is, before the key: $\sqrt{x}$. Otherwise, we get the wrong answer.

For all calculators, we can use the sequence. First, [1][5][x^2][=]. The answer then is 225, which we write down and then clear the calculator. Then we have the sequence: [8][x^2][+][2][2][5][=][$\sqrt{x}$][=] *Answer:* 17. On Type 1 calculators, the equal sign [=] may be omitted before the [$\sqrt{x}$]. On others, it must be used.

There is another way to solve this problem on any calculator. If we divide both terms, 15^2 and 8^2, by the quantity 8^2 and then multiply by 8^2, we can reduce the problem to the form: $(\sqrt{(\frac{15}{8})^2 + 1})\ (8)$
Then, for any calculator, we can use the sequence:

[1][5][÷][8][=][x^2][+][1][=][$\sqrt{x}$][×][8][=] *Answer:* 17.

Amazingly enough, on the simpler Type 1 calculators, the first two equal signs [=] may be omitted. On some calculators, these equal signs must be used.

Example 21

Find: $\sqrt{33^2 + 56^2}$.

Solution

If we use the final method in the preceding example, we reduce the problem to $(\sqrt{(\frac{33}{56})^2 + 1})\ (56)$. We get the answer: 64.99999. . . . The answer is approximate only because 33 ÷ 56 is an unending decimal. For Type 2 calculators we can use the sequence:

[3][3][x^2][+][5][6][x^2][=][$\sqrt{x}$][=] *Answer:* 65.

Example 22

Find: $\sqrt{29^2 - 20^2}$.

Solution

On Type 2 calculators, we can use the sequence:

[2][9][x^2][−][2][0][x^2][=][$\sqrt{x}$][=] *Answer:* 21.

This sequence cannot be used for Type 1 calculators. However, then we can use the formation:

$$\left(\sqrt{\left(\frac{29}{20}\right)^2 - 1}\right)(20)$$

Then we use the sequence:

[2][9][÷][2][0][=][x^2][−][1][=][$\sqrt{x}$][×][2][0][=]

Some calculators must use the two first equal signs; others need not use them.

Example 23

Find: $\sqrt{193^2 - 168^2}$

Solution

On the Type 2 calculator, we can use the sequence:

[1][9][3][x^2][−][1][6][8][x^2][=][$\sqrt{x}$][=] *Answer:* 95.

The answer is displayed even without the final equal sign.

For this example we could not use this sequence for calculator Type 1. Instead, we could change the problem to:

$$\left(\sqrt{\left(\frac{193}{168}\right)^2 - 1}\right)(168)$$

Then we can use: [1][9][3][÷][1][6][8][x^2][−][1][$\sqrt{x}$][×][1][6][8][=]. This sequence gives the answer: 94.999997. This sequence could not be used for Type 2 without two extra equal signs.

This problem could always be worked as follows: [1][6][8][x^2][=]; Then write down the answer, which is $168^2 = 28224$. Then use clear [C], and use the sequence: [1][9][3][x^2][−][2][8][2][2][4][$\sqrt{x}$][=] The answer, 95, is displayed without the final equal sign.

54-6 TRIGONOMETRIC FUNCTIONS

If a calculator provides values of sine, cosine, and tangent of angles, we can find the values of these functions for angles in degrees or radians.

Example 24

Find: sin 20°. (Type 2 calculator.)

Solution

Sequence: [2][0][sin][=] *Answer*: 0.34202 (rounded). The answer is usually displayed without the final equal sign.

Example 25

Find: cos 1. (Here the "1" means 1 radian.)

Solution

We first see that the change is made from degrees to radians. Then we have the sequence: [1][cos]. *Answer is displayed:* 0.5403.

For the values of cotangent, secant, and cosecant, we take the reciprocals of tangent, cosine, and sine, respectively.

Example 26

Find: sec 30°.

Solution

Be sure to have the calculator set for degrees instead of radians. Then we have the sequence: [3][0][cos][1/x]. *Answer*: 1.1547.

Example 27

Find: log sin 40°.

Solution

To provide for a positive mantissa, we have the sequence:

[4][0][sin][log][+][1][0][=] *Answer:* 9.8080675, or 9.8081.

In an example of this kind, we add 10 to make the mantissa positive.

Exercise 54-1

Compute the following with a hand calculator (multiplication sign: ×):

1. $28.3 + 17.2 - 12.8$ **2.** $32.5 - 9.8 - 17.3$ **3.** $14 - 8.5 - 12.6$

4. $18 + (11 - 6)$ **5.** $41 - (10 + 7)$ **6.** $21.4 - (13 - 5)$

7. $25.4 + 7.2 \times 3.5$ **8.** $43.4 - 6.2 \times 2.3$ **9.** $53 \div 3 \times 12$

10. $35 - 13 \div 4 \times 20$ **11.** $13 - 8 \times 9 \div 12$

12. $28 + 0.32 \times 0.25 \div 12$ **13.** $26 - 30 \div 5 \times 2 + 4$

14. $8 - 1.2 \div 2 \times 0.4 - 3.2$ **15.** 7×2.5^2

16. $24 - 16 \div 0.3 \times .12$ **17.** $3.2(0.6)^2$ **18.** $0.8 + (4.2)^2 - 3$

19. $12 - (1.3)^2 + 7$

20. $45 - (6.4)^2 - 8$

21. $3.6 + (1.4)^2 - 5$

22. $15 + (5.2)(0.5^2)$

23. $12 + (0.9^2)(1.5)^2$

24. $21^2 + 3.5^2 - 4$

25. $43^2 - 18^2 + 13^2$

26. $\sqrt{289} + \sqrt{121}$

27. $\sqrt{3025} - \sqrt{324}$

28. $\sqrt{32 + 15 - 8}$

29. $\sqrt{53.2 - 4.3 + 5}$

30. $\sqrt{60^2 + 11^2}$

31. $\sqrt{157^2 - 85^2}$

32. $\sqrt{26^2 - 6^2}$

33. $\sin 28°$

34. $\log \sin 28°$

35. $\log \cos 35°$

36. $\log \tan 32°$

37. $\cot 20°$

38. $\sec 70°$

39. $\csc 24°$

APPENDIX TABLES

Table A-1. Four-place mantissas of common logarithms

N	0	1	2	3	4	5	6	7	8	9
10	0000	0043	0086	0128	0170	0212	0253	0294	0334	0374
11	0414	0453	0492	0531	0569	0607	0645	0682	0719	0755
12	0792	0828	0864	0899	0934	0969	1004	1038	1072	1106
13	1139	1173	1206	1239	1271	1303	1335	1367	1399	1430
14	1461	1492	1523	1553	1584	1614	1644	1673	1703	1732
15	1761	1790	1818	1847	1875	1903	1931	1959	1987	2014
16	2041	2068	2095	2122	2148	2175	2201	2227	2253	2279
17	2304	2330	2355	2380	2405	2430	2455	2480	2504	2529
18	2553	2577	2601	2625	2648	2672	2695	2718	2742	2765
19	2788	2810	2833	2856	2878	2900	2923	2945	2967	2989
20	3010	3032	3054	3075	3096	3118	3139	3160	3181	3201
21	3222	3243	3263	3284	3304	3324	3345	3365	3385	3404
22	3424	3444	3464	3483	3502	3522	3541	3560	3579	3598
23	3617	3636	3655	3674	3692	3711	3729	3747	3766	3784
24	3802	3820	3838	3856	3874	3892	3909	3927	3945	3962
25	3979	3997	4014	4031	4048	4065	4082	4099	4116	4133
26	4150	4166	4183	4200	4216	4232	4249	4265	4281	4298
27	4314	4330	4346	4362	4378	4393	4409	4425	4440	4456
28	4472	4487	4502	4518	4533	4548	4564	4579	4594	4609
29	4624	4639	4654	4669	4683	4698	4713	4728	4742	4757
30	4771	4786	4800	4814	4829	4843	4857	4871	4886	4900
31	4914	4928	4942	4955	4969	4983	4997	5011	5024	5038
32	5051	5065	5079	5092	5105	5119	5132	5145	5159	5172
33	5185	5198	5211	5224	5237	5250	5263	5276	5289	5302
34	5315	5328	5340	5353	5366	5378	5391	5403	5416	5428
35	5441	5453	5465	5478	5490	5502	5514	5527	5539	5551
36	5563	5575	5587	5599	5611	5623	5635	5647	5658	5670
37	5682	5694	5705	5717	5729	5740	5752	5763	5775	5786
38	5798	5809	5821	5832	5843	5855	5866	5877	5888	5899
39	5911	5922	5933	5944	5955	5966	5977	5988	5999	6010
40	6021	6031	6042	6053	6064	6075	6085	6096	6107	6117
41	6128	6138	6149	6160	6170	6180	6191	6201	6212	6222
42	6232	6243	6253	6263	6274	6284	6294	6304	6314	6325
43	6335	6345	6355	6365	6375	6385	6395	6405	6415	6425
44	6435	6444	6454	6464	6474	6484	6493	6503	6513	6522
45	6532	6542	6551	6561	6571	6580	6590	6599	6609	6618
46	6628	6637	6646	6656	6665	6675	6684	6693	6702	6712
47	6721	6730	6739	6749	6758	6767	6776	6785	6794	6803
48	6812	6821	6830	6839	6848	6857	6866	6875	6884	6893
49	6902	6911	6920	6928	6937	6946	6955	6964	6972	6981
50	6990	6998	7007	7016	7024	7033	7042	7050	7059	7067
51	7076	7084	7093	7101	7110	7118	7126	7135	7143	7152
52	7160	7168	7177	7185	7193	7202	7210	7218	7226	7235
53	7243	7251	7259	7267	7275	7284	7292	7300	7308	7316
54	7324	7332	7340	7348	7356	7364	7372	7380	7388	7396
N	0	1	2	3	4	5	6	7	8	9

Table A-1. Four-place mantissas of common logarithms (*Continued*)

N	0	1	2	3	4	5	6	7	8	9
55	7404	7412	7419	7427	7435	7443	7451	7459	7466	7474
56	7482	7490	7497	7505	7513	7520	7528	7536	7543	7551
57	7559	7566	7574	7582	7589	7597	7604	7612	7619	7627
58	7634	7642	7649	7657	7664	7672	7679	7686	7694	7701
59	7709	7716	7723	7731	7738	7745	7752	7760	7767	7774
60	7782	7789	7796	7803	7810	7818	7825	7832	7839	7846
61	7853	7860	7868	7875	7882	7889	7896	7903	7910	7917
62	7924	7931	7938	7945	7952	7959	7966	7973	7980	7987
63	7993	8000	8007	8014	8021	8028	8035	8041	8048	8055
64	8062	8069	8075	8082	8089	8096	8102	8109	8116	8122
65	8129	8136	8142	8149	8156	8162	8169	8176	8182	8189
66	8195	8202	8209	8215	8222	8228	8235	8241	8248	8254
67	8261	8267	8274	8280	8287	8293	8299	8306	8312	8319
68	8325	8331	8338	8344	8351	8357	8363	8370	8376	8382
69	8388	8395	8401	8407	8414	8420	8426	8432	8439	8445
70	8451	8457	8463	8470	8476	8482	8488	8494	8500	8506
71	8513	8519	8525	8531	8537	8543	8549	8555	8561	8567
72	8573	8579	8585	8591	8597	8603	8609	8615	8621	8627
73	8633	8639	8645	8651	8657	8663	8669	8675	8681	8686
74	8692	8698	8704	8710	8716	8722	8727	8733	8739	8745
75	8751	8756	8762	8768	8774	8779	8785	8791	8797	8802
76	8808	8814	8820	8825	8831	8837	8842	8848	8854	8859
77	8865	8871	8876	8882	8887	8893	8899	8904	8910	8915
78	8921	8927	8932	8938	8943	8949	8954	8960	8965	8971
79	8976	8982	8987	8993	8998	9004	9009	9015	9020	9025
80	9031	9036	9042	9047	9053	9058	9063	9069	9074	9079
81	9085	9090	9096	9101	9106	9112	9117	9122	9128	9133
82	9138	9143	9149	9154	9159	9165	9170	9175	9180	9186
83	9191	9196	9201	9206	9212	9217	9222	9227	9232	9238
84	9243	9248	9253	9258	9263	9269	9274	9279	9284	9289
85	9294	9299	9304	9309	9315	9320	9325	9330	9335	9340
86	9345	9350	9355	9360	9365	9370	9375	9380	9385	9390
87	9395	9400	9405	9410	9415	9420	9425	9430	9435	9440
88	9445	9450	9455	9460	9465	9469	9474	9479	9484	9489
89	9494	9499	9504	9509	9513	9518	9523	9528	9533	9538
90	9542	9547	9552	9557	9562	9566	9571	9576	9581	9586
91	9590	9595	9600	9605	9609	9614	9619	9624	9628	9633
92	9638	9643	9647	9652	9657	9661	9666	9671	9675	9680
93	9685	9689	9694	9699	9703	9708	9713	9717	9722	9727
94	9731	9736	9741	9745	9750	9754	9759	9763	9768	9773
95	9777	9782	9786	9791	9795	9800	9805	9809	9814	9818
96	9823	9827	9832	9836	9841	9845	9850	9854	9859	9863
97	9868	9872	9877	9881	9886	9890	9894	9899	9903	9908
98	9912	9917	9921	9926	9930	9934	9939	9943	9948	9952
99	9956	9961	9965	9969	9974	9978	9983	9987	9991	9996
N	0	1	2	3	4	5	6	7	8	9

Table A-2. Natural trigonometric functions for decimal fractions of a degree

Deg.	Sin	Tan	Cot	Cos	Deg.	Deg.	Sin	Tan	Cot	Cos	Deg.
0.0	0.00000	0.00000	∞	1.0000	90.0	0.5	.07846	.07870	12.706	.9969	0.5
.1	.00175	.00175	573.0	1.0000	.9	.6	.08020	.08046	12.429	.9968	.4
.2	.00349	.00349	286.5	1.0000	.8	.7	.08194	.08221	12.163	.9966	.3
.3	.00524	.00524	191.0	1.0000	.7	.8	.08368	.08397	11.909	.9965	.2
.4	.00698	.00698	143.24	1.0000	.6	.9	.08542	.08573	11.664	.9963	.1
.5	.00873	.00873	114.59	1.0000	.5	5.0	0.08716	0.08749	11.430	0.9962	85.0
.6	.01047	.01047	95.49	0.9999	.4	.1	.08889	.08925	11.205	.9960	.9
.7	.01222	.01222	81.85	.9999	.3	.2	.09063	.09101	10.988	.9959	.8
.8	.01396	.01396	71.62	.9999	.2	.3	.09237	.09277	10.780	.9957	.7
.9	.01571	.01571	63.66	.9999	.1	.4	.09411	.09453	10.579	.9956	.6
1.0	0.01745	0.01746	57.29	0.9998	89.0	.5	.09585	.09629	10.385	.9954	.5
.1	.01920	.01920	52.08	.9998	.9	.6	.09758	.09805	10.199	.9952	.4
.2	.02094	.02095	47.74	.9998	.8	.7	.09932	.09981	10.019	.9951	.3
.3	.02269	.02269	44.07	.9997	.7	.8	.10106	.10158	9.845	.9949	.2
.4	.02443	.02444	40.92	.9997	.6	.9	.10279	.10334	9.677	.9947	.1
.5	.02618	.02619	38.19	.9997	.5	6.0	0.10453	0.10510	9.514	0.9945	84.0
.6	.02792	.02793	35.80	.9996	.4	.1	.10626	.10687	9.357	.9943	.9
.7	.02967	.02968	33.69	.9996	.3	.2	.10800	.10863	9.205	.9942	.8
.8	.03141	.03143	31.82	.9995	.2	.3	.10973	.11040	9.058	.9940	.7
.9	.03316	.03317	30.14	.9995	.1	.4	.11147	.11217	8.915	.9938	.6
2.0	0.03490	0.03492	28.64	0.9994	88.0	.5	.11320	.11394	8.777	.9936	.5
.1	.03664	.03667	27.27	.9993	.9	.6	.11494	.11570	8.643	.9934	.4
.2	.03839	.03842	26.03	.9993	.8	.7	.11667	.11747	8.513	.9932	.3
.3	.04013	.04016	24.90	.9992	.7	.8	.11840	.11924	8.386	.9930	.2
.4	.04188	.04191	23.86	.9991	.6	.9	.12014	.12101	8.264	.9928	.1
.5	.04362	.04366	22.90	.9990	.5	7.0	0.12187	0.12278	8.144	0.9925	83.0
.6	.04536	.04541	22.02	.9990	.4	.1	.12360	.12456	8.028	.9923	.9
.7	.04711	.04716	21.20	.9989	.3	.2	.12533	.12633	7.916	.9921	.8
.8	.04885	.04891	20.45	.9988	.2	.3	.12706	.12810	7.806	.9919	.7
.9	.05059	.05066	19.74	.9987	.1	.4	.12880	.12988	7.700	.9917	.6
3.0	0.05234	0.05241	19.081	0.9986	87.0	.5	.13053	.13165	7.596	.9914	.5
.1	.05408	.05416	18.464	.9985	.9	.6	.13226	.13343	7.495	.9912	.4
.2	.05582	.05591	17.886	.9984	.8	.7	.13399	.13521	7.396	.9910	.3
.3	.05756	.05766	17.343	.9983	.7	.8	.13572	.13698	7.300	.9907	.2
.4	.05931	.05941	16.832	.9982	.6	.9	.13744	.13876	7.207	.9905	.1
.5	.06105	.06116	16.350	.9981	.5	8.0	0.13917	0.14054	7.115	0.9903	82.0
.6	.06279	.06291	15.895	.9980	.4	.1	.14090	.14232	7.026	.9900	.9
.7	.06453	.06467	15.464	.9979	.3	.2	.14263	.14410	6.940	.9898	.8
.8	.06627	.06642	15.056	.9978	.2	.3	.14436	.14588	6.855	.9895	.7
.9	.06802	.06817	14.669	.9977	.1	.4	.14608	.14767	6.772	.9893	.6
4.0	0.06976	0.06993	14.301	0.9976	86.0	.5	.14781	.14945	6.691	.9890	.5
.1	.07150	.07168	13.951	.9974	.9	.6	.14954	.15124	6.612	.9888	.4
.2	.07324	.07344	13.617	.9973	.8	.7	.15126	.15302	6.535	.9885	.3
.3	.07498	.07519	13.300	.9972	.7	.8	.15299	.15481	6.460	.9882	.2
.4	.07672	.07695	12.996	.9971	85.6	.9	.15471	.15660	6.386	.9880	81.1
Deg.	Cos	Cot	Tan	Sin	Deg.	Deg.	Cos	Cot	Tan	Sin	Deg.

Table A-2. Natural trigonometric functions for decimal fractions of a degree (*Continued*)

Deg.	Sin	Tan	Cot	Cos	Deg.	Deg.	Sin	Tan	Cot	Cos	Deg.
9.0	0.15643	0.15838	6.314	0.9877	81.0	0.5	.2334	.2401	4.165	.9724	0.5
.1	.15816	.16017	6.243	.9874	.9	.6	.2351	.2419	4.134	.9720	.4
.2	.15988	.16196	6.174	.9871	.8	.7	.2368	.2438	4.102	.9715	.3
.3	.16160	.16376	6.107	.9869	.7	.8	.2385	.2456	4.071	.9711	.2
.4	.16333	.16555	6.041	.9866	.6	.9	.2402	.2475	4.041	.9707	.1
.5	.16505	.16734	5.976	.9863	.5	14.0	0.2419	0.2493	4.011	0.9703	76.0
.6	.16677	.16914	5.912	.9860	.4	.1	.2436	.2512	3.981	.9699	.9
.7	.16849	.17093	5.850	.9857	.3	.2	.2453	.2530	3.952	.9694	.8
.8	.17021	.17273	5.789	.9854	.2	.3	.2470	.2549	3.923	.9690	.7
.9	.17193	.17453	5.730	.9851	.1	.4	.2487	.2568	3.895	.9686	.6
10.0	0.1736	0.1763	5.671	0.9848	80.0	.5	.2504	.2586	3.867	.9681	.5
.1	.1754	.1781	5.614	.9845	.9	.6	.2521	.2605	3.839	.9677	.4
.2	.1771	.1799	5.558	.9842	.8	.7	.2538	.2623	3.812	.9673	.3
.3	.1788	.1817	5.503	.9839	.7	.8	.2554	.2642	3.785	.9668	.2
.4	.1805	.1835	5.449	.9836	.6	.9	.2571	.2661	3.758	.9664	.1
.5	.1822	.1853	5.396	.9833	.5	15.0	0.2588	0.2679	3.732	0.9659	75.0
.6	.1840	.1871	5.343	.9829	.4	.1	.2605	.2698	3.706	.9655	.9
.7	.1857	.1890	5.292	.9826	.3	.2	.2622	.2717	3.681	.9650	.8
.8	.1874	.1908	5.242	.9823	.2	.3	.2639	.2736	3.655	.9646	.7
.9	.1891	.1926	5.193	.9820	.1	.4	.2656	.2754	3.630	.9641	.6
11.0	0.1908	0.1944	5.145	0.9816	79.0	.5	.2672	0.2773	3.606	.9636	.5
.1	.1925	.1962	5.097	.9813	.9	.6	.2689	.2792	3.582	.9632	.4
.2	.1942	.1980	5.050	.9810	.8	.7	.2706	.2811	3.558	.9627	.3
.3	.1959	.1998	5.005	.9806	.7	.8	.2723	.2830	3.534	.9622	.2
.4	.1977	.2016	4.959	.9803	.6	.9	.2740	.2849	3.511	.9617	.1
.5	.1994	.2035	4.915	.9799	.5	16.0	0.2756	0.2867	3.487	0.9613	74.0
.6	.2011	.2053	4.872	.9796	.4	.1	.2773	.2886	3.465	.9608	.9
.7	.2028	.2071	4.829	.9792	.3	.2	.2790	.2905	3.442	.9603	.8
.8	.2045	.2089	4.787	.9789	.2	.3	.2807	.2924	3.420	.9598	.7
.9	.2062	.2107	4.745	.9785	.1	.4	.2823	.2943	3.398	.9593	.6
12.0	0.2079	0.2126	4.705	0.9781	78.0	.5	.2840	.2962	3.376	.9588	.5
.1	.2096	.2144	4.665	.9778	.9	.6	.2857	.2981	3.354	.9583	.4
.2	.2113	.2162	4.625	.9774	.8	.7	.2874	.3000	3.333	.9578	.3
.3	.2130	.2180	4.586	.9770	.7	.8	.2890	.3019	3.312	.9573	.2
.4	.2147	.2199	4.548	.9767	.6	.9	.2907	.3038	3.291	.9568	.1
.5	.2164	.2217	4.511	.9763	.5	17.0	0.2924	0.3057	3.271	0.9563	73.0
.6	.2181	.2235	4.474	.9759	.4	.1	.2940	.3076	3.251	.9558	.9
.7	.2198	.2254	4.437	.9755	.3	.2	.2957	.3096	3.230	.9553	.8
.8	.2215	.2272	4.402	.9751	.2	.3	.2974	.3115	3.211	.9548	.7
.9	.2233	.2290	4.366	.9748	.1	.4	.2990	.3134	3.191	.9542	.6
13.0	0.2250	0.2309	4.331	0.9744	77.0	.5	.3007	.3153	3.172	.9537	.5
.1	.2267	.2327	4.297	.9740	.9	.6	.3024	.3172	3.152	.9532	.4
.2	.2284	.2345	4.264	.9736	.8	.7	.3040	.3191	3.153	.9527	.3
.3	.2300	.2364	4.230	.9732	.7	.8	.3057	.3211	3.115	.9521	.2
.4	.2317	.2382	4.198	.9728	76.6	.9	.3074	.3230	3.096	.9516	72.1
Deg.	Cos	Cot	Tan	Sin	Deg.	Deg.	Cos	Cot	Tan	Sin	Deg.

Table A-2. Natural trigonometric functions for decimal fractions of a degree (*Continued*)

Deg.	Sin	Tan	Cot	Cos	Deg.	Deg.	Sin	Tan	Cot	Cos	Deg.
18.0	0.3090	0.3249	3.078	0.9511	72.0	0.5	.3827	.4142	2.414	.9239	0.5
.1	.3107	.3269	3.060	.9505	.9	.6	.3843	.4163	2.402	.9232	.4
.2	.3123	.3288	3.042	.9500	.8	.7	.3859	.4183	2.391	.9225	.3
.3	.3140	.3307	3.024	.9494	.7	.8	.3875	.4204	2.379	.9219	.2
.4	.3156	.3327	3.006	.9489	.6	.9	.3891	.4224	2.367	.9212	.1
.5	.3173	.3346	2.989	.9483	.5	23.0	0.3907	0.4245	2.356	0.9205	67.0
.6	.3190	.3365	2.971	.9478	.4	.1	.3923	.4265	2.344	.9198	.9
.7	.3206	.3385	2.954	.9472	.3	.2	.3939	.4286	2.333	.9191	.8
.8	.3223	.3404	2.937	.9466	.2	.3	.3955	.4307	2.322	.9184	.7
.9	.3239	.3424	2.921	.9461	.1	.4	.3971	.4327	2.311	.9178	.6
19.0	0.3256	0.3443	2.904	0.9455	71.0	.5	.3987	.4348	2.300	.9171	.5
.1	.3272	.3463	2.888	.9449	.9	.6	.4003	.4369	2.289	.9164	.4
.2	.3289	.3482	2.872	.9444	.8	.7	.4019	.4390	2.278	.9157	.3
.3	.3305	.3502	2.856	.9438	.7	.8	.4035	.4411	2.267	.9150	.2
.4	.3322	.3522	2.840	.9432	.6	.9	.4051	.4431	2.257	.9143	.1
.5	.3338	.3541	2.824	.9426	.5	24.0	0.4067	0.4452	2.246	0.9135	66.0
.6	.3355	.3561	2.808	.9421	.4	.1	.4083	.4473	2.236	.9128	.9
.7	.3371	.3581	2.793	.9415	.3	.2	.4099	.4494	2.225	.9121	.8
.8	.3387	.3600	2.778	.9409	.2	.3	.4115	.4515	2.215	.9114	.7
.9	.3404	.3620	2.762	.9403	.1	.4	.4131	.4536	2.204	.9107	.6
20.0	0.3420	0.3640	2.747	0.9397	70.0	.5	.4147	.4557	2.194	.9100	.5
.1	.3437	.3659	2.733	.9391	.9	.6	.4163	.4578	2.184	.9092	.4
.2	.3453	.3679	2.718	.9385	.8	.7	.4179	.4599	2.174	.9085	.3
.3	.3469	.3699	2.703	.9379	.7	.8	.4195	.4621	2.164	.9078	.2
.4	.3486	.3719	2.689	.9373	.6	.9	.4210	.4642	2.154	.9070	.1
.5	.3502	.3739	2.675	.9367	.5	25.0	0.4226	0.4663	2.145	0.9063	65.0
.6	.3518	.3759	2.660	.9361	.4	.1	.4242	.4684	2.135	.9056	.9
.7	.3535	.3779	2.646	.9354	.3	.2	.4258	.4706	2.125	.9048	.8
.8	.3551	.3799	2.633	.9348	.2	.3	.4274	.4727	2.116	.9041	.7
.9	.3567	.3819	2.619	.9342	.1	.4	.4289	.4748	2.106	.9033	.6
21.0	0.3584	0.3839	2.605	0.9336	69.0	.5	.4305	.4770	2.097	.9026	.5
.1	.3600	.3859	2.592	.9330	.9	.6	.4321	.4791	2.087	.9018	.4
.2	.3616	.3879	2.578	.9323	.8	.7	.4337	.4813	2.078	.9011	.3
.3	.3633	.3899	2.565	.9317	.7	.8	.4352	.4834	2.069	.9003	.2
.4	.3649	.3919	2.552	.9311	.6	.9	.4368	.4856	2.059	.8996	.1
.5	.3665	.3939	2.539	.9304	.5	26.0	0.4384	0.4877	2.050	0.8988	64.0
.6	.3681	.3959	2.526	.9298	.4	.1	.4399	.4899	2.041	.8980	.9
.7	.3697	.3979	2.513	.9291	.3	.2	.4415	.4921	2.032	.8973	.8
.8	.3714	.4000	2.500	.9285	.2	.3	.4431	.4942	2.023	.8965	.7
.9	.3730	.4020	2.488	.9278	.1	.4	.4446	.4964	2.014	.8957	.6
22.0	0.3746	0.4040	2.475	0.9272	68.0	.5	.4462	.4986	2.006	.8949	.5
.1	.3762	.4061	2.463	.9265	.9	.6	.4478	.5008	1.997	.8942	.4
.2	.3378	.4081	2.450	.9259	.8	.7	.4493	.5029	1.988	.8934	.3
.3	.3795	.4101	2.438	.9252	.7	.8	.4509	.5051	1.980	.8926	.2
.4	.3811	.4122	2.426	.9245	67.6	.9	.4524	.5073	1.971	.8918	63.1
Deg.	Cos	Cot	Tan	Sin	Deg.	Deg.	Cos	Cot	Tan	Sin	Deg.

Table A-2. Natural trigonometric functions for decimal fractions of a degree (*Continued*)

Deg.	Sin	Tan	Cot	Cos	Deg.	Deg.	Sin	Tan	Cot	Cos	Deg.
27.0	0.4540	0.5095	1.963	0.8910	63.0	0.5	.5225	.6128	1.6319	.8526	0.5
.1	.4555	.5117	1.954	.8902	.9	.6	.5240	.6152	1.6255	.8517	.4
.2	.4571	.5139	1.946	.8894	.8	.7	.5255	.6176	1.6191	.8508	.3
.3	.4586	.5161	1.937	.8886	.7	.8	.5270	.6200	1.6128	.8499	.2
.4	.4602	.5184	1.929	.8878	.6	.9	.5284	.6224	1.6066	.8490	.1
.5	.4617	.5206	1.921	.8870	.5	32.0	0.5299	0.6249	1.6003	0.8480	58.0
.6	.4633	.5228	1.913	.8862	.4	.1	.5314	.6273	1.5941	.8471	.9
.7	.4648	.5250	1.905	.8854	.3	.2	.5329	.6297	1.5880	.8462	.8
.8	.4664	.5272	1.897	.8846	.2	.3	.5344	.6322	1.5818	.8453	.7
.9	.4679	.5295	1.889	.8838	.1	.4	.5358	.6346	1.5757	.8443	.6
28.0	0.4695	0.5317	1.881	0.8829	62.0	.5	.5373	.6371	1.5697	.8434	.5
.1	.4710	.5340	1.873	.8821	.9	.6	.5388	.6395	1.5637	.8425	.4
.2	.4726	.5362	1.865	.8813	.8	.7	.5402	.6420	1.5577	.8415	.3
.3	.4741	.5384	1.857	.8805	.7	.8	.5417	.6445	1.5517	.8406	.2
.4	.4756	.5407	1.849	.8796	.6	.9	.5432	.6469	1.5458	.8396	.1
.5	.4772	.5430	1.842	.8788	.5	33.0	0.5446	0.6494	1.5399	0.8387	57.0
.6	.4787	.5452	1.834	.8780	.4	.1	.5461	.6519	1.5340	.8377	.9
.7	.4802	.5475	1.827	.8771	.3	.2	.5476	.6544	1.5282	.8368	.8
.8	.4818	.5498	1.819	.8763	.2	.3	.5490	.6569	1.5224	.8358	.7
.9	.4833	.5520	1.811	.8755	.1	.4	.5505	.6594	1.5166	.8348	.6
29.0	0.4848	0.5543	1.804	0.8746	61.0	.5	.5519	.6619	1.5108	.8339	.5
.1	.4863	.5566	1.797	.8738	.9	.6	.5534	.6644	1.5051	.8329	.4
.2	.4879	.5589	1.789	.8729	.8	.7	.5548	.6669	1.4994	.8320	.3
.3	.4894	.5612	1.782	.8721	.7	.8	.5563	.6694	1.4938	.8310	.2
.4	.4909	.5635	1.775	.8712	.6	.9	.5577	.6720	1.4882	.8300	.1
.5	.4924	.5658	1.767	.8704	.5	34.0	0.5592	0.6745	1.4826	0.8290	56.0
.6	.4939	.5681	1.760	.8695	.4	.1	.5606	.6771	1.4770	.8281	.9
.7	.4955	.5704	1.753	.8686	.3	.2	.5621	.6796	1.4715	.8271	.8
.8	.4970	.5727	1.746	.8678	.2	.3	.5635	.6822	1.4659	.8261	.7
.9	.4985	.5750	1.739	.8669	.1	.4	.5650	.6847	1.4605	.8251	.6
30.0	0.5000	0.5774	1.7321	0.8660	60.0	.5	.5664	.6873	1.4550	.8241	.5
.1	.5015	.5797	1.7251	.8652	.9	.6	.5678	.6899	1.4496	.8231	.4
.2	.5030	.5820	1.7182	.8643	.8	.7	.5693	.6924	1.4442	.8221	.3
.3	.5045	.5844	1.7113	.8634	.7	.8	.5707	.6950	1.4388	.8211	.2
.4	.5060	.5867	1.7045	.8625	.6	.9	.5721	.6976	1.4335	.8202	.1
.5	.5075	.5890	1.6977	.8616	.5	35.0	0.5736	0.7002	1.4281	0.8192	55.0
.6	.5090	.5914	1.6909	.8607	.4	.1	.5750	.7028	1.4229	.8181	.9
.7	.5105	.5938	1.6842	.8599	.3	.2	.5764	.7054	1.4176	.8171	.8
.8	.5120	.5961	1.6775	.8590	.2	.3	.5779	.7080	1.4124	.8161	.7
.9	.5135	.5985	1.6709	.8581	.1	.4	.5793	.7107	1.4071	.8151	.6
31.0	0.5150	0.6009	1.6643	0.8572	59.0	.5	.5807	.7133	1.4019	.8141	.5
.1	.5165	.6032	1.6577	.8563	.9	.6	.5821	.7159	1.3968	.8131	.4
.2	.5180	.6056	1.6512	.8554	.8	.7	.5835	.7186	1.3916	.8121	.3
.3	.5195	.6080	1.6447	.8545	.7	.8	.5850	.7212	1.3865	.8111	.2
.4	.5210	.6104	1.6383	.8536	58.6	.9	.5864	.7239	1.3814	.8100	54.1
Deg.	Cos	Cot	Tan	Sin	Deg.	Deg.	Cos	Cot	Tan	Sin	Deg.

Table A-2. Natural trigonometric functions for decimal fractions of a degree (*Continued*)

Deg.	Sin	Tan	Cot	Cos	Deg.	Deg.	Sin	Tan	Cot	Cos	Deg.
36.0	0.5878	0.7265	1.3764	0.8090	54.0	0.5	.6494	.8541	1.1708	.7604	0.5
.1	.5892	.7292	1.3713	.8080	.9	.6	.6508	.8571	1.1667	.7593	.4
.2	.5906	.7319	1.3663	.8070	.8	.7	.6521	.8601	1.1626	.7581	.3
.3	.5920	.7346	1.3613	.8059	.7	.8	.6534	.8632	1.1585	.7570	.2
.4	.5934	.7373	1.3564	.8049	.6	.9	.6547	.8662	1.1544	.7559	.1
.5	.5948	.7400	1.3514	.8039	.5	41.0	0.6561	0.8693	1.1504	0.7547	49.0
.6	.5962	.7427	1.3465	.8028	.4	.1	.6574	.8724	1.1463	.7536	.9
.7	.5976	.7454	1.3416	.8018	.3	.2	.6587	.8754	1.1423	.7524	.8
.8	.5990	.7481	1.3367	.8007	.2	.3	.6600	.8785	1.1383	.7513	.7
.9	.6004	.7508	1.3319	.7997	.1	.4	.6613	.8816	1.1343	.7501	.6
37.0	0.6018	0.7536	1.3270	0.7986	53.0	.5	.6626	.8847	1.1303	.7490	.5
.1	.6032	.7563	1.3222	.7976	.9	.6	.6639	.8878	1.1263	.7478	.4
.2	.6046	.7590	1.3175	.7965	.8	.7	.6652	.8910	1.1224	.7466	.3
.3	.6060	.7618	1.3127	.7955	.7	.8	.6665	.8941	1.1184	.7455	.2
.4	.6074	.7646	1.3079	.7944	.6	.9	.6678	.8972	1.1145	.7443	.1
.5	.6088	.7673	1.3032	.7934	.5	42.0	0.6691	0.9004	1.1106	0.7431	48.0
.6	.6101	.7701	1.2985	.7923	.4	.1	.6704	.9036	1.1067	.7420	.9
.7	.6115	.7729	1.2938	.7912	.3	.2	.6717	.9067	1.1028	.7408	.8
.8	.6129	.7757	1.2892	.7902	.2	.3	.6730	.9099	1.0990	.7396	.7
.9	.6143	.7785	1.2846	.7891	.1	.4	.6743	.9131	1.0951	.7385	.6
38.0	0.6157	0.7813	1.2799	0.7880	52.0	.5	.6756	.9163	1.0913	.7373	.5
.1	.6170	.7841	1.2753	.7869	.9	.6	.6769	.9195	1.0875	.7361	.4
.2	.6184	.7869	1.2708	.7859	.8	.7	.6782	.9228	1.0837	.7349	.3
.3	.6198	.7898	1.2662	.7848	.7	.8	.6794	.9260	1.0799	.7337	.2
.4	.6211	.7926	1.2617	.7837	.6	.9	.6807	.9293	1.0761	.7325	.1
.5	.6225	.7954	1.2572	.7826	.5	43.0	0.6820	0.9325	1.0724	0.7314	47.0
.6	.6239	.7983	1.2527	.7815	.4	.1	.6833	.9358	1.0686	.7302	.9
.7	.6252	.8012	1.2482	.7804	.3	.2	.6845	.9391	1.0649	.7290	.8
.8	.6266	.8040	1.2437	.7793	.2	.3	.6858	.9424	1.0612	.7278	.7
.9	.6280	.8069	1.2393	.7782	.1	.4	.6871	.9457	1.0575	.7266	.6
39.0	0.6293	0.8098	1.2349	0.7771	51.0	.5	.6884	.9490	1.0538	.7254	.5
.1	.6307	.8127	1.2305	.7760	.9	.6	.6896	.9523	1.0501	.7242	.4
.2	.6320	.8156	1.2261	.7749	.8	.7	.6909	.9556	1.0464	.7230	.3
.3	.6334	.8185	1.2218	.7738	.7	.8	.6921	.9590	1.0428	.7218	.2
.4	.6347	.8214	1.2174	.7727	.6	.9	.6934	.9623	1.0392	.7206	.1
.5	.6361	.8243	1.2131	.7716	.5	44.0	0.6947	0.9657	1.0355	0.7193	46.0
.6	.6374	.8273	1.2088	.7705	.4	.1	.6959	.9691	1.0319	.7181	.9
.7	.6388	.8302	1.2045	.7694	.3	.2	.6972	.9725	1.0283	.7169	.8
.8	.6401	.8332	1.2002	.7683	.2	.3	.6984	.9759	1.0247	.7157	.7
.9	.6414	.8361	1.1960	.7672	.1	.4	.6997	.9793	1.0212	.7145	.6
40.0	0.6428	0.8391	1.1918	0.7660	50.0	.5	.7009	.9827	1.0176	.7133	.5
.1	.6441	.8421	1.1875	.7649	.9	.6	.7022	.9861	1.0141	.7120	.4
.2	.6455	.8451	1.1833	.7638	.8	.7	.7034	.9896	1.0105	.7108	.3
.3	.6468	.8481	1.1792	.7627	.7	.8	.7046	.9930	1.0070	.7096	.2
.4	.6481	.8511	1.1750	.7615	49.6	.9	.7059	.9965	1.0035	.7083	.1
						45.0	0.7071	1.0000	1.0000	0.7071	45.0
Deg.	Cos	Cot	Tan	Sin	Deg.	Deg.	Cos	Cot	Tan	Sin	Deg.

Table A-3. Secants and cosecants for decimal fractions of a degree

Deg.	Sec	Csc	Deg.	Deg.	Sec	Csc	Deg.	Deg.	Sec	Csc	Deg.
0.0	1.0000		90.0	5.0	1.0038	11.474	85.0	10.0	1.0154	5.7588	80.0
.1	1.0000	572.96	.9	.1	1.0040	11.249	.9	.1	1.0157	5.7023	.9
.2	1.0000	286.48	.8	.2	1.0041	11.034	.8	.2	1.0161	5.6470	.8
.3	1.0000	190.99	.7	.3	1.0043	10.826	.7	.3	1.0164	5.5928	.7
.4	1.0000	143.24	.6	.4	1.0045	10.626	.6	.4	1.0167	5.5396	.6
.5	1.0000	114.59	.5	.5	1.0046	10.433	.5	.5	1.0170	5.4874	.5
.6	1.0001	95.495	.4	.6	1.0048	10.248	.4	.6	1.0174	5.4362	.4
.7	1.0001	81.853	.3	.7	1.0050	10.068	.3	.7	1.0177	5.3860	.3
.8	1.0001	71.622	.2	.8	1.0051	9.8955	.2	.8	1.0180	5.3367	.2
.9	1.0001	63.665	.1	.9	1.0053	9.7283	.1	.9	1.0184	5.2883	.1
1.0	1.0002	57.299	89.0	6.0	1.0055	9.5668	84.0	11.0	1.0187	5.2408	79.0
.1	1.0002	52.090	.9	.1	1.0057	9.4105	.9	.1	1.0191	5.1942	.9
.2	1.0002	47.750	.8	.2	1.0059	9.2593	.8	.2	1.0194	5.1484	.8
.3	1.0003	44.077	.7	.3	1.0061	9.1129	.7	.3	1.0198	5.1034	.7
.4	1.0003	40.930	.6	.4	1.0063	8.9711	.6	.4	1.0201	5.0593	.6
.5	1.0003	38.202	.5	.5	1.0065	8.8337	.5	.5	1.0205	5.0159	.5
.6	1.0004	38.815	.4	.6	1.0067	8.7004	.4	.6	1.0209	4.9732	.4
.7	1.0004	33.708	.3	.7	1.0069	8.5711	.3	.7	1.0212	4.9313	.3
.8	1.0005	31.836	.2	.8	1.0071	8.4457	.2	.8	1.0216	4.8901	.2
.9	1.0006	30.161	.1	.9	1.0073	8.3238	.1	.9	1.0220	4.8496	.1
2.0	1.0006	28.654	88.0	7.0	1.0075	8.2055	83.0	12.0	1.0223	4.8097	78.0
.1	1.0007	27.290	.9	.1	1.0077	8.0905	.9	.1	1.0227	4.7706	.9
.2	1.0007	26.050	.8	.2	1.0079	7.9787	.8	.2	1.0231	4.7321	.8
.3	1.0008	24.918	.7	.3	1.0082	7.8700	.7	.3	1.0235	4.6942	.7
.4	1.0009	23.880	.6	.4	1.0084	7.7642	.6	.4	1.0239	4.6569	.6
.5	1.0010	22.926	.5	.5	1.0086	7.6613	.5	.5	1.0243	4.6202	.5
.6	1.0010	22.044	.4	.6	1.0089	7.5611	.4	.6	1.0247	4.5841	.4
.7	1.0011	21.229	.3	.7	1.0091	7.4635	.3	.7	1.0251	4.5486	.3
.8	1.0012	20.471	.2	.8	1.0093	7.3684	.2	.8	1.0255	4.5137	.2
.9	1.0013	19.766	.1	.9	1.0096	7.2757	.1	.9	1.0259	4.4793	.1
3.0	1.0014	19.107	87.0	8.0	1.0098	7.1853	82.0	13.0	1.0263	4.4454	77.0
.1	1.0015	18.492	.9	.1	1.0101	7.0972	.9	.1	1.0267	4.4121	.9
.2	1.0016	17.914	.8	.2	1.0103	7.0112	.8	.2	1.0271	4.3792	.8
.3	1.0017	17.372	.7	.3	1.0106	6.9273	.7	.3	1.0276	4.3469	.7
.4	1.0018	16.862	.6	.4	1.0108	6.8454	.6	.4	1.0280	4.3150	.6
.5	1.0019	16.380	.5	.5	1.0111	6.7655	.5	.5	1.0284	4.2837	.5
.6	1.0020	15.926	.4	.6	1.0114	6.6874	.4	.6	1.0288	4.2527	.4
.7	1.0021	15.496	.3	.7	1.0116	6.6111	.3	.7	1.0293	4.2223	.3
.8	1.0022	15.089	.2	.8	1.0119	6.5366	.2	.8	1.0297	4.1923	.2
.9	1.0023	14.703	.1	.9	1.0122	6.4637	.1	.9	1.0302	4.1627	.1
4.0	1.0024	14.336	86.0	9.0	1.0125	6.3925	81.0	14.0	1.0306	4.1336	76.0
.1	1.0026	13.987	.9	.1	1.0127	6.3228	.9	.1	1.0311	4.1048	.9
.2	1.0027	13.654	.8	.2	1.0130	6.2546	.8	.2	1.0315	4.0765	.8
.3	1.0028	13.337	.7	.3	1.0133	6.1880	.7	.3	1.0320	4.0486	.7
.4	1.0030	13.035	.6	.4	1.0136	6.1227	.6	.4	1.0324	4.0211	.6
.5	1.0031	12.745	.5	.5	1.0139	6.0589	.5	.5	1.0329	3.9939	.5
.6	1.0032	12.469	.4	.6	1.0142	5.9963	.4	.6	1.0334	3.9672	.4
.7	1.0034	12.204	.3	.7	1.0145	5.9351	.3	.7	1.0338	3.9408	.3
.8	1.0035	11.951	.2	.8	1.0148	5.8751	.2	.8	1.0343	3.9147	.2
.9	1.0037	11.707	.1	.9	1.0151	5.8164	.1	.9	1.0348	3.8890	.1
5.0	1.0038	11.474	85.0	10.0	1.0154	5.7588	80.0	15.0	1.0353	3.8637	75.0
Deg.	Csc	Sec	Deg.	Deg.	Csc	Sec	Deg.	Deg.	Csc	Sec	Deg.

Table A-3. Secants and cosecants for decimal fractions of a degree (*Continued*)

Deg.	Sec	Csc	Deg.	Deg.	Sec	Csc	Deg.	Deg.	Sec	Csc	Deg.
15.0	1.0353	3.8637	75.0	20.0	1.0642	2.9238	70.0	25.0	1.1034	2.3662	65.0
.1	1.0358	3.8387	.9	0.1	1.0649	2.9099	.9	.1	1.1043	2.3574	.9
.2	1.0363	3.8140	.8	0.2	1.0655	2.8960	.8	.2	1.1052	2.3486	.8
.3	1.0367	3.7897	.7	0.3	1.0662	2.8824	.7	.3	1.1061	2.3400	.7
.4	1.0372	3.7657	.6	0.4	1.0669	2.8688	.6	.4	1.1070	2.3314	.6
.5	1.0377	3.7420	.5	0.5	1.0676	2.8555	.5	.5	1.1079	2.3228	.5
.6	1.0382	3.7186	.4	0.6	1.0683	2.8422	.4	.6	1.1089	2.3144	.4
.7	1.0388	3.6955	.3	0.7	1.0690	2.8291	.3	.7	1.1098	2.3060	.3
.8	1.0393	3.6727	.2	0.8	1.0697	2.8161	.2	.8	1.1107	2.2976	.2
.9	1.0398	3.6502	.1	0.9	1.0704	2.8032	.1	.9	1.1117	2.2894	.1
16.0	1.0403	3.6280	74.0	21.0	1.0711	2.7904	69.0	26.0	1.1126	2.2812	64.0
.1	1.0408	3.6060	.9	0.1	1.0719	2.7778	.9	.1	1.1136	2.2730	.9
.2	1.0413	3.5843	.8	0.2	1.0726	2.7653	.8	.2	1.1145	2.2650	.8
.3	1.0419	3.5629	.7	0.3	1.0733	2.7529	.7	.3	1.1155	2.2570	.7
.4	1.0424	3.5418	.6	0.4	1.0740	2.7407	.6	.4	1.1164	2.2490	.6
.5	1.0429	3.5209	.5	0.5	1.0748	2.7285	.5	.5	1.1174	2.2412	.5
.6	1.0435	3.5003	.4	0.6	1.0755	2.7165	.4	.6	1.1184	2.2333	.4
.7	1.0440	3.4799	.3	0.7	1.0763	2.7046	.3	.7	1.1194	2.2256	.3
.8	1.0446	3.4598	.2	0.8	1.0770	2.6927	.2	.8	1.1203	2.2179	.2
.9	1.0451	3.4399	.1	0.9	1.0778	2.6811	.1	.9	1.1213	2.2103	.1
17.0	1.0457	3.4203	73.0	22.0	1.0785	2.6695	68.0	27.0	1.1223	2.2027	63.0
.1	1.0463	3.4009	.9	0.1	1.0793	2.6580	.9	.1	1.1233	2.1952	.9
.2	1.0468	3.3817	.8	0.2	1.0801	2.6466	.8	.2	1.1243	2.1877	.8
.3	1.0474	3.3628	.7	0.3	1.0808	2.6354	.7	.3	1.1253	2.1803	.7
.4	1.0480	3.3440	.6	0.4	1.0816	2.6242	.6	.4	1.1264	2.1730	.6
.5	1.0485	3.3255	.5	0.5	1.0824	2.6131	.5	.5	1.1274	2.1657	.5
.6	1.0491	3.3072	.4	0.6	1.0832	2.6022	.4	.6	1.1284	2.1584	.4
.7	1.0497	3.2891	.3	0.7	1.0840	2.5913	.3	.7	1.1294	2.1513	.3
.8	1.0503	3.2712	.2	0.8	1.0848	2.5805	.2	.8	1.1305	2.1441	.2
.9	1.0509	3.2535	.1	0.9	1.0856	2.5699	.1	.9	1.1315	2.1371	.1
18.0	1.0515	3.2361	72.0	23.0	1.0864	2.5593	67.0	28.0	1.1326	2.1301	62.0
.1	1.0521	3.2188	.9	0.1	1.0872	2.5488	.9	.1	1.1336	2.1231	.9
.2	1.0527	3.2017	.8	0.2	1.0880	2.5384	.8	.2	1.1347	2.1162	.8
.3	1.0533	3.1848	.7	0.3	1.0888	2.5282	.7	.3	1.1357	2.1093	.7
.4	1.0539	3.1681	.6	0.4	1.0896	2.5180	.6	.4	1.1368	2.1025	.6
.5	1.0545	3.1515	.5	0.5	1.0904	2.5078	.5	.5	1.1379	2.0957	.5
.6	1.0551	3.1352	.4	0.6	1.0913	2.4978	.4	.6	1.1390	2.0890	.4
.7	1.0557	3.1190	.3	0.7	1.0921	2.4879	.3	.7	1.1401	2.0824	.3
.8	1.0564	3.1030	.2	0.8	1.0929	2.4780	.2	.8	1.1412	2.0757	.2
.9	1.0570	3.0872	.1	0.9	1.0938	2.4683	.1	.9	1.1423	2.0692	.1
19.0	1.0576	3.0716	71.0	24.0	1.0946	2.4586	66.0	29.0	1.1434	2.0627	61.0
.1	1.0583	3.0561	.9	0.1	1.0955	2.4490	.9	.1	1.1445	2.0562	.9
.2	1.0589	3.0407	.8	0.2	1.0963	2.4395	.8	.2	1.1456	2.0498	.8
.3	1.0595	3.0256	.7	0.3	1.0972	2.4300	.7	.3	1.1467	2.0434	.7
.4	1.0602	3.0106	.6	0.4	1.0981	2.4207	.6	.4	1.1478	2.0371	.6
.5	1.0608	2.9957	.5	0.5	1.0989	2.4114	.5	.5	1.1490	2.0308	.5
.6	1.0615	2.9811	.4	0.6	1.0998	2.4022	.4	.6	1.1501	2.0245	.4
.7	1.0622	2.9665	.3	0.7	1.1007	2.3931	.3	.7	1.1512	2.0183	.3
.8	1.0628	2.9521	.2	0.8	1.1016	2.3841	.2	.8	1.1524	2.0122	.2
.9	1.0635	2.9379	.1	0.9	1.1025	2.3751	.1	.9	1.1535	2.0061	.1
20.0	1.0642	2.9238	70.0	25.0	1.1034	2.3662	65.0	30.0	1.1547	2.0000	60.0
Deg.	Csc	Sec	Deg.	Deg.	Csc	Sec	Deg.	Deg.	Csc	Sec	Deg.

Table A-3. Secants and cosecants for decimal fractions of a degree (*Continued*)

Deg.	Sec	Csc	Deg.	Deg.	Sec	Csc	Deg.	Deg.	Sec	Csc	Deg.
30.0	1.1547	2.0000	60.0	35.0	1.2208	1.7434	55.0	40.0	1.3054	1.5557	50.0
.1	1.1559	1.9940	.9	.1	1.2223	1.7391	.9	.1	1.3073	1.5525	.9
.2	1.1570	1.9880	.8	.2	1.2238	1.7348	.8	.2	1.3093	1.5493	.8
.3	1.1582	1.9821	.7	.3	1.2253	1.7305	.7	.3	1.3112	1.5461	.7
.4	1.1594	1.9762	.6	.4	1.2268	1.7263	.6	.4	1.3131	1.5429	.6
.5	1.1606	1.9703	.5	.5	1.2283	1.7221	.5	.5	1.3151	1.5398	.5
.6	1.1618	1.9645	.4	.6	1.2299	1.7179	.4	.6	1.3171	1.5366	.4
.7	1.1630	1.9587	.3	.7	1.2314	1.7137	.3	.7	1.3190	1.5335	.3
.8	1.1642	1.9530	.2	.8	1.2329	1.7095	.2	.8	1.3210	1.5304	.2
.9	1.1654	1.9473	.1	.9	1.2345	1.7054	.1	.9	1.3230	1.5273	.1
31.0	1.1666	1.9416	59.0	36.0	1.2361	1.7013	54.0	41.0	1.3250	1.5243	49.0
.1	1.1679	1.9360	.9	.1	1.2376	1.6972	.9	.1	1.3270	1.5212	.9
.2	1.1691	1.9304	.8	.2	1.2392	1.6932	.8	.2	1.3291	1.5182	.8
.3	1.1703	1.9249	.7	.3	1.2408	1.6892	.7	.3	1.3311	1.5151	.7
.4	1.1716	1.9194	.6	.4	1.2424	1.6852	.6	.4	1.3331	1.5121	.6
.5	1.1728	1.9139	.5	.5	1.2440	1.6812	.5	.5	1.3352	1.5092	.5
.6	1.1741	1.9084	.4	.6	1.2456	1.6772	.4	.6	1.3373	1.5062	.4
.7	1.1753	1.9031	.3	.7	1.2472	1.6733	.3	.7	1.3393	1.5032	.3
.8	1.1766	1.8977	.2	.8	1.2489	1.6694	.2	.8	1.3414	1.5003	.2
.9	1.1779	1.8924	.1	.9	1.2505	1.6655	.1	.9	1.3435	1.4974	.1
32.0	1.1792	1.8871	58.0	37.0	1.2521	1.6616	53.0	42.0	1.3456	1.4945	48.0
.1	1.1805	1.8818	.9	.1	1.2538	1.6578	.9	.1	1.3478	1.4916	.9
.2	1.1818	1.8766	.8	.2	1.2554	1.6540	.8	.2	1.3499	1.4887	.8
.3	1.1831	1.8714	.7	.3	1.2571	1.6502	.7	.3	1.3520	1.4859	.7
.4	1.1844	1.8663	.6	.4	1.2588	1.6464	.6	.4	1.3542	1.4830	.6
.5	1.1857	1.8612	.5	.5	1.2605	1.6427	.5	.5	1.3563	1.4802	.5
.6	1.1870	1.8561	.4	.6	1.2622	1.6390	.4	.6	1.3585	1.4774	.4
.7	1.1883	1.8510	.3	.7	1.2639	1.6353	.3	.7	1.3607	1.4746	.3
.8	1.1897	1.8460	.2	.8	1.2656	1.6316	.2	.8	1.3629	1.4718	.2
.9	1.1910	1.8410	.1	.9	1.2673	1.6279	.1	.9	1.3651	1.4690	.1
33.0	1.1924	1.8361	57.0	38.0	1.2690	1.6243	52.0	43.0	1.3673	1.4663	47.0
.1	1.1937	1.8312	.9	.1	1.2708	1.6207	.9	.1	1.3696	1.4635	.9
.2	1.1951	1.8263	.8	.2	1.2725	1.6171	.8	.2	1.3718	1.4608	.8
.3	1.1964	1.8214	.7	.3	1.2742	1.6135	.7	.3	1.3741	1.4581	.7
.4	1.1978	1.8166	.6	.4	1.2760	1.6099	.6	.4	1.3763	1.4554	.6
.5	1.1992	1.8118	.5	.5	1.2778	1.6064	.5	.5	1.3786	1.4527	.5
.4	1.2006	1.8070	.4	.6	1.2796	1.6029	.4	.6	1.3809	1.4501	.4
.7	1.2020	1.8023	.3	.7	1.2813	1.5994	.3	.7	1.3832	1.4474	.3
.8	1.2034	1.7976	.2	.8	1.2831	1.5959	.2	.8	1.3855	1.4448	.2
.9	1.2048	1.7929	.1	.9	1.2849	1.5925	.1	.9	1.3878	1.4422	.1
34.0	1.2062	1.7883	56.0	39.0	1.2868	1.5890	51.0	44.0	1.3902	1.4396	46.0
.1	1.2076	1.7837	.9	.1	1.2886	1.5856	.9	.1	1.3925	1.4370	.9
.2	1.2091	1.7791	.8	.2	1.2904	1.5822	.8	.2	1.3949	1.4344	.8
.3	1.2105	1.7745	.7	.3	1.2923	1.5788	.7	.3	1.3972	1.4318	.7
.4	1.2120	1.7700	.6	.4	1.2941	1.5755	.6	.4	1.3996	1.4293	.6
.5	1.2134	1.7655	.5	.5	1.2960	1.5721	.5	.5	1.4020	1.4267	.5
.6	1.2149	1.7610	.4	.6	1.2978	1.5688	.4	.6	1.4044	1.4242	.4
.7	1.2163	1.7566	.3	.7	1.2997	1.5655	.3	.7	1.4069	1.4217	.3
.8	1.2178	1.7522	.2	.8	1.3016	1.5622	.2	.8	1.4093	1.4192	.2
.9	1.2193	1.7478	.1	.9	1.3035	1.5590	.1	.9	1.4118	1.4167	.1
35.0	1.2208	1.7434	55.0	40.0	1.3054	1.5557	50.0	45.0	1.4142	1.4142	45.0
Deg.	Csc	Sec	Deg.	Deg.	Csc	Sec	Deg.	Deg.	Csc	Sec	Deg.

Table A-4. Squares, square roots, cubes, cube roots

No.	Square	Square root	Cube	Cube root	No.	Square	Square root	Cube	Cube Root
0	0	0.000	0	0.000	50	2500	7.071	125,000	3.684
1	1	1.000	1	1.000	51	2601	7.141	132,651	3.708
2	4	1.414	8	1.260	52	2704	7.211	140,608	3.732
3	9	1.732	27	1.442	53	2809	7.280	148,877	3.756
4	16	2.000	64	1.587	54	2916	7.348	157,464	3.780
5	25	2.236	125	1.710	55	3025	7.416	166,375	3.803
6	36	2.449	216	1.817	56	3136	7.483	175,616	3.826
7	49	2.646	343	1.913	57	3249	7.550	185,193	3.849
8	64	2.828	512	2.000	58	3364	7.616	195,112	3.871
9	81	3.000	729	2.080	59	3481	7.681	205,379	3.893
10	100	3.162	1000	2.154	60	3600	7.746	216,000	3.915
11	121	3.317	1331	2.224	61	3721	7.810	226,981	3.936
12	144	3.464	1728	2.289	62	3844	7.874	238,328	3.958
13	169	3.606	2197	2.351	63	3969	7.937	250,047	3.979
14	196	3.742	2744	2.410	64	4096	8.000	262,144	4.000
15	225	3.873	3375	2.466	65	4225	8.062	274,625	4.021
16	256	4.000	4096	2.520	66	4356	8.124	287,496	4.041
17	289	4.123	4913	2.571	67	4489	8.185	300,763	4.062
18	324	4.243	5832	2.621	68	4624	8.246	314,432	4.082
19	361	4.359	6859	2.668	69	4761	8.307	328,509	4.102
20	400	4.472	8000	2.714	70	4900	8.367	343,000	4.121
21	441	4.583	9261	2.759	71	5041	8.426	357,911	4.141
22	484	4.690	10,648	2.802	72	5184	8.485	373,248	4.160
23	529	4.796	12,167	2.844	73	5329	8.544	389,017	4.179
24	576	4.899	13,824	2.884	74	5476	8.602	405,224	4.198
25	625	5.000	15,625	2.924	75	5625	8.660	421,875	4.217
26	676	5.099	17,576	2.962	76	5776	8.718	438,976	4.236
27	729	5.196	19,683	3.000	77	5929	8.775	456,533	4.254
28	784	5.292	21,952	3.037	78	6084	8.832	474,552	4.273
29	841	5.385	24,389	3.072	79	6241	8.888	493,039	4.291
30	900	5.477	27,000	3.107	80	6400	8.944	512,000	4.309
31	961	5.568	29,791	3.141	81	6561	9.000	531,441	4.327
32	1024	5.657	32,768	3.175	82	6724	9.055	551,368	4.344
33	1089	5.745	35,937	3.208	83	6889	9.110	571,787	4.362
34	1156	5.831	39,304	3.240	84	7056	9.165	592,704	4.380
35	1225	5.916	42,875	3.721	85	7225	9.220	614,125	4.397
36	1296	6.000	46,656	3.302	86	7396	9.274	636,056	4.414
37	1369	6.083	50,653	3.332	87	7569	9.327	658,503	4.431
38	1444	6.164	54,872	3.362	88	7744	9.381	681.472	4.448
39	1521	6.245	59,319	3.391	89	7921	9.434	704,969	4.465
40	1600	6.325	64,000	3.420	90	8100	9.487	729,000	4.481
41	1681	6.403	68,921	3.448	91	8281	9.539	753,571	4.498
42	1764	6.481	74,088	3.476	92	8464	9.592	778,688	4.514
43	1849	6.557	79,507	3.503	93	8649	9.644	804,357	4.531
44	1936	6.633	85,184	3.530	94	8836	9.695	830,584	4.547
45	2025	6.708	91,125	3.557	95	9025	9.747	857,375	4.563
46	2116	6.782	97,336	3.583	96	9216	9.798	884,736	4.579
47	2209	6.856	103,823	3.609	97	9409	9.849	912,673	4.595
48	2304	6.928	110,592	3.634	98	9604	9.899	941,192	4.610
49	2401	7.000	117,649	3.659	99	9801	9.950	970,299	4.626
50	2500	7.071	125,000	3.684	100	10,000	10.000	1,000,000	4.642

Answers to Selected Odd-Numbered Exercises

Exercise 1.1

1. 652, 1475; 1610; 1013; 1225; 808; 1231 **3.** 3674; 3805; 3863; 4158; 3573; 3879; 3601 **5.** 22610 **7.** 18520 **9.** 347764 **11.** 3146; 6831; 231; 1285; 3164 **13.** 3741; 949; 453; 2191; 267 **15.** 17 **17.** 68 **19.** 63 **21.** 32

Exercise 1.2

1. 140777; 344568; 287153; 428060; 450148 **3.** 3640472; 5498898; 4542531; 6285780; 4636989 **5.** 222222; 888888; 1244421; 44435556; 1030304 **7.** 30134 **9.** 10472 **11.** 100344 **13.** 104830 **15.** 437 **17.** 619 **19.** 762 **21.** 784 **23.** 568 **25.** 508 **27.** 706 **29.** 2909 (rem. 11) **31.** 12345679 **33.** 6247 **35.** 12960 ft **37.** 5,865,696,000,000 miles **39.** 7,500,000 **41.** 2100 tons **43.** 351 ohms **45.** 62 kwh

Exercise 2.1

1. (a) **3.** (a) **5.** (b) **7.** (a) **9.** (b) **11.** (b) **13.** (b) **15.** (a)

Exercise 2.2

1. $\frac{17}{3}$ **3.** $\frac{19}{2}$ **5.** $\frac{83}{5}$ **7.** $\frac{95}{6}$ **9.** $\frac{146}{3}$ **11.** $\frac{100}{7}$ **13.** $\frac{50}{3}$ **15.** $\frac{200}{3}$ **17.** $\frac{80}{9}$ **19.** $\frac{183}{32}$ **21.** $4\frac{3}{4}$ **23.** $8\frac{3}{5}$ **25.** $6\frac{5}{6}$ **27.** $5\frac{10}{11}$ **29.** $7\frac{4}{9}$ **31.** $6\frac{8}{13}$ **33.** $53\frac{1}{6}$ **35.** $66\frac{6}{7}$ **37.** 33 **39.** $312\frac{5}{16}$

Exercise 2.3

1. $\frac{1}{3}$ **3.** $\frac{5}{7}$ **5.** $\frac{3}{7}$ **7.** $\frac{7}{11}$ **9.** $\frac{3}{7}$ **11.** $\frac{7}{9}$ **13.** $\frac{5}{8}$ **15.** $\frac{5}{7}$ **17.** $\frac{4}{9}$ **19.** $\frac{3}{13}$ **21.** $\frac{3}{4}$ **23.** $\frac{5}{16}$ **25.** $\frac{5}{6}$ **27.** $\frac{3}{8}$ **29.** $\frac{5}{12}$ **31.** $\frac{21}{22}$ **33.** $\frac{18}{24}$; $\frac{16}{24}$; $\frac{9}{24}$; $\frac{14}{24}$ **35.** $\frac{70}{105}$; $\frac{90}{105}$; $\frac{84}{105}$; $\frac{56}{105}$

Exercise 2.4

1. 24 **3.** 60 **5.** 60 **7.** 36 **9.** 420 **11.** 1260 **13.** 3600 **15.** 5040 **17.** 5400 **19.** (a) 68; $49\frac{5}{16}$; $64\frac{1}{2}$; $108\frac{4}{5}$; $79\frac{5}{18}$; $90\frac{3}{5}$; $113\frac{1}{7}$; (b) $29\frac{1}{4}$; $14\frac{15}{16}$; $17\frac{11}{14}$; $57\frac{1}{5}$; $29\frac{11}{18}$; $34\frac{3}{5}$; 62 **21.** $13\frac{17}{42}$ **23.** $8\frac{43}{48}$ **25.** $8\frac{47}{72}$ **27.** $\frac{13}{40}$ **29.** $4\frac{5}{16}$ **31.** $8\frac{5}{12}$ **33.** $39\frac{19}{36}$ **35.** $43\frac{5}{12}$ **37.** $5\frac{4}{105}$ **39.** $9\frac{2}{15}$ **41.** $2\frac{16}{21}$ **43.** $2\frac{53}{72}$ **45.** $1\frac{3}{10}$

Exercise 2.5

1. $\frac{8}{35}$ **3.** $1\frac{5}{7}$ **5.** 8 **7.** $63\frac{3}{8}$ **9.** $8\frac{4}{5}$ **11.** $46\frac{1}{2}$ **13.** $86\frac{1}{3}$ **15.** $62\frac{3}{14}$ **17.** $1\frac{1}{5}$ **19.** $1\frac{1}{6}$

21. $6\frac{1}{2}$ **23.** $3\frac{7}{60}$ **25.** $\frac{25}{42}$ **27.** $127\frac{1}{2}$ **29.** $9\frac{89}{99}$ **31.** $\frac{14}{129}$ **33.** $1727\frac{509}{720}$ **35.** $81\frac{459}{700}$ **37.** $52\frac{436}{567}$ **39.** $1842\frac{17}{24}$ **41.** $\frac{9}{140}$ **43.** $5\frac{1}{4}$ **45.** $\frac{721}{1438}$ **47.** 800 **49.** $21\frac{7}{8}$ **51.** $4\frac{7}{8}$ **53.** $18\frac{3}{8}$ **55.** 5

Exercise 3.1

1. 0.5 **3.** 0.625 **5.** 0.3636... **7.** 3.1875 **9.** 6.875 **11.** 8.4 **13.** 7.833... **15.** 1.09375 **17.** $9.888\frac{8}{9}$ **19.** $8.666\frac{2}{3}$ **21.** 9.15625 **23.** 1.28125 **25.** 8.4375 **27.** $\frac{1}{4}$ **29.** $\frac{4}{25}$ **31.** $1\frac{1}{5}$ **33.** $\frac{1}{3}$ **35.** $8\frac{1}{20}$ **37.** $3\frac{1}{40}$ **39.** $15\frac{3}{40}$ **41.** $5\frac{3}{80}$ **43.** $8\frac{1}{200}$ **45.** $\frac{601}{2000}$ **47.** $11\frac{2}{25}$ **49.** $\frac{1}{700}$

Exercise 3.2

1, 1516.97 **3.** 295.535 **5.** 80683.18 **7.** 188.587 **9.** 758.9838 **11.** 41.06 **13.** 3550.53 **15.** 709.815 **17.** 188.968 **19.** 1.12086 **21.** 2.75 **23.** 1.61342 **25.** 0.93973 **27.** 0.57303 **29.** 1.36175 **31.** 221.6 mi

Exercise 3.3

1. 473530; 473500; 474000; 470000 **3.** 5081400; 5081000; 5080000; 5100000 **5.** 852.92; 852.9; 853; 850 **7.** 49.250; 49.25; 49.3; 49 **9.** 0.17397; 0.1740; 0.174; 0.17 **11.** 488540; 488500; 489000; 490000 **13.** 9270400; 9270000; 9270000; 9300000 **15.** 3.2581; 3.258; 3.26; 3.3 **17.** 0.0058404; 0.005840; 0.00584; 0.0058 **19.** 57.005; 57.00; 57.0; 57

Exercise 3.4

1. 440.16 **3.** 8.1528 **5.** 5.8875 **7.** 0.65504 **9.** 2.9172 **11.** 0.0015822 **13.** 0.85695 **15.** 28.771 **17.** 284.71 **19.** 0.058366 **21.** 0.043641 **23.** 1012.9 **25.** 0.080938 **27.** 0.0025787 **29.** 3622.6 **31.** 0.0084641 **33.** 747890 **35.** 1.3709 **37.** 29.633 **39.** 680.85 **41.** 0.000069191

Exercise 3.5 (No answers shown for Nos. 1–44)

45. 42.955 **47.** 0.036641

Exercise 3.6

1. 34.6; 41.3; 25.9; 58.3; 63.3; total: 223.4; Avg: 15.62 mi; $9.95 **3.** 25.18 **5.** 72.3 ft; 316 ft^2 **7.** 253.8 ft^2; 64.6 ft **9.** 4.6406 **11.** 0.14 in. **13.** 15.335 **15.** 10.88 **17.** 5.035 **19.** $19\frac{3}{4}$ **21.** 15 **23.** 113.64 **25.** 50.88 **27.** 308 ohms

Exercise 4.1

1. $\frac{7}{20}$; 0.35 **3.** $\frac{7}{160}$; 0.04375 **5.** $\frac{1}{800}$; 0.00125 **7.** $\frac{3}{8}$; 0.375 **9.** $\frac{1}{7}$; 0.14286 **11.** $1\frac{3}{4}$; 1.75 **13.** 10; 10 **15.** $\frac{1}{3}$; 0.3333 **17.** $\frac{3}{160}$; 0.01875 **19.** $\frac{1}{75}$; 0.01333 **21.** 80% **23.** 10.9375% **25.** 562.5% **27.** 2.5% **29.** 145.3% **31.** $66\frac{2}{3}$% **33.** $355\frac{5}{9}$% **35.** 510% **37.** 0.05% **39.** 60%

Exercise 4.2

1. $840 **3.** 335.4 miles **5.** $155.54 **7.** $561 **9.** $79.20 **11.** $0.36

13. 960 lb **15.** \$45500 **17.** \$1200 **19.** \$55,775 **21.** \$14,700; \$20,580; \$48,720 **23.** 7540 bushels

Exercise 4.3

1. 26.7% **3.** 32% **5.** 17.5% **7.** 2.19% **9.** 0.56% **11.** 150% **13.** 120% **15.** 240% **17.** 20.9%; 17.2%; 10.2%; 15.4%; 10.9%; 10.6%; 12.7%; 2.1%

Exercise 4.4

1. \$63.40 **3.** \$68.00 **5.** \$276.00 **7.** 340 lb **9.** \$764.00 **11.** \$435.00 **13.** \$1240 **15.** \$185 **17.** \$16,338

Exercise 4.5

1. \$8.50 **3.** 315 **5.** 15% **7.** \$620.00 **9.** 137.5% **11.** \$2394; \$2268; \$1890; \$1323; \$1071; \$1222.20; \$1512; \$919.80 **13.** 39,928 **15.** 5.5% **17.** $16\frac{2}{3}$%; $14\frac{2}{7}$% **19.** \$90; \$72 **21.** A: 0.289; B: 0.286 **23.** 0.0032% **25.** 226.2 **27.** 155.6 **29.** 105.1 **31.** 206.5; 131.5 **33.** 3780; 3220

Exercise 5.1

1. 235 **3.** 74.31 **5.** 42.7 **7.** 1.63 **9.** 0.403 **11.** 3.541 **13.** 19.71 **15.** 93.20 **17.** 8.36 **19.** 65.21 **21.** 5.541 **23.** 0.9354 **25.** 0.07797 **27.** 0.05727 **29.** 71.30 **31.** 30.60 **33.** 4.243 **35.** 3.464 **37.** 10.39 **39.** 4.899 **41.** 11.18 **43.** 10 **45.** 22.9 **47.** 26.83 **49.** 72.73

Exercise 6.1

1. $a + b + c$ **3.** $3x + 4y$ **5.** $7 + x$ **7.** $y - 8$ **9.** $15 + n$ **11.** $10 - x$ **13.** $a - b$ **15.** $\frac{m}{n}$ **17.** $\frac{(x + y)}{xy}$ **19.** $(a + b) - (x + y)$ **21.** $3(h - k)$ **23.** $(4a)(7b)$

Exercise 6.2

1. 59; 41; −62; 32; −25; −59 **3.** 4; −32; 16; −1; −43; 0 **5.** −57.604; 3.276; −57.6; −53.01; −41.52; 0.065

Exercise 6.3

1. 14; 3; 5; 64; −73; −27 **3.** 119; −91; −133; 0; 25; −74 **5.** −72; 212; −322; 322; −398; 1000 **7.** 1.72; 4.128; 6.289; −8.897; −6.463; −5.5322

Exercise 6.4

1. 12 **3.** 1 **5.** −16 **7.** −17 **9.** 9 **11.** −35 **13.** 46 **15.** 148

Exercise 6.5

1. −84 **3.** 40 **5.** −18.06 **7.** 2.88 **9.** −30.24 **11.** −96 **13.** −45 **15.** −60

17. $-\frac{32}{3}$ **19.** -180 **21.** 3402; 3844; -2254; 912; -0.00936; -2.4544; -0.015385

Exercise 6.6

1. 5 **3.** -8 **5.** 5 **7.** -16 **9.** -15 **11.** $\frac{2}{3}$ **13.** -2.6 **15.** $\frac{4}{3}$ **17.** $-\frac{11}{14}$ **19.** 117; -14; -23; 17; $+0.02$ **21.** $-\frac{7}{3}$; -1.5; 0.00211; -0.8; -224 **23.** -14 yd; -2 yd

Exercise 7.1

1. Adding: $8x$; $-11x^2$; $-st$; $2n$; $-3xy$; $4xy - 3x$; $-7x^2y - xy$; Subtracting: $-2x$; $3x^2$; $-9st$; $8n$; $3xy$; $4xy + 3x$; $-7x^2y + xy$ **3.** Adding: $8xy^3$; $-11st^2$; $-xy$; 0; $-m^2n$; $-18x^2y$; $-12xy - 12xy^2$; Subtracting: $-6xy^3$; $5st^2$; $-13xy$; $-6xyz$; $-7m^2n$; 0; $-12xy + 12xy^2$ **5.** 0 **7.** $-x^2y^3$ **9.** $6a + 6b + 7c - ab$ **11.** $5x + 6y - z + 9$ **13.** $y^2 - 2xy$ **15.** $2x^2 - 5xy + 11y^2$ **17.** $6x^2 - 2xy + 3y^2$ **19.** $4x^2 + 5nx - 6n^2$ **21.** $y^2 + 8xy - x^2 + 5x$ **23.** $-2x^2 + 3xy - 2y^2$ **25.** $5x^3 - x^2 + 9x$

Exercise 8.1

1. $2^3x^4y^2$ **3.** 5^4x^3yz **5.** $10^4x^2y^2z^5$ **7.** $6x^2$ **9.** $-6x^3y^4$ **11.** $-7ax^4y^3$ **13.** $18x^9$ **15.** $15abcd$ **17.** $-60x^6y^4z^6$ **19.** $96a^3m^4n^6$ **21.** $189x^6y^3z^8$ **23.** $-a^5b^4c^3$ **25.** $720abcde$ **27.** $12x^3 + 15x^2y - 6xy^2$ **29.** $8x^3yz - 10x^2y^2z - 14xy^3z$ **31.** $8a^4c^2 - 12a^3bc^3 + 20a^2b^2c^4 - 12ab^3c^6$ **33.** $8x^4y - 10x^3y^2 + 8x^2y^3 - 2xy^4 + 4x^2y + 2xy$ **35.** $-10x^5yz^4$ **37.** $2my$ **39.** -1 **41.** $-7x^2z$ **43.** $-6x^2y^3 + 5y^2 - 4xy$ **45.** $4r^2s^2t^3 - 3rs + 2t^2 - \frac{5}{2}r^3s^3t^4$

Exercise 8.2

1. $6x^2 - xy - 15y^2$ **3.** $6x^3 - 7x^2y + y^3$ **5.** $2x^3 - 2x^2y - xy^2 - 14x^2 + 14xy + 7y^2$ **7.** $x^4 + 6x^2y^2 + y^4 - 4x^3y - 4xy^3$ **9.** $6x^3 - 5x^2 - 14x - 5$ **11.** $4x^4 - 7x^3y + 2x^2y^2 + y^4$ **13.** $3x^4 - 13x^3 - 4x^2 - 5x - 6$ **15.** $12x^5 - 6x^4 - 25x^3 + 14x^2 + 12x - 8$ **17.** $x^5 - 4x^4 + 6x^3 - 5x^2 - x - 3$ **19.** $2x^3 - 9x^2 - 11x + 60$ **21.** $8x^3 + 27y^3$ **23.** $a^4 + 2a^3b - b^4 + a^2b^2$ **25.** $x^5 - 8x^4 - x^3 + 92x^2 - 60x - 144$ **27.** $x^5 + y^5$ **29.** $8x^5 - 8x^4 + 25x^2 - 26x - 15$

Exercise 8.3

1. $x + 5$ **3.** $3x - 2 + \dfrac{2}{x + 4}$ **5.** $x^2 - 4x - 2 - \dfrac{5}{2x - 1}$ **7.** $3x^2 - 6x + 7 - \dfrac{7}{x + 2}$ **9.** $4x^2 - x - 1 - \dfrac{3}{2x - 3}$ **11.** $3x^2 + x - 2 - \dfrac{6}{4x - 3}$ **13.** $3c + 2d - \dfrac{3d^2}{2c - 3d}$ **15.** $2v^3 - 4v^2 + 3v - 5$ **17.** $4x^3 + 5x^2 - x + \dfrac{2}{3} - \dfrac{\frac{7}{3}}{3x - 4}$ **19.** $6x^3 + 4x^2 + 7x - 3 - \dfrac{9}{3x - 2}$ **21.** $8x^3 + 10x^2 - 2x -$

$3 - \dfrac{12}{4x - 5}$ **23.** $2x^3 + 4x^2 + 5x - 3 - \dfrac{4}{4x - 5}$ **25.** $3x^3 - 4x - 3 - \dfrac{2}{4x - 3}$ **27.** $8x^3 - 6x^2 - 4x - 3 - \dfrac{4}{3x - 2}$ **29.** $3x^2 + 3xy + \dfrac{7}{2}y^2 + \dfrac{45}{2}\left(\dfrac{y^3}{2x - 3y}\right)$ **31.** $2c^2 + 3c + 1$ **33.** $4x^2 - 5xy - 3y^2$ **35.** $4x^2 + 3$

Exercise 8.4 (*Answers for 1, 5, 9, 13, 17*) (*R denotes remainder*)

1. $2x^2 - 4x - 1$, R: 1; $2x^2 - 8x + 11$, R: -9; $2x^2 - 2x - 1$, R: 0; $2x^2 + 3$, R: 11; $2x^2 - 10x + 23$, R: -44 **5.** $3x^3 - 4x^2 + 6x - 2$, R: -14; $3x^3 - 10x^2 + 20x - 28$, R: 16; $3x^3 - x^2 + 8x + 8$, R: 4; $3x^3 + 2x^2 + 16x + 40$, R: 108; $3x^3 + 5x^2 + 30x + 112$, R: 436 **9.** $3x^3 + x^2 + 2x + 1$, R: -6; $3x^3 - 5x^2 + 6x - 7$, R: 0; $3x^3 + 4x^2 + 9x + 17$, R: 27; $3x^3 - 8x^2 + 17x - 35$, R: 63; $3x^3 + 10x^2 + 41x + 163$, R: 645 **13.** $(x + 1)(x - 2)(x - 3)(x + 4)$ **17.** $x^6 - 2x^5 + 4x^4 - 8x^3 + 16x^2 - 32x + 64$

Exercise 9.1

1. 9 **3.** no solution **5.** 6 **7.** 2 **9.** 8 **11.** 12 **13.** 0 **15.** 4 **17.** -7 **19.** $\frac{7}{2}$ **21.** 2 **23.** 4.5

Exercise 9.2

1. 4 **3.** -2 **5.** $\frac{7}{3}$ **7.** no solution **9.** -5 **11.** 4 **13.** no solution **15.** $-\frac{9}{7}$ **17.** 0 **19.** $\frac{1}{3}$ **21.** 0 **23.** no solution **25.** 8 **27.** -4 **29.** 3

Exercise 9.3

1. $\frac{20}{9}$ **3.** 2.6 **5.** 12 **7.** $-\frac{20}{3}$ **9.** -2 **11.** 0.5 **13.** -1.5 **15.** -3 **17.** -4 **19.** identity **21.** 1.5 **23.** 0 **25.** no solution **27.** -7 **29.** 0.5 **31.** no solution

Exercise 10.1

1. $2x + 3y$ **3.** $(5x)(7y)$ **5.** $2x + 13$ **7.** $2(x + 5x)$ **9.** $\dfrac{x + y}{xy}$ **11.** $x + 5$; $2x + 5$ **13.** $4x$; $x - 5$; $4x - 5$ **15.** $x + 6x + 20$ **17.** $280x$ cents **19.** $n + 2$ **21.** even **23.** $50(8 - x)$ miles **25.** $3x - 5$; $2x + 2(3x - 5)$ **27.** $x - 50$; $x - 100$; $x - 150$

Exercise 10.2

1. 17; 51 **3.** 23; 92 **5.** rule: \$15.50; drawing set: \$23.50 **7.** rod: \$23.50; reel: \$16.50 **9.** visitors: 19; home: 33 **11.** 13 free throws, 34 field goals **13.** 176; 352; 292 **15.** width: 12 ft; length: 19 ft **17.** 7 in.; 11.5 in.; 15.5 in.

Exercise 10.3

1. 9; 36 **3.** 8.5; 32.5 **5.** 10; 2 **7.** 10; 36 **9.** 24; 48

Exercise 10.4

1. 66; 67; 68; 69 **3.** 37; 39; 41; 43; 45; 47 **5.** 440; 380; 320; 260 **7.** top: $9\frac{5}{8}$ in.; bottom: $18\frac{3}{8}$ in. **9.** 510; 430; 350

Exercise 10.5

1. 63 nickels; 26 dimes **3.** 33 nickels; 11 dimes; 28 quarters **5.** 70 at 65¢; 33 at $1.50 **7.** $33\frac{1}{3}$ of 95¢ kind; $66\frac{2}{3}$ of 50¢ kind **9.** 14 W's; 31 L's; 3 J's. **11.** 180 marks; 240 kroner; 80 gulden

Exercise 10.6

1. $7\frac{1}{3}$ hr **3.** $5\frac{15}{17}$ hr **5.** 54 mph **7.** tr: 450; car: 225 **9.** 1:30 PM **11.** 2100 miles

Exercise 10.7

1. 19; 52 **3.** 43; 65 **5.** 11; 41 **7.** lot: $4160; house: $33,640 **9.** 80; 75; 70; 65; 60 **11.** 1.12; 2.58 **13.** 2400; 47,600 **15.** 0.075; 0.120 **17.** $7\frac{1}{2}$; 30 **19.** $8\frac{2}{3}$ by $20\frac{1}{3}$ **21.** 42.5°; 52.5°; 85° **23.** 55.8; 55.8; 68.4 **25.** 84; 86; 88 **27.** 13 n.; 19 d.; 18 q. **29.** 191 children; 69 adults **31.** 35 at 70¢; 15 at $1.20 **33.** train: 600; car 360 **35.** A: 12; B: 14; C: 21; D: 11; E: 18

Exercise 11.1

1. $24x^4y^3z$ **3.** $-60m^4n^6$ **5.** $6x^5y^4$ **7.** $-168a^7b^3c^6$ **9.** $-540x^{10}y^7$ **11.** $-1728a^9b^3x^3yz$ **13.** $54r^6s^3t^2x^3y^6$

Exercise 11.2

1. $6x - 10y + 4z$ **3.** $20x - 28y + 12$ **5.** $-30n^3 + 54n$ **7.** $-6x^2y^2 + 8xy^3$ **9.** $-5x^5y + 5x^2y^3$ **11.** $4x^2y + 8x^2y^4z - 12x^5y^3$ **13.** $-8x^3y^2 - 12x^4y + 20x^2y^4 + 4x^2y$ **15.** $5x^5y^2z^3 + 4x^4y^3z^3 - x^2yz^5 + x^2yz^3$ **17.** $15h^3k - 9h^3k^2 - 12h^2k^2$ **19.** $-2a^2b^3c^4 + 9a^2b^3c - 4ab^4cx + 2abc$

Exercise 11.3

1. $4(x + 5)$ **3.** $8(3n - 1)$ **5.** $2n(3n - 16m)$ **7.** $10x^3(3x - 2)$ **9.** $4x^3y(3x^2 - 2y^2)$ **11.** $\pi h(R^2 - r^2)$ **13.** $3x(4x^2 + 3x - 1)$ **15.** $5y^2(4y^3 - 3y^2 - 2y - 1)$ **17.** $x(x^4 - x^3 + x^2 - x - 1)$ **19.** $3nx(2n + 4n^2x - 3n^3x^2 - 1)$ **21.** $(m + n)(m + n + 1)$ **23.** $(x - y)(x - y)(x - y + 1)$

Exercise 11.4

1. $x^2 - 49$ **3.** $16x^2 - 25y^2$ **5.** $25n^2 - 64a^2b^2$ **7.** $x^4 - 1$ **9.** $36y^8 - \frac{1}{9}$ **11.** $49a^2 - 81x^2y^2$ **13.** $400x^2 - 169y^2$ **15.** $64x^4y^6 - z^2$ **17.** $2500 - 1 = 2499$ **19.** $(80 + 2)(80 - 2) = 6400 - 4 = 6396$

Exercise 11.5

1. $(x - 6)(x + 6)$ **3.** $(h - 7)(h + 7)$ **5.** $(2x - 7y)(2x + 7y)$

7. $(4t - 1)(4t + 1)$ **9.** $(9y + 1)(9y - 1)$ **11.** $(14a^3 + 5x)(14a^3 - 5x)$ **13.** $(ab^2 - 3c^3)(ab^2 + 3c^3)$ **15.** $(x - \sqrt{2})(x + \sqrt{2})$ **17.** $(15n - \frac{1}{3})(15n + \frac{1}{3})$ **19.** $(5n - 0.4)(5n + 0.4)$ **21.** $(3x^2 - 0.2)(3x^2 + 0.2)$ **23.** $4(x + 4)(x - 4)$ **25.** $(x + y - z)(x + y + z)$ **27.** $(x - y - 4)(x + y + 4)$ **29.** $(x + y - 7)(x - y - 3)$

Exercise 11.6

1. $x^2 + 16x + 64$ **3.** $25x^2 + 70x + 49$ **5.** $16r^2 + 72rs + 81s^2$ **7.** $1 + 22x + 121x^2$ **9.** $100x^2 + 60x + 9$ **11.** $4x^6 + 20x^3 + 25$ **13.** $64y^2 + 8y + \frac{1}{4}$ **15.** $4x^2 + x + \frac{1}{16}$ **17.** $25x^2 + 2x + 0.04$ **19.** $4900 + 140 + 1$ **21.** $90000 - 600 + 1$ **23.** $(a + b)^2 + 2c(a + b) + c^2$ **25.** $25x^2 + 80x + 64$ **27.** $81x^2 + 72x + 16$ **29.** $16n^2 + 2n + \frac{1}{16}$ **31.** $b^4 - 6b^2x^3 + 9x^6$ **33.** $x^2 = 9a^2 + 48a + 64$ **35.** $x - 3 = 36$

Exercise 11.7

1. No square **3.** $(y + 5)^2$ **5.** $(3x - 8)^2$ **7.** No square **9.** $(6 - 5n^2)^2$ **11.** No square **13.** No square **15.** $(2x - \frac{1}{2})^2$ **17.** $(15n - 1)^2$ **19.** $(xy - \frac{3}{2}a)^2$

Exercise 11.8

1. $6x$; $(x + 3)^2$ **3.** $16n$; $(n - 8)^2$ **5.** $12x$; $(2x + 3)^2$ **7.** $10x$; $(5x + 1)^2$ **9.** $72x$; $(4x + 9)^2$ **11.** $60x$; $(6x + 5)^2$ **13.** $2x$; $(3x - \frac{1}{3})^2$ **15.** $6ab^2$; $(ab^2 - 3)^2$ **17.** x; $(x - \frac{1}{2})^2$ **19.** $3x$; $(3x - \frac{1}{2})^2$ **21.** $48x$; $(12 + 2x)^2$ **23.** $40F$; $(F + 20)^2$

*Further answers for Chapter 11 continued on page 859.

Exercise 12.1

1. $\frac{3x}{5y^2}$ **3.** $\frac{2cx^4}{3n^2z}$ **5.** $\frac{3xz}{4aby}$ **7.** $2x^2y$ **9.** $\frac{1}{3abx}$ **11.** -1 **13.** $x + 4$ **15.** $x - 1$ **17.** $\frac{(x - 1)^2}{(x^2 + x + 1)}$ **19.** $\frac{1}{(a + b)}$ **21.** $\frac{(R + r)}{(R - r)}$ **23.** $\frac{(3 - x)}{(6 + x)}$ **25.** $\frac{(x + 4)}{(2x + 4)}$ **27.** $\frac{y + 3}{2y(y + 6)}$ **29.** $\frac{3x(n + 6)}{2y(n + 3)}$ **31.** $\frac{2y(x - 4)}{3(x - 2)}$ **33.** $\frac{3(6n - 5)}{5(3n - 2)}$ **35.** $\frac{3c - 5}{2c(c - 2)}$ **37.** $\frac{a(x - 2)}{3b(3x - 4)}$ **39.** $\frac{u^2 + uv + v^2}{u(2u + v)}$ **41.** $\frac{a + b}{a - b + 1}$

43. not reducible

Exercise 12.2

1. $\frac{3cxy^2}{4ab}$ **3.** $\frac{ab^2c}{2x}$ **5.** $\frac{2}{5bx}$ **7.** $\frac{x}{5y}$ **9.** $\frac{5(x + 3)}{x(x + 5)}$ **11.** $\frac{x - 8}{x - 4}$ **13.** $\frac{2(x + 2)(x + 2)}{3(x - 6)(x + 4)}$ **15.** $\frac{3(a + 3)}{a + 9}$ **17.** $\frac{4x(2x + 1)}{2x - 1}$ **19.** $\frac{3(c + 2)}{3c + 4}$ **21.** $\frac{(x - 1)(x - 1)}{(x + 3)(x + 3)}$ **23.** $\frac{2(r + 3)}{2r + 3}$ **25.** $\frac{y + 8}{4(y + 2)}$ **27.** $\frac{(a + b)(a - b + 1)}{3(a^2 + ab + b^2)}$

Exercise 12.3

1. $\frac{14}{45}$ **3.** $\frac{(x + y)}{xy}$ **5.** $\frac{2x}{5}$ **7.** $\frac{6x}{21}$ **9.** $\frac{(7y + 35)}{12}$ **11.** $\frac{(3n - 2)}{8}$ **13.** $\frac{23 - 3r}{12}$
15. $\frac{51x - 76}{30}$ **17.** $\frac{2x^2 + 9y + 12x - 6xy}{3xy}$ **19.** $\frac{9x + 8x^2 - 6 - 18x^3}{6x^3}$
21. $\frac{4x - 29xy + 9y}{6xy}$ **23.** $\frac{r_2r_3 - r_1r_3 + r_1r_2}{r_1r_2r_3}$ **25.** $\frac{8}{x - 4}$ **27.** $\frac{5n^2 - n}{(n - 3)(n - 4)}$
29. $\frac{x^2 + 2x - 2}{(x - 3)(x + 3)}$ **31.** $\frac{x + 4}{x - 4}$ **33.** $\frac{-x - 12}{x(x + 3)}$ **35.** $\frac{20x^2 - 4x^3 + 15}{x(x - 5)}$
37. $\frac{-2x^3 - 6x^2 - 18x - 12}{x(x + 3)}$

Exercise 12.4

1. $\frac{4}{5}$ **3.** $3x - 2$ **5.** $4(4x - 3)$ **7.** $\frac{7x}{(5x + 12)}$ **9.** $\frac{(b + a)}{5}$ **11.** $\frac{4(x - 1)}{x(2x + 3)}$
13. $\frac{3 - 2x}{2x - 1}$ **15.** $\frac{28 - 20x}{15 - 3x}$ **17.** $2 - x$ **19.** $\frac{x + 2y}{x - 2y}$ **21.** $\frac{3x + 1}{3 - 2x}$

Exercise 13.1

1. 12 **3.** 20 **5.** $-\frac{35}{4}$ **7.** $\frac{9}{2}$ **9.** $\frac{32}{7}$ **11.** $\frac{105}{4}$ **13.** $-\frac{48}{17}$ **15.** $-\frac{24}{5}$ **17.** $-\frac{48}{5}$ **19.** $\frac{8}{3}$
21. −12 **23.** $-\frac{29}{8}$ **25.** \$36,000 ; \$45,000 **27.** 7.5; 12.5
29. \$75; \$7.50; \$45 **31.** 30 by 25.5

Exercise 13.2

1. −6 **3.** $\frac{7}{2}$ **5.** $-\frac{13}{4}$ **7.** $-\frac{8}{3}$ **9.** $-\frac{41}{15}$ **11.** 40 **13.** $\frac{110}{13}$ **15.** 62 **17.** 17 **19.** 6.5

Exercise 13.3

1. 11 **3.** 1 **5.** 5.25 **7.** 4.5 **9.** 0.5 **11.** −280 **13.** 600 **15.** 3792 **17.** 450
19. 283

Exercise 13.4

1. −4 **3.** −10 **5.** $\frac{22}{5}$ **7.** $\frac{8}{3}$ **9.** $\frac{10}{3}$ **11.** $-\frac{31}{3}$ **13.** 1.6 **15.** 7 **17.** 1 **19.** 2
21. $-\frac{19}{6}$ **23.** $\frac{8}{7}$ **25.** 2.8 **27.** −0.5 **29.** −9 **31.** 3.5 **33.** 2 **35.** −2.5

Exercise 13.5

1. $W = \frac{V}{LH}$ **3.** $r = \frac{d}{t}$ **5.** $h = \frac{V}{\pi r^2}$ **7.** $R = \frac{E - Ir}{I}$ **9.** $a = \frac{bf}{b - f}$
11. $M_1 = \frac{d^2F}{KM_2}$ **13.** $A = \frac{2S - NL}{N}$ **15.** $n = \frac{L - a + d}{d}$ **17.** $r = \frac{S - a}{S - L}$
19. $h = \frac{3V}{\pi r^2}$ **21.** $h = \frac{A - 2\pi r^2}{2\pi r}$ **23.** $Z_2 = \frac{Z_1Z_t}{Z_1 - Z_t}$ **25.** $W = \frac{V}{LH}$; 12.5
27. $r = \frac{d}{t}$; 52.5 mph **29.** $h = \frac{V}{\pi r^2}$; 7.5 **31.** $r = \frac{E - IR}{I}$; 0.5

33. $b = \dfrac{af}{a - f}$; 60 **35.** $M_1 = \dfrac{Fd^2}{KM_2}$; 20 **37.** $T = \dfrac{abc}{bc + ac + ab}$; $\dfrac{40}{3}$

Exercise 13.6

1. 12; 15 **3.** $\frac{51}{54}$ **5.** $\frac{24}{7}$ hr **7.** 2.4 hr **9.** 5 hr **11.** 2 hr, 21 min (approx.) **13.** 18 by 30; perimeter 2 less **15.** 2 **17.** 26 **19.**$2500 at 3%; $1500 at 7% **21.** $4800 at 5%; $3000 at 7% **23.** $9\frac{4}{19}$ gal **25.** 7.5 qt **27.** $523\frac{1}{3}$ **29.** 3 **31.** 150000; 12000 **33.** 180 miles **35.** 42

Exercise 13.7

1. 10 **3.** 4 **5.** −12 **7.** 2 **9.** −15 **11.** 5 **13.** 1 **15.** 24 **17.** 12 **19.** −7.8 **21.** $-4\frac{6}{7}$ **23.** 9 **25.** 1.9 **27.** 9 **29.** −2 **31.** $1\frac{7}{11}$

Exercise 14.1 (Answers indicate ordered pairs; some decimals)

1. 2; −1 **3.** −1; −2 **5.** 3; −4 **7.** $\frac{25}{11}$; $-\frac{7}{11}$ **9.** $\frac{4}{3}$; $-\frac{1}{3}$ **11.** 34; 25 **13.** 5; 0 **15.** −4; −3 **17.** $\frac{95}{68}$; $-\frac{45}{17}$ **19.** 1.86; 1.30 **21.** 31.5; 30.9 **23.** $\dfrac{bm + cn}{m^2 + n^2}$; $\dfrac{bn - cm}{m^2 + n^2}$

Exercise 14.2

1. 3; −2 **3.** 2; −1 **5.** −1; + 6 **7.** $-\frac{7}{11}$; $-\frac{79}{11}$ **9.** $-\frac{8}{5}$; $-\frac{62}{5}$ **11.** $\frac{2}{11}$; $-\frac{7}{11}$ **13.** −2; 5 **15.** −1; −4 **17.** $-\frac{11}{3}$; $-\frac{13}{3}$ **19.** $\frac{11}{13}$; $-\frac{5}{13}$ **21.** $\frac{38}{11}$; $-\frac{26}{11}$

Exercise 14.3

1. $\frac{6}{7}$; $-\frac{5}{7}$ **3.** $-\frac{40}{17}$; $-\frac{39}{17}$ **5.** $\frac{9}{23}$; $-\frac{29}{23}$ **7.** $\frac{165}{7}$; $\frac{83}{7}$ **9.** $-\frac{11}{16}$; $-\frac{207}{32}$ **11.** $-\frac{19}{11}$; $-\frac{6}{11}$

Exercise 14.4

1. 27 nickels; 46 quarters **3.** $5600 at 5%; $2800 at 7% **5.** car: 2.5 hr; 112.5 miles; train: 9.5 hr; 617 miles **7.** plane: $234\frac{2}{3}$ mph; wind: $21\frac{1}{3}$ mph **9.** 21.6 qt of 7.5%; 2.4 qt of 82.5% **11.** 52.5 volts; 32.5 volts

Exercise 14.5 (Some answers omitted here)

7. 0.5; 3; −0.5 **9.** 1; −4.2; −4.4 **15.** 3.100; 1.520; 0.632; 0.565 **17.** 0.1466; 2.503; −0.623; 0.728 **19.** −0.5; 2.75; 2.65; 3.75; −3.6 **21.** 2.685; 3.777; 4.404

Exercise 14.6

1. 27 nickels; 58 dimes; 8 quarters **3.** $6500 at 4%; $3400 at 7%; $2100 at 9% **5.** 444.4 cc of 4%; 222.2 cc of 28%; 333.3 cc of 60% **7.** boat: 45 miles; car 180 miles; train: 275 miles

Exercise 14.7

1. $\frac{37}{18}$; $-\frac{1}{6}$ **3.** 3; 0 **5.** inconsistent **7.** dependent **9.** 0; 3 **11.** inconsistent **13.** −12; −2; 9 **15.** −4; −1; −3

Exercise 15.3 (Answers approximate to nearest tenth)

1. 4; 3 **3.** 3; 1 **5.** −3; −1 **7.** −7; 4 **9.** −3; 6 **11.** 4.4; −1.3 **13.** −1; −4 **15.** −2; 4 **17.** −3.5; 0 **19.** −2.6; 1.9 **21.** −1.2; −11.2

Exercise 15.4

1. (a) $3x - 5y = -7$; (b) $2x + 5y = 17$; (c) $5x + 3y = 22$; (d) $x + 2y = -5$; (e) $3x - 5y = 30$; (f) $5x - 2y = -18$; (g) $x - 2y = 0$; (h) $x + 2y = 5$; (i) $2x + y = -5$; parallel: (a) and (e); (d) and (h); perpendicular: (a) and (c); (g) and (i) **3.** (a) $y = -2$; (b) $x = 4$; (c) $y = 5$ **5.** $5x - 4y = -9$; $2x + y = 12$; $x - 6y = 19$ **7.** Opposite slopes are equal: $-\frac{2}{3}$ and $\frac{3}{5}$

Exercise 16.1

1. x^9 **3.** a^5 **5.** n^{-7} **7,** x^5 **9.** $x^{\frac{31}{24}}$ **11.** 10^8 **13.** x^8 **15.** y^{5a-3} **17.** $10^{-0.85}$ **19.** x^6 **21.** n^4 **23.** y^5 **25.** n^{20} **27.** $y^{-3.326}$ **29.** $10^{-2.29}$ **31.** x^{a^2-a} **33.** $\frac{1024}{9}$ **35.** −16 **37.** 0.0001 **39.** $-y^{15}$ **41.** 10^{-12} **43.** x^{-1} **45.** 2^{2x^2} **47.** 10^{-6} **49.** $10^{0.89}$ **51.** $x^{12}y^{16}$ **53.** $256x^{16}$ **55.** $27x^6y^{-9}$ **57.** $4x^{-12}$ **59.** $64a^{-\frac{1}{2}}b$ **61.** $\dfrac{256x^8}{625y^{12}}$ **63.** $\dfrac{64n^3}{y^{-2}}$ **65.** $\dfrac{x^0y^2z}{a^3b^4c^2}$

Exercise 16.2

1. 1 **3.** 1 **5.** −1 **7.** −1 **9.** 8 **11.** −6 **13.** 10 **15.** −2 **17.** $-\dfrac{y^2}{x^3z}$ **19.** $\dfrac{a^5d}{b^4c}$ **21.** $-\dfrac{c}{9ab^3}$ **23.** $\dfrac{5b^4y^2}{2ac^3x^3z}$ **25.** $-ab$ **27.** $\dfrac{y-x}{xy}$ **29.** $\frac{25}{144}$ **31.** 0.1 **33.** $\frac{9}{200}$ **35.** $\frac{11}{3}$ **37.** $\frac{80}{81}$ **39.** −9 **41.** −4 **43.** 36 **45.** −16 **47.** 36 **49.** $-\frac{1}{18}$ **51.** $\frac{2}{45}$ **53.** −588

Exercise 16.3

1. $(ab)^{\frac{1}{2}}$ **3.** $(343)^{\frac{1}{3}}$ **5.** $(x^3 + y^2)^{\frac{1}{4}}$ **7.** $(x^2y^2)^{\frac{1}{3}}$ **9.** 4.76 **11.** $\frac{2}{3}$ **13.** $\frac{27}{125}$ **15.** 0.1 **17.** 8 **19.** $\frac{1}{8}$ **21.** −16 **23.** 5 **25.** 17 **27.** 25

Exercise 16.4

1. $\frac{1}{3}$ **3.** 64 **5.** x^{-2n} **7.** h^{2n-2} **9.** $1/81x^4$ **11.** $9y^4/x^3$ **13.** $\dfrac{4x^4}{9}$ **15.** $\frac{1}{4}$ **17.** $\frac{16}{5}$ **19.** $1/200y^3x^5$

Exercise 16.5

1. $(9.3)(10^7)$ **3.** $(6.3)(10^{-4})$ **5.** 10^{-7} **7.** $(5)(10^{-9})$ **9.** $(5.872)(10^{12})$ **11.** $(8.26)(10^7)$ **13.** $(5.29)(10^{-1})$ **15.** $(8.30)(10^1)$ **17.** $(7.38)(10^2)$ **19.** $(5.07)(10^{-4})$ **21.** $(4.05)(10^{-6})$

Exercise 17.1

1. x^4 **3.** $10^{2.8}$ **5.** 64 **7.** $4x^3$ **9.** $-4x^3$ **11.** $2x^2$ **13.** $3x^{10}$ **15.** $5a$ **17.** $\frac{5}{7}$ **19.** $2x^4$ **21.** $2x$ **23.** $x + y$ **25.** $n^{\frac{4}{3}}$ **27.** $-9x^3y^4z$ **29.** $x - 2$

Exercise 17.2

1. $4\sqrt{3}$ **3.** $9\sqrt{5}$ **5.** $3\sqrt{2}$ **7.** $\frac{1}{2}\sqrt{6}$ **9.** $3\sqrt{3}$ **11.** $\frac{21}{4}$ **13.** $\frac{3}{8}\sqrt{10}$ **15.** $\frac{4}{5}\sqrt{15}$ **17.** $2\sqrt[3]{2}$ **19.** $2\sqrt[4]{3}$ **21.** $3x\sqrt{2x}$ **23.** $2x^3y^3\sqrt{3x}$ **25.** $\frac{1}{x}\sqrt{7}$ **27.** $\frac{1}{x^2}\sqrt{3ax}$ **29.** $2x^2\sqrt[3]{2x}$ **31.** $3xy\sqrt[3]{3xy^2}$ **33.** 48 **35.** 60 **37.** $x^2\sqrt{x^2 + 1}$ **39.** 13

Exercise 17.3

1. 43.84 **3.** 74.62 **5.** −11.345 **7.** 58.34 **9.** 22.37 **11.** 1.945 **13.** 1.633 **15.** 83.84 **17.** 4.339 **19.** 27.91

Exercise 17.4

1. 37.95 **3.** 30 **5.** 144 **7.** 268.3 **9.** 272.2 **11.** 60 **13.** −31.205 **15.** 167.57 **17.** −98.38 **19.** −14.693 **21.** 2.679 **23.** 0.1255 **25.** (a) $\sqrt[6]{x^2y^3}$ (b) $\sqrt[12]{648}$ (c) $4\sqrt[4]{2}$ (d) $\sqrt[10]{112500}$

Exercise 17.5

1. $\frac{6\sqrt{2}}{5}$ **3.** $\frac{\sqrt{3}}{3}$ **5.** $\frac{3\sqrt{13}}{13}$ **7.** $\frac{2\sqrt{15}}{15}$ **9.** $\frac{2\sqrt{7}}{7}$ **11.** $\frac{9 + 3\sqrt{5}}{4}$ **13.** $16 + 8\sqrt{3}$ **15.** $\frac{6 + \sqrt{15}}{7}$ **17.** $2 - \sqrt{3}$ **19.** $7 + 4\sqrt{3}$ **21.** $\frac{8\sqrt{5} + 26\sqrt{3} + \sqrt{15} + 25}{122}$ **23.** $4\sqrt{2} - 3\sqrt{3}$ **25.** $\frac{4\sqrt{5} - \sqrt{3}}{11}$ **27.** $\frac{\sqrt{5} - 18}{11}$ **29.** $\frac{\sqrt{3} - 39}{66}$

Exercise 17.6

1. $4\sqrt{15}$ **3.** $25\sqrt{5}$ **5.** $10\sqrt{2}$ **7.** $5\sqrt{2}$ **9.** $\frac{9}{5}\sqrt{2}$ **11.** $4\sqrt{7}$ **13.** $\frac{1}{2}\sqrt[3]{3}$ **15.** $\frac{1}{16}\sqrt[4]{24}$ **17.** $4x^8$ **19.** $4x^3y^3\sqrt{6x}$ **21.** $\frac{2}{x^2}\sqrt{x}$ **23.** $\frac{1}{3ab^2}\sqrt{15ab}$ **25.** 3.34 **27.** 17.40 **29.** $6\sqrt{3} - 10$ **31.** $80 - 48\sqrt{3}$ **33.** $28 + 16\sqrt{3}$ **35.** $\frac{1}{2}(5\sqrt{3} - 9)$ **37.** $\frac{1}{4}(\sqrt{5} - 3)$ **39.** 1.57; 38.2 **41.** 0.962; 62.4

Exercise 18.1

1. ± 6 **3.** ± 12 **5.** $\pm\frac{7}{4}$ **7.** $\pm\sqrt{3}$ **9.** $\pm\sqrt{7}$ **11.** $\pm\frac{1}{4}\sqrt{14}$ **13.** $\pm\sqrt{-9}$ **15.** $\pm\sqrt{-16}$ **17.** $\pm\sqrt{-\frac{169}{6}}$ **19.** $\pm\frac{2}{3}\sqrt{5}$ **21.** $\pm\frac{5}{4}\sqrt{3}$ **23.** $h = \pm k$ **25.** $r = \sqrt{\frac{A}{\pi}}$ **27.** $r = \sqrt{\frac{3V}{\pi h}}$ **29.** $v = \sqrt{\frac{2K}{M}}$

Exercise 18.2

1. 3; −2 **3.** 5; −4 **5.** 5; −3 **7.** 3; 6 **9.** 3; 4 **11.** $\frac{5}{2}$; $\frac{5}{2}$ **13.** 0; $\frac{3}{5}$ **15.** 0; $\frac{9}{4}$ **17.** $\pm\frac{4}{3}$ **19.** Not factorable **21.** $\frac{5}{2}$; $-\frac{3}{2}$ **23.** $\frac{4}{3}$; $-\frac{3}{4}$

Exercise 18.3

1. 5; −1 **3.** 9; −1 **5.** 2; 10 **7.** 5; −2 **9.** 3; $\frac{1}{2}$ **11.** $\dfrac{2 \pm \sqrt{10}}{3}$ **13.** $\dfrac{3 \pm \sqrt{19}}{5}$ **15.** $\dfrac{5 \pm \sqrt{73}}{8}$ **17.** 0; $\frac{5}{3}$ **19.** Imaginary **21.** $4 \pm \sqrt{7}$ **23.** Imaginary

Exercise 18.4

1. 1; $\frac{7}{2}$ **3.** 1; $-\frac{2}{5}$ **5.** 7; $\frac{1}{2}$ **7.** $\frac{2}{3}$; $\frac{2}{3}$ **9.** $-\frac{5}{2}$; $-\frac{5}{2}$ **11.** $\dfrac{1 \pm 2\sqrt{2}}{7}$ **13.** $\dfrac{8 \pm \sqrt{-16}}{10}$ **15.** $\dfrac{6 \pm \sqrt{-36}}{18}$ **17.** $\pm\frac{2}{5}\sqrt{5}$ **19.** −2; $\frac{1}{4}$ **21.** 1; −5 **23.** $\dfrac{-6 \pm \sqrt{-44}}{10}$ **25.** $\dfrac{-6 \pm 2\sqrt{3}}{3}$ **27.** $\dfrac{3 \pm \sqrt{6}}{2}$ **29.** $\dfrac{-5 \pm \sqrt{37}}{6}$ **31.** $\dfrac{-3 \pm \sqrt{-23}}{8}$ **33.** $\dfrac{4 \pm \sqrt{-44}}{10}$ **35.** $\frac{9}{2}$; $\frac{8}{3}$ **37.** 2; −9 **39.** $\frac{4}{5}$

Exercise 18.5

1. 6; 14 **3.** 8; −9 **5.** 12 and 4; −14 and −12 **7.** 5 by 9 **9.** 60 by 90 **11.** 11.68 by 18.68 ft **13.** 1; 6 **15.** 45 mph; 30 mph **17.** 48 mph **19.** 5 sec **21.** 30 **23.** 18 by 24 in.

Exercise 18.6

1. 13 **3.** 6 **5.** 8 **7.** −2 **9.** −3 **11.** $\frac{16}{25}$ **13.** 1; 2 **15.** 2 **17.** 7 **19.** 4 **21.** 1; $\frac{13}{9}$ **23.** 7 **25.** 5; −3 **27.** None **29.** 512

Exercise 19.1

1. $2i$ **3.** $7i$ **5.** $13j$ **7.** $4jx$ **9.** $8iy^3$ **11.** $2i\sqrt{2}$ **13.** $3i\sqrt{3}$ **15.** $\frac{1}{2}i$ **17.** $\frac{1}{4}i\sqrt{6}$ **19.** $5nr^2i\sqrt{2nr}$ **21.** $10i$ **23.** $13j$ **25.** $11i$ **27.** −10 **29.** −5 **31.** −40 **33.** −18 **35.** −21 **37.** −7 **39.** $-j$ **41.** i **43.** −1

Exercise 19.2

1. $3 \pm 2i$ **3.** $-\frac{1}{4} \pm \frac{1}{4}\sqrt{7}i$ **5.** $2 \pm i\sqrt{5}$ **7.** $-3 \pm i$ **9.** $-2 \pm 2i\sqrt{3}$ **11.** $\frac{5}{2} \pm \frac{1}{2}i\sqrt{183}$ **13.** $2 \pm \frac{1}{2}i\sqrt{2}$ **15.** $10 \pm 20i$ **17.** $\pm 4i\sqrt{2}$ **19.** $\pm\frac{1}{2}i\sqrt{6}$ **21.** $\frac{3}{5} \pm \frac{1}{5}i$ **23.** $\frac{5}{2} \pm \frac{1}{2}i$

Exercise 19.3

1. $8 - i$ **3.** $-2 - 8i$ **5.** $-2 - 2j$ **7.** $7 - 2i$; $3 + j$; $-7 - 9i$; $-1 - j$; $10 + 0i$ **9.** $-4 - 2i$; $-3 - 2j$; $-7 + 0i$; $-2 + 13j$; $5 - 11j$ **11.** $0 - 2j$; $14 + 4j$; $6 - j7$; $3 - 8i$; $-5 - j$ **13.** $-14 + 0j$ **15.** $6 + 0j$ **17.** $12 + 0j$ **19.** $0 + 0j$

21. $0 - 6j$ **23.** $0 - j14$ **25.** $0 - 12j$ **27.** $0 - 8j$ **29.** $(a + bi) - (a - bi) = 0 + 2bi$ **31.** $x = 5$; $y = 11$

Exercise 19.4

1. $-12 + 15i$ **3.** $20 + 8i$ **5.** $-6 - 16j$ **7.** $-1 + 17j$ **9.** $-21 - j$ **11.** $55 + j$ **13.** $-21 - 20i$ **15.** $3 - 11i$ **17.** $-7 + 24j$ **19.** $-10 + 0i$ **21.** $5 - 21i$ **23.** $0 + 2j$ **25.** $60 + 32j$ **27.** $-24 + 10j$ **31.** (a) 29; (b) 58; (c) 41; (d) 25; (e) 65; (f) 72; (g) 81; (h) −16

Exercise 19.5

1. $\frac{11}{10} - \frac{13}{10}i$ **3.** $-\frac{9}{17} - \frac{2}{17}i$ **5.** $\frac{6}{5} - \frac{7}{5}i$ **7.** $-\frac{3}{17} + \frac{22}{17}j$ **9.** $\frac{5}{2} + \frac{1}{2}j$ **11.** $3 + j$ **13.** $-5 + 3i$ **15.** $\frac{8}{5} - \frac{1}{5}i$ **17.** $\frac{15}{29} + \frac{6}{29}j$ **19.** $-\frac{4}{5} - \frac{3}{5}j$ **21.** $-\frac{9}{5} + \frac{3}{5}j$ **23.** $-\frac{1}{10} - \frac{1}{5}j$ **25.** (a) $\frac{3}{13} - \frac{2}{13}j$ (c) $-i$

Exercise 19.6

1. $7 + 6i$ **3.** $-4 + 7i$ **5.** $9 - 6i$ **7.** $-6 - 2j$ **9.** $-4 + j$ **11.** $3 - 3j$ **13.** $0 + 10i$ **15.** $-1 + i$ **17.** $-7 + i$ **19.** $-8 - 3i$ **21.** $3 - 2i$ **23.** $-6 + i$ **25.** $-1 + 8i$ **27.** $0 + 8j$ **29.** $10 + 7j$ **31.** $11 + 9j$ **33.** $-5 + 14i$ **35.** $3 + 11j$ **37.** $5 + 15i$

Exercise 20.1

1. 3;−1 **3.** $\frac{5}{2}$; $\frac{5}{2}$ **5.** 2; $\frac{1}{3}$ **7.** $\frac{3}{2}$; $\frac{3}{2}$ **9.** 1; −7 **11.** $\dfrac{5 \pm \sqrt{-7}}{8}$ **13.** $\dfrac{7 \pm \sqrt{17}}{4}$ **15.** $\dfrac{1 \pm \sqrt{37}}{6}$ **17.** $\frac{4}{3}$; $\frac{4}{3}$

Exercise 20.2

[Abbrev.: re. (real); im. (imaginary); eq. (equal); un. (unequal); ra. (rational); ir. (irrational)]
1. 25; re; un; ra **3.** 4; re; un; ra; **5.** 81; re; un; ra **7.** 121; re; un; ra **9.** 0; re; eq; ra **11.** 169; re; un; ra **13.** 64; re; un; ra **15.** 9; re; un; ra **17.** 1849; re; un; ra **19.** 60; re; un; ir **21.** 20; re; un; ir **23.** −144; im; un **25.** −64; im; un **27.** −35; im; un **29.** 48; re; un; ir

Exercise 20.3

1. 9 **3.** $-\frac{1}{8}$ **5.** $\frac{9}{32}$ **7.** 1, −11 **9.** No real value of m

Exercise 20.4

1. $x^2 + 2x - 15 = 0$ **3.** $9x^2 - 9x + 2 = 0$ **5.** $x^2 + 3x = 0$
7. $x^2 - 4x + 1 = 0$ **9.** $4x^2 + 8x + 1 = 0$ **11.** $9x^2 + 36x + 31 = 0$
13. $x^3 - 7x + 6 = 0$
15. $2x^4 - 7x^3 - 2x^2 + 13x + 6 = 0$

Exercise 20.5

1. 0; 6; −2 **3.** −2; $1 \pm i\sqrt{3}$ **5.** ±3; ±3i **7.** 0; ±1; ±2 **9.** 1; ±2 **11.** 1; −2; −3 **13.** 1; $-1 \pm 2i$ **15.** 0; −1; 2; 3 **17.** −1; 2; ±3 **19.** 1; 2; $-\frac{3}{2} \pm \frac{3}{2}i$ **21.** −1; 2; $\frac{3}{2} \pm \frac{5}{2}i$ **23.** 0; ±2; 4; $\frac{1}{3}$

Exercise 20.6

1. −2; −4; 6 **3.** 2; 2; 6 **5.** 3; 4; −4; 6 **7.** 4; −4; −4; 6 **9.** 4; −4; −6; 12 **11.** −3; 4; −4; 6; −6

Exercise 21.1 (C: center; r: radius)

1. $C(0, 0)$; $r = 7$ **3.** $C(0, 0)$; $r = 1$ **5.** $C(0, 0)$; $r = \frac{9}{2}$ **7.** $x^2 + y^2 = 64$ **9.** $x^2 + y^2 - 6x + 2y = 39$ **11.** $C(3, 5)$; $r = 7$

Exercise 21.2 (Omit graphs)

Exercise 21.3 (Major axis)

1. 6 **3.** 10 **5.** 8

Exercises 21.4 (Omit graphs)

Exercise 21.5

1. (0, −3); (2.4, 1.8) **3.** Imaginary **5.** (4, 2); (−1.3, −3.3) **7.** (4.3, ±4.2) **9.** (±3.7, ±4.8) **11.** (4, ±3); (−5, 0) **13.** (3.3, 5.6); (−0.3, −1.6)

Exercise 21.6

1. $(\sqrt{5}, -2\sqrt{5})$; $(-\sqrt{5}, 2\sqrt{5})$ **3.** $(3 + i, -3 + i)$; $(3 - i, -3 - i)$ **5.** (−1, −3); $(\frac{13}{5}, \frac{3}{5})$ **7.** (3, 0); (−5, −4) **9.** (2, 3) **11.** (5, 0); (−3, 4) **13.** (4, ±3) **15.** $(\sqrt{5}, 4)$; $(-\sqrt{5}, -4)$ **17.** (3, 1); (−3, −1); $(i, -3i)$; $(-i, 3i)$ **19.** (1, 4); (1, −4) **21.** (0, 1); (8, 9) **23.** (−3, +4); (−3, −4) **25.** (−2, −3); (−2, 1) **27.** (±3, ±4); (±4, ±3) **29.** (−3, 0); $(\frac{12}{5}, \frac{9}{5})$

Exercise 22.1

1. $x < 3$ **3.** $x < \frac{2}{3}$ **5.** $x > \frac{5}{7}$ **7.** $x > -2$ **9.** $x > -\frac{3}{4}$ **11.** $x < 4$

Exercise 22.2 (Graphs not shown)

Exercise 22.3

1. $-2 < x < 1$ **3.** $-2 < x < 2$ **5.** $0 < x < 5$ **7.** $-4 < x < 1$ **9.** No solution **11.** $x < -2$; $x > 1$ **13.** $-2 < x < 3$

Exercise 23.1

1. $\frac{2}{3}$ **3.** $\frac{4}{5}$ **5.** $\frac{5}{12}$ **7.** 16 **9.** $\frac{1}{5280}$ **11.** 2; 2; 4 **13.** $\frac{1}{3}$; $\frac{1}{3}$; $\frac{1}{9}$ **15.** $\frac{1}{633600}$ **17.** 7.82; 11.38; 10.50; 19.33; 2.50; 2.69; 0.5

Exercise 23.2

1. 21 **3.** $\frac{54}{5}$ **5.** ±12 **7.** 54 **9.** 6 **11.** −2 **13.** $\frac{8}{3}$ **15.** 7.5 **17.** ±18 **19.** $\frac{35}{8}$ **21.** $\frac{8}{3}$ **23.** 84 **25.** $\frac{45}{4}$ **27.** 9 in.

Exercise 23.3

1. 64 in.; 80 in. **3.** $\frac{40}{3}$; $\frac{140}{3}$ **5.** 80; 80; 1840 lb **7.** 140; 120; 40; 60 **9.** 32; 20; 8 **11.** 150; 200; 250

Exercise 23.4

1. C = Kp **3.** L = Kw **5.** C = Kr **7.** I = Kprt **9.** d = Kt^2 **11.** C = Klwp

Exercise 23.5

1. t = K/r **3.** F = K/d^2 **5.** R = K/d^2 **7.** V = K/p **9.** L = K/$\sqrt{s}$

Exercise 23.6

1. 20 **3.** 2.5 **5.** $135 **7.** 800 lb **9.** 400 ft; 144 ft **11.** 33.75 **13.** 1080 gal **15.** 405 watts **17.** 53$\frac{1}{3}$ days **19.** 20 dynes **21.** Four times as much light **23.** 7.92 (speed = 50) **25.** 1200 BTU

Chapter 24 (No answers shown for Chapter 24. All graphs.)

Exercise 25.1

1. 462 in.^{2}; 89 in. **3.** 24.5 in.; 113 in. **5.** 18.4 in. **7.** 57.2 in.; 204.5 in.^{2} **9.** 32 in.; 128 in. **11.** $416 **13.** 75 ft **15.** 8100 ft^2; 0.186 acre (approx.) **17.** 0.24 acre (approx.) **19.** 43.2 ft^2 **21.** (a) 900 in.^{2}; (b) 864 in.^{2}; (c) 800; (d) 675; (e) 576; (f) 500; (g) 416; (h) 324; (i) 224; (j) 116; (k) 59; (l) 29.75

Exercise 25.2

1. 460.25 in.^{2} **3.** $30\sqrt{2}$, or 42.426 in.^{2} **5.** 16.4 in. **7.** 536.9 in.^{2} **9.** 34.6 in. **11.** 330 yd^2 **13.** 114 ft^2

Exercise 26.1

1. 70.84 in.^{2}; 41.27 in. **3.** 8.77 cm^2; 14.72 cm **5.** 615.8 yd^2; 126.2 yd **7.** 238.1 ft **9.** 28.2 miles **11.** 67.08 rd **13.** 9.33 ft **15.** (a) 18.67 in.; (c) 4.98 meters; (e) 6463 mm (g) 509.1 rd; (i) 7.778 yd; (k) 19.20 cm; (m) 4.243 in. **17.** (a) 140.3 in.^{2}; (c) 84.87 in.^{2}; (e) 2270 ft^2

Exercise 27.1

1. (a) 1385 in.^{2}; 131,95 in.; (c) 15.904 ft^2; 14.137 ft; (e) 61.28 ft^2; 27.75 ft; (g) 11.310 ft^2; 377 ft **3.** 6.179 **5.** 43.29 **7.** 775.7 **9.** 47,180 ft^2 **11.** 4335 ft^2 112.84ft **15.** 45.52 mph **17.** 16π; 20π; 28π; 36π **19.** 8.485 in. **21.** 16.97 in. **23.** 1.5 amp **25.** 6 in.; 44.67 in.^{2} **27.** 44π ft; 228π ft^2 **29.** 12π **31.** 17.4 in.^{2} **33.** 36.9 in.^{2} **35.** 9.13 in.^{2}

Exercise 28.1

1. 1829 in.^{3}; 923 in.^{2} **3.** 320 ft^3 **5.** 1056 in.^{2}; 5376 in.^{3} **7.** 188.5 ft^3 **9.** 4096 ft^3; 1238 ft^2 **11.** 1953.125 in.^{3}; 937.5 in.^{2} **13.** 12.25 in.; 1838.3 in.^{3} **15.** 292.5 in.^{2} **17.** 7560 gal; 62,899 lb **19.** 503.5 yd^3 **21.** 188.7 ft; 9,745,461,000,000 gal **23.** 216 **25.** 252 **27.** 3893(10^9)

Exercise 29.1

1. 34.48 **3.** 5.1 gal **5.** 15.96 in. **7.** 2641π in.^{2}; or 57.6 ft^2 **9.** first: 16.55 gal; second: 10.46 gal **11.** 125.7 ft^2 **13.** 1 ft^3; 8.5 ft^2; 6 ft^2; 192 ft^2 **15.** 52.36 ft^3 **17.** 14.28 **19.** 214.9 in.^{3}

Exercise 30.1

1. 1280 in.^{3}; 426.7 in.^{3} **3.** 3.81 in. **5.** 798.7 lb **7.** 297; 294.9 **9.** 10.72 in.^{3}; 3.72 in.^{3} less **11.** 9.83 ft^3 **13.** 25.46 in.^{3} **15.** 528 in.^{2} **17.** 11,068.2 lb **19.** 575.4 **23.** 250π **25.** 9 in.

Exercise 31.1 (Answers in order of items in problems)

1. 8 in.; 16π in.; 256π in.^{2}; 2144.7 in.^{3} **3.** 11.2 in.; 70.4 in.; 1576.3 in.^{2}; 588.5 in.^{3} **5.** 2.62 in.; 1.31 in.; 21.56 in.^{2}; 9.42 in.^{3} **7.** 9.486 in.; 4.743 in.; 282.6 in.^{2}; 446.8 in.^{3} **9.** 6 in.; 12 in.; 12π in.; 288π in.^{3} **11.** 4.5 in.; 9 in.; 9π in.; 121.5π in.^{3} **13.** 5 in.; 10 in.; 31.4 in.; 314.16 in.^{2} **15.** 6; 12; 12π; 144π **17.** 3.92 gal **19.** 12 in. **21.** 78.8 **23.** 4.99 **25.** 244.2 **27.** 512; 432; 707.7

Exercise 32.1

1. 83.82 **3.** 15.88 **5.** 68.58 **7.** 38.4 **9.** 5.504 **11.** 29.8 **13.** 30.48 **15.** 76.2 **17** 48.26 **19.** 7.77 **21.** 10.47 **23.** 486.16 **25.** 2.012 **27.** 192.4 **29.** 3.15 **31.** 36 mils; 0.9144 mm

Exercise 32.2

1. 29.26 **3.** 15.85 **5.** 114.3 **7.** 70.87 **9.** 83.7 **11.** 17.72 **13.** 22 **15.** 8.94

Exercise 32.3

1. 80.26 cm; 399.4 cm^2 **3.** 38.1 cm; 87.1 cm^2 **5.** 95.8 cm; 729.5 cm^2 **7.** 43.1 cm; 147.7 cm^2 **9.** 31.5 cm; 23.6 cm; 371.7 cm^2 **11.** 167.6 cm; 1756 cm^2 **13.** 25.1 m^2 **15.** 80.47 meters by 50.29 meters; perimeter: 261.52 meters; area: 4046.8 m^2 **17.** 34.65 in.; 72.5 in.^{2}

Exercise 32.4

1. 3933 cm^3; 1522 cm^2 **3.** 1635 cm^2; 4500 cm^3 **5.** 23,890 cm^3 **7.** 729 cm^2; 1853 cm^3 **9.** 128 **11.** 30.2 **13.** 73.2 in.^{3}

Exercise 32.5

1. 545 gm **3.** 880 lb **5.** 7.054 oz **7.** 14,176 mg **9.** 37.86 **11.** yes **13.** yes

15. yes **17.** 43.3 gm **19.** 9.36 yd **21.** 3.15 meters **23.** 11.35 cm^2 **25.** 0.224 km^2 **27.** 0.2 km **29.** 45.7 meters **31.** 28.4 gm **33.** 0.9144 meters **35.** 65 mg **37.** 329.2 meters **39.** 660,000 meters **41.** 625 kc **43.** 17,468 mph

Exercise 33.1 (Alternate odds)

1. 25; $\log_5 25 = 2$ **5.** 4096; $\log_8 4098 = 4$ **9.** 256; $\log_4 256 = 4$ **13.** $\frac{1}{81}$; $\log_3 \frac{1}{81} = -4$ **17.** 2; $\log_8 2 = \frac{1}{3}$ **21.** $\log_N y = x$ **25.** $\log_3 28 = x$ **29.** $\log_3 15.59 = 2.5$ **33.** $8^3 = 512$ **37.** $4^{-\frac{1}{2}} = \frac{1}{2}$ **41.** $b^c = A$ **45.** 3 **49.** 2401 **53.** 10

Exercise 33.2

1. $\log 85 = 1.9294$ **3.** $\log 2500 = 3.3979$ **5.** $\log 812 = 2.9096$ **7.** $\log 6 = 0.7782$ **9.** $\log 600 = 2.7782$ **11.** $6800 = 10^{3.8325}$ **13.** $9.15 = 10^{0.9614}$ **15.** $75 = 10^{1.8751}$ **17.** $24.8 = 10^{1.3945}$

Exercise 33.3

1. $1.86(10^5)$ **3.** $4.96(10^3)$ **5.** $5.8(10^{-4})$ **7.** $3.937(10^1)$ **9.** $9.5(10^{11})$ **11.** $5.0(10^{-1})$ **13.** 10^7 **15.** $7.3(10^{-9})$ **17.** $2.54(10^{-7})$ **19.** $3.3136(10^4)$ **21.** $5.893(10^{-7})$ **23.** (6.28)(eighteen zeros)

Exercise 33.4

1. 4.2355 **3.** 1.4983 **5.** 0.8432 − 1 **7.** 0.9299 − 1 **9.** 4.3139 **11.** 3.4914 **13.** 0.7832 − 2 **15.** 0.6911 − 3 **17.** 0.7782 **19.** 0.9484 − 1 **21.** 0.0253 **23.** 0.7559 − 3 **25.** 1.9863 **27.** 0.6990 − 1 **29.** 15,300 **31.** 32.7 **33.** 0.0564 **35.** 0.792 **37.** 9,550,000 **39.** 31.7 **41.** 0.00000417 **43.** 5.61

Exercise 33.5

1. 2.1021 **3.** 0.2261 − 2 **5.** 0.2842 − 1 **7.** 1.3608 **9.** 2.4080 **11.** 0.4750 − 1 **13.** 0.5362 **15.** 0.6228 − 1 **17.** 1.6956 **19.** 0.7035 **21.** 0.8419 − 2 **23.** 0.8951 − 1 **25.** 1.9503 **27.** 0.9909 − 2 **29.** 14,720 **31.** 0.01923 **33.** 253,700 **35.** 0.002915 **37.** 3284 **39.** 0.00004571 **41.** 5.249 **43.** 7.808

Exercise 34.1

1. 34.6 **3.** 310,000 **5.** 55,300 **7.** 0.0007073 **9.** 0.1965 **11.** 0.4748 **13.** 9.418 **15.** 44.68 **17.** −16,610,000 **19.** 3986 **21.** 412.1 **23.** 4080 **25.** 75.1

Exercise 34.2

1. 872.9 **3.** −53.81 **5.** 0.002839 **7.** 84,840 **9.** 621,000 **11.** $8.226(10^{-9})$ **13.** 0.00436 **15.** 3500 **17.** 26,260 **19.** 0.005181 **21.** 743.3 **23.** 0.0004649 **27.** 17.61 **29.** 62.49

Exercise 34.3

1. log 3 + log 11 **3.** log 15 + log 48 + log 37 **5.** log 3.2 + log 65 + log 23 **7.** log h − log k **9.** log 23 − log 46 **11.** log 6 + log 500 − log 24 − log 16 **13.** log 1440 **15.** log 0.12 **17.** log 12 **19.** log 280 **21.** log 300

Exercise 34.4

1. $5.96(10^7)$ **3.** 0.0635 **5.** $7.66(10^{-10})$ **7.** $9.02(10^{-6})$ **9.** $4.55(10^{13})$ **11.** $9.29(10^{10})$ **13.** $1.209(10^{11})$ **15.** 85,720 **17.** 9,742,000 **19.** 176.2 **21.** 1830 **23.** 52.28 **25.** 0.4464 **27.** 42.5

Exercise 34.5

1. 67.67 **3.** 838.5 **5.** 42.09 **7.** 0.1782 **9.** 6.645 **11.** 0.2864 **13.** 9.6 **15.** 6.14 **17.** 0.6286 **19.** 0.512 **21.** 30.24 **23.** 16.1 **25.** 0.0005236 **27.** 0.0000852 **29.** 39.17 **31.** 0.09036 **33.** 0.08998 **35.** 7.3

Exercise 34.6

1. log 2 + log 3 **3.** log 3 + 2 log 2 **5.** 4 log 3 **7.** 5 log 2 **9.** 3 log 3 + log 2 **11.** 1 − log 2 **13.** 1 + 2 log 2 **15.** 4 log 2 + 2 log 3 **17.** 4 log 2 + 3 log 3 **19.** 9 log 2 **21.** log 3 − log 2 **23.** 2 log 2 − 1 **25.** log 2 − log 3

Exercise 35.1

1. 483 **3.** 732.5 **5.** 944 **7.** 1720 **9.** 52.9 **11.** 60.9; 95.1 **13.** 354 ft **15.** 13.4 cm **17.** 463 in.^{3} **19.** $5.88(10^{12})$ **21.** 1.175 ft

Exercise 35.2

1. 3.2660 **3.** 1.6096 **5.** 2.1614 **7.** 2.122 **9.** 3.6912 **11.** 2.581 **13.** 3.6056 **15.** −0.0255 **17.** 4.4783 **19.** 0.7143 **21.** 5.5566

Exercise 35.3

1. 100 **3.** 351.7 **5.** 4.487 **7.** 0.2812 **9.** $\frac{2}{3}$ **11.** 3.402 **13.** 46.3 **15.** 1.738 **17.** 3.281 **19.** 2.512 **21.** 0.817 **23.** 0.7326 **25.** 0.3059 **27.** 7.581

Exercise 35.4

1. 3.457 **3.** 0.761 **5.** 7.30 **7.** 11.379 **9.** −0.320 **11.** 1.000 **13.** 64.2 **15.** 614 **17.** 741 **19.** 1.35 **21.** 0.232 **23.** 0.082

Exercise 36.1 (Answers not shown)

Exercise 36.2

A. $\frac{12}{13}$; $\frac{5}{13}$; $\frac{12}{5}$ **C.** $-\frac{3}{5}$; $\frac{4}{5}$; $-\frac{3}{4}$ **E.** $-\frac{12}{13}$; $\frac{5}{13}$; $-\frac{12}{5}$ **G.** $\frac{7}{25}$; $-\frac{24}{25}$; $-\frac{7}{24}$ **I.** 0.2425; 0.9701; 0.2500 **K.** 0.7071; 0.7071; 1 **M.** 0.7071; −0.7071; −1 **O.** 0.7809; 0.6247; 1.2500 **Q.** −0.8575; 0.5145; −1.6667 **S.** 0.5528; 0.8333; 0.6633

Exercise 37.1

1. 0.4242 **3.** 0.9532 **5.** 0.5362 **7.** 3.630 **9.** 0.7501 **11.** 0.5948 **13.** 1.5577 **15.** 0.3719 **17.** 0.02094 **19.** 0.9996 **21.** 0.02269 **23.** 22.02 **25.** 0.01047 **27.** 0.01047 **29.** 0.2136 **31.** 3.413 **33.** 1.0065 **35.** 3.7657 **37.** 14.7° **39.** 23.4° **41.** 35.8° **43.** 16.9° **45.** 28.23° **47.** 34.95°

Exercise 37.2

1. 19.6° **3.** 44.3° **5.** 29.7° **7.** 75.2° **9.** 71.16° **11.** 77.11° **13.** 23.58° **15.** 36.87° **17.** 0 **19.** 30°

Exercise 38.1 (Unknown side; sin; cos; tan; angle)

A. 10.82; 0.5547; 0.8321; 0.6667; 33.7° **C.** 6.32; 0.9035; 0.4286; 2.108; 64.6° **E.** 12.73; 0.7071; 0.7071; 1; 45° **G.** 5.57; 0.9375; 0.3480; 2.694; 69.6° **I.** 33.0; 0.5755; 0.8178; 0.7037; 35.1°

Exercise 38.2

1. 19.6° **3.** 44.3° **5.** 29.7° **7.** 75.2° **9.** 71.16° **11.** 77.11° **13.** 23.58° **5.** B =26.3°; a =84.98; c =94.81 **7.** A =25.94°; b =64.06°; b =43.16 **9.** A = 58°; B = 32°; c = 56.6 **11.** A = 33.56°; B = 56.44°; a = 29.85 **13.** B = 32.7°; a = 26.9; b = 17.3 **15.** B = 20.6°; a = 42.56; c = 45.48 **17.** A = 12.8°; b = 123.3; c = 126.4 **19.** A = 56.3°; B = 33.7°; a = 183 **21.** A = 21.5°; B = 68.5°; b = 390.8 **23.** B = 76.9°; a = 123.3; c = 544.1

Exercise 38.3

1. B = 68.1°; a = 9.97; c = 26.73 **3.** A = 51.2°; b = 96.48; c = 153.98 **5.** B = 31.8°; b = 19.96; c = 37.89 **7.** B = 16.3°; a = 1436.4; c = 1496.3 **9.** A = 8.2°; a = 3.401; c = 23.84 **11.** B = 19.8°; a = 59.73; c = 63.48 **13.** A = 20.08°; B = 69.92°; c = 771.9 **15.** A = 71.2°; B = 18.8°; a = 170.4

Exercise 38.4

1. 30.5 ft; 9.57 ft **3.** 142.6 ft **5.** 622 ft **7.** 17.5° **9.** 623 ft **11.** 75.8°; 23.3 ft **13.** 275 ft; 265 ft **15.** 1767 ft **17.** 155.4 ft **19.** 120.4 ft

Exercise 39.1 (Order: sine, cosine; tangent; cotangent)

1. 0.4019; −0.9157; −0.4390; −2.278 **3.** 0.2605; −0.9655; −0.2698; −3.706 **5.** −0.5707; −0.8211; 0.6950; 1.4388 **7.** −0.3665; −0.9304; 0.3939; 2.539 **9.** −0.9311; −0.3649; 2.552; 0.3919 **11.** −0.2656; 0.9641; −0.2754; −3.630 **13.** −0.4035; 0.9150; −0.4411; −2.267 **15.** −0.7694; 0.6388; −1.2045; −0.8302 **17.** 0.2990; 0.9542; 0.3134; 3.191 **19.** 0.9143; −0.4051; −2.257; −0.4431 **21.** −0.6225; 0.7826; −0.7954; −1.2572 **23.** −0.9737; −0.2278; 4.275; 0.2339 **25.** 12.27 **27.** −9.90 **29.** −11.79

Exercise 39.2

1. 204.9°; 335.1° **3.** 252.6°; 289.4° **5.** 64.4°; 295.6° **7.** 147.3°; 327.3° **9.** 117.9°; 297.9° **11.** 68.1°; 248.1°

Exercise 39.3 (Order: sin; cos; tan; cot; sec; csc. Given values omitted.)

1. $\frac{12}{13}$; $\frac{5}{12}$; $\frac{12}{5}$; $\frac{13}{12}$; $\frac{13}{5}$ **3.** $\frac{3}{5}$; $\frac{4}{5}$; $\frac{4}{3}$; $\frac{5}{4}$; $\frac{5}{3}$ **5.** 0.866; 0.577; 1.732; 1.155; 2 **7.** 0.816; 0.577; 0.707; 1.732; 1.225 **9.** 0.949; 0.316; 3; 0.333; 1.054 **11.** 0.96; 3.429; 0.292; 3.571; 1.042 **13.** 0.8575; 0.5145; 0.600; 1.944; 1.166 **15.** 0.745; 0.667; 1.118; 0.894; 1.342 **17.** $\frac{\sqrt{x^2-16}}{x}$; $\frac{4}{\sqrt{x^2-16}}$; $\frac{\sqrt{x^2-16}}{4}$; $\frac{x}{\sqrt{x^2-16}}$; $\frac{x}{4}$ **19.** hypotenuse $= \sqrt{x^2+25}$ **21.** unknown side $= \sqrt{9-4t^2}$ **23.** 0.577 **25.** 15°

Exercise 40.1 (No answers shown)

Exercise 40.2

1. 67° **3.** 48° **5.** 28° **7.** 26.3° **9.** 8.6° **11.** 45° **13.** cos **15.** $90° - \theta$ **17.** $90° - \alpha$ **19.** M **21.** $-30°$ **23.** 10° **25.** cot **27.** 60° **29.** tan

Exercise 41.1

1. 540° **3.** 90° **5.** 30° **7.** 1800° **9.** 270° **11.** $3\pi/4$ **13.** $\pi/10$ **15.** 4π

Exercise 41.2 (no answers shown for No. 17–43)

1. $\pi/4$ **3.** $5\pi/4$ **5.** $11\pi/6$ **7.** $-\pi/2$ **9.** $\frac{2}{3}$ **11.** 120π **13.** $3\pi/2$ **15.** $5\pi/8$

Exercise 41.3

1. 386 deg^2 **3.** 108.3 deg^2 **5.** 228 deg^2 **7.** 364 deg^2 **9.** 69.7 ft **11.** 67.6 in.

Exercise 42.1 (Order: sin; cos; tan; cot; sec; csc)

1. $\frac{\sqrt{3}}{2}$; $-\frac{1}{2}$; $-\sqrt{3}$; $-\frac{\sqrt{3}}{3}$; -2; $\frac{2\sqrt{3}}{3}$ **5.** $-\frac{\sqrt{2}}{2}$; $-\frac{\sqrt{2}}{2}$; 1; 1; $-\sqrt{2}$; $-\sqrt{2}$ **9.** 1; 0; not defined; 0; not defined; 1 **13.** same as No. 5 **17.** 1 **19.** 0 **21.** $2\sqrt{3}$ **23.** 1 **25.** undefined **27.** $\frac{3}{64}$ **29.** $-6\sqrt{3}$ **31.** $2+\sqrt{6}$ **33.** $\ln 3 = 1.0986$ **35.** 1.242

Exercise 42.2

1. 30°; 150° **3.** 0°; 90°; 180°; 270° **5.** 60°; 120°; 240°; 300° **7.** 30°; 90°; 150°; 210°; 270°; 330 **9.** 27°; 63°; 99°; 135°; 171°; 207°; 243°; 279°; 315°; 351° **11.** 30°; 150°; 210°; 330° **13.** 60°; 240° **15.** 10°; 50°; 70°; 110°; 130°; 170°; 190°; 230°; 250°; 290°; 310°; 350° **17.** 0; 180° **19.** 90°

Chapter 43. (No answers required for this chapter)

Exercise 44.1

1. $C = 74.7°$; $a = 28.2$; $c = 44.0$ **3.** $B = 23.2°$; $a = 68.9$; $b = 28.5$ **5.** $B = 70.6°$; $a = 27.1$; $c = 28.9$ **7.** $A = 59.5°$; $b = 19.46$; $c = 28.7$ **9.** $C = 37.3°$; $a = 27.7$; $b = 63.8$ **11.** $A = 49.2°$; $C = 62.6°$; $c = 62.2$ **13.** $A = 62.5°$; $B = 43.0°$; $a = 22.1$ **15.** $A = 47.9°$; $C = 19.4°$; $a = 60.3$ **17.** $B = 53.4°$; $C = 89.6°$; $c = 19.94$; *or* $B = 126.6°$; $C = 16.4°$; $c = 5.63$ **19.** $B = 49.9°$; $C = 95.1°$; $c = 41.67$; *or* $B = 130.1°$; $C = 14.9°$; $c = 10.76$ **21.** no solution **23.** 282 ft **25.** 189 ft

Exercise 44.2

1. $B = 71.6°$; $C = 57.5°$; $a = 36.8$ **3.** $A = 75.2°$; $C = 62.7°$; $b = 26.4$ **5.** $A = 48.3°$; $C = 27.4°$; $b = 67.5$ **7.** $A = 21.6°$; $B = 12.1°$; $c = 31.7$ **9.** $A = 43.9°$; $B = 60.1°$; $C = 76.1°$ **11.** $A = 52.6°$; $B = 56.4°$; $C = 71.0°$ **13.** $A = 43.5°$; $B = 31.1°$; $C = 105.4°$ **15.** $A = 122°$; $B = 22.7°$; $C = 35.3°$ **17.** $A = 25.2°$; $C = 131.2°$; $b = 15.9$ **19.** $B = 110.3°$; $C = 22.06°$; $a = 27.6$ **21.** 333 miles; 19.9° north of east **23.** 4.29 miles; 36.8° north of east **25.** 414 miles; 7° south of west **27.** 432 ft; 49.6° north of east **29.** 69.63°; 57.53°; 52.83° **31.** 66.4° with bank, upstream; 18 min

Exercise 45.1

1. $9.43\angle 32.0°$ **3.** $13.89\angle 210.3°$ **5.** $6.2\angle 121.1°$ **7.** $5.46\angle -76.9°$ **9.** $15.8\angle 18.4°$ **11.** $7.33\angle 10.4°$ **13.** $8.46\angle -6.24°$ **15.** $42.3\angle -1.9°$ **17.** $x = 70.44$; $y = 37.93$ **19.** $x = 13.62$; $y = 48.11$ **21.** $x = -5.15$; $y = 3.11$ **23.** $x = -136.05$; $y = -63.15$ **25.** $x = -38.54$; $y = 58.44$ **27.** $x = -6.832$; $y = -9.865$ **29.** $x = 4.54$; $y = -14.30$ **31.** $x = 12.76$; $y = -21.50$ **33.** 231.33 miles at 6.2° south of east **35.** 12.5 mph; 16.26° downstream **37.** 120.8 lb east; 105 lb north **39.** plane: 290.5 mph; air: 51.2 mph

Exercise 45.2

1. $56.7\angle 107.5°$ **3.** $59.5\angle 66.6°$ **5.** $48.09\angle 91°\,45'$ **7.** $73.96\angle -72.9°$ **9.** $14.88\angle -69.25°$ **11.** $46.05\angle -37.3°$ **13.** $19.11\angle 251.7°$ **15.** $39\angle 22.6°$ **17.** 104.4 lb at 64.5° north of west **19.** 75.5 lb at 23.4° south of west **21.** 39.1 lb at 31.6° north of west

Exercise 45.3

1. $13\angle 22.6°$; $13(\cos 22.6° + j \sin 22.6°)$ **3.** $2\angle 150°$; $2(\cos 150° + j \sin 150°)$ **5.** $15\angle 210°$; $-12.99 - j7.5$ **7.** $0 + j60$ **9.** $65\angle 120.5°$ **11.** $10\angle -60°$ **13.** $5(\cos 90° + j \sin 90°)$ **15.** $16\angle 240°$ **17.** $256\angle 0°$ **19.** $64\angle 270°$ **21.** $2\angle 0°$; $2\angle 120°$; $2\angle 240°$ **23.** $2\angle 60°$; $2\angle 150°$; $2\angle 240°$; $2\angle 330°$ **25.** 3 amp

Exercise 45.4

1. 28; 45; 53 **3.** 65; 72; 97 **5.** 119; 120; 169 **7.** 20; 48; 52 **9.** 75; 308; 317

Chapter 46. (Graphs of trigonometric functions; no answers shown)

Exercise 47.1

1. $-4; 0; 0; 9x^2 - 6x - 3; 4x^2 - 8x$ **3.** $9\pi; 49\pi; 16\pi; \pi; 4\pi r^2$ **5.** $3; 0; 4; -12; 3 - 2x^2 - x^4$ **7.** $x^2 + 7x - 1; 13; x^2 + 3x + 5; 31$ **9.** 2; 3 **11.** −1; −0.4163

Exercise 47.2

1. 47.5 mph **3.** −2° per hr **5.** 176 ft/sec **7.** 13π **9.** 390 **11.** 15 **13.** 740 **15.** 5.8 **17.** −1.5 **19.** 270

Exercise 47.3

1. 9 **3.** 20 **5.** 3 **7.** 6 **9.** 4 **11.** no limit **13.** no limit **15.** 0 **17.** $\frac{3}{4}$

Exercise 47.4

1. $2x$ **3.** $6x + 3$ **5.** $2x + 4$ **7.** $-6 - 2x$ **9.** $2t - 3$ **11.** $2t - 4$ **13.** $-6/x^2$ **15.** $12/(2 - 3x)^2$ **17.** 2 **19.** −15 **21.** $-\frac{4}{3}$ **23.** 120

Exercise 47.5

1. −3 **3.** 101 **5.** 72 **7.** $-\frac{2}{9}$ **9.** 3 **11.** 24π **13.** $12x^3 - 10x$ **15.** $80 - 4w$ **17.** $y'' = \frac{3}{4}x^{-\frac{1}{2}} - \frac{8}{9}x^{-\frac{5}{3}}$; $y''' = -\frac{3}{8}x^{-\frac{3}{2}} + \frac{40}{27}x^{-\frac{8}{3}}$ **19.** $6x - 4$; 2; −4 **21.** $2t + 5$; 1; 13 **23.** −1; 3 **25.** 10π **27.** −0.16

Exercise 48.1 (In Nos. 1–7, answers for s, v, a when $t = 1$ sec.)

1. 9; 4; 0 **3.** 5; 13; 8 **5.** 10; 1; −2 **7.** −11; −2; 14 **9.** $t = 1; s = 3; t = 5; s = -29$ **11.** 80; 96; 3; 224; $32\sqrt{14}$ **13.** 640; 640 ft/sec **15.** 1600 ft; 320 ft/sec; 1120 ft/sec **17.** 78 ft/sec

Exercise 48.2

1. 2; 0; 1; **3.** 9; −7; −5 **5.** $x - y = 3$ **7.** $y = 3$ **9.** $x + 2 = y$ **11.** (2, −7) **13.** $(\frac{5}{4}, \frac{1}{8})$ **15.** (−1, 15); (3, −17) **17.** (−2; 52); (3, −73) **19.** $3x + y = -1$ **21.** $-\frac{1}{4}$

Exercise 48.3

1. (2, −3), min **3.** (1, 5), max **5.** $(-\frac{5}{2}, -\frac{9}{4})$, min **7.** $(\frac{3}{2}, \frac{1}{2})$, min **9.** (1, 13), max; (5, −19), min **11.** (−2, 13), max; (2, −7), min **13.** (2, −4), max; (3, −5), min **15.** (−2, 26), max; (−4, 22), min **17.** $t = 3$; $s = 204$ **19.** 6

Exercise 48.4

1. 25 by 50 rd **3.** 12 by 15 yd **5.** $w = 12$; $l = 12$; $h = 6$ **7.** $4\sqrt{3}$ by $6\sqrt{3}$ **9.** Min: −30 volts; max: 60 volts **11.** 30.2π

Exercise 48.5

1. $dy = 15$; $\Delta y = 15.13554081$ **3.** $dy = -0.108$ **5.** 400 ft; 32 ft **7.** −6 watts **9.** 6.4π **11.** 144π **13.** 0.0882π **15.** 2.2 watts **17.** 0.0306π

Exercise 49.1

1. $20(4x - 3)^4$ **3.** $8(x^2 - 3x + 4)^7(2x - 3)$ **5.** $2(t^2 + 4t)(2t + 4)$ **7.** $5(4t - t^2)^4(4 - 2t)$ **9.** $4(1 - 3t - t^2)^3(-3 - 2t)$ **11.** $(4t - 1)(6t^2 - 3t)^{-\frac{2}{3}}$ **13.** $-8(5 - x^2 - x^3)^7(2x + 3x^2)$ **15.** $(\frac{1}{2})(x^3 + 3x^2 - 4)^{-\frac{1}{2}}(3x^2 + 6x)$ **17.** $\frac{5}{6}$ **19.** $\frac{3}{2}$ **21.** $\frac{3}{8}$

Exercise 49.2

1. $12x^2 - 6x - 20$ **3.** $4x^3 - 5x^4 + 6x^2 - 6x$ **5.** $10x^4 - 8x^3 - 66x^2 + 60x$ **7.** $(3 - t)^3(9t^2 - 11t^4)$ **9.** $(x + 4)^4(x - 2)^2(x + 2)(6 - 11x)$ **11.** $(2t^2 + 2t)(t^2 - 9)$ **13.** 20 **15.** 0 **17.** $\frac{124}{3}$ **19.** 28 **21.** $-\frac{3}{8}$

Exercise 49.3

1. $15(2x + 5)^{-2}$ **3.** $x^2(4x - 1)^{-\frac{1}{2}}$ **5.** $(6x^2 - 2x)(4x - 1)^{-\frac{3}{2}}$ **7.** $(2x^2 - 3)(x^{-4})(1 - x^2)^{-\frac{1}{2}}$ **9.** $(2x + 9)(3 - x^2)^{-\frac{3}{2}}$ **11.** $(y^2 - 4y)^{\frac{1}{2}}(y - 2)^{-1}$ **13.** $\frac{1}{40}$ **15.** $\frac{4}{5}$ **17.** $-\frac{5}{4}$ **19.** $-\frac{7}{27}$

Exercise 49.4

1. $-\frac{2}{5}$ **3.** $-\frac{5}{3}$ **5.** $-\frac{3}{5}$ **7.** $-\frac{7}{3}$ **9.** $-\frac{1}{2}$ **11.** $\frac{17}{20}$ **13.** $\frac{3}{5}$ **15.** $-\frac{3}{2}$ **17.** $-\frac{5}{7}$

Exercise 49.5

1. $\frac{9}{2}$ **3.** -10 **5.** 32 **7.** 0.45 ft/sec **9.** $81/16\pi$ ft/sec (approx. 1.6) **11.** $\frac{1}{540}\pi$ **13.** 8 cm^2/min **15.** 54.7 mi/hr **17.** 1.89 ft/sec **19.** 600 ergs/sec

Exercise 50.1

1. $6x + C$ **3.** $\dfrac{3x^4}{2} + C$ **5.** $q + C$ **7.** $\dfrac{5t^2}{2} + C$ **9.** $2t^4 + C$ **11.** $-3/x^2 + C$ **13.** $\frac{1}{4}t^4 - \frac{1}{2}t^2 + C$ **15.** $-\frac{1}{3}x^3 + 1/x + C$ **17.** $-2t^{-2} - 6t^{\frac{1}{2}} + C$ **19.** $x^4 - 4x^3 + 9x^2/2 + C$ **21.** $\frac{1}{3}x^3 + x^2 - 15x + C$ **23.** $-3/x^2 + C$ **25.** $-2x^2 - 6x^{-1} - x^3 - x^2/2 + 2x + C$ **27.** $x^3/3 - 3x^2/2 + C$ **29.** $y = 3x^2/2 - 5x + C$; parabola **31.** $4t^3/3 - 3t^2/2 + 5t + C$ **33.** $q = t^3 + 2t^2 + 3t + C$

Exercise 50.2

1. $y = x^2 + 5x - 6$ **3.** $y = x^3 - x^2 + x$ **5.** $y = 2x^4 - x^3 - 3x^2 - 13$ **7.** $y = x^4 - 4x^2 + 2x - 9$ **9.** $i = 2t^2 - 5t + 6$ **11.** $v = 8t + 14$; $s = 4t^2 + 14t - 4$ **13.** $v = 2400 - 32t$; $s = 2400t - 16t^2$; 90,000 ft **15.** 41.6 ft

Exercise 50.3

1. 21 **3.** 21 **5.** 30 **7.** $\frac{10}{3}$ **9.** 19.5 **11.** 8500 **13.** $-149\frac{5}{12}$ **15.** 3 **17.** $\frac{2}{9}$ **19.** $\frac{28}{3}$ **21.** $\frac{13}{6}$ **23.** 9

Exercise 50.4

1. $\frac{1}{8}(x^2 - 5)^4 + C$ **3.** $\frac{1}{28}(x^2 + 2)^7 + C$ **5.** $\frac{1}{8}(x^2 + 6x + 1)^4 + C$
7. $\frac{2}{9}(x^3 - 4)^6 + C$ **9.** $\frac{5}{3}(x^2 + 8x)^{\frac{3}{2}} + C$ **11.** $\frac{2}{15}(t^3 - 6t)^5 + C$
13. $-(4x - x^2)^{\frac{1}{2}} + C$ **15.** 78 **17.** $\frac{124}{9}$ **19.** $\frac{8}{3}$

Exercise 51.1

1. 12 **3.** 4 **5.** $\frac{32}{3}$ **7.** $\frac{16}{3}$ **9.** $21\frac{1}{12}$ **11.** $\frac{16}{3}$ **13.** 32 **15.** $\frac{8}{3}$ **17.** $\frac{16}{3}$ **19.** 9

Exercise 51.2

1. 32π **3.** $(\frac{256}{5})\pi$ **5.** $(\frac{128}{7})\pi$ **7.** $(\frac{512}{3})\pi$

Exercise 51.3

1. 96 in.-lb **3.** 62.5 in.-lb **5.** 15.7 in.-lb **7.** 5000 ft-lb **9.** 10,608 ft-lb
11. $181{,}710\pi$ ft-lb **13.** $80{,}870\pi$ ft-lb **15.** $117{,}150\pi$ ft-lb **17.** 16 ft **19.** 46
21. -10.5

Exercise 52.1

1. $5 \cos 5x$ **3.** $4 \sec^2 4t$ **5.** $2 \sec 2x \tan 2x$ **7.** $20 \cos 20t$ **9.** $-6x \cos (5 - 3x^2)$ **11.** $-50 \sin 50t$ **13.** $300 \cos 6t$ **15.** $15{,}080 \cos 377t$ **17.** $20 \cos \omega t$
19. $600 \cos 20t - 300 \sin 20t$ **21.** 1; $-\frac{1}{2}$; 0 **23.** 0; -3 **25.** 10 amp/sec
27. $-10 \sin 2t \cos^4 2t$ **29.** $-12 \cot^5 2t \csc^2 2t$ **31.** $-8 \cot 2t \csc^4 2t$
33. $t^2(3 \cos 2t - 2t \sin 2t)$ **35.** $-\frac{9}{4}$

Exercise 52.2

1. $\dfrac{2x}{x^2 + 5}$ **3.** $\dfrac{2x + 5}{x^2 + 5x - 2}$ **5.** $\dfrac{4}{4x - 7}(0.4343)$ **7.** $\dfrac{4}{x}$ **9.** $\dfrac{6x}{x^2 - 5}$
11. $-2t \tan t^2$ **13.** $-10 \cot 5t$ **15.** $12 \sec 3t \csc 3t$ **17.** $3/(5 - 3x)$
19. $1/x$ for all **21.** 1 **23.** $\ln x$ **25.** $-\frac{1}{2}$ **27.** (2, 0)

Exercise 52.3

1. $6e^{6x}$ **3.** $(2x + 3)e^{x^2+3x}$ **5.** $-2xe^{-x^2}$ **7.** $8te^{4t^2}$ **9.** $(2t - 2)e^{t^2-2t-3}$
11. $3(4^{3x})(\ln 4)$ **13.** $2(\cos 2t)e^{\sin 2t}$ **15.** $4(\sec^2 4x)e^{\tan 4x}$
17. $3t^2(1 - t)e^{-3t}$ **19.** $-(t + 2)(t^{-3})e^{-t}$ **21.** $5e^{5t} - 5e^{-5t}$
23. $e^x(5 \cos 5x + \sin 5x)$ **25.** $e^{-3t}(17 \cos 20t - 23 \sin 20t)$
27. $10e^{-20t}(5 \cos 30t + \sin 30t)$ **29.** $6e^{3x^2}(6x^2 + 1)$
31. $e^{-3t}(9t^2 - 12t + 2)$ **33.** 0 **35.** $-100e^{-20t}(17 \sin 30t + 7 \cos 30t)$
37. 1 **39.** $y = e$; $m = e$ **41.** $y = 1$; $m = 0$

Exercise 52.4

1. $\frac{1}{4} \sin 4x + C$ **3.** $(\frac{1}{10}) \sin^5 2x + C$ **5.** $(-\frac{1}{12}) \cos^4 3x + C$
7. $\frac{1}{2} \ln (x^2 - 3) + C$ **9.** $-\frac{1}{2} \ln (x^2 - 4x + 1) + C$ **11.** $\frac{1}{3} \ln \tan 3x$
13. $-\frac{1}{5}e^{-5x} + C$ **15.** $-\frac{1}{2}e^{\cos 2x} + C$ **17.** 0.8647 **19.** $\ln \sec + C$
21. $\frac{1}{4} \sec 4x + C$ **23.** $\frac{1}{3} \tan^3 x - \tan x + x + C$ **25.** 1 **27.** $\ln 2$ **29.** 2.3504

Exercise 52.5

1. 60; 42.4; 38.2 **3.** 35.4; 22.5; 1.11; 0.9 **5.** 367.7; 234.3

Exercise 53.1

1. 13031 **3.** 35113 **5.** 23305 **7.** 13743 **9.** 101011 **11.** 155 **13.** 177 **15.** 139 **17.** 1241 **19.** 243 **21.** 3031 **23.** 315; remainder: 1

Exercise 53.2

1. 23 **3.** 85 **5.** 181 **7.** 427 **9.** 100011 **11.** 101011 **13.** 1110001 **15.** 11000 **17.** 11111 **19.** 11101 **21.** 101000001 **23.** 1000 **25.** 110101 **27.** 1000001 **29.** 101011111 **31.** 10101; remainder: 110

Exercise 53.3

1. 731 **3.** 5771 **5.** 300,000 **7.** 1575 **9.** 3269 **11.** 1064 **13.** 14633 **15.** (a) 11111100; (b) 101110011; (c) 10001011111110; (d) 10000111110011

Exercise 53.4

1. (a) 275; (c) CAB; (e) 22F3 **2.** (a) 917; (c) 2213; (e) 3049 **3.** (a) F7; (b) ICE **4.** IC

*Chapter 11 answers continued from page 841.

Exercise 11.9

1. 16; $(x + 4)^2$ **3.** 64; $(x + 8)^2$ **5.** 225; $(x + 15)^2$ **7.** 1/4; $(y + \frac{1}{2})^2$ **9.** 25/4; $(n + \frac{5}{2})^2$ **11.** 0.16; $(y + 0.4)^2$ **13.** 49; $(2n + 7)^2$ **15.** 25; $(4y + 5)^2$ **17.** 49; $(6t + 7)^2$ **19.** 25/16; $(2x + \frac{5}{4})^2$ **21.** 49/36; $(3x + \frac{7}{6})^2$ **23.** 25/36; $(3x + \frac{5}{6})^2$

Exercise 11.10

1. $x^2 - 9x + 20$ **3.** $x^2 - 11x + 30$ **5.** $x^2 - 8x + 12$ **7.** $x^2 - x - 12$ **9.** $x^2 + x - 56$ **11.** $x^2 + 5x - 24$ **13.** $x^2 + 11x - 12$ **15.** $x^2 + 9x - 10$ **17.** $x^2 - 5x - 50$ **19.** $y^2 - 9y - 36$ **21.** $y^2 - 8y - 65$ **23.** $t^2 - 36$ **25.** $n^2 - 13n - 90$ **27.** $n^2 + 45n - 144$ **29.** $x^6 + 5x^3 - 36$

Exercise 11.11

1. $(x + 4)(x + 5)$ **3.** $(x + 2)(x + 9)$ **5.** $(x + 5)(x + 7)$ **7.** $(t + 5)(t - 3)$ **9.** $(t + 6)(t - 2)$ **11.** $(y - 11)(y + 1)$ **13.** $(n - 6)(n + 5)$ **15.** $(n + 10)(n - 9)$ **17.** $(x + 6)^2$ **19.** $(y + 15)(y - 4)$ **21.** $(y + 16)(y - 3)$ **23.** $(x + 4)(x + 16)$ **25.** $(y - 18)(y + 4)$ **27.** $(y - 36)(y + 2)$ **29.** $(x - 21)(x + 4)$

Exercise 11.12

1. $6x^2 + 19x + 15$ **3.** $12x^2 - x - 20$ **5.** $15x^2 - 14x - 8$ **7.** $20x^2 - 9x - 20$ **9.** $12y^2 - 28y + 15$ **11.** $35t^2 + t - 12$ **13.** $4n^2 - 25n + 6$ **15.** $30a^2 - 11a - 6$ **17.** $24 - 2xy - 15x^2y^2$ **19.** $12x^4 + 7x^2y - 12y^2$ **21.** $4n^2 - 65n + 16$ **23.** $6x^2 + 49x + 8$

Exercise 11.13

1. $(3x + 2)(x - 5)$ **3.** $(3n + 1)(n - 4)$ **5.** $(3c + 2)(2c - 3)$ **7.** $(5y - 6)(y - 1)$ **9.** $(5a - 6)(6a + 7)$ **11.** $(5n + 7)(2n - 3)$ **13.** $(3t - 4)^2$ **15.** not factorable **17.** $(3y - 8)^2$ **19.** not factorable **21.** $(3n + 8x)(2n - 3x)$ **23.** $(4k + 7)^2$ **25.** $(3n - 5)(12n - 5)$ **27.** $(4n + 7)(n + 7)$ **29.** not factorable.

Exercise 11.14

1. $(x - 2)(x^2 + 2x + 4)$ **3.** $(T + 10)(T^2 - 10T + 100)$ **5.** $(x^2 - 3)(x^4 + 3x^2 + 9)$ **7.** $(7t - 4)(49t^2 + 28t + 16)$ **9.** $(x - 1)(x^2 + x + 1)(x^6 + x^3 + 1)$ **11.** $(2y + 5)(4y^2 - 10y + 25)$ **13.** $(4n - 9)(16n^2 + 36n + 81)$ **15.** $[x + y - a + b][(x + y)^2 + (x + y)(a - b) + (a - b)^2]$

Exercise 11.15

1. $(x^2 + y^2)(x + y)(x - y)$ **3.** $(x - y)(x^6 + x^5y + x^4y^2 + x^3y^3 + x^2y^4 + xy^5 + y^6)$ **5.** not factorable **7.** not factorable **9.** $(x - 2)(x + 2)(x^2 + 2x + 4)(x^2 - 2x + 4)$

Exercise 11.16

1. $4(2x - 3)(2x + 3)$ **3.** $2(a - 6)(a + 6)$ **5.** $9x^2(x^2 + 9y^2)$ **7.** $(x^2 + 4)(x - 2)(x + 2)$ **9.** $x^6(x^2 + 1)(x - 1)(x + 1)$ **11.** $a^2(h - k)(h + k)$ **13.** $x(x^3 - 81)$ **15.** $5xy(y + 4)(y^2 - 4y + 16)$ **17.** $2n^3(5 - n)(25 + 5n + n^2)$ **19.** $x^3(x + 6)(x - 7)$ **21.** $4y^2(y + 3)(y - 4)$ **23.** $6n^2(n + 2)(n - 6)$ **25.** $3x^2(2x - 5y)(3x + 2y)$ **27.** $4n(3n + 4)^2$ **29.** $(a - 3b - x)(a - 3b + x)$ **31.** not factorable **33.** $(4x^2 - x + 1)(4x^2 + x + 1)$ **35.** $(2y - x + 3)(2y + x - 3)$ **37.** $(x - y + 4)(x + y - 2)$ **39.** $(x - 3y + 4)(x + 3y)$

Index